UNDERSTANDING

Statistics

GRAHAM UPTON AND IAN COOK

OXFORD UNIVERSITY PRESS

Oxford University Press, Great Clarendon Street, Oxford OX2 6DP

Oxford   New York
Athens   Auckland   Bangkok   Bogota   Bombay
Buenos Aires   Calcutta   Cape Town   Dar es Salaam   Delhi
Florence   Hong Kong   Istanbul   Karachi
Kuala Lumpur   Madras   Madrid   Melbourne
Mexico City   Nairobi   Paris   Singapore
Taipei   Tokyo   Toronto   Warsaw

and associated companies in
Berlin   Ibadan

Oxford is a trade mark of Oxford University Press

© Graham Upton and Ian Cook 1996

First published in 1996
Reprinted (with corrections) 1997

ISBN 0 19 914391 9   School and college edition
ISBN 0 19 914351 X   Bookshop edition

A CIP catalogue record for this book is available from the British Library.

Typeset by Tech Set Ltd., Gateshead, Tyne and Wear
Printed and bound in Great Britain by
Butler & Tanner Ltd, Frome and London

Contents

Preface

These days statistics are thrown at a cringing public from all sides and few people can escape them completely. Some pieces of statistical information are more interesting than others and some are more important. The well-educated citizen needs to develop a critical awareness and understanding of what is important and of what can be legitimately deduced from the deluge of statistical information. Even specialist research workers in disciplines as disparate as Linguistics and Sports Science need basic statistical ideas in order to understand, properly, the numerical information with which they will inevitably be confronted.

The purpose of this book is to present a wide range of essential statistical ideas in a simple and (we hope) enjoyable fashion. If a user of the book does not derive the tiniest bit of pleasure from some part of the book then we will be disappointed (but there will be no cash refunds).

The content of the book is dictated by the need to cover the relevant A-Level material. In these days of modularity and with a plethora of syllabi to choose from, there is a wide variety of material to be covered. We have tried to cover *all* the statistical material in Modules T1 to T4 of the ULEAC syllabus, Modules 2 and 6 of the AEB syllabus, Papers 4, 7, 8 and 9 of the NEAB syllabus, Modules S1 to S4 of the UCLES Modular syllabus, Papers 2 and 4 of the UCLES Linear Mathematics syllabus, and Paper 2 of the UCLES Linear Further Mathematics syllabus. Although the contents are dictated by school syllabi, the book is also suitable as a textbook for an introductory course at university level.

For some years we have run a course entitled 'Statistics at A-Level' for *teachers* of A-level statistics. We could not find a suitable course text since, in our opinion, existing books either presented the subject as a series of scarcely explained formulae, or used very advanced mathematics, or swamped the reader in oceans of prose. Some books also contained an inadequate number of exercises or were particularly short on examination questions. In *Understanding Statistics* we have tried to find a happy compromise in which formulae are simply derived, with the reasoning succinctly presented. There are over a thousand problems, with a great many coming from published A-level and AS-level examination papers. In addition, we have included many suggestions for practical work (both in and out of the classroom) and for calculator and computer practice.

We hope that in working through the exercises the readers get all the answers right first time. We also hope that we have got all the answers right first time! The numerical answers given in the back of the book are, of course, our responsibility and any errors are due to us and not to the examining boards.

We are very grateful to the examining boards listed below for permission to reproduce their questions. The source of each question is indicated by the corresponding initials at the end of the question.

Associated Examining Board [AEB]

Northern Examination and Assessment Board [NEAB], formerly the Joint Matriculation Board [JMB]

Oxford and Cambridge Schools Examination Board [O&C], which also gave permission to use questions from the examinations for the Mathematics in Education and Industry Project [MEI] and the School Mathematics Project [SMP]

University of Cambridge Local Examinations Syndicate [UCLES]

University of London Examinations and Assessment Council [ULEAC] formerly the University of London School Examinations Board [ULSEB]

University of Oxford Delegacy for Local Examinations [UODLE]

Welsh Joint Education Committee [WJEC]

In some cases it is appropriate to use only part of a question. This is indicated by (P) after the attribution. In Chapter 18 a few questions have been adapted to ask for a dispersion test: such questions are indicated with an (A). In Chapter 22, a few questions have been adapted to ask for Kendall's coefficient instead of Spearman's coefficient and a few have been adapted to ask for a Wilcoxon signed-rank test: these too are indicated with an (A).

The illustration of the Paris–Lyon train timetable is taken from R J Marey, *La Méthode Graphique* which was published in Paris in 1885. The graph by C J Minard of Napoleon's journey through Russia is taken from the same volume. These graphs, and many more, are reproduced in two wonderful books by Edward Tufte entitled *The Visual Display of Quantitative Information* and *Envisioning Information*. Both books are published by Graphics Press, Cheshire, Connecticut 06410. The diagram illustrating the use of Chernoff Faces was kindly made available to the authors by Dr Daniel Dorling of the University of Newcastle-upon-Tyne.

In conclusion, we would like to thank our patient editor, Don Manley, for his help and encouragement; Katherine Pate, the indefatigable copy editor, and Rob Fielding, the reader, for pointing out various howlers; our frustrated colleagues in the Department of Mathematics for waiting patiently in the queue at the laser printer; and, most of all, our understanding families for putting up with ever-growing piles of paper over more years than we care to remember. The errors that remain are, of course, due to ourselves.

GJGU
ITC
University of Essex
Colchester
November 1995

Glossary of notation

Inevitably some letters are used to denote different quantities in different contexts. However, this should cause no confusion since the context will make clear which definition is appropriate.

∞	Infinity.
$\sim$	'has distribution'.
$\approx$	'approximately equals'.
$^-$	'mean'.
$\hat{}$	'estimate'.
$'$	'complementary' (of events).
$\mid$	'conditional', so $A\mid B$ means 'A occurs given that B occurs'.
$!$	Factorial: $r! = r(r-1)(r-2)\cdots 1$.
$\cap$	The intersection of two events.
$\sum$	Summation: $\sum_{i=1}^{n} x_i = x_1 + \cdots + x_n$.
$\cup$	The union of two events.
a	Intercept of (estimated) regression line.
α (alpha)	Intercept of population regression line.
A_2	Factor for control chart for mean.
b	Slope of (estimated) regression line.
β (beta)	Slope of population regression line.
b	Number of blocks (Chapter 21).
B	The random variable corresponding to b.
$\mathrm{B}(n,p)$	The binomial distribution with parameters n and p.
c	A critical value.
cdf	The cumulative distribution function.
χ_d^2	Chi-squared distribution with d degrees of freedom.
$\mathrm{Cov}(X, Y)$	The covariance of the random variables X and Y.
d	Degrees of freedom (of t, F or χ^2 distributions).
d_i	A difference between paired values (or ranks).
$\bar{d}$	The mean difference of paired values.
D	Deviance, also known as residual sum of squares.
D_3, D_4	Factors for control charts for the range.
e	$2.718\,281\,828\ldots.$
$\mathrm{E}(X)$	Expected value or expectation of X.
E	An event.
E'	The complementary event to E.
E_i	An expected frequency.
f_j	The frequency with which the value x_j occurs.
$\mathrm{f}(x)$	The probability density function of X.
$\mathrm{F}(x)$	The cumulative distribution function of X.
$F_{u,v}$	The F-distribution with u and v degrees of freedom.
$\mathrm{G}(t), \mathrm{G}_X(t)$	The probability generating function (of X).
$\mathrm{G}'(t), \mathrm{G}''(t)$	First and second derivatives of $\mathrm{G}(t)$.
$\mathrm{H}_0, \mathrm{H}_1$	The null and alternative hypotheses.
I	Index of dispersion.
λ (lambda)	The parameter of a Poisson or exponential distribution.
L	Laspeyres price index.
L	Lower percentage point of a distribution.

m	The median.
M	The test statistic for a sign test.
mgf	The moment generating function.
$M(t)$, $M_X(t)$	The moment generating function (of X).
μ (mu)	The population mean.
n	The number of observations.
$n(E)$	The number of outcomes in the event E.
$\binom{n}{r}$	The number of distinct combinations of r objects chosen from n.
^nP_r	The number of distinct permutations of r objects chosen from n.
$N(\mu, \sigma^2)$	Normal distribution, mean μ, variance σ^2.
ν (nu)	Used in place of d, to denote degrees of freedom in some tables.
O_i	An observed frequency.
P	Paasche price index.
P	The sum of positive ranks.
p, P	Probability.
p_n, p_o	Prices in current and original years.
P_x	$P(X = x)$
$P(z)$	Used in place of $\Phi(z)$ by some tables.
pdf	Probability density function.
pgf	Probability generating function.
ϕ (phi)	Probability density function for $N(0, 1)$.
Φ	Cumulative distribution function for $N(0, 1)$.
π (pi)	$3.141\,592\,653\,59\ldots$.
q	$1 - p$, often the probability of a 'failure'.
Q	Sum of negative ranks.
Q	The minimum number of neighbour swaps.
q_n, q_o	Quantities in current and original years.
Q_1, Q_2, Q_3	The lower quartile, median and upper quartile.
$Q(z)$	Used by some tables to mean $1 - \Phi(z)$.
ρ (rho)	Population correlation coefficient.
r	The number of successes.
r	Product–moment correlation coefficient.
r_s	Spearman's rank correlation coefficient.
R	The rank-sum.
R, $\bar{R}$	The range, the average range.
s^2	($= \sigma_{n-1}^2$) An unbiased estimate of the population variance.
s	The square root of s^2.
S	The sample space.
S^2	The random variable corresponding to s^2.
S_{xx}, S_{xy}, S_{yy}	$\sum(x_i - \bar{x})^2$, $\sum(x_i - \bar{x})(y_i - \bar{y})$, $\sum(y_i - \bar{y})^2$
σ (sigma), σ^2	The population s.d., variance.
σ_n, σ_n^2	The sample s.d., variance (or the s.d., variance of a population of size n).
σ_{n-1}^2	($= s^2$) An unbiased estimate of the population variance.
τ (tau)	Kendall's rank correlation coefficient.
t	The number of treatments (Chapter 21).
t_d	A t-distribution with d degrees of freedom.
T	A random variable having a t-distribution.
T	The smaller of P and Q (Chapter 22).

U	An upper percentage point of a distribution.
$\text{Var}(X)$	The variance of X.
W	The sum of signed ranks.
$\bar{x}, \bar{X}$	The sample mean (value or random variable).
$\bar{\bar{x}}$	The mean of sample means.
x_i	An observed value.
$X, Y, \ldots$	Random variables.
X^2	The goodness-of-fit statistic.
X_c^2	The Yates-corrected version of X^2.
Z	A random variable having a N(0,1) distribution.
z	A test statistic: an observation on Z.

1 Summary diagrams and tables

She may look at it because it has pictures

Florence Nightingale, on a book of statistics that she had sent to Queen Victoria

One picture is worth ten thousand words

Frederick R Barnard

1.1 The purpose of Statistics

In most countries the biggest employer of statisticians is the government, which collects numerical information about all aspects of life. The information collected in the form of human statistics (such as the numbers out of work), of financial statistics (such as the rate of inflation), and on other aspects of life is regularly reported in newspapers and on the news. In addition to these **population** statistics, **sample** statistics are also reported. For example, market research agencies (e.g. Gallup) also collect numerical information which can dominate the news in advance of a general election.

As an example, a single issue of *The Times* contained the following:

- Drink–drive statistics (*Source:* The Government).
- Mothers' feelings about going back to work (sample statistics – *Source:* Gallup).
- Numbers of visitors to Britain subdivided by nationality (sample statistics – *Source:* International Passenger Survey).
- A breakdown of British Rail assets (*Source:* British Rail annual report).
- Pages of statistics on stocks and shares.
- Much more interesting pages of sports statistics!
- World weather statistics (*Source:* The Meteorological Office).

In the modern world we are inundated with statistics; the subject Statistics is concerned with trying to make sense of all this numerical information.

Project _____

Choose a single issue of a 'quality' newspaper and search for reports that include statistics. Try to decide what type of organisation collected the reported statistics.
Counting just one for all the sports reports, one for all the financial reports and one for all the weather information, how many different reports can you find in a single issue of the paper?
How many different organisations appear to have collected the statistics?

A good example of the purpose of Statistics is provided by the opinion poll. A poll is taken of a few thousand people. From the information that these people provide, remarkably accurate conclusions are drawn that refer to the entire population, which is many times greater in number.

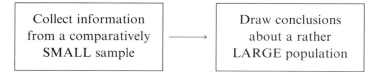

1.2 Variables and observations

The term **variable** refers to the description of the quantity being measured, and the term **observed value** or **observation** is used for the result of the measurement. Some examples are:

Variable	*Observed value*
Weight of a person	80 kg
Speed of a car	70 mph
Number of letters in a letter box	23
Colour of a postage stamp	Tyrian plum

If the value of a variable is the result of a random observation or experiment (e.g. the roll of a die), then the variable is called a **random variable**.

1.3 Types of data

The word 'data' is the plural of 'datum', which means a piece of information – so **data** are pieces of information. There are three common types of data: qualitative, discrete and continuous.

Qualitative data consist of descriptions using *names*. For example:

'Head' or 'Tail'
'Black' or 'White'
'Bungalow', 'House' or 'Castle'

Discrete data consist of numerical values in cases where we can make a list of the possible values. Often the list is very short:

1, 2, 3, 4, 5 and 6.

Sometimes the list will be infinitely long, as for example, the list:

0, 0.5, 1.0, 1.5, 2.0, 2.5, 3.0, 3.5, 4.0, . . .

Continuous data consist of numerical values in cases where it is not possible to make a list of the outcomes. Examples are measurements of physical quantities such as weight, height and time.

Note

◆ The distinction between discrete and continuous data is often blurred by the limitations of our measuring instruments. For example, when measuring people, we may record their heights to the nearest centimetre, in which case observations of a continuous quantity are being recorded using discrete values.

1.4 Tally charts and frequency distributions

Given below are the scores made in their final round by the 30 leading golfers in the 1992 Scottish Open golf championship:

62, 65, 63, 65, 70, 68, 65, 67, 67, 69, 70, 70, 70, 67, 68,
68, 66, 69, 74, 67, 68, 69, 69, 69, 71, 71, 72, 69, 72, 68

The data are discrete since the scores are all whole numbers. It is easy to see that most scores are around 70 or a little less. But it is not so easy to see which score was most common. A simple summary is provided by a **tally chart**.

The tally chart is constructed on a single 'pass' through the data. For each score a vertical stroke is entered on the appropriate row, with a diagonal stroke being used to complete each group of five strokes. This is much easier than going through the data counting the number of occurrences of a 62 and then repeating this for each individual score.

Notes
- ◆ Counting the tallies is made easy by using the 'five-bar gates'.
- ◆ If the tallies are equally spaced then the chart provides a useful graphical representation of the data.

The tally count for each outcome is called the **frequency** of that outcome. For example, the frequency of the outcome 65 was 3. The set of outcomes with their corresponding frequencies is called a **frequency distribution**, which can be displayed in a **frequency table**, as illustrated below:

Tally chart of the scores made in their final round by the 30 leading golfers in the 1992 Scottish Open

Score	Tallies
62	|
63	|
64	
65				
66	|
67					
68	ℍℍ
69	ℍℍ |
70					
71			
72			
73	
74	|

Final round score	62	63	64	65	66	67	68	69	70	71	72	73	74
Number of golfers	1	1	0	3	1	4	5	6	4	2	2	0	1

1.5 Stem-and-leaf diagrams

Tally charts become uncomfortably long if the range of possible values is very large, as with these individual scores from a low-scoring Sunday league cricket match:

22, 58, 12, 17, 4, 7, 26, 10, 13, 1, 39, 0, 1, 10, 6, 0, 11, 14, 1, 0

A convenient alternative is the **stem-and-leaf diagram**, in which the stem represents the most significant digit (i.e. the 'tens') and the leaves are the less significant digits (the 'units'). The following stem-and-leaf chart has been created following the order of the data:

```
0 | 4, 7, 1, 0, 1, 6, 0, 1, 0
1 | 2, 7, 0, 3, 0, 1, 4
2 | 2, 6
3 | 9
4 |
5 | 8
tens |        units
```

If the original stem-and-leaf diagram had been created on rough paper then a tidied version could have the leaves neatly ordered as shown below:

0	0, 0, 0, 1, 1, 1, 4, 6, 7
1	0, 0, 1, 2, 3, 4, 7
2	2, 6
3	9
4	
5	8
tens	*units*

These charts are sometimes presented with 'split' stems (for finer detail). This is illustrated below, with the units between 0 and 4 (inclusive) separated from the units between 5 and 9 (inclusive):

0	0, 0, 0, 1, 1, 1, 4
0	6, 7
1	0, 0, 1, 2, 3, 4
1	7
2	2
2	6
3	
3	9
4	
4	
5	
5	8
tens	*units*

It is now particularly easy to see that most players scored less than 15 and that the highest score of 58 was a long way clear of the rest.

Stem-and-leaf diagrams retain the original data information, but present it in a compact and more easily understandable way: this is the hallmark of an efficient data summary.

Stem-and-leaf diagrams can be used both with discrete data and with continuous data (treating the latter as though it were discrete). They are much easier to understand when the stem involves a power of ten, but other units may be employed if the stem would otherwise be too long or too short. It is often wise to provide an explanation (a **Key**) with the diagram.

Example 1

The internal phone numbers of a random selection of individuals from a large organisation are given below.

Summarise these numbers using a stem-and-leaf diagram.

3315, 3301, 2205, 2865, 2608, 2886, 2527, 3144, 2154, 2645, 3703, 2610, 2768, 3699, 2345, 2160, 2603, 2054, 2302, 2997, 3794, 3053, 3001, 2247, 3402, 2744, 3040, 2459, 3699, 3008, 3062, 2887, 2215, 2213, 3310, 2508, 2530, 2987, 3699, 3298, 2021, 3323, 2329, 2845, 2247, 3196, 3412, 2021

A quick glance at the data reveals that all the numbers begin with either a 2 or a 3, implying that they all lie between 2000 and 3999 (inclusive). Taking a stem with units of 100 would lead to a large diagram: instead, therefore, we work with units of 200, so that the leaves range between 0 and 199, inclusive. In this case the number 3315 is represented as a stem of '3200' and a leaf of '115' (so that 3200 + 115 = 3315). The resulting diagram (with unordered leaves) is as follows:

```
2000 | 154, 160, 54, 21, 21
2200 | 5, 145, 102, 47, 15, 13, 129, 47
2400 | 127, 59, 108, 130
2600 | 8, 45, 10, 168, 3, 144
2800 | 65, 86, 197, 87, 187, 45
3000 | 144, 53, 1, 40, 8, 62, 196
3200 | 115, 101, 110, 98, 123          Key:
3400 | 2, 12                            3200 | 115 = 3315
3600 | 103, 99, 194, 99, 99
```

The stem could also be labelled '20', '22', etc, with the caption 'hundreds'.

Example 2

The masses (in g) of a random sample of 20 sweets were as follows:

1.13, 0.72, 0.91, 1.44, 1.03, 1.39, 0.88, 0.99, 0.73, 0.91,
0.98, 1.21, 0.79, 1.14, 1.19, 1.08, 0.94, 1.06, 1.11, 1.01

Summarise these results using a stem-and-leaf diagram.

A quick scan reveals that the masses are all in the region of 1 g, so that an appropriate choice would be multiples of 0.1 for the stem and multiples of 0.01 for the leaves.

```
0.7 | 2, 3, 9        Key:
0.8 | 8              0.7 | 2 = 0.72
0.9 | 1, 9, 1, 8, 4
1.0 | 3, 8, 6, 1
1.1 | 3, 4, 9, 1
1.2 | 1
1.3 | 9
1.4 | 4
```

Exercises 1a

1 The numbers of absentees in a class over a period of 24 days were:

0, 3, 1, 2, 1, 0, 4, 0, 1, 1, 2, 3,
1, 0, 0, 2, 4, 6, 4, 2, 1, 0, 1, 1

By first drawing up a tally chart obtain a frequency table.

2 A bridge player keeps a note of the numbers of aces that she receives in successive deals. The numbers are:

0, 2, 3, 0, 0, 2, 1, 1, 0,
2, 3, 0, 1, 1, 2, 1, 0, 0

Draw up a tally chart and hence obtain a frequency table.

3 The numbers of eggs laid each day by 8 hens over a period of 21 days were:

6, 7, 8, 6, 5, 8, 6, 8, 6, 5, 6,
4, 7, 6, 8, 7, 5, 7, 6, 7, 5

Draw up a tally chart and hence obtain a frequency table.

4 For each potato plant, a gardener counts the numbers of potatoes whose mass exceeds 100 g. The results are:

8, 5, 7, 10, 8, 6, 5, 6, 4, 8,
10, 9, 8, 7, 3, 10, 11, 6, 9, 8

Obtain a frequency table.

5 A choirmaster keeps a record of the numbers turning up for choir practice. The numbers were:

25, 28, 32, 31, 31, 34, 28, 31, 29,
28, 32, 32, 30, 29, 29, 31, 28, 28

Obtain a frequency table.

6 The numbers of matches in a box were counted for a sample of 25 boxes. The results were:

51, 52, 48, 53, 47, 48, 50, 51, 50,
46, 52, 53, 51, 48, 49, 52, 50,
48, 47, 53, 54, 51, 49, 47, 51

Obtain a frequency table.

7 The marks obtained in a mathematics test marked out of 50 were:

35, 42, 31, 27, 48, 50, 24, 27,
21, 37, 41, 34, 12, 18, 27

Construct a stem-and-leaf diagram to represent the data.

8 A baker kept a count of the number of doughnuts sold each day.
The numbers were:

35, 47, 34, 46, 62, 41, 35, 47, 51,
56, 73, 38, 41, 44, 51, 45, 74

Construct a stem-and-leaf diagram to show the data.

9 The total scores in a series of basketball matches were:

215, 224, 182, 200, 229, 219,
209, 217, 195, 162, 210, 213,
204, 208, 197, 192, 187, 213

Construct a stem-and-leaf diagram to represent the above data.

10 The masses (in g) of a random collection of 16 pebbles are as follows:

17.4, 32.1, 24.4, 37.6, 51.0, 41.4,
19.9, 36.2, 41.3, 50.2, 37.7, 28.4,
26.3, 22.2, 33.5, 42.4

Summarise these data using a stem-and-leaf diagram.

1.6 Bar charts

The lengths of the rows of a tally chart or of a stem-and-leaf diagram provide an instant picture of the data. This picture is neatened by using bars whose lengths are proportional to the numbers of observations of each outcome (i.e. to the frequencies). In the resulting diagram, known as a **bar chart**, the bars may be either **horizontal** (like the tally chart) or **vertical**.

Notes
  ◆ Bar charts are easier to read if the width of the bars is different from the width of the gaps between the bars.
  ◆ It is not necessary to show the origin on the graph.

Example 3

Illustrate the golf scores of Section 1.4 (p.3) using a bar chart.

With lots of different values we use narrow bars centred on the values 62, 63, etc. The origin does not appear!

Vertical bar chart of the scores made in their final round by the 30 leading golfers in the 1992 Scottish Open

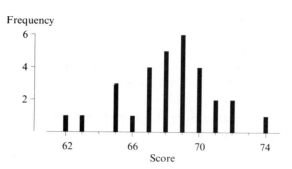

Example 4

A car salesman is interested in the colour preferences of his customers. For one type of car his records are as follows.

Blue	White	Red	Others
12	23	16	18

Represent these figures using a vertical bar chart.

With just four categories narrow bars would look silly! We therefore use wide bars separated by narrower gaps. The categories are not numerical so they could be arranged in any order. A sensible order is to arrange the single colours in descending order of observed frequency, ending with the 'Others' category.

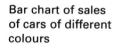

Bar chart of sales of cars of different colours

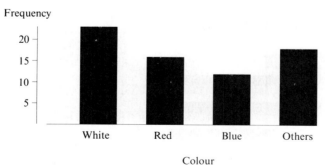

Practical _____

Roll an ordinary six-sided die 24 times, recording each outcome as it occurs (e.g. 3, 6, 2, 2, ...). Summarise the data using a tally chart and write down your frequency distribution. Compare your distribution with that of a neighbour. There may be large differences due to random variation! Combine the two sets of results and illustrate them with a vertical bar chart. Does it look as though your dice were fair?

Calculator practice

> *Graphical calculators can produce crude bar graphs. These are good enough to provide an idea of the data, but fail to indicate that discrete x-values are involved. Produce a diagram for the golf data in Section 1.4 on your calculator and compare it with our diagram.*

1.7 Multiple bar charts

When data occur naturally in groups and the aim is to contrast the variations within different groups, a **multiple bar chart** may be used. This consists of groups of two or more adjacent bars separated from the next group by a gap having, ideally, a different width to the bars themselves.

The diagram may be horizontal or vertical with the values either specified on the diagram or indicated using a standard axis.

▼

Example 5

The following data, taken from the *Monthly Bulletin of Statistics* published by the United Nations, show the 1970 and 1988 estimated populations (in millions) for five countries.
Illustrate the data using a multiple bar chart.

	France	Mexico	Nigeria	Pakistan	UK
1970	50	51	57	56	55
1988	55	82	104	105	57

The data show the differing rates of population growth of the two European countries and the three non-European countries and provide a graphic (literally!) illustration of a world problem. To increase visibility the countries are re-ordered in terms of their 1988 populations.

Populations of five countries in 1970 and in 1988 (figures are in millions)

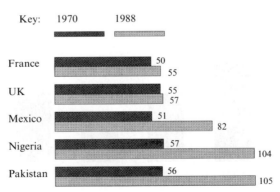

▲

1.8 Compound bars for proportions

In a compound bar chart the length of a complete bar signifies 100% of the population. The bar is subdivided into sections that show the relative sizes of components of the populations. By comparing the sizes of the subdivisions of two parallel compound bars, differences can be seen between the compositions of the separate populations. The populations need not be populations of living creatures – they could be, for example, the populations of nails in two builders' trucks!

Example 6

One consequence of the dramatic growth in population of the 'third world' countries is that a high proportion of the population of these countries is young and there are few old people. The United Nations publication *World Population Prospects* gives the following figures for 1990 populations:

	France	Mexico	Nigeria	Pakistan	UK
% under 15	20.2	37.2	48.4	45.7	18.9
% 15 to 64	66.0	59.0	49.2	51.6	65.6
% 65 and over	13.8	3.8	2.4	2.7	15.5

Illustrate these figures in an appropriate diagram.

The data are conveniently presented in percentage form and, since comparisons are intended, composite bar charts are appropriate. It is difficult to know in what order to present the countries: we have used increasing order of the youngest age group, since this appears on the left of the diagram.

Compound bars showing, for five countries, the proportions of the population in three age ranges

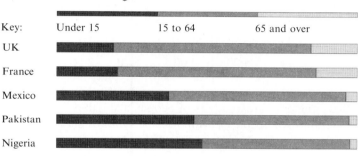

1.9 Population pyramids

A **population pyramid** is used in examining age distributions. A typical pyramid consists of two multiple bar charts (one for males and one for females) placed back to back, with the bars referring to different age categories. Here are two examples that contrast the age distribution of a European nation with that of an African nation. As well as showing the differing age structures, the greater life expectancy for females is evident in the European 'pyramid'.

Population pyramids for European and African nations showing age distribution and the division between sexes

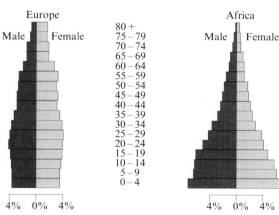

Note

♦ Each pyramid has the same area (representing 100% of the population): the contrast is between the age distributions rather than the population sizes.

1.10 Pictograms

Pictograms are imprecise bar charts, in which the bars are replaced by symbols illustrating the subject of the data. Their only virtue is variety!

Note

♦ Pictograms often misrepresent data by scaling both height *and* width in proportion to the data. For example, if sales of bacon are represented by pigs(!), then it is the *areas* of the pigs (or their *volumes*, if three-dimensional pictures are being used) that should be in proportion to the sales.

Example 7

The following table shows, for those living in houses, the average numbers of people per dwelling (in 1984) in various countries.

Represent the data using a suitable pictogram.

Average numbers per dwelling				
France	Mexico	Nigeria	Pakistan	UK
2.49	5.20	4.00	5.20	2.56

Since the data are numbers of people, a suitable diagram uses people as symbols. The symbols are all of the same size. Variations in the numbers are reflected by using different numbers of symbols. As a refinement, the possible effect of the living conditions is reflected in the expressions of the symbols!

France   2.49

UK   2.56

Nigeria   4.00

Mexico   5.20

Pakistan   5.20

Average number of people per dwelling

1.11 Pie charts

Pie charts are the circular equivalent of compound bar charts. The areas of the portions of the pie are in proportion to the quantities being represented. Occasionally you may see pies of different sizes; these indicate different population sizes. When drawn correctly the areas (and not the radii) will be in proportion to the differing population sizes.

Example 8

The European Community *Forest Health Report 1989* classifies trees by the extent of their defoliation (i.e. by their loss of leaves). Trees that are in good health have defoliation levels of between 0% and 10%. The following data show the proportions of conifers with various amounts of defoliation in France and the UK.

Illustrate the data using pie charts.

	Extent of defoliation			
	0%–10%	11%–25%	26%–60%	61%–100%
France	0.750	0.176	0.068	0.006
UK	0.358	0.303	0.250	0.089

The separate pies have the same size (since we are not concerned with the *quantities* of conifers in the two countries). The pie segments are shaded to assist with their visibility. The shading scale is chosen so that the colour is darker where there are more leaves (least defoliated).

It can be seen that a sizeable proportion of the UK conifers are heavily defoliated, whereas about three-quarters of the French conifers are in good health (0%–10% defoliation). However, the comparison is not quite fair since other information in the report shows that French conifers are rather younger than those in the UK.

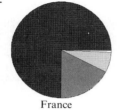

France

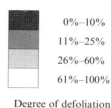

0%–10%
11%–25%
26%–60%
61%–100%

Degree of defoliation

UK

Pie charts comparing the amounts of defoliation of conifers in France and in the UK in 1989

Exercises 1b

1 The numbers of absentees in a class over a period of 24 days were:

    0, 3, 1, 2, 1, 0, 4, 0, 1, 1, 2, 3,
    1, 0, 0, 2, 4, 6, 4, 2, 1, 0, 1, 1

Construct a bar chart for the above data.

2 A bridge player keeps a note of the numbers of aces that she receives in successive deals. The numbers are:

    0, 2, 3, 0, 0, 2, 1, 1, 0,
    2, 3, 0, 1, 1, 2, 1, 0, 0

Construct a bar chart for the above data.

3 The numbers of eggs laid each day by 8 hens over a period of 21 days were:

    6, 7, 8, 6, 5, 8, 6, 8, 6, 5, 6,
    4, 7, 6, 8, 7, 5, 7, 6, 7, 5

Construct a bar chart for the above data.

4 For each potato plant, a gardener counts the numbers of potatoes whose mass exceeds 100 g. The results are:

    8, 5, 7, 10, 8, 6, 5, 6, 4, 8,
    10, 9, 8, 7, 3, 10, 11, 6, 9, 8

Construct a bar chart for the above data.

5 The numbers of goals scored in the first three divisions of the Football League on 4 February 1995 were:

    6, 5, 3, 3, 5, 3, 1, 2, 4, 2, 2, 5,
    1, 2, 2, 3, 8, 2, 4, 5, 3, 3, 0, 2,
    5, 0, 1, 0, 3, 0, 1, 2, 7, 1, 2

Construct a bar chart for the above data.

6 The shoe sizes of the members of a football team are:

    10, 10, 8, 11, 10, 9, 9, 10, 11, 9, 10

Represent the data on (i) a bar chart,
(ii) a compound bar chart, (iii) a pie chart.

7 A school recorded the numbers of candidates achieving the various possible grades in their A-level subjects.

    E: 21; D: 47; C: 69; B: 72; A: 53

Represent the data on (i) a bar chart,
(ii) a compound bar chart, (iii) a pie chart.

8 The proportions of males in the audiences of various sporting fixtures are as follows:
Football match 85%, Rugby match 70%, Tennis match 45%, Badminton match 40%, Gymnastics 35%.
Represent these findings using a compound bar chart.

9 Random samples of individuals aged 20–60 are interviewed in five regions of the country. The percentages of males and of females who are found to be in full-time employment are given in the following table.

	Male	Female
South-East	84	61
East Anglia	78	57
West Midlands	70	58
South-West	65	40
Scotland	63	36

Illustrate the data using a multiple bar chart.

10 A school recorded the numbers of candidates achieving the various possible grades in their A-level subjects.
For boys the figures were:

E: 14; D: 29; C: 42; B: 42; A: 21

For girls the figures were:

E: 7; D: 18; C: 27; B: 30; A: 32

Illustrate these results:
(i) using a compound bar chart,
(ii) using two pie charts,
(iii) using a multiple bar chart.

11 Construct a population pyramid from the following data, for the population of the United Kingdom in the middle of 1993.
Males:

All ages 28 474	Under 1 389	1–4 1603	5–14 3808

15–24 3965	25–34 4723	35–44 3904	45–59 5017	60–64 1374

65–74 2333	75–84 1117	85 and over 242

Females:

All ages 29 718	Under 1 370	1–4 1526	5–14 3609

15–24 3758	25–34 4572	35–44 3883	45–59 5054	60–64 1466

65–74 2836	75–84 1903	85 and over 740

Source: *Population Trends, No 78, Winter 1994*

12 The Registrar General's *Annual Reports* reveals the following figures concerning the marital status of men who married in the years 1872, 1931 and 1965:

	1872	1931	1965
Bachelor	86.3	91.7	88.5
Widower	13.7	7.6	4.9
Divorced	*	0.7	6.6

The figures are percentages, with * indicating a figure of less than 0.1%. Display the figures using:
(i) pie charts,
(ii) a multiple bar chart,
(iii) a compound bar chart.

13 Construct (i) a pictogram, (ii) a pie chart, to display the following data for attendance, in millions, at Museums and Galleries in 1993.
British Museum: 5.8; National Gallery: 3.9; Tate Gallery: 1.8; Natural History Museum: 1.7; Science Museum: 1.3
Source: *Social Trends, 1995*

14 Draw (i) a pictogram, (ii) a pie chart, (iii) a horizontal bar chart, to illustrate the following data.
Domestic air passengers, in thousands, at UK airports in 1993:
Heathrow: 6753; Glasgow: 2399; Edinburgh: 2155; Manchester: 2042; Belfast: 1629; Aberdeen: 1460; Gatwick: 1398; Birmingham: 778; Newcastle: 629; Stansted: 336; East Midlands: 265; Other: 4333; Total all airports: 24 177
Source: *Social Trends, 1995*

1.12 Triangles

When there are just *three* classes an alternative way of representing data uses points inside an equilateral triangle. Suppose the proportions in the three classes are denoted by x, y and z, where $x + y + z = 1$. The cases $x = 0$, $y = 0$ and $z = 0$ correspond to locations on, respectively, the left edge of the triangle, the right edge of the triangle and the bottom of the triangle, while the case $x = y = z = \frac{1}{3}$ corresponds to the centroid of the triangle. Thus, as z increases, so the location of the point representing the data becomes more distant from the base of the triangle. Corresponding statements apply to the values of x and y. The vertices correspond to cases where two of x, y and z are equal to zero. As an illustration the diagram shows the location of the point corresponding to $x = 0.45$, $y = 0.35$, $z = 0.2$.

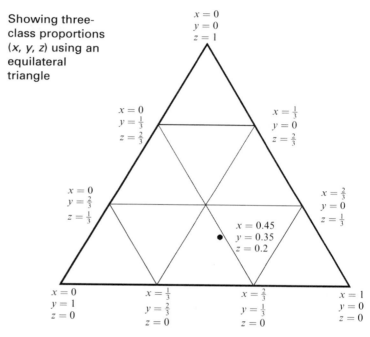

Showing three-class proportions (x, y, z) using an equilateral triangle

Triangular distributions can also be used to show change. This is illustrated in the context of the changes in support for the three main British political parties between the general elections of 1987 (start of arrow) and 1992 (arrow tip). Each arrow represents the aggregate change experienced by a group of constituencies. The general decline of the Liberal Democrats is apparent. The three kite-shaped regions indicate which party got the most votes.

Triangular representation of the change in support for the three main British political parties between the general elections of 1987 and 1992

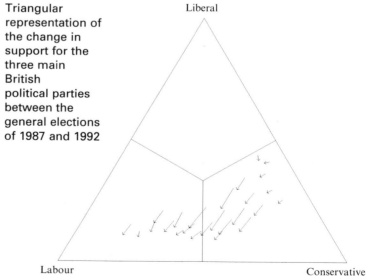

Note

♦ 'Triangular graph paper' is commercially available. Alternatively the (x, y, z)-'coordinates' can be translated into ordinary Cartesian (X, Y)-coordinates by using $X = 1 + x - y$ and $Y = z\sqrt{3}$.

Example 9

The percentages of sand, silt and clay in the samples of sediment taken
from the floor of an Arctic lake are given in the table below, together with
the water depth (in m) at the core site.

Illustrate these data in an appropriate diagram.

Sand	78	72	51	52	70	67	43	53	16	32	66	70	17
Silt	19	25	36	41	26	32	55	37	54	41	28	29	54
Clay	3	3	13	7	4	1	2	10	30	27	6	1	29
Depth	10	12	13	13	16	16	18	19	21	22	22	24	26

Sand	11	38	11	18	5	16	32	10	17	11	5	3	11
Silt	70	43	53	51	47	50	45	53	48	55	55	45	53
Clay	20	19	36	31	48	34	23	37	35	34	41	52	36
Depth	32	34	37	38	37	42	47	47	48	49	50	59	60

Sand	7	7	4	7	5	4	7	7	6	6	2	2	2
Silt	47	50	45	52	49	49	52	47	46	49	54	48	48
Clay	46	43	51	41	46	47	41	46	47	45	40	50	50
Depth	62	62	69	74	74	78	83	88	88	90	91	98	104

Since there are three components to each sample, a triangular diagram
is appropriate. It is evident that the composition varies with water
depth and the data circles have therefore, as an extra touch, been
shaded to indicate depth. The key shows the colours for a variety of
depths (the deeper the darker). Close to shore, in shallow water, there is
a far greater proportion of sand, but it gets less sandy as the water gets
deeper.

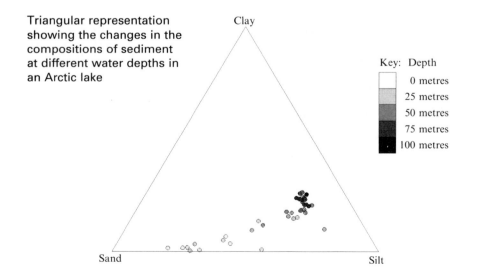

Triangular representation
showing the changes in the
compositions of sediment
at different water depths in
an Arctic lake

Key: Depth
  0 metres
  25 metres
  50 metres
  75 metres
  100 metres

Project _____

> *This is an exercise in visualisation!*
> *Determine which region of the political triangle corresponds to:*
> *(i) a Conservative victory,*
> *(ii) Conservatives in third place,*
> *(iii) the outcome: Conservative vote > Labour vote > Liberal vote.*
> *Draw lines on the triangle corresponding to:*
> *(i) no change in the Conservative vote,*
> *(ii) no change in the ratio of the Labour vote to the Conservative vote.*
> *Find out the latest election results for your local constituency and mark them on the triangle.*

1.13 Grouped frequency tables

The following data are the masses (in g) of 30 brown pebbles chosen at random from those on one area of a shingle beach:

> 3.4, 12.3, 7.5, 8.2, 8.6, 15.4, 6.9, 7.0, 2.9, 5.0,
> 13.5, 8.4, 9.9, 11.8, 4.6, 7.7, 3.8, 7.7, 8.6, 14.6,
> 4.3, 7.9, 9.1, 11.9, 17.4, 6.3, 8.7, 10.1, 5.1, 10.2

A bar chart of these data would look like a very old comb that had had an unfortunate accident! It is obviously sensible to work with ranges of values, which we call **classes**, rather than with the individual values. As a start we summarise the data (perhaps using a tally chart to help with the counting) in order to form a **grouped frequency table**:

Range of masses (g)	1.95– 3.95	3.95– 5.95	5.95– 7.95	7.95– 9.95	9.95– 11.95	11.95– 13.95	13.95– 15.95	15.95– 17.95
Frequency	3	4	7	7	4	2	2	1

Notes
- Inspecting the recorded data it appears that the measurements were made correct to the nearest 0.1 g. Thus pebbles with masses recorded as lying in the range 2 g–3.9 g have true masses lying in the range 1.95 g–3.95 g.
- The values 1.95, 3.95, ... , 15.95 are the **lower class boundaries (l.c.b.)** of their classes, while the values 3.95, 5.95, ..., 17.95 are the **upper class boundaries (u.c.b.)**. Therefore:
  - u.c.b. of one class = l.c.b. of the next class
  - class width = (u.c.b. − l.c.b.)
  In the example, each of the eight classes has width 2.
- Published tables frequently use the rounded figures in the grouped frequency table, and may give only the class mid-point or just one of the class boundaries (usually the l.c.b.). For example the pebble data might be reported thus:

Range of masses (nearest 0.1 g)	Frequency
2–	3
4–	4
6–	7
8–	7
10–	4
12–	2
14–	2
16–	1

Great care and some ingenuity is often needed to deduce the true class boundaries – this is, however, typical of published data!

♦ Many quantities that we measure are not really continuous, but are best treated as such. The following data consists of the advertised prices (in £, in 1992) of second-hand Ford Sierras all less than 3 years old.

> 8195, 4995, 9995, 9995, 8995, 8695, 5995, 5495, 7495, 7895,
> 7295, 8995, 8695, 8495, 7495, 8995, 4995, 7495, 4795, 4995,
> 4995, 8895, 5495, 6495, 5795, 5695, 5195, 5995, 7995, 7350,
> 12 395, 4995, 9495, 6495

These prices would be much easier to read with a 5 added! Although price in £ is not a continuous quantity (since all the prices are in whole numbers of pounds), the possible prices are so close together that it is sensible to treat it as such.

Price range (£)	4000–4999	5000–5999	6000–6999	7000–7999	8000–8999
Frequency	6	7	2	7	8

Price range (£)	9000–9999	10 000–10 999	11 000–11 999	12 000–12 999
Frequency	3	0	0	1

1.14 Difficulties with grouped frequencies

♦ **The value zero** For example, suppose the durations of phone calls are measured to the nearest minute. Then a call of duration '2 minutes' actually lasted for between 1.5 and 2.5 minutes – a range of one minute. Similarly, a call of duration '3 minutes' refers to a range of one minute. The same is true for every recorded phone call length *except* '0 minutes' which refers to calls of between 0 and 0.5 minutes in duration. The treatment of zero here (for a *continuous* variable) should be contrasted with that below.

♦ **Grouped discrete data** Suppose a test is marked out of 100 and it is decided to use the classes 0–24, 25–49, 50–74 and 75–100. Natural intermediate class boundaries are 24.5, 49.5 and 74.5. These boundaries lie 0.5 outside the stated ranges of the classes. In order to be consistent, this suggests using −0.5 and 100.5 as the two remaining boundaries, even though negative marks, and marks in excess of 100, are not feasible. The treatment of zero in this note differs from that in the previous note because the quantity being measured here is discrete and not continuous.

♦ **Age** Unlike almost every other variable, age is reported in *truncated* form. A person who claims to be 'aged 14' is actually aged at least 14.0, but has not yet reached 15.0.

Adolphe Quetelet (1796–1874) was a dominant force in Belgian science for 50 years. His job was as astronomer and meteorologist at the Royal Observatory in Brussels, but his fame was due to his work as a statistician and sociologist! He was one of the founders of the Royal Statistical Society (of London). He spent much time constructing tables and diagrams to show relationships between variables. He was interested in the concept of an 'average man' in the same way as today we talk of the 'average family'.

1.15 Histograms

Bar charts are not appropriate for data with grouped frequencies for ranges of values. A **histogram** is a diagram using rectangles to represent frequency. It differs from the bar chart in that the rectangles may have different widths, but the key feature is that, for each rectangle:

> **area** is proportional to **class frequency**

When all the class widths are equal, histograms are easy to construct, since then not only is area $\propto$ frequency, but also height $\propto$ frequency.

Note

♦ Some computer packages attempt to make histograms three-dimensional. Avoid these if you can, since the effect is likely to be misleading.

▼ ▼

Example 10

Use a histogram to display the following data, which refer to the heights of 5732 Scottish militia men. The data were reported in the *Edinburgh Medical and Surgical Journal* of 1817 and were analysed by Adolphe Quetelet (see above).

Height (ins)	64–65	66–67	68–69	70–71	72–73
Frequency	722	1815	1981	897	317

It appears that the heights were recorded to the nearest inch, so the class boundaries are 63.5, 65.5, 67.5, 69.5, 71.5 and 73.5. These define the locations of the sides of the rectangles while the heights are proportional to 722, 1815, etc.

Histogram of the heights of 5732 Scottish militia men in 1817

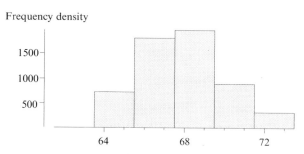

Height of Militia man (inches)

Notes

♦ The *y*-axis has been labelled *frequency density* rather than frequency because it is *area* which is proportional to frequency.
♦ Because the classes all have the same width, the vertical scale (frequency density) could be labelled 'Frequency per 2 inch height range'. However, these units have been omitted because they would confuse rather than inform anyone looking at the diagram!

▲ ▲

▼ ▼

Example 11

As glaciers retreat they leave behind rocks known as 'erratics' because they are of a different type to the normal rocks found in the area. In Scotland many of these erratics have been used by farmers in the walls of their fields. One measure of the size of these erratics is the cross-sectional area, measured in cm^2, visible on the outside of a wall. The areas of 30 erratics are given overleaf.

Provide an appropriate display of these data.

> 216, 420, 240, 100, 247, 128, 540, 594, 160, 286, 216, 448, 380, 509, 90,
> 156, 135, 225, 304, 144, 152, 143, 135, 266, 286, 154, 154, 386, 378, 160

A quick inspection of the data reveals that the values range from 90 to 594. Since many values are possible for cross-sectional area, it will be necessary to group the data and to portray it using a histogram. A good impression of the distribution of a set of values can usually be obtained by using between 5 and 15 classes. This suggests using classes of width 50, with 'natural' boundaries at 50, 100 and so on.

We use a tally chart to help with the counting and obtain:

Range of areas (cm^2)	50–99	100–149	150–199	200–249
Frequency	1	6	6	5

Range of areas (cm^2)	250–299	300–349	350–399	400–449
Frequency	3	1	3	2

Range of areas (cm^2)	450–499	500–549	550–599
Frequency	0	2	1

The resulting histogram shows that there is a long 'tail' of large values but no corresponding tail of small values: the distribution is said to be **skewed to the right** or **positively skewed**. This commonly happens when (as here) we are dealing with physical quantities that have no obvious upper bound, but cannot be negative.

Histogram of cross-sectional area of erratics, using classes of equal width

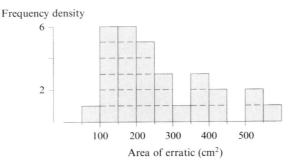

Notes
- ◆ The class boundaries are really at 49.5, 99.5, etc, and the histogram is plotted using these values. However, the axis is labelled (accurately) using less 'awkward' values. Of course, you might not have noticed!
- ◆ The internal 'boxes' serve to emphasise the relation between frequency and area and would not normally be shown.

The histogram shows that the typical erratic had a cross-sectional area of around 200 cm^2, and that some were much larger. With a bigger sample we would expect more or less steadily decreasing frequencies as the values of area increase. In order to eliminate the 'jagged' nature of this diagram, we could use wider categories for the larger values of cross-sectional area:

Range (cm^2)	50–99	100–149	150–199	200–249
Frequency	1	6	6	5

Range (cm^2)	250–299	300–449	450–599
Frequency	3	6	3

The histogram corresponding to the revised table has a reasonably smooth outline, with a more-or-less steady decrease from the peak. Effectively all that has happened is that a few of the 'boxes' have tumbled off local peaks into neighbouring troughs!

Smoothed histogram of cross-sectional area of erratics, using classes of unequal width

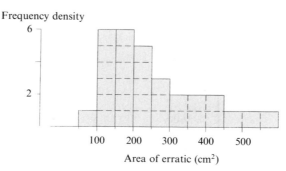

Notes
- The total area of the histogram is unaltered.
- The units on the *y*-axis are simply to enable the viewer to get an accurate impression of the relative heights of different parts of the histogram.

Example 12

The table below summarises the results of a 1992 assessment of the knowledge of the mathematics content of the National Curriculum by 7-year-olds.
Results were reported for 105 Local Education Authorities, with the figures in the table being the percentages of pupils who succeeded in attaining level 2 or better.
Illustrate these data in an appropriate fashion.

% reaching level 2	50–59	60–63	64–65	66–67	68–69	70–71
Number of LEAs	4	4	5	8	7	18

% reaching level 2	72–73	74–75	76–77	78–79	80–83
Number of LEAs	17	11	21	5	5

Assuming that the reported figures have been rounded to the nearest percentage point the class boundaries are 49.5, 59.5, 63.5, 65.5, ... , 79.5 and 83.5. Most classes have a width of 2 percentage points, but the classes at either end are wider. Taking 2 percentage points as being the 'standard' width, and recalling that it is *area* that is proportional to frequency, the height of the rectangle representing the final class frequency must be $\frac{5}{2}$, since this class is twice as wide as the standard class. Similarly, the heights of the first two classes will be $\frac{4}{5}$ and $\frac{4}{2}$, since their widths are respectively 5 times and 2 times the standard width.

An incorrect histogram in which, for the end classes, it is height rather than area that has been made proportional to frequency

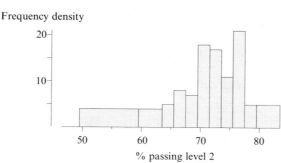

We can set out these calculations in a table as follows.

Class	Class width w	Frequency f	Frequency density $\dfrac{f}{w}$
50–59	10	4	0.40
60–63	4	4	1.00
64–65	2	5	2.50
66–67	2	8	4.00
68–69	2	7	3.50
70–71	2	18	9.00
72–73	2	17	8.50
74–75	2	11	5.50
76–77	2	21	10.50
78–79	2	5	2.50
80–83	4	5	1.25

The heights of the sections of the histogram are in proportion to the frequency densities given in the final column of the table.

The resulting correct histogram with its thin tails should be contrasted with the incorrect fat-tailed histogram in which no allowance has been made for the extra width of the end intervals.

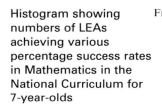

Histogram showing numbers of LEAs achieving various percentage success rates in Mathematics in the National Curriculum for 7-year-olds

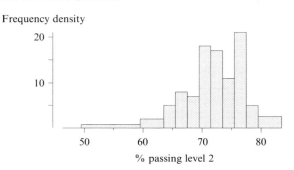

Note

◆ The most convenient scale for the *y*-axis is usually in terms of frequencies for the narrowest class width. In this example the indicated scale of 10 and 20 would correspond to frequencies of 10 and 20 in a 2% range of success rates.

Practical _____

How long can you get a 10p piece to spin on a flat surface?
Using a watch from which seconds can be read accurately, note the lengths of times of four spins. Your personal value will be the length of the longest spin.
Collect the personal values for the whole class and represent the data using a histogram.
Is the histogram roughly symmetrical, or is it skewed?
Was your personal value typical, or was it unusually short or long?

Calculator practice _____

Some 'bar charts' produced by graphical calculators are really histograms! These charts are only suitable for cases where the class widths are all the same. Use your calculator to reproduce the histogram of heights of militia men.

Exercises 1c

1 A market gardener plants 20 potatoes and weighs the potatoes obtained from each plant. The results, in g, are as follows:

   853, 759, 891, 923, 755, 885, 821, 911,
   789, 854, 861, 915, 784, 853, 891, 942,
   758, 867, 896, 835

Construct a frequency table with class boundaries at 750, 800, ..., 950.
Show the results in a histogram.

2 The lengths of 20 cucumbers were measured with the following results, in cm:

   29.3, 30.5, 34.0, 31.7, 27.8, 29.4, 32.6,
   33.4, 29.8, 29.8, 35.4, 36.3, 26.4, 38.8,
   37.5, 34.5, 28.6, 31.9, 27.6, 32.0

Construct a frequency table with class boundaries at 25.0, 27.0, ..., and draw a histogram for the data.

3 A consumers' association tests the lives of car batteries of a particular brand, with the following results, in completed months:

   45, 49, 55, 61, 47, 55, 63, 68, 58,
   51, 40, 46, 50, 51, 57, 58, 49, 44,
   65, 62, 53, 58, 43, 37, 48

Represent the data by a histogram.

4 The mileages travelled by delegates at a conference were as follows:

   38, 47, 22, 15, 71, 54, 43, 22, 79, 65,
   43, 33, 23, 12, 58, 63, 52, 32, 43, 48,
   21, 25, 27, 48, 55, 10, 23, 37, 47, 51

Represent the data by a histogram.

5 The masses (in g to the nearest g) of a random collection of offcuts taken from the floor of a carpenter's shop are summarised below:

0–19	20–39	40–59	60–99
4	17	12	6

Display the data using a histogram.

6 The marks gained in an examination are summarised below.

0–29	30–49	50–69	70–99
4	12	37	14

Represent the data using a histogram.

7 In 1993 the age distribution of the population of the UK (in thousands) was:

Total	Under 1	1–4	5–14
58 191	759	3129	7417

15–24	25–34	35–44	45–59
7723	9295	7787	10 070

60–64	65–74	75–84	Over 85
2839	5169	3020	982

Source: *Population Trends, No 78, Winter 1994*

Choosing a sensible upper limit (which you should state) for the top age category, construct a histogram showing the above data.

8 In 1991 the distribution of the age of a mother at the live birth of a child in the UK was (in thousands):

All	Under 20	20–24	25–29
699.2	52.4	173.4	248.7

30–34	35–39	Over 40
161.3	53.6	9.8

Source: *Population Trends, No 78, Winter 1994*

Making suitable assumptions, which you should state, construct a histogram showing the above data.

9 Quetelet (see earlier biography) analysed the following data, which give the heights (x mm) of potential French conscripts. Those with heights less than 157 cm were excused from military service.

Height range	Frequency
$1435 \leqslant x < 1570$	28 620
$1570 \leqslant x < 1597$	11 580
$1597 \leqslant x < 1624$	13 990
$1624 \leqslant x < 1651$	14 410
$1651 \leqslant x < 1678$	11 410
$1678 \leqslant x < 1705$	8780
$1705 \leqslant x < 1732$	5530
$1732 \leqslant x < 1759$	3190
$1759 \leqslant x < 1840$	2490

Plot these data on a histogram.

10 A survey of cars in a car park reveals the following data on the ages of cars:

<2 yrs	2–4 yrs	5–8 yrs	9–12 yrs
35	51	83	35

Represent the data by a histogram.

11 The lengths (in minutes, to the nearest minute) of the phone calls made between two teenagers are summarised in the table below.

0–4	5–9	10–14	15–19	20–29
2	7	15	18	5

Illustrate these data using a histogram.

1.16 Frequency polygons

The idea of the histogram is to give a visual impression of which values are likely to occur and which values are less likely. The 'chunky' outline of a histogram is not 'a thing of beauty' and an alternative exists *whenever the classes are all of equal width.* The **frequency polygon** is constructed as follows. For each class, locate the point with x-coordinate equal to the mid-point of the class and with y-coordinate corresponding to the class frequency. Successive points are then joined to form the polygon. In order to obtain a closed figure, extra classes with zero frequencies are added at either end of the frequency distribution.

Notes
- ◆ As with the histogram it is *area* that is proportional to frequency.
- ◆ The area of a frequency polygon equals that of the corresponding histogram.
- ◆ Since the frequency polygon is only used with classes of equal width, class frequencies provide a convenient scale for the y-axis.

Example 13

Illustrate the data on the heights of Scottish militia men (Example 10) using a frequency polygon.

Height (ins)	64–65	66–67	68–69	70–71	72–73
Frequency	722	1815	1981	897	317

After adding extra classes, having the same widths but zero frequencies, the data are now summarised in the following table. The addition of the end classes enables us to complete the frequency polygon.

Class mid-point (ins)	62.5	64.5	66.5	68.5	70.5	72.5	74.5
Frequency	0	722	1815	1981	897	317	0

Frequency polygon showing the heights of 5732 Scottish militia men in 1817

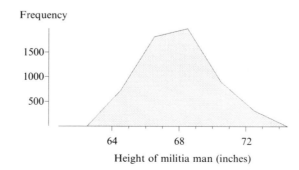

Calculator practice

> *Using the statistical draw mode, a graphical calculator produces a frequency polygon very easily – though the instructions may refer to 'a line graph'. Use your calculator to reproduce the previous polygon.*

1.17 Cumulative frequency diagrams

An alternative form of diagram provides answers to questions such as 'What proportion of the data have values less than x?'. In such a diagram, cumulative frequency on the y-axis is plotted against observed value on the x-axis. The result is a graph in which, as the x-coordinate increases, the y-coordinate cannot decrease.

With grouped data the first step is to produce a table of cumulative frequencies. These are then plotted against the corresponding upper class boundaries (u.c.b.). The successive points may be connected either by straight-line joins (in which case the diagram is called a **cumulative frequency polygon**) or by a curve (in which case the diagram is called an **ogive**).

▼   ▼

Example 14

In studying bird migration a standard technique is to put coloured rings around the legs of the young birds at their breeding colony. The source of a bird subsequently seen wearing coloured rings can therefore be deduced. The following data, which refer to recoveries of razorbills, consist of the distances (measured in hundreds of miles) between the recovery point and the breeding colony. Illustrate these data using a cumulative frequency polygon and estimate the distance exceeded by 50% of the birds.

Distance (miles) (x)	Frequency	Cumulative frequency
$x < 100$	2	2
$100 \leqslant x < 200$	2	4
$200 \leqslant x < 300$	4	8
$300 \leqslant x < 400$	3	11
$400 \leqslant x < 500$	5	16
$500 \leqslant x < 600$	7	23
$600 \leqslant x < 700$	5	28
$700 \leqslant x < 800$	2	30
$800 \leqslant x < 900$	2	32
$900 \leqslant x < 1000$	0	32
$1000 \leqslant x < 1500$	2	34
$1500 \leqslant x < 2000$	0	34
$2000 \leqslant x < 2500$	2	36

The cumulative frequency polygon shows that 50% of the razorbills had travelled more than 520 miles.

Cumulative frequency polygon of the distances travelled by razorbills between their breeding colony and their recovery point.

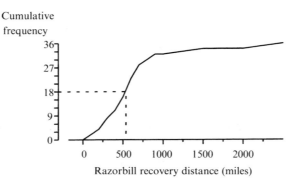

Razorbill recovery distance (miles)

Note

♦ If the recording inaccuracy (e.g. 'to the nearest mile') is small by comparison with the range of the data (2500 miles), there is no need to be over-particular about the end-points. The difference between a value plotted at $x = 99.5$ and $x = 100$ will not be visible!

Calculator practice

> *Write a routine for cumulating frequencies and use the line graph facility to draw a cumulative frequency diagram. Test it with the razorbill data.*

Step diagrams

A cumulative frequency diagram for ungrouped data is sometimes referred to as a **step polygon** or **step diagram** because of its appearance.

▼

Example 15

In a compilation of Sherlock Holmes stories, the 13 stories that comprise *The Return of Sherlock Holmes* have the following numbers of pages:

13.7, 15.5, 16.4, 12.8, 20.8, 13.7, 11.2, 13.7, 11.7, 15.0, 14.1, 14.8, 17.1

The lengths are given to the nearest tenth of a page.
Illustrate these data using a step diagram.

Treating the values as being exact, we use them as the boundaries in a cumulative frequency table. We first need to order the values:

11.2, 11.7, 12.8, 13.7, 13.7, 13.7, 14.1, 14.8, 15.0, 15.5, 16.4, 17.1, 20.8

The resulting table is therefore:

Story length, x	Cumulative frequency
$x < 11.2$	0
$11.2 \leqslant x < 11.7$	1
$11.7 \leqslant x < 12.8$	2
$12.8 \leqslant x < 13.7$	3
$13.7 \leqslant x < 14.1$	6
$14.1 \leqslant x < 14.8$	7
⋮	⋮
$20.8 \leqslant x$	13

Notice that the cumulative frequencies 'jump' at each of the observed values. It is this that gives rise to the vertical strokes in the diagram. The horizontal strokes represent the ranges given in the table.

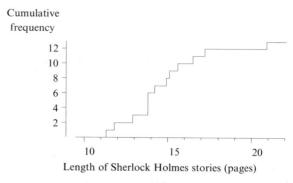

Step polygon of lengths of Sherlock Holmes stories

Exercises 1d

1 Students were asked to estimate the length (in mm) of a line. Their responses are summarised in the following table.

10–19	20–29	30–39	40–49	50–59	60–69
1	3	10	16	6	1

Represent the data using (i) a frequency polygon, (ii) a cumulative frequency polygon.

2 The lifetimes (in hours to the nearest hour) of bulbs in an advertising hoarding were recorded and are summarised in the following table.

650–699	700–749	750–799	800–849	850–899
1	7	18	9	2

Represent the data using (i) a frequency polygon, (ii) a cumulative frequency polygon.

3 The numbers of eggs laid each day by 8 hens over a period of 21 days were:

6, 7, 8, 6, 5, 8, 6, 8, 6, 5, 6,
4, 7, 6, 8, 7, 5, 7, 6, 7, 5

Display these results using a step diagram.

4 For each potato plant, a gardener counts the numbers of potatoes whose mass exceeds 100 g. The results are:

8, 5, 7, 10, 8, 6, 5, 6, 4, 8,
10, 9, 8, 7, 3, 10, 11, 6, 9, 8

Display these results using a step diagram.

5 The numbers of goals scored in the first three divisions of the Football League on 4 February 1995 were:

6, 5, 3, 3, 5, 3, 1, 2, 4, 2, 2, 5, 1, 2, 2, 3, 8, 2,
4, 5, 3, 3, 0, 2, 5, 0, 1, 0, 3, 0, 1, 2, 7, 1, 2

Display these results using a step diagram.

6 A survey of cars in a car park reveals the following data on the ages of cars:

< 2 yrs	2–4 yrs	5–8 yrs	9–12 yrs
35	51	83	35

Draw a cumulative frequency polygon.

7 In 1993 the age-distribution of the population of the UK (in thousands) was:

Total	Under 1	1–4	5–14
58 191	759	3129	7417

15–24	25–34	35–44	45–59
7723	9295	7787	10 070

60–64	65–74	75–84	Over 85
2839	5169	3020	982

Source: *Population Trends, No 78, Winter 1994*
Draw a cumulative frequency polygon.

8 In 1992 the distribution of the age of a mother at the live birth of a child in the UK was (in thousands):

All	Under 20	20–24	25–29
699.2	52.4	173.4	248.7

30–34	35–39	Over 40
161.3	53.6	9.8

Source: *Population Trends, No 78, Winter 1994*
Display the data using a cumulative frequency polygon.

Florence Nightingale (1820–1910) is best known for her work as a nurse during the Crimean War, where the soldiers called her 'The Lady with the Lamp'. She was a most efficient hospital administrator and compiled quantities of statistics in her drive for hospital reform: she standardised the reporting of deaths using 'Miss Nightingale's scheme for Uniform Hospital Statistics'. She was a great admirer of Quetelet's work and wrote an essay about it following his death. She described Statistics as 'the most important science in the whole world'. Naturally we wouldn't disagree!

1.18 Cyclic and circular data

Deaths (per thousand soldiers) from cholera etc, and from war wounds, in the hospital at Scutari between September 1854 and July 1855

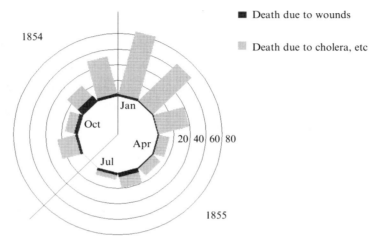

■ Death due to wounds

▨ Death due to cholera, etc

Much of the data collected by Florence Nightingale was concerned with the identification of causes of death and in tracing seasonal fluctuations in illnesses. She invented a circular diagram to portray the seasonal variation in deaths at the army hospital at Scutari, where she was working during the winter of 1854–55 in the middle of the Crimean War.

Wars are notoriously dangerous, yet, in January 1855, only 83 of the 3168 deaths in the hospital were due to wounds and injuries, the rest being largely the result of cholera and related diseases ('mitigable and preventible pestilences' in the words of Miss Nightingale). Florence Nightingale introduced improvements in the sanitary arrangements which took effect during March 1855 and were largely responsible for the subsequent abrupt reduction in deaths. Miss Nightingale returned from the war as a heroine.

Similar diagrams are useful with directional data. The figure shows the results of an experiment reported in the journal *Animal Behaviour*. Colorado potato beetles were collected and then let loose, one at a time, in the middle of an unfamiliar environment – a wheat field! The beetles showed a clear preference for a walk towards the north-west!

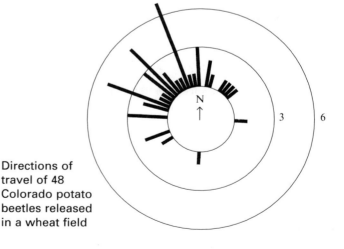

Directions of travel of 48 Colorado potato beetles released in a wheat field

1.19   Time series

Time-series graphs are probably the type of diagram most frequently encountered in newspapers. They are also possibly the most straightforward: time is plotted on the *x*-axis and the quantity of interest is plotted on the *y*-axis.

Cumulative sales of satellite dishes in the UK

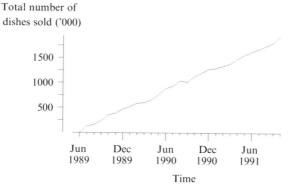

The figure shows the growth in total sales of satellite dishes in the United Kingdom between July 1989 and October 1991. The source is a report published in *The Times* in November 1991. The data probably consist of estimates rather than direct counts, since otherwise one would have to conclude that around 20 000 satellite dishes were returned to the shops in September 1990!

The straight-line joins between the successive values are useful here since they enable us to estimate the total sales at intermediate points in time. The implication of the (almost) relentless upward progress of the graph is that monthly sales of satellite dishes remained steady at an average of around 70 000 a month.

We will consider time series in more detail in Chapter 3.

Note

◆ Beware advertisements showing time series that rise rapidly! There are two possible explanations:
  1 The series was probably falling fast in the previous time period!
  2 The vertical scale may be exaggerated – check where 0 would occur.

1.20   Train timetables

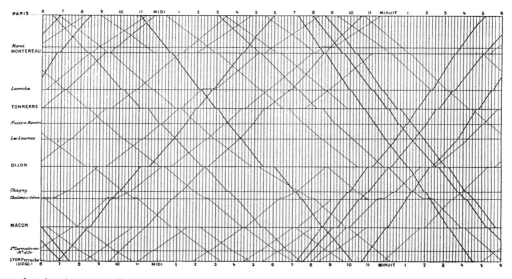

The passage of trains between Paris and Lyon in the 1880s

Train timetables are, of course, full of numbers (not always to be believed!). The figure is a reproduction of a classic graph dating from 1885 which illustrates the movements of trains between Paris and Lyon. The *y*-axis corresponds to the stations (with Paris at the top) and the *x*-axis to time. The positions of the stations on the *y*-axis reflect the inter-station distances: the whole graph being an example of a distance–time graph of the type that may be encountered in Mechanics.

Lines from top left to bottom right show the passage of trains in one direction, and lines from bottom left to top right represent trains going in the opposite direction. The diagonal lines appear to be curiously broken: this is because they include horizontal components representing periods where the trains are stopped at stations.

1.21 Moscow or bust!

Napoleon's march on Moscow
in the winter of 1812–13

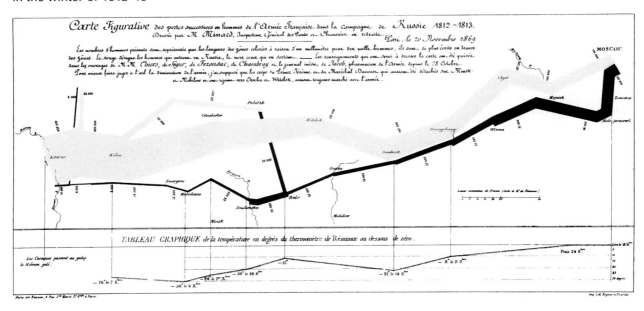

Another classic graph, produced in 1861 by Charles Minard, shows the progress of Napoleon's army in its advance on, and subsequent retreat from, Moscow. The graph shows a large army starting at the left of the picture – the size being represented by the width of the 'river' of troops. As the army progressed towards Moscow (on the right) some parts were detached on flanking missions (seen as tributaries of the main stream), but these detachments are not the main cause of the decline in the size of the army. That main cause is chronicled in the sub-graph at the bottom which shows the extraordinarily low temperatures that the army suffered. Eventually the army was forced to turn back. Of the 422 000 that initially crossed the Niemen river only 10 000 lived to tell the tale. The graph shows time, geography, temperature and army size in a single picture.

1.22 Scatter diagrams

The time-series graph of Section 1.19 was an example of using ordinary Cartesian coordinates to examine the relationship between two variables. Relationships between variables are particularly interesting since the variation in the values of one variable (x) may to some extent explain the variation in the other (y).

Time series data are ordered and their order is indicated on the plot by joining successive values. By contrast, in a scatter diagram, there is no order and the pairs of values are indicated by points (or crosses, or some other symbol).

Note
◆ We return to scatter diagrams in Chapter 20 – just 19 chapters to go!

Example 16

The following data relate soil erosion (in kg/day) to daily average wind velocity (in $\mathrm{km\,h^{-1}}$) in a region in the sandy plains of Rajasthan in India. Plot these data in an appropriate scatter diagram.

Wind velocity	13.5	13.5	14	15	17.5
Soil erosion	0	10	31	20	20

Wind velocity	19	20	21	22	23
Soil erosion	66	76	137	71	122

Wind velocity	25	25	26	27
Soil erosion	188	300	239	315

We begin with a straightforward diagram in which the x-coordinate indicates the daily average wind velocity and the y-coordinate shows the resulting estimated soil erosion.

Scatter diagram showing the relation between wind velocity and soil erosion in a sandy Indian plain

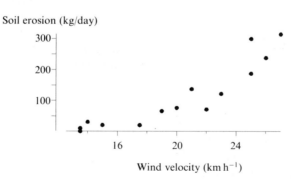

The statistician often needs to try many diagrams before finding the one that makes the most useful display of the data. Our first effort suggests that soil erosion increases dramatically as the wind velocity picks up, and that the relation between the two variables is not linear. One possibility is that the relationship is exponential: this can be examined by plotting the (natural) logarithm of the soil erosion against wind speed.

Revised scatter diagram
showing the relation
between wind velocity
and soil erosion in a
sandy Indian plain
using a logarithmic
scale for the *y*-axis

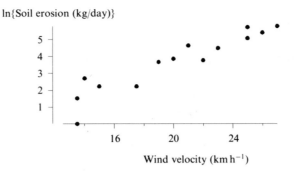

The revised diagram does appear to show a linear relation between
ln (soil erosion) and wind velocity.

▲ ─── ▲

Calculator practice ───────────────────────────────

*Graphical calculators are particularly effective for drawing scatter diagrams
of pairs of values. Reproduce the original scatter diagram from Example 16
and then investigate other ways of transforming the data so as to get scatter
diagrams that appear to be more linear in form.*

Computer project ───────────────────────────────

*Investigate how to produce scatter diagrams using a spreadsheet. As with
graphical calculators it is easy to experiment with transformations of one
or both of x and y with the aim of producing an approximate line on the
diagram.*

Exercises 1e ───────────────────────────────────────

1   The total amount of snow cover over Europe
and Asia during October for the years 1970 to
1979 is given (in millions of square kilometres)
in the table below.

1970	1971	1972	1973	1974
6.5	12.0	14.9	10.0	10.7

1975	1976	1977	1978	1979
7.9	21.9	12.5	14.5	9.2

Display this information on a suitable diagram.

2   The acidity of milk (the pH value, *y*) depends
upon the temperature at which it is stored ($x\,°C$).
Some experimental results are shown below.

x	4	24	38	40
y	6.85	6.63	6.62	6.57

x	40	60	70	78
y	6.52	6.38	6.32	6.34

Display this information on a suitable diagram.

3   The average numbers of deaths per 1000
population in Norway during the period 1750–
1850 are summarised below.

1750	1770	1790	1810	1830	1850
25.5	23.6	22.9	26.8	19.7	17.2

Display this information on a suitable
diagram.

4   A doctor records the number of patients that
he sees in the second week of four particular
months in the year. The numbers for 1994 and
1995 are as follows:

Month	Jan	Apr	Jul	Oct
No. seen ('94)	255	235	176	219
No. seen ('95)	215	207	139	243

Represent the data using a suitable
diagram.

5 The total value of goods (in thousands of £) produced by a manufacturer each quarter, together with the value of goods exported (in thousands of £) are given below.

1993	1st qtr	2nd qtr	3rd qtr	4th qtr
Production	238	316	297	286
Exports	57	89	94	82

1994	1st qtr	2nd qtr	3rd qtr	4th qtr
Production	211	297	241	270
Exports	63	82	108	103

1995	1st qtr	2nd qtr	3rd qtr	4th qtr
Production	224	289	285	228
Exports	76	97	114	91

Represent the data using a suitable diagram.

1.23 Contingency tables

When opinion poll organisations conduct their polls, they are not content to ask their interviewees a single question! Often by asking lots of questions they can obtain a greater understanding of the phenomenon that they are investigating. As a simple example, suppose a random sample of twenty individuals are asked whether they prefer coffee (C) or tea (T) and whether they prefer wine (W) or beer (B). Their answers are listed below.

(C,B) (T,W) (C,W) (C,W) (T,B) (C,B) (T,B) (C,W) (T,B) (C,B)
(C,W) (T,B) (T,B) (T,B) (C,B) (T,B) (C,W) (C,B) (T,W) (T,B)

The information is difficult to comprehend when presented as a list, but when cross-classified using a contingency table (which shows the frequencies with which each combination occurs), it is much easier to assimilate.

	Prefers coffee	Prefers tea
Prefers wine	5	2
Prefers beer	5	8

On the evidence of this very limited survey it appears that tea drinkers have a preference for beer rather than wine.

Note
- We return to contingency tables in Section 18.5 (p. 496).

1.24 Cartograms

Many data sets have a geographical component; for example, the birth rates in the countries of Africa. It would be nice to show such data on some form of map so that the numbers can be properly related to one another. Larger atlases often show 'misshapen' maps in an effort to illustrate this sort of data.

Rectangular cartogram of British parliamentary constituencies showing variation in turnout in the 1992 general election

Rectangle areas are proportional to ln(ln(Physical area))

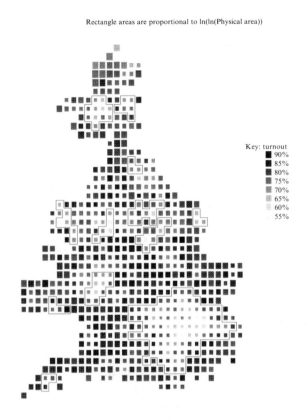

Key: turnout
■ 90%
■ 85%
■ 80%
■ 75%
▨ 70%
▨ 65%
▨ 60%
  55%

We show here an example of a 'rectangular **cartogram**' in which each rectangle represents a British parliamentary constituency. The size of each rectangle is an indicator of the area of the constituency, while the depth of colour is an indication of the turnout (the percentage of those able to vote who actually voted in the 1992 election). The major cities are indicated, with the importance of London (86 constituencies) being very evident. It is clear that turnout is lowest in inner-city constituencies.

1.25 Chernoff faces

Scatter diagrams show the relation between two variables; Chernoff faces attempt to show the relation between rather more! In the version illustrated here, apart from geographical variation (each face represents a British parliamentary constituency) we are presented with information about house prices (fat faces = high prices!), the extent of unemployment (sad faces = high unemployment), the extent of voting turnout (high turnout = big nose) and the percentage of those in work that were employed in service industries (larger percentages = bigger eyes). The information relates to 1990. The most noticeable feature is the high house prices in the South East.

Chernoff faces showing
various features of British
parliamentary
constituencies

The Changing Distribution of Voting, Housing, Employment and Industrial Compositions in Constituencies Between 1983 and 1987

Relative Social Indicators

Decreasing   **Increasing**

% Services Employees
% Electorate Voting
% Adult Employment
£ Mean Housing Price

Facial features are in proportion to the changes in the social and economic characteristics of the areas they represent.

Chapter summary

♦ The principal purpose of Statistics is the drawing of conclusions about large populations (human or otherwise) from comparatively small amounts of data.

♦ Diagrams are an effective way of conveying information.

♦ If only a small number of discrete values are possible, then the best approach is often to use a **tally chart**, followed by a summary in a **frequency table** and representation using a **bar chart**.

♦ If a large number of discrete values are possible, then the best approach is often to use a **stem-and-leaf diagram**, followed by a summary in a **grouped frequency table** and representation using a **histogram.**

♦ **Pie charts** and **compound bar charts** are useful when the features of interest are the relative sizes of the frequencies in alternative categories.

♦ When data are collected on two variables simultaneously, representation using either a **scatter diagram** or a **time-series** graph may be appropriate.

♦ There are many ways of portraying data. Whatever method is used, try to make it self-explanatory for the reader (and, if possible, interesting!).

Exercises 1f (Miscellaneous)

1 The total scores given in *The Times* for Welsh and Scottish Rugby matches on 4 February 1995 were:

> 44, 21, 23, 26, 24, 39, 56, 22, 28, 25, 63,
> 83, 42, 39, 24, 23, 38, 61, 44, 19, 31, 24,
> 24, 60, 45, 48, 39, 34, 50, 46, 53, 43, 43

Construct a stem-and-leaf diagram to display the above data.

2 The daughter of a market gardener plants twenty sunflower plants in her garden. When they are full grown, she measures them and records their heights in metres as follows:

> 1.60, 1.72, 2.23, 2.12, 1.70, 1.93, 1.69,
> 2.11, 1.99, 2.08, 2.11, 1.79, 2.01, 1.88,
> 1.93, 2.22, 1.92, 2.44, 1.87, 1.76

Summarise these data using a stem-and-leaf (!) diagram.

3 In 1665, a total of 97 308 people died in London (compared to just 9967 births). The principal cause of death was the plague, which accounted for 68 596 of the deaths. Show this information on a pie chart.

Of the deaths not due to the plague, the principal causes (according to the *Annual Bill of Mortality for London*, and using its spelling) were these:

Aged	1545
Ague and Feaver	5257
Chrisomes and Infants	1258
Consumption and Tissick	4808
Convulsion and Mother	2036
Dropsie and Timpany	1478
Griping in the Guts	1288
Spotted feaver and Purples	1929
Surfet	1251
Teeth and Worms	2614

Illustrate this information on a bar chart.

4 The infant mortality rate (IMR) in 1960 for various Latin American countries and the corresponding *per capita* mean kilocalorie intake for adults during 1959–61 are shown in the following table.

	IMR	Kcal
Mexico	67.7	2.58
Guatemala	92.8	1.97
Panama	42.9	2.37
Colombia	88.2	2.28
Peru	94.8	2.06
Argentina	60.7	3.22
Uruguay	47.4	3.03

Display this information on a suitable diagram.

5 One set of data that Quetelet (see earlier biography) analysed was concerned with the conviction rates of the French courts of assize during the period 1825–30.
The following table gives the total numbers accused and convicted during each year.

	Accused	Convicted
1825	7234	4594
1826	6988	4348
1827	6929	4236
1828	7396	4551
1829	7373	4475
1830	6962	4130

Plot these two time series on a single graph.

Calculate the conviction rate (number convicted as a proportion of the number accused) for each year.
Plot your results as a time series.
State your conclusions.

6 During a particular month a family spends £52.27 on meat, £23.10 on fruit and vegetables, £19.72 on drink, £12.41 on toiletries, £102.68 on groceries and £9.82 on miscellaneous items. These data are to be represented by a pie chart of radius 5 cm.
(*a*) Calculate, to the nearest degree, the angle corresponding to each of the above classifications. (DO NOT DRAW THE PIE CHART.)
The following month the family spends 20% more in total.
(*b*) Find the radius of a comparable pie chart to represent the data on this occasion. [ULEAC]

7 Telephone calls arriving at a switchboard are answered by the telephonist. The following table shows the time, to the nearest second, recorded as being taken by the telephonist to answer the calls received during one day.

Time to answer (to nearest second)	Number of calls
10–19	20
20–24	20
25–29	15
30	14
31–34	16
35–39	10
40–59	10

Represent these data by a histogram.

Give a reason to justify the use of a histogram to represent these data.

[ULEAC]

8 The following table shows the time to the nearest minute, spent reading during a particular day by a group of school children

Time	Number of children
10–19	8
20–24	15
25–29	25
30–39	18
40–49	12
50–64	7
65–89	5

(*a*) Represent these data by a histogram.
(*b*) Comment on the shape of the distribution.

[ULEAC]

2 General summary statistics

Lies, damned lies and statistics

Benjamin Disraeli

2.1 The purpose of summary statistics

Simple! The purpose of summary statistics is to replace a huge indigestible mass of numbers (the **data**) by just one or two numbers that, together, convey most of the essential information.

Well, perhaps it is not so simple, since this is a pretty stiff challenge! No single summary statistic can tell us all about a set of data. Different statistics emphasise different aspects of the data and it will not always be evident which aspect is more important. An example of the difficulties is provided by an interchange many years ago, in the Houses of Parliament. The MPs were debating the need for road signs in Wales to give directions in both Welsh and English. The discussion went something like this:

MP A: Since less than 10% of the population of Wales speak Welsh it is unnecessary to include directions in Welsh.

This seems like a pretty convincing statistic! But wait:

MP B: Over 90% of the area of Wales is inhabited by a population whose principal language is Welsh – directions in Welsh are essential.

It is easy to see why Disraeli was rather hard on Statistics! Both the above statements were essentially correct at the time (though the percentages are invented by the present authors): but they led to opposite inferences. Clearly we have to be careful to choose our summary statistics to be appropriate.

For **univariate** data (i.e. data concerned with a single quantity) there are two main types of summary statistic: **measures of location** and **measures of spread**. Measures of location answer the question 'What sort of size values are we talking about?'. Measures of spread answer the question 'How much do the values vary?'. Both are discussed in this chapter.

The main purpose of Statistics is to draw conclusions about a (usually large) **population** from a (usually small) **sample** of **observed values**: the **observations**. In this chapter we study various ways of providing numerical summaries of the observations.

2.2 The mode

The **mode** of a set of discrete data is the single value that occurs most frequently. This is the simplest of the measures of location, but is of limited use. If there are two such outcomes that occur with equal frequency then there is no unique mode and the data are described as being **bimodal**; if there are three or more such outcomes then the data are called **multimodal**. The associated adjective is 'modal', so we are sometimes asked to find the **modal value**.

Example 1

At the supermarket I buy 8 tins of soup. According to the information on the tins, four have mass 400 g, three have mass 425 g and one has mass 435 g. Find the mode.

_____

The mode is 400 g because 400 g is the most common value.

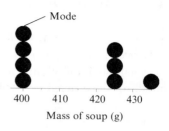

Example 2

After unpacking the shopping, I feel hungry and have soup for lunch. I choose one of the 400 g cans bought in the previous example. What is an appropriate description of the frequency distribution of the remaining 7 masses?

_____

There are now two modes, one at 400 g and one at 425 g: an appropriate description is 'bimodal'.

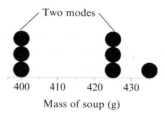

Calculator practice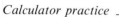

> *When used to plot a bar chart or frequency polygon, graphical calculators may also indicate the value of the mode and the associated frequency. Use a calculator to determine the mode of the soup tin data. What happens if the 435 g tin is replaced by a 425 g tin?*

Modal class

For continuous data (or for grouped discrete data) the mode exists only as an idea! When measured with sufficient accuracy, all observations on a continuous variable will be different: even if John and Jim both *claim* to be 1.8 metres tall, we can be certain that their heights will not be *exactly* the same. However, if we plot a histogram of a sample of men's heights, we will usually find that it has a peak in the middle: the class corresponding to this peak is called the **modal class**.

For **qualitative data** (in which items are described by the qualities they possess) we again refer not to a mode, but to a modal class. In the next example 'hair colour' is a qualitative variable.

Example 3

The hair colours and heights of a class of male university students are summarised below.
Determine the modal classes.

Hair colours			
Black	Brown	Blond	Red
63	53	9	3

Heights (m), x			
$1.4 \leqslant x < 1.7$	$1.7 \leqslant x < 1.9$	$1.9 \leqslant x < 2.0$	$2.0 \leqslant x < 2.2$
19	56	42	11

_____

For hair-colour the modal class is 'Black'. For height the modal class is '1.9–2.0 m' and not '1.7–1.9 m'. To see why, imagine drawing the histogram: the heights of the middle two rectangles would be in the proportion 28 to 42 (because the width of the second class is twice the width of the third class).

2.3 The median

The word 'median' is just a fancy name for 'middle'! After all the observations have been collected, they can be arranged in a row in order of magnitude, with the smallest on the left and the largest on the right (or vice versa). Values in the middle of the ordered row will therefore be intermediate in size and should give a good idea of the general size of the data. For example, suppose the observed values are 13, 34, 19, 22 and 16. Arranged in order of magnitude these become:

> 13, 16, **19**, 22, 34

The middle value, 19, is called the **median**.

When there are an even number of observations there are two middle values. By convention the median is then taken as their average. For the soup tins in Example 1, the values were:

> 400, 400, 400, **400**, **425**, 425, 425, 435

The value of the median is taken to be

$$\tfrac{1}{2}(400 + 425) = 412.5$$

In general, with n observed values arranged in order of size, the median is calculated as follows:

If n is odd and equal to $(2k + 1)$, say, then the median is the $(k + 1)$th ordered value.

If n is even and equal to $2k$, say, then the median is the average of the kth and the $(k + 1)$th ordered values.

Note

◆ A useful preliminary is to summarise the data using a stem-and-leaf diagram, since this immediately provides an ordering of the data.

Example 4

A chemistry professor has an accurate weighing machine and two sons, Charles and James, who are keen on playing conkers. One day, Charles and James collect some new conkers. On their return home, following a dispute over who has the best conkers, they use their father's balance to determine the weights of the conkers (in g). Their results are as follows:

> Charles: 31.4, 44.4, 39.5, 58.7, 63.6, 51.5, 60.0
> James: 60.1, 34.7, 42.8, 38.6, 51.6, 55.1, 47.0, 59.2

Which boy's collection of conkers has the higher median weight?

We first arrange each set of values in ascending order, and then highlight the central value(s):

> Charles: 31.4, 39.5, 44.4, **51.5**, 58.7, 60.0, 63.6
> James: 34.7, 38.6, 42.8, **47.0**, **51.6**, 55.1, 59.2, 60.1

The median for James is the average of 47.0 and 51.6, which is 49.3. This is less than the median for Charles, which is 51.5, so Charles's collection has the higher median weight.

Calculator practice _____

> *Some calculators will report the median value. You may need to delve deeply into the calculator manual in order to find the correct sequence of keystrokes. You should check that the calculator reports the correct value for the median both in the case of an even number of data items and in the case of an odd number. Use the data above as a check. (Remember that the calculator depends on its built-in instructions – these are not always correct!)*

2.4 The mean

This measure of location is often called the **average**, and can be used with both discrete and continuous data. The mean is equal to the sum of all the observed values divided by the total number of observations. Unlike the value of the mode, the value of the mean will usually not be equal to any one of the individual observed values. Thus the mean mass (in g) of the tins of soup is:

$$\frac{(400 + 400 + 400 + 400 + 425 + 425 + 425 + 435)}{8} = 413.75$$

It is time to introduce some algebra! Suppose that the data set consists of n observed values, denoted by $x_1, x_2, \ldots, x_n$. Then the sample mean, which is usually denoted by $\bar{x}$, is given by

$$\bar{x} = \frac{(x_1 + x_2 + \cdots + x_n)}{n} \tag{2.1}$$

One way of thinking about the mean is as the **centre of mass** when the observations are 'balanced' on the x-axis.

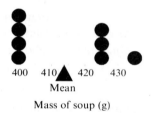

The sample mean viewed as a centre of mass

Calculator practice _____

> *Many calculators are capable of calculating the mean of a set of data using an appropriate sequence of key strokes.*
> *Determine how to do this using your own calculator.*

Exercises 2a _____

1 A school records the numbers of candidates achieving the various possible grades in their A-level subjects.

    E: 21; D: 47; C: 69; B: 72; A: 53

Find the modal class.

2 A survey of cars in a car park reveals the following data on the ages of cars:

< 2yrs	2–4 yrs	5–8 yrs	9–12 yrs
35	51	83	35

Determine the modal class.

3 In 1993 the age distribution of the population of the UK (in thousands) was:

Total	Under 1	1–4	5–14
58 191	759	3129	7417

15–24	25–34	35–44	45–59
7723	9295	7787	10 070

60–64	65–74	75–84	Over 85
2839	5169	3020	982

Source: *Population Trends, No 78, Winter 1994*

Determine the modal class.

4 Most people have more than the average number of legs!
Explain.

5 Eight athletes run 100 m. The times taken (in s) are:

   10.34, 10.68, 10.81, 11.02,
   11.35, 11.71, 11.82, 11.95

Find the average time taken.

6 A bridge player keeps a note of the numbers of aces that she receives in successive deals. The numbers are:

   0, 2, 3, 0, 0, 2, 1, 1, 0,
   2, 3, 0, 1, 1, 2, 1, 0, 0

Find (i) the mode, (ii) the mean, of the numbers of aces received.

7 A gardener classifies a potato having a mass over 100 g as being 'large'. The gardener grows a number of potato plants and, for each plant, he counts the number of large potatoes, obtaining the following results:

   8, 5, 7, 10, 8, 6, 5, 6, 4, 8,
   10, 9, 8, 7, 3, 10, 11, 6, 9, 8

Find (i) the mode, (ii) the mean, of the numbers of large potatoes.

8 The heights, in m, of 12 walnut seedlings, after twenty years' growth, were:

   4.3, 5.2, 4.1, 3.5, 5.2, 4.8,
   5.3, 4.8, 3.7, 4.1, 4.5, 5.0

Find the mean height.

9 A computer is programmed to generate 8 random numbers between −1 and +1.
The numbers generated are:

   0.269, −0.679, 0.507, −0.663,
   0.325, −0.960, 0.741, 0.484

Find the mean.

10 A student's bank balance at the end of each month was recorded in £. A negative quantity denotes an overdraft.
The figures were as follows:

   341.32,   97.53,   −57.44,   255.93,
     5.89,   −83.33,   152.81,   −23.11
  −105.73, −204.50, −150.46,   −85.39

Find her mean bank balance at the end of each month.

11 The weights, in kg, of the Cambridge Boat Race crew in 1995 were:

   90.7, 89.4, 93.4, 92.1, 82.6, 92.5, 94.4, 89.8

The weights of the Oxford crew were:

   86.9, 90.3, 94.8, 97.5, 89.6, 89.8, 91.9, 89.1

Find the mean weight of each crew and verify that the Oxford crew is heavier than the Cambridge crew by an average of 0.63 kg per man.

12 The mean of the following numbers is 20:

   20, 18, c, 24, 23, 13

Find the value of c.

13 The numbers of goals scored in the first three divisions of the Football League Championship on 4 February 1995 were:

   6, 5, 3, 3, 5, 3, 1, 2, 4, 2, 2, 5, 1, 2, 2, 3, 8, 2,
   4, 5, 3, 3, 0, 2, 5, 0, 1, 0, 3, 0, 1, 2, 7, 1, 2

Find (i) the mode, (ii) the mean, of the number of goals scored.

14 The heights of the Cambridge Boat Race crew in the 1995 race were:

   6 ft 3 in, 6 ft 5 in, 6 ft 3 in, 6 ft 4 in,
   6 ft 2 in, 6 ft 6 in, 6 ft 4 in, 6 ft 2 in

The heights of the Oxford crew were:

   6 ft 3 in, 6 ft 1 in, 6 ft 5 in, 6 ft 4 in,
   6 ft 5 in, 6 ft 3 in, 6 ft 3 in, 6 ft 2 in

Find the difference in their median heights.

15 The numbers of matches in a box were counted for a sample of 25 boxes. The results were:

   51, 52, 48, 53, 47, 48, 50, 51, 50,
   46, 52, 53, 51, 48, 49, 52, 50, 48,
   47, 53, 54, 51, 49, 47, 51

By constructing a tally chart, or otherwise, find the median number of matches in a box.

16 A record is kept of the number of patients attending each day at a medical practice. The numbers are:

   45, 41, 37, 48, 44, 29, 32, 43, 41, 37, 38,
   31, 43, 39, 35, 31, 42, 40, 35, 42, 35

Construct a stem-and-leaf diagram and hence find the median number of patients attending per day.

17 A baker keeps count of the number of doughnuts sold each day for three weeks. The numbers are:

> 35, 47, 34, 46, 55, 82, 41, 35, 47,
> 51, 56, 75, 38, 41, 44, 51, 45, 74

By constructing a steam-and-leaf diagram, or otherwise, find the median number of doughnuts sold per day.

18 The marks obtained in a mathematics test marked out of 50 were:

> 35, 42, 31, 27, 48, 50, 24, 27,
> 21, 37, 41, 34, 12, 18, 27

Find:

(i) the mean mark,

(ii) the median mark.

19 A choirmaster keeps a record of the numbers turning up for choir practice:

> 25, 28, 32, 31, 31, 34, 28, 31, 29,
> 28, 32, 32, 30, 29, 29, 31, 28, 28

(i) Find the mean number attending.

(ii) Determine the median number.

20 The shoe sizes of the members of a football team are:

> 10, 10, 8, 11, 10, 9, 9, 10, 11, 9, 10

Find:

(i) the mean shoe size,

(ii) the median shoe size,

(iii) the modal shoe size.

2.5 Advantages and disadvantages of the mode, mean and median

Advantages

- If a mode exists it is certain to have a value that was actually observed.
- The median can be calculated in some cases where the mean or mode cannot. For example, suppose 99 homing pigeons fly from A to B. The median time of flight can be calculated as soon as the 50th pigeon has arrived – we don't need to wait for the last exhausted traveller (who may never arrive!).

Disadvantages

- The mode may not be unique (because two or more values may be equally frequent).
- The mean may be significantly affected by the inclusion of a mistaken observation (e.g. a tin of soup misreported as having mass 4000 g) or of an unusual observation (e.g. the salary of the boss of a factory included with those of the factory workers).
- The statistical properties of the mode and the median are difficult to determine.

In practice much more use is made of the mean than of either of the other two measures of location.

Practical _____

> *How many four-legged pets does the typical family have?*
> *Use a tally chart to record the combined number of dogs, cats, hamsters, etc, for each member of your class.*
> *Determine the mean, median, and mode of these values. Which was easiest to calculate?*
> *An organisation wishes to estimate the total number of four-legged pets in your area.*
> *Which of your three statistics is likely to be most useful to them?*

2.6 Sigma (Σ) notation

Expressions such as $(x_1 + x_2 + \cdots + x_n)$ are tedious to write. We want to write 'Sum of the x-values', but this is not very mathematical – it doesn't contain Greek letters! Instead, therefore, we write

$$\sum_{i=1}^{n} x_i = x_1 + x_2 + \cdots + x_n \tag{2.2}$$

The Σ sign is the Greek equivalent of S and is pronounced 'sigma'. Confusingly, we shall also meet shortly the Greek equivalent of s which is also pronounced 'sigma', but looks quite different (σ) and has a very different statistical interpretation.

Notes
- In the shorthand formula the letter i is simply an index. Any letter could be used, but it must replace i everywhere it appears. For example:

$$\sum_{j=1}^{4} y_j = \sum_{i=1}^{4} y_i = \sum_{r=1}^{4} y_r = y_1 + y_2 + y_3 + y_4$$

- Changing the value of n results in a change in the terms being summed. For example:

$$\sum_{j=1}^{3} y_j = y_1 + y_2 + y_3 \quad \text{but} \quad \sum_{j=1}^{2} y_j = y_1 + y_2$$

Applications of sigma notation

Here are some further examples of the use of the Σ sign:

$$\sum_{r=1}^{3} r = 1 + 2 + 3 = 6$$

$$\sum_{s=2}^{4} s^2 = 2^2 + 3^2 + 4^2 = 29$$

$$\sum_{j=1}^{2} (2j + 5) = \{(2 \times 1) + 5\} + \{(2 \times 2) + 5\} = 16$$

$$\sum_{k=2}^{3} (k^2 + 6k) = \{2^2 + (6 \times 2)\} + \{3^2 + (6 \times 3)\} = 43$$

There are four particularly useful results that involve manipulation of the Σ sign:

$$\sum_{i=1}^{n} (x_i + y_i) = \sum_{i=1}^{n} x_i + \sum_{i=1}^{n} y_i \tag{2.3}$$

$$\sum_{i=1}^{n} cx_i = c \sum_{i=1}^{n} x_i \tag{2.4}$$

$$\sum_{i=1}^{n} c = nc \tag{2.5}$$

$$\sum_{i=1}^{n} x_i = \sum_{i=1}^{m} x_i + \sum_{i=m+1}^{n} x_i \tag{2.6}$$

In the above c is a constant and m is an integer such that $1 \leqslant m < n$. Result (2.5) in particular should be noted. It follows immediately from result (2.4) by putting all the x-values equal to 1. All four results are easily proved by writing out the various summations in full.

Notes

♦ Often the limits of the summation are obvious, in which case they may be dropped from the formula. For example, for the mean of n observations $x_1, x_2, \ldots, x_n$ we could write

$$\bar{x} = \frac{\Sigma x_i}{n}$$

♦ In ordinary text we write $\sum_{i=1}^{n} x_i$ instead of

$$\sum_{i=1}^{n} x_i$$

♦ As a shorthand, when the formulae are thick on the ground, the suffix may also be omitted:

$$\bar{x} = \frac{\Sigma x}{n} \tag{2.7}$$

Exercises 2b

1 It is given that $x_1 = 2$, $x_2 = 3$, $x_3 = 5$, $x_4 = 1$, $x_5 = 3$, $x_6 = 2$, $x_7 = 0$, $x_8 = 2$
   Verify that:

   (i) $\displaystyle\sum_{i=1}^{8}(x_i + 2) = \sum_{i=1}^{8} x_i + 16$

   (ii) $\displaystyle\sum_{i=1}^{8} x_i = \sum_{i=1}^{4} x_i + \sum_{i=5}^{8} x_i$

   (iii) $\displaystyle\sum_{i=1}^{8}(3x_i) = 3\sum_{i=1}^{8} x_i$

   (iv) $\displaystyle\sum_{j=1}^{8} x_j = \sum_{i=1}^{4} x_i + \sum_{k=5}^{8} x_k$

2 It is given that $x_1 = 2$, $x_2 = -3$, $x_3 = 0$, $x_4 = -1$, $y_1 = 3$, $y_2 = -2$, $y_3 = 10$, $y_4 = 2$
   Verify that:

   (i) $\displaystyle\sum_{i=2}^{4}(x_i + y_i) = \sum_{i=2}^{4} x_i + \sum_{i=2}^{4} y_i$

   (ii) $\displaystyle\sum_{i=1}^{4}(x_i y_i) = 10$

   (iii) $\displaystyle\left(\sum_{i=1}^{4} x_i\right)\left(\sum_{i=1}^{4} y_i\right) = -26$

3 Find:

   (i) $\displaystyle\sum_{j=1}^{8} j$

   (ii) $\displaystyle\sum_{j=1}^{8} j^2$

   (iii) $\displaystyle\sum_{j=1}^{8} (j-2)^2$

4 A set of data for 10 observations has $\Sigma x = 365$. Find the mean.

5 The summarised data for a set of observations is $n = 60$, $\Sigma y = 74\,344$.
   Find the mean value of y.

6 The results of 30 experiments to find the value of the acceleration due to gravity are summarised by $\Sigma g = 294.34$.
   Find the mean value.

7 Eight numbers have a mean of 16.
   Given that the first seven numbers have a total of 130, determine the value of the eighth number.

8 A set of 25 observations was found to have a mean of 15.2. It was subsequently found that one item of data had been wrongly recorded as 23 instead of 28.
   Find the revised value of the mean.

2.7 The mean of a frequency distribution

We have seen in Chapter 1 that data are often represented by a frequency distribution. For example, for the soup tins (see Section 2.4) we have:

Reported mass (g) x	400	425	435
Observed frequency f	4	3	1

The sum of the frequencies $(4 + 3 + 1)$ is equal to n, the total number of observations. The sum of the three products 4×400, 3×425 and 1×435 is equal to the sum of the eight observations, and so the mean mass is $\dfrac{1600 + 1275 + 435}{4 + 3 + 1} = 413.75$ g, as before. All that we have done is to collect together equal values of x.

So an alternative general formula for the sample mean is

$$\bar{x} = \frac{\displaystyle\sum_{j=1}^{m} f_j x_j}{\displaystyle\sum_{j=1}^{m} f_j} \tag{2.8}$$

where here the summation is over the m different values of x that were recorded. In the example, $m = 3$, $x_1 = 400$, $x_2 = 425$, $x_3 = 435$, $f_1 = 4$, $f_2 = 3$ and $f_3 = 1$.

Now $\sum_{j=1}^{m} f_j$ equals n, the total number of observations, so a simpler form for the previous formula is:

$$\bar{x} = \frac{1}{n} \Sigma f_j x_j$$

which we may write (more casually!) as $\dfrac{\Sigma fx}{n}$.

Calculator practice _____

> *Most calculators with statistical functions have some special key sequence for dealing with the input of grouped frequencies. Investigate how this can be done with your calculator and test the procedure using the soup tin data.*

2.8 The mean of grouped data

The formula for the mean of a frequency distribution can also be used to provide an estimate of the sample mean of a set of grouped data:

$$\bar{x} = \frac{\Sigma f_j x_j}{n}$$

In this case x_j is the **class mid-point** for the jth of m classes, f_j is the frequency for this class and $n = \sum_{j=1}^{m} f_j$. This is only an estimate of the actual sample mean since we do not know the individual sample values.

Notes
- ◆ The estimate is often referred to as the **grouped mean**.
- ◆ The difference in the values of the grouped mean and the true sample mean will usually be very small.
- ◆ It is usually much quicker to group a set of data and calculate the grouped .mean than to calculate the sample mean directly (unless a computer is taking the strain!).
- ◆ Sometimes we only have grouped data available!

Example 5

The following data summarise the distances travelled by a fleet of 190 buses before experiencing a major breakdown.

Distance ('000 miles) (d)	$d \leqslant 60$	$60 < d \leqslant 80$	$80 < d \leqslant 100$
Mid-point (x)	30	70	90
Frequency (f)	32	25	34

Distance ('000 miles) (d)	$100 < d \leqslant 120$	$120 < d \leqslant 140$	$140 < d \leqslant 220$
Mid-point (x)	110	130	180
Frequency (f)	46	33	20

Calculate the grouped mean of these data.

———————

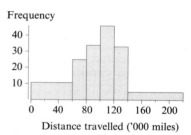

Frequency

Histogram of the distances travelled by a fleet of buses before breakdown

It is a good idea to draw a rough sketch of the data in order to 'get a feel' for the data. A glance at the (not so rough!) sketch suggests that the distribution has centre of mass at about 100 thousand miles. If our calculated answer is very different from this then it should be checked for a possible error.

Consider the 32 buses that travelled less than 60 000 miles before breaking down. Each breakdown occurred somewhere between 0 miles and 60 000 miles so a sensible estimate of the average distance travelled by these buses would be 30 000 miles. Hence an estimate of the total distance travelled by those 32 buses would be $32 \times 30\,000 = 960\,000$ miles. Repeating for each of the classes, our overall estimate of the total mileage is $\Sigma f_i x_i = 18\,720\,000$,

and hence the grouped mean, $\dfrac{18\,720\,000}{32}$, is about 98 500 miles.

Calculator practice ————————————————————————

Most statistical calculators can be used to calculate the mean of grouped data.
Test your calculator using the data given above.

Computer project ————————————————————————

It is easy to program a spreadsheet to calculate the mean of a set of grouped data.
Test your program using the previous data.

Exercises 2c ————————————————————————

1 A marine biologist is studying the population of limpets on a rocky coast. The numbers of a rare type of limpet that are found in 1 m square sections of the undercliff are summarised in the table below.

No. of limpets	0	1	2	3	4
No. of squares	73	19	5	2	1

Calculate the mean number of limpets per square metre of undercliff.

2 A proofreader reads through a 250-page manuscript. The numbers of mistakes found on each page are summarised in the table below.

No. of mistakes	0	1	2	3	4
No. of pages	61	109	53	23	4

Determine the mean number of errors found per page.

3 Construct a frequency distribution for the following data:

> 5, 7, 5, 3, 1, 4, 5, 4, 3, 2, 1, 3, 4, 5, 7, 6
> 8, 4, 3, 1, 5, 3, 5, 7, 3, 2, 4, 2, 6, 5, 2, 2

Find the mean and the median.

4 Construct a frequency distribution for the following data:

> 20, 30, 35, 25, 20, 30, 35, 25,
> 20, 30, 35, 40, 30, 35, 35, 25,
> 20, 40, 20, 25, 25, 30, 20, 20

Find the mean and the median.

5 A shop sells light bulbs. Mr Watt, the proprietor, makes a note one week of the wattage of the bulbs that he sells. At the end of the week he has noted the following:

> 100, 100, 100, 60, 100, 40, 150,
> 60, 100, 100, 100, 60, 40, 150,
> 100, 100, 100, 60, 100, 60

Construct a frequency distribution for the data and obtain (i) the mean, (ii) the median.

6 A sales representative records his daily mileage (in completed miles) for a period of 4 weeks:

> 153, 127, 142, 82, 91, 125, 113,
> 105, 93, 105, 88, 122, 96, 145,
> 136, 115, 107, 125, 98, 94

Group the data using class intervals of width 20, giving classes of 80–99, 100–119, etc. Find:
(i) the grouped mean,
(ii) the modal class.

7 A garage notes the mileages of cars brought in for a 15 000-mile service. The data is summarised in the following table.

Mileage ('000 miles)	14–	15–	16–	17–
No. of cars	8	15	13	9

Assuming that the upper limit of the final class is 17 999, find (i) an estimate for the mean, (ii) the modal class.

8 Each day, x, the number of diners in a restaurant was recorded and the following grouped frequency table was obtained.

x	16–20	21–25	26–30	31–35	36–40
No. of days	67	74	38	39	42

Using the above grouped data find:
(i) an estimate of the mean,
(ii) the modal class.

9 A die is rolled twenty times with the following results.

Outcome	1	2	3	4	5	6
Frequency	2	4	a	7	2	b

Given that the mean is 3.6, obtain the values of a and b.

10 Subsidies for loft insulation are offered to households whose net income is less than £25 000 per annum. Applicants for these subsidies classified their incomes as follows.

Annual income	No. of applicants
–£4999	3
£5000–£9999	17
£10 000–£14 999	31
£15 000–£19 999	28
£20 000–£24 999	16

Determine the value of the grouped mean.

2.9 Using coded values to simplify calculations

Consider the problem of finding the mean of the following values:

> 3001, 3003, 3005, 3005, 3007, 3007, 3007, 3009

We could calculate:

$$\frac{1}{8}\{3001 + 3003 + (2 \times 3005) + (3 \times 3007) + 3009\} = 3005.5$$

but this needs a calculator and lots of button pressing. It is much easier to calculate:

$$3000 + \frac{1}{8}\{1 + 3 + (2 \times 5) + (3 \times 7) + 9\} = 3005.5$$

As a second example, consider the problem of finding the mean of:

$$0.000\,01, \; 0.000\,03, \; 0.000\,05, \; 0.000\,05,$$
$$0.000\,07, \; 0.000\,07, \; 0.000\,07, \; 0.000\,09$$

We could calculate:

$$\frac{1}{8}\{0.000\,01 + 0.000\,03 + (2 \times 0.000\,05) + (3 \times 0.000\,07)$$
$$+ \, 0.000\,09\} = 0.000\,055$$

but it is much easier to calculate:

$$0.000\,01 \times \frac{1}{8}\{1 + 3 + (2 \times 5) + (3 \times 7) + 9\} = 0.000\,055$$

Both the examples above have used **coded** data. Algebraically, we replaced the observations $x_1, x_2, \ldots$ by the coded values $y_1, y_2, \ldots$ In the first example, $y_i = x_i - 3000$ and in the second example, $y_i = 100\,000 x_i$. The first example used a shift of location and the second a change of scale.

These two ideas may be combined. Suppose we want to find the mean of the following data:

$$10\,500, \; 11\,500, \; 12\,500, \; 12\,500, \; 13\,500, \; 13\,500, \; 13\,500, \; 14\,500$$

Writing $y = \dfrac{x - 10\,000}{500}$ we once again get the values 1, 3, 5, 5, 7, 7, 7 and 9 which have mean 5.5. Since $x = 10\,000 + 500y$, the mean of the x-values is:

$$10\,000 + (500 \times 5.5) = 12\,750$$

A general shift of location and change of scale is represented algebraically by the (linear) coding:

$$y = \frac{x - a}{b}$$

For convenience b is taken to be positive. When rewritten this expression gives:

$$x = a + by$$

and the mean, $\bar{x}$, is related to the mean, $\bar{y}$, of the coded values by

$$\bar{x} = a + b\bar{y}$$

In the first example $a = 3000$, $b = 1$; in the second $a = 0$, $b = \dfrac{1}{10\,000}$ and in the third $a = 10\,000$, $b = 500$.

Example 6

The jackets on display in the window of a men's outfitters have the following prices (in £):

49.95, 79.95, 79.95, 99.95, 139.95

Use a coding method to determine the average jacket price.

———

Let x be the displayed price. A useful coding is $y = x + 0.05$. The prices then become 50, 80, 80, 100, 140. The sum of the 5 y-values is 450, so $\bar{y} = 90$. Thus $\bar{x} = \bar{y} - 0.05 = 90 - 0.05 = 89.95$.
The average price is £89.95.

Example 7

A bus inspector notes the numbers of passengers on buses travelling on a certain route. He records the following values:

31, 45, 40, 38, 39, 42, 36, 38, 44, 39, 32, 32, 38

Using the coding $y = x - 30$, determine the mean of these data.

Taking the observed values to be x, the y-values are:

1, 15, 10, 8, 9, 12, 6, 8, 14, 9, 2, 2, 8

These are simple numbers that won't strain our powers of mental arithmetic! Their total is 104 and $n = 13$, so that $\bar{y} = \dfrac{104}{13} = 8$ and hence $\bar{x} = \bar{y} + 30 = 38$.

The mean number of passengers is 38.

Example 8

A manufacturer wished to test the accuracy of the '2000 ohm' resistors being produced by a machine. A random sample of 100 resistors was selected and their actual resistances were determined (correct to the nearest ohm). The results are shown in the table.

Determine the mean resistance of these resistors.

The values are clustered around the nominal value of 2000. A sensible coding is therefore provided by $y = x - 2000$, where x is the recorded resistance.

Resistance	Frequency
1995	1
1996	3
1997	5
1998	9
1999	19
2000	21
2001	16
2002	15
2003	4
2004	4
2005	2
2006	1

x	y	f	fy	Total
1995	−5	1	−5	
1996	−4	3	−12	
1997	−3	5	−15	
1998	−2	9	−18	
1999	−1	19	−19	−69
2000	0	21	0	
2001	1	16	16	
2002	2	15	30	
2003	3	4	12	
2004	4	4	16	
2005	5	2	10	
2006	6	1	6	90
Total		100		21

Notice the way that the negative values are summed separately from the positive values. The overall total is $90 - 69 = 21$ and so $\bar{y} = \dfrac{21}{100} = 0.21$. Since $\bar{x} = \bar{y} + 2000$, the mean resistance is 2000.21 ohms.

An alternative coding, that would avoid negative numbers, would be $y = x - 1995$.

Calculator practice

> *Compare the speed and accuracy of calculating the mean of the two sets of data given above using (i) the actual values and (ii) the coded values. You should find that using the coded values you work both more quickly and more accurately.*

Exercises 2d

1 Given that the numbers 3, 5, 6, 14 and 12 have mean 8, write down the mean of each of the following sets of numbers:
(i) 1003, 1005, 1006, 1014, 1012
(ii) 2.03, 2.05, 2.06, 2.14, 2.12
(iii) 1030, 1050, 1060, 1140, 1120

2 Find the mean, median and mode of the following observations:

> 1.000 000 002, 1.000 000 005,
> 1.000 000 006, 1.000 000 003,
> 1.000 000 009, 1.000 000 006,
> 1.000 000 005, 1.000 000 006,
> 1.000 000 003

3 The valuations (£x) of a collection of 12 antiques are reported as being:

> 600, 680, 1000, 750, 600, 850,
> 1000, 880, 1000, 650, 600, 1000

Use the coding $y = \frac{1}{10}(x - 600)$ to find the mean value of y and hence determine the mean valuation.

4 A choirmaster keeps a record of the numbers turning up for choir practice.

> 25, 28, 32, 31, 31, 34, 28, 31, 29,
> 28, 32, 32, 30, 29, 29, 31, 28, 28

Using a coding with each number reduced by 20, find the mean number turning up for choir practice.

5 The prices (£x) of pairs of shoes in the window display of a shoe shop are given below.

> 34.95, 44.95, 49.95, 69.95,
> 54.95, 64.95, 64.95, 54.95

(i) Using the coding $y = \frac{1}{5}(x + 0.05)$, determine the mean price.
(ii) Verify that the same result is obtained using the coding $y = x - 54.95$

6 The prices (in £) of various Indian dishes in a supermarket are given below.

> 1.99, 3.99, 2.99, 2.99, 1.99, 2.49, 1.99, 2.49

Using an appropriate coding, determine the mean price of these dishes.

7 The scores obtained by the leading 50 competitors in the first round of the 1995 US Masters are summarised below:

Score	66	67	68	69	70	71	72	73	74
Frequency	3	2	2	7	7	9	7	9	4

Using a suitable coding, find the mean of these scores.

8 The numbers of matches in a box were counted for a sample of 25 boxes. The results were:

> 51, 52, 48, 53, 47, 48, 50, 51, 50,
> 46, 52, 53, 51, 48, 49, 52, 50, 48,
> 47, 53, 54, 51, 49, 47, 51

Use a coding in which 40 is subtracted from each number to find the mean number of matches in a box.

Use a coding in which 50 is subtracted from each number (giving some negative values) to find the mean number of matches in a box.

Verify that your two answers are the same.

9 Records are kept for 18 days of the midday barometric pressure, in millibars.

> 1022, 1016, 1032, 1008, 998, 985,
> 993, 1004, 1009, 1011, 1015, 1020,
> 1007, 1001, 995, 993, 975, 972

Using a suitable coding, find the mean midday barometric pressure.

10 The gap, x mm, in a sample of spark plugs was measured with the following results:

> 0.81, 0.83, 0.81, 0.81, 0.82, 0.80, 0.81,
> 0.83, 0.84, 0.81, 0.82, 0.84, 0.80

Use the coding $y = 100x - 80$ to find the mean gap.

11 A garage notes the mileages of cars brought in for a 15 000-mile service. The data is summarised in the following table.

Mileage ('000 miles)	14–	15–	16–	17–
No. of cars	8	15	13	9

Taking the groups as having mid-points 14 500, ..., 17 500, and using the coding $y = \dfrac{1}{1000}(x - 14\,500)$, where x is the mileage, find the grouped mean.

12 Each day, x, the number of diners in a restaurant was recorded and the following grouped frequency table was obtained.

x	16–20	21–	26–	31–	36–40
No. of days	67	74	38	39	42

Using the coding $y = \dfrac{1}{5}(x - 18)$, where x is the number of diners, estimate the mean number of diners per day.

13 A set of data is summarised by $n = 8$, $\Sigma(x - 5) = 7.2$
Find the mean of x.

14 The annual salaries (x £'000) of the employees of a company are summarised in the following table.

Salary	Frequency
$5 \leqslant x < 10$	35
$10 \leqslant x < 15$	42
$15 \leqslant x < 20$	58
$20 \leqslant x < 30$	14
$30 \leqslant x < 50$	3
$50 \leqslant x < 100$	1

Use the coding $y = \dfrac{1}{5}(x - 7.5)$ to find the grouped mean salary.

15 A set of data is summarised by
$\Sigma_1^{12}(y + 0.5) = 0.234$
Find the mean of y.

16 A set of ten observations is such that
$\Sigma(2x + 3) = 427$
Find the mean of x.

17 Given that $n = 8$ and $\Sigma\{2(z + 3)\} = 752$, find the mean of z.

2.10 The median of grouped data

We begin by forming a cumulative frequency distribution. The median can then be estimated using linear interpolation (shown in this section) or, less accurately, by reading off a value from a cumulative frequency diagram (shown in the next section). In either case, with grouped data and n observations, it is customary to use $\dfrac{n}{2}$ in the calculations rather than $\dfrac{n+1}{2}$ (though the resulting difference is most unlikely to affect one's view of the data!).

▼

Example 9

Continuing with Example 5, we report the bus data using cumulative frequencies and upper class boundaries:

Distance ('000 miles)	$\leqslant 60$	$\leqslant 80$	$\leqslant 100$	$\leqslant 120$	$\leqslant 140$	$\leqslant 220$
Cumulative frequency	32	57	91	137	170	190

Estimate the distance exceeded by half the buses.

There are 190 buses and $\dfrac{190}{2} = 95$. Since 95 falls between 91 and 137, the median distance falls between 100 and 120 thousand miles. A total of $(137 - 91) = 46$ buses fall in this class. The median is estimated as:

$$100 + \frac{(95 - 91)}{(137 - 91)} \times (120 - 100) = 101.74 \text{ (to 2 d.p.)}$$

so the median is about 101 700 miles.

▲

Example 10

During 1983, motorists in Adelaide in South Australia were subject to random tests for alcohol consumption. Measurements of blood alcohol content (BAC) were made in units of mg of alcohol per 100 ml of blood.

BAC: Upper class boundary	15	25	35	45	65
Cumulative frequency	397	785	1083	1298	1580

BAC: Upper class boundary	95	125	155	205	400
Cumulative frequency	1793	1903	1951	1989	2003

Estimate the median BAC value.

———

One half of 2003 is 1001.5, which lies between 785 and 1083. The median therefore lies between 25 and 35. The estimated value is given by:

$$25 + \frac{(1001.5 - 785)}{(1083 - 785)} \times (35 - 25) = 32.27 \text{ (to 2 d.p.)}$$

The median blood alcohol content is estimated as being about 32 mg per 100 ml of blood.

Calculator practice ————————————————

Investigate whether your calculator is able to determine the median of grouped data. If there is not a preset sequence of key strokes available, then you may wish to write a short program to calculate the quantity.

2.11 Quartiles, deciles and percentiles

The median is a value that subdivides the ordered data into two halves. Further subdivision is also possible: the **quartiles** subdivide the data into quarters, the **deciles** provide a subdivision into tenths, and the **percentiles** provide a subdivision into hundredths. There are three quartiles: the **lower quartile**, Q_1, the median (Q_2), and the **upper quartile**, Q_3. The percentiles are simply called the 1st percentile, the 2nd percentile, and so on. The median is the 5th decile and the 50th percentile. A study of the values of the deciles or quartiles gives us an idea of the spread of the data, but an 'idea' is all we get and there is no need for great precision.

Grouped data

With grouped data, life is straightforward! In general, the rth percentile is the '$\left(\frac{rn}{100}\right)$th' observation. The median is therefore the '$\left(\frac{n}{2}\right)$th' observation

as in the previous section), while the quartiles are the '$\left(\frac{n}{4}\right)$th' and '$\left(\frac{3n}{4}\right)$th'

observations. We have used inverted commas as a reminder that interpolation will usually be needed (though it would be inappropriate to report the value obtained to any great accuracy).

Example 11

Determine the lower and upper quartiles and the 9th decile of the Adelaide motorists' data in Example 10:

BAC: Upper class boundary	15	25	35	45	65
Cumulative frequency	397	785	1083	1298	1580

BAC: Upper class boundary	95	125	155	205	400
Cumulative frequency	1793	1903	1951	1989	2003

Since the data are grouped we can use linear interpolation within the groups or we can attempt to read the figures off the cumulative frequency diagram. We first attempt to use the diagram.

Cumulative frequency diagram of blood alcohol content of Adelaide motorists in 1983

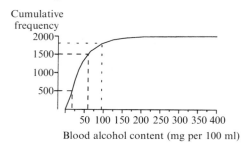

The lower and upper quartiles correspond to cumulative frequencies of $2003 \times 0.25 = 501$ and $2003 \times 0.75 = 1502$. Reading off the diagram (with considerable difficulty!) we find that these correspond to about 20 and 60. The 9th decile (the 90th percentile) corresponds to a cumulative frequency of $2003 \times 0.90 = 1803$ and, from the diagram, has a value of about 100.

We now use interpolation. For the lower quartile we have the estimate:

$$15 + \frac{(501 - 397)}{(785 - 397)} \times (25 - 15) = 18$$

and, for the upper quartile we have:

$$45 + \frac{(1502 - 1298)}{(1580 - 1298)} \times (65 - 45) = 59$$

For the 9th decile, the same approach gives:

$$95 + \frac{(1803 - 1793)}{(1903 - 1793)} \times (125 - 95) = 98$$

Ungrouped data

In Section 2.3 (p.38) the definition given for the median of ungrouped data was quite complicated! It does not look much better when expressed in another way: the median of ungrouped data is the '$\left(\frac{n}{2} + \frac{1}{2}\right)$th' observation, where the inverted commas serve as a reminder that interpolation may be needed.

Noting that $\frac{n}{2} = \frac{50n}{100}$, we now generalise this result and define the rth percentile of ungrouped data to be the '$\left(\frac{rn}{100} + \frac{1}{2}\right)$th' observation.

With this definition the lower and upper quartiles are, respectively, the '$\left(\frac{n}{4} + \frac{1}{2}\right)$th' and '$\left(\frac{3n}{4} + \frac{1}{2}\right)$th' ordered observations.

Note

◆ *There are no universally agreed formulae for any of these quantities (except for the median)*. However, since quartiles and percentiles are of limited use, this is not really a source of worry! There is no virtue in reporting values for the quartiles to great accuracy: they should be reported using at most one more decimal place than that given in the original data.

Example 12

The numbers of words in the first 18 sentences of Chapter 1 of *A Tale of Two Cities* by Charles Dickens are as follows:

118, 39, 27, 13, 49, 35, 51, 29, 68, 54, 58, 42, 16, 221, 80, 25, 41, 33

whilst the numbers of words in the first 17 sentences of Chapter 1 of *Not a Penny More, Not a Penny Less* by Jeffrey Archer are as follows:

8, 10, 15, 13, 32, 25, 14, 16, 32, 25, 5, 34, 36, 19, 20, 37, 19

Determine the median, quartiles and first decile for each data set.

———————

Rearranging the Dickens data in order of magnitude we get:

9				1										
13   16   25   27   29   33   35   39   41   42   49   51   54   58   68   80   118   221

2.7 | 6.3   0.5 | 0.5

18.7   29   41.5   58

Since $\frac{18}{2} + \frac{1}{2} = 9.5$ the median is the average of the 9th and 10th ordered observations, namely $\frac{1}{2}(41 + 42) = 41.5$.

Since $\frac{18}{4} + \frac{1}{2} = 5$, the lower quartile is the 5th observation, namely 29.

Since $\frac{3 \times 18}{4} + \frac{1}{2} = 14$, the upper quartile is 58.

For the first decile, we need the '$\left(\frac{10 \times 18}{100} + \frac{1}{2}\right)$th' observation. Since $\frac{10 \times 18}{100} + \frac{1}{2} = 2.3$, we need to interpolate between the 2nd and 3rd ordered observations, and the required value is:

$$16 + \{0.3 \times (25 - 16)\} = 18.7$$

The first decile is about 19.

Rearranging the Archer data in order we get:

2		1									0				
5   8   10   13   14   15   16   19   19   20   25   25   32   32   34   36   37

0.4 | 1.6   0.75 | 0.25   0 | 0

8.4   13.75   19   32

Since $\frac{17}{2} + \frac{1}{2} = 9$ the median is the 9th largest observation, namely 19.

Since $\frac{17}{4} + \frac{1}{2} = 4.75$, we must interpolate between the 4th and 5th ordered observations, getting $13 + \{0.75 \times (14 - 13)\} = 13.75$. The lower quartile is about 14.

Since $\frac{3 \times 17}{4} + \frac{1}{2} = 13.25$, we must interpolate between the 13th and 14th ordered observations. Since both of these are 32, the interpolated value will also be 32, which is therefore the value of the upper quartile.

For the first decile we need to calculate the value of $\frac{10 \times 17}{100} + \frac{1}{2}$, which is 2.2. Interpolating between the second and the third of the ordered observations we get $8 + \{0.2 \times (10 - 8)\} = 8.4$. The first decile is about 8.

The difference in writing styles is evident!

Computer project _____

> *Write a computer program to calculate quartiles, deciles and percentiles. If your computer has graphical capabilities then the program could be extended to display the cumulative frequency diagram along with indications of the locations of the quartiles. A well-written program should automatically scale the axes so that the diagram fills the screen.*

Exercises 2e _____

1 The gap, x mm, in a sample of spark plugs was measured with the following results:

> 0.81, 0.83, 0.81, 0.81, 0.82, 0.80, 0.81, 0.83, 0.84, 0.81, 0.82, 0.84, 0.80

Find the lower and upper quartiles for this data set.

2 The numbers of matches in a box were counted for a sample of 25 boxes. The results were:

> 51, 52, 48, 53, 47, 48, 50, 51, 50, 46, 52, 53, 51, 48, 49, 52, 50, 48, 47, 53, 54, 51, 49, 47, 51

Find the second and eighth deciles for this set of data.

3 Records are kept for 18 days of the midday barometric pressure, in millibars.

> 1022, 1016, 1032, 1008, 998, 985, 993, 1004, 1009, 1011, 1015, 1020, 1007, 1001, 995, 993, 975, 972

Find the values of the lower and upper quartiles.

4 Find the lower and upper quartiles and the 9th decile for the following data:

> 20, 30, 35, 25, 20, 30, 35, 25, 20, 30, 35, 40, 30, 35, 35, 25, 20, 40, 20, 25, 25, 30, 20, 20

5 Find the lower and upper quartiles and the 15th percentile for the following data:

> 5, 7, 5, 3, 1, 4, 5, 4, 3, 2, 1, 3, 4, 5, 7, 6, 8, 4, 3, 1, 5, 3, 5, 7, 3, 2, 4, 2, 6, 5, 2, 2

6 A baker keeps a count of the number of doughnuts sold each day for three weeks. The numbers are:

> 35, 47, 34, 46, 55, 82, 41, 35, 47, 51, 56, 75, 38, 41, 44, 51, 45, 74

By constructing a stem-and-leaf diagram, or otherwise, find the lower and upper quartiles of the number of doughnuts sold per day. Find also the 4th decile.

7 A garage notes the mileages of cars brought in for a 15 000-mile service. The data are summarised in the following table.

Mileage ('000 miles)	14–	15–	16–	17–
No. of cars	8	15	13	9

Find the lower and upper quartiles and the 5th and 20th percentiles.

8 Each day, x, the number of diners in a restaurant was recorded and the following grouped frequency table was obtained.

x	16–20	21–	26–	31–	36–40
No. of days	67	74	38	39	42

Treating x as though it is a continuous variable with class boundaries at 15.5, 20.5, 25.5, 35.5 and 40.5, find the lower and upper quartiles and the 2nd and 8th deciles.

9 In an investigation of delays at a roadworks, the times spent, by a sample of commuters, waiting to pass through the roadworks were recorded to the nearest minute. Shown below is part of a cumulative frequency table resulting from the investigation.

Upper class boundary	2.5	4.5	7.5	8.5	9.5
Cumulative number of commuters	0	6	21	48	97

Upper class boundary	10.5	12.5	15.5	20.5
Cumulative number of commuters	149	178	191	200

(a) For how many of the commuters was the time recorded as 11 minutes or 12 minutes?

(b) Estimate (i) the lower quartile, (ii) the 81st percentile, of these waiting times.

[ULEAC]

10

Volume (in litres) of petrol	Number of sales
5 or less	6
10 or less	20
15 or less	85
20 or less	148
25 or less	172
30 or less	184
35 or less	194
40 or less	200

The table gives an analysis of a random sample of 200 sales of unleaded petrol at a petrol station.

(*a*) Using scales of 2 cm to 5 litres on the horizontal axis and 2 cm to 20 sales on the vertical axis, draw a cumulative frequency curve for the data.

(*b*) Use your curve to estimate the median volume of unleaded petrol sales.

Unleaded petrol is sold for 52.3p per litre. Use your curve to estimate

(*c*) the 40th percentile of the value of unleaded petrol sales,

(*d*) the percentage of sales above £12. [ULSEB]

2.12 Range and interquartile range

The **range** of a set of numerical data is the difference between the highest and lowest values. It is the simplest possible measure of spread. It cannot be used with grouped data and it ignores the distribution of intermediate values. A single very large or very small value would give a misleading impression of the spread of the data. This happens with the Dickens data where the range $(221 - 13 = 208)$ gives a distorted impression because of the single unusually long sentence.

More useful, because it concentrates on the middle portion of the distribution, is the **interquartile range (IQR)** which is the difference between the upper and lower quartiles. The **semi-interquartile range** is sometimes quoted: it is half the interquartile range.

For the Dickens data of Example 12, the interquartile range is $Q_3 - Q_1 = 58 - 29 = 29$, and the semi-interquartile range is 14.5. The less variable Archer data has a semi-interquartile range equal to 9.1 (to 1 d.p.).

2.13 Box–whisker diagrams

Box–whisker diagrams present a simple picture of the data based on the values of the quartiles. They are also known as **boxplots**. The general form of a box–whisker diagram is shown in the diagram.

The general form of a box–whisker diagram

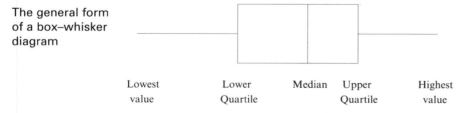

| Lowest value | Lower Quartile | Median | Upper Quartile | Highest value |

Box–whisker diagrams provide a particularly convenient way of comparing two distributions.

Note

◆ There is no agreed rule for determining the thickness of the box. When comparing samples a sensible procedure would be to make the box areas proportional to the sample sizes. The thicknesses of the boxes are therefore in proportion to the ratios of the respective sample sizes divided by the corresponding IQR.

Example 13

Use box–whisker diagrams to compare the sentence lengths of Dickens and Archer for the data of Example 12.

Following the previous note, we give the boxes areas in the ratio 18 to 17. The interquartile ranges were 29 and 18.25, so we use thicknesses in proportion to $\dfrac{18}{29}:\dfrac{17}{18.25}$. Note that this is an optional extra!

Box–whisker diagrams comparing the sentence lengths of two authors

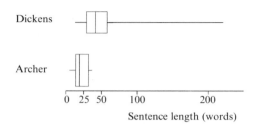

The difference in the distributions of the sentence lengths is very apparent.

Refined boxplots

A useful refinement of the simple box–whisker diagram highlights any unusually extreme data values (which are known as **outliers** and should be examined for possible transcription or other errors). In these **refined boxplots** the whiskers cannot exceed some specified length. A typical choice is 1.5 times the interquartile range (the length of the box). Thus the upper whisker extends at most from Q_3 to $Q_3 + 1.5(Q_3 - Q_1)$ and the lower whisker extends at most from Q_1 to $Q_1 - 1.5(Q_3 - Q_1)$. The whiskers reach these limits if there are outlier points: the outliers are then indicated individually using crosses.

Example 14

Draw a refined boxplot for the sentence lengths of Dickens given in Example 12.

For the sentences from *A Tale of Two Cities* we had $Q_3 = 58$ and $Q_1 = 29$. The interquartile range was $58 - 29 = 29$ and the maximum whisker length is therefore $1.5 \times 29 = 43.5$. The upper whisker is therefore curtailed at $58 + 43.5 = 101.5$ with the result that the sentences of lengths 118 and 221 are indicated as outliers.

Refined boxplot of the sentence lengths used by Dickens

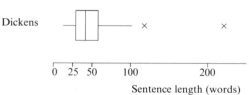

Project ——————————————————————————

Choose two of your own favourite authors and repeat the Dickens/Archer experiment. Try to choose authors whose styles you think may be different. Choose descriptive passages rather than passages of dialogue, but don't choose them because they seem to have particularly long (or short) sentences, or you will bias the results! Construct box–whisker diagrams or refined boxplots for each author.

Do there seem to be differences in the two distributions of sentence length?

Exercises 2f

1 The following data are the numbers of deaths of army officers caused by horse kicks, for the Prussian Army during the period 1875 to 1894. In order of size the numbers are:

> 3, 4, 5, 5, 6, 6, 7, 8, 9, 9, 10,
> 11, 11, 11, 12, 14, 15, 15, 17, 18

Find the range and interquartile range. Illustrate the data using a box–whisker diagram.

2 One year the numbers of academic staff (including part-time staff) in the various departments of the University of Essex (a small, friendly university) were as follows:

> 19.0, 15.7, 25.3, 28.0, 15.0, 10.0,
> 12.0, 10.3, 22.0, 24.8, 13.8, 25.9,
> 23.0, 21.3, 12.0, 11.0, 23.0

Find the range and interquartile range. Illustrate the data using a box–whisker diagram.

3 The record times (in hours) for marathon sessions of various games, as reported in the 1986 *Guinness Book of Records*, are as follows:

> Backgammon 151, Bridge 180, Chess 200, Darts 133, Draughts 108, Monopoly 660, Pool 300, Scrabble 153, Snooker 301, Table tennis 148, Tiddlywinks 300

Illustrate the data using a refined boxplot.

4 One year the numbers of undergraduates in the various departments of that friendly University of Essex were as shown below:

> 173, 166, 255, 225, 107, 146, 199, 107,
> 348, 327, 236, 390, 424, 252, 125, 161, 343

Illustrate the data using a refined boxplot.

5 The systolic blood pressures of 12 smokers and 12 non-smokers are as follows (in the standard units):

> Smokers:
> 122, 146, 120, 114, 124, 126,
> 118, 128, 130, 134, 116, 130

> Non-smokers:
> 114, 134, 114, 116, 138, 110,
> 112, 116, 132, 126, 108, 116

Contrast these two sets of data using side-by-side refined boxplots.

6 One year the number of overseas postgraduate students in the departments of a certain university (you know where!) were as given below:

> 13, 8, 34, 24, 26, 22, 14, 44,
> 0, 104, 19, 26, 9, 7, 41, 57, 6

Illustrate these data using a refined boxplot.

7 The numbers of students on degree schemes involving mathematics at a really excellent university are as follows:

> 22, 11, 5, 20, 15, 13, 3, 2, 5, 1, 2

Determine the range and interquartile range and illustrate the data using a refined boxplot.

8 The times (in s) taken for a group of experienced rats to run through a maze are to be compared with the times for a group of inexperienced rats. The data are:

> Experienced rats:
> 121, 137, 130, 128, 132, 127, 129,
> 131, 135, 130, 126, 120, 118, 125

> Inexperienced rats:
> 135, 142, 145, 156, 149, 134, 139,
> 126, 147, 152, 153, 145, 144

(i) Summarise the two data sets using stem-and-leaf diagrams.
(ii) Find the median and the upper and lower quartiles for each group of rats.
(iii) Plot the two sets of data on a single graph using boxplots.
    Comment on the results.

9 A random sample of size 500 was selected from the persons listed in a residential telephone directory of a county in Wales. The number of letters in each surname was counted and the distribution of the name-lengths is given in the table below.

Name-length	3	4	5	6	7	8	9	10	11
Frequency	4	31	103	124	111	63	38	19	7

(a) (i) Represent this distribution graphically.
    (ii) Find the median and the interquartile range of the distribution.
(b) The 500 persons in the sample are unlikely to be representative of the population of the county.
    Name one group of people which is likely to be under-represented.
(c) State, with a reason, whether the sample of surnames is likely to be representative of the surnames of the population of this county in Wales.   [NEAB]

10 A random sample of 51 people were asked to record the number of miles they travelled by car in a given week. The distances, to the nearest mile, are shown below.

67	76	85	42	93	48	93	46
52	72	77	53	41	48	86	78
56	80	70	70	66	62	54	85
60	58	43	58	74	44	52	74
52	82	78	47	66	50	67	87
78	86	94	63	72	63	44	47
57	68	81					

(a) Construct a stem and leaf diagram to represent these data.
(b) Find the median and the quartiles of this distribution.
(c) Draw a box plot to represent these data.
(d) Give one advantage of using (i) a stem and leaf diagram, (ii) a box plot, to illustrate data such as that given above.

[ULEAC]

11 The following table, extracted from *Welsh Social Trends*, shows the distribution of the number of persons per household in Wales in 1981.

Number of persons in household	1	2	3	4	5	6+
Percentage of households	21	31	18	18	8	4

(i) By plotting the cumulative percentage step polygon, or otherwise, determine the median and the semi-interquartile range of the number of persons per household.
(ii) State, giving your reason, whether the mean number of persons per household is greater than, equal to, or less than the median number.

[WJEC]

2.14 Deviations from the mean

Suppose we wish to summarise the following data:

0, 99, 99, 100, 100, 100, 100, 100, 101, 101, 200

This set of data has mean, median and mode equal to 100, lower quartile equal to 99, upper quartile equal to 101 and range equal to 200. The same is true for this second set of data:

0, 0, 99, 99, 100, 100, 100, 101, 101, 200, 200

However, this second set of data has four extreme observations, compared with only two in the first set. This extra variability can be quantified by calculating the differences between the observations and their mean:

Set 1: $-100,\quad -1,\quad -1,\quad 0,\quad 0,\ 0,\ 0,\ 0,\quad 1,\quad 1,\quad 100$
Set 2: $-100,\ -100,\quad -1,\ -1,\quad 0,\ 0,\ 0,\quad 1,\ 1,\quad 100,\ 100$

In each case the differences sum to zero. This always happens since, for a set of n observations $x_1, \ldots, x_n$ with sample mean $\bar{x}$, given by $n\bar{x} = \sum x_i$:

$$\sum_{i=1}^{n} (x_i - \bar{x}) = (x_1 - \bar{x}) + \cdots + (x_n - \bar{x})$$

$$= (x_1 + \cdots + x_n) - (\bar{x} + \cdots + \bar{x})$$

$$= \sum_{i=1}^{n} x_i - n\bar{x}$$

$$= n\bar{x} - n\bar{x}$$

$$= 0$$

2.15 The mean deviation

If we ignore the signs of the differences between the observations and their mean, and work with absolute values (moduli), then a natural measure of spread is provided by the average value of the differences. This is called the **mean deviation** or **mean absolute deviation** (MAD):

$$\frac{\Sigma |x_i - \bar{x}|}{n}$$

For the two sets of data in the previous section the mean deviations are $\frac{204}{11} = 18.5$ and $\frac{404}{11} = 36.7$. The extra variability of the second set leads to a larger value of the mean deviation.

Despite its apparent simplicity the mean deviation is little used because its use of absolute values makes the subsequent theory difficult. In fact you would be MAD to use it . . . .

2.16 The variance

An alternative is to work with the sum of the squares of the deviations from the mean:

$$\Sigma (x_i - \bar{x})^2 = (x_1 - \bar{x})^2 + \cdots + (x_n - \bar{x})^2$$

The more variation there is in the x-values, the larger will be the value of $\Sigma (x_i - \bar{x})^2$. However, the sum might be large simply because of the number of x-values, and some sort of average value is needed.

Dividing by n would seem natural, but (unfortunately!) there is a strong case for dividing instead by $(n - 1)$.

Using the divisor n

This is appropriate in two cases:

1 If the values $x_1, \ldots, x_n$ represent an entire population.
2 If the values $x_1, \ldots, x_n$ represent a sample from a population and we are interested in *the variation within the sample itself*.

In both cases the n observed values are all that interest us and the natural average squared deviation, denoted by σ_n^2, is given by

$$\sigma_n^2 = \frac{1}{n} \sum_{i=1}^{n} (x_i - \bar{x})^2 \tag{2.9}$$

The quantity σ_n^2 should be read as 'sigma n squared'.

If $x_1, \ldots, x_n$ represent a sample of data then σ_n^2 is called the **sample variance**, while if $x_1, \ldots, x_n$ represent the entire population then σ_n^2 is called the **population variance**.

An example of a case where the x-values refer to the entire population is where $x_1, \ldots, x_n$ represent the heights of *all* the children in a particular class in a school. If, for some reason, we are *only* interested in this class then σ_n^2 is appropriate.

Using the divisor $(n - 1)$

This is appropriate in the following case:

The values $x_1, \ldots, x_n$ represent a sample from a population and we are interested in estimating *the variation in the population*. The sample is important only because it gives information about the larger population.

For example, we might collect information about the heights of the children in a class so as to gain an impression of the distribution of the heights of children in corresponding classes nationwide.

In this case, a slight adjustment is made to the formula by dividing by $(n-1)$ instead of by n. We shall justify this in Section 8.6 (p.206). The revised quantity is sometimes denoted by σ_{n-1}^2, sometimes by $\hat{\sigma}^2$, but more commonly by s^2:

$$s^2 = \sigma_{n-1}^2 = \frac{1}{n-1} \sum_{i=1}^{n} (x_i - \bar{x})^2 \tag{2.10}$$

We will call this quantity the **unbiased estimate of the population variance** (and will use the s^2 notation).

Notes

- Since s^2 and σ_n^2 are positive multiples of a sum of squares:
  –they cannot have negative values,
  –they have units which are squares of the units of x.
- If s^2 or σ_n^2 is equal to zero then each of the x-values must be equal to the mean, $\bar{x}$, and therefore also equal to each other.
- Except when both are zero, $s^2 > \sigma_n^2$.
- Practising statisticians seldom use the divisor n because they are interested in drawing inferences about a population from a sample. The important questions are those about the unseen population rather than the particular sample observed.

Special note

- There is considerable variation from book to book, from exam board to exam board and from one set of statistical tables to another, concerning the names and symbols to be used for the two forms of variance formula introduced above.
  The quantity with divisor n, which we denote by σ_n^2, is denoted by $\hat{s}^2$ in one set of tables and by s^2 in another, both of which call it the 'sample variance'.
  Another set of tables uses S^2 for the same quantity and calls it the 'unadjusted variance'.
  A majority of tables use s^2 as we do, to denote the quantity with divisor $n-1$. There is also a general agreement that s^2 should be referred to as '**the unbiased estimate of the population variance**', although it too is often called the 'sample variance'.
  You should consult the formula sheet, tables, or syllabus for your exam board to be sure which formula you will be expected to use.

2.17 Calculating the variance

If $\bar{x}$ is an integer then the values of $(x_1 - \bar{x})^2, \ldots, (x_n - \bar{x})^2$ will be quite easy to calculate. However, $\bar{x}$ will usually be an awkward decimal and it is much easier to use the result:

$$\Sigma(x_i - \bar{x})^2 = \Sigma x_i^2 - \frac{(\Sigma x_i)^2}{n} \tag{2.11}$$

Hence:

$$\sigma_n^2 = \frac{\Sigma x_i^2}{n} - \frac{(\Sigma x_i)^2}{n^2}$$

and:

$$s^2 = \sigma_{n-1}^2 = \frac{1}{n-1} \left\{ \Sigma x_i^2 - \frac{1}{n}(\Sigma x_i)^2 \right\} = \frac{n}{n-1} \sigma_n^2$$

Notes

◆ In all important cases the quantities Σx_i^2 and $(\Sigma x_i)^2$ are *not* equal, since:

$$\Sigma x_i^2 = x_1^2 + x_2^2 + \cdots + x_n^2$$

whereas:

$$(\Sigma x_i)^2 = (x_1 + x_2 + \cdots + x_n)^2$$
$$= (x_1^2 + x_2^2 + \cdots + x_n^2) + 2(x_1 x_2 + x_1 x_3 + \cdots + x_{n-1} x_n)$$

◆ The proof of the result in Equation (2.11) requires some messy algebra:

$$\Sigma(x_i - \bar{x})^2 = \Sigma(x_i^2 - 2x_i\bar{x} + \bar{x}^2)$$
$$= \Sigma x_i^2 - 2\bar{x}\Sigma x_i + n\bar{x}^2$$
$$= \Sigma x_i^2 - 2\left(\frac{\Sigma x_i}{n}\right)\Sigma x_i + n\left(\frac{\Sigma x_i}{n}\right)^2$$
$$= \Sigma x_i^2 - \frac{1}{n}(\Sigma x_i)^2$$

◆ Another way of writing $\Sigma(x_i - \bar{x})^2$ is as $\Sigma x_i^2 - n\bar{x}^2$, but for numerical calculations it is usually more accurate to calculate $\frac{1}{n}(\Sigma x_i)^2$ than to calculate $n\bar{x}^2$. The latter form is more useful in algebraic manipulations.

2.18 The sample standard deviation

We define the sample standard deviation, σ_n, as being the square root of the sample variance, σ_n^2

$$\sigma_n = \sqrt{\frac{\Sigma x_i^2}{n} - \frac{(\Sigma x_i)^2}{n^2}} \qquad\qquad (2.12)$$

Notes

◆ The units of the standard deviation are the same as the units of x – i.e. if x is a number of apples then so is σ_n.
◆ In some books you may come across s, the square root of s^2, being described as the 'sample standard deviation'.
◆ The words 'standard deviation' are often abbreviated to s.d.

Calculator practice ────────────────────────────────

Calculators with statistical functions will calculate one or both of σ_n and $s (= \sigma_{n-1})$. You should check which statistic(s) your calculator provides. Usually the values of n, Σx and Σx^2 will have been calculated and stored in accessible memories in the process. You should be aware of where these quantities are stored and how they can be accessed.

▼───▼

Example 15

The nine planets of the solar system have approximate equatorial diameters (in thousands of km) as follows:

4.9, 12.1, 12.8, 6.8, 142.8, 120.0, 52.4, 49.5, 2.5

Determine the standard deviation of these diameters.

────────────

We begin by calculating $\Sigma x_i = (4.9 + \cdots + 2.5) = 403.8$ and $\Sigma x_i^2 = (4.9^2 + \cdots + 2.5^2) = 40\,374.6$. The mean diameter is $\frac{1}{9} \times 403.8 = 44.87$ thousand kilometres. Assuming that we are interested in these nine planets for their own sake rather than for what they may imply about planets elsewhere in the universe, we now calculate:

$$\sigma_n = \sqrt{\frac{\Sigma x_i^2}{n} - \frac{(\Sigma x_i)^2}{n^2}}$$

$$= \sqrt{\frac{40\,374.6}{9} - \frac{403.8^2}{9^2}}$$

$$= \sqrt{4486.066\,667 - 2013.017\,778}$$

$$= \sqrt{2473.048\,889}$$

$$= 49.73 \text{ (to 2 d.p.)}$$

The standard deviation of the equatorial diameters is about 50 thousand kilometres.

Notes

◆ The working is carried through to considerable accuracy to guard against **round-off errors** and against loss of significance, since the calculations often involve determining the relatively small difference between two large numbers. With inaccurate calculations

$$\frac{\Sigma x_i^2}{n} - \frac{(\Sigma x_i)^2}{n^2}$$

may *appear* to be negative due to loss of significant figures!
◆ When reporting results a reasonably accurate value (49.73) should be easily available, but the description (50) of the result should be as simple as possible. In most situations there will be little interest in the difference between 49.73 and 50!

Example 16

An office manager wishes to get an idea of the number of phone calls received by the office during a typical day. A week is chosen at random and the numbers of calls on each day of the (5-day) week are recorded. They are as follows:

15, 23, 19, 31, 22

Determine (i) the sample mean, (ii) the sample standard deviation, (iii) s^2, the unbiased estimate of the population variance.

(i)   For these data $\Sigma x = 110, \Sigma x^2 = 2560$. The mean is $\frac{110}{5} = 22$.
(ii) The calculation of the sample standard deviation takes a little longer:

$$\sigma_n = \sqrt{\frac{\Sigma x_i^2}{n} - \frac{(\Sigma x_i)^2}{n^2}}$$

$$= \sqrt{\frac{2560}{5} - \frac{110^2}{25}}$$

$$= \sqrt{28}$$

$$= 5.29 \text{ (to 2 d.p.)}$$

The sample standard deviation is about 5.

(iii) This requires the $(n - 1)$ divisor:

$$s^2 = \frac{1}{n - 1} \left\{ \Sigma x_i^2 - \frac{1}{n} (\Sigma x_i)^2 \right\}$$

$$= \frac{1}{4} \left(2560 - \frac{110^2}{5} \right)$$

$$= \frac{140}{4}$$

$$= 35$$

The unbiased estimate of the population variance, s^2, is equal to 35.

Approximate properties of the standard deviation

Providing the sample size is reasonably large and the data are not too skewed (i.e. there is not a long 'tail' of very large or very small values) it is possible to make the following approximate statements which are based on theory covered later in Chapter 12:

- About two-thirds of the individual observations will lie within one standard deviation of the sample mean.
- About 95% of the individual observations will lie within two standard deviations of the sample mean.
- Almost all the data will lie within three standard deviations of the sample mean.
- A useful check that your calculations have not gone hopelessly wrong is provided by noting that the standard deviation will usually be between a third and a sixth of the range.

These are *very approximate* statements which enable us to check our calculations. Because they are approximate we need not worry whether we are using s or σ_n (hurrah!).

As an example of their use, suppose that the observed data consists of values ranging between 0 and 30. We expect a mean of about 15 (since this is half-way between 0 and 30) and a standard deviation of between 5 and 10. If our calculations find a standard deviation of 4 then this should not worry us, but if we calculate a value of 40, then we will certainly have made a mistake.

The statements also enable us to draw inferences about the population from which the data has been sampled. As stated at the beginning of Chapter 1, this is the principal purpose of Statistics.

Example 17

Use the approximate properties of the standard deviation to make statements concerning the likely numbers of daily phone calls received by the office featured in Example 16.

_____

Here the sample is very small, so we cannot place too much reliance on our approximations.

The range of values observed was $31 - 15 = 16$, so we anticipate a mean of about $\frac{1}{2}(31 + 15) = 23$ and a standard deviation of between

$\frac{1}{3} \times 16 = 5.3$ and $\frac{1}{6} \times 16 = 2.6$. The calculated mean and standard deviation were 22 and 5.29. It is reasonable to assume that we have not made a mistake!

The office manager can conclude that on two-thirds of days the office will receive between $22 - 6 = 16$ and $22 + 6 = 28$ calls (there is no point in using great precision since these are only very crude approximations).

Assuming that the week sampled was typical, the office is unlikely ever to receive fewer than $22 - (3 \times 6) = 4$ calls, or more than $22 + (3 \times 6) = 40$ calls.

▲ _____ ▲

Exercises 2g

1 A card player notes the number of hearts that she receives during a sample of 5 random deals. The numbers are 3, 2, 4, 4, 1.
Find the sample mean and the sample standard deviation.
Find also the mean deviation.

2 The numbers of television licences bought at a particular Post Office on a sample of 5 randomly chosen weekdays were 15, 9, 23, 12, 17.
Find the mean and standard deviation of this sample.

3 The numbers of potatoes in a sample of 2 kg bags were 12, 15, 10, 12, 11, 13, 9, 14.
Find the mean and an unbiased estimate of the population variance.

4 During his entire life, Mr I Walton, a most unlucky angler, caught just six fish. Their masses, in kg, were 1.35, 0.87, 1.61, 1.24, 0.95, 1.87.
Find the mean and variance of this population.

5 A random sample of seven runner beans have lengths (in cm, to the nearest cm) given as 28, 31, 24, 33, 28, 32, 30.
Find the value of s.

6 In an experiment, a cupful of cold water is poured into a kettle and the time taken for the water to boil is noted. The experiment was conducted six times giving the following results (in seconds):

125, 134, 118, 143, 128, 131.

Find the value of s^2.

7 The midday temperature (in °C) was noted at an Antarctic weather station on every day of a particular week of the year. The results were $-25, -18, -41, -34, -25, -33, -27$. Treating these results as a population, find their mean and standard deviation.

8 A random sample has values summarised by $n = 8$, $\Sigma x = 671$, $\Sigma x^2 = 60\,304$.
Find the mean and the value of s^2.

9 Twenty observations of t are summarised by $\Sigma t = 23.16$, $\Sigma t^2 = 35.4931$.
Find the mean and the value of s^2.

10 A population is summarised by

$$\sum_{j=1}^{52} y_j = 3.751, \quad \sum_{j=1}^{52} y_j^2 = 0.535\,691$$

Find the mean and the value of s.

11 A random sample is summarised by $n = 13$, $\Sigma u = -27.3$, $\Sigma u^2 = 84.77$.
Find the mean and sample standard deviation of u.

12 A random sample has $n = 11$, $s = 1.4$ and $\Sigma x^2 = 50$.
Find $\bar{x}$.

13 A random sample of 15 observations has sample mean 11.2 and sample variance 13.4. One observation of 21.2 is judged to be unreliable.
Find the sample mean and the sample variance of the remaining 14 observations.

2.19 Variance and standard deviation for frequency distributions

When data have been summarised in the form:

'the value x_i occurs with frequency f_i'

the formulae for the variance need rewriting. With m distinct values of x, the formula for the sample variance, σ_n^2, becomes:

$$\sigma_n^2 = \frac{1}{n} \left\{ \sum_{j=1}^{m} f_j x_j^2 - \frac{1}{n} \left(\sum_{j=1}^{m} f_j x_j \right)^2 \right\} \qquad (2.13)$$

and the formula for the unbiased estimate of the population variance becomes:

$$s^2 = \frac{1}{n-1} \left\{ \sum_{j=1}^{m} f_j x_j^2 - \frac{1}{n} \left(\sum_{j=1}^{m} f_j x_j \right)^2 \right\} \qquad (2.14)$$

where n is the total of the individual frequencies:

$$n = \sum_{j=1}^{m} f_j$$

As before, the sample standard deviation is simply the square root of the sample variance.

The same revised formulae are used when working with grouped data. In this case the x-values are the mid-points of the class intervals and the f-values are the class frequencies. The value obtained will usually be a slight under-estimate of the true sample variance (or of the true value of s^2).

Example 18

Determine the variance of the marks obtained by 99 students which are summarised in the following grouped frequency table:

Mark range	10–19	20–29	30–39	40–49	50–59	60–69	70–79	80–89
Mid-point (x)	14.5	24.5	34.5	44.5	54.5	64.5	74.5	84.5
Frequency (f)	8	18	25	22	16	6	3	1

We start by calculating:
$\Sigma f_i x_i = (8 \times 14.5) + \cdots + (1 \times 84.5) = 3965.5$ and
$\Sigma f_i x_i^2 = (8 \times 14.5^2) + \cdots + (1 \times 84.5^2) = 182\,084.75$. Thus:

$$\sigma_n^2 = \frac{1}{99} \left(182\,084.75 - \frac{3965.5^2}{99} \right) \approx 234.79$$

and $\sigma_n = 15.32$ (to 2 d.p.).

A quick check suggests that the calculations are correct since the range of the mid-points is 70 and 15.32 lies comfortably inside the predicted range of $\frac{70}{6} = 11.7$ to $\frac{70}{3} = 23.3$.

The mean is $\frac{1}{99} \times 3965.5 = 40.06$ (to 2 d.p.). Suppose at this stage that the original frequency table is mislaid! Applying the approximate rules we deduce that about two-thirds of the data are in the interval between $(40 - 15) = 25$ and $(40 + 15) = 55$, whilst almost all the data are in the interval -6 (!) to 86.

Calculator practice ───────────────────────────

> *If your calculator is described as 'statistical', then it can probably be used to calculate the mean, standard deviation and variance of grouped data. Find out the correct sequence of buttons to press!*

Computer project ───────────────────────────

> *Computers love numbers! An advantage of a spreadsheet is that you can see what is happening: if you enter the wrong number it is likely to become obvious as the computer performs the calculations. Also, of course, it is easy to correct an error.*
>
> *Write a program to calculate the mean and variance for the data of the previous example. Revise the data by reducing all the marks by 20.*
> *What happens to the mean and variance?*
> *What happens if you now double the previous marks?*

2.20 Variance calculations using coded values

Earlier we introduced the general coding $y_i = \dfrac{x_i - a}{b}$, with $b > 0$. When rearranged, this gives:

$$x_i = a + by_i$$

This coding resulted in the mean, $\bar{y}$, of the coded values being related to the original mean, $\bar{x}$, by:

$$\bar{x} = a + b\bar{y}$$

so that:

$$x_i - \bar{x} = (a + by_i) - (a + b\bar{y}) = b(y_i - \bar{y})$$

Thus:

$$\sum_{i=1}^{n}(x_i - \bar{x})^2 = b^2 \sum_{i=1}^{n}(y_i - \bar{y})^2$$

If we denote the sample variance of the x-values by σ_x^2, and the sample variance of the y-values by σ_y^2, then on dividing the previous equation through by n on both sides we get:

$$\sigma_x^2 = b^2 \sigma_y^2$$

Notes

- Essentially the same formulae apply to grouped data. In this case the original x-values are the class mid-points.
- Writing s_x^2 and s_y^2 for the unbiased estimates of the population variances of the x-values and the y-values, the previous coding leads to the comparable result that:

$$s_x^2 = b^2 s_y^2$$

▼──▼

Example 19

The protein content of milk depends upon a cow's diet. The following observations are the percentages of protein in the milk produced by 25 cows fed on a diet of barley:

> 3.73, 3.33, 3.25, 3.11, 3.53, 3.73, 3.42, 3.57, 3.13, 3.27, 3.60, 3.26, 3.40, 3.24, 3.63, 3.15, 3.00, 3.28, 3.84, 3.57, 3.35, 3.24, 3.66, 3.50, 3.47

Calculate the mean, variance and standard deviation of these data.

───────────────

Calculation of the mean and variance of these data is simplified by using the coding $y = 100(x - 3)$, for which $a = 3$ and $b = \frac{1}{100}$. This results in the y-values:

73, 33, 25, 11, 53, 73, 42, 57, 13, 27, 60, 26, 40,
24, 63, 15, 0, 28, 84, 57, 35, 24, 66, 50, 47

For these y-values we have $\Sigma y_i = 1026$ so that $\bar{y} = \dfrac{1026}{25} = 41.04$.

Hence $\bar{x} = 3 + \dfrac{1}{100} \bar{y} = 3.4104$. The mean is 3.41 (to 2 d.p.).

We also have $\Sigma y_i^2 = 53\,814$, so that:

$$\Sigma (y_i - \bar{y})^2 = \Sigma y_i^2 - \frac{1}{n}(\Sigma y_i)^2 = 53\,814 - \frac{1026^2}{25} = 11\,706.96$$

and hence $\sigma_y^2 = \dfrac{1}{25} \times 11\,706.96 = 468.28$. Hence:

$$\sigma_x^2 = \left(\frac{1}{100}\right)^2 \times 468.28 = 0.046\,828$$

The variance is therefore 0.047 (to 3 d.p.).
The standard deviation is $\sqrt{0.046\,828} = 0.216$ (to 3 d.p).

Exercises 2h

1 The numbers of absentees in a class over a sample period of 24 days were:

0, 3, 1, 2, 1, 0, 4, 0, 1, 1, 2, 3,
1, 0, 0, 2, 4, 6, 4, 2, 1, 0, 1, 1

Find:
(i) the mean number of absentees,
(ii) the modal number of absentees,
(iii) the sample variance of the number of absentees.

2 The numbers of eggs laid each day by 8 hens over a period of 21 days were:

6, 7, 8, 6, 5, 8, 6, 8, 6, 5, 6,
4, 7, 6, 8, 7, 5, 7, 6, 7, 5

Find:
(i) the modal number of eggs laid per day,
(ii) the mean number of eggs laid per day,
(iii) the sample standard deviation of the number of eggs laid per day.

3 The shoe sizes of the members of a football team are:

10, 10, 8, 11, 10, 9, 9, 10, 11, 9, 10

Using the coding $y = x - 10$, where x is the shoe size, determine the variance of this population.

4 A choirmaster keeps a record of the numbers turning up for choir practice on a sample of 18 randomly chosen days.
The numbers are:

25, 28, 32, 31, 31, 34, 28, 31, 29,
28, 32, 32, 30, 29, 29, 31, 28, 28

Using the coding $y = x - 30$, where x is the observed number, determine the value of s_x^2.

5 A biased six-sided die is tossed 60 times giving the following results:

Side of die	1	2	3	4	5	6
Frequency	6	15	2	4	16	17

Without using a calculator (except for the final division), calculate the sample mean and the unbiased estimate of the population variance, showing your working clearly.

6 A driver keeps records of his average mileage per gallon, recording his findings to the nearest integer. His first 25 results are summarised below.

mpg	34	35	36	37	38	39
Frequency	2	4	10	6	2	1

Using the transformation $x = m - 34$, where m is the mpg, and without using a calculator (except for the final division), calculate the sample mean and the unbiased estimate of the population variance, showing your working clearly.

7 A random sample of values of x is:

> 20, 30, 35, 25, 20, 30, 35, 25, 20, 30, 35, 40,
> 30, 35, 35, 25, 20, 40, 20, 25, 25, 30, 20, 20

Using the coding $y = \dfrac{x - 35}{5}$, determine the unbiased estimate of the variance of the population.

8 The prices of a set of books, in £, are as follows:

> 12.95, 12.95, 12.95, 9.95, 16.95,
> 16.95, 16.95, 16.95, 14.95, 14.95

Use a suitable coding to determine the sample mean and the sample variance of these prices.

9 The gap, x mm, in a sample of spark plugs was measured with the following results:

> 0.81, 0.83, 0.81, 0.82, 0.80, 0.81, 0.81
> 0.83, 0.84, 0.81, 0.82, 0.84, 0.80

Use the coding $y = 100x - 80$ to find the unbiased estimate of the population variance of spark plug gaps.

10 A garage notes the mileages of cars brought in for a 15 000-mile service. The data is summarised in the following table.

Mileage ('000 miles)	14–	15–	16–	17–
No. of cars	8	15	13	9

Taking the groups as having mid-points 14 500, ..., 17 500, and using the coding $y = \dfrac{x - 14\,500}{1000}$, where x is the mileage, find the unbiased estimate of the population variance for these grouped data.

11 Each day, x, the number of diners in a restaurant was recorded and the following grouped frequency table was obtained.

x	16–20	21–25	26–30	31–35	36–40
No. of days	67	74	38	39	42

Using the coding $y = \dfrac{x - 18}{5}$, where x is the number of diners, find the value of s_x.

12 Records are kept for 18 days of the midday barometric pressure, in millibars.

> 1022, 1016, 1032, 1008, 998, 985,
> 993, 1004, 1009, 1011, 1015, 1020,
> 1007, 1001, 995, 993, 975, 972

Using a suitable coding, find the value of s.

13 The total scores in a series of basketball matches were:

> 215, 224, 182, 200, 229, 219,
> 209, 217, 195, 162, 210, 213,
> 204, 208, 197, 192, 187, 213

Using a suitable coding, find the sample mean and the sample variance.

14 The heights of a random sample of 100 Christmas trees taken from a field were measured with the following results, where h is the height of a tree in metres.

$0.5 < h \leqslant 1.0$	$1.0 < h \leqslant 1.5$	$1.5 < h \leqslant 2.0$
8	23	48

$2.0 < h \leqslant 2.5$	$2.5 < h \leqslant 3.0$
16	5

Find the grouped mean and the value of s for this set of grouped data.

15 A shopkeeper analyses his sales, in order to determine how much each customer spends. The amount spent is denoted by £c. The results are summarised below.

$0 < c \leqslant 10$	$10 < c \leqslant 15$	$15 < c \leqslant 20$
128	223	148

$20 < c \leqslant 30$	$30 < c \leqslant 50$
56	15

Find the grouped mean amount spent per customer.
Find also the value of s for these grouped data.

16 A machine tests the distance w, measured in thousands of km, that car tyres travel before the tyre wear reaches a critical amount. For a random sample of tyres, the results are summarised as follows.

$0 < w \leqslant 25$	$25 < w \leqslant 30$	$30 < w \leqslant 35$
12	23	48

$35 < w \leqslant 45$	$45 < w \leqslant 60$
15	3

Find the grouped mean for these data.
Find also the unbiased estimate of the population variance based on these grouped data.

17 To test their ability to perform tasks accurately, a class of chemistry students are asked to put precisely one kg of flour into a beaker. The class teacher then chooses six students at random and uses an extremely accurate balance (that records weights in milligrams) to determine the actual amounts of flour.

The results are:

1 000 007, 1 000 006, 999 992, 1 000 015, 999 998, 1 000 000

Obtain the sample mean and the value of s, giving your answers in milligrams, correct to two decimal places.

2.21 Symmetric and skewed data

If a population is approximately **symmetric** then in a sample of reasonable size the mean and median will have similar values. Typically their values will also be close to that of the mode of the population (if there is one!).

A population that is not symmetric is said to be **skewed**. A distribution with a long 'tail' of high values is said to be **positively skewed**, in which case the mean is usually greater than the mode or the median. If there is a long tail of low values then the mean is likely to be the lowest of the three location measures and the distribution is said to be **negatively skewed**

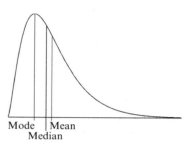

A positively skewed distribution

Various measures of skewness exist. One, known as **Pearson's coefficient of skewness**, is given by:

$$\frac{\text{mean} - \text{mode}}{\text{standard deviation}}$$

If the mode is not known, or if there is more than one, or if there is insufficient data for it to be reliably calculated, an alternative is:

$$\frac{3(\text{mean} - \text{median})}{\text{standard deviation}}$$

An alternative to Pearson's coefficient is the **quartile coefficient of skewness**:

$$\frac{Q_3 - 2Q_2 + Q_1}{Q_3 - Q_1} = \frac{(Q_3 - Q_2) - (Q_2 - Q_1)}{Q_3 - Q_1}$$

This coefficient takes values between -1 (when $Q_2 = Q_3$) and 1 (when $Q_2 = Q_1$). If the median (Q_2) lies midway between the two quartiles then this coefficient has value 0. It is positive if $(Q_3 - Q_2) > (Q_2 - Q_1)$ and negative if $(Q_3 - Q_2) < (Q_2 - Q_1)$.

Note

♦ It is feasible for one coefficient to have a negative value and another to have a positive value! Nowadays, professional statisticians rarely use any of these coefficients: they are included here only for syllabus reasons!

Example 20

The distribution of essay marks (out of 20) in a group of 80 students is as follows:

Mark, x	9	10	11	12	13	14	15	16	17	18
Frequency, f	1	1	4	11	17	19	15	8	3	1

Determine the values of the various measures of skewness.

The quartile coefficient is easily calculated. The quartiles are given by $Q_1 = 13$, $Q_2 = 14$, and $Q_3 = 15$ so that $(Q_3 - 2Q_2 + Q_1) = 0$ and the quartile coefficient is therefore 0.

Pearson's coefficient is, perhaps, more sensitive – but also requires more calculation. The sample mean is 13.8 and the sample standard deviation is 1.68, so, using the mean and the mode (14) we get -0.12, while when using the median we get -0.36.

There is some indication of negative skewness, but the three different formulae give (typically!) rather different values.

Exercises 2i

1 Calculate Pearson's coefficient of skewness:
   (i)   for a set of data having mean 15.0, mode 12.0 and standard deviation 3.1,
   (ii) for a set of data having mean 100, mode 112 and standard deviation 20,
   (iii) for a set of data having mean -0.9, median -1.1 and variance 0.9.

2 Calculate a measure of skewness for a set of data having 25th, 50th and 75th percentiles equal to, respectively, 14, 31 and 73.

3 The numbers of games of squash played in a given week by a random sample of university students were as follows:

No. of games	0	1	2	3	7
No. of students	42	11	2	1	1

Determine Pearson's coefficient of skewness using the mode.

4 Heights of Sitka spruce trees in a plantation, x m, are summarised below.

$x < 1.5$	$1.5 \leqslant x < 2$	$2 \leqslant x < 2.5$
11	58	74

$2.5 \leqslant x < 3$	$3 \leqslant x < 4$
41	6

Determine the quartile coefficient of skewness.

5 The cost (£x) of the purchases by 30 randomly chosen customers in a supermarket are summarised in the following table.

$x < 20$	$20 \leqslant x < 50$	$50 \leqslant x < 80$
3	19	5

$80 \leqslant x < 100$	$100 \leqslant x < 150$
2	1

Determine the quartile coefficient of skewness.

6 A random sample of 100 adults were asked to state which of the numbers 0, 1, 2, 5, 10, 50, 100 was the best approximation to the number of times that they had been to church in the previous year. Their replies are summarised below:

No. of times	0	1	2	5	10	50	100
No. of replies	72	13	7	2	3	2	1

Determine Pearson's coefficient of skewness using the mode.

7 The numbers of letters that are delivered to a particular house are recorded for 30 consecutive days (excluding Sundays).
On 4 days no letters are delivered, on 12 days 1 letter is delivered, on 6 days 2 letters are delivered, on 5 days 3 letters are delivered. On the remaining days more than 3 letters are delivered. Calculate a coefficient of skewness.

8 After an oil spill, local beaches are checked for oiled birds. To get an idea of the nature of the problem, the beaches are divided into 100 m stretches and the numbers of oiled birds are recorded separately for each stretch. Fifty of the records are summarised below.

   0, 1, 5, 2, 19, 47, 21, 8, 7, 4, 0, 1, 1, 0, 0, 0,
   0, 0, 0, 1, 3, 15, 11, 4, 3, 7, 2, 2, 0, 0, 0, 0,
   0, 1, 0, 0, 1, 0, 0, 1, 4, 6, 6, 0, 1, 2, 2, 0, 0, 0

   (i)   Determine the mean, median, standard deviation, lower quartile and upper quartile.
   (ii) Determine Pearson's coefficient of skewness using the mode.
   (iii) Determine Pearson's coefficient of skewness using the median.
   (iv) Determine the quartile coefficient of skewness.

9 A supermarket stocks 184 different types of wine. The prices (£x) are summarised in the following table.

$x < 2$	$2 \leqslant x < 3$	$3 \leqslant x < 4$
7	59	58

$4 \leqslant x < 5$	$5 \leqslant x < 8$	$x > 8$
37	15	8

Determine the value of a coefficient of skewness.

2.22 Standardising to a prescribed mean and standard deviation

By careful choice of the constants a and b, where $b > 0$, it is always possible to use the coding $y = \dfrac{x - a}{b}$ to transform the original x-values to new y-values having some predetermined mean and standard deviation (denoted by $\bar{y}$ and σ_y). Let the mean and standard deviation of the x-values be $\bar{x}$ and σ_x, respectively. The required values are:

$$b = \frac{\sigma_x}{\sigma_y}$$

and:

$$a = \bar{x} - b\bar{y}$$

so that the revised value y corresponding to an original value x is given by the equation

$$y = \bar{y} + \frac{\sigma_y}{\sigma_x}(x - \bar{x})$$

An equivalent expression, which presents the original and standardised values in a pleasingly symmetric form, is

$$\frac{y - \bar{y}}{\sigma_y} = \frac{x - \bar{x}}{\sigma_x}$$

These results may be easily obtained by using the formulae $\bar{x} = a + b\bar{y}$ and $\sigma_x^2 = b^2 \sigma_y^2$ obtained previously.

▼   ▼

Example 21

The mean and standard deviation of a set of exam marks were found to be 40.06 and 15.32, respectively. The school has a policy that all exams should have mean 50 and standard deviation 12. Determine the necessary transformation.

———————

Here x is the original exam mark and y is the required mark. The transformation required therefore has $b = \dfrac{1}{12} \times 15.32 = 1.277$ and $a = 40.06 - (1.277 \times 50) = -23.77$. An original mark of 80 is transformed to a new mark of $50 + \dfrac{(80 - 40.06)}{1.277}$ which is equal to 81 (to the nearest whole number).

▲   ▲

2.23 Calculating the combined mean and variance of several samples

Sometimes we have information in the form of the sample size, the sample mean, and the sample variance for each of several independent samples. We wish to amalgamate the information so as to discover the overall mean and variance of the combined set of data. We illustrate the calculations for the case of two samples having sample sizes n_1 and n_2, sample means $\bar{x}_1$ and $\bar{x}_2$ and sample variances $\sigma_{n_1}^2$ and $\sigma_{n_2}^2$.

The sum of the n_1 observed values in the first sample is $n_1\bar{x}_1$ and the sum of the n_2 observed values in the second sample is $n_2\bar{x}_2$, so that the overall sum of the two sets of observed values is $n_1\bar{x}_1 + n_2\bar{x}_2$. If we denote the overall mean by $\bar{x}$ and the combined sample size by n, then the overall mean is given by:

$$\bar{x} = \frac{(n_1\bar{x}_1 + n_2\bar{x}_2)}{n}$$

With k samples this formula generalises to:

$$\bar{x} = \frac{\displaystyle\sum_{j=1}^{k} n_j\bar{x}_j}{n} \tag{2.15}$$

where $\bar{x}_j$ is the mean of the jth sample, n_j is the size of the jth sample, and $n = \Sigma n_j$.

In order to calculate σ_n^2, the variance of the combined sample, it is necessary first to calculate the combined sum of squares of the observed values. The general formula for a sample variance for a single sample of size n having observations $x_1, \ldots, x_n$ and sample mean $\bar{x}$ is given by the equation:

$$\sigma_n^2 = \frac{1}{n}(\Sigma x_i^2 - n\bar{x}^2)$$

which can be rearranged in the form:

$$\Sigma x_i^2 = n\sigma_n^2 + n\bar{x}^2 = n(\sigma_n^2 + \bar{x}^2)$$

Thus, in the case of two samples, with sample variances $\sigma_{n_1}^2$ and $\sigma_{n_2}^2$, the total sum of squares, T, is given by:

$$T = \{n_1(\sigma_{n_1}^2 + \bar{x}_1^2)\} + \{n_2(\sigma_{n_2}^2 + \bar{x}_2^2)\}$$

For the case of k samples this generalises to:

$$T = \sum_{j=1}^{k} n_j(\sigma_{n_j}^2 + \bar{x}_j^2)$$

The variance of the combined sample is therefore:

$$\sigma_n^2 = \frac{1}{n}\left\{T - \frac{1}{n}\left(\sum_{j=1}^{k} n_j\bar{x}_j\right)^2\right\}$$

Note
- ◆ Equivalent formulae hold when working with unbiased estimates of the population variance. If these are denoted by $s_1^2, s_2^2, \ldots$, for the separate samples, and by s^2 for the combined sample, then by an argument equivalent to that given above:

$$s^2 = \frac{1}{n-1}\left\{\sum_{j=1}^{k}(n_j - 1)s_j^2 + \sum_{j=1}^{k} n_j\bar{x}_j^2 - \frac{1}{n}\left(\sum_{j=1}^{k} n_j\bar{x}_j\right)^2\right\}$$

Example 22

The birthweights of three groups of babies born at St George's Hospital in London between 1982 and 1984 are summarised in the table below. We wish to find the overall mean and standard deviation of the birthweights of the combined sample of 1001 babies.

Group	Number in group	Mean birthweight (g)	Standard deviation (g)
1	353	3353	427
2	401	3478	408
3	247	3587	440

It is much easier to work with coded data. Instead of using the values 3353, 3478 and 3587, we work with 53, 178 and 287, using the coding $y = x - 3300$. We first calculate the combined total coded birthweight (in g) as:

$$(353 \times 53) + (401 \times 178) + (247 \times 287) = 160\,976$$

The overall mean coded birthweight is therefore $\dfrac{160\,976}{1001} = 160.8$

Hence the mean birthweight of the combined set of babies is about $3300 + 161 = 3461$ g.

We next calculate the overall sum of squares of the 1001 coded birthweights:

$$\{353 \times (427^2 + 53^2)\} + \cdots + \{247 \times (440^2 + 287^2)\} = 212\,975\,405$$

Note that without the coding these calculations would have given an even more alarmingly large number!

Since the coding does not involve any scaling ($b = 1$), the standard deviations of the coded values are equal to those of the original values. The variance (in g^2) of the combined data set is therefore:

$$\frac{1}{1001} \left(212\,975\,405 - \frac{160\,976^2}{1001} \right) = 186\,901.1(\text{ to 1 d.p.})$$

and hence the standard deviation of the combined set is equal to $\sqrt{186\,901.1} = 432.3$ g (to 1 d.p.).

The combined data set of 1001 birthweights has a mean of approximately 3461 g and a standard deviation of approximately 432 g.

2.24 Combining proportions

Suppose that we are told that, in a certain population (consisting only of middle class and working class families), 54% of middle class families have a video recorder, whereas the proportion in working class families is just 14%. Without knowledge of the relative sizes of the two classes, all that we can say about the overall proportion of families that have a video recorder is that it lies in the range 14% to 54%.

Suppose we are also told that 63% of all families are middle class with the remainder being working class. We can now be more precise! Suppose there are n families in the population:

Class	Number of families		Proportion with a video recorder		Number with a video recorder
Middle	$0.63n$	$\times$	0.54	$=$	$0.3402n$
Working	$0.37n$	$\times$	0.14	$=$	$0.0518n$
Total	n	$\times$	?	$=$	$0.3920n$

The overall proportion with a video recorder is therefore 39.2%.

Exercises 2j

1 A set of data has mean 10 and variance 16.
   (i) Suppose 10 were added to each observation.
   Determine the mean and variance of the new set of data.
   (ii) Suppose instead that each observation was multiplied by 2.
   What now would be the mean and variance?

2 The mean and standard deviation of a set of marks are found to be 60 and 16. Devise a coding that results in revised marks having mean 50 and standard deviation 10.

3 In a Chemistry test the mean mark (out of 20) was 13 and the standard deviation of the marks was 1.5. The teacher uses coding to adjust the marks to have mean 50 and standard deviation 15. In the test Muggins got 0 and Einstein got 20. What are their coded marks?

4 In an English exam, Smith's original mark is 40 and Brown's mark is 60. After coding Smith's mark becomes 52 and Brown's mark becomes 84. If the mean and standard deviation of the original marks were 45 and 12, determine the mean and standard deviation of the coded marks.

5 A set of marks having mean 60 is coded, to have a mean of 50. Fred's original mark of 50 becomes 42 after coding. His twin sister, Freda, originally got 55.
   Find her coded mark.
   Given that the original marks had a standard deviation of 12.4, determine the standard deviation of the coded marks.

6 A class consists of 18 boys and 12 girls. In the Maths test the average mark is 56. If the mean mark obtained by the boys is 60, determine the mean mark obtained by the girls.

7 Sixty per cent of the students in a university are male. A survey of university students reveals that 58% of males support the Labour party, whereas only 53% of female students support the Labour party.
   What proportion of students in the university support the Labour party?

8 A typist spends 30% of her time typing letters, 40% of her time preparing accounts and 30% of her time working on legal documents. On average she makes 5 mistakes per hour when typing letters, 12 per hour working on accounts and 10 per hour with legal documents.
   Determine her mean number of mistakes per hour.

9 The windspeeds at 10 coastal locations have mean 25 knots and sample standard deviation 4 knots. The windspeeds at 20 inland locations have mean 20 knots and sample standard deviation 5 knots.
   Calculate the mean and sample standard deviation of the combined set of 30 windspeeds.

10 The amounts spent (in pence) by single customers eating at a restaurant varies between males and females. A random sample of 15 female customers finds a mean expenditure of 880, with $s_f = 146$, whereas a random sample of 12 male customers finds a mean expenditure of 1244, with $s_m = 211$.
   Determine the mean amount spent.
   Determine also the value of s for the combined sample of 27 customers.

11 In a certain country, the 20% of the population who are under 16 watch a mean of 3.2 hours of television a day, with a population standard deviation of 1.0 hours. The remainder of the population watch 2.6 hours on average, with a population standard deviation of 1.2 hours. Determine the mean and standard deviation for the entire population.

12 The mean height of a sample of 15 boys is 1.38 m and the mean height of a sample of 20 girls is 1.22 m.
Find the mean height of the combined sample of boys and girls. [ULEAC]

Chapter summary

♦ **Sigma notation:**

$$\sum_{i=1}^{n} x_i = x_1 + x_2 + \ldots + x_n$$

$$\Sigma c = nc$$

$$\Sigma(x + y) = \Sigma x + \Sigma y$$

♦ **Measures of location:**
- The **mode** is the single value that occurs most frequently (if there is one).
- The **mean** is the 'average value', denoted by $\bar{x}$.
  ○ For individual values, $x_1, \ldots, x_n$:
  $$\bar{x} = \frac{\Sigma x_j}{n}$$
  ○ When value x_j occurs with frequency f_j:
  $$\bar{x} = \frac{\Sigma f_j x_j}{\Sigma f_j}$$
  ○ For grouped data, x_j is the mid-point of class j.
- The **median** (Q_2) is the middle value of ordered values.
  ○ With $(2k + 1)$ observations the median is the $(k + 1)$th.
  ○ With $2k$ observations the median is the average of the kth and the $(k + 1)$th.
  ○ With grouped data (n observations) the median is calculated as the value of the '$\left(\frac{n}{2}\right)$th' observation, using linear interpolation.
- **Quartiles** (Q_1, Q_2 and Q_3) and **deciles** divide the ordered data into, respectively, quarters and tenths.

♦ **Measures of spread:**
- The **range** is the difference between the largest and smallest observations.
- The **interquartile range** (IQR) is the difference between the upper and lower quartiles.

- The **variance** has units that are the squares of the units of x.
  - If the observations comprise the entire population, or they represent a sample from a population and we are interested in *the variation within the sample itself*, then the variance is:

$$\sigma_n^2 = \frac{1}{n} \sum_{i=1}^n (x_i - \bar{x})^2$$

$$= \frac{1}{n} \left\{ \Sigma x_i^2 - \frac{1}{n} (\Sigma x_i)^2 \right\}$$

  If $x_1, \ldots, x_n$ constitutes the entire population then σ_n^2 is the **population variance**; otherwise it is the **sample variance**.

  - If the observations constitute a sample from a population, then the quantity referred to as the **unbiased estimate of the population variance**, is:

$$s^2 = \sigma_{n-1}^2 = \frac{1}{n-1} \sum_{i=1}^n (x_i - \bar{x})^2$$

$$= \frac{1}{n-1} \left\{ \Sigma x_i^2 - \frac{1}{n} (\Sigma x_i)^2 \right\}$$

  - If the data are summarised with x_j occurring with frequency f_j, where x_j may denote a class mid-point, then the sample variance is:

$$\sigma_n^2 = \frac{1}{n} \left\{ \Sigma f_j x_j^2 - \frac{1}{n} (\Sigma f_j x_j)^2 \right\}$$

  and the corresponding value of s^2 is given by:

$$s^2 = \frac{1}{n-1} \left\{ \Sigma f_j x_j^2 - \frac{1}{n} (\Sigma f_j x_j)^2 \right\}$$

  where $n = \Sigma f_j$.

- The **standard deviation** (s.d.) is the square root of the variance and has the same units as x.

- **Using coded values:**

  Using the coding $y = \frac{x-a}{b}$, with $b > 0$,

$$\bar{x} = a + b\bar{y}, \qquad \sigma_x^2 = b^2 \sigma_y^2, \qquad s_x^2 = b^2 s_y^2$$

- **Boxplots:**

  - **Box–whisker diagrams** indicate the least and greatest values together with the quartiles and the median.
  - **Refined boxplots** have whiskers whose lengths do not exceed 1.5 times the IQR. Outliers are indicated individually.

♦ **Skewness:**

Pearson's coefficient equals

$$\frac{(\text{mean}-\text{mode})}{\text{s.d.}} \quad \text{or} \quad \frac{3(\text{mean}-\text{median})}{\text{s.d.}}$$

The **quartile coefficient** equals

$$\frac{(Q_3 - 2Q_2 + Q_1)}{(Q_3 - Q_1)}$$

Exercises 2k (Miscellaneous)

1 Shirt sizes are given in multiples of $\frac{1}{2}$. The following data refer to the shirt sizes of a random sample of 250 adult males.

Size	14	$14\frac{1}{2}$	15	$15\frac{1}{2}$	16	> 16
Frequency	19	41	43	53	38	56

For these data calculate, where possible, the values of the mean, median, mode, range and variance.

2 For England and Wales, the percentages of households of various sizes, in 1993, were as follows:

1 person	27
2 people	35
3 people	16
4 people	15
5 people	5
6 or more people	2

Source: *Social Trends, 25*, 1995.

Find the modal class.
Represent the data by a suitable diagram.

3 The total scores in a series of basketball matches were:

215, 224, 182, 200, 229, 219,
209, 217, 195, 162, 210, 213,
204, 208, 197, 192, 187, 213

Use a stem-and-leaf diagram to find the median total score.

4 A market gardener sowed 20 sunflower seeds in each of 100 specially prepared seed trays. The number of seeds, n, that germinated in each of the trays was recorded. The values of n and their frequencies are summarised below.

No. germinating	20	19	18	17	16	15	< 15
No. of trays	53	25	12	6	3	1	0

(*a*) Exhibit the distribution of n using a line graph.
(*b*) Calculate the mean and the mode of the distribution of n.
(*c*) Calculate the overall proportion of seeds that germinated.

5 The numbers of households, in England, receiving local authority home help or home care services were tabulated, in thousands, against the age of the oldest client as follows:

Under 18	3.5
18–64	47.0
65–74	83.8
75–84	207.9
85 and over	143.6

Source: *Social Trends, 25*, 1995

Find the modal class.
Represent the data by a suitable diagram.

6 The midnight temperature is recorded, in °C. The figures for 25 Dec to 6 Jan are:

−3, −2, −5, 1, 3, 2, 2,
0, −4, −7, −8, −4, 2

Find:
(i) the mean temperature,
(ii) the median temperature,
(iii) the variance of the temperatures,
(iv) the standard deviation of the temperatures,
(v) the mean deviation of the temperatures.

7 A region is divided into a lattice of 100 one-metre square quadrats. The number of different plant species is determined for each quadrat, giving the results summarised below.

No. of species	4	5	6	7	8	9
No. of quadrats	1	2	5	9	8	15

No. of species	10	11	12	13	14	15
No. of quadrats	12	8	10	15	10	5

(i) For this set of data determine the mean, median, standard deviation, lower quartile and upper quartile.

(ii) Explain why Pearson's coefficient of skewness using the mode cannot be calculated. Calculate it using the median.

(iii) Determine the quartile coefficient of skewness.

8 The cumulative distribution of the ages (in years) of the employees of a company is given in the following table.

Age	< 15	< 20	< 30	< 40
Cum. Freq.	0	17	39	69

Age	< 50	< 60	< 65	< 100
Cum. Freq.	87	92	98	98

Find:
(i) the median age and the upper and lower quartiles,

(ii) the grouped mean and standard deviation for this population.

9 A grouped frequency distribution of the ages of 358 employees in a factory is shown in Table 1.

Age last birthday	16–20	21–25	26–30
Number of employees	36	56	58

Age last birthday	31–35	36–40	41–45
Number of employees	52	46	38

Age last birthday	46–50	51–60	61–
Number of employees	36	36	0

Table 1

Estimate, to the nearest month, the mean and the standard deviation of the ages of these employees.

Graphically, or otherwise, estimate

(a) the median and the interquartile range of the ages, each to the nearest month,

(b) the percentage, to one decimal place, of the employees who are over 27 years old and under 55 years old.   [ULSEB]

10 On September 1st the frequency distribution of the ages (in completed years) of the pupils in Forms 1–5 in a certain school is given in the following table:

Age (in completed years)	11	12	13	14	15
Frequency	111	119	150	159	161

(i) Draw the cumulative frequency polygon and estimate the median age of these pupils.

(ii) Calculate estimates for the mean and standard deviation of the ages of these pupils.

If, in addition, it is known that the mean and the standard deviation of the ages of the 100 pupils in Form 6 are 16.9 and 0.8 years respectively, find estimates for

(iii) the mean and standard deviation of the ages of all the pupils in the school,

(iv) the median age of all the pupils in the school.   [WJEC]

11 The number of seeds in each of 20 pods from a new variety of flower is summarised in Table 1.

No. of seeds per pod	3	4	5	6	7
No. of pods	2	3	1	4	2

No. of seeds per pod	8	9	10	11
No. of pods	5	2	0	1

Table 1

(a) Determine the median, the mode, the mean and, to 2 decimal places, the standard deviation of the number of seeds per pod for this sample of pods.

(b) Calculate unbiased estimates of the mean, and, to 3 decimal places, the variance of the number of seeds per pod for the population of all pods of this new variety of flower.

(c) Thirty more pods from the same variety of flower are chosen at random. This sample is found to have a mean of 7.2 seeds per pod and a standard deviation of 2.4 seeds per pod. From the combined sample of 50 results, obtain further unbiased estimates of the population mean and, giving your answer to 2 decimal places, the population variance.

(d) State, giving your reasons, which of the two sets of estimates would be expected to be closer to the true population values.   [ULSEB]

12 (i) At a university, a random sample of 100 students was taken and each student recorded his/her intake of milk (in ml) during a given day. The results are summarized in Table 1.

Milk intake	<25	25–	50–	100–	150–
No. of students	1	3	20	48	11

Milk intake	200–	300–	500–	700–	800–
No. of students	11	4	1	1	0

Table 1

(a) Draw a histogram to illustrate these data.

(b) Estimate the mean milk intake, explaining the limitations of your calculation.

(c) Draw a cumulative frequency curve to fit these data. From your curve, estimate, to the nearest 5 ml, the median intake of milk on that day for all students at this university.

(d) State, with reasons, whether you consider the mean or the median to be the more appropriate measure of the milk intake of students at the university on that day.

(ii) A measuring rule was used to measure the length of a rod of stated length 1 m. On 8 successive occasions the following results, in millimetres, were obtained.

999 1000 999 1002
1001 1000 1002 1001

Calculate unbiased estimates of the mean and, to 2 significant figures, the variance of the errors occurring when this rule is used for measuring a 1 m length. **[ULSEB]**

13 The table given below shows a grouped frequency distribution of the recorded heights, measured to the nearest centimetre, of 50 girls.

Height (cm)	102–105	106–107	108–109
No. of girls	14	16	10

Height (cm)	110–111	112–115
No. of girls	8	2

Find estimates of
(i) the median of the heights,
(ii) the upper quartile of the heights,
(iii) the proportion of the girls whose heights exceed 108.8 cm. **[WJEC]**

14 In each of the twenty Olympic Games this century, there has been a Men's Discus event. The distance of the Gold medal throw has ranged from 36.04 m in 1900 to 68.82 m in 1988. The total distance of the winning throws comes to 1082.64 m. What is the mean distance?

(a) The variance of the Men's Gold medal distances is $102.110 \, \text{m}^2$. Calculate the total of the squares of the distances.

(b) Women have had a Discus event in each of the 14 games since 1928. The mean of the Women's Gold medal throws is 56.34 m. Calculate the mean distance for Men's and Women's events combined.

(c) The total of the squares of the Women's Gold medal distances is 46 074.28. Use this information to calculate the variance of the distances for Men's and Women's events combined.

(d) Comment briefly on the validity of combining the two collections of data. **[UODLE]**

15 Summarised below are the values of the orders (to the nearest £) taken by a sales representative for a wholesale firm during a particular year.

Value of order (£)	Number of orders
less than 10	3
10–19	9
20–29	15
30–39	27
40–49	29
50–59	34
60–69	19
70–99	10
100 or more	4

(a) Using interpolation, estimate the median and the semi-interquartile range for these data.

(b) Explain why the median and the semi-interquartile range might be more appropriate summary measures for these data than the mean and standard deviation. **[ULEAC]**

16 The table below shows the age (at last birthday) at which women married in 1986 in England and Wales.

Age (in yrs)	16–20	21–24	25–29	30–34
Women (in tens of thousands)	6	12	8	3

Age (in yrs)	35–44	45–54	55–99
Women (in tens of thousands)	3	1	1

Draw a histogram and a cumulative frequency diagram to illustrate these data.

Hence estimate

(i)   the number of women who were aged 40 or over when they married,

(ii) the median age of marriage for women.
                                                      [O&C]

17 Measurements of the time intervals between successive arrivals of telephone calls at an office exchange were taken.

The first 100 time intervals were recorded and the following grouped frequency distribution was obtained.

Time interval (x mins)	Frequency
$0 < x \leqslant 0.5$	39
$0.5 < x \leqslant 1.0$	23
$1.0 < x \leqslant 2.0$	23
$2.0 < x \leqslant 3.0$	9
$3.0 < x \leqslant 6.0$	6

(i)   Draw a histogram to illustrate this distribution.

(ii) Calculate, showing your working, estimates for the mean and the standard deviation of the distribution.

(iii) Explain briefly which aspects of the data are measured by the mean and the standard deviation.   [JMB]

18 On September 1st 1992 the grouped frequency distribution of the ages (in completed years) of 1000 pupils aged under 16 in a comprehensive school was as given in the following table.

Age (in completed years)	11	12	13	14	15
Frequency	165	184	216	231	204

(i)   Calculate, to three significant figures, estimates for the mean and standard

deviation of the ages of these pupils on September 1st 1992.

(ii) Draw a cumulative frequency polygon and estimate, to three significant figures, the median age of the pupils on September 1st 1992.

(iii) Given in addition that there were 222 pupils aged 16 or over, estimate, to three significant figures, the median age of all the pupils in the school on September 1st 1992.
                                                      [NEAB]

19 The weekly consumption of cheese in ounces has been estimated for 50 participants in a nutrition study. The figures are given below.

3.89	3.80	4.01	3.84	3.91
4.16	3.98	3.87	3.97	4.04
3.96	4.12	4.05	4.03	4.02
4.07	3.90	4.03	3.94	3.91
3.97	3.91	3.98	4.05	4.03
4.16	4.09	4.13	4.07	4.00
4.06	3.97	4.07	3.90	3.91
4.02	4.20	4.11	3.99	4.02
4.01	4.01	4.05	4.18	3.99
4.21	3.77	3.96	3.84	3.83

(*a*) Construct a stem and leaf diagram of these data.

(*b*) Find the quartiles.

(*c*) Represent the data by a box and whisker plot.

(*d*) Using classes of common width and taking the first class to be 3.75–3.79, form a grouped frequency distribution from the data and represent this grouped distribution by a suitable histogram.

(*e*) Give a brief summary of the main features of the distribution of consumption of cheese by the 50 participants.   [O&C]

20 Auditem Ltd., an accounting firm, recorded the time, x minutes, to the nearest minute, taken to audit each account. The values of x below are those recorded for a random sample of accounts they have audited.

37	33	24	36	31	31	24	51
31	47	40	40	55	42	30	34
41	36	42	46	34	38	33	42
56	37	39	36	31	30	45	50
43	41	46	41	30	51	36	21
32	34	62	43	46	34	34	56
32	62	30					

(*a*) For these data,

(i)   construct a stem and leaf diagram,

(*continued*)

(ii) find the median and quartiles,

(iii) draw a box plot.

(b) Write down which of the mode, the median and the mean you would prefer to use as a representative value for these data. Justify your choice.   [ULEAC]

21 Give **one** advantage and **one** disadvantage of grouping data in a frequency table.

The table shows the trunk diameters, in centimetres, of a random sample of 200 larch trees.

Diameter (cm)	15–	20–	25–	30–	35–	40–50
Frequency	22	42	70	38	16	12

Plot a cumulative frequency curve of these data.

By use of this curve, or otherwise, estimate the median and the interquartile range of the trunk diameters of larch trees.

A random sample of 200 spruce trees yields the following information concerning their trunk diameters, in centimetres.

Minimum	Lower quartile	Median	Upper quartile	Maximum
13	27	32	35	42

Use this data summary to draw a second cumulative frequency curve on your graph.

Comment on any similarities or differences between the trunk diameters of larch and spruce trees.   [AEB 93]

22 The table shows the distribution of ages of school pupils in the United Kingdom in 1984.

Age in completed years	Number of pupils (1000)
2 to 4	887
5 to 10	4140
11	825
12 to 14	2631
15 to 16	1183
17 to 18	210

(a) What is the age range represented by the entry 11 in the table? Explain what is meant by the number 2631 in the table. How many pupils are represented in this table?

(b) Calculate an estimate of the mean age of pupils.

(c) For each of the classes calculate its frequency density and on graph paper draw a histogram of the data. Comment briefly on the distribution of ages of pupils.   [UODLE]

23 (a) Data are often presented in graphical form rather than in their raw state.

Give

(i) **one** reason for using graphical presentation,

(ii) **one** disadvantage of graphical presentation.

Explain briefly the difference **in use** between a *bar diagram* and a *histogram*.

(b) Electric fuses, nominally rated at 30 A, are tested by passing a gradually increasing current through them and recording the current, x amperes, at which they blow. The results of this test on a sample of 125 such fuses are shown in the following table.

Current (xA)	Number of fuses
$25 \leqslant x < 28$	6
$28 \leqslant x < 29$	12
$29 \leqslant x < 30$	27
$30 \leqslant x < 31$	30
$31 \leqslant x < 32$	18
$32 \leqslant x < 33$	14
$33 \leqslant x < 34$	9
$34 \leqslant x < 35$	4
$35 \leqslant x < 40$	5

Draw a histogram to represent these data.

For this sample calculate

(i) the median current,

(ii) the mean current,

(iii) the standard deviation of current.

A measure of the *skewness* (or asymmetry) of a distribution is given by

$$\frac{3(\text{mean} - \text{median})}{\text{standard deviation}}.$$

Calculate the value of this measure of skewness for the above data. Explain briefly how this skewness is apparent in the shape of your diagram.   [JMB]

24 In an attempt to devise an aptitude test for applicants seeking work on a factory's assembly line, it was proposed to use a simple construction puzzle. As an initial step in the evaluation of this proposal, the times taken to complete the puzzle by a random sample of 95 assembly line employees were observed with the following results.

Time to complete puzzle (seconds)	Number of employees
10–	5
20–	11
30–	16
40–	19
45–	14
50–	12
60–	9
70–	6
80–100	3

Draw a cumulative frequency diagram to represent these data. Hence, or otherwise, estimate the median and the interquartile range.

Calculate estimates of the mean and the standard deviation of this sample.

It is decided to grade the applicants on the basis of their times taken, as good, average or poor.

Method *A* states that the percentages of applicants in these grades are to be approximately 15%, 70% and 15% respectively. Estimate the grade limits.

Method *B* grades applicants as

good, if the time taken is less than (mean−standard deviation),

poor, if the time taken is more than (mean+standard deviation),

average, otherwise.

Compare methods *A* and *B* with respect to the percentages in each grade, and comment.

[JMB]

25 (a) Give an example of data for which the most appropriate measure of location might reasonably be
      (i) the mode,
      (ii) the median,
      (iii) the mean.
    (b) As part of a work study investigation for the Royal Mail, a daily record was made

for each of six days of the number of letters, *x*, delivered to each of the 175 private houses on a particular postal route. The table below summarises the results for the 1050 possible deliveries.

Number of letters delivered daily	Percentage of deliveries
0	13.2
1	26.7
2	18.9
3	15.8
4	10.5
5	5.2
6	2.6
7	1.1
$\geqslant 10$	6.0

Construct a suitable pictorial representation of these data.

Calculate the median and the interquartile range for the number of letters delivered daily.

For these data give **two** reasons why the interquartile range is a more appropriate measure of dispersion than the standard deviation.   [NEAB(P)]

26 The following table shows a grouped frequency distribution of the gross annual earnings of 110 employees at a certain factory in 1988.

Gross Earnings	Number of employees
up to £5000	14
above £5000 and up to £6000	25
above £6000 and up to £7000	30
above £7000 and up to £10 000	25
above £10 000 and up to £15 000	13
above £15 000	3

(a) Estimate graphically, or by calculation,
    (i) the median and the semi-interquartile range of the gross earnings of these 110 employees, giving **each** answer correct to the nearest £,   (*continued*)

(ii) the percentage of the employees whose gross earnings exceeded £9000, giving your answer to the nearest whole number.

(b) Explain why it is not possible, from the above data, to obtain reliable estimates of the mean and the standard deviation of the gross earnings.

(c) More detailed information on the gross earnings of the 110 employees in 1988 showed that the mean was £7470 and the standard deviation was £3550. During the year, the employees' union were negotiating for the mean gross earnings to be increased in 1989 to £8000. Find the constant percentage increase across the board that would meet this claim; give your answer correct to the nearest whole number.

If this percentage was granted find the value of the standard deviation of the employees' gross earnings in 1989.

Suppose, instead, that the employers decided to give each employee an extra £500 in 1989. In this case, find the mean and the standard deviation of the 110 employees' gross earnings in 1989.

[WJEC]

27 Over a period of four years a bank keeps a weekly record of the number of cheques with errors that are presented for payment. The results for 200 accounting weeks are as follows.

Number of cheques with errors (x)	Number of weeks (f)
0	5
1	22
2	46
3	38
4	31
5	23
6	16
7	11
8	6
9	2

($\Sigma fx = 706$; $\Sigma fx^2 = 3280$)

Construct a suitable pictorial representation of these data.

State the modal value and calculate the median, mean and standard deviation of the number of cheques with errors in a week.

Some textbooks measure the *skewness* (or asymmetry) of a distribution by

$$\frac{3(\text{mean} - \text{median})}{\text{standard deviation}},$$

and others measure it by

$$\frac{(\text{mean} - \text{mode})}{\text{standard deviation}}.$$

Calculate and compare the values of these two measures of skewness for the above data.

State how this skewness is reflected in the shape of your graph.   [AEB 90]

3 Special summary statistics

There is no cure for birth and death save to enjoy the interval

Soliloquies in England, George Santayana

3.1 Standardised birth and death rates

Birth rates are usually measured on an annual basis, by dividing the total number of births in the year by the total number of people in the population in the preceding year. If no account is taken of the nature of the population then we are said to have calculated a **crude** rate, whereas if some account is taken of the nature of the population then the rate is said to have been **standardised**.

Suppose there are two countries, A and B. Suppose that the proportion of the population who are women aged between 20 and 30 is 30% in one country, but 5% in the other. Suppose that the birth rate for these young women is an average of 0.8 children per person per year in *both* countries, while the birth rate for the remainder of the population is 0.05 children per year in *both* countries. In other words the birth rates are really the same, but the compositions of the two populations are very different.

Reference		Country A	Country B
1	Population size	N_A	N_B
2	Proportion of population that are women aged 20–30	0.30	0.05
$3 = 1 \times 2$	Number of women aged 20–30	$0.30\,N_A$	$0.05\,N_B$
4	Birth rate for the above	0.8	0.8
$5 = 3 \times 4$	Number of births to above	$0.24\,N_A$	$0.04\,N_B$
6	Proportion of population that are *not* women aged 20–30	0.70	0.95
$7 = 1 \times 6$	Number in rest of population	$0.70\,N_A$	$0.95\,N_B$
8	Birth rate for the above	0.05	0.05
$9 = 7 \times 8$	Number of births to above	$0.035\,N_A$	$0.0475\,N_B$
$10 = 5 + 9$	Total number of births	$0.275\,N_A$	$0.0875\,N_B$
$11 = 10 \div 1$	**Crude national birth rate**	**0.275**	**0.0875**

The result of failing to notice the difference in the compositions of the two countries is that the birth rates appear to be very different.

To avoid this problem, it is usual to report rates that have been standardised to some reference population.

Example 1

Comment on the following data which refer to deaths of white males in
the USA as a consequence of heart disease during the years 1968, 1970,
and 1972.

Age	Population (10^6)	Deaths (10^3)	Death rate (10^{-3})
	1968		
Younger	9.7	3.0	0.31
Middle-aged	19.5	73.3	3.76
Older	8.3	144.8	17.45
	1970		
Younger	9.7	2.9	0.30
Middle-aged	19.6	70.8	3.61
Older	8.6	140.6	16.35
	1972		
Younger	10.1	2.6	0.26
Middle-aged	19.7	68.9	3.50
Older	8.9	142.6	16.02

Calculate standardised death rates, based on the 1970 figures, and
comment on the results.

The table shows that:
* The population is increasing.
* The population designated as older increased by more than that
  designated as younger (so that the population has 'aged').
* The incidence of heart disease is decreasing over the years in all age groups.
* Heart disease is far more prevalent in the older group.

If we just look at the *numbers* of deaths, the decrease in heart disease
will be under-estimated because no account will have been taken of the
increasing population size. That is why we calculate death *rates*. The crude
death rate for 1968 is the total number of deaths, divided by the
population size. For 1968 this is:

$$\frac{221.1 \times 10^3}{37.5 \times 10^6} = 5.90 \times 10^{-3}$$

with the corresponding figures for 1970 and 1972 being 5.65×10^{-3} and
5.53×10^{-3} respectively. However, these crude rates do not take account
of the ageing of the population noted above.

To standardise the data, we arbitrarily assume that the breakdown of
the population was as in 1970 *throughout* the period. The proportions in
the three age groups in 1970 were 25.59%, 51.72%, and 22.69%. So the
standardised death rate for 1968 is calculated as:

$(0.2559 \times 0.31) + (0.5172 \times 3.76) + (0.2269 \times 17.45) = 5.98$

per thousand, with the 1972 figure being 5.51 per thousand.

After standardising we have a fall from 5.98 to 5.51 per thousand (a drop of 0.47 per thousand), whereas the crude figures diminished the drop to $5.90 - 5.53 = 0.37$ per thousand, by failing to allow for the ageing of the population.

▲ ▲

Exercises 3a

1   In Ruritania 80% of families are Royalist and the rest are Republican.

   In Transylvania the proportions are 40% Royalist and 60% Republican. Suppose that, in both countries, 10% of Royalists own horses, whereas only 1% of Republicans own horses.
   (i)   Show that the crude horse-owning rates are in the ratio 41 to 23.
   (ii) Suppose that the rates fall to 9% of Royalists and 0.5% of Republicans. Determine the new crude rates. Is the ratio still 41 to 23?

2   Farmer Giles has three sorts of pigs: Large White, Tamworth and Saddleback. Forty per cent of his pigs are Large White, and he has equal numbers of Tamworths and Saddlebacks.

   Farmer Ham has the same three types of pigs, but he has sixty per cent Saddlebacks and only ten per cent Tamworths.

   Farmer Giles does his utmost to look after his pigs, but (taken as a whole) their average litter size is only 16.3, compared with the average litter size of 16.4 for Farmer Ham's pigs, which live in much worse conditions.

   National records show that Large Whites have an average of 16.1 offspring per litter, compared with 15.4 for Tamworth and 16.9 for Saddlebacks.
   (i)   Show that, after taking account of pig types, the average litter size for Father Giles' pigs is greater than the national average.
   (ii) Show that, after taking account of pig types, the average litter size for Father Ham's pigs is smaller than the national average.

3   A country contains people of three different races (white, coloured and black). Nationally 95% of whites have received a full education, but the numbers for coloureds and blacks are just 35% and 5%, respectively.

   Two of the cities in this country are called Maytown and Jaytown. In Maytown, 10% of the population are white, 25% are coloured and the rest are black. The proportions for Jaytown are 8%, 30% and 62%.
   In which city would you expect to find the larger proportion of people that had had a full education?

3.2 The weighted mean and index numbers

A company gives all its employees a £1000 pay rise, and wants to know the consequent rise *in percentage terms* in its annual wage bill. To find the answer we need to know either all the original wages, or, both the number of employees and the original total wage bill. Suppose that we have the following information:

Staff type	Old wage (£'000)	New wage (£'000)	Number of employees
Junior	15	16	24
Middle	20	21	8
Senior	25	26	8

We begin by calculating the original total wage bill (in £'000) for the 40 employees. This was:

$$(24 \times 15) + (8 \times 20) + (8 \times 25) = 720$$

so the old mean wage was $\frac{720}{40} = 18$ thousand pounds. Note that the average wage was not a simple average of the three possible wages, but was **weighted** by the numbers of employees involved. Without any effort we have calculated a **weighted mean**! If we have a number of values (x) and associated weights (w), then the weighted average is:

$$\frac{\Sigma w_i x_i}{\Sigma w_i}$$

which is simply the usual formula $\dfrac{\Sigma f_i x_i}{\Sigma f_i}$ in disguise.

The new wage bill is:

$$(24 \times 16) + (8 \times 21) + (8 \times 26) = 760$$

so the overall percentage increase is:

$$\frac{(760 - 720)}{720} \times 100 = 5.56\%$$

For every £100 that the firm used to pay, it will now pay £105.56.

Suppose we wish to contrast this increase with that for another department that uses the same pay scales but has 21 Junior, 3 Middle and 1 Senior member of staff. The old total bill for this department was £400 000, while the new bill is £425 000. This is an increase of 6.25%: for every £100 that the firm used to pay it will now pay £106.25.

Reporting the changes by reference to the convenient yardstick of £100 makes for easy comparisons. The £ sign is redundant so far as the comparisons are concerned so that a summary using **index numbers** would be as shown in the following table:

	First department	Second department
Index for original wage	100	100
Revised index	105.56	106.25

Exercises 3b

1 A box of fruit contains ten apples, thirty bananas and forty lemons. If the average weight of the apples is 140 g, the average weight of the bananas is 130 g and the average weight of the lemons is 115 g, determine the average weight of the fruit in the box.

2 The mark awarded to a Mathematics student is a weighted average of the student's coursework and her exam mark, with the weights being 4 to 1 in favour of the exam mark.
If the student gets 53% in her exam and 62% in her coursework, determine her overall mark.

3 A sports shop stocks four grades of badminton racket. These sell at £15, £25, £30 and £50. Given that the shop has, respectively, 20, 12, 8 and 2 of the four types, determine the mean cost of the badminton rackets in the shop.

4 A particular type of postage stamp occurs in two versions. According to the stamp catalogue one version is worth 20p while the other is worth 60p. If the average value of this type of stamp is 23p and if there are 5 million of the cheaper version in circulation, determine how many there are of the dearer version.

5 (i) If the price of eggs rises by 10% and the old price index for eggs was 120, determine the new price index.
(ii) If the price of bacon rises by 8% and the old index was 116, determine the new index.
(iii) An 'all-day breakfast' consists of eggs costing 40p, bacon costing 35p, and other items costing 62p. The eggs and bacon are now subject to the price rises noted in parts (i) and (ii), while the other breakfast ingredients remain at their old price.
If the old price index for the breakfast was 100, determine the new price index.

6 The Jones family give regular parties. For an average party they buy 6 bottles of wine and 2 kg of a special cheese. The Smiths are trying to keep up with the Joneses, so they buy the same sorts of wine and cheese. However, for their parties, they buy an average of 7 bottles of wine and 1.8 kg of cheese.

In 1990 wine cost an average of £3.10 a bottle and cheese cost an average of £5.25 per kilogram. In 1995 both these prices had risen by £1.

Denoting the cost of a 1990 party by the index number 100, find, for each family, the 1995 index number.

7 The April 1995 version of the *Monthly Digest of Statistics* records the changes in the numbers of various types of road transport using indices.

Year 19–	Cars & taxis	Buses & coaches	Vans	Motor-cycles
87	147	126	131	108
88	157	134	145	97
89	171	140	160	96
90	173	142	161	90
91	173	149	168	87
92	173	142	166	73
93	173	142	164	68

(i) Convert these to indices with 100 corresponding to the 1987 value.
(ii) Plot these results on a time-series graph.
(iii) Which form of transport has shown the greatest percentage increase between 1987 and 1993?

3.3 Price indices

Indices are widely used as summaries of change in financial contexts by governments and industries. A common idea is to imagine buying a basket of commodities, which might be items of food, electrical goods, train tickets or a mixture of all of these! One index commonly quoted on the news refers to a 'basket' of shares and gives shareholders an immediate impression of the general behaviour of the stock market. Using an index enables us to obtain a quick and easy comparison of the *overall* price (or value) of the items in the basket at different points in time. However, there are several ways that this might be done.

The **Laspeyres price index**, L, fixes the contents of the 'basket' and monitors the subsequent price changes using the formula:

$$L = \frac{\Sigma p_n q_o}{\Sigma p_o q_o} \times 100 \qquad (3.1)$$

In this formula p refers to *price* and q to *quantity*, while the suffix 'o' refers to the original reference year (the **base year**) and the suffix 'n' to the current year. The summations are over the items in the 'basket'.

The **Paasche price index**, P, recognises that we live in a changing world! Today's restaurant customers may prefer prawn cocktail, where twenty years ago their predecessors would have chosen tomato soup. The Paasche index continually updates the contents of the 'basket', and asks how much today's basket would have cost in earlier years. Note that changes in the basket contents may also reflect changes in commodity prices: if a commodity's price increases then consumers will buy less of that commodity. The formula for Paasche's index is

$$P = \frac{\Sigma p_n q_n}{\Sigma p_o q_n} \times 100 \qquad (3.2)$$

Note

♦ Traditionally indices were reported as integers (with base 100). With increased se of computers more 'accurate' indices are reported: thus, at the time of writing, the 'FT all-share index' is reported as standing at 1098.98.

Example 2

Find the Laspeyres and Paasche price indices for the following 'basket of commodities'. The contents of the basket have changed with the passage of time because of changes in the habits of consumers (which may be partly in response to price changes). The base year is 1980.

Commodity	1980 Price (pence) p_o	1980 Quantity q_o	1990 Price (pence) p_n	1990 Quantity q_n
A	112	12	145	15
B	26	40	38	50
C	500	2	600	8

We first calculate
$$\Sigma p_n q_o = (145 \times 12) + (38 \times 40) + (600 \times 2) = 4460$$
and
$$\Sigma p_o q_o = (112 \times 12) + (26 \times 40) + (500 \times 2) = 3384$$
so that the Laspeyres price index is equal to $\dfrac{4460}{3384} \times 100 = 132$ (to the nearest whole number).

For Paasche's index we calculate:
$$\Sigma p_n q_n = (145 \times 15) + (38 \times 50) + (600 \times 8) = 8875$$
and
$$\Sigma p_o q_n = (112 \times 15) + (26 \times 50) + (500 \times 8) = 6980$$
so that the Paasche index equals $\dfrac{8875}{6980} \times 100 = 127$ (to the nearest whole number). Note that the Paasche index has an appreciably smaller value than the Laspeyres index because of the increased demand for commodity C, whose individual price rose by only 20% between 1980 and 1990.

3.4 Record prices and record earnings!

We often read that something has been sold for a record price. But does this mean that it has really become more valuable? Not necessarily!

In the sports pages we learn that a golfer has achieved record yearly earnings and that a footballer has been transferred for a record fee. Does this make them the greatest players ever? Again, the answer is not necessarily!

The underlying cause for all these statements is likely to be inflation. One way of judging the *real* value of something is to ask what else we could buy with the same amount of money. If this has increased, then the original item *has* increased in value.

One statistic often quoted is the **Retail Price Index** (the RPI), which is based on a weighted combination of the prices of a 'basket' of commodities. A commodity whose apparent value rises in step with the RPI is holding its value.

Example 3

The table below gives the January prices of Kellogg's Cornflakes for the years from 1985 to 1990 (as reported by *Shaw's Guide to Retail Prices*) together with the changing RPI values (with January 1985 taken as the base year).

Year	RPI (1985 = 100)	Cornflakes (Giant packet) Price (in p)
1985	100.0	86.5
1986	105.5	97
1987	109.6	97
1988	113.3	89
1989	121.7	96
1990	131.0	99

The RPI has risen from 100 to 131 over the 5-year period: this means that prices generally have risen by nearly one third during this period (this is referred to as **inflation**). Suppose that my salary has risen exactly in proportion to the RPI (if only that were so!). Would I be able to buy more cornflakes with my money in 1990 than I could in 1985?

———

The easiest way of answering this question is to convert the Cornflakes prices into indices, with the 1985 price being indexed to 100. This is accomplished by use of the formula $\frac{100p_n}{p_o}$. The ratio $\frac{100}{p_o}$ is $\frac{100}{86.5}$ which equals 1.156. Multiplying the prices for each year by 1.156 we get:

Year	RPI (1985 = 100)	Cornflakes price (1985 = 100)
1985	100.0	100.0
1986	105.5	112.1
1987	109.6	112.1
1988	113.3	102.9
1989	121.7	111.0
1990	131.0	114.5

During 1986 and 1987 Cornflakes were more expensive in real terms (indices greater than the RPI), but subsequently they became noticeably cheaper in real terms.

An alternative, but equivalent question, is to ask 'What is the *real* price of Cornflakes, given the cost of living?'. We can answer this by dividing the third column in the previous table by the second column, and multiplying by 100, to obtain:

Year	RPI (1985 = 100)	Cornflakes price (1985 = 100)	Cornflakes price in 1985 money
1985	100.0	100.0	100.0
1986	105.5	112.1	106.3
1987	109.6	112.1	102.3
1988	113.3	102.9	90.8
1989	121.7	111.0	91.2
1990	131.0	114.5	87.4

Note that the final column can be calculated directly from the original data, using, for example:

$$\frac{99 \times 100}{86.5 \times 131} \times 100 = 87.4$$

The general formula (using r to denote the RPI) is:

$$\frac{p_n r_0}{p_0 r_n} \times 100$$

Exercises 3c

1 A small 'shopping basket' consists of milk, bread and fish. In 1990 the basket contains 200 g of meat, 400 g of bread and 50 g of fish, with the prices per 100 g being, respectively, £1.20, £0.80 and £1.60. In 1995 the basket contains 150 g of meat, 300 g of bread and 80 g of fish, and the prices per 100 g are now £1.50, £1.00 and £1.50. Taking the index for 1990 as being 100, determine:
   (i) the Laspeyres index,
   (ii) the Paasche index.

2 If the price of apples rises by 5% during a period in which the retail price index rises from 111 to 118, have apples become relatively cheaper or more expensive?

3 An alcoholic consumes wine at £3 a bottle and beer at £1 a can. One week he consumes 4 bottles of wine and 20 cans of beer. Six months later the alcoholic is, amazingly, still alive and consumes 5 bottles of wine and 12 cans of beer, with the prices now being £3.50 and £1.10.

Taking an original index of 100, determine, for the second time period:
(i) the Laspeyres index and
(ii) the Paasche index,
giving your answers to one decimal place.

4 Indices of the total prices of various types of household goods or services are given in the table below.

Year	Furn-iture	Electrical goods	China etc.	DIY	TV hire & repair
1990	100	100	100	100	100
1992	109	103	109	112	100
1994	132	115	118	127	100

The total amounts spent (in millions of pounds) for 1990 were as follows: Furniture, 7.3; Electrical goods, 7.4; China etc, 2.1; DIY, 4.2 and TV hire and repair, 1.4.
(i) Show that the Laspeyres index for the 1992 total amount spent on these items is 107.0.
(ii) Find the corresponding Laspeyres index for 1994.

3.5 Time series

Government and industry are particularly concerned with forecasting the future. The basis of their forecasts of the future is what has happened in the past. For example, studying past records of the numbers of deaths in each month enables us to predict that next winter there will be more deaths than in the following summer. Similarly, the knowledge that there is almost always a rise in the number out of work in late summer (because of an influx of school leavers) will be taken into account in studying next year's figures.
   A time series is likely to be composed of a number of different components.

An underlying trend
Most financial series display an underlying upwards trend because of inflation (for example, the Cornflakes data of Example 3).

Periodicity

There may be regular cycles in the data. An obvious financial example is the pre-Christmas sales boom. There are also many examples in nature. If the periodicity is seasonal, then its effect can be allowed for: the government often reports **seasonally adjusted** figures, which are simply the raw figures from which the seasonal component has been subtracted.

▼   ▼

Example 4

An example of a naturally occurring set of periodic data is given in the table below, which shows the maximum numbers of common sandpipers recorded in Essex between January 1989 and December 1990 (as reported in the Essex bird reports for those years).
Illustrate the data in an appropriate fashion.

Year	Jan	Feb	Mar	Apr	May	Jun	Jul	Aug	Sep	Oct	Nov	Dec
1989	10	7	5	10	39	7	260	316	142	11	4	9
1990	9	7	8	24	87	6	243	214	87	11	9	10

———

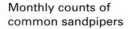

Monthly counts of common sandpipers

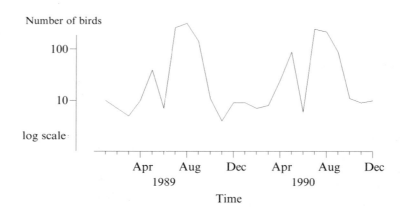

The numbers are mostly small, but contain a few high values. To make the graph fit into a smaller area it is sensible to use a logarithmic scale for the *y*-axis.

The autumn peak of birds flying south to winter in the warmer African climate is very apparent. By comparison, the northerly passage in spring almost passes unnoticed.

▲   ▲

Cycles

What happens at one point in time usually resembles what happened at the previous point in time. This results in a series of values which slowly vary without any fixed periodic pattern.

Physical time series may also demonstrate what might be called semi-periodicity. One of the best known is the '11-year' sunspot cycle: there is a huge increase in the numbers of sunspots roughly every 11 years, though successive peaks may be between 9 and 14 years apart. The precise reasons for either the variability in numbers, or the lack of precise periodicity remain to be established.

Example 5

A remarkable example of semi-periodicity from the animal kingdom concerns the numbers of lynx trapped in the neighbourhood of the Mackenzie River in Canada between the years 1821 and 1934. A portion of this data set is given in the table below. Illustrate the data using a time series graph.

Decade	0	1	2	3	4	5	6	7	8	9
1820s		269	321	585	871	1475	2821	3928	5943	4950
1830s	2577	523	98	184	279	409	2285	2685	3409	1824
1840s	409	151	45	68	213	546	1033	2129	2536	957
1850s	361	377	225	360	731	1638	2725	2871	2119	684

Once again we use a logarithmic scale for the *y*-axis to make the plot easier to view.

Variations in
numbers of
trapped lynx,
1821–1934

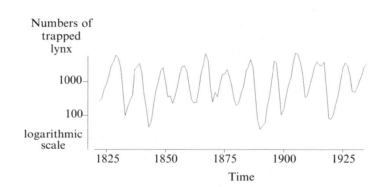

The reason for the cycle is not a variation in the amount of trapping, but a variation in the size of the lynx population. The Canadian lynx lives almost entirely on the snowshoe rabbit and the two populations have compensating cycles. When there are few lynx the rabbit population increases rapidly in the relatively predator-free conditions. Those lynx present therefore have lots to eat and their population increases fast. This results in the near extermination of the rabbits, and the subsequent starvation of the lynxes. Each complete cycle generally takes between 9 and 10 years.

Random variation

This is due to the cumulative effects of all the short-term unpredictable influences.

The analysis of time series data is very difficult: it has been said that if two statisticians are presented with the same time series they will come up with not just two, but three, separate explanations of the data! In this chapter we have room for just one approach.

Moving averages

Suppose we have ordered observations $x_1, x_2, x_3, \ldots$ These observations may be affected by all of the components listed earlier. However, we can highlight the underlying trend by looking at moving averages. Typical choices involve 3, 4 (for seasonal data), 5, 7 (for weekly data), or 12 (for monthly data) data points. The procedure is illustrated below for the cases 3, 5 and 7:

	Range of i	Value
Observation	1 to n	x_i
3-pt moving average	2 to $(n-1)$	$\frac{1}{3}(x_{i-1} + x_i + x_{i+1})$
5-pt moving average	3 to $(n-2)$	$\frac{1}{5}(x_{i-2} + x_{i-1} + x_i + x_{i+1} + x_{i+2})$
7-pt moving average	4 to $(n-3)$	$\frac{1}{7}(x_{i-3} + x_{i-2} + x_{i-1} + x_i + x_{i+1} + x_{i+2} + x_{i+3})$

▼ ▼

Example 6

The following data give the typical price (in US cents) of a pound of copper.

Year, 19–	70	71	72	73	74	75	76	77	78	79
Price	63	51	49	80	92	57	62	60	61	90

Year, 19–	80	81	82	83	84	85	86	87	88
Price	98	78	67	71	62	64	62	80	119

Determine, and plot, the 3-year and 7-year moving averages for these data.

———————

The calculations are simplified if running totals are kept, as illustrated below.

Year	Price	3-year total	3-year moving average	7-year total	7-year moving average
1970	63				
1971	51	163	54.3		
1972	49	$163 + 80 - 63 = 180$	60.0		
1973	80	$180 + 92 - 51 = 221$	73.7	454	64.9
1974	92	$221 + 57 - 49 = 229$	76.3	$454 + 60 - 63 = 451$	64.4
1975	57	$229 + 62 - 80 = 211$	70.3	$451 + 61 - 51 = 461$	65.9
1976	62	$211 + 60 - 92 = 179$	59.7	$461 + 90 - 49 = 502$	71.7
1977	60	$179 + 61 - 57 = 183$	61.0	$502 + 98 - 80 = 520$	74.3
1978	61	$183 + 90 - 62 = 211$	70.3	$520 + 78 - 92 = 506$	72.3
1979	90	$211 + 98 - 60 = 249$	83.0	$506 + 67 - 57 = 516$	73.7
1980	98	$249 + 78 - 61 = 266$	88.7	$516 + 71 - 62 = 525$	75.0
1981	78	$266 + 67 - 90 = 243$	81.0	$525 + 62 - 60 = 527$	75.3
1982	67	$243 + 71 - 98 = 216$	72.0	$527 + 64 - 61 = 530$	75.7
1983	71	$216 + 62 - 78 = 200$	66.7	$530 + 62 - 90 = 502$	71.7
1984	62	$200 + 64 - 67 = 197$	65.7	$502 + 80 - 98 = 484$	69.1
1985	64	$197 + 62 - 71 = 188$	62.7	$484 + 119 - 78 = 525$	75.0
1986	62	$188 + 80 - 62 = 206$	68.7		
1987	80	$206 + 119 - 64 = 261$	87.0		
1988	119				

After calculating the first total, each subsequent total is created from its predecessor by adding in the latest value and discarding the earliest value.

This method minimises the amount of calculation, but care is needed since any error would affect all subsequent calculations. It is therefore wise to check the last total by direct summation. In the example we can verify that $(62 + 80 + 119) = 261$ and that $(67 + 71 + 62 + \cdots + 119) = 525$. In the diagram the 7-year average is noticeably less erratic than the 3-year average. Both suggest a gradual overall rise in price.

3-year and 7-year moving averages of copper prices in US cents per lb

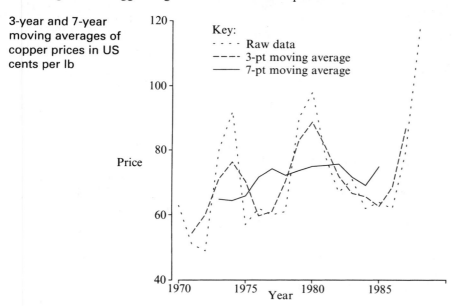

4-point and 12-point moving averages

Yearly data are often presented in the form of quarterly or monthly information. It is natural to take averages over an entire year, which requires averaging over four or twelve time points. The procedure is slightly different because 4 and 12 are even numbers! The result is that the middle time points do not correspond to times at which observations were taken. This is not a problem so far as plotting the data is concerned.

If averages are required that do correspond to the time points to which the data refer, then this can be achieved by averaging successive 4-point averages to produce a so-called **centred average**.

Example 7

The sales of shoes in a shop in successive quarters are shown below.

Year	Jan–Mar	Apr–Jun	Jul–Sep	Oct–Dec
1988	2100	1800	600	2460
1989	2140	2160	820	2300
1990	2460	2280	910	2800
1991	2580	2500	1130	2980

Treating each month as being equally long, calculate 4-point moving averages. Calculate also averages centred on the quarters.

We arrange the data as a single column with gaps between successive values. The averages (shown in italics) are produced following a succession of totals:

Observation	Sum of two observations	Sum of four observations	Average	Sum of two averages	Centred average
2100					
	2100 + 1800 = 3900				
1800					
	1800 + 600 = 2400	3900 + 3060 = 6960	*1740*		
600				3490	*1745*
	600 + 2460 = 3060	2400 + 4600 = 7000	*1750*		
2460				3590	*1795*
	2460 + 2140 = 4600	3060 + 4300 = 7360	*1840*		
2140				3735	*1867.5*
	4300	4600 + 2980 = 7580	*1895*		
2160				3750	*1875*
	2980	4300 + 3120 = 7420	*1855*		
820				3790	*1895*
	3120	2980 + 4760 = 7740	*1935*		
2300				3900	*1950*
	4760	3120 + 4740 = 7860	*1965*		
2460				3952.5	*1976.25*
	4740	4760 + 3190 = 7950	*1987.5*		
2280				4100	*2050*
	3190	4740 + 3710 = 8450	*2112.5*		
910				4255	*2127.5*
	3710	3190 + 5380 = 8570	*2142.5*		
2800				4340	*2170*
	5380	3710 + 5080 = 8790	*2197.5*		
2580				4450	*2225*
	5080	5380 + 3630 = 9010	*2252.5*		
2500				4550	*2275*
	3630	5080 + 4110 = 9190	*2297.5*		
1130					
	4110				
2980					

In the table, the fourth column gives the 4-point averages and the final
column gives the corresponding centred averages. The centred averages
rise in each successive period, indicating a consistent rise in the underlying
sales pattern.

Periodicity in shoe sales

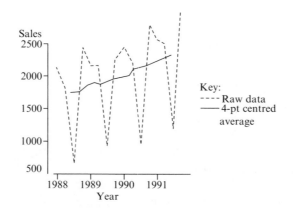

Estimating periodic effects

In the previous section we saw how to compute an average value (a_i, say) corresponding to a time point i for which there was an observed value x_i. There are two simple ways of evaluating the extent of any periodic effects:

1 The difference $x_i - a_i$

2 The ratio $\frac{x_i}{a_i}$

It is not easy to choose which of these to use. If a plot of the data suggests that in each cycle the amplitudes of the changes are comparable, then $x_i - a_i$ should be used. This is called the **additive model**. On the other hand, if in successive cycles the amplitudes appear to be changing in size in proportion to the general level (for example, as a consequence of inflation or an increasing population size), then a **multiplicative model** is appropriate and $\frac{x_i}{a_i}$ should be used.

▼ ▼

Example 8

Estimate the periodic effects for the shoe sales data of Example 7
(i) using the additive model, and (ii) using the multiplicative model.

(i) We begin by calculating the differences:

Sales	Centred average	Difference
x_i	a_i	$x_i - a_i$
600	1745	−1145
2460	1795	665
2140	1867.5	272.5
2160	1875	285
820	1895	−1075
2300	1950	350
2460	1976.25	483.75
2280	2050	230
910	2127.5	−1217.5
2800	2170	630
2580	2225	355
2500	2275	225

We now arrange the calculated differences in a table that shows the years and periods to which the differences relate.

Year	Jan–Mar	Apr–Jun	Jul–Sep	Oct–Dec
1988			−1145	665
1989	272.5	285	−1075	350
1990	483.75	230	−1217.5	630
1991	355	225		
Total	1111.25	740	−3437.5	1645
Average	*370.4*	*246.7*	*−1145.8*	*548.3*

There is little difference between quarters 1, 2 and 4, but sales in quarter 3 are considerably depressed. Since the amplitudes of the differences do not appear to be changing greatly with time, this additive model is probably satisfactory.

(ii) The calculations for the multiplicative model follow a similar pattern.

Sales x_i	Centred average a_i	Ratio $\dfrac{x_i}{a_i}$
600	1745	0.344
2460	1795	1.370
2140	1867.5	1.146
2160	1875	1.152
820	1895	0.433
2300	1950	1.179
2460	1976.25	1.245
2280	2050	1.112
910	2127.5	0.428
2800	2170	1.290
2580	2225	1.160
2500	2275	1.099

The nature of the periodic effect is displayed in the following table:

Year	Jan–Mar	Apr–Jun	Jul–Sep	Oct–Dec
1988			0.344	1.370
1989	1.146	1.152	0.433	1.179
1990	1.245	1.112	0.428	1.290
1991	1.160	1.099		
Total	3.551	3.363	1.205	3.839
Average	*1.184*	*1.121*	*0.402*	*1.280*

We see that the sales in the third quarter are estimated as being only about 40% of what might otherwise have been expected. There is little difference amongst the other quarters.

Predicting the future

The last three sections have been concerned with the underlying trend and the seasonal effects. Combining these ideas we can estimate the future – but we should not estimate too far ahead because our prediction errors are likely to get ever larger!

In Chapter 20 we will discuss methods for fitting straight lines through a set of data points. If there appears to be an underlying linear trend over the most recent data points, then, if we are prepared to sacrifice some accuracy, a simple procedure is to estimate the underlying trend by joining with a straight line the first and last of the recent centred averages. This line can then be extended forward and, using the periodic increments or multipliers, estimated values for future periods can be obtained.

Example 9

Predict the quarterly shoe sales for the first two quarters of 1992 (i) using the additive model, and (ii) using the multiplicative model from Example 8.

(i) There were 16 sales figures in the original data set. The centred average for the third quarter of 1988 was 1745, while that for the second quarter of 1991 (11 data points further on) was 2275. A straight-line fit implies that the change per data point is given by $\frac{1}{11}(2275 - 1745) = 48.182$. Starting with 2275 we then obtain the subsequent trend values by successive additions as shown below. The 'adjustments' shown in the table are the estimated seasonal effects found earlier.

Year	Period	Trend value	Adjustment	Estimated	True
1991	Apr–Jun	2275			
	Jul–Sep	2323.182	−1145.8	1177	1130
	Oct–Dec	2371.364	548.3	2920	2980
1992	Jan–Mar	2419.546	370.4	2790	
	Apr–Jun	2467.728	246.7	2714	

The estimates are reported to the same accuracy as the original data. The small differences between the estimates for the final two quarters of 1991 and the true values given in the original data suggests that the additive model provides a good description of the data. The predictions should be quite reliable.

(ii) After the trend values have been calculated the remaining calculations for the predictions based on the multiplicative model (using the multipliers estimated previously) are shown below.

Year	Period	Trend value	Multiplier	Estimated	True
1991	Apr–Jun	2275			
	Jul–Sep	2323.182	0.402	930	1130
	Oct–Dec	2371.364	1.280	3040	2980
1992	Jan–Mar	2419.546	1.184	2860	
	Apr–Jun	2467.728	1.121	2770	

The relatively large error in the prediction of the sales for Jul–Sep 1991 suggests that these predictions may be less reliable than those from the additive model. However, the estimates for the first two quarters of 1992 are very similar to those obtained by the additive model in part (i).

▲ ▲

Exercises 3d

1 A factory making shoes works three shifts a day. The output, in hundreds of pairs of shoes, during one week is shown in the table below.

	Day				
Shift	1	2	3	4	5
I	16.4	16.2	17.0	17.2	17.4
II	15.1	15.8	15.9	15.9	16.5
III	14.1	14.5	15.1	15.4	15.7

Display the data on a time series graph. Calculate the 3-point moving averages and display these on the same graph.

2 The numbers of meals served during a three-week period in a college canteen, open five days a week, were as shown in the table below.

Week	Mon	Tue	Wed	Thu	Fri
1	182	217	253	211	138
2	165	212	231	210	131
3	154	198	232	195	127

(i) Plot the data, together with a suitable moving average.

(ii) Estimate the numbers of meals required on each day of the following week.

3 The numbers of current TV licences in the United Kingdom are given, for the months of 1994, in the following table. The numbers reported are in thousands and all values are reduced by 19 million. The figure '224' therefore represents 19 224 000 TV licences!

Jan	Feb	Mar	Apr	May	Jun
224	305	524	442	463	509

Jul	Aug	Sep	Oct	Nov	Dec
577	602	649	667	626	667

Source: *Monthly Digest of Statistics*, April 1995

Plot these data on a time series graph together with the corresponding 5-point moving averages.

4 A householder's quarterly electricity bills, in £, over a period of 3 years, were as follows.

Year	Q1	Q2	Q3	Q4
1	112	137	161	154
2	133	147	188	161
3	163	184	219	201

(i) Plot the data together with an appropriate moving average.
(ii) Estimate the bill for each of the next three quarters.

5 The numbers of unemployed in the United Kingdom are given for selected months in the following table. The figures given in the table are the numbers of thousands in excess of 2 million, so that the figure 711 corresponds to 2 711 000 unemployed.

Year	Feb	May	Aug	Nov
1992	711	708	845	864
1993	1042	917	960	769
1994	841	653	638	423

Source: *Monthly Digest of Statistics*, April 1995

Plot these data on a time series graph.
(i) Is there evidence of periodicity?
(ii) Determine the 4-point (annual) moving average corresponding to the observed time points.
Is there evidence of a linear trend?
(iii) Assuming a linear trend starting in February 1993, estimate the number of unemployed in February 1995. (The actual number was 2 459 000.)

6 The numbers of marriages in Scotland for the quarter years starting from the fourth quarter of 1991 up to the fourth quarter of 1994 are given (in thousands) in the table below.
(i) Is there evidence of periodicity?
(ii) Determine the 4-point (annual) moving average corresponding to the observed time points.
Is there evidence of a trend?
(iii) Estimate the number of marriages in the first quarter of 1995.

Year	Q1	Q2	Q3	Q4
1991				7.0
1992	4.5	10.1	13.5	6.9
1993	4.2	9.7	12.8	6.6
1994	3.9	9.1	12.1	6.4

Source: *Monthly Digest of Statistics*, April 1995

Computer project

It is easy to see from the huge tables of calculations shown above that the calculations associated with the analysis of time series are long and tedious. However, they are also simple and repetitive. This is precisely where computers come into their own.
Use a spreadsheet to repeat the calculations in the previous examples.
You can probably get it to draw your graphs as well!

Chapter summary

♦ **Standardised birth and death rates** are the rates referred to a standard population with specified numbers in the various age categories.

♦ The **weighted mean** of values $x_1, x_2, \ldots$, with weights $w_1, w_2, \ldots$, is given by:

$$\frac{\Sigma w_i x_i}{\Sigma w_i}$$

♦ The **index number** for some value (e.g. sales figures) is the ratio of this value to a reference value (usually a corresponding value from a previous time point). The ratio is usually multiplied by 100.

♦ There are several price indices. Denoting price by p, and quantity by q, with suffixes n and o referring to a new year and an old year,

  • The **Laspeyres price index**, L, is given by:

  $$L = \frac{\Sigma p_n q_o}{\Sigma p_o q_o} \times 100$$

  • The **Paasche price index**, P, is given by:

  $$P = \frac{\Sigma p_n q_n}{\Sigma p_o q_n} \times 100$$

♦ The **3-point moving average** corresponding to the item x_i in a time series of n points is (for $2 \leqslant i \leqslant n - 1$):

$$\frac{(x_{i-1} + x_i + x_{i+1})}{3}$$

Similar results apply to other moving averages having odd numbers of time points.

♦ The **4-point moving average** corresponding to the item x_i in a time series of n points is the average of:

$$\frac{(x_{i-2} + x_{i-1} + x_i + x_{i+1})}{4} \quad \text{and} \quad \frac{(x_{i-1} + x_i + x_{i+1} + x_{i+2})}{4}$$

Similar results apply to other moving averages having even numbers of time points.

♦ **Periodic effects** may be estimated using the average (over all items in the time series that refer to the given period) of the difference $x_i - a_i$ (additive model) or the ratio $\frac{x_i}{a_i}$ (multiplicative model), where a_i is the moving average value corresponding to the observed x_i.

Exercises 3e (*Miscellaneous*)

1 A teacher is introducing the concept of weighted averages and index numbers. She uses as her example the cost of a typical breakfast for one person. Her typical breakfast consists of egg, bacon, bread, butter and tea. She has the following information available.

Item	Quantity	Cost 1986	Cost 1990
Bacon	1 lb	£0.88	£1.60
Butter	$\frac{1}{2}$ lb	£0.42	£1.00
Egg	1 dozen	£0.36	£1.20
Bread	1 loaf	£0.35	£0.65
Tea	$\frac{1}{4}$ lb	£0.44	£1.08

(*a*) One student suggested that a reasonable estimate for the cost of a typical breakfast in 1986 would be obtained by adding together all the costs in that column and dividing the total by 5. His teacher was not impressed. Put forward **two** criticisms of this proposed method.

(*b*) The teacher went on to say that the sort of quantities consumed in this typical breakfast were:

$\frac{1}{10}$ lb of bacon, $\frac{1}{30}$ lb of butter, 1 egg, 2 slices of bread ($=\frac{1}{10}$ of a loaf), $\frac{1}{100}$ lb of tea.

Calculate a realistic cost for such a meal in 1986.

(*c*) By 1990 the costs of the items had risen to that shown in the 1990 column. Assuming the same consumption as in 1986, calculate an index number for the cost of a typical breakfast in 1990 using 1986 as a base.   [UODLE]

2 The table below shows part of the calculation of a simple unweighted index of wage costs for a firm that employs semi-skilled workers, skilled workers, clerical staff and supervisory staff. Copy the table, filling in the missing values indicated by dots.

	DATA Weekly wage rates (£)		
	1980	1984	1988
Semi-skilled	52	72	82
Skilled	79	93	110
Clerical	58	71	76
Supervisory	88	126	160

	CALCULATION Wage rates relative to 1980 (1980=100)		
	1980	1984	1988
Semi-skilled	100	138.46	157.69
Skilled	100	117.72	•
Clerical	100	•	•
Supervisory	100	•	•
Index (overall average index of wage costs (1980=100)	100	130.44	•

What drawback does this index have in indicating the firm's wage costs in these three years?

In 1980, the firm employed 30 semi-skilled workers, 40 skilled workers, 20 clerical staff and 10 supervisory staff. Calculate the firm's total weekly wage bill for these workers in 1980. Calculate also these totals for 1984 and 1988, assuming the numbers of workers remained the same. Use these totals to construct an index, based on 1980=100, for 1984 and 1988.

Do you consider this new index to be an improvement on the original unweighted index? Why?

What further information would you wish to have in judging whether the new index is satisfactory? Supposing this information were available, describe briefly how you would use it in index construction. Discuss whether there would still be any drawbacks in your procedure.   [O&C]

3 Most students at a certain university either live in the university's own halls of residence or rent accommodation in the neighbourhood. This rented accommodation is broadly divided into three types (*A*, *B*, *C*) by the university's accommodation office. The table below shows part of the calculation of an index, based on 1982 = 100, of accommodation using the 'Laspeyres method of base-year weighting'; copy the table, filling in the missing values indicated by dots.   (*continued*)

DATA			
	Weekly rents (£) 1982 1986 1990		Numbers of students 1982
Halls	19	23	29	2780
Rented *A*	26	30	39	1668
Rented *B*	32	38	46	1034
Rented *C*	34	40	50	412
			Total = 5894

CALCULATION			
	1982	1986	1990
Halls	19×2780 =52820	23×2780 =63940	•
Rented *A*	26×1668 =43368	•	•
Rented *B*	•	•	•
Rented *C*	•	•	•
Total	•	169752	•
Index	100.00	118.47	•

What drawback does this method of calculation have in providing an index of accommodation cost over the period 1982/1986/1990? How can this be overcome?

Suppose that, in each year from 1982 onwards, student grants had been increased by 5% compared with the previous year. Use the index you have calculated to judge whether this increase has kept pace with the cost of accommodation over the period 1982/1986/1990.   [O&C]

4 The Table shows the quarterly sales of beer at a supermarket in thousands of litres for the four years 1989 to 1992.

Year	Q_1	Q_2	Q_3	Q_4
1989	49	56	84	60
1990	53	62	90	65
1991	58	62	95	71
1992	63	67	99	70

(a) Plot the time series.

(b) Calculate four-quarter moving averages and plot them on the same graph as the time series. *Note: You are not required to find centred moving averages.*

(c) Comment on what your graph shows.
        [O&C]

5 The following time series is for an index of sales of a certain product.

		Quarter			
		1	2	3	4
	1985	104	114	124	107
	1986	109	126	138	117
Year	1987	125	137	150	130
	1988	136	148	155	138
	1989	144	160	170	151

Plot the time series on a graph. Give an approximate indication on your graph of what you think will happen to the series in the first two quarters of 1990. What assumptions are implicit in your approximate predictions?

Does it appear that there is any cyclical (as opposed to seasonal) component in this series? Assuming that there is not, use suitable moving averages to find the underlying trend. Plot the trend values on your graph.   [O&C]

6 Explain briefly the purpose of time series analysis. Write down the simple additive model and define each term used in your model.

The sales of a product, in thousands, which show a marked seasonal variation are given below for May 1985 to April 1988.

	Sales		
Year	Jan/Feb	Mar/Apr	May/June
1985			10.3
1986	6.8	9.0	10.9
1987	7.0	9.5	11.6
1988	7.3	9.9	

	Sales		
Year	July/Aug	Sept/Oct	Nov/Dec
1985	12.8	20.3	32.4
1986	14.4	23.1	35.4
1987	15.3	24.5	37.4
1988			

(*continued*)

Corresponding 6 point centred moving averages are

| 15.3 | 15.5 | 15.9 | 16.4 | 16.6 | 16.7 |
| 16.8 | 16.9 | 17.1 | 17.4 | 17.6 | 17.6 |

As sales manager in charge of this product it is your responsibility to make sales forecasts. Assuming that the general rate of increase of sales is maintained, estimate graphically the trend values for each of the next two two-month periods of 1988.

Find the seasonal component for the period July/August and estimate the sales that will occur in the period July/August 1988.

Explain briefly how you would make use of the residuals from your calculations, (DO NOT CALCULATE THE RESIDUALS).

[ULSEB]

7 A market stallholder sells clothes on three days a week – Tuesday (T), Friday (F), and Saturday (S). Her takings, £x, over a five week period were as follows

Week	1			2		
Day	T	F	S	T	F	S
x	196	210	343	267	274	336

Week	3			4		
Day	T	F	S	T	F	S
x	168	279	315	160	258	310

Week	5		
Day	T	F	S
x	154	242	312

(a) Plot the data together with a suitable moving average.
(b) On one of the 15 days a nearby clothes stall was closed. Suggest which day this was.
(c) On another day the stallholder overslept and opened late. Suggest which day this was.
(d) Forecast the takings on Tuesday of week 6. Indicate how you have made the forecast and discuss whether or not your method would be suitable for forecasting the takings on Tuesday of week 26. [ULSEB]

8 What is the purpose of time series analysis? Explain briefly the decomposition of a time series.

The table below shows the number of houses sold per quarter by an estate agent.

Year	Houses Sold			
	1st qtr	2nd qtr	3rd qtr	4th qtr
1987	160	300	275	180
1988	200	325	310	215
1989	225	350	320	240
1990	245	370	–	–

Estimate the sales for each of the 3rd and 4th quarters of 1990.
(You may assume that the trend line through the moving averages can be represented by

Trend$=230.67+6.33t$,

where t takes the values 1,2..., and $t = 1$ for the 3rd quarter of 1987.)

[ULSEB]

9 The following table records the distribution in 1987, classified by sex and by age group, of the population of England and Wales and of the North of England. It also records the number of deaths for each sex and age group in the North of England.

	Population (thousands)			
	England and Wales		North of England	
Age	Males	Females	Males	Females
Under 1	339.9	324.4	20.3	19.4
1–14	4477.4	4243.7	277.6	262.9
15–24	4095.0	3932.2	241.9	238.4
25–44	7126.1	7052.9	434.3	425.0
45–64	5321.4	5457.2	337.7	347.7
65–74	1989.6	2499.0	122.6	155.8
75 & over	1143.1	2240.7	64.1	129.1
Totals	24492.5	25750.1	1498.5	1578.3

	Number of deaths North of England	
Age	Males	Females
Under 1	203	149
1–14	44	49
15–24	194	72
25–44	582	358
45–64	4247	2635
65–74	5946	4129
75 & over	7477	11359
Totals	18693	18751

Source: *HMSO, Mortality Statistics, 1987*

(continued)

(i) *Using the figures for the North of England only*, draw, on graph paper, a population pyramid showing the age distribution of males and females.

(ii) The mean age of males in the North of England is 36.6 years with standard deviation 22.0 years. The mean age of females in the North of England is 39.8 years with standard deviation 23.7 years. Referring to these statistics and to your population pyramid, write a brief comparison of the age distributions of males and females in the North of England.

(iii) Obtain an estimate of the death rate for females in the North of England standardised on the age distribution of females in England and Wales.

The death rates for England and Wales are 11.4 per thousand for males and 11.1 per thousand for females. The death rate for males in the North of England, standardised on the age distribution of males in England and Wales is 12.8 per thousand.

Comment on these **four** death rates.　　[JMB]

4 Data sources

It is a capital mistake to theorize before one has data

The Memoirs of Sherlock Holmes, Sir Arthur Conan Doyle

One way of classifying data is in terms of the data collector: data that one has collected oneself is sometimes referred to as **primary data**, whereas data that has already been collected by somebody else may be called **secondary data**. The advantages and disadvantages of these two types of data are summarised in the table below:

Type	Advantage	Disadvantage
Primary data	We collected it so we understand any curiosities that it contains.	We collected it so there were probably not very many observations.
Secondary data	There will often be a very large number of observations.	We did not collect the data, so it may not really answer the question that interests us.

In this short chapter:

(a) we discuss methods of obtaining primary data, emphasising the possible pitfalls,

(b) we list the principal sources of secondary (national) data.

Later, in Chapter 13, we discuss strategies for attempting to ensure that the data collected are representative of the population of interest.

4.1 Data collection by observation

This has been the standard method of data collection for millennia. The famous theories, such as Newton's Theory of Gravitation and Einstein's Theory of Relativity, all have their roots in numerical data collected by careful observation. On a more mundane level, decisions concerning local traffic flow (e.g. 'Would it help to replace the crossroads by a roundabout?') are based on observations of flow made by video cameras or teams of observers.

The collection of data of a scientific nature (e.g. physical, chemical, biological data) relies almost exclusively on observation. However, in the last two centuries there has been increasing interest in the social sciences (e.g. sociology, politics, economics) for which other methods of data collection are relevant.

4.2 Methods of data collection by questionnaire (or survey)

The most common method for collecting social science data is by means of a **questionnaire** which consists of a series of questions concerning the facts of someone's life or their opinions on some subjects. The recipient of a questionnaire is usually referred to as the **respondent**.

There are three principal methods of collecting the data using a questionnaire:

1 Face-to-face interview
2 By post
3 By phone

The face-to-face interview

In this case the interviewer and the respondent communicate directly, either in a street interview (in which the interviewer selects passers-by for interview) or in an interview in the respondent's home.

Advantages

◆ **Complex structure** The structure of the questionnaire (e.g. 'If answer is "Yes" then go to question 23c') can be relatively complicated, since only the interviewer needs to understand it.

◆ **Consistency** If the interviewer does the writing, then the questionnaire will be completed in a consistent fashion.

◆ **Help** If the respondent has difficulty understanding a question, then the interviewer is available to give an explanation.

◆ **Response rate** The **response rate** is defined as the number of interviews completed divided by the number attempted. Assuming that the interviewer is friendly, this is likely to be quite high (say 70%).

Disadvantages

◆ **Expense** The procedure uses up a lot of time for each interviewer. There may also be costs associated with the travelling between respondents.

◆ **Bias** Although the questionnaire is completed in a consistent fashion, this consistency may contain bias (e.g. the interviewer consistently misinterprets an answer, or gives misleading guidance).

◆ **Lack of anonymity** A respondent may refuse to answer questions because of being embarrassed by the presence of the interviewer.

The 'postal' questionnaire

Here we mean any questionnaire that is given out for self-completion and return by (anonymous) respondents. An example would be a questionnaire about school food consisting of questions on two sides of paper, to be returned by 'posting' in a box at the end of a lunchtime.

The principal advantage of this method of gathering information is:

◆ **Economy** Since no interviewer is required, it is a cheap method of collecting data.

However, set against this advantage is a major disadvantage:

◆ **Non-response** The response rate (measured as the proportion of questionnaires that are returned) can be very low indeed (e.g. 10%) and is rarely greater than 50%. This low level of response is a problem because the replies received are unlikely to be representative of those of the population as a whole. People who take the trouble to fill in and return a questionnaire are not typical (it is well known that 'apathy reigns O.K.')! If the response rate is very low then the replies may be seriously misleading.

The telephone interview

Telephone interviews are occasionally used by market research organisations as a cheaper alternative to face-to-face interviews in the case of short questionnaires. A major problem with a telephone interview (apart from the would-be respondent putting the phone down!) is that it is difficult to relate the information obtained to the population as a whole, because the people interviewed will not be representative.

In Britain in the 1990s, only about 85% of households possess a phone, while about 25% of domestic subscribers are ex-directory. Consequently only about two-thirds of British households are listed in the telephone directory.

When this problem is compounded with that of non-response it is easy to see that the reliability of telephone surveys is rather doubtful. If you read in a newspaper that a telephone survey has unearthed some interesting new facts about British society, then our advice would be to take this information with a pinch of salt.

4.3 Questionnaire design

To ask someone a series of questions might seem to be a ridiculously simple task, but this is certainly not the case. It is easy accidentally to create unanswerable questions, while small changes to the wording can make a difference to the answer obtained. Even the order of questions needs careful thought.

Some poor questions

1 Do you think that boys or girls have the better dress sense or is it simply the influence of their parents?
   Unanswerable! This 'question' is at least two questions and is unlikely to be understood by anyone (including its author!).

2 Does your family watch a lot of television?
   Unanswerable! Some family members may be TV addicts, whereas others scarcely ever watch. Also 'a lot' is not a well-defined quantity.

3 Do you think that Statistics is:
   (a) a very interesting subject,
   (b) an interesting subject,
   (c) quite an interesting subject?
   A biased set of choices.

4 Are you alive?
   Not worth asking! Avoid questions that will be answered the same way by everyone (or almost everyone).

5 I am going to ask you about the Monarchy. Bertrand Russell once said ... [something long and rambling taking several minutes to read]. Do you agree?
   Avoid long questions – the respondent will forget what the question is about.

6 You are against the death penalty, aren't you?
This is a leading question – the respondent is being pressurised into saying 'Yes'. The defence counsel would object!

7 What do you think of the verisimilitude of this simulacrum?
Avoid unfamiliar words.

8 Are you aged
   (a) over 30,
   (b) under 21,
   (c) under 18?
When giving a range of alternatives make sure that they are non-overlapping and include all possibilities.

9 When they are not playing at home, Arsenal are not a good side at scoring goals. Do you agree or disagree?
Avoid double (or multiple) negatives – some respondents will misunderstand the question.

10 Please don't be embarrassed by this question: do you pick your nose?
But for the preamble, many respondents would have answered the question without worry. Don't invite respondents not to respond!

11 Where were you on March 7th?
Unless this question is asked soon afterwards, it is unlikely to get a response! Questions about the distant past are likely to require the respondent to guess.

12 Are you a communist?
Since communists are rather out of fashion at present, some supporters of communism are unlikely to own up. Respondents tend to give 'socially acceptable' answers.

Some good questions

The best questions are probably those that have been used in surveys conducted by market research or other organisations that specialise in asking questions. From their experience they will know which questions work well. A large public library may be able to help with this.

♦ Books on survey methods may contain example questionnaires.

♦ The 'quality' newspapers may report questions asked in national surveys by an organisation such as Gallup.

♦ The survey organisations themselves may publish questionnaire details.

Most good questions are *short* and *simple*.
The same applies to questionnaires!

The order of questions

Two general rules are:

♦ Start with easy questions.
   This encourages the respondent to participate.

♦ Ask general questions (e.g. 'How satisfied are you with school lunches?') first, and specific questions (e.g. 'What do you dislike most about school lunches?') afterwards.

> *This is to avoid the 'satisfaction' question being influenced by the subsequent 'dislike' question.*

Some questions occur naturally before others. For example, if one were investigating a respondent's history, it would be natural to begin with questions about childhood before questions about middle-age.

Question order and bias
The order in which questions are asked can influence a respondent's reply. Contrast:

1 Do you intend to be an organ donor?
2 Did you know that dozens of people die each year because there are not enough organ donors?

with:

1 Did you know that dozens of people die each year because there are not enough organ donors?
2 Do you intend to be an organ donor?

Filtered questions
Many questionnaires have what might be described as 'miss-out sections' (flagged by statements such as 'If NO then go to Q24'). Thus a question such as:

> How much money did you earn last week?

should not precede:

> Were you employed last week?

since, if the answer to the second question is 'No', then the first question should not be asked (it should be **filtered** out).

Open and closed questions
An **open question** is one in which there are no suggested answers:

> What is your opinion of the Prime Minister?

The advantage of this type of question is that the respondent can choose precisely how to answer. The disadvantage is that every respondent may answer in a different way, making it difficult to summarise the data obtained.

A **closed question** is one in which there is a prescribed set of alternative answers:

> How do you think the present Government compares with others that we have had?
> Is it (i) above average, (ii) average, (iii) below average?

With a closed question the respondent may find difficulty because none of the alternatives offered is found to be suitable. However, this problem will not arise if all possibilities are covered (as in this example).

The order of answers for closed questions
We noted earlier that question order can affect the responses obtained. The same is true of the alternative answers provided for closed questions.

◆   There is a bias towards the left-hand answer in 'postal' questionnaires.
    Because the respondent reads from left to right and may get bored before reaching the right-hand answers.

4 Data sources 111

◆ There is a bias towards the right-hand answer in face-to-face interviews. *Because this is the last answer read out and is therefore the one that the respondent remembers most easily.*

◆ If there is a sequence of similar questions the respondent is likely to develop a 'habit' and answer each the same way. *So it is a good idea to vary the questions – this also makes the questionnaire more interesting.*

The pilot study

Before using a questionnaire it is essential to make sure that it 'works'. Are there any ambiguous questions? Are there closed questions that cause trouble because a possibility has been overlooked? Are there any questions that you have forgotten to ask? The **pilot study** uses the entire questionnaire with a small number of people who need not be chosen in any scientific way. The aim is simply to find and overcome any difficulties *before* using the real questionnaire.

4.4 National surveys

National censuses

The best known of all English Censuses was conducted under the orders of William the Conqueror and is better known as the *Domesday Book*.

The first modern British census took place in 1801 and there have been further censuses every 10 years since then. Attempting to get information from *every* household in the country is a formidable task. More than 120 000 people were employed in connection with the 1991 census, which cost about £135 million. The completed census returns, which are kept underground, occupy 12 miles of shelving!

In the 1991 census there were seven general household questions:

◆ What type of property?   ◆ How many rooms?
◆ Owned or rented?   ◆ Is there a bath?
◆ Is there an inside toilet? ◆ Any central heating?
◆ How many cars?

There were also about 20 questions asked of every member of the household. These included questions about migration ('Where did you live 12 months ago?'), health, employment and professional qualifications. About 98.5% of all returns answered all the questions.

Government surveys

The Government runs a number of regular large-scale surveys in order to monitor all aspects of life in Britain. Similar surveys are conducted in most developed countries. The most important of the British government surveys are probably the following:

1 The General Household Survey
    This survey has been conducted annually since 1971. Interviews are conducted with about 10 000 households each year (a 70% response rate). The subjects covered include population, housing, education, health and employment.

2 The Labour Force Survey

This concentrates on employment, including items such as hours of work. Inaugurated in 1973, it is the largest of the government surveys and currently interviews some 60 000 households every three months. It has a response rate of about 80%.

3 The National Food Survey

Each year information is collected from around 7500 households. Each household keeps a record for a 7-day period of all food bought, how much the various foodstuffs cost and details of the meals eaten. Unsurprisingly the response rate (in terms of completed weekly records) is only 50%.

4 The Family Expenditure Survey

This annual survey was instigated in 1957 and has a response rate of around 70% despite its demands on the participants. Around 7000 households complete a fortnightly diary that details expenditure on items such as food and drink, clothing, and heating. The results are used to calculate the Retail Prices Index (the RPI).

Other national surveys in Britain

An independent research organisation called SCPR (Social and Community Planning Research) carries out an annual survey, *The British Social Attitudes Survey*. This survey, as its title suggests, examines people's attitudes on all manner of subjects, and reports the outcomes of each year's survey in a *Yearbook*. A useful feature of this publication is that it includes the original questionnaire as an appendix.

There are three *National Child Development Studies*. These started in 1946, 1958 and 1970, respectively, and all have the same form. Each study concentrates on individuals who were born in a specific week of the year concerned. Family records were collected at birth and individual records have been collected at regular intervals thereafter. Of the 17 133 individuals included in the 1958 study, around 70% were traced during the 1991 re-interviews. Participants in these studies get a birthday card each year. Following individuals over these long periods of time reveals fascinating details: for example, the eating patterns in 1991 were related to the level of education of the mother, which was a detail recorded in 1958.

The ESRC (Economic and Social Research Council) is currently funding a rather different household survey known as the *British Household Panel Study* which is run by a group of researchers based at Essex University. The crucial difference lies in the word *panel*. The previous surveys we have mentioned all use fresh samples of households each year. A **panel study** revisits the *same* households each year, so that household changes can be studied directly. The study started in 1991 with a panel of 5000 households and the findings are to be reported in a series of books entitled (provisionally) *British Households Today*.

4.5 Government publications

The Central Statistical Office (CSO) publishes a large number of compilations of data gleaned from the surveys described in the previous section. The *Monthly Digest of Statistics* contains articles on varied topics, while the *Annual Abstract of Statistics* provides a yearly overview of the statistical state of the nation.

Family Spending is self-explanatory (taking its data from the Family Expenditure Survey), while *Economic Trends*, *Social Trends* and *Regional Trends* are specialist publications slanted at particular topics.

A cheap summary is provided by the annual *Key Data* publication, which provides numerical information on topics of general interest (such as the numbers of households with video recorders, and the changes in average family size during the century).

4.6 The Data Archive

You may wonder what happens to all the data after they have been collected and some initial analysis has taken place. The answer is that they are kept very safe in a number of different forms (e.g. magnetic tape) suitable for reading into a computer. The Data Archive, which is funded by the ESRC (the Economic and Social Research Council), is at Essex University. All the data for the surveys mentioned in this chapter, as well as geographical and historical data and data relating to thousands of other surveys (from various countries), are stored in the Archive. The Archive not only stores the data but also makes it available to researchers. From time to time data sets are prepared for use in schools. Enquiries should be directed to The Director, The Data Archive, University of Essex, Colchester, Essex, CO4 3SQ.

5 Probability

Probable impossibilities are to be preferred to improbable possibilities

Aristotle

5.1 Relative frequency

Suppose we roll a die and are interested in the outcome '6'. To get some idea of how likely the outcome is, we roll the die repeatedly. Here are the first 30 rolls:

2 4 4 1 2   3 2 4 3 1   4 5 6 4 3
2 3 6 2 4   3 4 2 2 5   4 6 5 3 3

After 10 rolls we have had no '6's and might think that getting a '6' is impossible! However, as the number of rolls increases so '6's begin to appear: after 30 rolls, we have had 3 '6's – a **relative frequency** of $\frac{3}{30} = 0.1$.

What will happen as we increase the number of rolls? The answer is that the number of '6's will increase, but the proportion of '6's (the relative frequency) will stabilise. The limiting value of this relative frequency is called the **probability**. So, if all six sides of the die are equally likely (which is the case for a *fair die*), then the limit of the relative frequency will be $\frac{1}{6}$ and we will say that the probability of a '6' is $\frac{1}{6}$.

5.2 Preliminary definitions

♦ A **statistical experiment** is one in which there are a number of possible outcomes and we have no way of predicting which outcome will actually occur. Sometimes the experiment may have already taken place, but we remain ignorant of the outcome.

♦ The **sample space**, often denoted by S, is the set of all possible outcomes of the experiment.
  ● The use of the word *sample* in the definition of S is an unfortunate historical accident – it does *not* refer to a sample of observations.

♦ An **event** is any set of possible outcomes of the experiment (thus an event is a subset of S). When rolling a die we might be interested in events such as 'getting an even number' or 'getting a number greater than 3'.

♦ A **simple event** is an event consisting of a single outcome. When rolling a die the simple events are '1', '2', etc.

Example 1

Many board games require the rolling of an ordinary six-sided die. The possible outcomes are 1, 2, 3, 4, 5, 6. Before the die is rolled we cannot predict the outcome – so this is an example of a statistical experiment.

In our new notation the six values are the six possible simple events and the sample space is $S = \{1, 2, 3, 4, 5, 6\}$. An example of a simple event is 'the outcome is a 6'. Examples of events are 'the outcome is an even number' and 'the outcome is a number less than 4'.

5.3 The probability scale

Assigned to the event E is a number, known as the probability of the event E, which takes a value in the range 0 to 1 (inclusive). The number is denoted by $P(E)$. In addition to satisfying:

$$0 \leqslant P(E) \leqslant 1$$

the value of $P(E)$ is chosen so that:

> If E is impossible, then $\quad P(E) = 0$
> If E is certain to occur, then $P(E) = 1$

Intermediate values of $P(E)$ have natural interpretations:

> $P(E) = 0.5 \quad \longrightarrow \quad E$ is as likely to occur as not to occur
> $P(E) = 0.001 \quad \longrightarrow \quad E$ is very unlikely
> $P(E) = 0.999 \quad \longrightarrow \quad E$ is highly likely

Example 2

Suppose we toss an ordinary coin. Define the events A and B to be:

> A: The coin comes down heads.
> B: The coin explodes in a flash of green light.

We can reasonably assume that $P(A) = \frac{1}{2}$ and that $P(B) = 0$.

5.4 Probability with equally likely outcomes

Suppose that the sample space, S, consists of $n(S)$ possible outcomes, and suppose that each is *equally likely*. Suppose that the number of outcomes in the event E is $n(E)$. Then $P(E)$, the probability that the event E occurs, is given by the equation:

$$P(E) = \frac{n(E)}{n(S)} \tag{5.1}$$

This clearly satisfies the requirement that $0 \leqslant P(E) \leqslant 1$.

Example 3

A fair die is tossed. The event A is defined as 'the number obtained is a multiple of 3'.
Determine $P(A)$.

Here, the sample space S consists of the outcomes $\{1, 2, 3, 4, 5, 6\}$, so that $n(S) = 6$. The outcomes corresponding to A are $\{3, 6\}$, so $n(A) = 2$. Thus
$P(A) = \dfrac{n(A)}{n(S)} = \dfrac{2}{6} = \dfrac{1}{3}$.

Example 4

Two fair coins are tossed. The event A is defined to be 'exactly one head is tossed'.
Determine $P(A)$.

Consider the coin that is tossed first. This coin is equally likely to give a head (H) or a tail (T). Suppose it gives a head. The second coin is now tossed. This coin is also equally likely to give a head or a tail, so, if the first coin was a head there are two equally likely sequences: HH and HT. On the other hand, if the first coin gave a tail then the equally likely sequences are TH and TT. Since the first coin was equally likely to give a head as a tail, the four outcomes HH, HT, TH, TT, which make up the sample space, S, are equally likely. The event A corresponds to the outcomes HT, TH. Thus $n(A) = 2$, $n(S) = 4$ and so $P(A) = \dfrac{n(A)}{n(S)} = \dfrac{2}{4} = \dfrac{1}{2}$.

Exercises 5a

1 An unbiased die is thrown.
Find the probability that:
(i) the score is even,
(ii) the score is at least two,
(iii) the score is at most two,
(iv) the score is divisible by 3.

2 A box contains 4 red balls, 6 green balls and 5 yellow balls. A ball is drawn at random.
Find the probability that:
(i) the ball is green,
(ii) the ball is red,
(iii) the ball is not yellow.

3 A card is drawn at random from a pack.
Find the probability that:
(i) the card is a Spade,
(ii) the card is an Ace,
(iii) the card is the Ace of Spades,
(iv) the card is a 'court card' (King, Queen or Jack).

4 A computer produces a 4-digit random number in the range 0000 to 9999 inclusive, in such a way that all such numbers are equally likely.
Find the probability that:
(i) the number is at least 1000,
(ii) the number lies between 1000 and 5000 inclusive,
(iii) the number is 4321,
(iv) the number ends in 0,
(v) the number begins and ends with 1.

5 A disc carries the numbers 1 and 2 on its faces. It is thrown with a fair die. The score is the sum of the two numbers that show.
Find the probability that:
(i) the score is at least 4,
(ii) the score is at most 6.

6 A bag contains 30 balls. The balls are numbered 1, 2, 3,..., 30. A ball is drawn at random.
Find the probability that the number on the ball:
(i) is divisible by 3,
(ii) is not divisible by 3,
(iii) is divisible by 4,
(iv) is a prime (2, 3, 5,...),
(v) differs from 10 by less than 5,
(vi) differs from 25 by more than 6.

7 Two unbiased dice, one red and one green, are thrown and the separate scores are observed. Represent the result as (r, g), where r and g are the scores on the red and green dice respectively.
Give a reason why there are 36 of these simple events.
Hence find the probability that:
(i) a double six is obtained,
(ii) a double (any score) is obtained,
(iii) the sum of the two scores is 4,
(iv) the sum of the two scores is 5,
(v) the score on the red die is 3 more than the score on the green die,
(vi) both scores are divisible by 3.

8 I have 14 coins in my purse. There are two 1p coins, three 2p coins, four 5p coins and five 10p coins. I choose a coin at random.
Find the probability that:
(i) it is a 2p coin,
(ii) it is worth at least 5p,
(iii) it is worth less than 3p,
(iv) it is worth at least 1p,
(v) it is worth at least 20p.

5.5 The complementary event, E'

An event E either occurs or it does not! We cannot have events 'half-occurring'. Each of the possible equally likely outcomes therefore corresponds to the event occurring or to the event not occurring. If $n(E)$ is the number of outcomes for which E occurs and $n(S)$ is the size of the sample space, then $n(S) - n(E)$ is the number of outcomes corresponding to the event 'E does not occur', which is called the **complementary event**, and is denoted by E'. Thus:

$$P(E') = \frac{n(S) - n(E)}{n(S)} = 1 - \frac{n(E)}{n(S)} = 1 - P(E)$$

This result:

$$P(E') = 1 - P(E) \tag{5.2}$$

or its equivalent:

$$P(E) = 1 - P(E')$$

often enables us to simplify calculations.

Notes

- The complementary event is sometimes denoted by $\bar{E}$, $C(E)$ or E^c.
- $(E')' = E$, since if E' does not occur then E occurs and vice versa.

Example 5

We toss a red die and a blue die. Both dice are fair.
We wish to find $P(A)$, where A is the event 'the total of the numbers shown by the two dice exceeds 3'.

We begin by finding $n(S)$, the number of possible outcomes in the sample space. There are six equally likely outcomes for the red die.
Whichever of these outcomes arises, there will also be six equally likely outcomes for the blue die. In all, therefore, there are thirty-six equally likely outcomes: $n(S) = 36$. We can see this easily on a diagram which can also be used to show the possible totals of the two dice:

```
                    Red die
              1   2   3   4   5   6
          1   2   3   4   5   6   7
          2   3   4   5   6   7   8
Blue      3   4   5   6   7   8   9
die       4   5   6   7   8   9   10
          5   6   7   8   9   10  11
          6   7   8   9   10  11  12
```

The complementary event, A', is the event that 'the total of the two dice does not exceed 3'. Now whereas there are lots of outcomes for which A occurs, there are very few for which A' occurs and it is easy to count them: the (red die, blue die) possibilities are $\{(1,1), (1,2), (2,1)\}$. Hence $n(A') = 3$. The 33 remaining outcomes in the diagram correspond to the event A.

Now, since all the outcomes are equally likely:

$$P(A') = \frac{n(A')}{n(S)} = \frac{3}{36} = \frac{1}{12}$$

But $P(A) = 1 - P(A')$, so that $P(A) = \frac{11}{12}$.

Note

♦ The fact that the dice were coloured does not affect $P(A)$. It simply makes it easier to describe what is happening. All that is required is some method of distinguishing the dice, and this is *always* possible, even if the dice are described as being identical! We could refer to the dice as being rolled one after the other, or being rolled by different people, or being rolled at the same time from different starting points in the die shaker.

John Venn (1834–1923) was a Cambridge lecturer whose speciality was logic. His major work, *The Logic of Chance*, was published in 1866. It is best known today for the introduction of the diagrams that now bear his name. Venn had a general interest in all branches of Statistics and a letter that he wrote in 1887 to the editor of the influential journal *Nature* stimulated an explosion of interest in the mathematical theory of Statistics.

5.6 Venn diagrams

A Venn diagram is a simple representation of the sample space, that is often helpful in seeing 'what is going on'. Usually the sample space is represented by a rectangle, with individual regions within the rectangle representing events.

It is often helpful to imagine that the actual areas of the various regions in a Venn diagram are in proportion to the corresponding probabilities. However, there is no need to spend a long time drawing these diagrams – their use is simply as a reminder of what is happening.

Venn diagrams illustrating the events E and E'

5.7 Unions and intersections of events

Suppose A and B are two events associated with a particular statistical experiment. We now consider the events denoted by $A \cup B$ and $A \cap B$, which are defined as follows:

$A \cup B$	'A **or** B'	At least one of A and B occurs.
$A \cap B$	'A **and** B'	Both A and B occur.

Notes

♦ $A \cup B$ includes the possibility that both A *and* B occur.
♦ In set notation:
   $A \cup B$ is called the **union** of A and B,
   $A \cap B$ is called the **intersection** of A and B.

The number of outcomes in A is $n(A)$ and the number of outcomes in B is $n(B)$. Also a total of $n(A \cap B)$ outcomes is in both A and B. The outcomes in $A \cup B$ include all those in A and all those in B but no others. However, if we simply add together $n(A)$ and $n(B)$ we will overstate the number in $A \cup B$ because we will have counted those in $A \cap B$ twice.

Hence:

$$n(A \cup B) = n(A) + n(B) - n(A \cap B)$$

Dividing throughout by $n(S)$ we get:

$$P(A \cup B) = P(A) + P(B) - P(A \cap B) \tag{5.3}$$

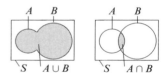

Venn diagrams illustrating the events $A \cup B$ and $A \cap B$

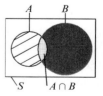

Outcomes in $A \cap B$ are counted twice

Notes

♦ In ordinary English, the phrase '*A or B*' usually means one of *A* and *B*, but *not* both. However, in probability questions '*A or B*' does not rule out the possibility of both occurring. The ambiguity disappears if set notation is used.

♦ From Equation (5.3) (or from the Venn diagram) it can be seen that $P(A \cup B)$ must be at least as great as the greater of $P(A)$ and $P(B)$.

♦ Similarly, $P(A \cap B)$ cannot be greater than the lesser of $P(A)$ and $P(B)$.

Example 6

Each month a mail-order firm awards a 'Star Prize' to a randomly chosen shopper. The firm uses the following procedure. It first chooses eight shoppers at random. The names of these eight shoppers are put into a hat. A guest celebrity then draws the name of the lucky winner of the 'Star Prize' from the hat and the other seven shoppers are awarded consolation prizes.

One month the first of the eight shoppers was a male living in the south of England. The complete list of those chosen was:

	Shopper number							
	1	2	3	4	5	6	7	8
Male (M) or Female (F)	M	F	F	F	M	F	F	F
North (N) or South (S)	S	S	N	S	N	S	S	N

The events *A* and *B* are defined by:

    A: The winner of the 'Star Prize' is male.
    B: The winner of the 'Star Prize' lives in the south.

(i)   Define, in words, the events $A \cap B$ and $A \cup B$.
(ii) Determine the probabilities of these events.

(i)   The event $A \cap B$ is the event: 'the winner of the "Star Prize" is a male living in the south'.
    The event $A \cup B$ is the event: 'the winner of the "Star Prize" is either a male, or lives in the south (or both)'.

(ii) The situation is illustrated in the Venn diagram, with the eight simple events, which are all equally likely, making up the sample space, *S*, being identified by numbers. Note that not all sets are egg-shaped! It can be seen that only the first of the eight simple events corresponds to $A \cap B$.

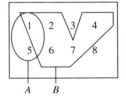

The following table provides a comprehensive list of the various events:

Event (E)	Simple events in E	$n(E)$	$P(E)$
Sample space, S	1, 2, 3, 4, 5, 6, 7, 8	8	1
A	1, 5	2	$\frac{2}{8} = \frac{1}{4}$
B	1, 2, 4, 6, 7	5	$\frac{5}{8}$
$A \cap B$	1	1	$\frac{1}{8}$
$A \cup B$	1, 2, 4, 5, 6, 7	6	$\frac{6}{8} = \frac{3}{4}$

As a check note that:

$$P(A \cup B) = P(A) + P(B) - P(A \cap B) = \tfrac{2}{8} + \tfrac{5}{8} - \tfrac{1}{8} = \tfrac{6}{8}$$

The probabilities of the two events are $\frac{1}{8}$ and $\frac{3}{4}$.

Example 7

For the sample space S it is known that $P(A) = 0.5$, $P(B) = 0.6$.
Determine the minimum and maximum possible values of $P(A \cap B)$.
Illustrate each case using a Venn diagram.

Substituting into Equation (5.3) we have:

$$P(A \cup B) = 0.5 + 0.6 - P(A \cap B) = 1.1 - P(A \cap B)$$

Since $P(A \cup B)$ cannot exceed 1, the minimum value of $P(A \cap B)$ is 0.1.
When $P(A \cap B) = 0.1$, $P(A \cup B)$ takes its maximum value $= 1$ and $A \cup B$ is
the whole of S. The smaller of $P(A)$ and $P(B)$ is $P(A)$, so the maximum
value for $P(A \cap B)$ is 0.5, in which case $A \cap B$ is the whole of A.

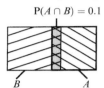

$P(A \cap B) = 0.1$

B A

$P(A \cap B) = 0.5$

B A

Example 8

Interviews with 18 people revealed that 5 of the 8 women and 8 of
the 10 men preferred drinking coffee to tea.
Determine the probability that the first person interviewed was either a
woman or someone who preferred coffee to tea.

In the absence of any information to the contrary we begin by assuming
that each of the 18 people was equally likely to have been the first to be
interviewed. If we were to guess who was first interviewed, there would be
a probability of $\frac{1}{18}$ that we would guess correctly.
    We define the events W: 'the person was a woman' and C: 'the person
preferred coffee to tea'. We *assume* that the interviewing was done at
random, so that each of the 18 people has probability $\frac{1}{18}$ of being the first
to be interviewed. The question as stated is slightly ambiguous (questions
often are!) – we assume it requires calculation of $P(W \cup C)$.
    Now $P(W) = \frac{8}{18}$, $P(C) = \frac{13}{18}$, and
$P(W \text{ and } C) = P$ (the person was a woman who preferred coffee to tea)
        $= P(W \cap C) = \frac{5}{18}$.
Hence:

$$P(W \text{ or } C) = \frac{8}{18} + \frac{13}{18} - \frac{5}{18} = \frac{16}{18} = \frac{8}{9}$$

so the probability that the first person interviewed was either a woman or
someone who preferred coffee to tea is $\frac{8}{9}$.

	Prefers coffee	Prefers tea	Total
Women	$\underline{5}$	$\underline{3}$	8
Men	$\underline{8}$	2	10
Total	13	5	18

An easy way of seeing the answer in this case is by totalling the underlined
numbers in the table, and expressing the total as a proportion of the
overall total (18).

5.8 Mutually exclusive events

Events A, B, ..., M, are said to be **mutually exclusive** if the occurrence of one of them implies that none of the others can occur. If D and E are two mutually exclusive events then $P(D \cap E) = 0$.

Note

◆ All simple events are mutually exclusive.

The addition rule

If the events A and B are mutually exclusive, then Equation (5.3) simplifies, since $P(A \cap B) = 0$, to give:

$$P(A \cup B) = P(A) + P(B) \tag{5.4}$$

which is known as the **addition rule**.

Note

◆ The addition rule *only* applies to mutually exclusive events.

Example 9

An Irish rugby club contains 40 players, of whom 7 are called O'Brien, 6 are called O'Connell, 4 are called O'Hara, 8 are called O'Neill and there are 15 others. The 40 players draw lots to decide who should be captain of the first team. Determine the probability that the captain of the first team is:
(i)   called either O'Brien or O'Connell,
(ii) is not called either O'Hara or O'Neill.

The sample space consists of the 40 players, each of whom is equally likely to be selected as captain. Denote the event that 'the captain is an O'Brien' by the symbol B, with C, H and N denoting the other events. The events B, C, H and N are mutually exclusive, since a player cannot have two surnames.

(i) $P(B \text{ or } C) = P(B \cup C) = P(B) + P(C) = \frac{7}{40} + \frac{6}{40} = \frac{13}{40}$

The probability that the captain is called O'Brien or O'Connell is $\frac{13}{40}$.

(ii) $P(\text{Neither } H \text{ nor } N) = 1 - P(H \text{ or } N) = 1 - \{P(H) + P(N)\}$
$$= 1 - \{\tfrac{4}{40} + \tfrac{8}{40}\} = 1 - \tfrac{12}{40} = \tfrac{28}{40} = \tfrac{7}{10}$$

The probability that the captain is not called O'Hara or O'Neill is $\frac{7}{10}$. The question can also be answered by making a table showing the possibilities:

	O'Brien	O'Connell	O'Hara	O'Neill	Others
Number	7	6	4	8	15
Satisfy part (i)	*	*			
Satisfy part (ii)	*	*			*

From the table we see immediately that there are 13 players satisfying the requirements of (i) and 28 satisfying (ii), so the probabilities are $\frac{13}{40}$ and $\frac{28}{40}$, respectively.

5.9 Exhaustive events

Two events are said to be **exhaustive** if it is certain that at least one of them occurs. For example, when rolling a die it is certain that at least one of the events A: 'the number obtained is either 1, 2, 3 or 5' and B: 'the number obtained is even' will occur. In this example, if a 2 is obtained then both A and B occur. If the events A and B are exhaustive then:

$$P(A \cup B) = 1 \qquad\qquad (5.5)$$

Notes

- ◆ Any event A and its complement, A', are both exhaustive and mutually exclusive:

  $$P(A \cup A') = 1, \qquad P(A \cap A') = 0$$

- ◆ The events $A, B, \ldots, N$ are said to be exhaustive if it is certain that at least one of them occurs:

  $$P(A \text{ or } B \text{ or } \ldots \text{ or } N) = P(A \cup B \cup \ldots \cup N) = 1$$

  Thus the simple events that make up the sample space, S, are mutually exclusive and exhaustive.

Exercises 5b

1 A fair die is thrown. Events A, B, C, D are defined as follows:

A: The score is even.
B: The score is divisible by 3.
C: The score is not more than 2.
D: The score exceeds 3.
Verify that:
$P(A) + P(B) = P(A \cup B) + P(A \cap B)$
Find:
(i) $P(A')$, (ii) $P(B')$, (iii) $P(C')$, (iv) $P(D')$
(v) Identify two pairs of events that are mutually exclusive, and verify the addition rule in each case.
(vi) Identify three events that are exhaustive.
(vii) Find $P(A \cup B \cup C)$.
(viii) Find $P(C \cap D)$.

2 Two fair dice, one red and one green, are thrown and the separate scores are observed. The outcome is denoted by (r, g), where r and g are the scores on the red and green dice respectively.
Represent these outcomes on a 6×6 grid, with r-axis horizontal and g-axis vertical. Events A, B, C are defined as follows:

A: The score on the red die exceeds the score on the green die.
B: The total score is six or more.
C: The score on the red die does not exceed 4.

(i) Identify on your diagram the sets corresponding to A, B, C.
(ii) Verify that:
$P(A) + P(B) = P(A \cup B) + P(A \cap B)$
(iii) Verify that:
$P(A) + P(C) = P(A \cup C) + P(A \cap C)$
(iv) Identify a pair of events that are exhaustive.
(v) Find $P(A')$, $P(B')$, $P(C')$.
(vi) Find $P(A' \cup B)$, $P(A \cap B')$, $P(B \cup C)$, $P(B' \cap C')$, $P(B' \cup C')$.

3 A man tosses two fair dice. One is numbered 1 to 6 in the usual way. The other is numbered 1, 3, 5, 7, 9, 11. The events A to E are defined as follows:
A: Both dice show odd numbers.
B: The number shown by the normal die is the greater.
C: The total of the two numbers shown is greater than 10.
D: The total is less than or equal to 4.
E: The total is odd.

(a) Determine the probability of each event.
(b) State which pairs of events (if any) are exclusive and which (if any) are exhaustive.

4 For the sample space S it is given that:
$P(A) = 0.5$, $P(A \cup B) = 0.6$,
$P(A \cap B) = 0.2$.
Find:
(i) $P(B)$,
(ii) $P(A' \cap B)$,
(iii) $P(A \cap B')$,
(iv) $P(A' \cap B')$.

5 For the sample space S it is given that:
$P(A' \cap B) = \frac{3}{7}$, $P(A \cap B') = \frac{2}{7}$, $P(A' \cap B') = \frac{1}{7}$.
Find:
(i) $P(A \cap B)$,
(ii) $P(A)$,
(iii) $P(B)$.

6 For the sample space S it is given that:
$P(B \cap C) = 0$, $P(A \cap B) = \frac{1}{20}$, $P(A \cap C) = \frac{2}{5}$,
$P(A) = \frac{3}{5}$, $P(B) = \frac{3}{20}$, $P(C) = \frac{13}{20}$.
Sketch a corresponding Venn diagram and
indicate $A \cap B$ and $A \cap C$.
Find:
(i) $P(A' \cap B)$,
(ii) $P(A' \cap C)$,
(iii) $P(A \cup B)$,
(iv) $P(B \cup C)$,
(v) $P(A' \cap B' \cap C')$.

7 A card is drawn at random from a normal
pack of 52 cards. Events A, B, C, D are defined
as follows:
A: The card drawn is either a Queen or a
Heart.
B: The card drawn is a black King.
C: The card drawn is either an Ace or a King
or a Queen or a Jack.
D: The card drawn is a Spade.
Find:
(i) $P(A)$,   (ii) $P(B)$,   (iii) $P(C)$,   (iv) $P(D)$,
(v) $P(A \cap D)$,   (vi) $P(A \cup D)$,   (vii) $P(A \cup B)$,
(viii) $P(C \cap D)$,   (ix) $P(C \cup D)$,
(x) $P(B \cap D)$,   (xi) $P(B \cup D)$.

8 A survey of 1000 people revealed the following
voting intentions.

	Women	Men	Total
Con	153	130	283
Lab	220	194	414
LibDem	157	146	303
Total	530	470	1000

A person is chosen at random from the sample.
Find the probability that the person chosen:
(i)   intends to vote Conservative,
(ii) is a woman intending to vote Labour,
(iii) is either a woman or intends to vote
Conservative,
(iv) is neither a man nor intends to vote
LibDem,
(v)   is a man and intends to vote either LibDem
or Labour.

5.10 Probability trees

Probability trees are diagrams that help us to see what is happening!
Consider the following problem. A fair coin is tossed three times. Determine
P(exactly two heads are obtained).

Each time we toss the coin the number of distinguishable outcomes
increases:

After first toss	Either H or T
After second toss	The sequence of outcomes must be HH, HT, TH or TT
After third toss	Either HHH, HHT, HTH, HTT, THH, THT, THH or TTT

The same possibilities are represented more simply (and we are less likely to miss out one of the possibilities!) in a tree diagram in which the final column lists the entire sample space.

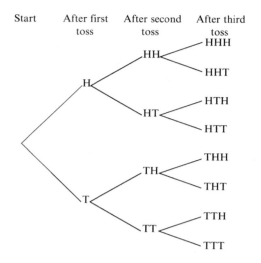

Each of the eight sequences is equally likely to occur. Three sequences include exactly two heads (HHT, HTH, THH) and so the probability of obtaining exactly two heads is $\frac{3}{8}$.

Consider the new problem.

A man tosses a 5p coin, a 10p coin and a 20p coin.
Determine P(exactly two heads are obtained).

Essentially the same tree diagram does the trick:

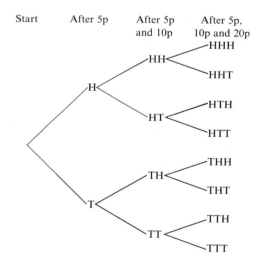

The probability is again equal to $\frac{3}{8}$.

Consider one final problem.

A woman tosses three coins.
Determine P(exactly two heads are obtained).

Once again we use the same tree diagram:

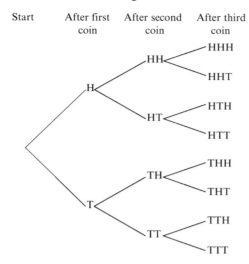

This time the tree has been labelled 'After first coin', 'After second coin' and 'After third coin'. We can think of the three coins as being tossed one after the other, so as to identify which coin is which. More mischievously, we can imagine having written the words 'First coin', 'Second coin' and 'Third coin' on the coins before tossing them. The required probability is again $\frac{3}{8}$.

Although all three problems refer to coin tosses, they describe different physical situations that are all equivalent in terms of their probability structure. This is an example of a general principle: most probability problems can be translated into problems concerning either the tossing of (possibly bent) coins, or the drawing of coloured balls from boxes! The setters of probability problems do their best to disguise this fact!

5.11 Sample proportions and probability

So far the probability to be associated with an event has been expressed in terms of the numbers of simple events in a sample space in which all the possible outcomes are equally likely. An alternative view of probability is a consequence of the general idea that a sample of observations gives information about the population from which it is derived. The bigger the sample, the more reliable is the information.

We have to adapt this approach when the outcomes in the sample space are no longer equally likely. For example, if we are interested in the probability that a bent penny comes down heads, then an obvious approach is to toss the penny a number of times (our sample) and see what proportion of the time a head is obtained:

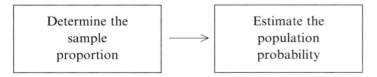

As the sample size increases, so the observed sample proportion of occasions on which the event E occurs will vary. However, the variations will generally decrease in magnitude, and we expect that the observed sample proportion will approach a value that we will take to be the probability of E and will denote by P(E).

Consider the following two situations:

Experiment	Event
A fair die is tossed.	*A*: A 6 is obtained.
A car is chosen at random.	*B*: The car is white.

For event *A* it seems reasonable that if we were to roll a fair die a huge number of times then 'obviously' the event *A* would occur on about one-sixth of occasions: $P(A) = \frac{1}{6}$. There is no need to do any real sampling – we need only think about it!

For event *B*, however, there is no alternative to real sampling. To have any idea of the value of $P(B)$, we need to examine a large sample of cars to find out what (roughly) is the proportion of cars that are white.

Project _____

So, what is the probability of the event, B, that a randomly chosen car is white? To answer this, you need to go to a convenient nearby road and count cars, keeping a tally of the number that are white. To see how the sample proportion stabilises as the sample size increases, complete the following table for your results:

Number of cars n	*Number of white cars* w	*Sample proportion* $p = \dfrac{w}{n}$
2		
5		
10		
50		
100		
200		
500		

You may wish to stop before seeing 500 cars, if the road is not a busy one!

Your best estimate of $P(B)$ *is simply your final value for p. Compare the value that you get with the values obtained by others in your class (who, hopefully, all observed different sets of cars). Decide on a class estimate for* $P(B)$.

Computer project _____

Computers are a good source of so-called 'random numbers'. For now, all we need to know about these numbers is that, if the random-number generator is set to produce numbers between 0 and 1, and is working correctly, then exactly 10% of the random numbers will have values less than 0.1. In probability terms, if E is the event 'number less than 0.1', then, theoretically, $P(E) = 0.1$.

Write a computer program to produce a table of the form shown for the car project above. Since the computer is doing the counting the table can go on a little longer – a final sample size of 10 000 should suffice!

Because of random variation, you should not expect always to see exactly 1000 'successes', but the theory discussed later in Chapter 14 suggests that you are likely to obtain between 940 and 1060 'successes', corresponding to an estimate of $P(E)$ *in the range 0.094 to 0.106.*

Calculator practice _____

> Many calculators also have an inbuilt random-number generator which generates random numbers between 0 and 1.
>
> If your calculator is programmable, then you could write a short program to simulate the rolling of a six-sided fair die. A random number between 0 and $\frac{1}{6}$ would correspond to 1, a number between $\frac{1}{6}$ and $\frac{2}{6}$ would correspond to 2, and so on.
>
> Use such a program to simulate 6000 rolls of a die and to count the numbers of 1's, 2's and so on. If the random-number generator is fair you should nearly always get between 900 and 1100 of each of the six outcomes.

5.12 Unequally likely possibilities

The results so far have been obtained while considering equally likely simple events. However, this restriction is artificial and Equations (5.2) to (5.5) hold equally well for unequally likely events.

▼

Example 10

The events A and B are such that $P(A) = 0.4$, $P(B') = 0.3$ and $P(A \cap B) = 0.2$.
Determine (i) $P(A \cup B)$, (ii) $P(A' \cap B')$.

_____

(i) Since $P(B') = 0.3$, $P(B) = 1 - 0.3 = 0.7$

$$P(A \cup B) = P(A) + P(B) - P(A \cap B)$$

$$P(A \cup B) = 0.4 + 0.7 - 0.2 = 0.9$$

(ii) By inspection of a Venn diagram we can see that
$P(A' \cap B') = 1 - P(A \cup B)$. Thus $P(A' \cap B') = 0.1$.

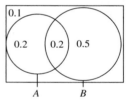

▲

5.13 Physical independence

The coin-tossing examples of Sections 5.4 (p. 115) and 5.10 (p. 123) are examples of situations in which the separate components (e.g. toss of one coin and toss of another coin) are physically independent events. By **physical independence** we mean that the outcome of one component (e.g. the first toss) can have no possible influence on the outcome of any other component (e.g. the second toss).

The multiplication rule

If A and B are two events *relating to physically independent situations* then:

$$P(A \cap B) = P(A) \times P(B) \tag{5.6}$$

More generally, if A, B, ..., N *all relate to physically independent situations* (for example, N separate tosses of a coin) then:

$$P(A \cap B \cap \cdots \cap N) = P(A) \times P(B) \times \cdots \times P(N) \tag{5.7}$$

This very useful result is known as the **multiplication rule**.

Example 11

A bent penny has probability 0.8 of coming down heads when it is tossed.
The penny is tossed six times.

What is the probability that it shows heads on every occasion?

_____

The six tosses are physically independent – there is no way that the
outcome of one of the tosses can affect the outcomes of the other tosses.
Therefore:

$$
\begin{aligned}
\text{P(6 heads)} &= \text{P('Head on first toss' } and \text{ 'Head on second toss'} \\
&\quad \cdots and \text{ 'Head on sixth toss')} \\
&= \text{P('Head on first toss')} \times \text{P('Head on second toss')} \times \\
&\quad \cdots \times \text{P('Head on sixth toss')} \\
&= 0.8 \times 0.8 \times \cdots \times 0.8 \\
&= 0.8^6 \\
&= 0.262 \text{ (to 3 d.p.)}
\end{aligned}
$$

The probability of getting 6 heads with the bent penny is just over a
quarter.

Example 12

A computer system consists of a keyboard, a monitor and the computer
itself. The three parts are manufactured separately. From past experience
it is known that, on delivery, the probability that the monitor works
correctly is 0.99, the probability that the keyboard works correctly is 0.98
and the probability that the computer works correctly is 0.95. What is the
probability that:
(i) the entire system works correctly,
(ii) exactly two of the components work correctly?

_____

Define the events M, K and C as follows:

   M: The monitor works correctly.
   K: The keyboard works correctly.
   C: The computer works correctly.

In part (i) we want P(M *and* K *and* C) = P($M \cap K \cap C$). Since the parts are
manufactured separately the three events refer to physically independent
manufacturing processes and therefore:

$$
\begin{aligned}
\text{P}(M \cap K \cap C) = \text{P}(M) \times \text{P}(K) \times \text{P}(C) &= 0.99 \times 0.98 \times 0.95 \\
&= 0.922 \text{ (to 3 d.p.)}
\end{aligned}
$$

To answer part (ii) we have to examine a number of possibilities. The
situation of interest is one in which just one of the three components is
working incorrectly (or not working at all!). This may be the monitor or it
may be the keyboard or it may be the computer. Writing it all out in
words would be dreadfully tedious, so we use the union/intersection
notation. The event of interest is:

$$E = E_1 \cup E_2 \cup E_3$$

where:

$$E_1 = (M' \cap K \cap C), \qquad E_2 = (M \cap K' \cap C), \qquad E_3 = (M \cap K \cap C')$$

Here, for example, M' is the complement of the event M, in other words the event: The monitor does not work correctly.

The events E_1, E_2 and E_3 are mutually exclusive, so, using the addition rule:

$$P(E) = P(E_1) + P(E_2) + P(E_3)$$

Because of physical independence:

$$P(E_1) = P(M' \cap K \cap C) = P(M') \times P(K) \times P(C)$$

and, since $P(M') = 1 - P(M)$, we finally get:

$$P(E) = \{1 - P(M)\} \times P(K) \times P(C) + P(M) \times \{1 - P(K)\} \times P(C)$$

$$+ P(M) \times P(K) \times \{1 - P(C)\}$$

$$= (0.01 \times 0.98 \times 0.95) + (0.99 \times 0.02 \times 0.95)$$

$$+ (0.99 \times 0.98 \times 0.05)$$

$$= 0.009\,31 + 0.018\,81 + 0.048\,51$$

$$= 0.076\,63$$

$$= 0.077 \text{ (to 3 d.p.)}$$

If the above solution seems rather daunting, then a probability tree will be very welcome:

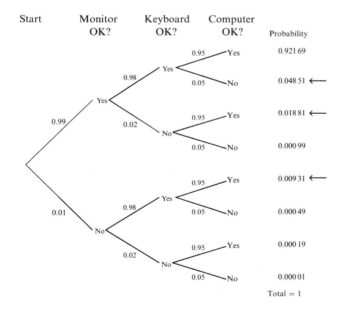

The final column gives the products of the probabilities of the corresponding branches.

Exercises 5c

1 Two fair dice, each having faces numbered 1, 1, 2, 2, 2, 3 are thrown.
Draw up a probability tree.
Hence find the probability that:
(i)   the total score is 4,
(ii) the total score is less than 4,
(iii) the total score exceeds 4,
(iv) at least one die shows 2.

2 Two fair dice, each having faces numbered 1, 1, 1, 1, 2, 2 are thrown.
Draw up a probability tree for the scores.
Find the probability that:
(i)   the total score is 2,
(ii) the total score is 3,
(iii) the total score is 4.
A third similar die is thrown.
Add this to your probability tree and hence find the probability that:
(iv) the total score is 4,
(v)   the total score is 5.

3 A man travels to work each day by train for three days. Each day the probability that the train is late is 0.1.
Find the probability that his train to work is late on at most two occasions.

4 The probability that a biased coin comes down heads is 0.4. It is tossed three times.
Find the probability of:
(i)   exactly two heads,
(ii) at least two heads.

5 Family A has two daughters and one son.
Family B has three daughters and one son.
Family C has two daughters and two sons. One child is chosen at random from each family.
Draw up a probability tree.
Find the probability that:
(i)   3 girls are chosen,
(ii) at least 2 girls are chosen,

(iii) no girls are chosen,
(iv) a girl is chosen from A and the other two are of opposite sex to one another.

6 A child is allowed a lucky dip from each of three boxes. One box contains 10 chocolates and 15 mints, one box contains 8 apples and 4 oranges, and the third box contains 7 (plastic) dinosaurs and 3 (plastic) turtles. Events *A*, *B*, *C* are defined as follows:
A: The child gets a chocolate and a dinosaur.
B: The child gets a mint or a turtle (or both).
C: The child gets an apple.
Find (i) P(*A*), (ii) P(*B*), (iii) P($A \cap C$),
(iv) P($B \cup C$), (v) P($A \cap B$), (vi) P($A \cup B$).

7 A woman travels to work by car. There are three roundabouts on the road. The probability that she is delayed at the first roundabout is 0.3. The corresponding figures for the second and third roundabouts are 0.5 and 0.7 respectively.
Find the probability that:
(i)   she is delayed at only one roundabout,
(ii) she is delayed at 2 or more roundabouts.

8 Two chess grand masters, Xerxes and Yorick, play a tournament of 3 games. Past experience of games between these two players suggests that the results of successive games are independent of one another and that, for each game:
P(Xerxes wins) = $\frac{1}{4}$
P(Yorick wins) = $\frac{1}{5}$
P(draw)   = $\frac{11}{20}$
Determine the probabilities of each of the following events:
A: Xerxes wins all three games.
B: Exactly two games are drawn.
C: Yorick wins at least one game.
D: Xerxes wins more games than Yorick wins.

5.14 Orderings

Consider the following problem.

Four markers are arranged in a line. The markers are labelled A, B, C and D. Assuming that all possible arrangements are equally likely, determine the probability that the markers are in the order ABCD.

A systematic (alphabetical) list of the possible arrangements is:

ABCD	ABDC	ACBD	ACDB	ADBC	ADCB
BACD	BADC	BCAD	BCDA	BDAC	BDCA
CABD	CADB	CBAD	CBDA	CDAB	CDBA
DABC	DACB	DBAC	DBCA	DCAB	DCBA

In all there are 24 possible orderings of the markers. Since each ordering is equally likely, the required probability is $\frac{1}{24}$.

The problem with this sort of approach is that frequently the number of elementary events in the sample space is so large that we may miss a few out! What is needed is a formula that allows us to count the possibilities *without* actually making a list. This formula can be deduced from study of the table of possibilities given above. There are 4 possibilities for the first marker. Suppose that this is A (the possible orderings are those in the first row of the table). There are then 3 possible candidates for the second marker (B, C or D). Suppose that B is second. Then there are 2 possibilities for third place (C or D), with whichever is left being in last place. We see that there are 24 possible orderings because $4 \times 3 \times 2 \times 1 = 24$.

In general, therefore, if there were n objects, the number of possible orderings would be:

$$n \times (n-1) \times (n-2) \times \cdots \times 3 \times 2 \times 1$$

This is tedious to write out, so we use the notation:

$$n! = n \times (n-1) \times (n-2) \times \cdots \times 3 \times 2 \times 1$$

The quantity $n!$ is read as 'n **factorial**'.

Notes
   ◆ $(n+1)! = (n+1) \times n!$
   ◆ For convenience, 0! is defined to be equal to 1.

Calculator practice

Check out the values of 0!, 1!, 2! and so on on your calculator.
What happens when you try to calculate 99!?
Why does this happen?
What is the largest value of n for which your calculator can calculate n!?
You could also try to calculate 4.5! More expensive calculators will give a value of about 52.3, whereas cheaper or older calculators will refuse to give an answer. If your calculator does give an answer, then you might like to plot the values of, say, 3.9!, 4!, 4.1!, ..., 5!, 5.1! on graph paper. Do you get a smooth curve?

Example 13

A supermarket uses a code to identify each product that it stocks. The code consists of an ordering (without repetition) of the letters A–E, followed by an ordering (without repetition) of the numbers 1–6. How many different codes can be formed?

The number of orderings of 5 objects is $5! = 5 \times 4! = 5 \times 24 = 120$. The number of orderings of 6 objects is $6! = 6 \times 5! = 720$. Since each ordering of the letters can be associated with any one of the 720 orderings of the numbers, there are a total of $120 \times 720 = 86\,400$ different codes.

Orderings of similar objects

Consider the following problem.

> Four markers are arranged in a line. The markers are labelled A, B, A and B. Determine the number of distinguishable orderings.

The only change from the previous situation is that the marker labelled C is now labelled A, while D has become B. Making the appropriate adjustments to the previous table, we get:

ABAB	ABBA	AABB	AABB	ABBA	ABAB
BAAB	BABA	BAAB	BABA	BBAA	BBAA
AABB	AABB	ABAB	ABBA	ABAB	ABBA
BABA	BAAB	BBAA	BBAA	BAAB	BABA

There is now a lot of repetition! The only distinguishable arrangements are **AABB, ABBA, ABAB, BAAB, BABA** and **BBAA**. The reduction comes about because A followed by C and C followed by A now give an identical result (A followed by A). This halves the number of distinguishable orderings. A similar halving results from the replacement of D by B.

The general rule is as follows:

If there are n objects, consisting of a of one type, b of a second type, and so on, then the number of *distinct* arrangements of the objects in a line is:

$$\frac{n!}{a!b!\ldots}$$

▼   ▼

Example 14

The four letters of the word COOK are arranged in a line.
(i) How many distinct arrangements are there?
(ii) If an arrangement is chosen at random, what is the probability that the two Os are consecutive?

———————

(i) There are 4 letters, consisting of 1 C, 2 Os and 1 K. The number of arrangements is therefore:

$$\frac{4!}{1!1!2!} = \frac{24}{1 \times 1 \times 2} = 12$$

There are 12 possible arrangements of the letters in the word COOK.
(ii) We require the two Os to be consecutive. Imagine that they are glued together. We then have only three items to arrange in order: C, OO and K. The number of possible orderings is $3! = 6$. Thus 6 of the 12 possible arrangements of the letters in the word COOK involve a double O: the required probability is $\frac{6}{12} = \frac{1}{2}$.
In this question the number of orderings is small enough that we could write them all out. Life is not always that easy, however, as the next example shows!

▲   ▲

Example 15

Five chairs are arranged in a line. Five boys are to be seated on the chairs.
If Alfred and Bruce sit next to each other then a fight is sure to start.

(i)   How many possible arrangements are there if there are no restrictions
      on the seating arrangements?
(ii) If the boys are assigned seats at random, what is the probability that
      Alfred and Bruce are not sitting next to one another?

———————

(i)   There are 5! = 120 possible arrangements, all equally likely.
(ii) The easy way to answer this is to consider the complementary event
      'a fight starts'! Imagine that Alfred and Bruce are 'glued' together in
      the order AB. There are now 4 'objects' (boys or doubleboys) to be
      arranged in order.

      There are then 4! = 24 possible arrangements of the objects. There
      are a further 24 possible arrangements with Albert and Bruce 'glued'
      in the order BA. In all, therefore, there are 48 unsatisfactory
      arrangements and therefore 120 − 48 = 72 satisfactory arrangements.
      The probability that Albert and Bruce are not sitting next to each
      other is therefore $\frac{72}{120} = \frac{3}{5}$.

C   A══B   E   D

One possible arrangement

Arrangements of n objects in a circle are more restrictive because there are n
possible 'starting points' for the circle. Denoting the directions North, South,
East and West by the letters N, S, E and W, the familiar clockwise ordering
NESW could also be represented as ESWN, SWNE or WNES, depending
upon one's starting point. The general rule for objects arranged in a circle is
as follows.

The number of arrangements of n objects arranged in a circle is equal to the
corresponding number of arrangements on a line, divided by n.

Note

♦ If the circle can be 'turned over', so that clockwise and anticlockwise
  arrangements are indistinguishable, the number of arrangements is equal to
  the corresponding number of arrangements on a line, divided by $2n$ rather
  than n.

Example 16

If the five chairs of the previous example are now arranged in a circle,
what is the probability that Albert and Bruce are not sitting next to each
other?

———————

The number of equally likely distinct arrangements is now $\frac{120}{5} = 24$. The
number of AB arrangements is now $\frac{24}{4} = 6$, and the number of BA
arrangements is also 6, so the total number of unsatisfactory arrangements
is 12. The probability that Albert and Bruce do not sit next to each other
is therefore $\frac{1}{2}$, somewhat smaller than before.

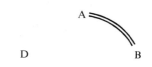

One possible arrangement

Exercises 5d

1 Six children, Alice, Brenda, Caroline, David, Edward and Frank, stand in a line.
How many different orders are possible?
They stand in random order.
Find the probability that:
(i) the three girls are next to each other,
(ii) Brenda and Frank are next to each other,
(iii) Caroline and David are not next to each other.

2 The five letters of the word UPTON are arranged in a line.
How many different arrangements are possible?
The letters are arranged in a random order.
Find the probability that:
(i) the two vowels are next to each other,
(ii) the two vowels are not next to each other.
(iii) either the letters NOT appear next to each other and in that order or the letters UP appear next to each other and in that order (or both).

3 A hand of cards consists of all 13 Hearts from an ordinary pack.
In how many different orders can they be arranged?
The cards are arranged in random order.
Find the probability that:
(i) the Ace is first and the King is last,
(ii) the Ace and King, in either order, are the first two cards,
(iii) either the Ace is first or the King is last or both,
(iv) the Ace is somewhere in front of the King.

4 I empty out my purse. There are four 1p coins, three 2p coins, two 5p coins and one 10p coin.
Assume that coins of the same value are indistinguishable from each other.
In how many different ways can the 10 coins be arranged in a line?
In how many of these ways are the three 2p coins all next to each other?
The coins are arranged in a line in random order.
Find the probability that:
(i) the two 5p coins are not next to each other,
(ii) the 10p coin has a 5p coin next to it on either side.

5 Thirteen counters, 4 red, 4 green, 3 blue and 2 yellow, are arranged in order in a line. The counters are identical except for their colour.
Find the number of distinguishable orderings.
The counters are arranged in random order.
Find the probability that:
(i) the 4 green counters are all next to each other,
(ii) all counters of the same colour are next to each other.

6 Find the number of different arrangements of the six letters in the word ELEVEN in which
(i) all three letters E are consecutive,
(ii) the first letter is E and the last letter is N.
[UCLES]

7 Six novels, labelled *A, B, C, D, E, F*, have to be arranged in order of merit for a literary prize.
Find the total number of different ways in which this can be done.
Suppose that the novels are arranged in random order.
Find the probability that:
(i) *F* is first,
(ii) *A* is last,
(iii) *C* is first and *D* is second,
(iv) *D* comes immediately after *C*,
(v) either *B* or *E* (or both) appear in the first two places.

8 The six children, Alice, Brenda, Caroline, David, Edward and Frank, now stand in a circle.
Distinguishing between clockwise order and anticlockwise order, find the number of different possible orders.
Find the probability that:
(i) the three girls are next to each other,
(ii) Brenda and Frank are next to each other,
(iii) Caroline and David are not next to each other.

5.15 Permutations and combinations

Consider the following problem.

A pack of 52 playing cards (all different) is shuffled. Determine the probability that the top card in the pack is the Ace of Spades, the next is the Ace of Hearts and the next is the Ace of Diamonds.

Now any one of the 52 cards could have been at the top of the pack. This leaves 51 cards, any one of which might have been next. Similarly, there are 50 possibilities for the third card. There are therefore a total of $52 \times 51 \times 50 = 132\,600$ possibilities for the first three cards in order. Only one of these corresponds to the event described, so the probability of that event is $\dfrac{1}{132\,600}$.

The number of *ordered* arrangements of r objects chosen from a collection of n objects, is denoted by $^{n}P_{r}$ (read as '**n p r**' or '**n perm r**') and each ordering is called a **permutation** of the selected objects.

The value of $^{n}P_{r}$ is given by:

$$^{n}P_{r} = n \times (n-1) \times \cdots \times (n-r+1) \tag{5.8}$$

Note that there are r terms in the expression on the right of this equation. An alternative expression, using factorials, is:

$$^{n}P_{r} = \frac{n!}{(n-r)!}$$

Using Equation (5.8) with $n = 52$ and $r = 3$, we get $^{52}P_{3} = 52 \times 51 \times 50 = 132\,600$, as before.

Consider now the slightly different problem.

A pack of 52 playing cards (all different) is shuffled. Determine the probability that the top three cards in the pack are the Ace of Spades, the Ace of Hearts and the Ace of Diamonds.

This problem differs from the previous one in that *the order in which the cards appear is irrelevant*. There are $3! = 6$ possible orders for three cards, so the number of *distinguishable* groups of three cards, chosen from 52, is the number of ordered possibilities ($132\,600$) divided by 6 giving the answer $22\,100$. The probability that the first three cards are the three aces is therefore $\dfrac{1}{22\,100}$.

The number of *unordered* arrangements of r objects selected from a collection of n objects, is denoted by either $^{n}C_{r}$ or $\dbinom{n}{r}$. In this book we use the second form which is that used in all modern advanced statistical texts. In either case the formula is read as either '**n c r**' or '**n choose r**'. Each collection of selected objects is a **combination**.

The general formula for $\dbinom{n}{r}$ is:

$$\binom{n}{r} = \frac{^{n}P_{r}}{r!} = \frac{n \times (n-1) \times \cdots \times (n-r+1)}{r \times (r-1) \times \cdots \times 1} = \frac{n!}{(n-r)!\,r!} \tag{5.9}$$

It should be noted that the fraction:

$$\frac{n \times (n-1) \times \cdots \times (n-r+1)}{r \times (r-1) \times \cdots \times 1}$$

has r terms in both the numerator and the denominator.

Using Equation (5.9) with $n = 52$ and $r = 3$, we get:

$$\binom{52}{3} = \frac{52 \times 51 \times 50}{3 \times 2 \times 1}$$

$$= 22\,100$$

Notes

◆ $\dbinom{n}{r} = \dbinom{n}{n-r}$; $\dbinom{n}{0} = \dbinom{n}{n} = 1$

◆ Some calculators have buttons for calculating permutations and combinations.

◆ Combinations occur naturally in the context of the binomial expansion, since:

$$(a + b)^n = \sum_{r=0}^{n} \binom{n}{r} a^r b^{n-r}$$

Example 17

A woman is planting rose bushes. She has eight different bushes, each with a different colour flower, and she will plant five of the bushes in her back garden.

How many different choices does she have?

———

Order matters here, so the number of possible arrangements is:

$$^8P_5 = \frac{8!}{5!} = 8 \times 7 \times 6 = 336$$

Example 18

A pack of cards is shuffled and a 'hand' of 13 randomly chosen cards is dealt to one card player.

How many possible hands can that player receive?

———

In this case the order in which the player receives the cards is irrelevant. The number of possible hands is therefore:

$$\binom{52}{13} = \frac{52 \times 51 \times 50 \times \cdots \times 41 \times 40}{13 \times 12 \times 11 \times \cdots \times 2 \times 1}$$

$$\approx 6.35 \times 10^{11}$$

There are about 635 thousand million possible hands!

Example 19

At the beginning of a game show a contestant is allowed a five-second glimpse of a table on which is placed a fluffy toy and four other objects (all different). At the end, the contestant is asked to name as many of the objects as possible.

(i)   How many different combinations of objects might be named?

(ii) What proportion include the fluffy toy?

———

(i) The contestant may name 0, 1, 2, 3, 4 or 5 of the objects. The total number of combinations is therefore:

$$\binom{5}{0} + \binom{5}{1} + \binom{5}{2} + \binom{5}{3} + \binom{5}{4} + \binom{5}{5}$$

$$= 1 + 5 + 10 + 10 + 5 + 1 = 32$$

(ii) Given that the fluffy toy is named, the contestant may name up to four of the remaining objects. The total number of combinations including the fluffy toy is therefore:

$$\binom{4}{0} + \binom{4}{1} + \binom{4}{2} + \binom{4}{3} + \binom{4}{4} = 1 + 4 + 6 + 4 + 1 = 16$$

The proportion of the combinations that include the fluffy toy is therefore $\frac{16}{32} = \frac{1}{2}$.

Note

◆ An alternative approach to part (i) is to argue that each of the five objects can either be 'chosen' or 'not chosen'. There are therefore 2 possibilities for each of 5 objects, so the total number of combinations is $2^5 = 32$. In part (ii) the number of possibilities is reduced to $2^4 = 16$, and so the required proportion is $\frac{16}{32} = \frac{1}{2}$.

Exercises 5e

1 A delegation of 3 students is to be chosen from a class of 15.
In how many ways can this be done?
The class consists of 10 girls and 5 boys.
(i) If two of the delegates are to be girls and the other is to be a boy, in how many ways can this be done?
(ii) If the delegation is to include at least one boy and at least one girl, in how many ways can this be done?

2 How many different hands of 13 cards, drawn from an ordinary pack, are there that contain 6 Spades, 4 Hearts, 2 Diamonds and 1 Club?
How many hands contain 6 from one suit, 4 from another, 2 from another and one from the fourth suit?

3 In the state of Utopia, the alphabet contains 25 letters. A car registration number consists of two **different** letters of the alphabet followed by an integer n such that $100 \leqslant n \leqslant 999$. Find the number of possible car registration numbers. [UCLES(P)]

4 A nursery school teacher has 4 apples, 3 oranges, and 2 bananas to share among 9 children, with each child receiving one fruit. Find the number of different ways in which this can be done. [UCLES]

5 A code consists of blocks of ten digits, four of which are zeros and six of which are ones; e.g. 1011011100. Calculate the number of such blocks in which the first and last digits are the same as each other. [UCLES]

6 A computer terminal displaying text can generate 16 different colours numbered 1 to 16. Any one of colours 1 to 8 may be used as 'background colour' on the screen, and any one of colours 1 to 16 may be used as 'text colour'; however, selecting the same colour for background and text renders the text invisible so this combination is not used. Find the number of different usable combinations of background colour and text colour. [UCLES]

7 Find the number of ways in which 4 questions can be chosen from the 7 questions in an examination paper, assuming that the order in which the questions are chosen is not relevant. [UCLES]

8 Prizes are to be awarded to four different members of a group of eight people. Find the number of ways in which the prizes can be awarded
(i) if there is a 1st prize, a 2nd prize, a 3rd prize and a 4th prize,
(ii) if there are two 1st prizes and two 2nd prizes. [UCLES]

9 Twelve horses run in a race. The published results list the horses finishing first, second and third. Assuming there are no dead-heats, find the number of different possible published results.   [UCLES]

10 A party of 12 people is to make a journey in 3 cars, with 4 people in each car. Each car is driven by its owner. Find the number of ways in which the remaining 9 people may be allocated to the cars. (The arrangement of people within a particular car is not relevant.)   [UCLES]

11 The digits of the number 314152 are rearranged so that the resulting number is odd. Find the number of ways in which this can be done.   [UCLES]

12 A school is asked to send a delegation of six pupils selected from six badminton players, six tennis players and five squash players. No pupil plays more than one game. The delegation is to consist of at least one, and not more than three, players drawn from each game. Giving full details of your working, find the number of ways in which the delegation can be selected.   [UCLES(P)]

5.16 Sampling with replacement

This is easy! The situation is one of physical independence and we can use the addition and multiplication rules and probability trees. Here is a typical problem.

> A pack of cards consists of the Queens of Spades, Hearts, Diamonds and Clubs together with the Ace, King and Jack of Spades. The pack is shuffled and a card is chosen at random. After its identity has been noted, the card is replaced in the pack, which is again shuffled. This is repeated on two further occasions. Determine the probability that a Queen is chosen on only one occasion.

On each occasion the probability that a Queen is chosen is $\frac{4}{7}$. Using Q to denote a Queen and R to denote one of the other cards, the possibilities that include exactly one Queen are RRQ, RQR and QRR. For each of these possibilities, the probability is the product of $\frac{3}{7}$, $\frac{3}{7}$ and $\frac{4}{7}$, so the overall probability is:

$$3 \times \left(\frac{3}{7}\right)^2 \times \frac{4}{7} = \frac{108}{343}$$

which is about 0.315 (to 3 d.p.).

5.17 Sampling without replacement

Consider the following problem.

> A pack of cards consists of the Queens of Spades, Hearts, Diamonds and Clubs together with the Ace, King and Jack of Spades. The pack is shuffled and three cards are chosen at random.
> Determine the probability that just one of the three cards is a Queen.

This is similar to the previous problem, but in this case the cards must be different, whereas in the previous case the same card might have been selected on more than one occasion.

In our new problem the order of selection is again unimportant and we are therefore concerned with combinations rather than permutations. The number of distinct combinations of three cards chosen from seven cards is:

$$\binom{7}{3} = \frac{7 \times 6 \times 5}{3 \times 2 \times 1} = 35$$

These are listed systematically in the following table using the shorthand of
A, K and J for the Ace, King and Jack and with S, H, D and C representing
the four queens.

When making lists it is important to work systematically (or we will get
hopelessly lost!). In this case we work alphabetically:

ACD	ACH	ACJ	ACK	ACS	ADH	ADJ
ADK	ADS	AHJ	AHK	AHS	AJK	AJS
AKS	CDH	CDJ	CDK	CDS	CHJ	CHK
CHS	CJK	CJS	CKS	DHJ	DHK	DHS
DJK	DJS	DKS	HJK	HJS	HKS	JKS

The 12 outcomes corresponding to the event of interest are underlined.

For each of the $\binom{4}{1} = 4$ possible selections of a Queen there are

$\binom{3}{2} = 3$ possible selections of two other cards from the three available. The

total number of possibilities is the product $\binom{4}{1} \times \binom{3}{2} = 4 \times 3 = 12$. The

probability of the event of interest is:

$$\frac{\binom{4}{1} \times \binom{3}{2}}{\binom{7}{3}} = \frac{12}{35}$$

This problem is a simple example of a general type illustrated by the
following.

A box contains a total of N balls. The balls are of k different types. There
are N_1 balls of type 1, N_2 of type 2, and so on $\left(\sum_{i=1}^{k} N_i = N\right)$. A random
sample of n balls is taken from the box *without replacement*.

What is the probability that the sample contains exactly n_1 balls of type 1,
n_2 of type 2, and so on $\left(\sum_{i=1}^{k} n_i = n\right)$? The order of selection is unimportant.

In this case an outcome consists of an unordered collection of n balls. The
total number of outcomes in the sample space is the number of ways in which
a random sample of n balls can be selected from a group of N balls. This is

just the number of ways of choosing n from N, which is $\binom{N}{n}$.

The number of ways of choosing n_1 balls from the N_1 balls of type 1, is

$\binom{N_1}{n_1}$. Whichever of these selections occurs there are also $\binom{N_2}{n_2}$ selections

of balls of type 2, and so on. The total number of selections corresponding to
the required event (i.e. the total number of outcomes) is therefore:

$$\binom{N_1}{n_1} \times \binom{N_2}{n_2} \times \cdots \times \binom{N_k}{n_k}$$

The probability of simultaneously choosing n_1 from N_1, n_2 from N_2, and so
on, is therefore:

$$\frac{\binom{N_1}{n_1} \times \binom{N_2}{n_2} \times \cdots \times \binom{N_k}{n_k}}{\binom{N}{n}}$$

Note

◆ The amount of thought required for this sort of problem can be minimised as follows! Write down in a row the numbers of each of the different types in the population (i.e. $N_1, N_2, \ldots, N_k$, which sum to N). In a row below these write down the corresponding numbers that are required for the sample (including zeros). These are the numbers $n_1, n_2, \ldots, n_k$ which sum to n. With suitably placed brackets we have the required numerator while $\begin{pmatrix} N \\ n \end{pmatrix}$ provides the denominator.

Example 20

A committee of five is chosen by drawing lots from a group of eight men and four women.
Determine the probability that the committee contains exactly three men.

Since nobody can be chosen more than once, selection is without replacement. An outcome consists of an unordered group of three people. We now suspend thought and simply identify the values of the parts of N and n. We have $N = 12$, $n = 5$, $N_1 = 8$, $N_2 = 4$, $n_1 = 3$ and $n_2 = 2$. Hence:

$$\frac{\begin{pmatrix} N_1 \\ n_1 \end{pmatrix} \times \begin{pmatrix} N_2 \\ n_2 \end{pmatrix}}{\begin{pmatrix} N \\ n \end{pmatrix}} = \frac{\begin{pmatrix} 8 \\ 3 \end{pmatrix} \times \begin{pmatrix} 4 \\ 2 \end{pmatrix}}{\begin{pmatrix} 12 \\ 5 \end{pmatrix}}$$

$$= \frac{8 \times 7 \times 6}{3 \times 2 \times 1} \times \frac{4 \times 3}{2 \times 1} \times \frac{5 \times 4 \times 3 \times 2 \times 1}{12 \times 11 \times 10 \times 9 \times 8}$$

$$= 56 \times 6 \times \frac{1}{792}$$

$$= \frac{14}{33}$$

The probability that the committee contains exactly three men is $\frac{14}{33}$, which is 0.424 to 3 decimal places.

Example 21

A notorious gang of outlaws contains five gunfighters called Smith, four called Jones and one called Cassidy. In a gunfight, three of the gang are killed. Assuming that each gunfighter had the same probability of being killed, what is the probability that the three killed in the gunfight all had different names?

This time the outcomes are unordered groups of three outlaws. There are three types of outlaw: Smith, Jones and Cassidy. The numbers of these are 5, 4 and 1 (total 10), while the numbers required are 1, 1 and 1 (total 3). Hence the required probability is:

$$\frac{\begin{pmatrix} 5 \\ 1 \end{pmatrix} \times \begin{pmatrix} 4 \\ 1 \end{pmatrix} \times \begin{pmatrix} 1 \\ 1 \end{pmatrix}}{\begin{pmatrix} 10 \\ 3 \end{pmatrix}} = \frac{5 \times 4 \times 1}{120}$$

$$= \frac{1}{6}$$

The probability that the three ex-gunfighters had different names is $\frac{1}{6}$.

Example 22

Three letters are chosen at random (without replacement) from the word STATISTICS.

What is the probability that:

(i) they are all the same,
(ii) they are all consonants,
(iii) they are all different,
(iv) exactly two are the same?

_____

The sample space consists of all possible unordered selections of 3 letters from the 10 letters S, T, A, T, I, S, T, I, C and S. The number of outcomes is the number of ways of choosing 3 letters from 10 letters (ignoring the repetition of the letters), and is therefore $\binom{10}{3} = 120$.

(i) One of these selections is SSS and another is TTT. These are the only outcomes that consist of three letters all the same and so the required probability is $\frac{2}{120} = \frac{1}{60}$.

(ii) There are seven consonants and three vowels in STATISTICS. The number of ways of choosing three consonants and no vowels is $\binom{7}{3} \times \binom{3}{0} = 35$. Hence the probability of this event is $\frac{35}{120} = \frac{7}{24}$.

(iii) Part (i) was simple because it was easy to spot that there were only two possible outcomes. Part (ii) was easy because the letters were split into two types. But this part is more difficult because there are 5 types of letter (S, T, A, I and C) to consider and – worse still – we need to consider these three at a time.

Since $\binom{5}{3} = 10$, there are 10 different types of outcomes to consider. These are (ignoring order) STI, STA, STC, SIA, SIC, SAC, ITA, ITC, IAC and TAC. The number of ways of obtaining, in some order, an outcome of type STI, is the number of ways of obtaining an S, times the number of ways of obtaining a T, times the number of ways of obtaining an I. This is:

$$\binom{3}{1} \times \binom{3}{1} \times \binom{2}{1} = 18$$

The table below shows the numbers of possibilities for all ten types of outcome.

Outcome type	STI	STA	STC	SIA	SIC	SAC	TIA	TIC	TAC	IAC	Total
Number of possibilities	18	9	9	6	6	3	6	6	3	2	68

The probability that the three chosen letters are all different is $\frac{68}{120} = \frac{17}{30}$.

(iv) This question can be answered by enumeration, though care is needed to make sure that all the possibilities have been noted. The method proceeds as before. Thus, for an outcome of type SSI the number of possibilities is:

$$\binom{3}{2} \times \binom{2}{1} = 6$$

However, it is simpler to recognise that we have already done the hard work! The three letters will either be all the same, all different or will have two letters the same and one different. We have calculated that there are 2 combinations in which the letters are all the same and 68 in which they are all different. By subtraction, therefore, the number of combinations in which one letter occurs exactly twice is 50 and the probability required is $\frac{50}{120}$, i.e. $\frac{5}{12}$.

▲ ▲

Exercises 5f

1 There are ten bottles arranged in a random order on a shelf. Five are green, three are blue and two are yellow. Two bottles are knocked off the shelf.
Determine the probability that:
(i) both bottles are green,
(ii) both bottles are the same colour,
(iii) the bottles are of different colours.

2 A class of 100 students comprises a group of 40 people called 'idiots' and a group of 60 called 'complete idiots'. A sample of three students is selected at random from the class. Determine the probability that the sample contains more 'complete idiots' than 'idiots'.

3 A bag of fruit contains 5 apples, 8 oranges and 3 pears. Three fruit are chosen, at random and without replacement, from the bag.
Find the probability that:
(i) no apples are chosen,
(ii) all the chosen fruit are different,
(iii) exactly one apple is chosen,
(iv) exactly two apples are chosen,
(v) three apples are chosen,
(vi) two apples and one orange are chosen.

4 A man is taking 12 shirts with him on a flight. He takes 4 formal shirts and 8 casual shirts, of which 3 are long-sleeved and 5 are short-sleeved. He splits his shirts randomly between his two cases, putting 6 shirts in each case. One of his cases is lost.
Find the probability that he has lost:
(i) exactly three formal shirts,
(ii) more than two formal shirts,
(iii) all his long-sleeved casual shirts.

5 A committee consists of 5 people: Anne, Bridget, Charles, Diana and Edward. Two members are to be chosen at random to be Chair and Vice-chair.
In how many different ways can these offices be filled?

Find the probability that:
(i) both the members chosen are men,
(ii) both are women,
(iii) the Chair is a woman and the Vice-chair is a man,
(iv) the Chair is a man and the Vice-chair is a woman,
(v) the two are of opposite sex.

6 Manjula has the following coins in her purse: eight 1p coins, three 2p coins, four 5p coins, two 10p coins and four 20p coins. In the dark she drops three coins.
Find the probability that:
(i) each of the coins lost is worth 5p or more,
(ii) the total value of the three coins is 3p,
(iii) the total value of the three coins is less than 7p,
(iv) all three coins have the same value.

7 A club committee consists of 2 married couples, 3 single women and 5 single men. Four members are to be chosen at random from the 12 members of the committee to form a delegation to represent the club at a conference. Find the probabilities that the delegation will consist of
(i) 4 single men,
(ii) 2 men and 2 women.   [JMB(P)]

8 A bag contains 5 red balls, 3 blue balls and 2 white balls. Four balls are drawn at random without replacement from the bag. Calculate the probability that the four balls drawn contain at least one of each colour.   [WJEC]

9 A choir has 7 sopranos, 6 altos, 3 tenors and 4 basses. At a particular rehearsal, three members of the choir are chosen at random to make the tea.
(i) Find the probability that all three tenors are chosen.
(ii) Find the probability that exactly one bass is chosen.   [UCLES(P)]

Chapter summary

♦ The **probability** of the event E is denoted by $P(E)$.

♦ The event 'E does not occur' is the **complementary event** and is denoted by E'.

$$P(E') = 1 - P(E)$$

♦ The event 'At least one of events A or B occurs' is the **union** of events A and B and is denoted by $A \cup B$.

♦ The event 'Both A and B occur' is the **intersection** of events A and B and is denoted by $A \cap B$.

♦ $P(A \cup B) = P(A) + P(B) - P(A \cap B)$

♦ If events A and B are **mutually exclusive** then:

$$P(A \cap B) = 0$$
$$P(A \cup B) = P(A) + P(B): \text{the } \textbf{addition rule.}$$

♦ If events A and B are **exhaustive** then:

$$P(A \cup B) = 1$$
$$P(A) + P(B) = 1 + P(A \cap B)$$

♦ If events A and B are **mutually exclusive and exhaustive** then:

$$P(A \cap B) = 0$$
$$P(A) + P(B) = 1$$

♦ If events A and B are **independent** then:

$$P(A \cap B) = P(A) \times P(B): \text{the } \textbf{multiplication rule.}$$

♦ *Orderings*

  • **Factorials** $n! = n \times (n-1) \times \cdots \times 1; \quad 0! = 1$
    The number of ways that n distinct objects can be arranged in order is $n!$

  • If a set of n objects comprises a objects of one type, b of another type, etc, then

    the number of distinct orderings is $\dfrac{n!}{a!b!\ldots}$

  • When r objects are chosen from a group of n unlike objects, with ordering being important, the number of distinct **permutations** is:

    $$^nP_r = n \times (n-1) \times \cdots \times (n-r+1) = \frac{n!}{(n-r)!}$$

  • When the order of drawing is unimportant, the number of possible collections (**combinations**) of r objects drawn from n is:

    $$^nC_r = \binom{n}{r} = \frac{n \times (n-1) \times \cdots \times (n-r+1)}{r \times (r-1) \times \cdots \times 1} = \frac{n!}{r!(n-r)!}$$

    $$\binom{n}{r} = \binom{n}{n-r}; \quad \binom{n}{0} = \binom{n}{n} = 1$$

> - The **probability of selecting** n_1 items of type 1, n_2 items of type 2, etc, from a population containing N_1 of type 1, N_2 of type 2, etc, when the order of selection is unimportant, is:
>
> $$\frac{\binom{N_1}{n_1} \times \binom{N_2}{n_2} \times \cdots}{\binom{N}{n}}$$
>
> where $N = \Sigma N_i$ and $n = \Sigma n_i$.

Exercises 5g (Miscellaneous)

1 In Ruritania all the cars are made by a single firm and vary only in their colouring. Six different colours are available (including 'communist red'). The same numbers of cars are painted in each of the six colours. Assuming that, when travelling on the roads, the colours of the cars occur in random order, determine the probability that:
  (a) the first six cars to pass Rudolf are all of different colours,
  (b) the second car is the same colour as the first, but the next 5 are all of different colours to their predecessors,
  (c) the first two cars have different colours, the third is the same colour as one or other of the first two, and the next four cars are all of different colours to their predecessors,
  (d) at least 8 cars pass Rudolf before all 6 colours are encountered.
  (e) none of the first six cars to pass Rudolf were painted in 'communist red'.

2 The National Lottery brochure claims that the chance of matching 6 different numbers from 1 to 49 are 1 in 13 983 816 and that the chance of matching 3 numbers out of the 6 is about 1 in 57. The order in which numbers occur is unimportant.
  Demonstrate that the quoted figures are correct.

3

	Small	Medium	Large
White	40	35	20
Blue	25	30	15
Cream	10	20	5

Table 1

Table 1 shows the distribution by size and colour of shirts in a batch of 200. A shirt is to be selected at random from the batch.
Calculate the probability that it will be
  (a) small,
  (b) either blue or white.
Two shirts are to be selected at random, without replacement, from the large shirts. Calculate, to 4 decimal places, the probability that
  (c) both shirts will be white,
  (d) one shirt will be white and one will be cream.   [ULSEB]

4 In the Upper Sixth Statistics class there are two boys and four girls, while in the Lower Sixth Statistics class there are four boys and six girls. Two different pupils are chosen at random from **each** of the two classes. Calculate the probabilities that the four chosen consist of
  (i) two boys from the Upper Sixth and two girls from the Lower Sixth,
  (ii) two boys and two girls.   [WJEC]

5 A book has 60 pages. The letter 'e' is the last letter on 15 of the pages, and the letters 's', 't' and 'd' are the last letters on 12, 9 and 6, respectively, of the pages. The last letters on each of the other pages are all different from each other and none is 'e', 's', 't' or 'd'. One page out of the 60 pages is chosen at random and the last letter is observed. This process is carried out two more times. Find
  (i) the probability that the letters obtained are 't', 'e', 'e' in that order,
  (ii) the probability that the letters obtained are 't', 'e', 'e' in any order.
A page is chosen at random and then a second different page is chosen at random. Find
  (iii) the probability that at least one of the two pages ends with the letter 's',
  (iv) the probability that the pages have the same last letter as each other.   [UCLES(P)]

6 In a computer game played by a single player, the player has to find, within a fixed time, the path through a maze shown on a computer screen. On the first occasion that a particular player plays the game, the computer shows a simple maze, and the probability that the player succeeds in finding the path in the time allowed is $\frac{3}{4}$. On subsequent occasions, the maze shown depends on the result of the previous game. If the player succeeded on the previous occasion, the next maze is harder, and the probability that the player succeeds is one half of the probability of success on the previous occasion. If the player failed on the previous occasion, a simple maze is shown and the probability of the player succeeding is again $\frac{3}{4}$. The player plays three games.

(i) Show that the probability that the player succeeds in all three games is $\frac{27}{512}$.

(ii) Find the probability that the player succeeds in exactly one of the games.

(iii) Find the probability that the player does not have two consecutive successes.

[UCLES(P)]

7 Four girls, Amanda, Beryl, Clare, and Dorothy, and three boys, Edward, Frank and George, stand in a queue in random order.

(i) Find the probability that the first two in the queue are Amanda and Beryl, in that order.

(ii) Find the probability that either Frank is first or Edward is last (or both).

(iii) Find the probability that no two girls stand next to each other.

(iv) Find the probability that all four girls stand next to each other. [UCLES(P)]

8 A class of twenty pupils consists of 12 girls and 8 boys. For a discussion session four 'officers' are to be chosen at random as 'Chairman', 'Recorder', 'Proposer' and 'Opposer'. Find, giving your answers correct to three significant figures,

(i) the probability that all four officers are girls,

(ii) the probability that two officers are girls and two are boys,

(iii) the probability that the Proposer and the Opposer are both girls. [UCLES(P)]

9 Each of a set of 26 cards is marked with one of the letters A to Z so that each card carries a different letter of the alphabet. Three of these cards are drawn at random. Find the number of different selections that can be made

(i) if the cards are drawn without replacement, and the order in which the cards are drawn is disregarded,

(ii) if the cards are drawn with replacement and the order in which the cards are drawn is taken into account. [UCLES]

6 Conditional probability

The theory of probabilities is at bottom nothing but common sense reduced to calculus

Laplace

The probability that we associate with the occurrence of an event is always likely to be influenced by the information that we have available. Suppose, for example, that I see a man lying motionless on the grass in a nearby park and am interested in the probability of the event 'the man is dead'. In the absence of other information a reasonable guess might be that the probability is one in a million. However, if I have just heard a shot ring out, and a suspicious-looking man with a smoking revolver is standing nearby then the probability would be rather higher!

6.1 Notation

We write:

$$P(B|A)$$

to mean the probability that the event B occurs (or has occurred) given the information that the event A occurs (or has occurred).

The quantity $B|A$ is read as 'B **given** A' and $P(B|A)$ is described as a **conditional probability** since it refers to the probability that B occurs (or has occurred) *conditional* on the event that A occurs (or has occurred).

▼

Example 1

A statistician has two coins, one of which is fair, while the other is double-headed. She chooses one coin at random and tosses it. The events A_1, A_2 and B are defined as follows:

A_1: The fair coin is chosen.
A_2: The double-headed coin is chosen.
B: A head is obtained.

Determine the values of $P(B|A_1)$ and $P(B|A_2)$.

———

If the fair coin is tossed then the probability of a head is $\frac{1}{2}$: $P(B|A_1) = \frac{1}{2}$.
If the double-headed coin is tossed then the probability is 1: $P(B|A_2) = 1$.

▲

Shortly we will relate $P(B|A)$ to the unconditional probabilities of the events A, B, $A \cap B$, but first we look at an example that involves equally likely simple events.

▼

Example 2

An electronic display is equally likely to show any of the digits $1, \ldots, 8, 9$. Determine the probability that it shows a prime number (i.e. one of 2, 3, 5 and 7):
(i) given no knowledge about the number,
(ii) given the information that the number is odd.

———

Let B be the event 'a prime number' and A be the event 'an odd number'.
Thus $A \cap B$ is the event 'an odd prime number'.

(i) Since there are nine possible outcomes, $n(S) = 9$. Since there are four
outcomes corresponding to the event of interest, $n(B) = 4$. Since the
outcomes are all equally likely, $\mathrm{P}(B) = \dfrac{n(B)}{n(S)} = \frac{4}{9}$.

(ii) Given the information that the number is odd, we know that it must
be one of the $n(A)$ numbers 1, 3, 5, 7 and 9. Initially, each of these
outcomes was equally likely. The knowledge that one of them has
occurred does not make their chances of occurrence unequal. Of these
five possible outcomes, three (3, 5 and 7) are prime. These outcomes
are the simple events corresponding to the event $A \cap B$. Thus,

$$\mathrm{P}(B|A) = \frac{n(A \cap B)}{n(A)} = \tfrac{3}{5}.$$

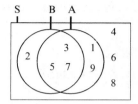

The previous example illustrated, for a particular case, the result that, for
equally likely simple events:

$$\mathrm{P}(B|A) = \frac{n(A \cap B)}{n(A)}$$

If we divide both the numerator and the denominator of the right-hand side
of this equation by $n(S)$, we obtain:

$$\mathrm{P}(B|A) = \frac{\mathrm{P}(A \cap B)}{\mathrm{P}(A)} \tag{6.1}$$

This result is always true (provided A is a possible event!) and is not confined
to equally likely events. We can illustrate the result using Venn diagrams.

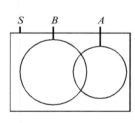   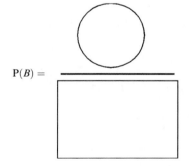   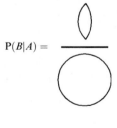

$$\mathrm{P}(B) = \underline{\hspace{3cm}} \qquad \mathrm{P}(B|A) = \underline{\hspace{2cm}}$$

Knowing that A has occurred means that we can ignore all of the sample
space except for that part occupied by the event A. The part of A in which B
also occurs is the part denoted by $A \cap B$, and Equation (6.1) is seen to be a
simple statement about proportions.

Rearranging the previous equation, we get:

$$\mathrm{P}(A \cap B) = \mathrm{P}(A) \times \mathrm{P}(B|A) \tag{6.2}$$

Reversing the roles of A and B:

$$\mathrm{P}(B \cap A) = \mathrm{P}(B) \times \mathrm{P}(A|B)$$

Since $A \cap B$ and $B \cap A$ are descriptions of the same event, the intersection of
A and B, we have:

$$\mathrm{P}(A \cap B) = \mathrm{P}(B \cap A)$$

and hence:

$$\mathrm{P}(A \cap B) = \mathrm{P}(A) \times \mathrm{P}(B|A) = \mathrm{P}(B) \times \mathrm{P}(A|B) \tag{6.3}$$

The generalised multiplication rule

For three events, repeated application of Equation (6.3) gives:

$$P(A \cap B \cap C) = P(A) \times P(B|A) \times P(C|A \cap B) \qquad (6.4)$$

from which the extension to larger numbers of events is clear.

Since $(A \cap B \cap C)$ is the same as, for example, $(B \cap C \cap A)$, another of many equivalent expressions for $P(A \cap B \cap C)$ is:

$$P(A \cap B \cap C) = P(B) \times P(C|B) \times P(A|B \cap C)$$

6.2 Statistical independence

Two events A and B are said to be **statistically independent** if knowledge that one occurs does *not* alter the probability that the other occurs. Formally, if A and B are two statistically independent events with non-zero probabilities, then:

- $P(A|B) = P(A)$
- $P(B|A) = P(B)$
- $P(A \cap B) = P(A) \times P(B)$

Notes

- Any one of the above three equations is enough to guarantee independence of A and B (assuming that both have non-zero probability of occurrence).
- Physically independent events are always statistically independent.
- The words 'statistically' and 'physically' are often omitted and events are simply referred to as being 'independent'.
- Exclusive events with positive probability cannot be independent.

▼ ▼

Example 3

Two events A and B are such that $P(A) = 0.5$, $P(B) = 0.4$ and $P(A|B) = 0.3$.
(i)   State whether the events are independent.
(ii) Find the value of $P(A \cap B)$.

———

(i)   The events A and B are *not* independent since $P(A) \neq P(A|B)$
(ii) $P(A \cap B) = P(B) \times P(A|B) = 0.4 \times 0.3 = 0.12$

▲ ▲

▼ ▼

Example 4

Two events A and B are such that $P(A) = 0.7$, $P(B) = 0.4$ and $P(A|B) = 0.3$. Determine the probability that neither A nor B occurs.

———

It is not obvious how to answer this! One way is to 'doodle', by writing down the probabilities of things we do know! So, from Equation (6.3):

$$P(A \cap B) = P(B) \times P(A|B) = 0.4 \times 0.3 = 0.12$$

From Equation (5.3) we can now obtain $P(A \cup B)$:

$$P(A \cup B) = P(A) + P(B) - P(A \cap B) = 0.7 + 0.4 - 0.12 = 0.98$$

But, looking at a Venn diagram:

$$P(\text{neither } A \text{ nor } B) = 1 - P(A \cup B)$$

The required probability is therefore $1 - 0.98 = 0.02$

———

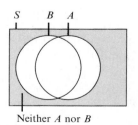

Neither A nor B

An alternative approach, more algebraic in nature, begins by organising the information in a table of probabilities of the joint events $A \cap B$, $A \cap B'$, $A' \cap B$, $A' \cap B'$, with the required value, $P(A' \cap B')$ being set equal to x.

	B	B'	Total
A			0.7
A'		x	0.3
Total	0.4	0.6	1.0

which gives:

	B	B'	Total
A	$0.1 + x$	$0.6 - x$	0.7
A'	$0.3 - x$	x	0.3
Total	0.4	0.6	1.0

In obtaining the second table we have used, for example, the fact that:

$$P(B) = P(B \cap A) + P(B \cap A')$$

We also know that $P(A|B) = 0.3$. Hence $\dfrac{0.1 + x}{0.4} = 0.3$. Multiplying both sides by 0.4, we get $0.1 + x = 0.3 \times 0.4$, so that $x = 0.12 - 0.1 = 0.02$, as before.

The same approach could be adopted using the Venn diagram shown.

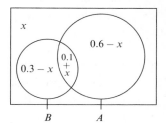

Example 5

The following contingency table (see Section 1.23, p. 31) gives information on two aspects of the habitats of some tropical lizards for a sample of 207 habitats.

		Perch diameter (cm)		Total
		$\leqslant 10$	> 10	
Perch	> 1.5	64	22	86
height (m)	$\leqslant 1.5$	86	35	121
Total		150	57	207

Suppose that one of the 207 habitat locations in the sample is chosen at random. Determine, correct to 2 decimal places, the probability that:
(i)   the perch diameter is greater than 10 cm,
(ii) the perch diameter is greater than 10 cm, given the information that the perch height is more than 1.5 m,
(iii) the perch height is more than 1.5 m,
(iv) the perch height is more than 1.5 m, given that the perch diameter is greater than 10 cm.

———

Define the events A and B as follows:

 A: The perch diameter is greater than 10 cm.
 B: The perch height is more than 1.5 m.

We can read the answers direct from the table.

(i) $P(A) = \frac{57}{207} = 0.28$ (to 2 d.p.)

(ii) $P(A|B) = \frac{22}{86} = 0.26$ (to 2 d.p.)

(iii) $P(B) = \frac{86}{207} = 0.42$ (to 2 d.p.)

(iv) $P(B|A) = \frac{22}{57} = 0.39$ (to 2 d.p.)

Since $P(A) \approx P(A|B)$ and $P(B) \approx P(B|A)$ it appears that perch height and perch diameter are approximately independent.

Example 6

A person is chosen at random from the population. Let A be the event 'the person is female' and let B be the event 'the person is aged at least 80'. Suppose that $P(A) = 0.5$, $P(B) = 0.1$ and $P(A|B) = 0.7$. Let the event C be defined by $C = A \cap B'$.

(i)   Describe the event C in real terms.

(ii) Determine $P(A|B')$.

———————

This question is much easier to answer when written in English!

In a certain population, 50% are female, 10% are aged at least 80 and 70% of these aged people are female.

(i)   The event C is 'a female aged less than 80'.

(ii) We need to find the probability that someone aged under 80 is female. A simple approach is to form a table. It may also help to give the population a definite size, N, say. The number of females aged 80 or over is therefore $0.7 \times 0.1N = 0.07N$. The remainder of the table is filled by subtraction.

	<80	$\geqslant 80$	
Males	$0.47N$	$0.03N$	$0.50N$
Females	$0.43N$	$0.07N$	$0.50N$
	$0.90N$	$0.10N$	N

The proportion of females amongst those aged under 80 is therefore $\dfrac{0.43N}{0.90N} = \frac{43}{90}$. So $P(A|B') = \frac{43}{90}$, which is just less than $\frac{1}{2}$.

Exercises 6a ————————————————————————————————

1 Given that $P(A) = 0.4$, $P(B) = 0.7$, $P(A \cap B) = 0.2$, find (i) $P(A|B)$, (ii) $P(A'|B)$, (iii) $P(A|B')$, (iv) $P(A'|B')$.

2 Given that $P(A) = 0.8$, $P(A|B) = 0.8$, $P(A \cap B) = 0.5$, find (i) $P(B)$, (ii) $P(B|A)$, (iii) $P(A \cup B)$, (iv) $P(A|A \cup B)$, (v) $P(A \cap B|A \cup B)$, (vi) $P(A \cap B|B')$, (vii) $P(A \cap B|A)$.

3 Given that $P(C \cap D) = \frac{1}{4}$, $P(C|D) = \frac{1}{3}$, $P(D|C) = \frac{3}{5}$, find (i) $P(C)$, (ii) $P(D)$, (iii) $P(C|D')$, (iv) $P(C|C \cup D)$.

4 Given that $P(A) = 0.8$, $P(B) = 0.7$, $P(C) = 0.6$, $P(A|B) = 0.8$, $P(C|B) = 0.7$, $P(A \cap C) = 0.48$, determine whether:

(i) A and B are independent,

(ii) A and C are independent,

(iii) B and C are independent.

5 Given that C and D are independent and that $P(C|D) = \frac{2}{3}$, $P(C \cap D) = \frac{1}{3}$, find (i) $P(C)$, (ii) $P(D)$.

6 Given that $P(B) = \frac{4}{5}$, $P(C) = \frac{2}{3}$, $P(A|B) = \frac{1}{2}$, $P(B|A) = \frac{3}{4}$, $P(C|A \cap B) = \frac{1}{3}$, find (i) $P(A \cap B)$, (ii) $P(A \cap B \cap C)$, (iii) $P(A)$, (iv) $P(A \cap B|C)$.

7 Three ordinary unbiased six-sided dice, one red, one green and one blue, are thrown simultaneously. Events R, G, S and T are defined as follows:

R: The score on the red die is 3.

G: The score on the green die is 2.

S: The sum of the scores on the red and the green dice is 4.

T: The total score for the three dice is 5.

Find the following probabilities:

(a) $P(R \cap G)$, (b) $P(S|R)$, (c) $P(R|S)$, (d) $P(R \cup G)$, (e) $P(T)$, (f) $P(S|T)$.

8 On the sunny tropical island of Utopia, one quarter of the large number of adult inhabitants are male and the remainder are female. The island's tourist welcoming committee consists of six individuals drawn at random from the adult inhabitants of the island.   *(continued)*

Determine the probability that:
(a) exactly one committee member is male,
(b) all the committee members are female,
(c) at least five committee members are female,
(d) all the committee members are of the same sex,
(e) all the committee members are female, given that it is known that they are all of the same sex.

9 A box contains 5 red balls and 3 white balls. A second box contains 4 red balls and 4 white balls. Two balls are drawn at random from the first box and placed in the second box. One ball is then drawn at random from the 10 balls now in the second box.
Determine the probability that this ball is red.

10 The Green Hand gang used to consist of 12 individuals, of whom 8 were called Smith and 4 were called Jones. One bad year, they fell foul of a rival gang and every month one member of the Green Hand gang was 'eliminated' at random.
Determine the probability of each of the following events:
A: Exactly three of the first five eliminated were named Jones.
B: The last two to be eliminated were named Smith.
Determine also $P(A|B)$ and $P(B|A)$.

11 The events A and B are such that $P(A) = \frac{2}{5}$, $P(B) = \frac{1}{6}$ and $P(A \cup B) = \frac{13}{30}$. Show that A and B are neither mutually exclusive nor independent. [WJEC]

12 A box contains ten objects of which 1 is a red ball, 2 are white balls, 3 are red cubes and 4 are white cubes. Three objects are drawn at random from the box, in succession and without replacement. Events B and R are defined as follows:
B: Exactly two of the objects drawn are balls.
R: Exactly one of the objects drawn is red.
Show that $P(B) = \frac{7}{40}$ and calculate $P(R)$, $P(B \cap R)$, $P(B \cup R)$ and $P(B|R)$. [UCLES]

13 A box contains 25 apples, of which 20 are red and 5 are green. Of the red apples, 3 contain maggots and of the green apples, 1 contains maggots. Two apples are chosen at random from the box. Find, in any order,
(i) the probability that both apples contain maggots,

(ii) the probability that both apples are red and at least one contains maggots,
(iii) the probability that at least one apple contains maggots, given that both apples are red,
(iv) the probability that both apples are red given that at least one apple is red. [UCLES]

14 A golfer observes that, when playing a particular hole at his local course, he hits a straight drive on 80 per cent of the occasions when the weather is not windy but only on 30 per cent of the occasions when the weather is windy. Local records suggest that the weather is windy on 55 per cent of all days.
(i) Show that the probability that, on a randomly chosen day, the golfer will hit a straight drive at the hole is 0.525.
(ii) Given that he fails to hit a straight drive at the hole, calculate the probability that the weather is windy. [JMB]

15 The events A and B are such that
$P(A') = \frac{3}{4}$,
$P(A|B) = \frac{1}{3}$,
$P(A \cup B) = \frac{2}{3}$,
where A' denotes the event "A does not occur". Find
(i) $P(A)$,
(ii) $P(A \cap B)$,
(iii) $P(B)$,
(iv) $P(A|B')$,
where B' denotes the event "B does not occur". Determine whether A and B are independent. [Answers may be given as fractions in their lowest terms.] [O&C]

16 A game is played with an ordinary six-sided die. A player throws this die, and if the result is 2, 3, 4 or 5, that result is the player's score. If the result is 1 or 6, the player throws the die a second time and the sum of the two numbers resulting from both throws is the player's score. Events A and B are defined as follows:
A: the player's score is 5, 6, 7, 8 or 9;
B: the player has two throws.
Show that $P(A) = \frac{1}{3}$.
Find (i) $P(A \cap B)$, (ii) $P(A \cup B)$, (iii) $P(A|B)$, (iv) $P(B|A')$. [UCLES]

17 A bag contains 4 red counters and 6 green counters. Four counters are drawn at random from the bag, without replacement. Calculate the probability that
(i)   all the counters drawn are green,
(ii) at least one counter of each colour is drawn,
(iii) at least two green counters are drawn,
      (iv) at least two green counters are drawn, given that at least one of each colour is drawn.

State with a reason whether or not the events 'at least two green counters are drawn' and 'at least one counter of each colour is drawn' are independent.   [UCLES]

18 A bag contains 5 white balls and 3 red balls. Two players, A and B, take turns at drawing one ball from the bag at random, and balls drawn are not replaced. The player who first gets two red balls is the winner, and the drawing stops as soon as either player has drawn two red balls. Player A draws first. Find the probability
(i)   that player A is the winner on his second draw,
(ii) that player A is the winner, given that the winning player wins on his second draw,
(iii) that neither player has won after two draws, given that A draws a red ball on his first draw.   [UCLES]

19 For married couples the probability that the husband has passed his driving test is $\frac{7}{10}$ and the probability that the wife has passed her driving test is $\frac{1}{2}$. The probability that the husband has passed, given that the wife has passed, is $\frac{14}{15}$. Find the probability that, for a randomly chosen married couple, the driving test will have been passed by
(a) both of them,
(b) only one of them,
(c) neither of them.

If two married couples are chosen at random, find the probability that only one of the husbands and only one of the wives will have passed the driving test.   [ULSEB]

20 Write down an expression involving probabilities for $P(B|A)$, the probability of event B given that event A occurs.
Alison and Brenda play a tennis match in which the first player to win two sets wins the match. In tennis no set can be drawn. The probability that Alison wins the first set is $\frac{1}{3}$; for sets after the first, the probability that Alison wins the set is $\frac{3}{5}$ if she won the preceding set, but is only $\frac{1}{4}$ if she lost the preceding set.
With the aid of a suitable diagram, or otherwise, determine the probability that
(i)   the match lasts for just two sets,
(ii) Alison wins the match given that it lasts for just two sets,
(iii) Alison wins the match,
(iv) Alison wins the match given that it goes to three sets,
(v)   if Alison wins the match, then she does so in two sets.   [JMB(P)]

6.3 Mutual and pairwise independence

If the events A, B, C, ... , M, each having non-zero probability are **mutually independent** then their probabilities and the probabilities of their intersections satisfy all possible equations of the general form:

$$P(E \cap F \cap \cdots \cap K) = P(E) \times P(F) \times \cdots \times P(K)$$

including:

$$P(A \cap B \cap C \cap \cdots \cap M) = P(A) \times P(B) \times P(C) \times \cdots \times P(M) \qquad (6.5)$$

If the events A, B, C, ..., M, each having non-zero probability are **pairwise independent** then their probabilities and the probabilities of their intersections satisfy all possible equations of the type:

$$P(E \cap F) = P(E) \times P(F) \qquad (6.6)$$

where E and F are any pair of the events.

Mutual independence clearly implies pairwise independence, but the reverse is untrue, as Example 7 illustrates.

Note

   ◆ A set of physically independent events will be both mutually independent and pairwise independent.

Example 7

Two fair coins are tossed and the events A, B and C are defined as follows:

   A: The first coin shows a head.
   B: The second coin shows a head.
   C: The two coins show different faces.

Demonstrate that A, B and C are pairwise independent but not mutually independent.

The outcomes corresponding to the various events of interest are summarised in the following table:

Event	Outcomes	Probability	Event	Outcomes	Probability
A	(H,H), (H,T)	$\frac{1}{2}$	B	(H,H), (T,H)	$\frac{1}{2}$
C	(H,T), (T,H)	$\frac{1}{2}$	$A \cap B$	(H,H)	$\frac{1}{4}$
$A \cap C$	(H,T)	$\frac{1}{4}$	$B \cap C$	(T,H)	$\frac{1}{4}$
$A \cap B \cap C$		0			

Thus:

$$P(A \cap B) = \tfrac{1}{4} = P(A) \times P(B)$$
$$P(A \cap C) = \tfrac{1}{4} = P(A) \times P(C)$$
$$P(B \cap C) = \tfrac{1}{4} = P(B) \times P(C)$$

Thus the events A, B and C display pairwise independence. However:

$$P(A \cap B \cap C) = 0 \neq P(A) \times P(B) \times P(C)$$

so the three events are *not* mutually independent.

6.4 The total probability theorem

Consider the following problem (already considered in Example 1).

   A statistician has a fair coin and a double-headed coin. She chooses one of the coins at random and tosses it.

   Determine the probability that she obtains a head.

We can illustrate this situation with a probability tree:

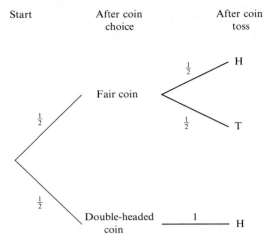

The total probability that she obtains a head is $\frac{3}{4}$, the sum of the two branches of the tree that end with the outcome 'Head'. (This probability could also be deduced by noting that the two coins have four sides between them and that three of the four equally likely sides are heads.)

The total probability theorem states (in mathematical language) that the whole is the sum of its parts. A simple illustration of the general idea is provided by the following.

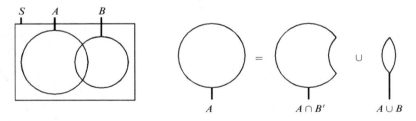

Translating the diagram into probability statements that use the fact that $A \cap B$ and $A \cap B'$ are mutually exclusive, we have:

$$P(A) = P(A \cap B) + P(A \cap B')$$
$$= \{P(B) \times P(A|B)\} + \{P(B') \times P(A|B')\}$$

In this case A consists of just two 'slices', $A \cap B$ and $A \cap B'$.
The result generalises easily to m 'slices' as follows. Suppose that $B_1, B_2, \ldots, B_m$ are m *mutually exclusive and exhaustive* events in the sample space S. Let A be some other event. A formal statement of the total probability theorem is that, for these events:

$$P(A) = \sum_{i=1}^{m} P(A \cap B_i) \qquad (6.7)$$

or equivalently using Equation (6.3):

$$P(A) = \sum_{i=1}^{m} P(B_i) \times P(A|B_i) \qquad (6.8)$$

Example 8

Of those students who do well in Physics, 80% also do well in Mathematics. Of those who do not do well in Physics, only 30% do well in Mathematics. If 40% do well in Physics, what proportion do well in Mathematics?

Define the events A, B_1 and B_2 as follows:

A: Does well in Mathematics.
B_1: Does well in Physics.
B_2: Does not do well in Physics.

The information given tells us that $P(A|B_1) = 0.8$, $P(A|B_2) = 0.3$ and $P(B_1) = 0.4$. From the latter we can deduce that $P(B_2) = 0.6$. The events B_1 and B_2 are mutually exclusive and exhaustive, so, using Equation (6.8):

$$P(A) = \{P(B_1) \times P(A|B_1)\} + \{P(B_2) \times P(A|B_2)\}$$
$$= (0.4 \times 0.8) + (0.6 \times 0.3)$$
$$= 0.50$$

Thus half the students do well in Mathematics.

Example 9

Here is an example involving both balls being drawn from a box and coins being tossed! Suppose that a box contains 3 balls numbered, respectively, 0, 1 and 2. A ball is drawn at random from the box and is found to have the number n, say. We now toss n coins.
What is the probability that we get exactly one head?

We begin by drawing a probability tree and we define events:

A: Exactly one head is obtained.
B_i: The ball chosen is numbered i, where $i = 0, 1$, or 2.

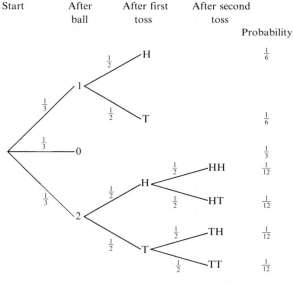

As the diagram shows, $P(B_0) = P(B_1) = P(B_2) = \frac{1}{3}$, while $P(A|B_0) = 0$, $P(A|B_1) = \frac{1}{2}$ and $P(A|B_2) = (\frac{1}{2} \times \frac{1}{2}) + (\frac{1}{2} \times \frac{1}{2}) = \frac{1}{2}$.
The total probability of the event A is given by:

$$P(A) = \{P(B_0) \times P(A|B_0)\} + \{P(B_1) \times P(A|B_1)\} + \{P(B_2) \times P(A|B_2)\}$$
$$= (\tfrac{1}{3} \times 0) + (\tfrac{1}{3} \times \tfrac{1}{2}) + (\tfrac{1}{3} \times \tfrac{1}{2})$$
$$= 0 + \tfrac{1}{6} + \tfrac{1}{6}$$
$$= \tfrac{1}{3}$$

The probability that we get exactly one head is $\frac{1}{3}$.

Example 10

A car is made in three versions: 2-door, 4-door and hatchback. The proportions of the three types made are 25%, 40% and 35% respectively. Each version of the car has either a 1400 cc engine or a 1600 cc engine. Of the 2-door version, 70% have 1400 cc engines. The proportions for the 4-door and hatchback versions are 40% and 35% respectively.

In a publicity stunt the car makers choose an owner at random to receive a prize of free car servicing for the lifetime of the car. Determine the probability that the owner's car has a 1600 cc engine.

———————

Define the events A, B_1, B_2 and B_3 as follows:

A: Owner's car has a 1600 cc engine.
B_1:   Owner's car is the 2-door version.
B_2:   Owner's car is the 4-door version.
B_3:   Owner's car is the hatchback version.

The events B_1, B_2 and B_3 are mutually exclusive and exhaustive, so using Equation (6.8):

$$P(A) = \{P(B_1) \times P(A|B_1)\} + \{P(B_2) \times P(A|B_2)\} + \{P(B_3) \times P(A|B_3)\}$$
$$= (0.25 \times 0.3) + (0.4 \times 0.6) + (0.35 \times 0.65)$$
$$= 0.5425$$

The probability that the owner's car has a 1600 cc engine is approximately 54%.

Exercises 6b

1 A vehicle insurance company classifies drivers as A, B or C according to whether or not they are a good risk, a medium risk or a poor risk with regard to having an accident. The company estimates that A constitutes 30% of drivers that are insured and B constitutes 50%. The probability that a class A driver will have one or more accidents in any 12 month period is 0.01, the corresponding values for B and C being 0.03 and 0.06 respectively.

(a) Find the probability that a motorist, chosen at random, is assessed as a class C risk and will have one or more accidents in a 12 month period.

(b) Find the probability that a motorist, chosen at random, will have one or more accidents in a 12 month period.

(c) The company sells a policy to a customer and within 12 months the customer has an accident. Find the probability that the customer is a class C risk.

(d) If a policy holder goes 10 years without an accident and accidents in each year are independent of those in other years, show that the probabilities that the policy holder belongs to each of the classes can be expressed, to 2 decimal places, in the ratio 2.71 : 3.69 : 1.08.   [AEB 90]

2 The events A and B are such that $P(A) = x + 0.2$, $P(B) = 2x + 0.1$, $P(A \cap B) = x$.

(a) Given that $P(A \cup B) = 0.7$, find the value of x and state the values of $P(A)$ and $P(B)$.

(b) Verify that the events A and B are independent.

The events A and C are mutually exclusive, $P(A \cup B \cup C) = 1$ and $P(B|C) = 0.4$.

(c) Find the values of $P(B \cap C)$ and $P(C)$.

(d) Giving a reason, state whether or not the events B and C are independent.   [ULSEB]

3 For the two events A and B, $P(A|B) = \frac{5}{11}$, $P(A \cup B) = \frac{9}{10}$, $P(B) = x$.

(a) Write $P(A \cap B)$ in terms of x and hence show that
$$P(A) = \frac{9}{10} - \frac{6x}{11}.$$
It is given that $P(A \cap B) = 2P(A \cap B')$.

(b) Find an equation for x.

(c) Deduce that $x = \frac{11}{15}$.

For the two events A and B and a third event C,
$$P(A \cup B \cup C) = 1,$$
A and C are mutually exclusive,
B and C are independent.

(d) Taking $P(B \cap C) = y$, form an equation for y and hence show that $P(C) = \frac{3}{8}$.

(e) Find the value of $P(A \cup C)$.   [ULSEB]

4 (i) Events A and B are such that $P(A) = \frac{2}{5}$, $P(B) = \frac{1}{4}$ and $P(A \cup B) = \frac{11}{20}$. Determine whether or not the events A and B are

    (*a*) independent,

    (*b*) mutually exclusive.

A third event C is such that $P(A \cup C) = \frac{7}{10}$, $P(B \cup C) = \frac{3}{4}$ and $P(A \cap C) = 2P(B \cap C)$.

    (*c*) Find $P(C)$ and determine whether or not the events B and C are independent.

(ii) A biased die is constructed so that each of the numbers 3 and 4 is twice as likely to occur as the numbers 1, 2, 5 and 6. Find

    (*a*) the probability of throwing a 4,

    (*b*) the probability of throwing a 4, given that the throw is greater than 2.

Two such dice are thrown.

    (*c*) Find the probability that the sum of the numbers thrown is 7.   [ULSEB]

5 (i) The events A and B are such that $P(A|B) = \frac{7}{10}$, $P(B|A) = \frac{7}{15}$ and $P(A \cup B) = \frac{3}{5}$. Find the values of

    (*a*) $P(A \cap B)$,

    (*b*) $P(A' \cap B)$.

(ii) A hand of four cards is to be drawn without replacement and at random from a pack of fifty two playing cards. Giving your answer in each case to three significant figures, find the probabilities that this hand will contain

    (*a*) four cards of the same suit,

    (*b*) either two aces and two kings *OR* two aces and two queens.   [ULSEB]

6 (a) Use the fact that $P(A) = P(A \cap B) + P(A \cap B')$ to show that $P(A'|B) = 1 - P(A|B)$.

(b) It is given that events A and B have non-zero probabilities, and that $P(A|B) = P(A)$.

    (i) Show that $P(B|A) = P(B)$.

    (ii) Use the result in (a) to show that $P(A'|B) = P(A')$.

    (iii) Given also that $P(B) \neq 1$, show also that $P(A|B') = P(A)$ and $P(A'|B') = P(A')$.

The Reverend Thomas Bayes (1701–61) was a Nonconformist minister in Tunbridge Wells, Kent. He was elected a Fellow of the Royal Society in 1742. The theorem (described below) that bears his name has led to the development of an approach to statistics that runs parallel to much of the material in later chapters of this book. This approach is referred to as 'Bayesian Statistics' and its advocates are referred to as 'Bayesians'. Ironically, the theorem was contained in an essay that did not appear until after his death and was largely ignored at the time.

6.5 Bayes' theorem

In introducing the idea of conditional probability we effectively asked the question:

Given that event B has occurred in the past, what is the probability that event A will occur?

We now consider the following 'reverse' question.

Given that the event A has just occurred, what is the probability that it was preceded by the event B?

As an example, consider the following problem.

A statistician has a fair coin and a double-headed coin. She chooses one of the coins at random and tosses it. She obtains a head.
Determine the probability that the coin that she tossed was double-headed.

We have looked at this situation before. We found that the total probability of a head was made up of a contribution of $\frac{1}{2} \times 1$ from the double-headed coin and $\frac{1}{2} \times \frac{1}{2}$ from the fair coin, giving a total probability of $\frac{3}{4}$. Two-thirds of this total is associated with the selection of the double-headed coin (because $\frac{1}{2} \times 1$ equals $\frac{1}{2}$, which is two-thirds of $\frac{3}{4}$). Expressing this in different words, on two-thirds of the occasions that a head is obtained the double-headed coin has been tossed. The required probability is therefore $\frac{2}{3}$.

If you found the last paragraph difficult to follow, fear not! We now develop a general result, beginning with a restatement of Equation (6.3):

$$P(A) \times P(B|A) = P(B) \times P(A|B)$$

Dividing through by $P(A)$ we get:

$$P(B|A) = \frac{P(B) \times P(A|B)}{P(A)} \tag{6.9}$$

Suppose that, instead of a single event, B, there were m alternative previous events that could have happened, namely, $B_1, B_2, \ldots, B_m$. Assume that, as was the case with the total probability theorem, these events are mutually exclusive and exhaustive. From Equation (6.9):

$$P(B_j|A) = \frac{P(B_j) \times P(A|B_j)}{P(A)}$$

and, on substituting for $P(A)$ using Equation (6.8), we get **Bayes' theorem**:

$$P(B_j|A) = \frac{P(B_j) \times P(A|B_j)}{\sum_{i=1}^{m}\{P(B_i) \times P(A|B_i)\}} \tag{6.10}$$

You may not believe it, but this is not as bad as it looks – the denominator is, after all, simply $P(A)$.

Note

♦ In Equation (6.10) it should be noted that the numerator, $P(B_j) \times P(A|B_j)$, is one of the terms in the sum in the denominator, $\sum_{i=1}^{m}\{P(B_i) \times P(A|B_i)\}$.

▼ ▼

Example 11

A statistician has a fair coin and a double-headed coin. She chooses one of the coins at random and tosses it. She obtains a head. Determine the probability that the coin that she tossed was double-headed.

This is the problem that we answered rather long-windedly at the beginning of this section! We now use a formal approach using Bayes' theorem. We define the events A, B_1 and B_2 as follows:

A: A head is obtained.
B_1: The fair coin is chosen.
B_2: The double-headed coin is chosen.

We want $P(B_2|A)$ and we know the following probabilities: $P(B_1) = \frac{1}{2}$, $P(B_2) = \frac{1}{2}$, $P(A|B_1) = \frac{1}{2}$, $P(A|B_2) = 1$. Using Bayes' theorem we have:

$$P(B_2|A) = \frac{P(B_2) \times P(A|B_2)}{P(B_1) \times P(A|B_1) + P(B_2) \times P(A|B_2)}$$

$$= \frac{\frac{1}{2} \times 1}{(\frac{1}{2} \times \frac{1}{2}) + (\frac{1}{2} \times 1)} = \frac{\frac{1}{2}}{\frac{1}{4} + \frac{1}{2}} = \frac{2}{3}$$

The good thing about Bayes' theorem is that (once the events have been carefully defined!) we do not need to think!

▲ ▲

Example 12

According to a firm's internal survey, of those employees living more than 2 miles from work, 90% travel to work by car. Of the remaining employees, only 50% travel to work by car. It is known that 75% of employees live more than 2 miles from work.
Determine:
(i)   the overall proportion of employees who travel to work by car,
(ii) the probability that an employee who travels to work by car lives more than 2 miles from work.

————

Define the events A, B_1 and B_2 as follows:

> A: Travels to work by car.
> B_1: Lives more than 2 miles from work.
> B_2: Lives not more than 2 miles from work.

The events B_1 and B_2 are mutually exclusive and exhaustive, with $P(B_1) = 0.75$, $P(B_2) = 0.25$, $P(A|B_1) = 0.9$ and $P(A|B_2) = 0.5$.
(i)   From the total probability theorem:

$$P(A) = \{P(B_1) \times P(A|B_1)\} + \{P(B_2) \times P(A|B_2)\}$$
$$= (0.75 \times 0.9) + (0.25 \times 0.5) = 0.675 + 0.125 = 0.8$$

so 80% of employees travel to work by car.
(ii) From Bayes' theorem:

$$P(B_1|A) = \frac{P(B_1) \times P(A|B_1)}{P(A)} = \frac{0.75 \times 0.9}{0.8} = 0.843\,75$$

so the probability that an employee, who travels to work by car, lives more than 2 miles from work, is about 0.84 (to 2 d.p.).

————

An alternative approach involves constructing the following table from the information in the question:

	More than 2 miles	Not more than 2 miles	Total
Travels by car	67.5	12.5	80.0
Does not travel by car	7.5	12.5	20.0
Total	75.0	25.0	100.0

The entries are percentages of the workforce. The first entry, 67.5%, is obtained by calculating the value corresponding to 90% of the 75% who live more than 2 miles from work (using $0.90 \times 0.75 = 0.675$).
(i)   The answer is the first row total, 80%.
(ii) The answer is the proportion of the first row that are contained in the top left cell of the table, namely $\dfrac{67.5}{80} = 0.84$ (to 2 d.p.).

Example 13

A box contains three coins. Two coins are fair, but the third coin is double-headed. A coin is chosen at random and tossed.
(i)   Determine the probability that a head is obtained.
(ii) If a head is obtained, determine the probability that it was the double-headed coin that was tossed.

————

We provide three alternative answers. One uses the formality of Bayes' theorem, one uses a probability tree and the last uses a 'common-sense' approach. The first is the recommended answer!

Define the events A, B_1 and B_2 as follows:

> A: A head is obtained.
> B_1: A fair coin was tossed.
> B_2: The double-headed coin was tossed.

The events B_1 and B_2 are mutually exclusive and exhaustive, with $P(B_1) = \frac{2}{3}$ and $P(B_2) = \frac{1}{3}$. Also $P(A|B_1) = \frac{1}{2}$ and $P(A|B_2) = 1$.

(i) $P(A) = \{P(B_1) \times P(A|B_1)\} + \{P(B_2) \times P(A|B_2)\}$

$$= \left(\tfrac{2}{3} \times \tfrac{1}{2}\right) + \left(\tfrac{1}{3} \times 1\right) = \tfrac{1}{3} + \tfrac{1}{3} = \tfrac{2}{3}$$

The probability of obtaining a head is $\frac{2}{3}$.

(ii) $P(B_2|A) = \dfrac{P(B_2) \times P(A|B_2)}{P(A)} = \dfrac{\frac{1}{3}}{\frac{2}{3}} = \tfrac{1}{2}$

Given that a head is obtained, the probability that it was the double-headed coin that was tossed is $\frac{1}{2}$.

We can see the various possibilities quite easily using a probability tree.

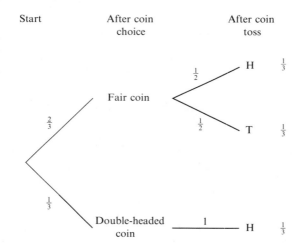

In this case there are three alternative outcomes, all with probability $\frac{1}{3}$. Two correspond to getting a head, hence $P(\text{Head}) = \frac{2}{3}$. Of these two outcomes one corresponds to the case where the double-headed coin was tossed and hence the second of the required probabilities is $\frac{1}{2}$.

An alternative argument is as follows. The three coins have six sides between them. The side actually seen is equally likely to be any of the six. Since four of the sides are heads, the probability of obtaining a head is $\frac{4}{6} = \frac{2}{3}$. Since two of the four heads are on the double-sided coin, the probability that it was this coin that was tossed is $\frac{2}{4} = \frac{1}{2}$. This type of argument is perfectly acceptable *when it is correct*! However, it is easy to go wrong – it is safer to follow the formulae!

Exercises 6c

1 It is given that B_1 and B_2 are mutually exclusive and exhaustive, and $P(A|B_1) = 0.3$, $P(A|B_2) = 0.4$, $P(B_1) = 0.4$.
Find (i) $P(B_1|A)$, (ii) $P(B_2|A)$, (iii) $P(B_1|A')$, (iv) $P(B_2|A')$.

2 It is given that $P(A) = 0.3$, $P(B) = 0.2$, $P(C) = 0.5$, $P(A \cap B) = 0$, $P(B \cap C) = 0$, $P(C \cap A) = 0$. It is also given that $P(D|A) = 0.1$, $P(D|B) = 0.4$, $P(D|C) = 0.6$.
Find (i) $P(A|D)$, (ii) $P(A'|D)$, (iii) $P(A|D')$, (iv) $P(A'|D')$.

3 A bag contains 7 white balls and 3 black balls. A white box contains 5 green balls and 2 red balls. A black box contains 3 green balls and 1 red ball. A ball is taken at random from the bag, and if this ball is white, a ball is taken at random from the white box. If it is black a ball is taken at random from the black box. Given that the ball taken from the box is red, determine the probability that the box is coloured white.

4 A factory has three machines making large numbers of components. 10% of the components made by machine I are faulty. The corresponding figures for machines II and III are 5% and 1% respectively. The proportions of the total output produced by machines I, II and III are 50%, 30% and 20% respectively.
(i) A randomly chosen component is found to be faulty.
   Find the probability that it was made by machine I.
   Find also the probability that it was not made by machine II.
(ii) A randomly chosen component is found not to be faulty.
   Find the probability that it was made by machine I.

5 Suppose that on one-third of the days of the year some rain falls on my garden. Suppose also that when it rains there is a probability of 0.7 that my barometer will be indicating rain, but when it does not rain there is a probability of 0.1 that my barometer nevertheless indicates rain.
(a) Determine the probability that, on a randomly chosen day of the year, my barometer indicates rain.
(b) Given that my barometer is indicating rain, determine the probability that it is actually raining.

6 In an examination, the probabilities of three candidates, Aloysius, Bertie and Claude, solving a certain problem are $\frac{4}{5}$, $\frac{3}{4}$ and $\frac{2}{3}$, respectively. Calculate the probability that the examiner will receive from these candidates:
(a) one, and only one, correct solution,
(b) not more than one correct solution,
(c) at least one correct solution.
Given that the examiner receives exactly one correct solution, determine the probability that this solution was provided by Bertie.

7 A test for a particular disease has the following characteristics. If someone has the disease the probability of a positive test is 90%, and the probability of a negative result (a 'false negative') is 10%. If someone does not have the disease the probability of a positive test (a 'false positive') is 20%, and the probability of a negative result is 80%. The proportion of the population that has the disease is denoted by p. A person is chosen at random from the population and tested. Given that the result of the test is positive, find, in terms of p, the probability P that the person has the disease.
Verify that when $p = 0.05$ the value of P is a little less than 20%.
Sketch the graph of P against p and comment on the results.

8 In a television game show, a contestant chooses one of three doors and receives the prize behind the door. The three doors are gold, silver and black. Behind one of the doors there is a Ferrari, and there is nothing behind the other two doors. From the contestant's point of view it is equally likely to be behind any one of the three doors. The contestant chooses the gold door, and the presenter, who knows where the Ferrari is, then opens one of the other two doors (say the black door) and shows that there is nothing behind it. He then says to the contestant 'It is now between gold and silver, and I will allow you to change your choice if you wish to.' The problem is to decide whether the contestant can improve her chances of winning by changing her choice to silver.
Suppose that in the case when the Ferrari is behind the gold door, the probability that the presenter opens the black door is p. In the other two cases he has no choice. Given that the presenter has opened

(*continued*)

the black door and shown that there is nothing behind it, show that the probability that the Ferrari is behind

the silver door is $\dfrac{1}{(1+p)}$. By considering the values of this probability for $p = 0$, $\frac{1}{2}$ and 1, determine whether the contestant can improve her chances of winning by changing her choice.

9   Three prisoners, A, B and C, are in separate cells. Prisoner A is told that two of the three prisoners are to be executed and the third is to be set free, but it is not permitted for him to be told his own fate. In the absence of any more information, prisoner A estimates the probability that he will be set free as $\frac{1}{3}$.

Prisoner A now says to his jailer 'At least one of the other two prisoners is bound to be executed, so you will not be giving anything away if you tell me the name of one of them who is to be executed'. After a little thought the jailer says 'All right, B is to be executed.' Prisoner A now re-estimates his probability of being set free as $\frac{1}{2}$, since either C or A will be set free!
To analyse this situation, suppose that, in the case when B and C are to be executed, and A is to be set free, the probability that the jailer says 'B is to be executed' is p. In the other two cases the jailer has no choice.
Show that the probability that A is to be set free, given that the jailer has said 'B is to be executed', is $\dfrac{p}{(1+p)}$.
Consider the value of this expression as p takes the values 0, $\frac{1}{2}$ and 1 and comment on the results.

10 Four machines A, B, C and D produce respectively 30%, 30%, 15% and 25% of the total number of items from a factory. The percentages of defective output of these machines are 1%, $1\frac{1}{2}$%, 3% and 2% respectively. Given that an item is to be selected at random from the total output, find the probability that the item will be defective. An item is selected at random and is found to be defective. Find the probability that the item was produced by machine A.   [ULSEB(P)]

11 In a simple model of the weather in October, each day is classified as either fine or rainy. The probability that a fine day is followed by a fine day is 0.8. The probability that a rainy day is followed by a fine day is 0.4. The probability that 1 October is fine is 0.75.

(i)   Find the probability that 2 October is fine and the probability that 3 October is fine.
(ii) Find the conditional probability that 3 October is rainy, given that 1 October is fine.
(iii) Find the conditional probability that 1 October is fine, given that 3 October is rainy.   [UCLES]

12 (a) Give an equation involving probabilities which is equivalent to the statement
      (i)   the events L and M are *mutually exclusive*,
      (ii) the events L, M and N are *exhaustive*.
If the events L, M and N are mutually exclusive as well as being exhaustive, write down an equation relating $P(L)$, $P(M)$ and $P(N)$.

    (b) A city Passenger Transport Executive (PTE) carries out a survey of the commuting habits of its city centre workers. The PTE discovers that 40% of commuters travel by bus, 25% travel by train and the remainder use private vehicles.
Of those who travel by bus, 60% have a journey of less than 5 miles and 30% have a journey of between 5 and 10 miles.
Of those who travel by train, 30% have a journey of between 5 and 10 miles and 60% have a journey of more than 10 miles.
Of those who use private vehicles, 20% travel less than 5 miles, with the same percentage travelling more than 10 miles. By organising the above information in a suitable table or diagram, or otherwise, determine the probability that a commuter chosen at random
      (i)   travels by bus for a journey of less than 5 miles,
      (ii) has a journey of more than 10 miles,
      (iii) travels by bus or has a journey of more then 10 miles,
      (iv) uses a private vehicle given that the commuter travels between 5 and 10 miles.   [JMB(P)]

13 State in words the relationship between two events E and F when
(a) $P(E \cap F) = P(E).P(F)$,
(b) $P(E \cap F) = 0$.
Given that $P(E) = \frac{1}{3}$, $P(F) = \frac{1}{2}$, $P(E' \cap F) = \frac{1}{2}$, find
(c) the relationship between E and F,
(d) the value of $P(E|F)$,
(e) the value of $P(E' \cap F')$.   (*continued*)

A boy always either walks to school or goes by bus. If one day he goes to school by bus, then the probability that he goes by bus the next day is $\frac{7}{10}$. If one day he walks to school, then the probability that he goes by bus the next day is $\frac{2}{5}$. Given that he walks to school on a particular Tuesday, draw a tree diagram and hence find the probability that he will go to school by bus on Thursday of that week.

Given that the boy walks to school on both Tuesday and Thursday of that week, find the probability that he will also walk to school on Wednesday.

[You may assume that the boy will not be absent from school on Wednesday or Thursday of that week.] [ULSEB]

Chapter summary

♦ $P(B|A)$ denotes the probability that the event B occurs (or has occurred) given the information that the event A occurs (or has occurred).
$P(B|A)$ is known as the **conditional probability** of B **given** A.

♦ $P(B|A) = \dfrac{P(A \cap B)}{P(A)}$

♦ $P(A \cap B) = P(A) \times P(B|A) = P(B)P(A|B)$

♦ **Statistical independence**: each of the following statements implies all the others.
● Events A and B are statistically independent.
● $P(A|B) = P(A)$
● $P(B|A) = P(B)$
● $P(A \cap B) = P(A) \times P(B)$ – the **multiplication rule**.

♦ **Physically independent** events are statistically independent.

♦ **The total probability theorem**:
If events $B_1, B_2, \ldots, B_n$ are mutually exclusive and exhaustive, then:
$P(A) = \Sigma P(A \cap B_i) = \Sigma\{P(B_i) \times P(A|B_i)\}$

♦ **Bayes' theorem**:
$$P(B_j|A) = \frac{P(B_j) \times P(A|B_j)}{\Sigma\{P(B_i) \times P(A|B_i)\}}$$

Exercises 6d (Miscellaneous)

1 The probability that a particular man will survive the next twenty-five years is 0.6, and independently, the probability that the man's wife will survive the next twenty-five years is 0.7. Calculate the probability that in twenty-five years' time
(i) only the man will be alive,
(ii) at least one will be alive. [WJEC]

2 (a) Explain what you understand by the following terms in relation to probability:
(i) mutually exclusive events,
(ii) independent events.

You may illustrate your answer with reference to experiments if you wish.

(continued)

(b) In an ordinary pack of 52 playing cards one card is selected. What is the probability of the card being:
  (i)   a heart;
  (ii) the queen of hearts;
  (iii) a red queen?

Consider the events of selecting a heart, a queen, a red card. Then for each of the possible pairs of events state whether the events are independent, mutually exclusive or otherwise, giving in each case the reasons for your answer.   [UODLE]

3 Show that, for any two events E and F,

$$P(E \cup F) = P(E) + P(F) - P(E \cap F).$$

Express in words the meaning of $P(E|F)$. Given that E and F are independent events, express $P(E \cap F)$ in terms of $P(E)$ and $P(F)$, and show that E' and F are independent.
In a college, 60 students are studying one or more of the three subjects Geography, French and English. Of these, 25 are studying Geography, 26 are studying French, 44 are studying English, 10 are studying Geography and French, 15 are studying French and English, and 16 are studying Geography and English.
Write down the probability that a student chosen at random from those studying English is also studying French.
Determine whether or not the events "studying Geography" and "studying French" are independent.
A student is chosen at random from all 60 students. Find the probability that the chosen student is studying all three subjects.   [ULSEB]

4 Students in a class were given two statistics problems to solve, the second of which was harder than the first. Within the class $\frac{5}{6}$ of the students got the first one correct and $\frac{7}{12}$ got the second one correct. Of those students who got the first one correct, $\frac{3}{5}$ got the second one correct. One student was chosen at random from the class.
Let A be the event that the student got the first problem correct and B be the event that the student got the second one correct.
(a) Express in words the meaning of $A \cap B$ and of $A \cup B$.
(b) Find $P(A \cap B)$ and $P(A \cup B)$.

(c) Given that the student got the second problem right, find the probability that the first problem was solved correctly.
(d) Given that the student got the second problem wrong, find the probability that the first problem was solved correctly.
(e) Given that the student got the first problem wrong, find the probability that the student also got the second problem wrong.   [ULSEB]

5 Two porcelain factories A and B produce cheap china cups in equal numbers. If closely examined, a cup from A will be found flawless with probability $\frac{3}{4}$, but one from B with probability $\frac{1}{2}$. Jim picks up two cups from a batch in a shop. The shopkeeper says that all the cups in the batch come from the same factory, but the batch is equally likely to come from factory A or from factory B.
(i)   What is the probability that the first cup Jim examines is flawless?
(ii) Given that the first cup is flawless, what is the conditional probability that the batch came from factory A?
(iii) Unfortunately, Jim drops the second cup before he can examine it. Given that the first cup was flawless, find the probability that the second cup was also flawless before the accident.   [SMP]

6 A high jumper estimates the probabilities that she will be able to clear the bar at various heights, on the basis of her experience in training. These are given in the table:

Height	Probability of success at each attempt
1.60 m	1
1.65 m	0.6
1.70 m	0.2
1.75 m	0

In a competition she is allowed up to three attempts to clear the bar at each height. If she succeeds, the bar is raised by 5 cm and she is allowed three attempts at the new height; and so on. It is assumed that the result of each attempt is independent of all her previous attempts.
(i) Show that the probability that she will be successful at 1.65 m is 0.936.

(continued)

(ii) Calculate the probability that, if she is successful at 1.65 m, she will not be successful at 1.70 m.

Hence find the probabilities that, in the competition, the height she jumps will be recorded as

(*a*) 1.60 m, (*b*) 1.65 m, (*c*) 1.70 m.   [SMP]

7 In a sales campaign, a petrol company gives each motorist who buys their petrol a card with a picture of a film star on it. There are 10 different pictures, one each of 10 different film stars, and any motorist who collects a complete set of all 10 pictures gets a free gift. On any occasion when a motorist buys petrol, the card received is equally likely to carry any one of the 10 pictures in the set.

(i) Find the probability that the first four cards the motorist receives all carry different pictures.

(ii) Find the probability that the first four cards received result in the motorist having exactly three different pictures.

(iii) Two of the ten film stars in the set are X and Y. Find the probability that the first four cards received result in the motorist having a picture of X or Y (or both).

(iv) At a certain stage the motorist has collected nine of the ten pictures. Find the least value of n such that

P(at most n more cards are needed to complete the set) > 0.99.   [UCLES]

8 The staff employed by a college are classified as academic, administrative or support. The following table shows the numbers employed in these categories and their sex.

	Male	Female
Academic	42	28
Administrative	7	13
Support	26	9

A member of staff is selected at random.

A is the event that the person selected is‘ female.

B is the event that the person selected is academic staff.

C is the event that the person selected is administrative staff.

($\bar{A}$ is the event not A, $\bar{B}$ is the event not B, $\bar{C}$ is the event not C.)

(a) Write down the values of
(i) P(A),
(ii) P(A ∩ B),
(iii) P(A ∪ $\bar{C}$),
(iv) P($\bar{A}$|C).

(b) Write down one of the events which is
(i) not independent of A,
(ii) independent of A,
(iii) mutually exclusive of A.
In each case justify your answer.

(c) Given that 90% of academic staff own cars, as do 80% of administrative staff and 30% of support staff,
(i) what is the probability that a staff member selected at random owns a car?
(ii) A staff member is selected at random and found to own a car. What is the probability that this person is a member of the support staff?   [AEB 91]

9 On one of his travels, Gulliver landed on an island inhabited by equal numbers of two groups of people – the Veracians who always told the truth and the Confusians who answered questions truthfully with probability $\frac{2}{3}$, independently for each question. The Veracians and Confusians were indistinguishable with regard to features, dress, etc. One afternoon Gulliver was lost on the island and on meeting a local inhabitant asked the following two questions:

"Is it night or day?"

"Which is the way to the nearest town?"

(i) Find the probability that the answer to the first question was correct.

(ii) Given that the answer to the first question was correct, calculate the conditional probability that the answer to the second question was correct.   [JMB]

10 Three children, A, B and C, are not good at keeping secrets. The probability that A will tell any secret to any other child is p. Similarly, for B and C the probabilities of telling secrets are q and r respectively. A knows a secret. Firstly A meets B, then A meets C and finally B meets C.

(continued)

(i) Show that, after these three meetings, the probability that B knows the secret and C does not is $p(1-p)(1-q)$.

It is known that $p = \frac{1}{2}$, $q = \frac{1}{3}$, $r = \frac{1}{4}$.

(ii) Show that the probability that, after these three meetings, both B and C know the secret is $\frac{19}{48}$.

(iii) After these three meetings a fourth child D, who never tells secrets, first meets B and then meets C. Find the probabilities that, after these five meetings,

(a) A, B and C know the secret and D does not,

(b) all four children know the secret,

(c) just three of the children know the secret.

[Assume independence of events throughout. Answers may be left as fractions in their lowest terms.]   [O&C]

7 Probability distributions and expectations

I am giddy: expectation whirls me round. The imaginary relish is so sweet that it enchants my sense

Troilus and Cressida, William Shakespeare.

This chapter is concerned with discrete random variables. Recall that a variable is described as being a **random variable** if its value is the result of a random observation or experiment, and that **discrete** implies that a list of its possible numerical values could be made. Here are some examples:

Discrete random variable	Possible values
The number obtained when rolling a fair six-sided die	1, 2, 3, 4, 5, 6
The number of heads obtained when four fair coins are tossed	0, 1, 2, 3, 4
The amount (in £) won in a lottery having prizes of 50p, £5 and £50	0, 0.5, 5, 50
The net gain (in £) from buying a 25p ticket in the above lottery	−0.25, 0.25, 4.75, 49.75
The number of rainy days in May	0, 1, ..., 31
The number of heads obtained when a single fair coin is tossed once	0, 1
The number of tosses of a fair coin until a head is obtained	1, 2, 3, ... (no limit!)

In each case the possible outcomes can be written down as a list of numerical values. These values do not have to be positive, nor do they have to be integers. Usually, but not always, the list is limited to just a few values.

7.1 Notation

We write:

RANDOM VARIABLES as e.g. X, Y, Z
observed values as e.g. x, y, z

This leads to a statement such as:

$$P(X = x) = \tfrac{1}{4}$$

which should be read as:

The probability that the random variable X takes the value x is $\tfrac{1}{4}$.

We can link this statement to the probability of an event by defining the event A as 'the random variable X takes the particular value x', so $P(A) = \tfrac{1}{4}$.

To simplify formulae we will often replace the cumbersome $P(X = x)$ by the simpler P_x.

7.2 Probability distributions

Suppose we roll a biased die which has sides numbered 1 to 6. Define the random variable X to be 'the number showing on the top of the die'. We know two things:

1 The observed value of X must be 1, 2, 3, 4, 5 or 6.
2 On a given roll the random variable X can only take *one* of those values.

These correspond to statements that the six outcomes are both exhaustive and mutually exclusive, hence:

$$P_1 + P_2 + \cdots + P_6 = 1$$

Generalising, for a discrete random variable X that can take only the distinct values $x_1, x_2, \ldots, x_m$:

$$\sum_{i=1}^{m} P_{x_i} = 1 \qquad\qquad (7.1)$$

The sizes of $P_{x_1}, P_{x_2}, \ldots,$ show how the total probability of 1 is *distributed* amongst the possible values of X. The most likely value for X will be the one with the highest probability. This is analogous to a frequency distribution, and, since the quantities are probabilities, the values $P_{x_1}, P_{x_2}, \ldots,$ are said to define a **probability distribution**.

Example 1

Tabulate the probability distribution of the number of heads obtained when a fair coin is tossed twice.

Let X be the random variable 'the number of heads obtained'. The possible values are 0, 1 and 2. The simplest way of finding the required probabilities is to use a probability tree from which we obtain the required table.

Start	After first toss	After second toss	Number of heads
		HH	2
	H		
		HT	1
		TH	1
	T		
		TT	0

Number of heads, x	0	1	2
P_x	$\frac{1}{4}$	$\frac{1}{2}$	$\frac{1}{4}$

Jean-le-Rond d'Alembert (1717–83) was found abandoned as a new-born infant near the church of Saint Jean-le-Rond in Paris. The gendarme who discovered the baby chose the name of the church for the baby. Despite this inauspicious beginning the boy did well! At the age of 24 he was admitted to the French Academy (the equivalent of the British Royal Society). He is best known for his work on kinetics in connection with what is now known as d'Alembert's principle. However, probabilists recall his name best because his answer to the previous example was wrong! He argued falsely that, since there were three possibilities, each probability must be $\frac{1}{3}$!

Practical

d'Alembert's error was to assume that the three possibilities were equally likely. To verify that they are not, toss two coins a total of twenty times. Draw up a tally chart of the number of heads (0, 1 or 2) obtained.
Do you believe d'Alembert? If he had seen the combined results from your class, he would have spotted his error!

The probability function

For many situations it will not be necessary to make a list of all *m* probabilities, in order to specify the probability distribution, because some simple all-embracing formula (sometimes called the **probability function**) can be found.

Example 2

Obtain a formula for the probability distribution of the random variable *X* defined as 'the result of rolling a fair six-sided die'.

_____

Each of the six possible values for *X* has probability $\frac{1}{6}$, so we can write:

$$P_x = \frac{1}{6} \qquad (x = 1, 2, \ldots, 6)$$

Illustrating probability distributions

As always in statistics, it is a good idea to draw pictures whenever possible. Since a discrete random variable can only take discrete values, a bar chart is appropriate, with the *y*-axis measuring probability.

Example 3

The random variable *X* is defined as 'the sum of the scores shown by two fair six-sided dice'.
Tabulate the probability distribution of *X* and draw an appropriate diagram.

_____

We begin by drawing up a table showing the 36 possible outcomes, all of which (since the dice are fair) are equally likely:

First die

		1	2	3	4	5	6
	1	2	3	4	5	6	7
	2	3	4	5	6	7	8
Second	3	4	5	6	7	8	9
die	4	5	6	7	8	9	10
	5	6	7	8	9	10	11
	6	7	8	9	10	11	12

Entries in table are the sums of the numbers shown by the two dice.

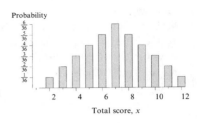

By inspection of the table (look along the NE-SW diagonals!) we can see that, of the 36 equally likely possibilities, there is just 1 possibility leading to the outcome $X = 2$, so $P_2 = \frac{1}{36}$. The most likely value for X is 7, which has probability $\frac{6}{36} = \frac{1}{6}$.

The full distribution is tabulated below.

Value of x	2	3	4	5	6	7	8	9	10	11	12
P_x	$\frac{1}{36}$	$\frac{2}{36}$	$\frac{3}{36}$	$\frac{4}{36}$	$\frac{5}{36}$	$\frac{6}{36}$	$\frac{5}{36}$	$\frac{4}{36}$	$\frac{3}{36}$	$\frac{2}{36}$	$\frac{1}{36}$

Exercises 7a

In Questions **1–9**, find the set of possible values of the random variable X, and draw up a table showing P_x (where $P_x = P(X = x)$) for each value of x. The distributions will be used in Exercises 7c, 7d, 7e.

1 A box contains 3 red marbles and 5 green marbles. Two marbles are taken at random without replacement, and X is the number of green marbles obtained.

2 A box contains 3 red marbles and 5 green marbles. Two marbles are taken at random with replacement, and X is the number of green marbles obtained.

3 A fair coin has the number '1' on one face and the number '2' on the other. The coin is thrown with a fair die and X is the sum of the scores.

4 In a raffle, 20 tickets are sold and there are two prizes. One ticket number is drawn at random and the corresponding ticket earns a £10 prize. A second, different, ticket number is drawn at random, and the corresponding ticket earns a £3 prize. The prize earned by a particular one of the original 20 tickets is £X.

5 Two cards are drawn at random (without replacement) from a pack of playing cards, and X is the number of Hearts obtained.

6 A fair die is thrown and X is the reciprocal of the score (i.e. 'one over the score').

7 Two fair dice, one red and the other green, are thrown and X is the score on the red die minus the score on the green die.

8 Two fair dice, one red and the other green, are thrown and X is the positive difference in the scores, (i.e. the modulus of the random variable in the preceding example).

9 Packets of 'Hidden Gold' cornflakes are sold for £1.20 each. One in twenty of the packets contains a £1 coin. A shopper buys two packets and £X is the net cost of the two packets.

10 Which of the following experiments give a discrete random variable? (You are not asked to find any probabilities.)
(i) A book is chosen at random from a shelf with 50 books and its author noted.

(continued)

(ii)   A book is chosen at random from a shelf with 50 books and the number of pages noted.

(iii)   A book is chosen at random from a shelf with 50 books and the fifth letter on the tenth page is noted.

(iv)   A pupil is chosen at random from a particular class and the pupil's name is noted.

(v)   A pupil is chosen at random from a particular class and the pupil's height is recorded to the nearest inch.

(vi)   The number of cars passing a given point on the road between 0800 and 0900 tomorrow.

(vii)   The colour of the first car to pass a given point after 0900 tomorrow.

(viii)   The time, after 0900 tomorrow, recorded to the nearest second, at which the telephone first rings in the local Town Hall.

(ix)   A point is chosen at random in the x–y plane and the distance from the origin is recorded to the nearest mm.

Estimating probability distributions

As we noted in Section 5.1 (p. 114), probabilities can be thought of as being the limiting values of relative frequencies. If we concentrate on a single outcome, such as obtaining a six when a die is rolled, and plot relative frequency against number of rolls, then we get a graph such as the following, which was obtained using the random number generator of a computer to simulate the rolling of a die.

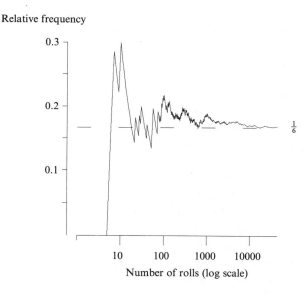

Note how the 'wiggles' die away as the number of rolls increases and the relative frequency becomes increasingly close to its limiting value of $\frac{1}{6}$. A summary of the results for all six outcomes is given below:

Number of rolls	Relative frequency of					
	1	2	3	4	5	6
36	0.222	0.083	0.167	0.167	0.139	0.222
216	0.185	0.139	0.167	0.162	0.185	0.162
1296	0.156	0.171	0.168	0.159	0.176	0.170
7776	0.161	0.164	0.173	0.160	0.168	0.173
46 656	0.168	0.164	0.167	0.168	0.165	0.168
Target	$\frac{1}{6}$	$\frac{1}{6}$	$\frac{1}{6}$	$\frac{1}{6}$	$\frac{1}{6}$	$\frac{1}{6}$

As the sample size (the number of rolls) increases, so the relative frequencies converge ever more closely on the theoretical probabilities, and the observed distribution of the possible outcomes converges on the theoretical probability distribution.

Computer project ————————————————————————

> *Write your own computer (or calculator) program to simulate the rolling of a die. To convert a random number r having a value between 0 and 1 into the score d on a die, set $d = 1 + INT(6 * r)$, where INT is a function that truncates a decimal to an integer (e.g. $INT(5.8) = 5$). Examine the relative frequencies as you increase the sample size – if they are all converging on $\frac{1}{6}$ then the program works!*

Practical ————————————————————————

> *This practical needs a little care, since it involves drawing pins! Take ten(unsquashed) drawing pins and drop them on to a flat surface. Count the number of drawing pins, x, that land with their point in the air. Repeat the experiment twenty times.*
> *Draw a bar chart of the results and determine the relative frequency of the outcome x = 5.*
> *Combine your results with four other members of the class and recalculate the relative frequency.*
> *What do you suppose is the numerical value of P_5?*

Project ————————————————————————

> *Car number plates provide a useful guide to the age of cars. The relationship between number plate and age is not perfect, of course, since some owners have 'cherished' number plates that they transfer from old cars to new ones. Furthermore, two cars with number plates of the same 'age' can differ in age by as much as 365 days (in a leap year!). Nevertheless, a rough indication of the age distribution of cars can be obtained. A convenient way of avoiding most of the problems of deciding on the age of a car is to define the random variable X to be 'the age of the car in completed years as indicated by the registration'. Thus cars registered in the current year correspond to $X = 0$.*
> *A number of interesting questions now arise! Is it the case that the age distribution of the cars on a dual carriageway (company cars, speeding executives, etc) is the same as that for cars in the supermarket car park (shoppers using older(?) 'second' cars). Does the age distribution vary according to the time of day? This could be the case if the 'first car' leaves early for work and the 'second car' leaves later for the shops. You will be able to think of other possibilities. Several hundred observations will be required before any differences can be detected with confidence.*

The cumulative distribution function

This is an alternative function for summarising a probability distribution.
It provides a formula for $P(X \leqslant x)$ in place of that for $P(X = x)$.

Example 4

Obtain the cumulative distribution function for the random variable X defined as 'the result of rolling a fair six-sided die'.

The following formula does the trick:

$$P(X \leqslant x) = \begin{cases} 0 & x < 1 \\ \frac{1}{6}m & m \leqslant x < (m+1); m = 1, 2, \ldots, 5 \\ 1 & x \geqslant 6 \end{cases}$$

since, for example:

$$P(X \leqslant 3) = P(X = 1) + P(X = 2) + P(X = 3) = \frac{1}{6} + \frac{1}{6} + \frac{1}{6} = \frac{3}{6}$$

7.3 Some special discrete probability distributions

The two most important discrete distributions are the binomial and Poisson distributions, which will be discussed in Chapters 9 and 10. We now look at some others.

The discrete uniform distribution

Here the random variable X is equally likely to take any of k values $x_1, x_2, \ldots, x_k$, so that the distribution is summarised by:

$$P_{x_i} = c \qquad (i = 1, 2, \ldots, k)$$

where c is a constant. Can c take any value that we please? Certainly not! Equation (7.1) specified that the probabilities must sum to 1 and so c is determined by the necessity that:

$$c + c + \cdots + c = 1$$

which implies that $c = \dfrac{1}{k}$. The distribution is properly specified by:

$$P_{x_i} = \frac{1}{k} \qquad (i = 1, 2, \ldots, k)$$

Example 5

The most familiar example occurs when X is defined as 'the score obtained when a fair six-sided die is rolled'. In this case $k = 6$. The distribution is tabulated below:

Value of X	1	2	3	4	5	6
Probability	$\frac{1}{6}$	$\frac{1}{6}$	$\frac{1}{6}$	$\frac{1}{6}$	$\frac{1}{6}$	$\frac{1}{6}$

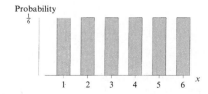

James Bernoulli (1654–1705) was a member of an extremely talented Swiss family. The most famous family members were James, his brother John and his nephews Nicholas and Daniel – though there were seven Bernoullis who would deserve a mention in a mathematician's *Who's Who!* James was 21 when he graduated (in Theology) from the University of Basel. He returned to the university as a lecturer (in Physics) when he was 29 and became a Professor of Mathematics at the age of 33. His principal work, *Ars Conjectandi* (The Art of Conjecture), was a treatise on probability.

The Bernoulli distribution

The Bernoulli distribution is very simple! It refers to a random variable X that can take only the values 0 and 1:

$$P_0 = 1 - p \qquad P_1 = p$$

An example of the random variable X is 'the number of heads obtained on a single toss of a bent coin', where the probability of a head is p. The importance of this simple distribution will become apparent in Chapter 9.

7.4 The geometric distribution

We can again use coin-tossing as an illustration. Suppose that we have a bent penny with $P(\text{Head}) = p$ and $P(\text{Tail}) = 1 - p$, with $0 < p < 1$. This time we embark on a succession of tosses and define the random variable X to be the number of tosses up to and including the first head (a 'Success').

Evidently $P_1 = p$, since this is the probability of an immediate head. For X to be equal to 2 we must obtain a tail on the first toss and a head on the second toss. Thus:

$$
\begin{aligned}
P_2 &= P(\text{Tail then Head}) \\
&= P(\text{Tail}) \times P(\text{Head}) \qquad \text{physically independent events} \\
&= (1 - p)p
\end{aligned}
$$

Similarly, for X to be equal to x, we must obtain a sequence of $(x - 1)$ tails followed by a head. Each tail occurs with probability $(1 - p)$ so that we get the general result:

$$P_x = (1 - p)^{x-1}p \qquad (x = 1, 2, \ldots) \tag{7.2}$$

This general result, which holds for all positive integer values of x, defines a **geometric distribution**.

For a fair penny, $p = \frac{1}{2}$. In a recent five-match test series the English test captain lost the first four tosses, but won the fifth. Using Equation (7.2) we see that the probability of his having to wait this long for a win during his next sequence of tosses is:

$$\left(1 - \tfrac{1}{2}\right)^4 \tfrac{1}{2} = \tfrac{1}{32}$$

Notes

- The distribution is called *geometric* because the successive probabilities, p, $(1 - p)p$, $(1 - p)^2 p$, ... form a **geometric progression** with first term p and common ratio $(1 - p)$.

◆ Writing q for $(1-p)$, and noting that $0 < q < 1$:

$$\sum_{x=1}^{\infty} P_x = p(1 + q + q^2 + q^3 + \cdots)$$

$$= p\,\frac{1}{1-q} \qquad \text{sum to infinity of a geometric progression}$$

$$= 1 \qquad \text{since } q = 1 - p$$

This shows that the total probability being distributed is equal to 1 as required. It also proves that a success will occur eventually – if you have just had 3000 failures, don't worry! Providing $0 < p < 1$, you will get a success eventually (if you don't die of exhaustion first ...).

◆ The distribution is sometimes written as:

$$P_x = (1-p)^x p \qquad (x = 0, 1, 2, \ldots)$$

Cumulative probabilities

To calculate $P(X \leqslant x)$ we note that this means that at least one of the first x trials must have been a success. The complement to this event is that all x trials were failures. If the probability of a failure is $(1-p)$ then the probability of x failures is $(1-p)^x$. Writing q for $(1-p)$ we have

$$P(X \leqslant x) = 1 - q^x \qquad (7.3)$$

Similarly:

$$P(X < x) = 1 - q^{x-1}$$
$$P(X > x) = q^x$$
$$P(X \geqslant x) = q^{x-1}$$

Note

◆ We can also prove the result in Equation (7.3) as follows:

$$P(X \leqslant x) = P(X = 1) + P(X = 2) + \cdots + P(X = x)$$
$$= p + pq + \cdots + pq^{x-1}$$
$$= p(1 + q + \cdots + q^{x-1})$$

The bracketed terms are a geometric series with sum $\dfrac{1 - q^x}{1 - q}$. Since $p = (1 - q)$, this establishes the given result.

Example 6

Only 1% of the vehicles leaving a motorway are prepared to give lifts to hitch-hikers. George Nerdowell arrives at a motorway exit and sticks out his thumb. Determine the probability that at least four vehicles fail to stop for him (i.e. that he doesn't get a lift until at least vehicle 5).

———

George will keep his thumb stuck out until he obtains a lift. So each vehicle is either a 'Success' (with probability, p, equal to 0.01), or a 'Failure' (with probability, q, equal to 0.99). Let X be the number of vehicles up to and including the vehicle that gives George a lift. The question requires us to calculate $P(X > 4)$:

$$P(X > 4) = q^4 = 0.99^4 = 0.961 \text{ (to 3 d.p.)}$$

The probability that at least four vehicles fail to stop for him is about 0.96.

A paradox!

Assuming that $0 < p < 1$, all geometric distributions have a similar shape: an infinite sequence of ever smaller probabilities. The rate of decline in the size of the probabilities depends upon the value of p, but the mode (the most probable value) is at $x = 1$ in each case.

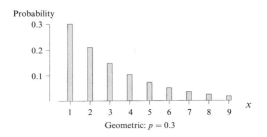

Geometric: $p = 0.3$

The practical consequences of this result are, to say the least, surprising! Suppose, for example, that I decide that I will stand outside my house until I see a red sports car. Clearly I may have to stand there for a long time, since red sports cars are not all that common. Consider therefore the following question: 'What is the most probable number of cars that pass my house up to and including the red sports car?'. The situation is geometric, with the value of p being rather small. Nevertheless, the previous result still holds and the answer to the question is that the most probable number of cars is just 1!

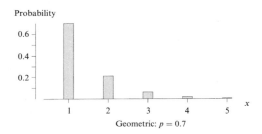

Geometric: $p = 0.7$

Note

◆ This result can easily be misinterpreted! The probability of the specific outcome 1 is $P_1 = p$, so $P(X > 1) = 1 - p$. The rarer the event of interest is (i.e. the smaller the value of p is), the more likely is the observed value to be greater than 1. Nevertheless, 1 remains the most probable single value.

Practical _____

Roll a normal six-sided die repeatedly until a 6 is obtained. Record the number of tosses required. Repeat a further 9 times. Pool your results with those of your neighbours, or with those for the entire class. You should find, as predicted, that the mode is at 1 – though some people may have had to roll as many as 20 times in order to get a 6!

Practical _____

A pack of cards is required for this exercise. Begin by shuffling the pack so that the cards are in a random order and draw a card at random from the pack. Replace the card, shuffle once again and repeat the procedure. Continue with this process until a 'Success' is obtained. The event to be called a 'Success' could be 'a Spade' ($p = \frac{1}{4}$), 'an Ace' ($p = \frac{1}{13}$) or whatever is of interest. However, it is inadvisable to choose a very rare event such as 'the

Ace of Spades' $(p = \frac{1}{52})$, unless there is a great deal of time available!
Record the number of cards required (x).

Repeat the entire procedure a number of times and pool your results with those of the rest of the class. Compare the class relative frequencies with the theoretical probabilities. There should be reasonable agreement, particularly for low values of x. In Chapter 18 we shall look at ways of testing how good the agreement is.

Exercises 7b

1 A book has pages numbered 1 to 300. A page is chosen at random and X is the last digit of the page number.
(i) Find the probability distribution of X.
(ii) The first digit of the page number is Y. Does Y have a discrete uniform distribution? Find the distribution of Y.

2 In Ludo it is necessary to throw a six with a single fair die in order to start. The number of throws needed to obtain the first six is N. Find the probability distribution of N. Find (i) $P(N > 6)$, (ii) $P(N \leqslant 5)$.

3 Determine the distribution of X, where X is the random variable denoting the number of sixes obtained on a single throw of a fair die.

4 The random variable X has a distribution which is both uniform and Bernoulli. Describe the distribution.

5 The random variable X is the number of heads obtained when a fair coin is thrown twice. Show that the distribution of X is (i) not uniform, (ii) not Bernoulli, (iii) not geometric.

6 Packets of 'Hidden Gold' cornflakes are sold for £1.20 each. One in twenty of the packets contains a £1 coin.
(i) A shopper goes on buying packets until a packet containing £1 is obtained. Find the probability distribution of the number of packets bought.
(ii) A shopper buys a single packet. Find the probability distribution of the number of coins obtained.

7.5 Expectations

In Chapter 2 we saw that the mean of a frequency distribution is given by:

$$\bar{x} = \frac{1}{n}\Sigma f_j x_j$$

where f_j is the frequency with which the value x_j occurred, n is the total number of observations, and the summation is over all the values of x_j. The formula can be rewritten as:

$$\bar{x} = \sum \left\{ x_j\left(\frac{f_j}{n}\right) \right\}$$

which emphasises that, in the summation, each value of x is multiplied by its relative frequency.

As the sample size increases, what happens to $\bar{x}$? We can get an idea by re-examining the die-rolling results (see Section 7.2, p. 171) and plotting the value of $\bar{x}$ against the number of rolls.

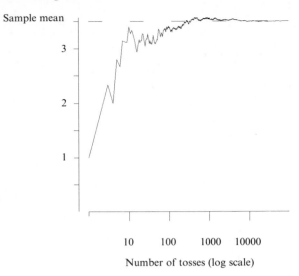

Number of tosses (log scale)

We can see that, as in the case of a relative frequency, after some initial oscillations it appears to be settling down to some value. We call this limiting value the **expectation** of X and denote it by $E(X)$. We can think of it as the long-term average value of X. Whilst $\bar{x}$ is approaching $E(X)$, each relative frequency is approaching the corresponding population probability. To determine the value of $E(X)$, therefore, we calculate:

$$E(X) = \Sigma x P_x \qquad\qquad (7.4)$$

where the summation is over all possible values of X. Because of its derivation as the limiting value of the sample mean we see that:

$E(X)$ is the population mean value of X.

Note

♦ The expectation of X does not have to be equal to an integer, nor does it have to be one of the possible values for X. We do not require these features for a sample mean, so there is no reason to require them for a population mean.

Example 7

The random variable X can only take the values 2 and 5. Given that the value 5 is twice as likely as the value 2, determine the expectation of X.

———————

Suppose we denote the probability that X equals 2 by p. Then the probability that X equals 5 is $2p$. Since these are the only possible values for X, the sum of their probabilities is 1: $p + 2p = 1$. Since $3p = 1$ it follows that $p = \frac{1}{3}$. The expectation of X is therefore given by:

$$E(X) = (2 \times P_2) + (5 \times P_5)$$
$$= \left(2 \times \tfrac{1}{3}\right) + \left(5 \times \tfrac{2}{3}\right)$$
$$= \tfrac{2}{3} + \tfrac{10}{3}$$
$$= 4$$

The random variable X has a long-term average value of 4, though no individual values of X will be equal to that value.

Example 8

Determine the expectation of the random variable X, which has probability distribution given below.

Value of X	0	1	2	3
Probability	0.3	0.4	0.2	0.1

$$E(X) = (0 \times P_0) + (1 \times P_1) + (2 \times P_2) + (3 \times P_3)$$

$$= 0 + 0.4 + 0.4 + 0.3$$

$$= 1.1$$

The expectation of X is 1.1.

Practical

Roll a die four times, recording your results using a tally chart. Calculate the sample mean. Compare your results with other members of the class. You should find that almost everyone has a sample mean between 2 and 5.

Now roll the die a further thirty-six times and calculate the sample mean for the combined set of forty values. How variable are people's results now? You should find that most people have obtained values in the range 3 to 4. As the sample size increases so the sample mean becomes less likely to deviate far from 3.5.

Calculate a sample mean for the entire class.

Expected value or expected number

Sometimes either '**expected value**' or '**expected number**' is used in place of 'expectation' – these are all synonyms for one another. Whichever term is used, the numerical value that is being sought can be thought of as being *the long-term average value*.

Note

◆ This is just one of several places where Statistics has 'borrowed' a word from the ordinary English vocabulary but subtly altered its meaning – the 'expected value of X', using the statistical meaning of the phrase, does not have to be a value of X that is actually 'expected', using the everyday interpretation of the word 'expected'!

Example 9

In a multiple-choice paper, each question is followed by four alternative answers. The candidate is asked to ring one of these answers. If the answer ringed is correct, then the candidate gains 3 marks, but if the answer is incorrect the candidate loses 1 mark. Determine the expected value of the mark gained per question by the candidate if:

(i)   the candidate chooses an answer at random,

(ii) the candidate knows that one of the incorrect answers is incorrect and chooses at random from the remaining three possibilities.

Comment on the results in each case.

Let X be the number of marks gained.

(i)   The probability distribution of X is:

$$P_3 = \tfrac{1}{4} \qquad P_{-1} = \tfrac{3}{4}$$

so that:

$$E(X) = \{3 \times (\tfrac{1}{4})\} + \{(-1) \times (\tfrac{3}{4})\} = \tfrac{3}{4} - \tfrac{3}{4} = 0$$

The examination marking scheme has been designed so that the expected mark obtained by someone who knows nothing and guesses every question will be zero.

(ii)   The revised probability distribution, after the elimination of one of the incorrect answers is:

$$P_3 = \tfrac{1}{3} \qquad P_{-1} = \tfrac{2}{3}$$

so that:

$$E(X) = \{3 \times (\tfrac{1}{3})\} + \{(-1) \times (\tfrac{2}{3})\} = 1 - \tfrac{2}{3} = \tfrac{1}{3}$$

Since $E(X)$ is greater than 0, if one or more of the possibilities can be eliminated as being certainly incorrect, then there will be an advantage in guessing the answer. If the candidate were to guess, under these conditions, the answers to lots of questions, then the average gain would be one-third of a mark per question.

Exercises 7c

Find the expectation of the random variables in each of Questions **1–14** below which are based on Questions **1–9** of Exercises 7a and Questions **1–5** of Exercises 7b.

1   A box contains 3 red marbles and 5 green marbles. Two marbles are taken at random without replacement, and X is the number of green marbles obtained.

2   A box contains 3 red marbles and 5 green marbles. Two marbles are taken at random with replacement, and X is the number of green marbles obtained.

3   A fair coin has the number '1' on one face and the number '2' on the other. The coin is thrown with a fair die and X is the sum of the scores.

4   In a raffle, 20 tickets are sold and there are two prizes. One ticket number is drawn at random and the corresponding ticket earns a £10 prize. A second, different ticket number is drawn at random, and the corresponding ticket earns a £3 prize. The prize earned by a particular one of the original 20 tickets is £X.

5   Two cards are drawn at random (without replacement) from a pack of playing cards, and X is the number of Hearts obtained.

6   A fair die is thrown and X is the reciprocal of the score (i.e. 'one over the score').

7   Two fair dice, one red and the other green, are thrown and X is the score on the red die minus the score on the green die.

8   Two fair dice, one red and the other green, are thrown and X is the positive difference in the scores (i.e. the modulus of the random variable in the preceding example).

9   Packets of 'Hidden Gold' cornflakes are sold for £1.20 each. One in twenty of the packets contains a £1 coin. A shopper buys two packets and £X is the net cost of the two packets.

10   A book has pages numbered 1 to 300. A page is chosen at random: X is the last digit of the page number and Y is the first digit of the page number.

11   In Ludo it is necessary to throw a six with a single fair die in order to start. The number of throws needed to obtain the first six is N.

12 X is the random variable denoting the number of sixes obtained on a single throw of a fair die.

13 The random variable X has a distribution which is both uniform and Bernoulli.

14 The random variable X is the number of heads obtained when a fair coin is thrown twice.

15 A breakfast cereal company gives away one of a set of six cards inside each cereal packet. A family already has five of the set and decides to buy packets, one at a time, until they obtain the sixth card. Using the geometric distribution, together with the result that:

$$1 + 2x + 3x^2 + 4x^3 + \cdots = (1 - x)^{-2}$$

show that the family must buy on average a further six packets.

Another family, that has none of the cards, decides to buy packets one at a time until they have collected the complete set of cards. Given that each packet costs 85p, show that they will have to spend, on average, just under £12.50.

Expectations of functions of random variables

We have seen that, essentially, $E(X)$ is the long-term average value of the random variable X. In the same way $E(X^2)$ is the long-term average value of X^2, $E(X^3 + 2X)$ is the long-term average value of $X^3 + 2X$ and so on. For a general function, $g(X)$, the value of $E[g(X)]$ is calculated using:

$$E[g(X)] = \Sigma g(x)P_x \qquad (7.5)$$

where the summation is over all possible values of X.

Note
- In general:

  $$E[g(X)] \neq g(E[X])$$

  So that, for example, we will usually find that:

  $$E(X^2) \neq [E(X)]^2$$

Example 10

Calculate the expected value of $\dfrac{1}{X}$, where X is the value obtained from rolling a fair six-sided die.

$$E\left(\frac{1}{X}\right) = \left(\frac{1}{1}\right) \times P_1 + \overline{\left(\frac{1}{2}\right)} \times P_2 + \cdots + \left(\frac{1}{6}\right) \times P_6$$
$$= \tfrac{1}{6} + \tfrac{1}{12} + \cdots + \tfrac{1}{36}$$
$$= \tfrac{49}{120}$$

The expected value of $\dfrac{1}{X}$ is about 0.4 and is *not* simply the reciprocal of $E(X)$ (which would have given the answer $\tfrac{2}{7} \approx 0.3$).

Example 11

The discrete random variable X is equally likely to take the values $0°$, $45°$ or $90°$. Determine the expected value of $\sin(X)$.

$$E[\sin(X)] = \sin(0°) \times P_{0°} + \sin(45°) \times P_{45°} + \sin(90°) \times P_{90°}$$
$$= (0 \times \tfrac{1}{3}) + (\tfrac{1}{2}\sqrt{2} \times \tfrac{1}{3}) + (1 \times \tfrac{1}{3})$$
$$= 0.569 \text{ (to 3 d.p.)}$$

The expected value of $\sin(X)$ is approximately 0.57.

Exercises 7d

Find $E(X^2)$ for Questions **1–9** below, which have been seen earlier in Exercises 7a and 7c.

1 A box contains 3 red marbles and 5 green marbles. Two marbles are taken at random without replacement, and X is the number of green marbles obtained.

2 A box contains 3 red marbles and 5 green marbles. Two marbles are taken at random with replacement, and X is the number of green marbles obtained.

3 A fair coin has the number '1' on one face and the number '2' on the other. The coin is thrown with a fair die and X is the sum of the scores.

4 In a raffle, 20 tickets are sold and there are two prizes. One ticket number is drawn at random and the corresponding ticket earns a £10 prize. A second, different, ticket number is drawn at random, and the corresponding ticket earns a £3 prize. The prize earned by a particular one of the original 20 tickets is £X.

5 Two cards are drawn at random (without replacement) from a pack of playing cards, and X is the number of Hearts obtained.

6 A fair die is thrown and X is the reciprocal of the score (i.e. 'one over the score').

7 Two fair dice, one red and the other green, are thrown and X is the score on the red die minus the score on the green die.

8 Two fair dice, one red and the other green, are thrown and X is the positive difference in the scores (i.e. the modulus of the random variable in the preceding example).

9 Packets of 'Hidden Gold' cornflakes are sold for £1.20 each. One in twenty of the packets contains a £1 coin. A shopper buys two packets and £X is the net cost of the two packets.

10 The random variable Y has distribution given by $P(Y = 0) = \frac{1}{4}$, $P(Y = 1) = \frac{1}{4}$, $P(Y = 4) = \frac{1}{2}$. Find (i) $E(Y)$, (ii) $E(Y^2)$, (iii) $E(\sqrt{Y})$, (iv) $E(Y - 1)$.

11 The random variable Z has distribution given by $P(Z = -1) = \frac{1}{5}$, $P(Z = 3) = \frac{2}{5}$, $P(Z = 8) = \frac{2}{5}$. Find (i) $E(Z)$, (ii) $E(Z^2)$, (iii) $E(|Z|)$, (iv) $E(\sqrt{Z + 1})$.

7.6 The variance

We have seen that, as a sample gets larger and larger so its properties will generally come increasingly to resemble those of the corresponding population. In particular we have:

Sample	Population
Relative frequency, $\frac{f_j}{n}$ $\longrightarrow$	Probability, P_{x_j}
Sample mean, $\bar{x}$ $\longrightarrow$	Population mean, $E(X)$
to these we now add:	
Sample variance, σ_n^2 $\longrightarrow$	Population variance, $\text{Var}(X)$

One formula for the population variance is:

$$\text{Var}(X) = E(X^2) - \{E(X)\}^2 \tag{7.6}$$

An alternative form, which is shown in Chapter 8 to be equivalent is:

$$\text{Var}(X) = E[\{X - E(X)\}^2] \tag{7.7}$$

Since probabilities and squared real quantities are never negative, we can deduce immediately that:

$$\text{Var}(X) \geqslant 0$$

The link between Equation (7.7) and the corresponding expression for the sample variance is given in the note below.

Notes

- In practice, the word 'population' is often omitted and we simply write 'the variance of X'.
- The formula for the sample variance can be written as:

$$\sigma_n^2 = \sum \left\{ (x_j - \bar{x})^2 \left(\frac{f_j}{n} \right) \right\}$$

As n increases, so $\dfrac{f_j}{n}$ approaches P_{x_j} and $\bar{x}$ approaches $\mathrm{E}(X)$. As the sample turns into the entire population the formula becomes:

$$\mathrm{Var}(X) = \Sigma \{x_j - \mathrm{E}(X)\}^2 P_{x_j}$$

where the summation is over all possible values of X. Comparing this expression with that for $\mathrm{E}[g(X)]$, we arrive at Equation (7.7).

Example 12

The random variable X has probability distribution given by:

Value of X	2	5
Probability	0.4	0.6

Determine the variance of X.

We first calculate the expectation of X:

$$\mathrm{E}(X) = (2 \times 0.4) + (5 \times 0.6) = 3.8$$

We next calculate $\mathrm{E}(X^2)$:

$$\mathrm{E}(X^2) = (2^2 \times 0.4) + (5^2 \times 0.6) = 16.6$$

Finally, using Equation (7.6), we get:

$$\mathrm{Var}(X) = \mathrm{E}(X^2) - \{\mathrm{E}(X)\}^2 = 16.6 - 3.8^2 = 2.16$$

Example 13

The random variable X has probability distribution given by:

Value of X	2	3	4
Probability	p	p	$1 - 2p$

Show that X has variance equal to $p(5 - 9p)$.

We first calculate the expectation of X:

$$\mathrm{E}(X) = (2 \times p) + (3 \times p) + \{4 \times (1 - 2p)\} = 4 - 3p$$

We next calculate $\mathrm{E}(X^2)$:

$$\mathrm{E}(X^2) = (2^2 \times p) + (3^2 \times p) + \{4^2 \times (1 - 2p)\} = 16 - 19p$$

Finally, using Equation (7.6), we get:

$$\mathrm{Var}(X) = \mathrm{E}(X^2) - \{\mathrm{E}(X)\}^2 = (16 - 19p) - (4 - 3p)^2$$
$$= 5p - 9p^2 = p(5 - 9p)$$

as required.

Example 14

The random variable X has probability distribution given by:

Value of X	2	3	4	5
Probability	0.1	0.2	0.3	0.4

Determine the expectation and variance of X.

We first calculate the expectation of X:

$$E(X) = (2 \times 0.1) + (3 \times 0.2) + (4 \times 0.3) + (5 \times 0.4) = 4$$

Since $E(X)$ is an integer this is a rare example in which it makes sense to use Equation (7.7):

$$\begin{aligned}E[\{X - E(X)\}^2] &= \{(2 - 4)^2 \times 0.1\} + \{(3 - 4)^2 \times 0.2\}\\ &\quad + \{(4 - 4)^2 \times 0.3\} + \{(5 - 4)^2 \times 0.4\}\\ &= 0.4 + 0.2 + 0 + 0.4\\ &= 1\end{aligned}$$

Thus $\mathrm{Var}(X) = 1$.

Using the simpler Equation (7.6) we must first calculate $E(X^2)$:

$$E(X^2) = (2^2 \times 0.1) + (3^2 \times 0.2) + (4^2 \times 0.3) + (5^2 \times 0.4) = 17$$

Since $E(X) = 4$:

$$\mathrm{Var}(X) = E(X^2) - \{E(X)\}^2 = 17 - 4^2 = 1$$

as before.

Example 15

The random variable X has the Bernoulli distribution:

$$P_0 = 1 - p \qquad P_1 = p$$

Find the expectation and variance of X.

Finding $E(X)$ is straightforward:

$$E(X) = \{0 \times (1 - p)\} + \{1 \times p\} = p$$

To determine the variance of X we use Equation (7.6) and first find $E(X^2)$:

$$E(X^2) = \{0^2 \times (1 - p)\} + \{1^2 \times p\} = p$$

Hence:

$$\begin{aligned}\mathrm{Var}(X) &= E(X^2) - \{E(X)\}^2\\ &= p - (p)^2\\ &= p(1 - p)\end{aligned}$$

A random variable having a Bernoulli distribution with parameter p has mean p and variance $p(1 - p)$.

Example 16

This example is algebraically demanding and should be omitted on an initial read of the chapter. However, the results obtained should be noted.

The random variable X has a geometric distribution:

$$P_x = (1 - p)^{x-1} p \qquad (x = 1, 2, \ldots)$$

Show that **the geometric distribution with parameter p has expectation $\dfrac{1}{p}$ and variance $\dfrac{1-p}{p^2}$**.

_____

We will now provide two alternative methods of solution that involve manipulating series. The first uses the geometric series and the second uses the binomial series. Neither is straightforward.

Method 1: Geometric series

We use the result that:

$$1 + q + q^2 + q^3 + \cdots = \frac{1}{1-q} = \frac{1}{p}$$

Now the mean, $E(X)$, is given by:

$$\begin{aligned}
E(X) &= (1 \times p) + (2 \times pq) + (3 \times pq^2) + (4 \times pq^3) + \cdots \\
&= (1 - q) + 2(1 - q)q + 3(1 - q)q^2 + 4(1 - q)q^3 + \cdots \\
&= 1 - q + 2q - 2q^2 + 3q^2 - 3q^3 + 4q^3 - 4q^4 + \cdots \\
&= 1 + q + q^2 + q^3 + \cdots \\
&= \frac{1}{1-q} = \frac{1}{p}
\end{aligned}$$

We now calculate $E(X^2)$:

$$\begin{aligned}
E(X^2) &= (1^2 \times p) + (2^2 \times pq) + (3^2 \times pq^2) + \cdots \\
&= (1 - q) + 4(1 - q)q + 9(1 - q)q^2 + \cdots \\
&= 1 - q + 4q - 4q^2 + 9q^2 - 9q^3 + \cdots \\
&= 1 + 3q + 5q^2 + 7q^3 + \cdots
\end{aligned}$$

At this point we become very clever indeed! Watch carefully! We add on, and also take away, the quantity $(1 + q + q^2 + q^3 + \cdots)$ which we know to be equal to $\dfrac{1}{p}$:

$$\begin{aligned}
E(X^2) &= 1 + 3q + 5q^2 + 7q^3 + \cdots \\
&= 2 + 4q + 6q^2 + 8q^3 + \cdots - (1 + q + q^2 + q^3 + \cdots) \\
&= 2(1 + 2q + 3q^2 + 4q^3 + \cdots) - \frac{1}{p}
\end{aligned}$$

Where have we seen something like $(1 + 2q + 3q^2 + 4q^3 + \cdots)$ before? In the expression for $E(X)$: it is $E(X)$ divided by $(1 - q)$. So:

$$E(X^2) = \left\{ 2 \times \frac{E(X)}{1-q} \right\} - \frac{1}{p}$$

and, since $E(X) = \dfrac{1}{p} = \dfrac{1}{1-q}$, we find that:

$$E(X^2) = \frac{2}{p^2} - \frac{1}{p}$$

Finally:

$$\text{Var}(X) = \text{E}(X^2) - \{\text{E}(X)\}^2$$

$$= \left(\frac{2}{p^2} - \frac{1}{p}\right) - \left(\frac{1}{p}\right)^2$$

$$= \frac{1}{p^2} - \frac{1}{p}$$

$$= \frac{1-p}{p^2}$$

As required, we have shown that $\text{Var}(X) = \dfrac{1-p}{p^2}$.

———

Method 2: Binomial series

We use the following binomial series:

$$(1 + q)^n = 1 + nq + \frac{n(n-1)q^2}{2!} + \cdots$$

Replacing q by $-q$ we get:

$$(1 - q)^n = 1 - nq + \frac{n(n-1)q^2}{2!} - \cdots$$

The restrictions on these results are that q should be less than 1 in magnitude. In the present context it is useful to write $q = 1 - p$, so that $p = 1 - q$. Hence:

$$p^n = 1 - nq + \frac{n(n-1)q^2}{2!} - \cdots$$

The particular cases that will be needed are those corresponding to $n = -2$ and $n = -3$:

$$(1 - q)^{-2} = p^{-2} = 1 + 2q + 3q^2 + 4q^3 + \cdots$$
$$(1 - q)^{-3} = p^{-3} = 1 + 3q + 6q^2 + 10q^3 + \cdots$$

We now return to the actual problem! The question refers to 'the mean and variance of X'. Since X is a random variable we are being asked for $\text{E}(X)$ and $\text{Var}(X)$. We begin with $\text{E}(X)$:

$$\text{E}(X) = \{1 \times p\} + \{2 \times (1-p)p\} + \{3 \times (1-p)^2 p\}$$
$$+ \{4 \times (1-p)^3 p\} + \cdots$$
$$= (1 + 2q + 3q^2 + 4q^3 + \cdots) \times p$$
$$= p^{-2} \times p$$
$$= \frac{1}{p}$$

In order to calculate $\text{Var}(X)$ we must first calculate $\text{E}(X^2)$:

$$\text{E}(X^2) = \{1^2 \times p\} + \{2^2 \times (1-p)p\} + \{3^2 \times (1-p)^2 p\}$$
$$+ \{4^2 \times (1-p)^3 p\} + \cdots$$
$$= (1 + 4q + 9q^2 + 16q^3 + \cdots) \times p$$

At this point, knowing the answer is a distinct advantage! We break up the series into two pieces:

$$E(X^2) = \{(1 + 2q + 3q^2 + 4q^3 + \cdots) \times p\}$$
$$\qquad + \{(1 + 3q + 6q^2 + \cdots) \times 2pq\}$$
$$\qquad = (p^{-2} \times p) + (p^{-3} \times 2pq)$$
$$\qquad = \frac{1}{p} + \frac{2q}{p^2}$$
$$\qquad = \frac{\{p + 2(1 - p)\}}{p^2}$$
$$\qquad = \frac{2 - p}{p^2}$$

Finally we can put it all together!

$$Var(X) = E(X^2) - \{E(X)\}^2$$
$$\qquad = \frac{2 - p}{p^2} - \frac{1}{p^2}$$
$$\qquad = \frac{1 - p}{p^2}$$

which is the required result.

▲ ▲

Project

The nature of this project, which involves the use of a telephone directory, will depend on your location. In a country district, you should choose some large town from those covered by the directory; in a city, you should choose some well-defined large sub-area. Open the directory at random and start at the top of the left-hand page (unless it is an advertisement!). Count the number of subscribers until the first that you encounter with a number in your selected town (or sub-area). The reason for choosing a large town is so that your counting does not take too long! Record your value. Repeat the process until you have a total of 50 values.

Summarise your data using a bar chart (if most values are small) or a histogram (if most values are large).

Does your data look as though it could be described by a geometric distribution?

Calculate the sample mean, $\bar{x}$. Recall that the population mean is equal to $\frac{1}{p}$, assuming a geometric distribution. Deduce an estimate of the proportion of subscribers in the directory that reside in your chosen town or sub-area. (This is a rather involved way of estimating this proportion – how else might you have estimated it?)

7.7 The standard deviation

The standard deviation of a random variable is simply the square root of its variance.

▼ ▼

Example 17

The discrete random variable X has probability distribution given by:

$$P_x = \begin{cases} kx^3 & (x = 1, 2, 3) \\ 0 & \text{otherwise} \end{cases}$$

Determine, correct to 3 decimal places, the values of:
(i) the constant k,
(ii) the expectation of X,
(iii) the standard deviation of X.

(i) In order to find the value of k we use the fact that the probabilities sum to 1. Thus:

$$k(1^3 + 2^3 + 3^3) = 1$$

The sum of the left-hand side is $36k$ and hence $k = \frac{1}{36}$.

(ii) The expectation is $E(X)$, which is given by:

$$E(X) = (1 \times k) + (2 \times 8k) + (3 \times 27k) = (1 + 16 + 81)k$$
$$= 98k = \frac{98}{36}$$
$$= 2.722 \text{ (to 3 d.p.)}$$

(iii) To calculate the standard deviation we must first calculate the variance. Since $E(X)$ is not an integer we use Equation (7.6) and begin by calculating $E(X^2)$:

$$E(X^2) = (1^2 \times k) + (2^2 \times 8k) + (3^2 \times 27k) = (1 + 32 + 243)k$$
$$= 276k = \frac{276}{36}$$

Hence:

$$\text{Var}(X) = E(X^2) - \{E(X)\}^2 = \frac{276}{36} - \left(\frac{98}{36}\right)^2 \approx 0.2562$$

The standard deviation of X is therefore $\sqrt{0.2562} = 0.506$ (to 3 d.p.).

Note

♦ In order to achieve a desired accuracy of 3 decimal places, intermediate calculations have used fractions wherever practicable and have otherwise worked with extended accuracy so as to reduce round-off errors.

Exercises 7e

Find the variance and standard deviation of X in each of Questions **1**–**9** below, which have been seen earlier in Exercises 7a, 7c and 7d.

1 A box contains 3 red marbles and 5 green marbles. Two marbles are taken at random without replacement, and X is the number of green marbles obtained.

2 A box contains 3 red marbles and 5 green marbles. Two marbles are taken at random with replacement, and X is the number of green marbles obtained.

3 A fair coin has the number '1' on one face and the number '2' on the other. The coin is thrown with a fair die and X is the sum of the scores.

4 In a raffle, 20 tickets are sold and there are two prizes. One ticket number is drawn at random and the corresponding ticket earns a £10 prize.

A second, different, ticket number is drawn at random, and the corresponding ticket earns a £3 prize. The prize earned by a particular one of the original 20 tickets is £X.

5 Two cards are drawn at random (without replacement) from a pack of playing cards, and X is the number of Hearts obtained.

6 A fair die is thrown and X is the reciprocal of the score (i.e. 'one over the score').

7 Two fair dice, one red and the other green, are thrown and X is the score on the red die minus the score on the green die.

8 Two fair dice, one red and the other green, are thrown and X is the positive difference in the scores (i.e. the modulus of the random variable in the preceding example).

9 Packets of 'Hidden Gold' cornflakes are sold for £1.20 each. One in twenty of the packets contains a £1 coin. A shopper buys two packets and £X is the net cost of the two packets.

10 Show that if X has a discrete uniform distribution on the integers $1, 2, \ldots, n$ then:

$$E(X) = \tfrac{1}{2}(n+1)$$
$$\operatorname{Var}(X) = \tfrac{1}{12}(n^2 - 1)$$

11 The random variable X has a geometric distribution with variance 6.
Find $E(X)$.

12 The random variable Y has a Bernoulli distribution with standard deviation $\frac{3}{10}$.
Find the possible values of the expectation of Y.

13 A woman removes the labels from 3 tins of tomato soup and from 4 tins of peaches. She sends the labels off to the manufacturers in order to win herself a huggable teddy bear. Delighted with the prospect of the forthcoming bear, she forgets to mark the tins, which, devoid of their labels, then appear identical. The next week she is entertaining guests and requires a tin of peaches. She chooses tins at random, opening each in turn until a tin of peaches has been located. Let p_r be the probability that it is the rth tin that first contains peaches.
Determine the values of p_1, p_2, p_3 and p_4. Determine the expectation and variance of the number of tins that are opened.

7.8 Greek notation

All branches of Mathematics display a liking for Greek symbols and Statistics is no exception! Conventionally the symbols μ (pronounced 'mu') and σ (a lower-case 'sigma') are reserved for the population mean and population standard deviation, with σ^2 being used for the variance. Thus, when studying the random variable X we may write:

$$E(X) = \mu \tag{7.8}$$
$$\operatorname{Var}(X) = \sigma^2 \tag{7.9}$$

Note

♦ A useful guide as to whether, for a random variable X, we have calculated μ and σ^2 incorrectly, is provided by noting the following:
   ♦ The population mean, μ, must have a value lying between the smallest and largest possible values for X.
   ♦ If the range of possible values of X is finite then it usually has a magnitude of between 3σ and 6σ.

▼

Example 18

Verify that the calculations in Example 17 seem reasonable.

———

In the previous example we calculated the expectation of X as being approximately 2.7 which does lie between 1 and 3, the extreme values that are possible for X. There is therefore no obvious indication that we calculated $E(X)$ incorrectly.

The range of the possible values of X is $3 - 1 = 2$. This is about 4 times 0.5, our calculated standard deviation, so it provides no suggestion of an incorrect calculation.

▲

Chapter summary

◆   A **random variable** is denoted by a CAPITAL letter, e.g. X; an observed value by a lower-case letter, e.g. x.

◆   The **probability** of the value x, $P(X = x)$, is denoted by P_x.

◆   Probabilities **sum to 1**. For a discrete random variable that can take values $x_1, x_2, \ldots, x_n$

$$P_{x_1} + \cdots + P_{x_n} = 1$$

◆   **Expectation** (population mean):

$$E(X) = \Sigma x P_x \qquad E[g(X)] = \Sigma g(x) P_x$$

◆   **Variance**: $\mathrm{Var}(X) = E(X^2) - \{E(X)\}^2$

◆   *Special distributions*:
  ● **Bernoulli**:

$$P_0 = 1 - p \qquad P_1 = p$$

$$E(X) = p \qquad \mathrm{Var}(X) = p(1 - p)$$

  ● **Geometric** $(0 < p < 1, q = 1 - p)$:

$$P_x = q^{x-1} p \qquad x = 1, 2, \ldots$$

$$P(X \leqslant x) = 1 - q^x \qquad P(X < x) = 1 - q^{x-1}$$

$$P(X > x) = q^x \qquad P(X \geqslant x) = q^{x-1}$$

$$E(X) = \frac{1}{p} \qquad \mathrm{Var}(X) = \frac{q}{p^2}$$

Exercises 7f (Miscellaneous)

1 A typically nutty statistician performs the following experiment. He first tosses a fair tetrahedron whose sides are numbered 0, 1, 2 and 3. When it lands, three sides are visible. Let n be the number on the fourth side. The statistician now tosses n unbiased coins.
   Let X be the number of heads obtained.
   (a) Obtain the probability distribution of X.
   (b) Show that X has expectation $\frac{3}{4}$. Find the variance of X.
   (c) Suppose that on a particular occasion the statistician has obtained exactly one head. Determine the probability that, on that occasion, $n = 2$.

2 Peter sets out with two 50p coins and three 10p coins in his pocket. When he comes to pay for goods at a shop, he finds that two of the coins are missing. Find
   (i) the probability that he can pay for £1 worth of goods,
   (ii) the expected total value of the money lost.
   [Assume that each coin is equally likely to be lost.]   [SMP]

3 An unbiased six-sided die has numbers 1, 1, 1, 2, 2, 3 printed on its faces. It is thrown twice.
   (i) By drawing a tree diagram, or otherwise, find the probability that the total score is 4.
   (ii) Find the expected value of the total score.   [SMP]

4 In the first trial of a random experiment the probability of a successful outcome is $\frac{2}{5}$. In the second trial the probability of a successful outcome will be $\frac{1}{5}$ if the outcome of the first trial was successful, and will be $\frac{4}{5}$ if the outcome of the first trial was not successful. Determine the probability distribution and the mean value of the number of successful outcomes that will be obtained in the first two trials of the random experiment.   [WJEC]

5 The probability distribution of a discrete random variable X is given by

$$P(X = r) = kr, \quad r = 1, 2, 3, \ldots, n,$$

where k is a constant. Show that

$$k = \frac{2}{n(n+1)}$$

and find, in terms of n, the mean of X.   [JMB]

6 An experiment is carried out with three coins. Two of the coins are fair, so that the probability of obtaining a 'head' on any throw is $\frac{1}{2}$, while the third coin is biased so that the probability of obtaining a 'head' on any throw is $\frac{1}{4}$.
The three coins are thrown, and events A and B are defined as follows:

A occurs if all three coins show the same result;
B occurs if the biased coin shows a 'head'.

Find (i) $P(A)$, (ii) $P(A \cup B)$, (iii) $P(A' \cap B)$.

The random variable N denotes the number of 'heads' showing as a result of the experiment being carried out. Tabulate the probability distribution of N, and hence or otherwise calculate $E(N)$.   [UCLES]

7 A circular card is divided into 3 sectors scoring 1, 2, 3 and having angles 135°, 90°, 135° respectively. On a second circular card, sectors scoring 1, 2, 3 have angles 180°, 90°, 90°, respectively. Each card has a pointer pivoted at its centre. After being set in motion, the pointers come to rest independently in random positions. Find the probability that

(i) the score on each card is 1,
(ii) the score on at least one of the cards is 3.

The random variable X is the larger of the two scores if they are different, and their common value if they are the same.
Show that $P(X = 2) = \frac{9}{32}$.
Show that $E(X) = \frac{75}{32}$ and find $Var(X)$.   [UCLES]

8 (a) A regular tetrahedron is a solid with four faces, all identical. What is the probability, if it is tossed in the air, that it will land on a given face?

A dice is made by numbering the four faces 1, 2, 3, 4. The random variable X represents the number on the face on which the tetrahedron lands. The mean of X is 2.5. Calculate the variance of X.

(b) A circular disc is marked with a '1' on one side and a '2' on the other. Y is the random variable representing the number showing on the visible face of the disc when it is tossed and lands. Demonstrate that the mean and variance of Y are $1\frac{1}{2}$ and $\frac{1}{4}$ respectively.

(c) The disc and the dice are tossed together. The sum of the outcomes is recorded. Z is the random variable representing independent sums of X and Y. Write down the probability distribution for Z, and use this to calculate its mean and variance.   [UODLE(P)]

9 A woman is waiting for a taxi to come into view so that she can hail it. Explain why the Geometric distribution is appropriate to model the number of vehicles she sees up to and including the first taxi.

If 5% of the vehicles in the locality are taxis:

(a) write down the mean number of vehicles up to and including the first taxi;

(b) calculate, giving your answer to three significant figures, the probability that the first taxi is the 6th vehicle to come into view;

(c) calculate the probability that the first taxi is among the first 6 vehicles she sees.   [UODLE]

8 Expectation algebra

Oft expectation fails, and most oft there where most it promises

All's Well That Ends Well, William Shakespeare

In this chapter we derive some very useful results that are concerned with transformations of a single random variable and with combining information on several random variables. Where appropriate we shall use the simplifying notation of P_x for $P(X = x)$. We begin with an example that foreshadows some of these results.

Example 1

Suppose that the discrete random variable X has probability distribution given by:

$$P_0 = P_1 = 0.4 \qquad P_2 = 0.2$$

The random variable Y is defined by $Y = 2X - 1$.
Determine the mean and variance of X and of Y.
Comment on the results.

———

The simplest approach is to make a table of the probabilities and the possible values for X and Y:

Probability	0.4	0.4	0.2
Value of X	0	1	2
Value of $Y = 2X - 1$	-1	1	3

$$E(X) = (0 \times 0.4) + (1 \times 0.4) + (2 \times 0.2) = 0.8$$
$$E(Y) = \{(-1) \times 0.4\} + (1 \times 0.4) + (3 \times 0.2) = 0.6$$

In order to obtain the variances, we first calculate $E(X^2)$ and $E(Y^2)$:

$$E(X^2) = (0^2 \times 0.4) + (1^2 \times 0.4) + (2^2 \times 0.2) = 1.2$$
$$E(Y^2) = \{(-1)^2 \times 0.4\} + (1^2 \times 0.4) + (3^2 \times 0.2) = 2.6$$

Hence:

$$Var(X) = 1.2 - 0.8^2 = 1.2 - 0.64 = 0.56$$
$$Var(Y) = 2.6 - 0.6^2 = 2.6 - 0.36 = 2.24$$

Comparing the values obtained we find:

$$E(Y) = E(2X - 1) = 2E(X) - 1$$
$$Var(Y) = Var(2X - 1) = 2^2 \times Var(X)$$

These connections between $E(X)$ and $E(Y)$ and between $Var(X)$ and $Var(Y)$ are not coincidental, but are examples of general results that we derive later in the chapter.

8.1 $E(X + a)$ and $Var(X + a)$

Suppose that the random variable X refers to the distance (in cm) between the top of a person's head and ground level. A group of three people is illustrated below.

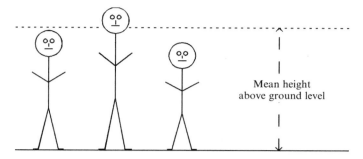

The three people now stand on a platform that is 20 cm high, as shown below.

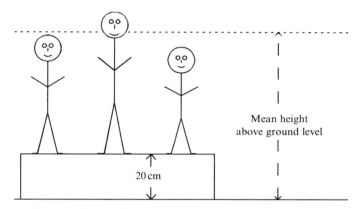

Since all are standing on the platform, the value of X for each person has increased by 20 cm and therefore the average value of X has increased by 20 cm. However, the new values of X are no more variable than the old ones, since the individual differences from the mean height are the same as they were previously. Generalising these results to a platform of height a cm we have the results:

$$E(X + a) = E(X) + a \qquad (8.1)$$
$$\mathrm{Var}(X + a) = \mathrm{Var}(X) \qquad (8.2)$$

Suppose instead that the people now stand in a pit that is 50 cm deep.

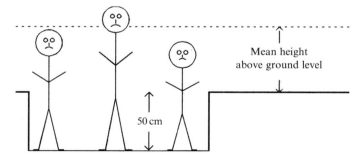

Each X value has been reduced by 50 cm, so the average X value is reduced by that amount. However, the variability of the values of X is again unaffected. Generalising to a pit of depth a cm we have:

$$E(X - a) = E(X) - a \qquad (8.3)$$
$$\mathrm{Var}(X - a) = \mathrm{Var}(X) \qquad (8.4)$$

We now prove these results algebraically, for a discrete random variable. Let $Y = X + a$, where a is any constant (either positive or negative). We want $E(Y)$ and $Var(Y)$.
Now:

$$E(Y) = E(X + a) = \Sigma(x + a)P_x$$

where the summation is over all possible values of X. Thus:

$$
\begin{aligned}
E(X + a) &= \Sigma x P_x + \Sigma a P_x \\
&= E(X) + a\Sigma P_x \qquad \text{by definition of E}(X) \\
&= E(X) + a \qquad\quad\ \ \text{since } \Sigma P_x = 1
\end{aligned}
$$

which proves the first result.

For the second result we use the alternative definition of variance given by Equation (7.7):

$$Var(Y) = E[\{Y - E(Y)\}^2]$$

Substituting $(X + a)$ for Y and $\{E(X) + a\}$ for $E(Y)$ we obtain:

$$
\begin{aligned}
Y - E(Y) &= (X + a) - \{E(X) + a\} \\
&= X - E(X)
\end{aligned}
$$

Hence:

$$
\begin{aligned}
Var(X + a) &= E[\{X - E(X)\}^2] \\
&= Var(X)
\end{aligned}
$$

which proves the second result.

Example 2

A lottery ticket costs 10p. There are 10 000 tickets for the lottery, which has a top prize of £100 and 9 runner-up prizes of £10 each. Determine the expected gain or loss resulting from the purchase of a ticket.

Let X be the random variable denoting the amount (in £) won by a ticket. Assuming all the tickets are sold, the probability distribution of X is given by:

Value of X	100	10	0
Probability	0.0001	0.0009	0.9990

Hence:

$$E(X) = (100 \times 0.0001) + (10 \times 0.0009) + (0 \times 0.999) = 0.01$$

Since the lottery ticket costs £0.10, we want the expectation of $Y = X - 0.10$. Using the general result:

$$E(Y) = E(X - 0.10) = E(X) - 0.10 = -0.09$$

On average, therefore, the purchase of a ticket will result in a loss of 9p. Of course this need not deter us! We can probably afford to gamble on losing 10p, even with these very unfavourable circumstances, for the slight chance of being £99.90 better off.

8.2 E(*aX*) and Var(*aX*)

For simplicity, suppose that $a = 2$ and that each of the people illustrated in the original diagram was one of a pair of identical twins, remarkably adept at gymnastics – as illustrated below.

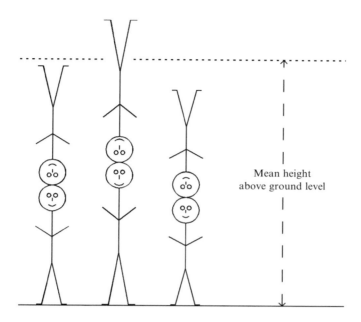

Mean height above ground level

Let the random variable Y be the distance from the top of the twin's feet to ground level. Obviously, for each pair of twins, $Y = 2X$, where X was the height of one of the twins. So the average of the Y-values must be double that of the X-values. The general result, true for any random variable and for any constant a (positive or negative) is:

$$\mathrm{E}(aX) = a\mathrm{E}(X) \tag{8.5}$$

More generally, for any function of X, g(X) say, we can write:

$$\mathrm{E}[a\mathrm{g}(X)] = a\mathrm{E}[\mathrm{g}(X)] \tag{8.6}$$

This result follows immediately from Equation (7.5) (p. 181).

It is obvious from the figure that the Y-values are a great deal more variable than the X-values, but to find out exactly how much more variable requires some algebra.

Let $Y = aX$ and denote $\mathrm{E}(X)$ by μ, so that:

$$\mathrm{Var}(X) = \mathrm{E}[(X - \mu)^2] \text{ and } \mathrm{E}(Y) = a\mu.$$

Then:

$$
\begin{aligned}
\mathrm{Var}(Y) &= \mathrm{E}[\{Y - \mathrm{E}(Y)\}^2] \\
&= \mathrm{E}[(aX - a\mu)^2] \\
&= \mathrm{E}[a^2(X - \mu)^2] \\
&= a^2\mathrm{E}[(X - \mu)^2] \qquad \text{from } \mathrm{E}[a\mathrm{g}(X)] = a\mathrm{E}[\mathrm{g}(X)] \text{ above} \\
&= a^2\mathrm{Var}(X)
\end{aligned}
$$

The general result is therefore that:

$$\mathrm{Var}(aX) = a^2\mathrm{Var}(X) \tag{8.7}$$

8.3 $E(aX + b)$, $Var(aX + b)$ and $E[g(X) + h(X)]$

Combining the results of the previous two sections we have the result that, if a and b are any two constants, then:

$$E(aX + b) = E(aX) + b \qquad \text{by Equation (8.1)} \qquad (8.8)$$
$$= aE(X) + b \qquad \text{by Equation (8.5)} \qquad (8.9)$$

and also:

$$Var(aX + b) = Var(aX) \qquad \text{by Equation (8.2)} \qquad (8.10)$$
$$= a^2 Var(X) \qquad \text{by Equation (8.7)} \qquad (8.11)$$

For the final result we use the basic definition of expectation:

$$E[g(X) + h(X)] = \Sigma\{g(x) + h(x)\}P_x$$
$$= \Sigma g(x)P_x + \Sigma h(x)P_x$$

Hence:

$$E[g(X) + h(X)] = E[g(X)] + E[h(X)] \qquad (8.12)$$

Example 3

At a fairground there is the following game. The player pays 20p in order to toss three coins. The stall-holder pays the player (in pence) 10 times the number of heads that the player obtains.
Determine the mean and variance of the player's net loss.

Let X be the random variable indicating the number of heads obtained. We require the mean and variance of the net loss which (in pence) is given by $Y = 20 - 10X$.

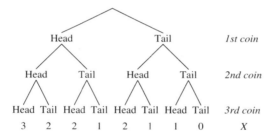

Assuming that the coins are fair, each of the 8 alternatives has probability $\frac{1}{8}$, so that the probability distribution for X is:

Value of X	0	1	2	3
Probability	$\frac{1}{8}$	$\frac{3}{8}$	$\frac{3}{8}$	$\frac{1}{8}$

We can calculate $E(X)$ from its definition, or, more simply, we can note that since the distribution is symmetric, $E(X)$ will be equal to the mid-range, which in this case is $\frac{3}{2}$. Also:

$$E(X^2) = (0^2 \times \tfrac{1}{8}) + (1^2 \times \tfrac{3}{8}) + (2^2 \times \tfrac{3}{8}) + (3^2 \times \tfrac{1}{8}) = \tfrac{24}{8} = 3$$

Hence:

$$Var(X) = 3 - (\tfrac{3}{2})^2 = \tfrac{3}{4}$$

Using the general results we therefore get:

$$E(Y) = E(20 - 10X) = 20 - 10E(X) = 20 - 10 \times (\tfrac{3}{2}) = 5$$

which implies an average net loss of 5p a go, and:

$$Var(Y) = Var(20 - 10X) = (-10)^2 Var(X) = 100 \times \tfrac{3}{4} = 75$$

Alternative expressions for Var(X)

In the last chapter we gave two alternative expressions for Var(X), namely
$E(X^2) - \{E(X)\}^2$ and $E[\{X - E(X)\}^2]$, which we now show to be equivalent.
Denote $E(X)$ by μ.

$$
\begin{aligned}
E[\{X - E(X)^2\}] &= E[(X - \mu)^2] \\
&= E(X^2 - 2\mu X + \mu^2) \\
&= E(X^2) + E(-2\mu X + \mu^2) \qquad \text{by Equation (8.12)} \\
&= E(X^2) - 2\mu E(X) + \mu^2 \qquad \text{by Equation (8.9)} \\
&= E(X^2) - 2\mu^2 + \mu^2 \\
&= E(X^2) - \mu^2 \\
&= E(X^2) - \{E(X)\}^2
\end{aligned}
$$

Exercises 8a

1 Given that $E(X) = 4$, Var(X) = 2, find
(i) $E(3X + 6)$, (ii) Var($3X + 6$), (iii) $E(6 - 3X)$,
(iv) Var($6 + 3X$).

2 Given that $E(X) = 3$, Var(X) = 4, find the
expectation and variance of (i) $X - 2$,
(ii) $3X + 1$, (ii) $2 - 3X$.

3 Given that $E(Y) = \frac{3}{4}$ and Var(Y) $= \frac{1}{4}$, find
$E(\frac{1}{3}Y)$ and Var($\frac{1}{3}Y$).

4 Given that the expectation of X is -2, and the
standard deviation of X is 9, find
(i) $E\{(X + 2)^2\}$,
(ii) $E(X^2)$,
(iii) $E\{(X - 1)(X + 3)\}$.

5 Given that $E(3Y + 2) = 8$ and
Var($4 - 2Y$) = 12, find the expected value and
the variance of Y.

6 Given that $E(Z) = 0$, Var(Z) = 1 and
$Y = 3Z - 4$, find $E(Y)$ and Var(Y).

7 The random variable U has mean 10 and
standard deviation 5. The random variable V is
defined by $V = \frac{1}{2}(U + 5)$. Find the mean and
standard deviation of V.

8 It costs £30 to hire a car for the day, and there
is a mileage charge of 10p per mile. The
distance travelled in a day has expectation
200 miles and standard deviation 20 miles.
Find the expectation and standard deviation of
the cost per day.

9 Given that $E(X) = \mu$ and Var(X) $= \sigma^2$, find
two pairs of values for the constants a and b
such that $E(aX + b) = 0$ and Var($aX + b$) = 1.

10 Given that $E(X) = \mu$, Var(X) $= \sigma^2$ and a is a
constant, show that:
$$
E\{(X - a)^2\} = (\mu - a)^2 + \sigma^2
$$
Hence show that, as a varies, $E\{(X - a)^2\}$ is
least when $a = \mu$ and find the least value.

11 The random variable T has mean 5 and
variance 16.
Find two pairs of values for the constants c
and d such that $E(cT + d) = 100$ and
Var($cT + d$) = 144.

12 Find $E(2S - 6)$ and Var($2S - 6$), where S is the
score resulting from a single throw of an
unbiased die.

13 The random variable Y takes the values $-1, 0, 1$
with probabilities $\frac{1}{4}, \frac{1}{2}, \frac{1}{4}$, respectively.
Find $E(10Y + 10)$ and Var($10Y + 10$).

8.4 Expectations of functions of more than one variable

We now quote a result that is rather obvious, but is surprisingly difficult to
prove, namely that for two random variables, X and Y:

$$
E(X + Y) = E(X) + E(Y) \tag{8.13}
$$

Combining this result with those from the previous sections we have the more general results:

$$E(aX + bY + c) = aE(X) + bE(Y) + c \tag{8.14}$$

$$E[g(X) + h(Y)] = E[g(X)] + E[h(Y)] \tag{8.15}$$

These results can be extended to the case of more than two random variables. For example:

$$E(R + S + T + U) = E(R) + E(S) + E(T) + E(U) \tag{8.16}$$

Var($X + Y$)

From the definition of variance we can write:

$$Var(X + Y) = E[(X + Y)^2] - \{E(X + Y)\}^2$$

We know that the second term on the right-hand side is equal to $\{E(X) + E(Y)\}^2$, so our problems centre on the first term:

$$E[(X + Y)^2] = E(X^2 + 2XY + Y^2)$$
$$= E(X^2) + 2E(XY) + E(Y^2)$$

since $E(R + S + T) = E(R) + E(S) + E(T)$.

Hence:

$$Var(X + Y) = \{E(X^2) + 2E(XY) + E(Y^2)\} - \{E(X) + E(Y)\}^2$$
$$= \{E(X^2) - E(X)^2\} + 2\{E(XY) - E(X)E(Y)\}$$
$$+ \{E(Y^2) - E(Y)^2\}$$
$$= Var(X) + 2\{E(XY) - E(X)E(Y)\} + Var(Y)$$

The quantity $\{E(XY) - E(X)E(Y)\}$ is called the **covariance** of X and Y, written $Cov(X, Y)$, for short. So we have:

$$Var(X + Y) = Var(X) + 2Cov(X, Y) + Var(Y)$$

We discuss covariance later, in Chapter 20.

A specially simple case occurs when $Cov(X, Y) = 0$. The most common reason for this simplification is that X and Y are independent random variables (i.e. knowing the value of one tells us nothing about the value of the other). Hence, if X and Y are independent, then:

$$Var(X + Y) = Var(X) + Var(Y) \tag{8.17}$$

Combining this result with Equation (8.11) we get the more general result that, if X and Y are independent, then:

$$Var(aX + bY + c) = a^2 Var(X) + b^2 Var(Y) \tag{8.18}$$

Once again, these results can be extended to cases involving more than two random variables. For example, if R, S, T and U are all mutually independent, then:

$$Var(R + S + T + U) = Var(R) + Var(S) + Var(T) + Var(U) \tag{8.19}$$

Notes
- Although, if X and Y are independent, $E(XY) = E(X)E(Y)$, it does not follow that if $E(XY) = E(X)E(Y)$ then X and Y are necessarily independent.
- A particular case of Equation (8.18) that should be noted is:

$$Var(X - Y) = Var(X) + Var(Y) \tag{8.20}$$

Example 4

Two fair six-sided dice are rolled. One die has its sides numbered
$0, 0, 0, 1, 1, 2$; the other die has its sides numbered $2, 2, 3, 3, 4, 4$.
Determine the mean and variance of Z, the total of the numbers shown by
the dice.

———————

Let X and Y be the numbers shown by the two dice.
We are interested in $Z = X + Y$. We require $E(Z)$ and $Var(Z)$.

We can use the result $E(Z) = E(X) + E(Y)$. Also, since the two dice are
independent of one another, $Var(Z) = Var(X) + Var(Y)$.

For X we have the probability distribution:

$$P(X = 0) = \tfrac{3}{6} \qquad P(X = 1) = \tfrac{2}{6} \qquad P(X = 2) = \tfrac{1}{6}$$

Hence:

$$E(X) = (0 \times \tfrac{3}{6}) + (1 \times \tfrac{2}{6}) + (2 \times \tfrac{1}{6}) = \tfrac{4}{6} = \tfrac{2}{3}$$

Also:

$$E(X^2) = (0^2 \times \tfrac{3}{6}) + (1^2 \times \tfrac{2}{6}) + (2^2 \times \tfrac{1}{6}) = 1$$

so that:

$$Var(X) = E(X^2) - \{E(X)\}^2 = 1 - (\tfrac{2}{3})^2 = \tfrac{5}{9}$$

For Y we have the probability distribution:

$$P(Y = 2) = P(Y = 3) = P(Y = 4) = \tfrac{1}{3}$$

By symmetry $E(Y) = 3$. Also:

$$E(Y^2) = (2^2 \times \tfrac{1}{3}) + (3^2 \times \tfrac{1}{3}) + (4^2 \times \tfrac{1}{3}) = \tfrac{29}{3}$$

so that:

$$Var(Y) = E(Y^2) - \{E(Y)\}^2 = \tfrac{29}{3} - 3^2 = \tfrac{2}{3}$$

Thus:

$$E(Z) = E(X) + E(Y) = \tfrac{2}{3} + 3 = \tfrac{11}{3}$$

and:

$$Var(Z) = Var(X) + Var(Y) = \tfrac{5}{9} + \tfrac{2}{3} = \tfrac{11}{9}$$

———————

An alternative (long!) approach

A check on the above results is provided by tackling the distribution of Z
head on! Again let X and Y be the numbers shown by the two dice. Since
X and Y are independent, the probability of the outcome (say) $X = 0$ and
$Y = 2$ is the product of their separate probabilities: $\tfrac{3}{6} \times \tfrac{1}{3} = \tfrac{3}{18} = \tfrac{1}{6}$. A
convenient summary of the 9 possible value combinations is as follows:

		Probabilities Second die					Totals Second die		
		2	3	4			2	3	4
	0	$\tfrac{3}{18}$	$\tfrac{3}{18}$	$\tfrac{3}{18}$		0	2	3	4
First die	1	$\tfrac{2}{18}$	$\tfrac{2}{18}$	$\tfrac{2}{18}$	First die	1	3	4	5
	2	$\tfrac{1}{18}$	$\tfrac{1}{18}$	$\tfrac{1}{18}$		2	4	5	6

The distribution for Z is therefore:

Value of Z	2	3	4	5	6
Probability	$\tfrac{3}{18}$	$\tfrac{5}{18}$	$\tfrac{6}{18}$	$\tfrac{3}{18}$	$\tfrac{1}{18}$

Hence:

$$E(Z) = (2 \times \tfrac{3}{18}) + (3 \times \tfrac{5}{18}) + (4 \times \tfrac{6}{18}) + (5 \times \tfrac{3}{18}) + (6 \times \tfrac{1}{18})$$
$$= \tfrac{66}{18} = \tfrac{11}{3}$$

as before.

Also:

$$E(Z^2) = (2^2 \times \tfrac{3}{18}) + \cdots + (6^2 \times \tfrac{1}{18}) = \tfrac{264}{18} = \tfrac{44}{3}$$

So that:

$$\mathrm{Var}(Z) = E(Z^2) - \{E(Z)\}^2 = \tfrac{44}{3} - (\tfrac{11}{3})^2 = \tfrac{11}{9}$$

▲ _____ ▲

Exercises 8b

1 The independent random variables X and Y are such that $E(X) = 5$, $E(Y) = 7$, $\mathrm{Var}(X) = 3$ and $\mathrm{Var}(Y) = 4$. Determine the mean and variance of the random variables U, V and W defined by:
$$U = 2X, \quad V = X + Y, \quad W = X - 2Y$$

2 It is given that $E(X) = 3$, $\mathrm{Var}(X) = 16$, $E(Y) = 4$, $\mathrm{Var}(Y) = 9$ and that X and Y are independent.
Find:
(i) $E(X + Y)$, (ii) $\mathrm{Var}(X + Y)$, (iii) $E(4X - 3Y)$, (iv) $\mathrm{Var}(4X - 3Y)$, (v) $E(\tfrac{1}{4}X + \tfrac{1}{3}Y)$, (vi) $\mathrm{Var}(\tfrac{1}{4}X + \tfrac{1}{3}Y)$.

3 It is given that X_1 and X_2 are independent, and $E(X_1) = E(X_2) = \mu$, $\mathrm{Var}(X_1) = \mathrm{Var}(X_2) = \sigma^2$.
Find $E(\bar{X})$ and $\mathrm{Var}(\bar{X})$, where $\bar{X} = \tfrac{1}{2}(X_1 + X_2)$.

4 It is given that $E(X) = -5$, $\mathrm{Var}(X) = 25$, $E(Y) = 8$, $\mathrm{Var}(Y) = 9$, and that X and Y are independent.
Find:
(i) $E(X^2)$ and $E(Y^2)$, (ii) $E(3X^2 + 4Y^2)$.

5 H is the number of heads obtained when an unbiased coin is thrown and S is the score obtained when an unbiased die is thrown. The random variable X is defined by $X = 2H - 6S$.
Find the expectation and variance of X.

6 A bank has two branches in Camchester. The number of customers on a Monday at the High Street branch has mean 100 and standard deviation 15. The number of customers on a Monday at the Station Road branch has mean 50 and standard deviation 20.
Find the mean and standard deviation of the total number of customers at both branches on a Monday. State any assumption that you need to make in order to be able to answer the question.

$E(X_1 + X_2)$ and $\mathrm{Var}(X_1 + X_2)$

An important application of the previous results concerns the case where the random variables X and Y are replaced by two variables, X_1 and X_2 which are independent, but have identical probability distributions. In other words X_1 and X_2 share the properties that:

$$P(X_1 = x) = P(X_2 = x) \quad \text{(for all values of } x\text{)}$$
$$E(X_1) = E(X_2)$$
$$\mathrm{Var}(X_1) = \mathrm{Var}(X_2)$$

Denoting the common value of the population mean by μ, and the common variance by σ^2, then, using Equations (8.3) and (8.17), we have:

$$E(X_1 + X_2) = \mu + \mu = 2\mu$$

and:

$$\mathrm{Var}(X_1 + X_2) = \mathrm{Var}(X_1) + \mathrm{Var}(X_2) = \sigma^2 + \sigma^2 = 2\sigma^2$$

The difference between $2X$ and $X_1 + X_2$

Suppose that each of the random variables X, X_1 and X_2 has mean μ and variance σ^2. Gathering the previous results together we have:

$$E(2X) = 2E(X) = 2\mu \qquad E(X_1) + E(X_2) = 2\mu$$

which is what we would expect. However, the results for the variances are not so accommodating:

$$\mathrm{Var}(2X) = 2^2\mathrm{Var}(X) = 4\sigma^2 \qquad \text{but} \qquad \mathrm{Var}(X_1 + X_2) = 2\sigma^2$$

Why is there a difference? To see the answer, consider the acrobats once again.

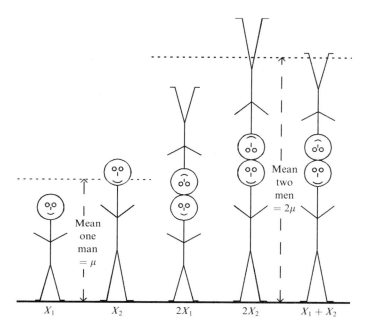

Suppose that the observed value of X_1 is less than μ: then $2X_1$ must certainly be less than 2μ. Likewise, if the observed value of X_2 is greater than μ, then $2X_2$ must be greater than 2μ.

However, on some occasions that X_1 is smaller than μ, X_2 will be larger than μ. Whenever this happens, the total of the values of X_1 and X_2 is likely to be quite close to 2μ. In this case therefore there is an opportunity for central values that does not exist in the previous case – hence the distribution is less variable.

Algebraically, we can observe the difference by recalling that:

$$\mathrm{Var}(X + Y) = \mathrm{Var}(X) + 2\{E(XY) - E(X)E(Y)\} + \mathrm{Var}(Y)$$

In the case of $\mathrm{Var}(2X)$ we have in effect that $Y = X$. The central term is therefore $2\{E(X^2) - E(X)E(X)\}$ which is just $2\mathrm{Var}(X)$, so that:

$$\mathrm{Var}(X + X) = \mathrm{Var}(X) + 2\mathrm{Var}(X) + \mathrm{Var}(X) = 4\mathrm{Var}(X) = 4\sigma^2$$

However, if we put $X = X_1$ and $Y = X_2$ then the central term is $2\{E(X_1 X_2) - E(X_1)E(X_2)\}$ which is the covariance of X_1 and X_2. Since X_1 and X_2 are independent of one another, this covariance is equal to zero. Hence:

$$\mathrm{Var}(X_1 + X_2) = \mathrm{Var}(X_1) + 0 + \mathrm{Var}(X_2) = 2\sigma^2$$

Practical ————————————————————————————

In order to verify that there really is a difference between $2X$ and $X_1 + X_2$, we can perform two simple experiments using dice.

1   Roll an ordinary die 25 times. On each roll double the score before recording it on a tally chart.
Calculate the values of the sample mean and variance.

2   Roll a pair of dice 25 times. On each roll record the total of the two dice on a second tally chart.
Calculate the values of the sample mean and variance.

Verify that the two sample means are about equal, whereas the first sample variance is about twice the second.
To see why this has occurred draw a barchart of the outcomes of the first experiment and superimpose (using a different colour or different shading) the bar chart for the second experiment.

▼ —————————————————————————————————— ▼

Example 5

The independent random variables X_1 and X_2 each have the probability distribution: $P_2 = 0.4$, $P_3 = 0.6$
Determine the values of (i) $Var(X_1)$, (ii) $Var(2X_1)$, (iii) $Var(X_1 + X_2)$.

——————————

For each X-variable we have:

$$E(X) = (2 \times 0.4) + (3 \times 0.6) = 2.6$$
$$E(X^2) = (2^2 \times 0.4) + (3^2 \times 0.6) = 7.0$$

Hence:

$$Var(X) = E(X^2) - \{E(X)\}^2 = 7.0 - 2.6^2 = 0.24$$

We can now answer the various questions:
(i) $Var(X_1) = 0.24$
(ii) $Var(2X_1) = 2^2 Var(X_1) = 4 \times 0.24 = 0.96$
(iii) $Var(X_1 + X_2) = Var(X_1) + Var(X_2) = 2 \times 0.24 = 0.48$

We can also obtain the final result by considering the distribution of $Y = X_1 + X_2$ directly:

$$P(Y = 4) = 0.4^2 = 0.16, \qquad P(Y = 6) = 0.6^2 = 0.36$$

and hence:

$$P(Y = 5) = 1 - 0.16 - 0.36 = 0.48$$

Hence:

$$E(Y) = (4 \times 0.16) + (5 \times 0.48) + (6 \times 0.36) = 5.2$$
$$E(Y^2) = (4^2 \times 0.16) + (5^2 \times 0.48) + (6^2 \times 0.36) = 27.52$$

and so:

$$Var(X_1 + X_2) = Var(Y) = E(Y^2) - \{E(Y)\}^2 = 27.52 - 5.2^2 = 0.48$$

▲ —————————————————————————————————— ▲

8.5 The expectation and variance of the sample mean

Suppose we take a total of m samples, each of n independent observations, on the random variable X. Each sample will have a first observation, a second observation, and so on. Denote the jth observation in the ith sample by x_{ij}. The observations are summarised in the following table:

Sample number	1st observation		jth observation		nth observation
1	x_{11}	$\cdots$	x_{1j}	$\cdots$	x_{1n}
2	x_{21}	$\cdots$	x_{2j}	$\cdots$	x_{2n}
i	x_{i1}	$\cdots$	x_{ij}	$\cdots$	x_{in}
m	x_{m1}	$\cdots$	x_{mj}	$\cdots$	x_{mn}

Consider the first observations of all the samples that we take: x_{11}, $x_{21}, \ldots, x_{m1}$. These values vary because of random variation. For the ith sample, x_{i1} can be thought of as an observation on the random variable 'the first observation' which we denote by X_1. In the same way we can define a further $(n-1)$ random variables: $X_2, X_3, \ldots, X_n$. Since the observations are independent and are all observations of the same underlying random variable X, the n random variables $X_1, \ldots, X_n$ are independent and identically distributed, with their common distribution being that of X.

Denote the sample mean for the ith sample by $\bar{x}_i$. Because of random variation the values of $\bar{x}_1, \bar{x}_2, \ldots, \bar{x}_m$ will also vary. There is therefore yet another lurking random variable, namely 'the sample mean', which we will denote by $\bar{X}$. Evidently:

$$\bar{X} = \frac{1}{n}(X_1 + X_2 + \cdots + X_n)$$

$$= \frac{1}{n}X_1 + \frac{1}{n}X_2 + \cdots + \frac{1}{n}X_n$$

Suppose that $\mathrm{E}(X) = \mu$ and $\mathrm{Var}(X) = \sigma^2$. Using the result concerning expectations of sums of random variables, we have:

$$\mathrm{E}(\bar{X}) = \mathrm{E}\left(\frac{1}{n}X_1 + \frac{1}{n}X_2 + \cdots + \frac{1}{n}X_n\right)$$

$$= \frac{1}{n}\mathrm{E}(X_1) + \frac{1}{n}\mathrm{E}(X_2) + \cdots + \frac{1}{n}\mathrm{E}(X_n)$$

$$= \frac{1}{n}\mu + \frac{1}{n}\mu + \cdots + \frac{1}{n}\mu$$

$$= n\left(\frac{1}{n}\mu\right)$$

$$= \mu$$

The expectation of the sample mean is therefore the population mean – a pleasing result.

Since the random variables $X_1, X_2, \ldots, X_n$ are mutually independent:

$$\text{Var}(\bar{X}) = \text{Var}\left(\frac{1}{n}X_1 + \frac{1}{n}X_2 + \cdots + \frac{1}{n}X_n\right)$$

$$= \text{Var}\left(\frac{1}{n}X_1\right) + \text{Var}\left(\frac{1}{n}X_2\right) + \cdots + \text{Var}\left(\frac{1}{n}X_n\right)$$

$$= \left(\frac{1}{n}\right)^2 \text{Var}(X_1) + \left(\frac{1}{n}\right)^2 \text{Var}(X_2) + \cdots + \left(\frac{1}{n}\right)^2 \text{Var}(X_n)$$

$$= \left(\frac{1}{n}\right)^2 \sigma^2 + \left(\frac{1}{n}\right)^2 \sigma^2 + \cdots + \left(\frac{1}{n}\right)^2 \sigma^2$$

$$= n\left(\frac{1}{n}\right)^2 \sigma^2$$

$$= \frac{\sigma^2}{n}$$

This is an important result because, for $n > 1$, this tells us that the sample mean is much less variable than are the individual observations. We can also see that the variance decreases as n increases, so that the sample mean is more and more likely to be close to the population mean as the sample size increases.

Example 6

The discrete random variable X has probability distribution $P_x = \dfrac{4-x}{10}$

for $x = 0, 1, 2, 3$.
Determine the variance of the sample mean (i) when the sample size is 2, (ii) when the sample size is 16.

In tabular form the probability distribution of X is as follows:

x	0	1	2	3
P_x	$\frac{4}{10}$	$\frac{3}{10}$	$\frac{2}{10}$	$\frac{1}{10}$

The probabilities sum to 1, so it seems that we have interpreted the formula correctly! To answer the question we must first obtain the variance of a single observation on X. Now:

$$E(X) = (0 \times 0.4) + (1 \times 0.3) + (2 \times 0.2) + (3 \times 0.1) = 1.0$$
$$E(X^2) = (0^2 \times 0.4) + (1^2 \times 0.3) + (2^2 \times 0.2) + (3^2 \times 0.1) = 2.0$$

so that:

$$\text{Var}(X) = E(X^2) - \{E(X)\}^2 = 2.0 - (1.0)^2 = 1.0$$

From the general formula for a sample of size n we therefore have the answers (i) $\text{Var}(\bar{X}) = \frac{1}{2}$, and (ii) $\text{Var}(\bar{X}) = \frac{1}{16}$.

Note

- The square root of the variance of the sample mean, $\dfrac{\sigma}{\sqrt{n}}$, is often called the **standard error of the mean**, or simply the **standard error**. The same terms may be used for the corresponding sample value $\dfrac{s}{\sqrt{n}}$.

Exercises 8c

1 A random variable has expectation 12 and standard deviation 3. A sample of 81 observations is taken.
Find the expectation and variance of the sample mean.

2 An unbiased die is thrown 100 times and the score is observed.
Find the expectation and standard error of the mean score.

3 An unbiased die is thrown until the first six is obtained. The number X of throws needed to obtain the first six is observed. The process is repeated 50 times, giving 50 observations of X. Denoting the sample mean by $\bar{X}$, find the mean and standard error of $\bar{X}$.

4 A random variable Y takes the value 1 and 10, with probabilities p and $1 - p$ respectively. 200 observations of Y are taken and the sample mean is $\bar{Y}$.
Find expressions for the mean and variance of $\bar{Y}$.

5 The mean weight of a soldier may be taken to be 90 kg, and the standard deviation may be taken to be 10 kg. 250 soldiers are on board an aircraft.
Find the expectation and variance of their average weight. State any assumption necessary.

Hence, or otherwise, find the mean and standard deviation of the total weight of the soldiers.

6 A random variable V has mean 150 and standard deviation 2. A random sample of n observations of V is taken.
Find the smallest value of n such that the standard error of the sample mean is less than 0.1.

7 A random variable R has mean 12 and variance 3. A random sample of n observations of R is taken.
Find the smallest value of n such that the expected value of the sample total exceeds 1000, and find the variance of the sample total for this value of n.

8 A computer program generates, with equal probabilities, one of the three numbers 0, 1 or 2. The variables X, Y and Z result from three independent runs of the program. If m is the mean of X, Y and Z, calculate the mean and variance of m.
If M is the median of X, Y and Z show that $P(M = 0) = \frac{7}{27}$. Deduce the values of $P(M = 2)$ and of $P(M = 1)$. Hence determine the mean and variance of M.
If U is the largest of X, Y and Z calculate the mean and variance of U. [SMP]

8.6 The unbiased estimate of the population variance

In Chapter 2 we introduced the quantity s^2, given, for a sample of n observations $x_1, x_2, \ldots, x_n$, by the formula:

$$s^2 = \frac{1}{n - 1} \Sigma(x_i - \bar{x})^2$$

The phrase 'unbiased estimate' simply means that the corresponding random variable has expectation equal to σ^2, the population variance:

$$E\left[\frac{1}{n - 1} \Sigma(X_i - \bar{X})^2\right] = \sigma^2$$

where, as in the previous section, X_i is the random variable 'the ith observation' and $\bar{X}$ is the random variable 'the sample mean'. We now set about proving this result.
For simplicity, we will write:

$$Y_i = X_i - \bar{X}$$

and we note that, since both X_i and $\bar{X}$ have mean μ, $E(Y_i) = \mu - \mu = 0$, for all i. We wish to show that:

$$\frac{1}{n-1} E\left(\Sigma Y_i^2\right) = \sigma^2$$

Now:

$$E(\Sigma Y_i^2) = \Sigma E(Y_i^2)$$
$$= \Sigma\left[\mathrm{Var}(Y_i) + \{E(Y_i)\}^2\right]$$
$$= \Sigma \mathrm{Var}(Y_i) \qquad \text{since } E(Y_i) = 0$$

To find $\mathrm{Var}(Y_i)$, we return to the definitions of Y_1 and $\bar{X}$ and write

$$Y_1 = X_1 - \frac{1}{n}(X_1 + X_2 + \cdots + X_n)$$
$$= X_1\left(1 - \frac{1}{n}\right) - \frac{1}{n}(X_2 + \cdots + X_n)$$

We now use the fact that $X_1, X_2, \ldots, X_n$ are independent random variables, together with the result that, for a constant a, $\mathrm{Var}(aX) = a^2\mathrm{Var}(X)$, to write:

$$\mathrm{Var}(Y_1) = \sigma^2\left(1 - \frac{1}{n}\right)^2 + \left(-\frac{1}{n}\right)^2(\sigma^2 + \cdots + \sigma^2)$$
$$= \frac{\sigma^2}{n^2}\{(n-1)^2 + (n-1)\}$$
$$= \frac{\sigma^2}{n^2}(n-1)\{(n-1) + 1\}$$
$$= \frac{(n-1)\sigma^2}{n}$$

Substituting this result, which will also hold for Y_2, Y_3, $\ldots$, we get:

$$E(\Sigma Y_i^2) = \sum\left\{\frac{(n-1)\sigma^2}{n}\right\}$$
$$= n\left\{\frac{(n-1)\sigma^2}{n}\right\}$$
$$= (n-1)\sigma^2$$

Hence:

$$E\left[\frac{1}{n-1}\Sigma Y_i^2\right] = \sigma^2$$

as required.

We have finally justified our original statement that the random variable corresponding to s^2, with its $(n-1)$ divisor, has expectation σ^2.

8.7 $E(t^X)$ – the probability generating function

In many branches of mathematics there are problems that can be solved by a variety of different methods. Using one method a problem can appear very difficult whereas using another method it turns out to be quite easy. In Statistics, sometimes the use of a probability generating function can make a hard problem much simpler.

By convention, the **probability generating function (pgf** for short) is denoted by G(t), and is defined for the discrete random variable X by the relation:

$$G(t) = E(t^X) = \sum\{t^x P(X = x)\} \tag{8.21}$$

where the summation is over all possible values of the discrete random variable X. The immediate question that occurs to one on looking at this definition is 'What on earth is t?' The rather mysterious answer is that 't is a variable whose powers can be thought of as labels for the probabilities'. We shall see that setting t equal to 1 will often be very convenient!

One use of the pgf is as an alternative method of obtaining the mean and variance of a probability distribution. The method entails differentiating G(t) with respect to t. Differentiating a quantity involving a Σ sign is not a problem, but on this first occasion we will write things out at length. Suppose the possible values for X are $x_1, x_2, \ldots$, so that:

$$G(t) = t^{x_1}P(X = x_1) + t^{x_2}P(X = x_2) + \cdots$$

Differentiating with respect to t, we get:

$$\frac{dG(t)}{dt} = x_1 t^{x_1-1}P(X = x_1) + x_2 t^{x_2-1}P(X = x_2) + \cdots$$
$$= \sum\{xt^{x-1}P(X = x)\}$$

where, as usual, the summation is over all possible values of X.

The notation $\dfrac{dG(t)}{dt}$ for the derivative is cumbersome, so we use the alternative notation G$'(t)$, with G$''(t)$ denoting the second derivative with respect to t. Using the usual notation, we get:

$$G'(t) = \Sigma\{xt^{x-1}P_x\}$$

with the second derivative with respect to t becoming:

$$G''(t) = \Sigma\{x(x-1)t^{x-2}P_x\}$$

It is now time to choose our convenient value (1) for t, obtaining:

$$G'(1) = \Sigma\{xP_x\}$$
$$G''(1) = \Sigma\{x(x-1)P_x\}$$

The right-hand sides of these two equations are simply expectations:

$$G'(1) = E(X) \tag{8.22}$$
$$G''(1) = E[X(X-1)] = E(X^2) - E(X) \tag{8.23}$$

Hence:

$$\text{Var}(X) = G''(1) + G'(1) - \{G'(1)\}^2 \tag{8.24}$$

Notes
- Putting $t = 1$, we get G$(1) = 1$.
- In strict pure mathematical terms, the 'function' is G, rather than G(t) (which is the image of t under G). In Statistics, however, it is standard to use image notation for generating functions.

Example 7

The random variable X has probability distribution given by
$P(X = 1) = P(X = 2) = P(X = 4) = \frac{1}{3}$.
Obtain the probability generating function for X.

From the definition of G(t) we have:

$$G(t) = \frac{1}{3}t^1 + \frac{1}{3}t^2 + \frac{1}{3}t^4 = \frac{t(1 + t + t^3)}{3}$$

Example 8

The random variable X has probability distribution given by
$P(X = 1.5) = P(X = 2.5) = \frac{1}{2}$.
Obtain the probability generating function for X.

From the definition of $G(t)$ we have:

$$G(t) = \frac{1}{2} t^{1.5} + \frac{1}{2} t^{2.5}$$

Now $t^{1.5}$ looks pretty frightening, but it isn't really! In fact this is just an unusual way of writing $t^{\frac{3}{2}} = \sqrt{t^3} = t\sqrt{t}$. Similarly, $t^{2.5} = t^2\sqrt{t}$, and so:

$$G(t) = \frac{1}{2} t(1 + t)\sqrt{t}$$

Example 9

The random variable Y has pgf $G(t)$ given by:

$$G(t) = k(1 + t^2 + t^5)$$

Determine: (i) the value of k, (ii) the probability distribution of Y, (iii) the expectation of Y.

(i)   To find k we use the fact that $G(1) = 1$:

$$k(1 + 1^2 + 1^5) = k(1 + 1 + 1) = 3k = 1$$

giving $k = \frac{1}{3}$.

(ii) The only possible values for Y are indicated by the indices of the powers of t in $G(t)$. Remembering that $t^0 = 1$, we see that the only possible values are 0, 2 and 5. Each of t^0, t^2 and t^5 has coefficient $k(= \frac{1}{3})$, and so the probability distribution for Y is:

Value of Y	0	2	5
Probability	$\frac{1}{3}$	$\frac{1}{3}$	$\frac{1}{3}$

(iii) $E(Y)$ could be obtained directly from the probability distribution, but it is easier here to use Equation (8.22). The derivative of $G(t)$ is given by:

$$G'(t) = k(0 + 2t + 5t^4)$$

Substituting for k and putting t equal to 1, we get:

$$E(Y) = G(1) = \frac{1}{3}(0 + 2 + 5) = \frac{7}{3}$$

The expectation of Y is $\frac{7}{3}$.

Example 10

Calculating the expectation and variance of the Bernoulli distribution was very simple (see Example 15 of Chapter 7), but it is equally simple to use the pgf.

The Bernoulli random variable X has distribution:

$$P_0 = 1 - p \qquad P_1 = p$$

so its pgf, $G(t)$ is given by:

$$G(t) = t^0(1 - p) + t^1 p = 1 - p + pt$$

Hence $G'(t) = p$ and $G''(t) = 0$ so that:

$$E(X) = G'(1) = p$$
$$\text{Var}(X) = G''(1) + G'(1) - \{G'(1)\}^2 = 0 + p - p^2 = p(1 - p)$$

Example 11

Obtaining the expectation and variance of the geometric distribution by direct manipulation of series was distinctly tricky (see Example 16 of Chapter 7). We now obtain the same results using the pgf.

The geometric random variable X has distribution:

$$P_x = q^{x-1}p \qquad (x = 1, 2, \ldots)$$

where $q = 1 - p$, so its pgf is given by:

$$G(t) = \Sigma t^x q^{x-1} p = tp\Sigma(tq)^{x-1}$$

The possible values for x are 1, 2, ... so the right-hand side becomes $1 + tq + (tq)^2 + (tq)^3 + \cdots$ which is a geometric progression with sum equal to $\dfrac{1}{1 - tq}$ (taking $|t| < \dfrac{1}{q}$, so that $|qt| < 1$). The pgf of the geometric distribution is therefore:

$$G(t) = \frac{tp}{1 - tq}$$

Differentiating with respect to t and simplifying we get:

$$G'(t) = \frac{p}{(1 - tq)^2}$$

Hence, putting t equal to 1, $G'(1) = \dfrac{p}{(1 - q)^2}$. But $q = 1 - p$ and $G'(1) = E(X)$ so we have that the expectation of X is equal to $\dfrac{1}{p}$.

The second derivative is easier to obtain, since t only appears in the denominator of $G'(t)$:

$$G''(t) = \frac{(-2)(-q)p}{(1 - tq)^3}$$

Setting t equal to 1, this gives:

$$G''(1) = \frac{2qp}{(1-q)^3} = \frac{2qp}{p^3} = \frac{2q}{p^2}$$

Hence:

$$\begin{aligned}
\text{Var}(X) &= G''(1) + G'(1) - \{G'(1)\}^2 \\
&= \frac{2(1-p)}{p^2} + \frac{1}{p} - \frac{1}{p^2} \\
&= \frac{2 - 2p + p - 1}{p^2} \\
&= \frac{1-p}{p^2}
\end{aligned}$$

Pgf for the sum of random variables

Suppose U and V are two independent random variables, and suppose that the random variable W is defined by $W = U + V$. Let the probability generating functions associated with U, V and W be denoted by $G_U(t)$, $G_V(t)$ and $G_W(t)$, respectively. Then:

$$\begin{aligned}
G_W(t) &= E(t^W) &&\text{by definition} \\
&= E(t^{U+V}) \\
&= E(t^U \times t^V) \\
&= E(t^U)E(t^V) &&\text{since } U \text{ and } V \text{ are independent} \\
&= G_U(t)G_V(t)
\end{aligned}$$

This simple result obviously extends to more than two independent random variables. It comes in particularly handy when dealing with n independent identically distributed random variables. Let the random variable S be defined by:

$$S = \sum_{i=1}^{n} X_i$$

where $X_1, \ldots, X_n$ are independent and identically distributed random variables, each with pgf $G_X(t)$. The pgf for S is simply:

$$\begin{aligned}
G_S(t) &= E(t^S) = E(t^{\Sigma X_i}) = E(t^{X_1} t^{X_2} \ldots t^{X_n}) \\
&= E(t^{X_1})E(t^{X_2}) \ldots E(t^{X_n}) = \{G_X(t)\}^n
\end{aligned}$$

From this expression, on differentiating both sides with respect to t, we obtain:

$$G_S'(t) = n\{G_X(t)\}^{n-1} G_X'(t)$$

Setting $t = 1$, we see that $G_S'(1) = nG_X'(1)$, since $G_X(1) = 1$, and hence we have shown once again that:

$$E(S) = E(\Sigma X_i) = nE(X)$$

Being able to obtain this result using probability generating functions may be reassuring, but is not very exciting since we knew it already! However, we also have the entire probability distribution of S encapsulated in $G_S(t)$, since $P(S = s)$ is, by definition of a pgf, equal to the coefficient of t^s in the series expansion of $G_S(t)$.

Example 12

Determine the probability that the sum of ten independent Bernoulli random variables is equal to exactly 7.

A Bernoulli random variable, X, has pgf $G_X(t) = (1 - p) + pt$. Denoting the sum of ten such variables by S, we have:

$$G_S(t) = \{(1 - p) + pt\}^{10}$$

$$= (1 - p)^{10} + \binom{10}{1}(1 - p)^9 pt + \binom{10}{2}(1 - p)^8 (pt)^2$$

$$+ \binom{10}{3}(1 - p)^3 (pt)^7 + \cdots + (pt)^{10}$$

using the binomial expansion. The probability that S equals 7 is the coefficient of t^7, which is $\binom{10}{3}(1 - p)^3 p^7$.

Exercises 8d

1 The random variable X takes values 1 and 2 with probabilities $\frac{1}{4}$ and $\frac{3}{4}$ respectively.
Show that the pgf of X is $\frac{1}{4}t(1 + 3t)$.
The random variable S is the sum of four independent observations of X.
Find the pgf of S and hence find (i) $P(S = 6)$, (ii) $E(S)$ and $Var(S)$.

2 The random variable X has pgf
$G_X(t) = \frac{1}{8}(1 + t)^3$.
Find the set of possible values for X and the corresponding probabilities.
Use the pgf to find $E(X)$ and $Var(X)$.

3 The random variable R has pgf
$$G_R(t) = \frac{(1 + t^2)^2}{4t^2}.$$
Find the set of possible values for R and the corresponding probabilities.

4 An unbiased die is thrown and S is the score obtained.
Show that the pgf of S is $\frac{1}{6}(t + t^2 + \cdots + t^6)$.
Hence find the mean and variance of S.

5 The random variable V takes values $-1, 1$ with equal probabilities.
Find the pgf of V.
A sample of ten observations of V is taken and T denotes the sample total.
Find the pgf of T and hence find
(i) $P(T = 4)$, (ii) $P(T = 5)$, (iii) $E(T)$, (iv) $Var(T)$.

6 The random variable X has probability generating function $G(t)$.
Show that:
(i) $2X$ has pgf $G(t^2)$,
(ii) $-X$ has pgf $G(t^{-1})$,
(iii) $X + 3$ has pgf $t^3 G(t)$,
(iv) aX, where a is a constant, has pgf $G(t^a)$,
(v) $aX + b$, where a and b are constants, has pgf $t^b G(t^a)$.
(vi) Deduce from (v) that:
$E(aX + b) = aE(X) + b$
$Var(aX + b) = a^2 Var(X)$

7 The discrete random variable X has probability distribution given by
$$P(X = r) = \frac{k}{r!}, \qquad r = 0, 1, 2, \ldots,$$
where k is a constant.
(i)   Find the value of k.
(ii) Find the probability generating function of X and hence or otherwise obtain the mean and variance of X.
(iii) $X_1, X_2, \ldots, X_n$ are independent observations of X, and $Y = \sum_{i=1}^{n} X_i$. Write down the probability generating function of Y and use it to show that
$$P(Y = r) = \frac{e^{-n} n^r}{r!}, \qquad r = 0, 1, 2, \ldots.$$
In the case $n = 2$, find $P(Y = 3 | Y > 0)$, giving your answer correct to three decimal places.   [UCLES(P)]

8 The discrete random variable X has a uniform distribution on the integers $1, 2, 3, \ldots, n$. Find, in either order, $E(X)$ and the probability generating function $G_X(t)$.

A die with four faces is numbered 1 to 4, a die with 6 faces is numbered 1 to 6 and a die with 12 faces is numbered 1 to 12. The three dice, each of which is unbiased, are thrown and Y denotes the sum of the numbers on the lowermost faces. In any order,

(i) find $E(Y)$,

(ii) show that

$$G_Y(t) = \frac{t^3(1 - t^4)(1 - t^6)(1 - t^{12})}{288(1 - t)^3},$$

(iii) show that $P(Y = 6) = \frac{5}{144}$,

(iv) find $P(Y \leqslant 6)$.   [UCLES]

9 The discrete random variable X has mean 4 and probability generating function given by

$$G(t) = \frac{t^3}{(a - bt)},$$

where a and b are constants.

(i) Find the values of a and b.

(ii) By considering the series expansion of $G(t)$, find the set of possible values of X, and show that, if r is one of these values, $P(X = r) = (\frac{1}{2})^{r-2}$.

(iii) Calculate the variance of X.

(iv) The discrete random variable Y is the sum of two independent observations of X. Write down the probability generating function of Y and hence, or otherwise, obtain $P(Y = 10)$.   [UCLES]

10 The discrete random variable X is defined by

$$P(X = r) = \left(\frac{1}{2}\right)^{r+1}, \qquad r = 0, 1, 2, \ldots.$$

Find the probability generating function of X. Hence or otherwise find $E(X)$ and $Var(X)$.

The discrete random variable Y is independent of X and is defined by

$$P(Y = r) = k\left(\frac{1}{2}\right)^r(r + 1), \quad r = 0, 1, 2, \ldots,$$

where k is a positive constant. Find the probability generating function of Y and determine the value of k.

The discrete random variable Z is defined by $Z = X + Y$. Find

(i) $E(Z)$,

(ii) $P(Z = 6)$.   [UCLES]

11 In each round of a quiz a contestant can answer up to 3 questions. Each correct answer scores 1 point and allows the contestant to go on to the next question. A wrong answer scores nothing and the contestant is allowed no further questions in that round once a wrong answer is given. If all three questions are answered correctly a bonus point of 1 is scored, making a total of 4 for the round. For a certain contestant, A, the probability of giving a correct answer to any question in any round is $\frac{2}{3}$. The random variable X_r is the number of points scored by A during the rth round. Write down the probability generating function of X_r and find the mean and variance of X_r.

Write down an expression for the probability generating function of $X_1 + X_2$ and find the probability that A has a total score of 4 at the end of two rounds.

Write down an expression for the probability generating function of $X_2 - X_1$. Find the probability that A scores equally in rounds one and two, and hence or otherwise find the probability of A scoring more in round two than in round one.   [UCLES]

12 The discrete random variable X has a geometric distribution defined by

$$P(X = r) = p(1 - p)^{r-1}, \qquad r = 1, 2, 3, \ldots.$$

(i) Find the probability generating function of X.

(ii) Find $E(X)$.

Each of n players, where $n > 2$, tosses a fair coin. If the result of any player's toss is different from that of all the remaining players, then that player wins the game. Otherwise all the players toss again until one player wins. Given that X is the number of tosses each player makes, up to and including the one on which the game is won, find $P(X = r)$ in terms of r and n.

Show that $E(X) = \dfrac{2^{n-1}}{n}$.

Given that $n = 7$, find the least number k such that $P(X \leqslant k) > 0.5$.   [UCLES]

13 (a) State, with reasons, which of the following do not give probability generating functions. For the remainder, find the expectation of the corresponding random variable.

(i) $f_1(t) = \dfrac{t^4}{(1 - t)}$,   *(continued)*

(ii) $f_2(t) = \dfrac{3t^4}{(1 + 2t)}$,

(iii) $f_3(t) = -\dfrac{\ln\left(1 - \frac{1}{2}t\right)}{\ln 2}$.

(b) The independent random variables X and Y have probability generating functions given by:

$$G_X(t) = \frac{\frac{3}{4}t}{1 - \frac{1}{4}t}, \qquad G_Y(t) = \tfrac{1}{2}t^2(1 + t).$$

Find:

(i) $P(3X - 2Y = 12)$,

(ii) $E(3X - 2Y)$.   [UCLES]

14 The values of the probability generating functions of the independent discrete random variables U and V are $G(z)$ and $H(z)$ respectively. The discrete random variable W, the sum of U and V, has probability generating function K, where $K(z) = G(z)H(z)$. Prove that

(i) $E(W) = E(U) + E(V)$,

(ii) $\mathrm{Var}(W) = \mathrm{Var}(U) + \mathrm{Var}(V)$.

When any brass drawing-pin is dropped onto the floor it has a probability of 0.4 of coming to rest point up. The corresponding probability for any plastic-coated drawing-pin is 0.3. A secretary drops a collection of 200 brass drawing-pins and 300 plastic-coated drawing-pins onto the floor. Assuming that the positions of rest of the 500 drawing-pins are independent of one another, determine the mean and variance of the total number of drawing-pins that come to rest point up.   [UCLES]

Chapter summary

♦ **Expectations and variances of functions of X:**

$$E(X + a) = E(X) + a$$
$$E(aX) = aE(X)$$
$$E[ag(X)] = aE[g(X)]$$
$$E[g(X) + h(X)] = E[g(X)] + E[h(X)]$$
$$\mathrm{Var}(X + a) = \mathrm{Var}(X)$$
$$\mathrm{Var}(aX) = a^2\mathrm{Var}(X)$$

♦ **Expectations of combinations of random variables:**

$$E(X + Y) = E(X) + E(Y)$$
$$E(aX + bY + c) = aE(X) + bE(Y) + c$$
$$E[g(X) + h(Y)] = E[g(X)] + E[h(Y)]$$
$$E(R + S + T + U) = E(R) + E(S) + E(T) + E(U)$$

♦ **Variances of combinations of *independent* random variables:**

$$\mathrm{Var}(aX + bY + c) = a^2\mathrm{Var}(X) + b^2\mathrm{Var}(Y)$$
$$\mathrm{Var}(X - Y) = \mathrm{Var}(X) + \mathrm{Var}(Y)$$
$$\mathrm{Var}(R + S + T + U) = \mathrm{Var}(R) + \mathrm{Var}(S) + \mathrm{Var}(T) + \mathrm{Var}(U)$$

♦ **Multiples and sums of random variables:**
Combinations of identically distributed random variables having mean μ and variance σ^2

$$E(2X) = 2\mu \quad \text{and} \quad E(X_1) + E(X_2) = 2\mu$$
$$\mathrm{Var}(2X) = 4\sigma^2 \quad \text{but} \quad \mathrm{Var}(X_1 + X_2) = 2\sigma^2$$

♦ **Expectation and variance of sample mean:**

$$E(\bar{X}) = \mu \qquad \mathrm{Var}(\bar{X}) = \frac{\sigma^2}{n}$$

♦ **Probability generating functions**:
$$G(t) = E(t^X) = \Sigma t^x P_x$$
$$E(X) = G'(1)$$
$$Var(X) = G''(1) + G'(1) - \{G'(1)\}^2$$

● Sum of n independent identically distributed random variables

If $S = \Sigma X_i$ then $G_S(t) = \{G_X(t)\}^n$

Exercises 8e (Miscellaneous)

1 At a certain institution, students living off-campus travel to campus on foot (0, 30%), by bicycle (2, 20%), by car (4, 30%) or by bus (6, 20%). The bracketed figures indicate the numbers of wheels of the mode of transport and the percentage of students involved.

(a) Let X be the number of wheels utilised by a randomly chosen student.
Determine $E(X)$ and $Var(X)$.

(b) Let S be the total number of wheels utilised by two randomly chosen students who travel independently of one another. Determine the probability distribution of S.
State the mean and variance of S.

(c) A third student is now randomly chosen. This student travels independently of the two previously chosen students. Let the total number of wheels utilised by the three students be denoted by T.
Show that $P(T = 4) = 0.117$.
Find $P(S = 2|T = 4)$.

(d) Let W be the event that at least one of the three students walks.
Find $P(T = 4|W)$ and $P(W|T = 4)$.

2 (i) Six fuses, of which two are defective and four are good, are to be tested one after another in random order until both defective fuses are identified. Find the probability that the number of fuses that will be tested is
(a) three,
(b) four or fewer.

(ii) A random variable R takes the integer value r with probability p(r) where

$$p(r) = kr^3, \qquad r = 1, 2, 3, 4,$$
$$p(r) = 0, \qquad \text{otherwise.}$$

Find
(a) the value of k, and display the distribution on graph paper,
(b) the mean and the variance of the distribution,
(c) the mean and variance of $5R - 3$.
[ULSEB]

3 A darts player practises throwing a dart at the bull's-eye on a dart board. Independently for each throw, her probability of hitting the bull's-eye is 0.2. Let X be the number of throws she makes, up to and including her first success.

(a) Find the probability that she is successful for the first time on her third throw.
(b) Write down the distribution of X, and give the name of this distribution.
(c) Find the probability that she will have at least 3 failures before her first success.
(d) Show that the mean value of X is 5. (You may assume the result $\sum_{r=1}^{\infty} rq^{r-1} = \dfrac{1}{(1-q)^2}$ when $|q| < 1$.)

On another occasion the player throws the dart at the bull's-eye until she has two successes. Let Y be the number of throws that she makes up to and including her second success. Given that $Var(X) = 20$, determine the mean and the variance of Y, and find the probability that $Y = 4$.
[ULSEB]

4 A random variable X takes values $-1, 0, +1$ with probabilities $p, q, 2p$, respectively, and can take no other values.

(i) Express q in terms of p.
(ii) Find, in terms of p, the expected value and standard deviation of X.
(iii) If X_1 and X_2 are independent random variables each having the same distribution as X, find the probability distribution of $Y = X_1 + X_2$ and find $E(Y)$, giving your answers in terms of p.
[UCLES]

5 A coin and a six-faced die are thrown simultaneously. The random variable X is defined as follows:

If the coin shows a head,

then X is the score on the die.

If the coin shows a tail,

then X is twice the score on the die.

Find the expected value, μ, of X and show that $P(X < \mu) = \frac{7}{12}$.

Show that $\mathrm{Var}(X) = \frac{497}{48}$.

The experiment is repeated and the sum of the two values obtained for X is denoted by Y.

Find $P(Y = 4)$ and $E(Y)$. [UCLES]

6 The random variable X takes values $-2, 0, 2$ with probabilities $\frac{1}{4}, \frac{1}{2}, \frac{1}{4}$ respectively. Find $\mathrm{Var}(X)$ and $E(|X|)$.

The random variable Y is defined by $Y = X_1 + X_2$, where X_1 and X_2 are two independent observations of X. Find the probability distribution of Y.

Find $\mathrm{Var}(Y)$ and $E(Y + 3)$. [UCLES]

7 The probability of there being X unusable matches in a full box of Surelite matches is given by

$P(X = 0) = 8k$, $P(X = 1) = 5k$,

$P(X = 2) = P(X = 3) = k$, $P(X \geqslant 4) = 0$.

Determine the constant k and the expectation and variance of X.

Two full boxes of Surelite matches are chosen at random and the total number Y of unusable matches is determined. Calculate $P(Y > 4)$, and state the values of the expectation and variance of Y. [UCLES]

8 Alfred and Bertie play a game, each starting with cash amounting to £100. Two dice are thrown. If the total score is 5 or more then Alfred pays £x, where $0 < x \leqslant 8$, to Bertie. If the total score is 4 or less then Bertie pays £$(x + 8)$ to Alfred. Show that the expectation of Alfred's cash after the first game is £$\frac{1}{3}(304 - 2x)$.

Find the expectation of Alfred's cash after six games.

Find the value of x for the game to be fair, i.e. for the expectation of Alfred's winnings to equal the expectation of Bertie's winnings. Given that $x = 3$, find the variance of Alfred's cash after the first game. [UCLES]

9 The binomial distribution

To be or not to be: that is the question

Hamlet, William Shakespeare

It is not well known that Hamlet played cricket. He was captain of the local side and his problem was that when it came to calling heads or tails at the start of the match, he was rarely correct – just twice in the last test series of six matches against the Visigoths. The consequences were disastrous: whole families wiped out ... He wondered what was the probability of being that unlucky?

Of course, the binomial distribution had not been discovered then! If it had been, then Hamlet would have known that, for 6 independent trials, with the probability of a success being $\frac{1}{2}$ for each trial, the probability of calling correctly just twice was $\binom{6}{2}\left(\frac{1}{2}\right)^6 = \frac{15}{64}$ (which is about $\frac{1}{4}$, so he wasn't really all that unlucky after all!).

9.1 Derivation

The essential elements of Hamlet's problem, for which we shall develop a general formula, are:

♦ A fixed number, n, of independent trials.
♦ Each trial results in either a 'success' or a 'failure'.
♦ The probability of success, p, is the same for each trial.

We have already encountered problems of this type for which a tree diagram helps.

▼   ▼

Example 1

Determine the probability of getting 2 heads in 3 tosses of a bent coin which has $P(\text{Head}) = \frac{1}{5}$.

The tree of possible outcomes is as follows.

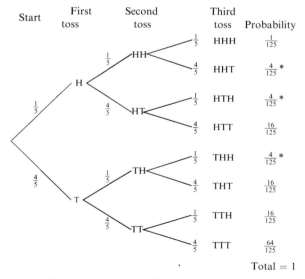

	First toss	Second toss	Third toss	Probability
			HHH	$\frac{1}{125}$
			HHT	$\frac{4}{125}$ *
			HTH	$\frac{4}{125}$ *
			HTT	$\frac{16}{125}$
			THH	$\frac{4}{125}$ *
			THT	$\frac{16}{125}$
			TTH	$\frac{16}{125}$
			TTT	$\frac{64}{125}$

Total = 1

There are three possible sequences (indicated by a *) that lead to the outcome 'exactly 2 heads'. Each sequence has probability $\frac{4}{125}$. Hence the total probability of obtaining exactly 2 heads is $\frac{12}{125}$.

Exercises 9a

1 A cube has the letter 'A' on four faces and the letter 'B' on the remaining two faces. It is thrown three times.

   Draw an appropriate tree diagram and find the probability that the number of A's obtained is (i) 0, (ii) 1, (iii) 2, (iv) 3.

   Find also the probability that the number of B's obtained is (v) 0, (vi) 1, (vii) 2, (viii) 3.

2 Four players each have a pack of cards and, after shuffling each pack, they each turn over the top card of their pack.

   By drawing an appropriate tree diagram, find the probability that the number of Hearts obtained is (i) 0, (ii) 2, (iii) 4.

3 A coin is tossed at the start of each cricket match in a series of 4 Test matches. One captain tosses and the other calls 'Heads' or 'Tails', at random. Find the probability that the toss is called correctly (i) exactly once, (ii) exactly twice.

   Suppose the caller always calls 'Heads'. Does this alter the probabilities?
   Give a reason for your answer.

4 Use a tree diagram to determine the probability of getting exactly two sixes when three fair dice are rolled one after another.

   Would it make any difference to the probability if:

   (i)   all the dice were rolled at once,
   (ii) instead of rolling three different dice, the same die was rolled each time?

5 A woman is trying to light a bonfire. She has only four matches left in her matchbox. Given that ten per cent of matches break when struck, determine the probability that:
   (i)   all four matches will break when struck,
   (ii) at least one match will not break when struck.
   (Assume that all four matches are struck in each case.)

6 Every thousandth visitor to an exhibition is given a voucher for £50.
   Assuming that 65% of the visitors to the exhibition are male, find the probability that, out of the first five to be given a voucher, exactly three are male.

Using a tree diagram is only feasible when the number of trials, n, is small. Otherwise we need a formula!

   Look back at Example 1. There were three possible sequences leading to the desired outcome, and each sequence had the same probability. The answer we calculated was, in effect:

$$P(\text{exactly 2 heads}) = (\text{Number of sequences}) \times \{P(\text{Head})\}^2 \times \{P(\text{Tail})\}^1$$
$$= 3 \qquad\qquad \times (\tfrac{1}{5})^2 \qquad \times (\tfrac{4}{5})^1$$
$$= \tfrac{12}{125}$$

This approach works every time.

▼   ▼

Example 2

Suppose a (rich) gambler has a biased coin for which the probability of a head is 0.55. He tosses the coin 8 times.
What is the probability of his getting 6 heads?
———————

Using the method above, we get:

$$P(\text{6 heads in 8 tosses}) = (\text{Number of sequences}) \times \{P(\text{Head})\}^6 \times \{P(\text{Tail})\}^2$$
$$= (\text{Number of sequences}) \times (0.55)^6 \qquad \times (0.45)^2$$

In this case the number of sequences leading to the desired result happens to be 28, so:

$$P(\text{6 heads in 8 tosses}) = 28 \times (0.55)^6 \times (0.45)^2 = 0.157 \text{ (to 3 d.p.)}$$

▲   ▲

The essential question is, for the general case of n independent trials, 'How many sequences in a probability tree lead to exactly r successes?' To answer this question, note that the r successes can be the result of any combination of r of the n trials. In Chapter 5 we introduced the notation $\binom{n}{r} = \dfrac{n \times (n-1) \times \ldots \times (n-r+1)}{r \times (r-1) \times \ldots \times 1}$ to represent the number of ways of choosing r out of n: the number of sequences is therefore $\binom{n}{r}$. (The number '28' used in Example 2 above is $\binom{8}{6}$.)

To illustrate this approach, consider the following problem:

> A marksman fires 10 times at a target.
> Assuming that the outcomes of the shots are independent of one another, and that each shot has probability 0.96 of being a 'bull', determine the probability that the marksman obtains exactly 9 bulls.

In this problem each shot is either a 'success' (a bull) or a 'failure'. Denoting the number of bulls obtained by X, we require $P(X = 9)$, which for convenience we will write as P_9. Since the number of sequences leading to exactly 9 bulls is $\binom{10}{9}$ ($= 10$), we obtain:

$$P_9 = 10(0.96)^9(0.04)^1 = 0.277 \text{ (to 3 d.p.)}$$

What would have happened if the marksman's probability of obtaining a bull had been 0.92, instead of 0.96? To find out we simply replace 0.96 by 0.92 and 0.04 by 0.08, to get:

$$P_9 = 10(0.92)^9(0.08)^1 = 0.378 \text{ (to 3 d.p.)}$$

As the marksman's probability of a bull changes, so we change the values in the formula. If his probability of getting a bull had been p, then we would have had:

$$P_9 = 10p^9(1-p)^1$$

The generalisation is clear:

> **The probability of obtaining r successes out of n independent trials, when for each trial the probability of a success is p, is:**
> $$P_r = \binom{n}{r}p^r(1-p)^{n-r} \qquad (9.1)$$

This result, which provides the definition of the **binomial distribution**, makes no assumptions about the size of r and is therefore valid for all values of r from 0 to n inclusive.

Notes

♦ Remember $\binom{n}{r} = \binom{n}{n-r}$, $\binom{n}{0} = \binom{n}{n} = 1$ and $p^0 = 1$.

♦ The quantity $1 - p$ is often written as q.

♦ Write q for $(1 - p)$ and consider the **binomial expansion** of $(q + p)^n$, which is:

$$(q + p)^n = q^n + \binom{n}{1}q^{n-1}p^1 + \binom{n}{2}q^{n-2}p^2 + \cdots + \binom{n}{n-1}q^1p^{n-1} + p^n$$

The probabilities $P_0, P_1, \ldots, P_n$ are the successive terms in this expansion. Since $q + p = 1$ this confirms that the sum of the binomial probabilities is 1.

♦ The most usual error in calculating a binomial probability is to forget that, in order for there to be *exactly* r successes, there must also be $n - r$ failures. The $(1 - p)^{n-r}$ factor must not be omitted from the formula!

- The binomial distribution can be used as a model for sampling *with replacement* from a population of any size.
- Only if a (finite) population is very large, can the binomial distribution be used as a model for sampling *without replacement*.

Example 3

According to a motoring magazine, in Britain, Japanese cars account for 5% of the cars on the road. Whilst held up in a traffic jam I occupy my time by examining the cars racing past on the other side of the road. Assuming that the magazine is correct, determine the probability that, of the first 50 cars that pass me, 4 are Japanese.

Each car is either Japanese (a 'success') or not Japanese. Assuming that the traffic jam is not immediately outside a car manufacturing plant, the 50 cars can be assumed to be a random sample of the cars on the road. The population of cars is sufficiently large for us to use the binomial distribution.

The number of trials, n is 50, since 50 cars are examined. The probability of a 'success', p, is 0.05 and the value of r is 4. Hence the required probability is:

$$\binom{50}{4}(0.05)^4(0.95)^{46} = 0.136 \text{ (to 3 d.p.)}$$

Example 4

Four cards are drawn at random from an ordinary pack of 52 cards. Determine the probability that precisely three are Spades (i) if the four are drawn *without* replacement, (ii) if the four are drawn one-at-a-time *with* replacement.

(i) The pack contains 13 Spades and 39 other cards. The probability that the first card drawn is a Spade is $\frac{13}{52} = \frac{1}{4}$. However, the probability of the second card drawn being a Spade depends upon the outcome of the first card drawn. If the first is a Spade then the probability of the second being a spade is $\frac{12}{51}$, whereas if the first is not a Spade then the probability of the second being a Spade is $\frac{13}{51}$. This situation was discussed in Section 5.17 (p. 138) and the required probability is:

$$\frac{\binom{13}{3} \times \binom{39}{1}}{\binom{52}{4}} = \frac{286 \times 39}{270\,725} = 0.041 \text{ (to 3 d.p.)}$$

(ii) In this case, for each draw the probability of getting a Spade is constant at $\frac{13}{52} = \frac{1}{4}$. The binomial distribution is now appropriate and the probability is:

$$\binom{4}{3}\left(\frac{1}{4}\right)^3\left(\frac{3}{4}\right)^1 = \frac{3}{64} = 0.047 \text{ (to 3 d.p.)}$$

The probability obtained in the 'with replacement' case is appreciably larger than that obtained in the 'without replacement' case.

Exercises 9b

1 The number of successes in n independent trials is X. The probability of a success in each trial is p.
  (i) Given that $n = 10$, $p = \frac{1}{4}$, find $P(X = 3)$.
  (ii) Given that $n = 8$, $p = \frac{3}{4}$, find $P(X = 6)$.
  (iii) Given that $n = 12$, $p = \frac{1}{3}$, find $P(X \leqslant 3)$.
  (iv) Given that $n = 11$, $p = \frac{4}{5}$, find $P(X \geqslant 9)$.
  (v) Given that $n = 7$, $p = \frac{1}{2}$, find
     $P(3 \leqslant X \leqslant 5)$.

2 Five per cent of bluebells (confusingly) have white flowers. The remainder have blue flowers. Determine the probability that a random sample of ten bluebell plants includes exactly one with white flowers.

3 In a telephone poll 22% of the respondents believed in astrology and 78% did not. Assuming that the same proportions apply to the whole population, find the probability that in a random sample of 10 people, less than 20% believe in astrology.
Comment on the validity of the extrapolation from the poll to the population.

4 There are 15 students in a class.
Assuming that each student is equally likely to have been born on any day of the week, find the probability that three or fewer were born on a Monday.
Find also the probability that four or more were born on a Tuesday.

5 Two parents each have the gene for cystic fibrosis. For each of their children, the probability of developing cystic fibrosis is $\frac{1}{4}$. If there are four children, find the probability that exactly two develop cystic fibrosis.

6 A pair of dice is thrown 20 times.
Find the probability of getting a double six at least 3 times.

7 When the Romans decimated a population they lined up the men and executed every tenth man. Six brothers stood in random places in the line. Find the probability that:
  (i) none were executed,
  (ii) four or more escaped execution.

8 A large box contains a mixture of three different types of bolt, in equal numbers. Another box contains the nuts for the bolts. Each nut only fits a bolt of the same type. A nut and a bolt are chosen at random and checked to see if they match (i.e. they are of the same type). The process is repeated 12 times. Find the probability that more than 4 matches are obtained.

9 Driving to work I have to negotiate three sets of traffic lights. I have observed that each of these shows green for 0.45 of the time, red for 0.45 of the time and amber for the remaining time.
Assuming that the colours of the traffic lights are independent of one another and of the time at which I reach them, determine the probability that exactly two of the lights force me (a law-abiding citizen) to stop (by showing either amber or red).

10 The characters in a film are classed as being either 'Good', 'Bad', or 'Ugly'. The proportions in these classes are, respectively, 0.4, 0.4 and 0.2. Seven of the characters have red hair. Assuming that class and hair colour are independent, determine the probability that exactly two of these characters are 'Ugly'.

Project

Cars provide a very convenient set of easily collected data with which to test how well a binomial model works. Suppose we define a success to be 'a number plate in which the last digit is a 3, a 6 or a 9'. Assuming that all 10 digits from 0 to 9 are equally likely, the probability of a success is therefore 0.3. In a sample of five cars the probability of observing, for example, 3 successes, should be:

$$\binom{5}{3}(0.3)^3(0.7)^2 = 0.132 \text{ (to 3 d.p.)}$$

Does it really work out like this? To find out, we noted the last digits of a sequence of 200 cars that passed by.

The first car that passed was an old banger. Its number was GJG 1944, which ends in a '4' – an immediate failure. After 15 cars had passed our records (the last numbers) looked like this:

Groups of five cars	4, 9, 1, 7, 4	3, 5, 5, 6, 0	4, 4, 2, 0, 8
Numbers of successes	1	2	0

When all the data had been collected the numbers of successes that we had obtained were as follows:

1,2,0,0,0   2,0,3,1,0   1,1,2,2,4   0,1,1,1,2
3,2,0,1,2   0,1,1,0,1   2,3,4,2,1   0,1,2,1,1

Our next step was to summarise the values using a tally chart, so that we could subsequently tabulate the results as follows:

Number of successes	0	1	2	3	4	5
Observed frequency	10	15	10	3	2	0
Relative frequency	0.250	0.375	0.250	0.075	0.050	0.000
Theoretical probability	0.168	0.360	0.309	0.132	0.028	0.002

All in all the model has not done too badly! The largest observed proportion corresponds to the case '1' as predicted, and the fact that we never observed a '5' should not be a surprise.

This project can easily be varied. For example, the value of n need not be five. Similarly, the value of p can easily be altered: for example, we could set p to be 0.15 by defining a 'success' to be a number plate ending in some number between '00' and '14', inclusive – ignoring single digit number plates.

Decide on values for n and p and collect some of your own. If class results are pooled together, then (assuming that the class members obtained independent samples) the agreement with the theoretical model should be even better.

Practical _____

We have seen that coin-tossing provides a simple example of a binomial situation. If a fair coin is tossed six times, and the random variable X denotes the number of heads obtained, then:

$$P_r = \binom{6}{r} (0.5)^r (0.5)^{6-r} \qquad r = 0, 1, \ldots, 6$$

$$= \binom{6}{r} (0.5)^6$$

the probabilities of the various values of r are:

Outcome, r	0	1	2	3	4	5	6
Probability (to 3 d.p.)	0.016	0.094	0.234	0.313	0.234	0.094	0.016

Toss a coin six times and record the number of heads, r, that you obtain. Repeat a further nine times and compare the relative frequencies for your outcomes with the theoretical probabilities. The resemblance is unlikely to be close since ten observations is a very small sample.

Combine your results with those of the rest of the class to get a larger amount of data. You should find the overall class results closely resemble those predicted by the binomial distribution.

In Chapter 18 we shall see how to use precise methods to test the goodness of fit between a theoretical distribution and an observed set of data.

Practical

Take out one suit from a pack of cards. Shuffle these cards and choose one card at random. Replace the card and repeat a further four times. Record the number of times (out of the five) that you obtain a court card (a Jack, Queen or King). For example, suppose the original card is a 7, and the next four are respectively 9, 3, King and 7. A court card has occurred on just one of the five occasions, so the outcome of this experiment is a '1'.

Repeat the entire process to obtain a total of twenty observations, each with a value between 0 and 5, inclusive. Combine your results with those of your neighbours in class and calculate the relative frequencies of the outcomes in your combined sample of results. Calculate the theoretical probabilities for this situation and compare them with your relative frequencies.

In the above experiment the cards were replaced after their values had been noted. Repeat the experiment without replacing the cards. This is most easily done by choosing five cards from the collection of thirteen and noting the number of court cards.

Compare your results with those obtained previously.

Calculator practice

If your calculator has a random number generator facility and can be set to show a fixed number of decimal places, then the generator can be tested for randomness as follows. Set the display to show (say) 6 decimal places. Generate a random number and count the number of 9s that it contains. The number of 9s is an observation from a binomial distribution with $n = 6$ and $p = 0.1$. Repeat the process a further 99 times, summarise your results using a tally chart and compare the observed proportions of the outcomes 0 to 6 with those predicted by the binomial model. If they appear to be very different then either something (your calculations?) is wrong or you have been very unlucky!

Computer project

With a computer it is easy to write a program that will both generate the random numbers (as in the previous calculator project) and count the number of occurrences of 9s (or anything else that takes one's fancy). The advantage of the computer is that it can do this a really large number of times. Advanced programmers will arrange for the output to list both the observed proportions and the theoretical probabilities. If a graphical display is available then pictures can also be created.

9.2 Notation

To save having to write: 'The random variable X has a binomial distribution. There are n independent trials. The probability of a "success" is p for each trial', we write:

$$X \sim \mathrm{B}(n, p)$$

Here, the symbol $\sim$ means 'has distribution' and 'B' is used as a shorthand for 'binomial'.

The quantities n and p are called the **'parameters'** of the distribution; they are the quantities whose values are required in order to specify the distribution completely.

9.3 'Successes' and 'failures'

Some good news! It does not matter which of the two possible outcomes we think of as being a 'success' – the calculations will be the same. For example, suppose I play a game of chance with an opponent and suppose my probability of winning is p. Obviously my 'successes' are my opponent's 'failures' and vice versa.

	Me	Opponent
P(success)	p	$q = 1 - p$
Number of successes	r	$n - r$

▼   ▼

Example 5

In Example 3, when we required the probability of observing four Japanese cars in a random sample of 50 cars, we defined a 'success' to be 'a Japanese car'. Suppose instead that we define a 'success' to be 'a *non*-Japanese car'. Thus n is 50, the probability of a 'success', p, is 0.95 and the value of r, the required number of 'successes', is now 46. The required probability is:

$$P_{46} = \binom{50}{46} (0.95)^{46}(0.05)^4 = 0.136 \text{ (to 3 d.p.)}$$

which is the same value that we obtained previously.

▲   ▲

9.4 The shape of the distribution

The shape of the binomial distribution depends upon the value of p. When $p = \frac{1}{2}$, this means that a 'success' is just as likely as a 'failure'. So, for example, the probability of obtaining two 'successes' (and hence $n - 2$ 'failures') is equal to the probability of obtaining two 'failures' (and hence $n - 2$ 'successes'); when $p = \frac{1}{2}$ the binomial distribution is symmetric. For other values of p the distribution is asymmetric (skewed), with a mode near np (see the examples in the diagram).

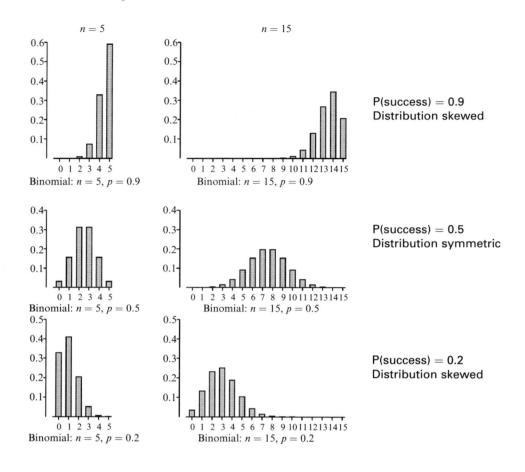

P(success) = 0.9
Distribution skewed

P(success) = 0.5
Distribution symmetric

P(success) = 0.2
Distribution skewed

Project _____

An important requirement for town planning is a knowledge of the type of traffic that uses the principal roads. In particular, planners need to know the proportion of vehicles that are cars (as opposed to lorries, vans, etc).

Suppose that you are part of the local transport authority. Go to a busy road junction and record the identities of the vehicles that pass you. For every car that passes record a 1, and for every other vehicle record a 0. Set out your records carefully in groups of ten vehicles, so that it might look like this:

1111110111   1100111011   1000111110....

Continue until you have recorded the identities of 200 vehicles. Now count the numbers of cars in each group of 10 vehicles. In the example above, the three groups contained 9, 7 and 6 cars respectively. Summarise your data, initially using a tally chart, and then using a frequency distribution, which might look like this:

Number of cars in group (r)	10	9	8	7	6
Number of groups with this number of cars	8	5	4	2	1
Relative frequencies	0.40	0.25	0.20	0.10	0.05

Now count the total number of cars that you have observed. In the example above, this is:

$$(10 \times 8) + (9 \times 5) + (8 \times 4) + (7 \times 2) + (6 \times 1) = 177$$

Finally determine your personal estimate of p, the probability of a vehicle being a car. In the example this is $\frac{177}{200} = 0.885$. You can now calculate the theoretical probabilities of seeing 10, 9, 8, ... cars in a group of 10 vehicles, based on your personal estimate of p. In our example the binomial model predicts that the proportion of groups consisting of all cars would be:

$$\binom{10}{10}(0.885)^{10}(0.115)^0 = 0.295 \text{ (to 3 d.p.)}$$

while the proportion with nine cars will be:

$$\binom{10}{9}(0.885)^9(0.115)^1 = 0.383 \text{ (to 3 d.p.)}$$

and so on.

An extension to this project is to compare (using, for example, a paired bar chart) the theoretical probabilities with your observed relative frequencies. If your relative frequencies for the small values of r are unexpectedly large then this suggests that lorries (and buses!) travel in convoys.

Naturally your personal estimate of p will not be very accurate. Therefore, as a final task, the comparisons may be repeated using the pooled data from the entire class. (For this to be sensible, the observations taken by each class member should be independent of one another, e.g. at slightly different times at the same road junction.) Does it appear that the binomial model is appropriate, or do 'non-cars' (and cars) appear to cluster?

9.5 Calculating binomial probabilities

The calculation of binomial probabilities can be simplified by noting a relation between them. We have:

$$P_r = \left(\frac{n}{1}\right)\left(\frac{n-1}{2}\right)\cdots\left(\frac{n-r+1}{r}\right)p^r q^{n-r}$$

and, hence, replacing r by $r-1$ in the above, for $r > 0$:

$$P_{r-1} = \left(\frac{n}{1}\right)\left(\frac{n-1}{2}\right)\cdots\left(\frac{n-r+2}{r-1}\right)p^{r-1} q^{n-r+1}$$

We can rewrite P_r as follows:

$$P_r = \left(\frac{n-r+1}{r}\right)\left(\frac{p}{q}\right)\left\{\left(\frac{n}{1}\right)\left(\frac{n-1}{2}\right)\cdots\left(\frac{n-r+2}{r-1}\right)p^{r-1} q^{n-r+1}\right\}$$

$$= \left(\frac{p}{q}\right)\left(\frac{n-r+1}{r}\right)P_{r-1}$$

We have therefore deduced a convenient **recurrence formula** for calculating a set of binomial probabilities relatively quickly:

$$P_r = \left(\frac{p}{q}\right)\left(\frac{n-r+1}{r}\right)P_{r-1} \tag{9.2}$$

Notes

- ♦ The recurrence formula is very efficient, but be warned! Once an error has been introduced (e.g. by pressing the wrong button on the calculator!) *all* the subsequent probabilities will be in error. It is therefore advisable to check the last value calculated by comparing it with the value obtained from Equation (9.1).
- ♦ It is usually best to begin by calculating one of the *largest* individual probabilities. This will be associated with a value of r near to np. Sometimes, therefore, the recurrence relation (9.2) will need to be rearranged:

$$P_{r-1} = \left(\frac{q}{p}\right)\left(\frac{r}{n-r+1}\right)P_r \tag{9.3}$$

Example 6

The random variable X has a binomial distribution with parameters $n = 20$ and $p = 0.4$.
Determine $P(X \leqslant 3)$.

Since $np = 8$, we commence by calculating $P(X = 3)$, and then use Equation (9.3). The recurrence formula includes a term, $\dfrac{q}{p}$, that is independent of r. When using a calculator with a memory, the first step is to calculate and store the value of this quantity. In the present case the ratio $\dfrac{0.6}{0.4}$ is equal to 1.5. We now calculate:

$$P_3 = \binom{20}{3}(0.4)^3(0.6)^{17} = 0.012\,349\,7$$

$$P_2 = 1.5 \times \frac{3}{18} \times P_3 \quad = 0.003\,087\,4$$

$$P_1 = 1.5 \times \frac{2}{19} \times P_2 \quad = 0.000\,487\,5$$

$$P_0 = 1.5 \times \frac{1}{20} \times P_1 \quad = 0.000\,036\,6$$

The required probability is the sum of the above which is equal to 0.016 (to 3 d.p.).

Computer project ——————————————————————

Computer programmers wanting a really stiff problem are invited to write a program to do the following:

    1 *Read in values of n and p provided by the user.*
    2 *Print out the values of r and P_r for all cases where $P_r > 0.001$.*

The above is harder than it looks because $\binom{n}{r}$ will often be very large, while $p^r q^{n-r}$ will be correspondingly very small. The solution is not to calculate either quantity! Instead, calculate their logarithms and convert back to probabilities after adding the logarithms. It is also efficient to start with a value of r near to np and to work outwards from this value using Equations (9.2) and (9.3) or their logarithmic equivalents, stopping when the probabilities become negligible.

Exercises 9c ——————————————————————

1 The random variable X has a B(6, 0.2) distribution.
Determine P_4.

2 $X \sim B(n, 0.6)$.
Given that $P_0 = 0.0256$, determine the value of n.

3 Starting by finding P_0, use the recurrence relation for a binomial distribution with parameters n and p to calculate (either directly or by programming your calculator) P_1, P_2, P_3, P_4, P_5:
    (i) in the case $n = 5$, $p = \frac{1}{5}$,
    (ii) in the case $n = 9$, $p = \frac{1}{4}$.

4 Starting by finding P_n, use the recurrence relation for a B(n, p) distribution to calculate (either directly or by programming your calculator) $P_{n-1}, P_{n-2}, P_{n-3}$:
(i) in the case $n = 6$, $p = 0.7$,
(ii) in the case $n = 10$, $p = 0.9$.

5 Find P_8 for a B$(25, \frac{1}{3})$ distribution. Use the recurrence relation to find P_7 and P_9.

6 Sally tosses a fair coin 3 times and Parminder tosses a fair coin 2 times. Give a reason why the total number of heads obtained by both of them has a binomial distribution with $n = 5$, $p = \frac{1}{2}$.

Generalise this result to state the distribution of the sum of two independent random variables having binomial distributions with the same value of p and values of n equal to n_1 and n_2 respectively.

7 For a B(20, 0.3) distribution, show that $P_4 < P_5 < P_6$ and $P_6 > P_7 > P_8$.
Hence determine the mode of the distribution.

8 A binomial distribution has parameters n and p.
Show that if $P_r = P_{r-1}$ then $r = (n+1)p$.
Show also that if $P_r > P_{r-1}$ then $r < (n+1)p$, and if $P_r < P_{r-1}$ then $r > (n+1)p$.

Suppose that $(n+1)p$ is not an integer and that i is the integer part of $(n+1)p$ (so that $i < (n+1)p < i+1$).
Deduce that the mode of the distribution is i.
Deduce also that if $r \leqslant i$ then $P_r > P_{r-1}$, (i.e. P_r increases as r increases) and if $r > i$ then $P_r < P_{r-1}$ (i.e. P_r decreases as r increases).
For the case where $(n+1)p$ is an integer i, deduce that $P_i = P_{i-1}$ and that the distribution has two modes $i - 1$ and i.
Deduce also that P_r increases as r increases if $r < i - 1$ and decreases if $r > i$.

9.6 Tables of binomial distributions

Tables of binomial distributions may take any of the following forms:

(i) tables of P$(X = r)$,
(ii) tables of P$(X \geqslant r)$,
(iii) tables of P$(X \leqslant r)$.

Tables vary from book to book (tables of the third form are given in the Appendix, p.617). The tables recommended or supplied by the various examination boards also vary. You should be careful to become familiar with the tables relevant to your examination: there is no point in wasting time in calculating a value if it is in the tables supplied!

Here is an extract from a table (of the third type) that gives the values of P$(X \leqslant r)$:

n	r	p 0.10
5	0	0.5905
	1	0.9185
	2	0.9914
	3	0.9995
	4	
	5	

The table gives the cumulative probabilities correct to four decimal places. For any combination of n and p, once a value greater than 0.999 95 has been reached it will be shown in the table as a blank.

In the fragment of table given we see that, if $X \sim$ B$(5, 0.1)$, then P$(X \leqslant 2) = 0.9914$.

In order to use the tables to find probabilities for individual values of r, or for other types of inequality, we need the following relations:

$$P(X < r) = P(X \leqslant r - 1)$$
$$P_r = P(X = r) = P(X \leqslant r) - P(X \leqslant r - 1)$$
$$P(X > r) = 1 - P(X \leqslant r)$$
$$P(X \geqslant r) = 1 - P(X \leqslant r - 1)$$

Notes

- It is easy to confuse $P(X < r)$ with $P(X \leqslant r)$, so questions should always be read very carefully.
- It is important to be familiar with the tables that are available for your examinations. The tables given in this book are just a convenience!
- In this book the values of p in the table range from 0.05 to 0.5 only. For values of p greater than 0.5 we must interchange the definitions of success and failure. Thus, if $Y \sim B(n, q)$ and $q = 1 - p$, then:

$$P(X = r) = P(Y = n - r)$$

so that, for example:

$$P(X \geqslant r) = P(Y \leqslant n - r)$$

- Inevitably the tables do not provide for *every* combination of n and p. If a value is required that cannot be obtained directly from the tables then, if possible, it should be calculated from the formula rather than by interpolation in the tables (since this may not be very accurate).

Example 7

Given that $X \sim B(5, 0.1)$ find (i) $P(X < 2)$, (ii) $P(X = 2)$, (iii) $P(X > 2)$, (iv) $P(X \geqslant 2)$.

(i) $P(X < 2) = P(X \leqslant 1) = 0.9185$
(ii) $P(X = 2) = P(X \leqslant 2) - P(X \leqslant 1) = 0.9914 - 0.9185 = 0.0729$
(iii) $P(X > 2) = 1 - P(X \leqslant 2) = 1 - 0.9914 = 0.0086$
(iv) $P(X \geqslant 2) = 1 - P(X \leqslant 1) = 1 - 0.9185 = 0.0815$

Example 8

Given that $X \sim B(5, 0.9)$, determine (i) $P(X \geqslant 2)$, (ii) $P(X \leqslant 2)$.

In this case the value of p is 0.9, which is greater than 0.5, so we must work instead with Y, where $Y \sim B(5, 0.1)$.
(i) $P(X \geqslant 2) = P[Y \leqslant (5 - 2)] = P(Y \leqslant 3) = 0.9995$
(ii) $P(X \leqslant 2) = P(Y \geqslant 3) = 1 - P(Y \leqslant 2) = 1 - 0.9914 = 0.0086$

Exercises 9d

Use the tables that you will use in your examination, or the tables in the Appendix at the back of this book, to answer the following questions.

1 Given that $X \sim B(8, 0.3)$, find (i) $P(X \leqslant 4)$, (ii) $P(X > 6)$.

2 Given that $X \sim B(10, 0.4)$, find (i) $P(X \geqslant 7)$, (ii) $P(X = 6)$, (iii) $P(X < 5)$.

3 Given that $X \sim B(15, 0.7)$, find (i) $P(X \geqslant 9)$, (ii) $P(X \leqslant 11)$.

4 Given that $X \sim B(12, 0.6)$, find $P(5 \leqslant X \leqslant 8)$.

5 When serving at tennis, the probability that Holly Hitter gets the first service in court is 30%. If the first service is a fault (i.e. does not go in court), there is a second service and the probability that the second service goes in court is 90%. Find the probability that out of 20 first services more than 10 go in court.

Show that the probability of a double fault (i.e. neither service goes in court) is 0.07.

Find the probability that out of 20 (combined) services more than 3 are double faults.

6 University student Joe Sleepwell often misses 9 o'clock lectures through oversleeping. The probability that he oversleeps is 0.4.
Find the probability that, in a nine-week term, with two 9 o'clock lectures each week, he misses more than half of them.

9.7 The expectation and variance of a binomial distribution

We want to find the values of $E(X)$ and $Var(X)$, where $X \sim B(n,p)$. It is a little difficult to calculate these quantities using the formula for $P(X = r)$. Instead we note that:

$$X = Y_1 + Y_2 + \cdots + Y_n$$

where Y_i is the number of successes (0 or 1) obtained on the ith trial. Now $Y_1, Y_2, \ldots$ are independent Bernoulli random variables of the type studied in Chapter 7. We found there that a Bernoulli random variable with parameter p has expectation equal to p and variance equal to $p(1-p)$. Combining this information with that from Chapter 8 on the expectations of sums of random variables, we have:

$$\begin{aligned} E(X) &= E(Y_1 + Y_2 + \cdots + Y_n) \\ &= E(Y_1) + E(Y_2) + \cdots + E(Y_n) \\ &= (p + p + \cdots + p) \\ &= np \end{aligned}$$

Similarly, writing q for $(1-p)$:

$$\begin{aligned} Var(X) &= Var(Y_1) + Var(Y_2) + \cdots + Var(Y_n) \\ &= pq + pq + \cdots + pq \\ &= npq \end{aligned}$$

Thus:

a random variable having a B(n,p) distribution has expectation np and variance npq.

Later results (Section 14.4) show that (if n is reasonably large, and p and q are not too small) on about 95% of occasions the observed value of a random variable X having a B(n, p) distribution will lie in the range: mean ± 2 standard deviations, i.e. in the range $np \pm 2\sqrt{npq}$.

Note
- The Bernoulli distribution is really a special case of the binomial distribution in which $n = 1$.

Example 9

A very lazy candidate has done no revision for his multiple-choice statistics exam and guesses the answer to each of the 40 questions. Given that each question offers four alternative answers, only one of which is correct, determine the mean and variance of X, the number of correct answers obtained.

———

The probability, p, that the candidate guesses the correct answer to a question is $\frac{1}{4}$. The situation is binomial, since, for each question, the candidate is either correct or not correct. Thus $X \sim B(40, \frac{1}{4})$. Hence:

$$E(X) = np = 40 \times 0.25 = 10$$

and:

$$Var(X) = npq = E(X) \times 0.75 = 7.5$$

Exercises 9e

1 Determine the expectation and variance of a binomial random variable X for which $n = 50$ and $p = 0.2$.

2 Two boys are throwing skimmers. The probability that a skimmer thrown by Alec will hop 5 times (a success!) is 0.2, whereas for Bill the probability is 0.1. Both boys throw 10 skimmers. Determine:
  (i) the expectation and variance of the number of successes obtained by Alec,
  (ii) the mean and variance of the number of successes obtained by Bill,
  (iii) the mean and variance of the total number of successes obtained by the two boys.

3 A die is thrown 10 times. Let X be the number of sixes obtained.
  Find μ, the expected number of sixes.
  Find also $Var(X)$ and $P(X < \mu)$.

4 A motorist making a regular journey to work finds that she is delayed at a particular level crossing once in five journeys, on average. Using a binomial model, find the expected number of journeys that are delayed at the level crossing in a month when she makes 22 journeys to work, and find also the probability that she is delayed on fewer than 4 occasions.

  Comment on the appropriateness of the binomial model.

5 Published articles in medical journals indicate that, on average, 35 out of 100 patients having a lumbar puncture will suffer SSH ('Severe Spinal Headache'). Twelve patients are given a lumbar puncture.

Using a binomial model, find the expected number of patients who will suffer SSH, and find also the standard deviation.

Find the probability that four or more of the twelve patients will suffer SSH.

6 The random variable X has a binomial distribution with mean 12 and variance 3. Find $P(X \geqslant 14)$.

7 It is given that $Y \sim B(9, p)$ and that the standard deviation of Y is $\frac{9}{10}$.
  Find the possible values of p.
  For each value of p find $P(Y = 4)$, giving 3 significant figures in your answers.

8 The Post Office claims that 92% of first-class letters are delivered the next day after posting. A company selects 20 letters at random from a large batch of first-class letters in order to determine the number X that were delivered the next day.
  Find the expectation μ and the standard deviation σ of X.
  Find $P(|X - \mu| < 1)$, and $P(|X - \mu| < 2\sigma)$.

9 The random variable X is such that $X \sim B(n, p)$. It is known that $\dfrac{Var(X)}{E(X)} = 0.3$ and that X has mean 10.5.
  Determine the values of n and p.

10 The random variable X is such that $X \sim B(n, 0.5)$.
  Determine the smallest value of n for which the ratio of the standard deviation of X to the mean of X is less than 1 to 10.

9.8 The probability generating function

In Section 8.7 (p.206) we denoted the probability generating function (pgf) of a random variable X taking integer values by $G(t)$, where:

$$G(t) = E(t^X) = \Sigma P_x t^x = P_0 + P_1 t + P_2 t^2 + P_3 t^3 + \cdots$$

If $X \sim B(n, p)$ then:

$$G(t) = q^n + \binom{n}{1} q^{n-1} pt + \binom{n}{2} q^{n-2} p^2 t^2 + \cdots$$

$$= q^n + \binom{n}{1} q^{n-1}(pt) + \binom{n}{2} q^{n-2}(pt)^2 + \cdots$$

$$= (q + pt)^n$$

Alternatively, using the Σ notation:

$$G(t) = \sum_{x=0}^{n} \binom{n}{x} p^x q^{n-x} t^x$$

$$= \sum_{x=0}^{n} \binom{n}{x} (pt)^x q^{n-x}$$

$$= (q + pt)^n$$

Example 10

Suppose that $X \sim B(n, p)$, that $Y \sim B(m, p)$ (so that this second distribution has the same success probability) and that X and Y are independent. Use probability generating functions to determine the distribution of the random variable Z, where $Z = X + Y$.

———

Denote the generating functions of X, Y and Z by $G_X(t)$, $G_Y(t)$ and $G_Z(t)$, respectively. Then:

$$
\begin{aligned}
G_Z(t) &= E(t^Z) \\
&= E(t^{X+Y}) \\
&= E(t^X t^Y) \\
&= E(t^X)E(t^Y) \qquad \text{since } X \text{ and } Y \text{ are independent} \\
&= G_X(t)G_Y(t) \\
&= (q + pt)^n (q + pt)^m \\
&= (q + pt)^{(m+n)}
\end{aligned}
$$

We can recognise this as the pgf of a binomial random variable with parameters $(m + n)$ and p. We have therefore shown that **the sum of two independent binomial random variables having the same value of p also has a binomial distribution**.

   The result is not really news! Effectively Z refers to a sequence of n trials followed by a sequence of m trials. Since p is constant throughout, this just amounts to a sequence of $m + n$ trials.

9.9 The negative binomial distribution

No, we are not going to introduce you to negative probabilities! This distribution is appropriate for a variant of the situations for which a geometric distribution (Section 7.4, p. 174) was appropriate.

Suppose we examine a sequence of outcomes, each of which can be a 'success' with probability p or a 'failure' with probability $1 - p$. We define the random variable X to be the number of outcomes considered up to and including the r th 'success'.

If the r th success occurs on the x th outcome then this implies that the first $(x - 1)$ outcomes included precisely $(r - 1)$ successes. This is the ordinary binomial situation, with probability:

$$\binom{x-1}{r-1} p^{r-1} (1-p)^{(x-1)-(r-1)}$$

We also require that the x th outcome was a success (which has probability p of occurring). Multiplying these probabilities together we get the formula for the negative binomial distribution, namely:

$$P_x = \binom{x-1}{r-1} p^r (1-p)^{x-r} \qquad x = r, (r+1), \ldots$$

We see that the negative binomial distribution has two parameters, r and p. The special case where r equals 1 corresponds to the number of outcomes up to the first event. The distribution becomes:

$$P_x = p(1-p)^{x-1}$$

which is the geometric distribution of Section 7.4. The geometric distribution has expectation $\dfrac{1}{p}$ and variance $\dfrac{1-p}{p^2}$. If we think of the more general negative binomial variable as a sum of r independent geometric variables then it is evident that the negative binomial has expectation $\dfrac{r}{p}$ and variance $\dfrac{r(1-p)}{p^2}$.

Example 11

Cricket captains toss a coin at the beginning of each game. Determine the probability that a captain will require at least 11 guesses before obtaining his third correct guess.

The required probability is 1 minus the probability that he requires between three and ten guesses inclusive. This is:

$$1 - \sum_{x=3}^{10} \binom{x-1}{3-1} \left(\frac{1}{2}\right)^3 \left(1 - \frac{1}{2}\right)^{x-3} = 1 - \sum_{x=3}^{10} \frac{(x-1)(x-2)}{2} \left(\frac{1}{2}\right)^x$$

$$= 1 - \left(\tfrac{1}{8} + \tfrac{3}{16} + \tfrac{6}{32} + \cdots + \tfrac{36}{1024}\right)$$

$$= \tfrac{7}{128}$$

Example 12

A crisp manufacturer randomly allocates prizes worth £10 000 in five bags of crisps, chosen at random from a batch of 5 million.
On average, how many bags will be bought before four of the prize bags have been bought?

We ignore the fact that p changes slightly as bags of crisps are bought and opened! Here $r = 4$ and $p = 0.000\,001$ so the (not unexpected!) answer is $\dfrac{4}{0.000\,001}$, which is 4 million.

Chapter summary

♦ **The binomial distribution** applies to situations in which each outcome is either a 'success' or a 'failure'. If n independent trials each have probability p of being a success and X denotes the number of successes, then:
  - $X \sim \mathrm{B}(n, p)$
  - $\mathrm{P}(X = r) = P_r = \binom{n}{r} p^r q^{n-r} \quad r = 0, 1, \ldots, n$ where $q = (1 - p)$
  - $\mathrm{E}(X) = np \qquad \mathrm{Var}(X) = npq$
  - $\mathrm{G}(t) = (q + pt)^n$
  - Recurrence relation: for $1 \leqslant r \leqslant n - 1$, $P_r = \left(\dfrac{p}{q}\right)\left(\dfrac{n - r + 1}{r}\right) P_{r-1}$

♦ **The negative binomial distribution** is concerned with the number of outcomes required until the rth success is observed.
  - $P_x = \binom{x - 1}{r - 1} p^r (1 - p)^{x-r} \qquad x = r, (r + 1), \ldots$
  - $\mathrm{E}(X) = \dfrac{r}{p}, \ \mathrm{Var}(X) = \dfrac{r(1 - p)}{p^2}$

Exercises 9f (Miscellaneous)

1 The germination of cactus seeds is not easy. From experience Mr Thorn, the expert cactus grower, knows that on average only 40% germinate. An intrepid collector returns from a very dry desert with six seeds of a previously unknown type of cactus.
  (i) Determine the probability that only 1 seed germinates.
  (ii) Determine the most likely number of germinating seeds.

2 A lorry carrying a large number of boxes of eggs is involved in an accident. Each box contains six eggs. After the accident the contents of a random sample of 100 boxes are examined and the numbers of broken eggs (x) are recorded. The numbers of boxes (n) containing various numbers of broken eggs are given in the table below.

x	0	1	2	3	4	5	6
n	31	37	22	7	2	1	0

From this frequency distribution estimate p, the proportion of broken eggs.

Calculate, correct to 1 d.p., the expected frequencies to be expected from a binomial distribution having this value of p.
(Expected frequencies are obtained by multiplying the theoretical probabilities by the sample size.)

3   A company produces electrical components, some of which are defective. The proportion of defectives is usually low, but if the proportion reached 10% then the company would want to know that this had happened in order to adjust the machine. A random sample of n components is therefore examined.

Given that the proportion of defectives currently being produced is indeed equal to 10%, determine an expression, in terms of n, for the probability that the sample contains no defectives.

Denoting this probability by P_0, determine the smallest value of n for which P_0 is less than 5%.

4   There is room for 53 passengers on flight ZJ142. Tickets are sold at £130. If a ticket is sold, but the would-be passenger is unable to fly, the airline pays back none of the money to the passenger. From past experience it is known that only 95% of ticket purchasers actually fly. If a ticket is sold to a passenger, but no seat is available then there is an average cost to the airline of £200.

By calculating the net revenue to an airline that results from its selling N tickets, for values of N from 53 to 57, determine the value of N that maximises the expected revenue.

5   On average, it is found that the failure rate for germination of geranium seeds, sold in packets of ten, is 0.8 seeds per packet. Find
   (*a*) the variance of the number of seeds per packet that fail to germinate,
   (*b*) the probability, to 3 decimal places, that a packet chosen at random will contain more than one seed that fails to germinate.

[ULSEB]

6   A boojum is a rare mammal which inhabits tropical seas and spends most of the time under the water. An ecological expedition suspects that there is a boojum in a certain area and attempts to obtain photographic evidence. The technique used is such that if a boojum is present, the probability that it will be visible on any particular photograph is $\frac{1}{4}$. A member of the expedition takes six photographs. If there is a boojum in the area, show that the probability that it will not be visible on any of the photographs is about 0.178.

Five members of the expedition independently search the area, each taking six photographs by the same technique. What is the probability that at least two of them will succeed in photographing the boojum?   [SMP]

7   A reader of a magazine enters for a competition in the magazine, in which the competitors have to choose the correct answers to a number of questions. There are five suggested answers for each question, but the reader is completely unskilful and selects an answer at random to each question, so that, for each question, the probability of choosing the right answer is $\frac{1}{5}$. For a competition with 12 questions, find the probability of the reader getting more than 3 correct answers, giving three decimal places in your answer.

[UCLES(P)]

8   A rifle competition is entered by teams of four people. Each person in a team fires one shot at the target. The table below shows the number of points for the number of hits by a team.

Number of hits	0	1	2	3	4
Number of points	0	2	4	8	16

For a particular team, each member of the team independently has probability 0.7 of hitting the target. Find:
   (i)   the probability of the team hitting the target r times, for each $r = 0, 1, 2, 3, 4$;
   (ii) the team's expected points score;
   (iii) the team's most likely points score;
   (iv) the probability that the team scores more than 6 points.   [O&C]

9   In each trial of a random experiment the probability that the event A will occur is 0.6 and the probability that the event B will occur is 0.5. It is known that A and B are independent.

   (i)   Calculate the probability that at least one of A and B will occur in a single trial.
   (ii) Using the tables provided, or otherwise, find the probability that in twenty independent trials the event A will occur exactly twelve times; give your answer correct to three decimal places.

[WJEC]

10 (*a*) A class of 16 pupils consist of 10 girls, 3 of whom are left-handed, and 6 boys, only one of whom is left-handed. Two pupils are to be chosen at random from the class to act as monitors. Calculate the probabilities that the chosen pupils will consist of

    (i)   one girl and one boy,

    (ii) one girl who is left-handed and one boy who is left-handed,

    (iii) two left-handed pupils,

    (iv) at least one pupil who is left-handed.

(*b*) The probability of a manufactured item being defective is 0.1. A batch consisting of a very large number of the items is inspected as follows. A random sample of five items is chosen. If this sample contains no defective item then the batch is accepted, while if the sample includes 3 or more defectives it is rejected. If the sample includes either 1 or 2 defectives then a second random sample of five items is chosen from the batch. The batch will be accepted if this second sample contains no defective item and will be rejected otherwise. Calculate, correct to three decimal places, the probabilities that

    (i)   the first sample will result in the batch being accepted,

    (ii) the first sample will result in the batch being rejected,

    (iii) the second sample will be necessary and will result in the batch being accepted.   [WJEC]

10 The Poisson distribution

In the United States there is more space where nobody is than where anybody is. That is what makes America what it is

The Geographical History of America, Gertrude Stein

10.1 The Poisson process

In a Poisson distribution (capital P because the distribution is named after Siméon Poisson, of whom more anon) the random variable is a count of events occurring *at random* in regions of time or space. 'At random' here has a very particular and strict definition: the occurrences of the events are required to be distributed through time or space so as to satisfy the following:

♦ Whether or not an event occurs at a particular point in time or space is independent of what happens elsewhere.
♦ At all points in time the probability of an event occurring within a small fixed interval of time is the same. This also applies to the occurrence of events in small regions of space.
♦ There is no chance of two events occurring at precisely the same point in time or space.

Events that obey these requirements are said to be described by a **Poisson process**. Typically, in a spatial Poisson process, there appear to be haphazardly arranged clusters of points as well as wide-open spaces.

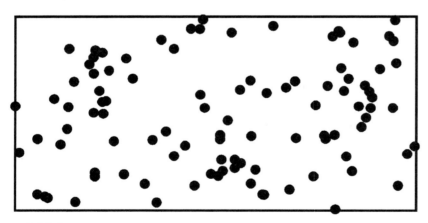

A spatial Poisson process

Examples of real-life Poisson processes are the following:

♦ The points in time at which a given piece of radioactive substance emits a charged particle.
♦ The points in space occupied by the micro-organisms in a random sample of well-stirred water taken from a pond.

A Poisson distribution describes the probabilities of the associated counts:

♦ The number of particles emitted in a minute by the radioactive substance.
♦ The number of micro-organisms in 1 ml of pond water.

There are many other situations in which a Poisson distribution provides a good approximation for small periods of time or for regions of space that are not too large:

- The number of phone calls received on a randomly chosen day.
- The number of cars passing in a randomly chosen five-minute period on a road with no traffic lights or long queues (assuming such a road exists!).
- The number of currants in a randomly chosen currant bun.
- The number of accidents in a factory during a randomly chosen week.
- The number of typing errors on a randomly chosen page of a manuscript.
- The number of daisies in a randomly chosen square metre of playing field.

Two somewhat bloodthirsty classic examples that appear prominently in older Statistics books are:

- The numbers of bomb craters in equal-sized areas of wartime London.
- The numbers of deaths of cavalrymen caused by horse kicks. These data were collected with military precision each year for each of the various Prussian army corps!

The locations of 65 pine saplings growing in a square region of side 5.7 m are typical of what we see in nature. The trees were not deliberately planted in this way and nature's pattern appears entirely haphazard. Looking at the figure, we cannot deduce where the saplings in the neighbouring plots will be found. This is a real-life example of an approximate spatial Poisson process.

The locations of 65 pine saplings

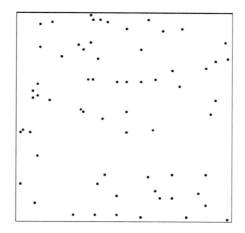

It would be wrong to imagine that all spatial patterns can be attributed to a Poisson process! Two alternatives are provided by the locations of the churches and schools of Norwich. Both diagrams show a rough outline of Norwich and its immediate surroundings. The churches are more densely packed near the centre of the city than in the outskirts. This variation in the rate of occurrence contradicts the constant rate assumption of the Poisson process and reflects the smaller size of Norwich at the time that the majority of the churches were built.

Examples of regular and clustered patterns

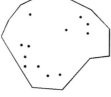

Norwich schools

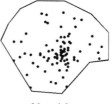

Norwich churches

The Norwich schools are not arranged randomly but are rather regularly placed through the city suburbs so that no children live too far from school. If a new school were built the locations of existing schools would be taken into account, so the independence property of the Poisson process is violated.

1500   1600   1700   1800   1900

Dates of accession of rulers

An example of a time-series of events that is an approximate realisation of a Poisson process is (surprisingly!) provided by considering the dates of accession of new rulers of England and Wales. Although the overall rate of occurrence of these 'events' appears to be constant, an outsider, with no knowledge of history, would be hard pressed to find a pattern. If you doubt this, then forget your own knowledge of history and try to predict the pattern for the 15th century from that shown – if it is even vaguely correct then you probably cheated!

The way in which a Poisson distribution relates to a Poisson process is shown in the diagram. The top figure shows a grid laid over the map of pine sapling locations and the bottom figure shows the resulting counts of the numbers of trees in each grid square. If the trees are truly randomly located, then these will be observations from a Poisson distribution.

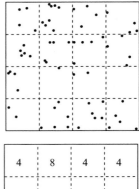

The link between a Poisson process and a set of Poisson counts

Calculator practice ────────────────────────────────

If you have a graphical calculator (or a computer with a suitable graphics package) then you might like to write a short program to display a spatial point process. All that is needed is to draw symbols at locations (x, y), where x is chosen (using the random number generator) at random between 0 and the largest value that will fit on the display screen, and y is chosen similarly. This was the method used to produce the diagram at the start of the chapter.

10.2 The form of the distribution

The formula for a Poisson distribution involves one of the 'magic numbers' of mathematics, the number e = 2.718 28 . . .

$$P_r = \mathrm{P}(X = r) = \frac{\lambda^r \mathrm{e}^{-\lambda}}{r!} \qquad r = 0, 1, 2, \dots \qquad (10.1)$$

where λ, pronounced 'lambda', is a positive number. For $r = 0$ this becomes:

$$P_0 = \mathrm{e}^{-\lambda}$$

since $0! = 1$ and $\lambda^0 = 1$ for all values of λ.

We shall prove later that for a Poisson distribution:

$$\mathrm{E}(X) = \mathrm{Var}(X) = \lambda$$

One way of defining the value of e is via the expression:

$$\mathrm{e}^c = 1 + \frac{c}{1!} + \frac{c^2}{2!} + \frac{c^3}{3!} + \frac{c^4}{4!} + \cdots$$

so that:

$$e = e^1 = 1 + \frac{1}{1} + \frac{1}{2} + \frac{1}{6} + \frac{1}{24} + \cdots$$

We can use the definition of e^c to verify that the probabilities of the Poisson distribution do indeed sum to 1:

$$P_0 + P_1 + P_2 + \cdots = \left(1 + \frac{\lambda}{1!} + \frac{\lambda^2}{2!} + \cdots\right)e^{-\lambda}$$
$$= e^\lambda e^{-\lambda}$$
$$= 1$$

Note

♦ Almost all scientific calculators have a button marked e^x which enables easy calculation of these probabilities without the need to remember the precise value of e. A useful recurrence relationship for the Poisson probabilities is given later in Section 10.4.

Example 1

Between 6 p.m. and 7 p.m., Directory Enquiries receives calls at the rate of 2 per minute.
Assuming that the calls arrive at random points in time, determine the probability that:
(i)   4 calls arrive in a randomly chosen minute,
(ii) 6 calls arrive in a randomly chosen two-minute period.

Since calls arrive at random points in time, a Poisson process is being described.
(i)   Let X be the number of calls that arrive in a randomly chosen minute. Since the mean number of calls per 1-minute period is 2, we put $\lambda = 2$. Hence:

$$P(X = 4) = \frac{2^4 e^{-2}}{4!} = 0.090 \text{ (to 3 d.p.)}$$

(ii) Let Y be the number of calls that arrive in a randomly chosen two-minute period. The mean number of calls per two-minute period is 4, so we put $\lambda = 4$. Hence:

$$P(Y = 6) = \frac{4^6 e^{-4}}{6!} = 0.104 \text{ (to 3 d.p.)}$$

Example 2

In a certain disease a small proportion of the red blood corpuscles display a tell-tale characteristic. A test consists of taking a random sample of 2 ml of a person's blood and counting the number of distinctive corpuscles. A count of five or more is taken to be an indication that the person has the disease. Mrs Wretched has the disease: the mean number of distinctive corpuscles in her blood is 1.6 per ml.
Determine the probability that a randomly chosen sample of 2 ml of her blood will contain five or more of the distinctive corpuscles.
Does the test appear to be a good one?

Let X be the number of distinctive corpuscles in 2 ml of Mrs Wretched's blood. On average, 2 ml of her blood contains 3.2 distinctive corpuscles. Assuming that the corpuscles are haphazardly distributed through her blood, the random variable X has a Poisson distribution with $\lambda = 3.2$. We want $P(X \geqslant 5)$. Since there is no upper bound to the possible values for X, it is 'infinitely' simpler to calculate the probability of the complementary event, $(X \leqslant 4)$:

$$P(X \leqslant 4) = P_0 + \cdots + P_4$$

$$= \left(1 + \frac{(3.2)^1}{1!} + \frac{(3.2)^2}{2!} + \frac{(3.2)^3}{3!} + \frac{(3.2)^4}{4!}\right)e^{-3.2}$$

$$= 0.781 \text{ (to 3 d.p.)}$$

Hence: $P(X \geqslant 5) = 1 - 0.781 = 0.219$

Hence the probability that 2 ml of Mrs Wretched's blood contains five or more of the distinctive corpuscles is 0.219 (to 3 d.p.). There is a chance of nearly 80% that the test will fail to suggest that Mrs Wretched has the disease – the test is not very good.

Exercises 10a

In Questions **1–5**, the random variable X has a Poisson distribution with mean λ.

1 Given that $\lambda = 2$, find (i) $P(X = 0)$,
(ii) $P(X = 1)$, (iii) $P(X = 2)$, (iv) $P(X \leqslant 2)$,
(v) $P(X \geqslant 2)$.

2 Given that $\lambda = 0.5$, find (i) $P(X < 3)$,
(ii) $P(2 \leqslant X \leqslant 4)$, (iii) $P(X \geqslant 3)$.

3 Given that $\lambda = 5$, find (i) $P(X = 5)$,
(ii) $P(X < 5)$, (iii) $P(X > 5)$.

4 Given that $\lambda = 1.4$ find $P(X = 1, 3 \text{ or } 5)$.

5 Given that $\lambda = 2.1$ and $P(X = r) = 0.1890$, find the value of r.

6 The number of currants in a randomly chosen currant bun can be modelled as a random variable having a Poisson distribution with mean 5.6.
Find the probability that a randomly chosen currant bun contains (i) fewer than 4 currants, (ii) more than 4 currants.

7 The number of accidents in a randomly chosen week at a factory can be modelled by a Poisson distribution with mean 0.7.
Find the probability that there are more than two accidents in a randomly chosen week.

8 The number of emergency calls received by a Gas Board in a randomly chosen day can be modelled by a Poisson distribution with mean 3.4.
Find the probability that, in a randomly chosen day, the number of emergency calls received is 5 or more.

9 Buttercups are randomly distributed across a playing field with the probability of a randomly chosen square metre of the field containing precisely r buttercups being:

$$\frac{2^r e^{-2}}{r!}, \qquad r = 0, 1, 2, \ldots$$

Determine the probability that a randomly chosen region of area 0.5 square metres contains precisely one buttercup.

10.3 The properties of a Poisson distribution

The expectation

We asserted in the last section that, if X has a Poisson distribution with parameter λ, then $E(X) = \lambda$. We now prove this result.

$$E(X) = \sum_{r=0}^{\infty} r P_r$$

$$= \sum_{r=0}^{\infty} r \frac{\lambda^r e^{-\lambda}}{r!}$$

$$= \left(0 \times \frac{\lambda^0 e^{-\lambda}}{0!}\right) + \left(1 \times \frac{\lambda^1 e^{-\lambda}}{1!}\right) + \left(2 \times \frac{\lambda^2 e^{-\lambda}}{2!}\right) + \left(3 \times \frac{\lambda^3 e^{-\lambda}}{3!}\right) + \cdots$$

$$= \left(0 + \lambda + \frac{\lambda^2}{1!} + \frac{\lambda^3}{2!} + \frac{\lambda^4}{3!} + \cdots\right) e^{-\lambda}$$

$$= \lambda \left(1 + \frac{\lambda^1}{1!} + \frac{\lambda^2}{2!} + \frac{\lambda^3}{3!} + \cdots\right) e^{-\lambda}$$

$$= \lambda e^{\lambda} e^{-\lambda}$$

$$= \lambda$$

A random variable having a Poisson distribution with parameter λ has expectation λ.

The variance

The usual method of calculating the population variance is by using the equation:

$$\mathrm{Var}(X) = E(X^2) - \{E(X)\}^2$$

However, because of the form of the Poisson distribution, with its denominator of $r! = r(r-1)(r-2)\cdots$, it is easier to work with $E[X(X-1)]$ using the fact that:

$$E[X(X-1)] = E(X^2 - X) = E(X^2) - E(X)$$

Thus:

$$\mathrm{Var}(X) = E[X(X-1)] + E(X) - \{E(X)\}^2$$

We therefore need to calculate $E[X(X-1)]$:

$$E[X(X-1)] = \sum_{r=0}^{\infty} r(r-1) P_r$$

$$= \sum_{r=0}^{\infty} r(r-1) \frac{\lambda^r e^{-\lambda}}{r!}$$

$$= \left(0 + 0 + \lambda^2 + \frac{\lambda^3}{1!} + \frac{\lambda^4}{2!} + \frac{\lambda^5}{3!} + \cdots\right) e^{-\lambda}$$

$$= \lambda^2 \left(1 + \frac{\lambda^1}{1!} + \frac{\lambda^2}{2!} + \frac{\lambda^3}{3!} + \cdots\right) e^{-\lambda}$$

$$= \lambda^2 e^{\lambda} e^{-\lambda}$$

$$= \lambda^2$$

Hence:

$$\mathrm{Var}(X) = \lambda^2 + \lambda - (\lambda)^2 = \lambda$$

A random variable having a Poisson distribution with parameter λ has variance λ.

The shape of a Poisson distribution

When $\lambda < 1$ the distribution has mode at $x = 0$ and is very skewed. As λ increases so the distribution takes on a more symmetrical appearance. Note that the diagrams are truncated – in each case it is *possible* for the Poisson random variable to take a larger value than those shown. However, although the range of possible values is infinite, the results given later in Section 12.12 (p. 343) suggest that in practice 95% of values will lie between $\lambda - 2\sqrt{\lambda}$ and $\lambda + 2\sqrt{\lambda}$ (i.e. in the range: mean $\pm$ 2 standard deviations).

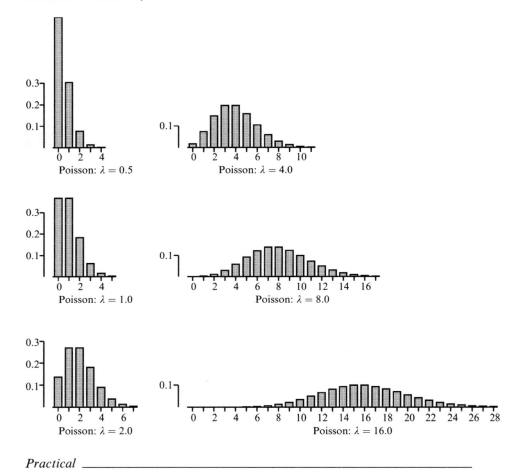

Poisson: $\lambda = 0.5$

Poisson: $\lambda = 4.0$

Poisson: $\lambda = 1.0$

Poisson: $\lambda = 8.0$

Poisson: $\lambda = 2.0$

Poisson: $\lambda = 16.0$

Practical

A chess board and a tube of Smarties™ (or something similar) are required for this project. The idea is to toss the Smarties one at a time on to the chess board and, when all are on the board (you may need several attempts!) to count the numbers of Smarties in each of the 64 squares. Providing the board is reasonably large and you didn't cheat, the arrangement of the Smarties should approximate a spatial Poisson process.

Check by drawing a bar chart of the 64 observations and calculating their mean and variance, which should be reasonably similar.

Project _____

> *Providing a road is not so busy that queues of traffic form, traffic flow may be modelled by a Poisson process. To investigate this, choose a reasonably busy road and count the numbers of cars (or lorries, or bicycles, or whatever) that pass in a particular direction in a period of one minute. Repeat for a complete half-hour. It will be easier to work in pairs, with one person counting and the other timing and recording. Choose a dry day and warm clothing!*
> *Represent your results using a bar chart.*
> *Calculate the sample mean and variance.*
>
> *If the stream of cars does form a Poisson process then the mean and variance should be quite similar. If the variance is much larger than the mean then this will suggest that there is appreciable clumping of the cars due to slower cars holding up faster ones or to the presence of a nearby roundabout or traffic light.*

Computer project _____

> *Write a computer program to generate a sequence of 10 000 random numbers, each with a value between 0 and 1. Test each number to see if its value is less than 0.002 (a 'success'). If the mth number in the sequence is a 'success' record the value of m. You should end up with about 20 'successes'. (Why?).*
> *Now draw a line 10 inches long on a piece of paper, with one end corresponding to the start of your simulation. Suppose your first 'success' occurred on the 876th random number. Illustrate this by placing a dot 0.876 inches (approximately!) from the start of the line. Suppose now that the second 'success' is obtained on the 83rd subsequent random number. This will be illustrated by a second dot at a point 0.959 inches (since 876 + 83 = 959) from the start of the line. Continue to add subsequent 'successes' in this way.*
>
> *When all the 'successes' have been marked you will have an illustration of a linear Poisson process somewhat similar to that shown earlier. Computer wizards, whose computers have graphical capabilities, may wish to get the computer to draw this display.*

Calculator practice _____

> *If you have a graphical calculator with programming facilities then the previous computer project can be carried out (more slowly!) on the calculator. You could arrange for each line of the screen to correspond to a single 'time sequence'. When the program stops, your display of alternative time sequences will be indistinguishable from a realisation of a spatial Poisson process.*

10.4 Calculation of Poisson probabilities

Since:

$$P_r = \frac{\lambda^r \mathrm{e}^{-\lambda}}{r!}$$

and, for $r \geqslant 1$:

$$P_{r-1} = \frac{\lambda^{r-1} \mathrm{e}^{-\lambda}}{(r-1)!}$$

we can write:

$$P_r = \frac{\lambda}{r} \left\{ \frac{\lambda^{r-1} e^{-\lambda}}{(r-1)!} \right\}$$

$$= \frac{\lambda}{r} P_{r-1}$$

This gives us a convenient **recurrence formula** for calculating a set of Poisson probabilities relatively quickly by starting with $P_0 = e^{-\lambda}$.

Example 3

The random variable X has a Poisson distribution with mean 1.7. Determine $P(X > 3)$.

_____

It is much simpler to calculate first the probability of the complementary event, $X \leqslant 2$. We start with P_0 and use the recurrence formula:

$$P_0 = e^{-1.7} \quad (= 0.182\ 683\ 5)$$

$$P_1 = \frac{1.7}{1} P_0 \quad (= 0.310\ 562\ 0)$$

$$P_2 = \frac{1.7}{2} P_1 \quad (= 0.263\ 977\ 7)$$

$$P_3 = \frac{1.7}{3} P_2 \quad (= 0.149\ 587\ 4)$$

In practice, using a calculator with several memories, virtually all the calculations could take place on the machine in a single sequence of operations. The intermediate values would be summed in one of the memories and are given here to an excessive number of decimal places to emphasise that premature approximation could lead to substantial inaccuracy in the final answer.

$$P(X > 3) = 1 - (P_0 + P_1 + P_2 + P_3)$$

$$= 1 - 0.906\ 810\ 6$$

$$= 0.093 \text{ (to 3 d.p.)}$$

Calculator practice _____

Write a short program on your calculator to reproduce the calculations given above.

Exercises 10b _____

In Questions **1–7**, the random variable X has a Poisson distribution with mean λ.

1 Given that $\lambda = 2$, calculate P_0 and use the recurrence relation to calculate P_r for $r = 1, 2, 3, 4$. Check by calculating P_4 directly.

2 Given that $\lambda = 5$, calculate P_4 directly and use the recurrence relation to calculate P_r for $r = 5, 6, 7$.

3 Given that $\lambda = 0.5$, calculate P_3 directly and use the recurrence relation (backwards) to calculate P_r for $r = 2, 1, 0$. Check by calculating P_0 directly.

4 Find λ, given that $P_4 = 2P_5$.

With this value of λ find $\dfrac{P_6}{P_8}$.

5   Given that $\lambda = 0.5$, give a reason why:

$$P_0 > P_1 > P_2 > \ldots$$

Give a reason why this result holds for any λ such that $0 < \lambda < 1$.

6   Given that $\lambda = 10$, show that $P_9 = P_{10}$.
Give a reason why P_r increases as r increases from 0 to 9, and decreases as r increases from 10.

Generalise this result to the case where λ is any integer.

7   Given that λ lies between the successive integers n and $n + 1$, show that P_r increases as r increases from 0 to n and thereafter P_r decreases as r increases from n and deduce that the most probable value of X is n.

10.5 Tables for Poisson distributions

As with tables of the binomial distribution, tables for Poisson distributions may occur in a variety of forms. The tables provided in the Appendix (p. 618) are tables of $P(X \leqslant r)$, for various values of λ. Our tables give probabilities correct to four decimal places. Cumulative probabilities exceeding 0.999 95 are omitted from the table. A (rearranged) extract from the table is given below:

| λ | r | | | | | | | | |
	0	1	2	3	4	5	6	7	8
1.4	0.2466	0.5918	0.8335	0.9463	0.9857	0.9968	0.9994	0.9999	

This table refers to the case $\lambda = 1.4$ and shows $P(X \leqslant r)$ for $r = 0, 1, \ldots, 7$. Thus $P(X \leqslant 4) = 0.9857$. There is no entry corresponding to $r = 8$ since the cumulative probability exceeds 0.999 95 and is therefore 1.0000 to 4 d.p.

In order to use the tables to find probabilities for individual values of r, or for other types of inequality, we need the following relations:

$$P(X < r) = P(X \leqslant r - 1)$$
$$P_r = P(X = r) = P(X \leqslant r) - P(X \leqslant r - 1)$$
$$P(X > r) = 1 - P(X \leqslant r)$$
$$P(X \geqslant r) = 1 - P(X \leqslant r - 1)$$

Notes
- ◆ It is easy to confuse $P(X < r)$ with $P(X \leqslant r)$, so questions should always be read very carefully!
- ◆ Inevitably the tables do not provide for *every* value of λ. If a probability is required for a value of λ that is not included in the tables then, if possible, it should be calculated from the formula rather than by interpolation in the tables.
- ◆ It is important to be familiar with the tables that are available for your examinations. The tables given in this book are just a convenience!

Example 4

Tadpoles are scattered randomly through a pond at the rate of 14 per litre.
A random sample of 0.1 litre is examined.
What is the probability that it will contain more than 3 tadpoles?

Assuming a Poisson distribution (since the tadpoles are distributed at
random in space) with mean 1.4 per 0.1 litre, we require:

$$1 - (P_0 + P_1 + P_2 + P_3)$$

which is:

$$1 - P(X \leqslant 3) = 1 - 0.9463$$

and so the probability that the sample contains more than 3 tadpoles is
0.054 (to 3 d.p.).

Example 5

The random variable Y has a Poisson distribution with mean 1.4. Determine
the probability that Y takes a value greater than 4, but less than 7.

The question is asking for $P_5 + P_6$.
Using the cumulative tables we calculate this as:

$$P(Y \leqslant 6) - P(Y \leqslant 4) = 0.9994 - 0.9857$$

and so the probability that Y takes a value greater than 4, but less than 7,
is 0.014 (to 3 d.p.).

Example 6

Use tables of cumulative Poisson probabilities to determine $(3 < X \leqslant 7)$,
where X has a Poisson distribution with mean 1.4.

The question requires $P_4 + P_5 + P_6 + P_7$, which we calculate as:

$$P(X \leqslant 7) - P(X \leqslant 3) = 0.9999 - 0.9463.$$

The required probability is 0.054 (to 3 d.p.).

Exercises 10c

In Questions **1–5**, the random variable X has a
Poisson distribution with mean λ. Use the tables
that you will use in your examination, or the tables
in the Appendix at the back of this book, to answer
the following questions.

1 Given that $\lambda = 3$ find (i) $P(X \leqslant 5)$, (ii) $P(X < 7)$.

2 Given that $\lambda = 0.9$ find (i) $P(X \geqslant 3)$,
(ii) $P(X > 4)$.

3 Given that $\lambda = 1.2$ find (i) $P(2 < X < 5)$,
(ii) $P(2 \leqslant X \leqslant 5)$.

4 Find λ given that $P(X \leqslant 5) = 0.9896$.

5 Find λ given that $P(X > 4) = 0.0527$.

6 Weak spots occur at random in the
manufacture of a certain cable at an average
rate of 1 per 100 metres.
If X represents the number of weak spots in
100 metres of cable, write down the distribution
of X.
Lengths of this cable are wound onto drums.
Each drum carries 50 metres of cable.
Find the probability that a drum will have 3 or
more weak spots.
A contractor buys five such drums.
Find the probability that two have just one
weak spot and the other three have none.

[AEB(P)91]

Siméon Denis Poisson (1781–1840) was a French mathematician who is variously described as lively and extremely hard-working. His principal interest lay in aspects of mathematical physics. His major work on probability was entitled *Researches on the probability of criminal and civil verdicts*. In this long book (over 400 pages) only about one page is devoted to the derivation of the distribution that bears his name! Poisson derived the distribution as a limiting form of the binomial (see below). He is quoted as having said that 'Life is good for only two things: to study mathematics and to teach it'. No comment!

10.6 The Poisson approximation to the binomial

If X has a binomial distribution with parameters n and p, and if n is large and p is near 0, then the distribution of X is closely approximated by a Poisson distribution with mean np.

Notes
- The approximation should only be used when it is not feasible to calculate the required probability exactly.
- The usual guidelines for the use of the approximation are that n should be greater than 50 and that p should be less than 0.1. These are not strict rules. All that can be said with confidence is that:
  - the smaller p, the better.
  - the larger n, the better.

Derivation of the result

We show first that, if p is small, binomial probabilities tend to Poisson probabilities as n increases.

For the binomial distribution with probability of success p and n trials:

$$P_r = \binom{n}{r} p^r q^{n-r}$$

where $q = 1 - p$. The recurrence relation for this distribution was given by Equation (9.2) (p. 225) as:

$$P_r = \frac{(n-r+1)}{r}\left(\frac{p}{q}\right)P_{r-1} = \frac{\left(1 - \frac{r-1}{n}\right)}{r}\left(\frac{np}{q}\right)P_{r-1}$$

We now set $np = \lambda$, and rewrite the relation as:

$$P_r = \frac{\left(1 - \frac{r-1}{n}\right)}{q}\frac{\lambda}{r}P_{r-1}$$

We are now ready to go to the limit! Imagine that n has become huge (equal to N perhaps!) so that:

$$\left(1 - \frac{r-1}{N}\right) \approx 1$$

Since the product of N and p is still equal to λ, p is now very small (equal $\frac{\lambda}{N}$). It follows that:

$$q = (1 - p) \approx 1$$

so that:

$$P_r \approx \frac{\lambda}{r} P_{r-1}$$

with equality in the limit.

In the limit, then:

$$P_r = \frac{\lambda}{r} P_{r-1}$$

which is precisely the recurrence formula used in Section 10.4 (p. 244). Successive applications of the formula give us:

$$P_1 = \frac{\lambda}{1} P_0$$

$$P_2 = \frac{\lambda}{2} P_1 = \frac{\lambda^2}{2!} P_0$$

$$P_3 = \frac{\lambda}{3} P_2 = \frac{\lambda^3}{3!} P_0$$

and so on.

Evidently:

$$P_r = \frac{\lambda^r}{r!} P_0$$

Since $\Sigma P_r = 1$:

$$\sum_{r=0}^{\infty} \left\{ \frac{\lambda^r}{r!} P_0 \right\} = 1$$

$$P_0 \sum \frac{\lambda^r}{r!} = 1$$

$$P_0 e^{\lambda} = 1$$

$$P_0 = e^{-\lambda}$$

Hence:

$$P_r = \frac{\lambda^r}{r!} e^{-\lambda}$$

which is the familiar form of the Poisson distribution. Thus for large n and small p we can expect a Poisson probability to be a good approximation to the corresponding binomial probability.

Example 7

The discrete random variable X has a binomial distribution with $n = 60$ and $p = 0.02$.
Determine $P(X = 1)$ (i) exactly, (ii) using a Poisson approximation.

(i)   The exact binomial probability is given by:

$$P(X = 1) = \binom{60}{1} (0.02)^1 (0.98)^{59} = 0.364 \text{ (to 3 d.p.)}$$

(ii) Since n is quite large and p is small, we can expect the Poisson approximation to be quite accurate. Setting $\lambda = np = 60 \times 0.02 = 1.2$, we have:

$$P(X = 1) \approx \frac{1.2}{1!} e^{-1.2} = 0.361 \text{ (to 3 d.p.)}$$

The approximation is indeed an accurate one.

Example 8

Past experience suggests that 0.4% of peaches show signs of mildew on arrival at the market. Occasionally, if storage conditions are faulty, the proportion of mildewed peaches may be much higher than this. Assuming that the conditions of individual fruit are independent of one another, and that the proportion of mildewed peaches is the usual 0.4%, determine the probability that a carton of 250 individually packed peaches contains more than three that show signs of mildew. What conclusions would you draw if a randomly chosen carton was found to contain 5 mildewed peaches?

This is a binomial situation. The parameters are $n = 250$ and $p = \dfrac{0.4}{100} = 0.004$ (not 0.4 which corresponds to 40%, and would mean that nearly half were mildewed!). The question asks for 'more than three', which means 4, 5, ..., 250. It is much easier to consider the complementary event that there are 0, 1, 2 or 3 mildewed peaches. Although it is feasible to calculate the required probability directly, using the binomial distribution, it is much easier to use the Poisson approximation with $\lambda = np = 250 \times 0.004 = 1$. The individual probabilities are tabulated below:

No. of peaches	0	1	2	3	3 or less
Exact binomial prob.	0.3671	0.3686	0.1843	0.0612	0.9813
Poisson approximation	0.3679	0.3679	0.1839	0.0613	0.9810

There is about a 2% chance that there are more than three peaches showing signs of mildew.

If a randomly sampled carton was found to contain five mildewed peaches then this would strongly suggest that the storage conditions had been faulty.

Practical _____

This is another practical involving the rolling of dice. Each member of the class should roll a die twice (or two dice once) and report a 'success' if two sixes are obtained. The total number of successes for the class should be recorded and the exercise repeated twenty times so as to give twenty observations from a binomial distribution with parameters n (the class size) and $p = \frac{1}{36}$.
Compare the observed relative frequencies with the exact binomial probabilities.
Calculate the approximating Poisson probabilities and compare them with the exact values.

Practical _____

Ordinary packs of playing cards are required for this practical. The event of interest is that a single card drawn from a pack is the Ace of Spades. Each class member should have 26 attempts at striking lucky, with the card chosen being returned to the pack, and the pack being shuffled between attempts.
Since $n = 26$ and $p = \frac{1}{52}$, the approximating Poisson distribution has $\lambda = \frac{1}{2}$. About 60% of the class should not see the Ace of Spades at all, but about 9% (where do these percentages come from?) should see the Ace more than once.

Exercises 10d

In Questions **1–3** use the Poisson approximation to find the required probability concerning the random variable X which has a binomial distribution with parameters n and p.

1 Given that $n = 40$, $p = 0.1$, find (i) $P(X \leqslant 3)$, (ii) $P(X \geqslant 3)$.

2 Given that $n = 100$, $p = 0.02$, find (i) $P(X \geqslant 2)$, (ii) $P(X < 4)$.

3 Given that $n = 55$, $p = \frac{1}{11}$, find $P(3 \leqslant X \leqslant 6)$.

4 Screws are packed in boxes of 200. For each screw the probability that it is faulty is 0.4%. Using a suitable approximation, find the probability that a box contains at most two faulty screws.

5 For a beginner taking photos, the probability that a photo is 'useless' is 0.1 and the probability that it is 'brilliant' is 0.05. The beginner takes 72 photos.
Use a suitable approximation to find the probability that:
(i) at least 3 photos are brilliant,
(ii) at most 3 photos are useless.

6 The proportion of red sports cars is 1 per 200 cars in the country as a whole. There are 500 cars in a car park.
Assuming these to be a random sample from the population, use a Poisson approximation to determine the probability that there are exactly 5 red sports cars in the car park.

7 A rare type of error in the printing of postage stamps is such that in a random sample of 1000 stamps there will be on average 2 stamps displaying the error. Using a Poisson approximation, calculate the probability of there being exactly one stamp displaying the error in a random sample of 100 stamps.
For a sample of this size, state the mean and variance of the number of stamps displaying the error.

8 Thirty digits are taken at random from a table of random numbers. Find the exact binomial probabilities of obtaining (i) one seven, (ii) two sevens, (iii) three sevens. Find the same probabilities using the Poisson approximation and compare the results.

9 A charity runs a prize draw every week, and each person who buys a ticket has a chance of 1 in 1000 of winning the prize. A contributor buys a ticket each week for 50 weeks.
Using a suitable approximation, find:
(i) the probability that she wins at least one prize
(ii) the smallest number of weeks that she must buy tickets in order that the probability of winning at least one prize exceeds 0.9.

10 A machine produces resistors of which 99% are up to standard. They are packed in boxes each containing 200 resistors. Using a suitable approximation, find the probability that a randomly chosen box contains at least 198 resistors that are up to standard.

11 A hockey team consists of 11 players. It may be assumed that, on every occasion, the probability of any one of the regular members of the team being unavailable for selection is 0.15, independently of all the other members. Calculate, giving three significant figures in your answers, the probability that, on a particular occasion,
(i) exactly one regular member is unavailable,
(ii) more than two regular members are unavailable
Taking the probability that more than 3 regular members are unavailable as 0.07, write down, for a season in which 50 matches are played, the expected value of the number of matches for which more than 3 regular members are unavailable.
Use a suitable Poisson distribution to find an approximation for the probability that, in the course of a season, more than 3 regular players will be unavailable at most twice.   [UCLES]

10.7 Sums of independent Poisson random variables

If X and Y are independent Poisson random variables with parameters λ and μ, respectively, then the random variable Z, defined by $Z = X + Y$, is a Poisson random variable with parameter $\lambda + \mu$.

A direct proof of this result is tedious (though a proof using probability generating functions is given in Section 10.9), but the result is obvious once we consider the Poisson process background to the distribution.

A Poisson distribution refers to counts of 'events' scattered at random in time or space. Suppose we have a collection of m red objects and n green objects which are all identical apart from their colour. We scatter these objects at random over a square region of area A. Focusing on the red objects alone we see a spatial Poisson process with rate m per unit area. Likewise, focusing on the green objects, we see a random arrangement with a rate of n per unit area. A colour-blind person would see a combined set of randomly distributed objects at a rate of $(m + n)$ objects per unit area.

Mixing together two random patterns produces another random pattern!

Example 9

An observer is standing beside a road. Both cars and lorries pass the observer at random points in time. On average there are 300 cars per hour, while the mean time between lorries is five minutes. Determine the probability that exactly 6 vehicles pass the observer in a one-minute period.

———————

Since the question refers to 'random points in time' a Poisson distribution is appropriate both for cars and for lorries. The mean rate for lorries is 12 an hour, so the combined rate is 312 vehicles per hour, which corresponds to 5.2 vehicles per minute. The required probability is therefore:

$$\frac{(5.2)^6 e^{-5.2}}{6!} = 0.151 \text{ (to 3 d.p.)}$$

10.8 The National Lottery

This provides an approximate space–time Poisson process! If we draw up a grid having 49 columns and a row corresponding to each lottery draw, then, on colouring the chosen balls, we get an arrangement of coloured balls that has most of the properties of a Poisson process and leads to precisely the pattern of apparent clusters and 'spaces' that one would expect.

Note
  ◆ The major difference from a real Poisson process is the restriction that there are exactly seven coloured balls on each row. A lottery in which the number of balls drawn was itself random would be *really* exciting!

Project ——————————————————————————————

> *Brighten up your life with the National Lottery! Draw up a 49-column grid and, adding a new row for each lottery draw, colour in the balls that are drawn using a different colour for each ball drawn.*
>
> *Each of the seven colours provides an approximately random spatial pattern, as does each aggregate of colours.*

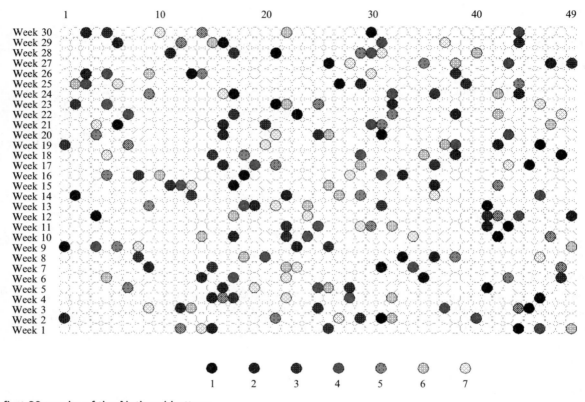

The first 30 weeks of the National Lottery

10.9 The probability generating function

The probability generating function (pgf) of a Poisson distribution with parameter λ is given by:

$$G(t) = E(t^X)$$

$$= t^0 P_0 + t^1 P_1 + t^2 P_2 + \cdots$$

$$= \left(1 \times \frac{\lambda^0 e^{-\lambda}}{0!}\right) + \left(t \times \frac{\lambda^1 e^{-\lambda}}{1!}\right) + \left(t^2 \times \frac{\lambda^2 e^{-\lambda}}{2!}\right) + \cdots$$

$$= \left(1 + \frac{t\lambda}{1!} + \frac{(t\lambda)^2}{2!} + \cdots\right) e^{-\lambda}$$

$$= e^{t\lambda} e^{-\lambda}$$

$$= e^{-\lambda(1-t)}$$

Alternatively, using Σ notation:

$$\begin{aligned} \mathrm{G}(t) &= \mathrm{E}(t^X) \\ &= \sum_{r=0}^{\infty} t^r \frac{\lambda^r \mathrm{e}^{-\lambda}}{r!} \\ &= \mathrm{e}^{-\lambda} \sum_{r=0}^{\infty} \frac{(t\lambda)^r}{r!} \\ &= \mathrm{e}^{-\lambda} \mathrm{e}^{t\lambda} \\ &= \mathrm{e}^{-\lambda(1-t)} \end{aligned}$$

Obviously, if we change the Poisson parameter to μ, all that happens is that the pgf becomes $\mathrm{e}^{-\mu(1-t)}$. Any pgf of this type must be the pgf of a Poisson distribution – we could prove this by expanding the pgf as a power series in t and looking at the coefficient of t^r, which is P_r. Thus, if we find that a distribution has a pgf given by $\mathrm{e}^{-\text{BANANA}(1-t)}$, we can deduce that it is a Poisson distribution with parameter equal to BANANA.

Consider the random variable Z, defined by $Z = X + Y$, where X and Y are independent Poisson random variables with parameters λ and μ, respectively. The pgf of Z is given by:

$$\begin{aligned} \mathrm{E}(t^Z) &= \mathrm{E}[t^{(X+Y)}] \\ &= \mathrm{E}(t^X t^Y) \\ &= \mathrm{E}(t^X)\mathrm{E}(t^Y) \qquad \text{since } X \text{ and } Y \text{ are independent} \\ &= \mathrm{e}^{-\lambda(1-t)} \mathrm{e}^{-\mu(1-t)} \\ &= \mathrm{e}^{-(\lambda+\mu)(1-t)} \end{aligned}$$

But this is the pgf for a Poisson random variable with parameter $(\lambda + \mu)$. We have therefore confirmed our previous assertion that the sum of two independent Poisson random variables is another Poisson random variable.

Note

◆ The *difference* of two independent Poisson random variables does *not* have a Poisson distribution (since negative values will be possible).

Chapter summary

◆ The Poisson distribution refers to the counts of items that occur at **random points in time or space** (a Poisson process).

◆ **Distribution**: $P_r = \dfrac{\lambda^r \mathrm{e}^{-\lambda}}{r!}$ $r = 0, 1, 2, \ldots$

◆ **Mean and variance**: $\mathrm{E}(X) = \mathrm{Var}(X) = \lambda$

◆ **Recurrence relation**: For $r \geqslant 1$, $P_r = \dfrac{\lambda}{r} P_{r-1}$

◆ **Approximation of binomial**: A binomial distribution with parameters n (large) and p (small) may be approximated by a Poisson distribution with parameter np.

◆ **Additivity**: If X and Y are independent Poisson random variables, with parameters λ and μ respectively, then $X + Y$ has a Poisson distribution with parameter $\lambda + \mu$.

◆ The **pgf** of a Poisson random variable with parameter λ is $\mathrm{e}^{-\lambda(1-t)}$.

Exercises 10e (Miscellaneous)

1 The numbers of emissions per minute from two radioactive objects A and B are independent Poisson variables with means 0.65 and 0.45 respectively.
Find the probabilities that:
(i) in a period of three minutes there are at least three emissions from A,
(ii) in a period of two minutes there is a total of less than four emissions from A and B together.

2 The number of customers per hour entering a jeweller's shop has a Poisson distribution. For the first hour after opening the mean is 0.7 per hour and for the next three hours the mean is 1.3 per hour.
Find the probability that there are between 4 and 6 (inclusive) customers entering the shop in the first four hours.

3 In a particular form of cancer, deformed blood corpuscles occur at random at the rate of 10 per 1000 corpuscles.
Use an appropriate approximation to determine the probability that a random sample of 200 corpuscles taken from a cancerous area will contain no deformed corpuscles.
How large a sample should be taken in order to be 99% certain of there being at least one deformed corpuscle in the sample?

4 The number of telephone calls arriving per minute at a small telephone exchange has a Poisson distribution with mean 2.25. Find, correct to three decimal places, the probability that
(i) exactly 2 calls arrive in a minute,
(ii) more than 4 calls arrive in a period of 2 minutes.   [WJEC]

5 The numbers of emissions per minute from two radioactive substances, A and B, are independent and have Poisson distributions with means 2.8 and 3.25, respectively.
Find, correct to three decimal places, the probabilities that in a period of one minute there will be
(i) exactly three emissions from A,
(ii) one emission from one of the two substances and two emissions from the other substance.   [WJEC]

6 Independently for each page of a printed book the number of errors occurring has a Poisson distribution with mean 0.2. Find, correct to three decimal places, the probabilities that
(i) the first page will contain no error,
(ii) four of the first five pages will contain no error,
(iii) the first error will occur on the third page.   [WJEC]

7 (i) The discrete random variable X has probability function given by

$$p(x) = \begin{cases} (\tfrac{1}{2})^x & x = 1, 2, 3, 4, 5, \\ C & x = 6, \\ 0 & \text{otherwise,} \end{cases}$$

where C is a constant.
Determine the value of C and hence the mode and arithmetic mean of X.

(ii) A process for making plate glass produces small bubbles (imperfections) scattered at random in the glass, at an average rate of four small bubbles per $10\,\text{m}^2$.
Assuming a Poisson model for the number of small bubbles, determine to 3 decimal places, the probability that a piece of glass $2.2\text{m} \times 3.0\text{m}$ will contain

(a) exactly two small bubbles,
(b) at least one small bubble,
(c) at most two small bubbles.

Show that the probability that five pieces of glass, each 2.5m by 2.0m will all be free of bubbles is e^{-10}.
Find, to 3 decimal places, the probability that five pieces of glass, each 2.5m by 2.0m, will contain a total of at least ten small bubbles.   [ULSEB]

8 Serious delays on a certain railway line occur at random, at an average rate of one per week. Show that the probability of at least four serious delays occurring during a particular 4-week period is 0.567, correct to 3 decimal places. Taking a year to consist of thirteen 4-week periods, find the probability that, in a particular year, there are at least ten of these 4-week periods during which at least 4 serious delays occur.
Given that the probability of at least one serious delay occurring in a period of n weeks is greater than 0.995, find the least possible integer value of n.   [UCLES]

9 In a certain country it is known that 35% of the adult population have some knowledge of a foreign language. If 10 adults are chosen at random from this population, find the probability that

(i) at least one of those chosen will have some knowledge of a foreign language,

(ii) at most three of those chosen will have some knowledge of a foreign language.

For one particular foreign language, only a very small proportion r% of the adult population have some knowledge of it. It is required to select n adults at random, where n is chosen so that the probability of obtaining at least one adult having some knowledge of the language is to be 0.99, as nearly as possible. Use a suitable Poisson approximation to show that $n \approx \dfrac{460.5}{r}$.

For the case when $r = \frac{1}{2}$ and $n = 921$, find the probability that precisely four adults in the sample will have some knowledge of the language. [UCLES]

10 A biased cubical die is such that the probability of any one face landing uppermost is proportional to the number on that face. Thus, if X denotes the score obtained in one throw of this die,

$$\mathrm{P}(X = r) = kr, \qquad r = 1, 2, 3, 4, 5, 6,$$

where k is a constant.

(i) Find the value of k.

(ii) Show that $\mathrm{E}(X) = 4\frac{1}{3}$, and find $\mathrm{Var}(X)$. This die is thrown 80 times, and the scores are noted. Use an appropriate Poisson distribution to estimate the probability of at least four "ones" being scored. [UCLES]

11 The number of telephone calls X made by a daughter D to her mother in each week has a Poisson distribution with mean 2, whilst the number of telephone calls Y made by her brother B in each week has a Poisson distribution with mean 1. Show that

$$(n + 1)\mathrm{P}(X = n + 1) = 2\mathrm{P}(X = n)$$

and

$$(n + 1)\mathrm{P}(Y = n + 1) = \mathrm{P}(Y = n),$$

$n = 0, 1, 2 \ldots$.

Assuming that X and Y are independent, find the probability, to 2 decimal places, that in a given week,

(a) neither B nor D makes a call,

(b) B and D make an equal number of calls not exceeding 2 calls each,

(c) B makes less than 4 calls, but makes more calls than D. [ULSEB]

12 (a) The independent Poisson random variables X and Y have means of 2.5 and 1.5, respectively. Obtain the mean and variance of the random variables
(i) $X - Y$, (ii) $2X + 5$.
For each of these random variables give a reason why the distribution is not Poisson.

(b) A car salesman receives £60 commission for each *new* car that he sells and £40 for each *used* car that he sells. The weekly number of *new* cars that he sells has a Poisson distribution with mean 3 and, independently, the weekly number of *used* cars that he sells has a Poisson distribution with mean 2.

(i) Determine the probability that he sells more than two *new* cars in a week.

(ii) Determine the probability that he sells no more than one car in a week.

(iii) Determine the probability that his commission in a week is exactly £100.

(iv) Calculate the mean and standard deviation of the salesman's weekly commission. [JMB]

13 Independently for each page the number of typing errors per page in the first draft of a novel has a Poisson distribution with mean 0.4.

(a) Calculate, correct to five decimal places, the probabilities that

(i) a randomly chosen page will contain no error,

(ii) a randomly chosen page will contain 2 or more errors,

(iii) the third of three randomly chosen pages will be the first to contain an error.

(b) Write down an expression for the probability that each of n randomly chosen pages will contain no error. Hence find the largest n for which there is a probability of at least 0.1 that each of the n pages contains no error.

Independently for each page the number, Y, of typing errors per page in the first draft of a Mathematics textbook also has a Poisson distribution. *(continued)*

(c) Given that $P(Y = 2) = 2P(Y = 3)$
  (i) find $E(Y)$,
  (ii) show that $P(Y = 5) = 4P(Y = 6)$.

(d) One page is chosen at random from the first draft of the novel and one page is chosen at random from the first draft of the Mathematics textbook. Calculate, correct to three decimal places, the probability that exactly one of the two chosen pages will contain no error.   [WJEC]

14 In a double-sampling scheme an initial sample of 50 items is taken from the large batch under investigation and the number m of defectives is noted. If $m = 0$, the whole batch is accepted without further testing. If $m = 1$ or 2, a second sample, this time of 100 items, is taken from the batch and the number n of defectives noted. The whole batch is now accepted if $m + n \leqslant 4$. In all other cases the batch is rejected. For a batch with 1% defective items, use suitable Poisson approximations to estimate
(i) the probability of the batch being accepted,
(ii) the expected number of items sampled. [SMP]

15 In a certain area, the probability of a randomly selected cow dying from 'mad cow' disease is 0.04.
(i)   Calculate the probability that in a random sample of 18 cows exactly 2 will die from the disease.
(ii) Find the probability that in a random sample of 20 cows more than 2 will die from the disease.
(iii) Find the probability that in a random sample of 50 cows between 2 and 5 (inclusive) will die from the disease.
(iv) When a random sample of n cows is taken, the probability that at least one cow will die from the disease exceeds 0.99. Find the smallest value of n.
(v)   Use a distributional approximation to find the probability that in a random sample of 150 cows fewer than 7 will die from the disease.   [WJEC]

16 Manufactured articles are packed in boxes each containing 200 articles, and, on average, $1\frac{1}{2}\%$ of all articles manufactured are defective. A box which contains 4 or more defective articles is substandard. Using a suitable approximation, show that the probability that

a randomly chosen box will be substandard is 0.353, correct to three decimal places.
A lorry-load consists of 16 boxes, randomly chosen. Find the probability that a lorry-load will include at most 2 boxes that are substandard, giving three decimal places in your answer.
A warehouse holds 100 lorry-loads. Show that, correct to two decimal places, the probability that exactly one of the lorry-loads in the warehouse will include at most 2 substandard boxes is 0.06.   [UCLES]

17 A randomly chosen doctor in general practice sees, on average, one case of a broken nose per year and each case is independent of other similar cases.
(i)   Regarding a month as a twelfth part of a year,
      (a) show that the probability that, between them, three such doctors see no cases of a broken nose in a period of one month is 0.779, correct to three significant figures,
      (b) find the variance of the number of cases seen by three such doctors in a period of six months.
(ii) Find the probability that, between them, three such doctors see at least three cases in one year.
(iii) Find the probability that, of three such doctors, one sees three cases and the other two see no cases in one year.   [UCLES]

18 (a) A box contains 12 golf balls, 3 of which are substandard. A random sample of 4 balls is selected, *without replacement*, from the box. The random variable R denotes the number of balls in the box that are substandard. Deduce that

$$P(R = r) = \frac{\binom{3}{r}\binom{9}{4-r}}{\binom{12}{4}},$$

and state the sample space for R.
Determine the probability that the random sample of 4 golf balls contains
(i) no substandard balls,
(ii) fewer than 2 substandard balls.

(*continued*)

(b) A large bin contains 5000 used golf balls, 1500 of which are defective. The random variable X denotes the number of defective balls in a random sample of 20 balls selected, without replacement, from the bin. Explain why X may be approximated as a binomial variate with parameters 20 and 0.3.
Using this binomial model, calculate the probability that this sample contains

(i) fewer than 5 defective balls,
(ii) at least 7 defective balls.

(c) The random variable Y denotes the number of defective golf balls in a sample of 2000, selected at random from a batch of 200 000 of which 3250 are defective. Completely specify an approximate distribution for Y, other than a binomial distribution. [JMB]

11 Continuous random variables

I've measured it from side to side: 'Tis three feet long, and two feet wide

The Thorn, William Wordsworth

Chapters 7 to 10 have focused on discrete random variables: quantities whose values are unpredictable but for which a list of the possible values can be made. Continuous random variables differ in that no such list is feasible, though the range of values can be described. Here are some examples:

Continuous random variable	Possible range of values
The height of a randomly chosen 18-year-old male student	1.3 m to 2.3 m
The true mass of a '1 kg' bag of sugar	990 g to 1010 g
The time interval between successive earthquakes of magnitude >7 on the Richter scale	Any (positive) time

The measurements all refer to **physical** quantities. The number of distinct values is limited only by the inefficiency of our measuring instruments. Since there are an uncountable number of possible values that a continuous random variable might take, the probability of any particular value is zero. Instead, we assign probabilities to (arbitrarily small) ranges of values.

If a continuous random variable is measured rather inaccurately, then we will treat it as though it is a discrete random variable.

Age of randomly chosen male (aged under 100) measured to nearest 10 years.	$\longrightarrow$	Treat as a discrete random variable with 11 categories.

Conversely, if a discrete random variable has a great many possible outcomes, then we may treat it as though it is a continuous random variable

Mark on exam paper (An integer in the range 0 to 100.)	$\longrightarrow$	Treat as a continuous variable on the interval [0, 100].

Because of this easy transition between the two types of variable, the ideas of expectation and the formulae interrelating expectations that were derived in Chapters 7 and 8 carry over to continuous variables – more of this anon!

11.1 Histograms and sample size

In Chapter 1 the histogram was introduced as being the appropriate method for displaying data on a continuous variable. The crucial part of the instructions for drawing a histogram was that *area should be proportional to frequency*. We now develop that idea by requiring that:

> area should be proportional to *relative* frequency.

As an illustration we consider some data concerning the geyser known as 'Old Faithful', which is situated in Yellowstone National Park in Wyoming, USA. This geyser is a great tourist attraction because of the regularity of its eruptions of steam. In August 1985 the geyser was watched continuously for a fortnight, with the times between its eruptions being recorded to the nearest minute. The first 50 times are shown below.

80	71	57	80	75		77	60	86	77	56
81	50	89	54	90		73	60	83	65	82
84	54	85	58	79		57	88	68	76	78
74	85	75	65	76		58	91	50	87	48
93	54	86	53	78		52	83	60	87	49

Using classes of width 10 minutes (with class boundaries at 39.5, 49.5, etc.) we can represent these data using a histogram. There are just two observations out of the 50 that fall in the 39.5–49.5 class, so the relative frequency density for that class is $\frac{2}{50} = 0.04$. The histogram looks like this:

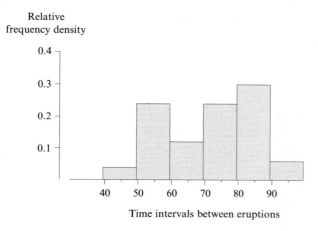

The histogram has a fairly chunky appearance! We now increase the sample size to 100 observations (twice the original size) and illustrate the combined set of data using classes of width 5 minutes (half the original size). With the same vertical scale, the area of the shaded region is the same as before.

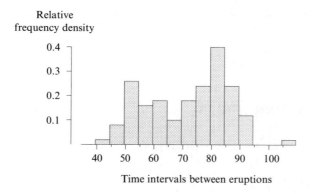

Finally, we add in a further 150 observations, raising the total to 250, and now illustrate it using classes that are one-fifth of the original width, so that the area of the shaded region remains the same once again.

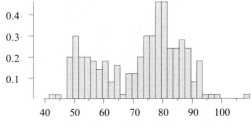

Time intervals between eruptions

Comparing the sequence of histograms it is easy to see that, as we increase the amount of data, so we increase the precision with which we can see the outline of the histogram. What would happen if we had not 250 observations, but 2500 or 25 000? There would still be the odd bit of random variation, but it seems likely that the dominating effect would be of a histogram with a remarkably smooth outline. With a very large sample (assuming 'Old Faithful' was still working faithfully!) we might obtain a diagram that appeared to be outlined by a smooth curve something like this:

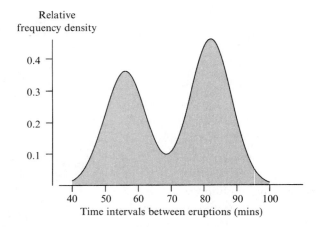

Time intervals between eruptions (mins)

It is clear that 'Old Faithful' behaves in a rather odd fashion! The periods between eruptions are either short (around 50–60 minutes) or long (around 75–90 minutes), with durations of around 66 minutes being rather unusual.

11.2 The probability density function, f

The data from 'Old Faithful' suggested a general result: as we allow the sample size to increase (with correspondingly narrower class intervals), the outline of a histogram will usually converge on a smooth curve. The areas of the individual sections of a histogram represent relative frequencies. We know that as the sample size increases so sample relative frequencies approach the corresponding population probabilities. The area of any section under the curve therefore represents a probability.

Probability density

$f(x)$

a b

x

Probability that X lies between a and b

When the curve is close to the x-axis the probability associated with a unit range of x is small, whereas when the curve is distant from the axis, the probability is much larger. The height of the curve represents the rate at which probability is accumulated as we move along the x-axis. The curve is the graph of the **probability density function**, written as **pdf** for short and the function is usually denoted by the letter f.

Properties of the pdf

We already know what these are!

1 Since we cannot have negative probabilities, the graph of f cannot dip below the x-axis:

$$f(x) \geqslant 0 \tag{11.1}$$

2 The probability of X taking a value in the interval (a, b) is given by the corresponding area. Since the area between any curve and the x-axis is given by the integral of that curve with respect to x, we therefore have:

$$P(a < X < b) = \int_a^b f(x)\,dx \tag{11.2}$$

3 The total of a set of relative frequencies is, by definition, equal to 1. The same is true for probabilities. The total area between the graph of $f(x)$ and the x-axis is therefore 1.

$$\int_c^d f(x)\,dx = 1 \tag{11.3}$$

where the limits of the integral are the end-points of the interval for which f is non-zero.

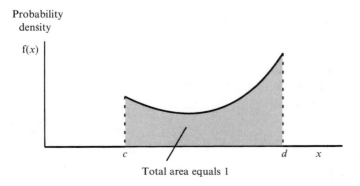

Probability density

$f(x)$

c d x

Total area equals 1

Notes

- Suppose k is somewhere between c and d, and let a be just less than k and let b be just greater than k. As a and b get closer to k, the value of the integral in Equation (11.2) approaches zero, so in the limit, $P(X = k) = 0$. This is an entirely general result and implies that:
  – we need not be fussy about whether we write $P(X < k)$ or $P(X \leqslant k)$, since the two have the same value.
- If f has a unique maximum when $x = M$, then M is called the **mode**.
  – Often the mode can be located by examination of a sketch of f(x).
- The function f measures probability *density*, not probability. Although f(x) usually has values less than 1, this need not be the case. For example:

$$f(x) = \begin{cases} 2 & 0 < x < \tfrac{1}{2} \\ 0 & \text{otherwise} \end{cases}$$

defines a proper probability density function that integrates to 1.
- Problems involving probability density functions often require the calculation of areas. Instead of using calculus it is often much quicker to use standard geometric results. In particular:

  A triangle of height h and base b has area $\tfrac{1}{2}hb$

  A trapezium with parallel sides of lengths k and l at a distance d apart has area $\tfrac{1}{2}d(k + l)$.

Example 1

The continuous random variable X has probability density function given by:

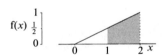

$$f(x) = \begin{cases} \tfrac{1}{2}x & 0 < x < 2 \\ 0 & \text{otherwise} \end{cases}$$

Determine $P(X > 1)$.

———

The statement that f$(x) = 0$ 'otherwise' merely emphasises that attention may safely be restricted to the interval (0, 2).

Glancing at the diagram we can see that the area corresponding to $P(X > 1)$ is greater than half of the total area between f(x) and the x-axis, so that the required probability will be greater than 0.5. If our calculations give a value smaller than 0.5 then we must have made an error (possibly in the diagram!).

Method 1: Calculus
The required probability is:

$$\int_1^2 \frac{x}{2}\,dx = \left[\frac{x^2}{4}\right]_1^2 = \frac{4 - 1}{4} = \frac{3}{4}$$

So $P(X > 1) = 0.75$, which, as anticipated, is considerably greater than 0.5.

———

Method 2: Geometry
The required probability is given by the area of a trapezium having parallel sides of lengths $\tfrac{1}{2}$ and 1 at a distance apart of $2 - 1 = 1$. The area corresponding to $P(X > 1)$ is therefore equal to:

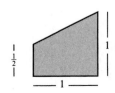

$$\tfrac{1}{2} \times 1 \times (\tfrac{1}{2} + 1) = \tfrac{3}{4}$$

confirming the result found by calculus.

Example 2

The continuous random variable Y has probability density function given by:

$$f(y) = \begin{cases} |y - 1| & 0 < y < 2 \\ 0 & \text{otherwise} \end{cases}$$

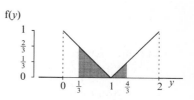

Determine $P(\frac{1}{3} < Y < \frac{4}{3})$.

This time the diagram suggests that the answer will be less than 0.5.

Method 1: Calculus

We begin by rewriting the formula for the pdf so that the | signs are not needed:

$$f(y) = \begin{cases} 1 - y & 0 < y \leqslant 1 \\ y - 1 & 1 < y < 2 \\ 0 & \text{otherwise} \end{cases}$$

We now use the result that:

$$P(\tfrac{1}{3} < Y < \tfrac{4}{3}) = P(\tfrac{1}{3} < Y \leqslant 1) + P(1 < Y < \tfrac{4}{3})$$

We therefore need:

$$\int_{\frac{1}{3}}^{1} (1 - y)\,dy + \int_{1}^{\frac{4}{3}} (y - 1)\,dy = \left[y - \frac{y^2}{2} \right]_{\frac{1}{3}}^{1} + \left[\frac{y^2}{2} - y \right]_{1}^{\frac{4}{3}}$$

$$= \left\{ \left(1 - \frac{1}{2} \right) - \left(\frac{1}{3} - \frac{1}{18} \right) \right\} + \left\{ \left(\frac{16}{18} - \frac{4}{3} \right) - \left(\frac{1}{2} - 1 \right) \right\}$$

$$= \left(\frac{1}{2} - \frac{5}{18} \right) + \left(\frac{-8}{18} - \frac{-1}{2} \right)$$

$$= \frac{4}{18} + \frac{1}{18} = \frac{5}{18}$$

The probability that Y takes a value between $\frac{1}{3}$ and $\frac{4}{3}$ is $\frac{5}{18}$, or 0.278 (to 3 d.p.).

Method 2: Geometry

In this case we have two triangles. At $y = \frac{1}{3}$, $f(y) = \frac{2}{3}$, so the left-hand triangle has both height and base equal to $\frac{2}{3}$ and therefore has area equal to:

$$\tfrac{1}{2} \times \tfrac{2}{3} \times \tfrac{2}{3} = \tfrac{2}{9}$$

At $y = \frac{4}{3}$, $f(y) = \frac{1}{3}$, so this triangle has area equal to:

$$\tfrac{1}{2} \times \tfrac{1}{3} \times \tfrac{1}{3} = \tfrac{1}{18}$$

The total area of the two triangles is therefore $\frac{2}{9} + \frac{1}{18} = \frac{5}{18}$, which agrees with the answer obtained by calculus.

Example 3

The continuous random variable X has pdf given by:

$$f(x) = \begin{cases} kx^2 & 1 < x < 3 \\ 0 & \text{otherwise} \end{cases}$$

Determine (i) the value of the constant k, (ii) $P(X < 2)$.

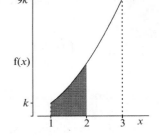

We will not attempt a geometric solution in this case since a curve is involved.

(i)   To find k we use the fact that f integrates to 1:

$$\int_1^3 kx^2 \, dx = \left[\frac{kx^3}{3} \right]_1^3$$
$$= \frac{k}{3}(27 - 1)$$
$$= \frac{26k}{3}$$

Since we know that the integral is equal to 1, it follows that $k = \frac{3}{26}$.

(ii) $$P(X < 2) = \int_1^2 kx^2 \, dx$$
$$= \left[\frac{kx^3}{3} \right]_1^2$$
$$= \frac{k}{3}(8 - 1)$$
$$= \frac{7k}{3}$$
$$= \frac{7}{26}$$

The probability that X takes a value less than 2 is $\frac{7}{26}$, or 0.269 (to 3 d.p.).

Calculator practice

If you have a calculator that can perform numerical integration then you should check that you know the appropriate instructions. Practice by checking that the answers to Examples 1–3 are correct.

Exercises 11a

In Questions **1–6**, the continuous random variable X has pdf f and k is a constant.

1   Given that:

$$f(x) = \begin{cases} kx^2 + \frac{1}{6} & 0 < x < 3 \\ 0 & \text{otherwise} \end{cases}$$

sketch the graph of f and find (i) the value of k, (ii) $P(X < 1)$, (iii) $P(X > 2)$.

2   Given that:

$$f(x) = \begin{cases} kx^2 & -2 < x < 2 \\ 0 & \text{otherwise} \end{cases}$$

sketch the graph of f and find (i) the value of k, (ii) $P(X > 1)$, (iii) $P(|X| < 1)$, (iv) $P(-1 < X < 0)$.

3 Given that:

$$f(x) = \begin{cases} \frac{1}{2}x & 1 < x < k \\ 0 & \text{otherwise} \end{cases}$$

sketch the graph of f and find (i) the value of k, (ii) $P(\frac{3}{2} < X < 2)$, (iii) $P(X > 3)$, (iv) $P(2 < X < 3)$, (v) the mode.

4 Given that:

$$f(x) = \begin{cases} 1 - kx & 0 < x < 2 \\ 0 & \text{otherwise} \end{cases}$$

Sketch the graph of f and find
(i) the value of the constant k,
(ii) $P(X \leqslant 1)$.

5 Given that:

$$f(x) = \begin{cases} 10k & -0.05 < x < 0.05 \\ 0 & \text{otherwise} \end{cases}$$

sketch the graph of f and find
(i) the value of k, (ii) $P(X > 0.1)$,
(iii) $P(X < 0.025)$.

6 Given that:

$$f(x) = \begin{cases} kx(6 - x) & 2 < x < 5 \\ 0 & \text{otherwise} \end{cases}$$

sketch the graph of f and find
(i) the value of k, (ii) the mode, m,
(iii) $P(X < m)$.

7 For each of the following functions, state, giving a reason, whether or not there is a value of the constant k for which the function can be a pdf. Sketch graphs may help. You do not have to find the value of k.

(i) $f(x) = \begin{cases} kx & -1 < x < 1 \\ 0 & \text{otherwise} \end{cases}$

(ii) $f(x) = \begin{cases} kx^2 & -1 < x < 2 \\ 0 & \text{otherwise} \end{cases}$

(iii) $f(x) = \begin{cases} 1 + kx & 0 < x < 3 \\ 0 & \text{otherwise} \end{cases}$

(iv) $f(x) = \begin{cases} 4 + x^2 & -k < x < k \\ 0 & \text{otherwise} \end{cases}$

8 A garage is supplied with petrol once a week. Its volume of weekly sales, X, in thousands of gallons, is distributed with probability density function $f(x)$ given by:

$$f(x) = \begin{cases} kx(1 - x)^2 & 0 < x < 1 \\ 0 & \text{otherwise} \end{cases}$$

Determine the value of the constant k.

Determine an expression for the probability that the sales are less than c hundred gallons.

Determine the value of this probability for each of $c = 7$, 7.5 and 8.

Hence, or otherwise, determine an appropriate capacity for the garage's tank, if it is to have a probability of only about 0.05 of being exhausted in a given week.

11.3 The cumulative distribution function, F

The cumulative distribution function is often referred to as the **distribution function** or, more simply, as the **cdf**. The graph of a cdf may be thought of as the limiting form of the cumulative frequency polygon (Section 1.17, p. 23). The function is defined by:

$$F(x) = P(X \leqslant x) = P(X < x) \tag{11.4}$$

and is related to the function f by:

$$F(b) = \int_{-\infty}^{b} f(x)\,dx \tag{11.5}$$

The lower limit of the integral is given as $-\infty$, but is in effect the smallest attainable value of X. In each of the following three diagrams, $P(X < a) = 0$ and $P(X > b) = 0$. The area under the graph of each density function is equal to 1.

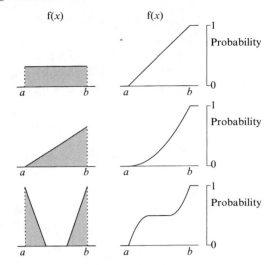

Notes

♦ Since it is always impossible to have a value of X smaller than $-\infty$ or larger than ∞:

    ◆ $F(-\infty) = 0$

    ◆ $F(\infty) = 1$

    (Strictly $F(-\infty)$ means 'the limiting value of $F(x)$ as x approaches $-\infty$' and $F(\infty)$ is similarly defined.)

♦ As x increases so $F(x)$ either increases or remains constant, but never decreases. The third diagram shows that this also applies to cases where f is discontinuous.

♦ F is a continuous function, even if f is discontinuous.

♦ Useful relations are:

$$P(c < X < d) = F(d) - F(c) \tag{11.6}$$

$$P(X > x) = 1 - F(x) \tag{11.7}$$

Example 4

The continuous random variable X has pdf f(x), given by:

$$f(x) = \begin{cases} 1 & 2 < x < 3 \\ 0 & \text{otherwise} \end{cases}$$

Determine the form of $F(x)$.

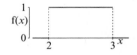

For $b \leq 2$, $F(b) = 0$, since there is no chance that X will take a value less than or equal to 2. Similarly, for $b \geq 3$, $F(b) = 1$, since it is certain that X takes a value less than 3.

For $2 \leq b \leq 3$, we use the definition:

$$F(b) = P(X \leq b) = \int_2^b 1 \, dx$$

$$= \left[x \right]_2^b$$

$$= (b - 2)$$

The form of $F(x)$ is therefore:

$$F(x) = \begin{cases} 0 & x \leq 2 \\ x - 2 & 2 \leq x \leq 3 \\ 1 & x \geq 3 \end{cases}$$

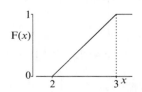

Example 5

The continuous random variable X has pdf given by:

$$f(x) = \begin{cases} 2k(x-1) & 1 < x < 2 \\ k(4-x) & 2 < x < 4 \\ 0 & \text{otherwise} \end{cases}$$

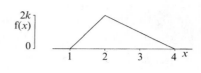

where k is a constant.

Determine (i) the value of k, (ii) the form of F(x).

Either geometry or calculus could be used to answer both parts of the question. The simplest procedure is to use geometry to find the first answer and calculus to find the second.

(i) To find the value of k we use the fact that the total area of the region between the graph of the probability density function and the x-axis is equal to 1.

The sketch reveals that the region of interest is a triangle. Since $f(x) = 2k$ at $x = 2$, and since the triangle has base equal to $(4-1) = 3$, the triangle has area $\frac{1}{2} \times 2k \times 3 = 3k$. The area is known to be equal to 1 and therefore $k = \frac{1}{3}$.

(ii) $F(x) = 0$ for $x \leqslant 1$ and $F(x) = 1$ for $x \geqslant 4$, since X only takes values in the interval 1 to 4. We consider the two intervals $[1, 2]$ and $[2, 4]$ separately since f has a different form in each interval.

For $1 \leqslant b \leqslant 2$:

$$F(b) = \int_1^b 2k(x-1)\,dx$$

$$= \left[2k\frac{(x-1)^2}{2} \right]_1^b$$

$$= k(b-1)^2$$

$$= \frac{(b-1)^2}{3}$$

In particular, $P(X < 2) = F(2) = \frac{1}{3}$.

For $2 \leqslant b \leqslant 4$, we write:

$$F(b) = P(X \leqslant b) = P(X \leqslant 2) + P(2 \leqslant X \leqslant b)$$

$$= \tfrac{1}{3} + P(2 \leqslant X \leqslant b)$$

We therefore need $P(2 \leqslant X \leqslant b)$:

$$P(2 \leqslant X \leqslant b) = \int_2^b k(4-x)\,dx$$

$$= \left[-k\frac{(4-x)^2}{2} \right]_2^b$$

$$= -\frac{k}{2}\{(4-b)^2 - (4-2)^2\}$$

$$= \frac{1}{6}\{4 - (4-b)^2\}$$

Hence, for $2 \leqslant b \leqslant 4$:

$$F(b) = \frac{1}{3} + \frac{4 - (4 - b)^2}{6}$$

$$= \frac{6 - (4 - b)^2}{6}$$

As a check, note that F(4) does indeed equal the maximum value, 1, and that F(2) equals $\frac{1}{3}$, the value obtained previously. The complete description of F(x) is therefore:

$$F(x) = \begin{cases} 0 & x \leqslant 1 \\ \dfrac{(x-1)^2}{3} & 1 \leqslant x \leqslant 2 \\ 1 - \dfrac{(4-x)^2}{6} & 2 \leqslant x \leqslant 4 \\ 1 & x \geqslant 4 \end{cases}$$

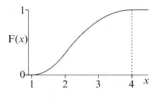

The median, *m*

The **median**, *m*, is the value that bisects the distribution in the sense that *X* is equally likely to be smaller or larger than *m*. Hence:

$$\int_{-\infty}^{m} f(x)\, dx = \int_{m}^{\infty} f(x)\, dx = 0.5 \qquad (11.8)$$

Notes

♦ If the graph of f is symmetric about the line $x = x_0$, then $m = x_0$.
♦ **Percentiles** and **quartiles** are defined similarly. For example, the 90th percentile is the solution of F(x) = 0.90, while the upper quartile is the solution of F(x) = 0.75.

Example 6

The continuous random variable *X* has pdf given by:

$$f(x) = \begin{cases} k(3 - x) & 1 < x < 2 \\ k & 2 < x < 3 \\ k(x - 2) & 3 < x < 4 \\ 0 & \text{otherwise} \end{cases}$$

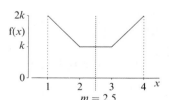

where *k* is a constant.
Determine the median.

The diagram shows that the pdf is symmetric about the line $x = 2.5$, which implies that the median is 2.5. We do not need to know the value of *k*.

Example 7

The continuous random variable X has pdf given by:

$$f(x) = \begin{cases} k - 5 + x & 5 < x < 6 \\ 0 & \text{otherwise} \end{cases}$$

where k is a constant.
Determine the median.

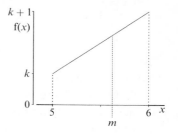

Method 1: Calculus

This time the pdf is not symmetric. We must begin by determining the value of k, which we do by using the fact that the total area is 1. The area is given by:

$$\int_5^6 (k - 5 + x)\, dx = \left[(k - 5)x + \frac{x^2}{2} \right]_5^6$$

$$= \left\{ 6(k - 5) + \frac{6^2}{2} \right\} - \left\{ 5(k - 5) + \frac{5^2}{2} \right\}$$

$$= (k - 5) + 18 - \frac{25}{2}$$

$$= k + \frac{1}{2}$$

Since $(k + \frac{1}{2}) = 1$ it follows that $k = \frac{1}{2}$ and so:

$$f(x) = x - \tfrac{9}{2} \qquad \text{for } 5 < x < 6$$

To find the median, m, we need to solve $F(m) = \frac{1}{2}$. Hence m is the solution of:

$$\frac{1}{2} = \int_5^m \left(x - \frac{9}{2} \right) dx$$

$$= \left[\frac{x^2}{2} - \frac{9x}{2} \right]_5^m$$

$$= \left(\frac{m^2}{2} - \frac{9m}{2} \right) - \left(\frac{25}{2} - \frac{45}{2} \right)$$

$$= \frac{m^2 - 9m + 20}{2}$$

Rearranging we get:

$$m^2 - 9m + 19 = 0$$

which has solution:

$$m = \frac{9 \pm \sqrt{81 - 76}}{2}$$

We need the root to be between 5 and 6 (since this is the range of possible values for X) and so it is the larger root, $\frac{1}{2}(9 + \sqrt{5})$, that is relevant. To three significant figures, the median is 5.62.

Method 2: Geometry

The region of interest is a trapezium with parallel sides of lengths k and $(k + 1)$. The 'distance apart' is $(6 - 5) = 1$, and so the area is:

$$\frac{1}{2} \times 1 \times \{k + (k + 1)\} = k + \frac{1}{2}$$

Since this area must be 1, we again conclude that $k = \frac{1}{2}$.

To find m we proceed in a similar way, by considering the smaller trapezium having parallel sides corresponding to the cases $x = 5$ and $x = m$. These sides have lengths k and $k - 5 + m$, so the area of this trapezium is:

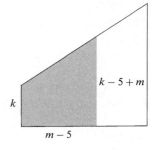

$$\frac{1}{2} \times (m - 5) \times \{k + (k - 5 + m)\} = \frac{1}{2}(m - 5)(2k - 5 + m) = \frac{1}{2}(m - 5)(m - 4)$$

We wish to choose m so that this area $= \frac{1}{2}$. Therefore, m is the solution of:

$$(m - 5)(m - 4) = 1$$

which on rearranging gives:

$$m^2 - 9m + 19 = 0$$

as before.

Example 8

The continuous random variable X has pdf given by:

$$f(x) = \begin{cases} ax & 1 < x \leqslant 3 \\ c(4 - x) & 3 \leqslant x < 4 \\ 0 & \text{otherwise} \end{cases}$$

where a and c are constants.
Determine the values of a and c, and also the median, m.

Method 1: Calculus

We begin by noting the duplicate definition at $x = 3$. This implies that:

$$f(3) = 3a = c$$

We replace c by $3a$ in the subsequent calculations.

As usual we begin by using the fact that the total area must equal 1. This total area is given by:

$$\int_1^3 ax \, dx + \int_3^4 3a(4 - x) \, dx = \left[a\frac{x^2}{2}\right]_1^3 + \left[-3a\frac{(4 - x)^2}{2}\right]_3^4$$

$$= a\left(\frac{9}{2} - \frac{1}{2}\right) + \left\{0 - \left(-3a \times \frac{1}{2}\right)\right\}$$

$$= 4a + \frac{3a}{2}$$

$$= \frac{11a}{2}$$

and we therefore conclude that $a = \frac{2}{11}$ (and hence that $c = \frac{6}{11}$).

A glance at the diagram shows that the median is less than 3. We therefore solve the equation:

$$\int_1^m ax \, dx = \frac{1}{2}$$

Now:

$$\int_1^m ax\,dx = \left[a\frac{x^2}{2}\right]_1^m = \frac{am^2}{2} - \frac{a}{2} = \frac{a(m^2-1)}{2} = \frac{m^2-1}{11}$$

Thus m is the solution of:

$$\frac{m^2-1}{11} = \frac{1}{2}$$

Multiplying through by 11 and rearranging we get:

$$m^2 = 1 + \frac{11}{2} = 6.5$$

and, since m is evidently positive:

$$m = \sqrt{6.5} = 2.55 \text{ (to 3 s.f.)}$$

————

Method 2: Geometry

The total area is made up of a trapezium corresponding to the region between $x = 1$ and $x = 3$ and a triangle for the remainder. The parallel sides of the trapezium have lengths a and $3a$, with the 'distance apart' being $(3 - 1) = 2$. The triangle has height $3a$ and base $(4 - 3) = 1$, so the combined area is:

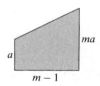

$$\frac{1}{2} \times (3 - 1) \times (a + 3a) + \frac{1}{2} \times 3a \times (4 - 3) = 4a + \frac{3}{2}a = \frac{11}{2}a$$

giving $a = \frac{2}{11}$, as before.

Seeing that the median is less than 3, we need only consider the trapezium bounded by $x = 1$ and $x = m$, with sides of length a and ma, respectively. This trapezium therefore has area:

$$\frac{1}{2} \times (m - 1) \times (a + ma) = \frac{1}{2}a(m - 1)(1 + m)$$

$$= \frac{1}{11}(m - 1)(m + 1) = \frac{1}{11}(m^2 - 1)$$

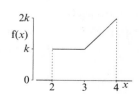

and so, as before, m is the solution of:

$$\frac{m^2-1}{11} = \frac{1}{2}$$

Example 9

The continuous random variable X has pdf given by:

$$f(x) = \begin{cases} k & 2 < x < 3 \\ k(x - 2) & 3 < x < 4 \\ 0 & \text{otherwise} \end{cases}$$

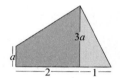

Determine the value of the median, m.

————

Method 1: Calculus

We begin by finding the value of k. The total area, corresponding to the total probability of 1, is given by:

$$\int_2^3 k\,dx + \int_3^4 k(x-2)\,dx = \Big[kx\Big]_2^3 + \left[\frac{k(x-2)^2}{2}\right]_3^4$$

$$= (3k - 2k) + \left(\frac{4k}{2} - \frac{k}{2}\right)$$

$$= k + \frac{3k}{2}$$

$$= \frac{5k}{2}$$

Since $\frac{5}{2}k = 1$, we see that $k = \frac{2}{5}$.

The diagram shows that the median is greater than 3. We can confirm this by noting that $P(X < 3)$ is given by:

$$\int_2^3 k\,dx = \Big[kx\Big]_2^3 = (3k - 2k) = k = \frac{2}{5}$$

The value of m therefore satisfies the equation:

$$\int_3^m k(x-2)\,dx = \tfrac{1}{2} - \tfrac{2}{5} = \tfrac{1}{10}$$

The integral is equal to:

$$\left[\frac{k(x-2)^2}{2}\right]_3^m = \frac{k(m-2)^2}{2} - \frac{k}{2} = \frac{k\{(m^2 - 4m + 4) - 1\}}{2}$$

$$= \frac{m^2 - 4m + 3}{5}$$

and hence m is the solution of:

$$\frac{m^2 - 4m + 3}{5} = \frac{1}{10}$$

or:

$$2(m^2 - 4m + 3) = 1$$

which simplifies to:

$$2m^2 - 8m + 5 = 0$$

and has solution:

$$m = \frac{8 \pm \sqrt{8^2 - 4 \times 2 \times 5}}{2 \times 2} = 2 \pm \sqrt{1.5}$$

Since $m > 3$, we need the larger solution: the median is 3.22 (to 2 d.p.).

———————

Method 2: Geometry

In this case the region of interest consists of a rectangle of sides k and 1 together with a trapezium with parallel sides of length k and $2k$, and 'distance apart' 1. The total area is therefore:

$$(k \times 1) + \frac{1}{2} \times 1 \times (k + 2k) = k + \frac{3}{2}k = \frac{5}{2}k$$

from which we (very quickly!) have found that, since the total area must be 1, the value of k must be $\frac{2}{5}$.

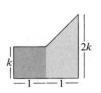

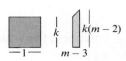

Substituting for k, we note that the area of the rectangle is $\frac{2}{5}$, which is less than $\frac{1}{2}$. It follows that the median, m, must exceed 3. We need to add to the rectangle a (thin!) trapezium with parallel sides of length k and $k(m-2)$, and 'distance apart' $(m-3)$.

The thin trapezium has area $\frac{1}{2} \times (m-3) \times \{k + k(m-2)\}$. We need this to have area $\left(\frac{1}{2} - \frac{2}{5}\right) = \frac{1}{10}$, and hence m is the solution of:

$$\tfrac{1}{2} \times (m-3) \times \{k(m-1)\} = \tfrac{1}{5} \times (m-3) \times (m-1) = \tfrac{1}{10}$$

Multiplying through by 10, this becomes:

$$2(m-3)(m-1) = 1$$

which finally simplifies to the equation obtained previously:

$$2m^2 - 8m + 5 = 0$$

from which we found that the median was 3.22 (to 2 d.p.).

Exercises 11b

In Questions **1–10**, the pdf and cdf of the continuous random variable X are f and F respectively, and k is a constant.

1 Given that:
$$f(x) = \begin{cases} kx + \frac{1}{4} & 1 < x < 3 \\ 0 & \text{otherwise} \end{cases}$$
sketch the graph of f.
Find: (i) the value of k, (ii) $F(x)$,
(iii) $P(X > 2)$, (iv) the median of X.

2 Given that:
$$f(x) = \begin{cases} -kx & -2 < x < 0 \\ kx & 0 < x < 2 \\ 0 & \text{otherwise} \end{cases}$$

sketch the graph of f.
Find: (i) the value of k, (ii) the median of X,
(iii) $F(x)$, (iv) $P(0 < X < 1)$.

3 Given that:
$$f(x) = \begin{cases} \frac{1}{2} & 1 < x < 2 \\ k & 2 < x < 4 \\ 0 & \text{otherwise} \end{cases}$$

sketch the graph of f.
Find:
(i) the value of k,
(ii) $F(x)$,
(iii) the tenth percentile of X,
(iv) the eightieth percentile of X.

4 It is given that:
$$f(x) = \begin{cases} k(x+2) & -2 < x < 0 \\ \frac{1}{2}k(3-x) & 0 < x < 2 \\ 0 & \text{otherwise} \end{cases}$$
Find:
(i) the value of k, (ii) $F(x)$
(iii) $P(-1 < X < 1)$, (iv) $P(1 < X < 3)$.

5 Given that:
$$f(x) = \begin{cases} kx^2 & -2 < x < 1 \\ 0 & \text{otherwise} \end{cases}$$
sketch the graph of f and find:
(i) the value of k, (ii) $F(x)$, (iii) the mode,
(iv) the median of X.

6 Given that:
$$f(x) = \begin{cases} k(x+2)^2 & 0 < x < 1 \\ 0 & \text{otherwise} \end{cases}$$
sketch the graph of f and find:
(i) the value of k, (ii) $F(x)$,
(iii) the median of X.

7 Given that:
$$f(x) = \begin{cases} kx^2 - c & 1 < x < 2 \\ 0 & \text{otherwise} \end{cases}$$
sketch the graph of f.
(i) Find, in terms of k, the maximum possible value for the constant c.
(ii) With this value for c, find the corresponding value of k.
(iii) With these values for k and c, find $F(x)$.

8 Given that:
$$f(x) = \begin{cases} 2x + k & 3 < x < 4 \\ 0 & \text{otherwise} \end{cases}$$
sketch the graph of f and find:
(i) the value of k, (ii) $F(x)$,
(iii) the lower and upper quartiles of X.

9 Given that:
$$f(x) = \begin{cases} k & 0 < x < 1 \\ 4k & 1 < x < 3 \\ 0 & \text{otherwise} \end{cases}$$
sketch the graph of f and find:
(i) the value of k, (ii) $F(x)$,
(iii) the difference between the median and the fifth percentile of X.

10 Given that:
$$f(x) = \begin{cases} 2(1 - x) & 0 < x < k \\ 0 & \text{otherwise} \end{cases}$$
sketch the graph of f and find:
(i) the value of k, (ii) $F(x)$,
(iii) the median of X.

11 At Wetville, the proportion of the sky covered in cloud, S, has pdf f given by:

$$f(s) = \begin{cases} k(3 + s) & 0 < s < 1 \\ 0 & \text{otherwise} \end{cases}$$
Sketch the graph of f and find:
(i) the value of k, (ii) $F(s)$, (iii) $P(S > 0.5)$,
(iv) the median of S.

12 The time, T hours, required to erect a type of wooden garden shed has pdf f given by:
$$f(t) = \begin{cases} kt^2 & 5 < t < 8 \\ 0 & \text{otherwise} \end{cases}$$
Sketch the graph of f and find:
(i) the value of k, (ii) $F(t)$,
(iii) the probability that it takes between 6 hours and 7 hours 20 minutes to erect a shed.

13 The continuous random variable Z has pdf f given by:
$$f(z) = \begin{cases} a + bz & 0 < z < 1 \\ 0 & \text{otherwise} \end{cases}$$
It is given that $F(0.5) = 0.6$.
(a) Determine the values of a and b.
(b) Determine the median and the mode of Z.

11.4 Expectation and variance

The formula for the sample mean, $\bar{x}$, can be written as:
$$\bar{x} = \sum x_j \left(\frac{f_j}{n} \right)$$
where $\frac{f_j}{n}$ is the relative frequency of the value x_j. As the sample increases in size, so, for a *discrete* random variable, $\frac{f_j}{n}$ converges on the corresponding population probability. However, for a *continuous* random variable, a probability can only be associated with a *range* of values, for example:
$$P\left[\left(x - \frac{\delta x}{2}\right) < X < \left(x + \frac{\delta x}{2}\right)\right] \approx f(x) \times \delta x$$
since the thin rectangle that approximates this probability has width δx and height f(x).

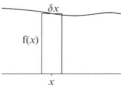

The analogue of $\sum x_j \left(\frac{f_j}{n} \right)$ is therefore $\Sigma x \, f(x) \delta x$, where the latter summation is over a huge number of values of x, each separated by a small amount δx. As we decrease the size of δx so the value of $\Sigma x \, f(x) \delta x$ converges on $\int x\,\mathrm{f}(x)\,\mathrm{d}x$. Hence, for a continuous variable X, the population mean is the expectation $\mathrm{E}(X)$ given by:

$$\mathrm{E}(X) = \int_{-\infty}^{\infty} x\,\mathrm{f}(x)\,\mathrm{d}x \tag{11.9}$$

The limits of the integral are given as $-\infty$ and ∞, but are in effect the largest and smallest attainable values of X. By a similar argument:

$$\mathrm{E}[\mathrm{g}(X)] = \int_{-\infty}^{\infty} \mathrm{g}(x)\mathrm{f}(x)\,\mathrm{d}x \tag{11.10}$$

In particular:

$$\mathrm{E}(X^2) = \int_{-\infty}^{\infty} x^2\,\mathrm{f}(x)\,\mathrm{d}x \tag{11.11}$$

and:

$$\mathrm{Var}(X) = \mathrm{E}[(X - \mu)^2] = \int_{-\infty}^{\infty} (x - \mu)^2\,\mathrm{f}(x)\,\mathrm{d}x \tag{11.12}$$

though it is usually easier to calculate $\mathrm{Var}(X)$ using:

$$\mathrm{Var}(X) = \mathrm{E}(X^2) - \{\mathrm{E}(X)\}^2$$

All the results of Chapter 8 continue to hold:

$$\mathrm{E}(X + a) = \mathrm{E}(X) + a$$
$$\mathrm{E}(aX) = a\mathrm{E}(X)$$
$$\mathrm{E}[a\mathrm{g}(X)] = a\mathrm{E}[\mathrm{g}(X)]$$
$$\mathrm{E}[\mathrm{g}(X) + \mathrm{h}(X)] = \mathrm{E}[\mathrm{g}(X)] + \mathrm{E}[\mathrm{h}(X)]$$

$$\mathrm{Var}(X + a) = \mathrm{Var}(X)$$
$$\mathrm{Var}(aX) = a^2\mathrm{Var}(X)$$

$$\mathrm{E}(X + Y) = \mathrm{E}(X) + \mathrm{E}(Y)$$
$$\mathrm{E}(aX + bY + c) = a\mathrm{E}(X) + b\mathrm{E}(Y) + c$$
$$\mathrm{E}[\mathrm{g}(X) + \mathrm{h}(Y)] = \mathrm{E}[\mathrm{g}(X)] + \mathrm{E}[\mathrm{h}(Y)]$$
$$\mathrm{E}(R + S + T + U) = \mathrm{E}(R) + \mathrm{E}(S) + \mathrm{E}(T) + \mathrm{E}(U)$$

For independent random variables we also have:

$$\mathrm{Var}(aX + bY + c) = a^2\mathrm{Var}(X) + b^2\mathrm{Var}(Y)$$
$$\mathrm{Var}(R + S + T + U) = \mathrm{Var}(R) + \mathrm{Var}(S) + \mathrm{Var}(T) + \mathrm{Var}(U)$$

Most are easy to prove. For example, consider $\mathrm{E}[\mathrm{g}(X) + \mathrm{h}(X)]$, where $\mathrm{g}(X)$ and $\mathrm{h}(X)$ are two arbitrary functions of a continuous random variable X having pdf f:

$$\mathrm{E}[\mathrm{g}(X) + \mathrm{h}(X)] = \int_{-\infty}^{\infty} \{\mathrm{g}(x) + \mathrm{h}(x)\}\mathrm{f}(x)\,\mathrm{d}x$$

$$= \int_{-\infty}^{\infty} \mathrm{g}(x)\mathrm{f}(x)\,\mathrm{d}x + \int_{-\infty}^{\infty} \mathrm{h}(x)\mathrm{f}(x)\,\mathrm{d}x$$

$$= \mathrm{E}[\mathrm{g}(X)] + \mathrm{E}[\mathrm{h}(X)]$$

Note

◆ If f is symmetric about the line $x = c$ and X has expectation $\mathrm{E}(X)$, then $\mathrm{E}(X) = c$. The median is also c, of course.

Example 10

The continuous random variable X has pdf given by:

$$f(x) = \begin{cases} \dfrac{3(1-x^2)}{4} & -1 < x < 1 \\ 0 & \text{otherwise} \end{cases}$$

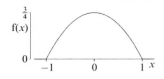

Determine $E(X)$ and $Var(X)$.

Denoting these quantities by μ and σ^2, respectively, determine the probability that an observed value of X has a value in the interval $(\mu - \sigma, \mu + \sigma)$.

———

We see from the sketch that f is symmetric about the line $x = 0$, hence $\mu = E(X) = 0$.

To calculate $Var(X)$ we need $E(X^2)$:

$$E(X^2) = \int_{-1}^{1} x^2 \times \frac{3(1-x^2)}{4} \, dx$$

$$= \frac{3}{4} \int_{-1}^{1} (x^2 - x^4) \, dx$$

$$= \frac{3}{4} \left[\frac{x^3}{3} - \frac{x^5}{5} \right]_{-1}^{1}$$

$$= \frac{3}{4} \left[\left(\frac{1}{3} - \frac{1}{5} \right) - \left\{ \frac{(-1)}{3} - \frac{(-1)}{5} \right\} \right]$$

$$= \frac{3}{2} \left(\frac{1}{3} - \frac{1}{5} \right)$$

$$= \frac{3}{2} \times \frac{2}{15}$$

$$= \frac{1}{5}$$

Hence $\sigma^2 = Var(X) = E(X^2) - \{E(X)\}^2 = \frac{1}{5}$.

Note that, because of symmetry, we could instead have calculated:

$$E(X^2) = 2 \int_{0}^{1} \frac{3}{4} x^2 (1 - x^2) \, dx$$

$$= \frac{3}{2} \int_{0}^{1} (x^2 - x^4) \, dx$$

$$= \frac{3}{2} \left[\frac{x^3}{3} - \frac{x^5}{5} \right]_{0}^{1}$$

$$= \frac{3}{2} \left(\frac{1}{3} - \frac{1}{5} \right)$$

$$= \frac{3}{2} \times \frac{2}{15}$$

$$= \frac{1}{5}$$

which is slightly quicker!

The probability of an observed value of X having a value in the interval $(\mu - \sigma, \mu + \sigma)$ is therefore:

$$P(\mu - \sigma < X < \mu + \sigma) = P\left(-\frac{1}{\sqrt{5}} < X < \frac{1}{\sqrt{5}}\right)$$

$$= \frac{3}{4}\int_{-\frac{1}{\sqrt{5}}}^{\frac{1}{\sqrt{5}}} (1 - x^2)\, dx$$

$$= \frac{3}{2}\int_{0}^{\frac{1}{\sqrt{5}}} (1 - x^2)\, dx \qquad \text{by symmetry}$$

$$= \frac{3}{2}\left[x - \frac{x^3}{3}\right]_{0}^{\frac{1}{\sqrt{5}}}$$

$$= \frac{3}{2}\left(\frac{1}{\sqrt{5}} - \frac{1}{3} \times \frac{1}{5\sqrt{5}}\right)$$

$$= \frac{3}{2} \times \frac{15 - 1}{15\sqrt{5}}$$

$$= \frac{3 \times 14}{2 \times 15\sqrt{5}}$$

$$= \frac{7}{5}\sqrt{\frac{1}{5}}$$

$$= 0.626 \text{ (to 3 d.p.)}$$

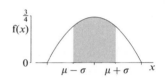

The required probability is 0.626, which is illustrated in the sketch.

Example 11

The continuous random variable X has pdf given by:

$$f(x) = \begin{cases} \dfrac{2(x + 1)}{3} & 0 < x < 1 \\ 0 & \text{otherwise} \end{cases}$$

Determine $E(X)$ and $Var(X)$.
Determine also the probability that two independent observed values of X both have values below the mean.

———————

Since $f(x)$ is not symmetric we need to carry out the integration:

$$E(X) = \int_{0}^{1} x \times \frac{2(x + 1)}{3}\, dx$$

$$= \frac{2}{3}\int_{0}^{1} (x^2 + x)\, dx$$

$$= \frac{2}{3}\left[\frac{x^3}{3} + \frac{x^2}{2}\right]_{0}^{1}$$

$$= \frac{2}{3}\left(\frac{1}{3} + \frac{1}{2}\right)$$

$$= \frac{2}{3} \times \frac{5}{6}$$

$$= \frac{5}{9}$$

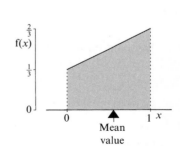

A glance at the sketch suggests that the shaded area *would* balance on a fulcrum placed at $x = \frac{5}{9}$.

In Section 2.18 (p. 63) we noted that, as a guide, the range of a random variable usually takes a value between 3σ and 6σ, where σ is the standard deviation. Since the range here is equal to 1 it follows that we expect that σ will lie between $\frac{1}{3}$ and $\frac{1}{6}$, and hence that $\text{Var}(X)$ $(=\sigma^2)$ will lie between $\frac{1}{9}$ and $\frac{1}{36}$ (i.e. between 0.111 and 0.028).

In order to calculate the variance we need $E(X^2)$:

$$E(X^2) = \int_0^1 x^2 \times \frac{2(x+1)}{3}\, dx$$

$$= \frac{2}{3}\int_0^1 (x^3 + x^2)\, dx$$

$$= \frac{2}{3}\left[\frac{x^4}{4} + \frac{x^3}{3}\right]_0^1$$

$$= \frac{2}{3}\left(\frac{1}{4} + \frac{1}{3}\right)$$

$$= \frac{2}{3} \times \frac{7}{12} = \frac{7}{18}$$

Hence:

$$\text{Var}(X) = E(X^2) - \{E(X)\}^2 = \frac{7}{18} - \left(\frac{5}{9}\right)^2 = \frac{63}{162} - \frac{50}{162} = \frac{13}{162}$$

Since $\frac{13}{162} = 0.080$ (to 3 d.p.) which lies comfortably in the anticipated range of $(0.028, 0.111)$, we have no indication of having made an error.

The probability that an observed value of X is less than the mean is given by:

$$P\left(X < \frac{5}{9}\right) = \int_0^{\frac{5}{9}} \frac{2(x+1)}{3}\, dx$$

$$= \frac{2}{3}\left[\frac{x^2}{2} + x\right]_0^{\frac{5}{9}}$$

$$= \frac{2}{3}\left(\frac{25}{162} + \frac{5}{9}\right)$$

$$= \frac{2}{3} \times \frac{115}{162} = \frac{115}{243}$$

The probability that two independent observed values of X are both smaller than the mean is therefore (using the multiplication rule) $\left(\frac{115}{243}\right)^2 = 0.224$ (to 3 d.p.).

Example 12

The continuous random variable X has pdf given by:

$$f(x) = \begin{cases} x & 0 < x < 1 \\ 2 - x & 1 < x < 2 \\ 0 & \text{otherwise} \end{cases}$$

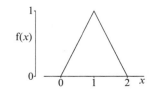

Determine $E(X)$ and $\text{Var}(X)$.

Denoting these quantities by μ and σ^2, respectively, determine the probability that an observed value of X lies in the interval $(\mu - \sigma, \mu + \sigma)$.

We see from the sketch that $f(x)$ is symmetric about the line $x = 1$, so $\mu = E(X) = 1$.

We now need $E(X^2)$ and must take care, since the form of $f(x)$ depends upon the value of x:

$$E(X^2) = \int_0^1 x^2 x \, dx + \int_1^2 x^2(2 - x) \, dx$$

$$= \int_0^1 x^3 \, dx + \int_1^2 (2x^2 - x^3) \, dx$$

$$= \left[\frac{x^4}{4}\right]_0^1 + \left[\left(2 \times \frac{x^3}{3}\right) - \frac{x^4}{4}\right]_1^2$$

$$= \frac{1}{4} + \left\{\left(2 \times \frac{2^3}{3}\right) - \frac{2^4}{4}\right\} - \left\{\left(2 \times \frac{1}{3}\right) - \frac{1}{4}\right\}$$

$$= \frac{1}{4} + \frac{16}{3} - \frac{16}{4} - \frac{2}{3} + \frac{1}{4}$$

$$= \frac{14}{3} - \frac{14}{4}$$

$$= \frac{7}{6}$$

Hence:

$$\sigma^2 = \text{Var}(X) = \frac{7}{6} - 1^2 = \frac{1}{6}$$

Calculation of the probability that an observed value of X lies in the interval $(\mu - \sigma, \mu + \sigma)$ is made easier by exploiting the symmetry of $f(x)$ and calculating the equivalent quantity $2P(\mu - \sigma < X < \mu)$:

$$P(\mu - \sigma < X < \mu) = \int_{\mu-\sigma}^{\mu} f(x) \, dx$$

$$= \int_{1-\frac{1}{\sqrt{6}}}^{1} x \, dx$$

$$= \left[\frac{x^2}{2}\right]_{1-\frac{1}{\sqrt{6}}}^{1}$$

$$= \frac{1}{2}\left[1 - \left(1 - \frac{1}{\sqrt{6}}\right)^2\right]$$

$$= \frac{1}{2}\left[1 - \left(1 - 2\frac{1}{\sqrt{6}} + \frac{1}{6}\right)\right]$$

$$= \frac{1}{2}\left(2\frac{1}{\sqrt{6}} - \frac{1}{6}\right)$$

$$= 0.3249 \text{ (to 4 d.p.)}$$

The probability that an observed value of X falls in the interval $(\mu - \sigma, \mu + \sigma)$ is therefore $2 \times 0.3249 = 0.650$ (to 3 d.p.).

Exercises 11c

In Questions **1–8** the continuous random variable X has pdf f.

1 It is given that:
$$f(x) = \begin{cases} \frac{1}{2}x & 0 < x < 2 \\ 0 & \text{otherwise} \end{cases}$$
Find (i) $E(X)$, (ii) $E(X^2)$, (iii) $Var(X)$.
Find also (iv) $P[X < E(X)]$.

2 It is given that:
$$f(x) = \begin{cases} \frac{1}{4}(3x^2 - 6x + 4) & 0 < x < 2 \\ 0 & \text{otherwise} \end{cases}$$
Find (i) $E(X)$ and (ii) $Var(X)$.

3 It is given that:
$$f(x) = \begin{cases} \frac{1}{4} & 0 < x < 1 \\ \frac{1}{2} & 1 < x < 2 \\ \frac{1}{4} & 2 < x < 3 \\ 0 & \text{otherwise} \end{cases}$$
Find (i) $E(X)$ and (ii) $Var(X)$.

4 It is given that:
$$f(x) = \begin{cases} \frac{2}{3}x & 0 < x < 1 \\ \frac{1}{3} & 2 < x < 4 \\ 0 & \text{otherwise} \end{cases}$$
Find (i) $E(X)$ and (ii) $Var(X)$.
Find also (iii) the median of X,
(iv) $P[X < E(X)]$, (v) $E(X^3)$.

Two independent observations of X are taken.
Find (vi) the probability that one of the observations exceeds the mean and the other is less than the median.

5 Given that:
$$f(x) = \begin{cases} 4 & 0 < x < \frac{1}{4} \\ 0 & \text{otherwise} \end{cases}$$
sketch the graph of f and find (i) $E(X)$,
(ii) $E(2X + 4)$, (iii) $Var(X)$, (iv) $Var(2X + 4)$.

6 Given that:
$$f(x) = \begin{cases} 2x & 0 < x < 1 \\ 0 & \text{otherwise} \end{cases}$$
sketch the graph of f.
The random variable Y is defined by
$Y = 4X + 2$.
Find (i) the expectation of Y,
(ii) the variance of Y.

7 Given that:
$$f(x) = \begin{cases} kx^3 & 2 < x < 3 \\ 0 & \text{otherwise} \end{cases}$$
sketch the graph of f and find
(i) the value of k, (ii) $E(X)$, (iii) $Var(X)$.

The independent continuous random variables X_1 and X_2 have the same distribution as X.
Find (iv) $E(X_1 - X_2)$, (v) $Var(X_1 - X_2)$.

8 Given that:
$$f(x) = \begin{cases} 1 & 0 < x < 1 \\ 0 & \text{otherwise} \end{cases}$$
Determine (i) the mean of X,
(ii) the variance of X.

The random variable Y is defined as the sum of 12 independent random variables that each has the same distribution as X.
Determine (iii) the mean of Y,
(iv) the variance of Y.

9 The amount of chemical, W g, produced by a reaction, has pdf given by:
$$f(w) = \begin{cases} k\{4 - (4 - w)^2\} & 3 < w < 6 \\ 0 & \text{otherwise} \end{cases}$$
Sketch the graph of f and find
(i) the value of k, (ii) $E(W)$, (iii) $P[W > E(W)]$.

10 The random variable X has probability density function given by:
$$f(x) = \begin{cases} x & 0 \leqslant x \leqslant 1 \\ k - x & 1 \leqslant x \leqslant 2 \\ 0 & \text{otherwise} \end{cases}$$
Find k.
Find also the mean, μ, and show that the variance, σ^2, is equal to $\frac{1}{6}$.

Determine the probability that a future observation lies in the interval $(\mu - \sigma, \mu)$.

11 The amount of cloud cover is measured on a scale from 0 to 1. In this country a reasonable model for the amount of cloud cover, X, at midday during the spring is provided by the probability density function f, defined below.
$$f(x) = \begin{cases} 8(x - 0.5)^2 + c & 0 \leqslant x \leqslant 1 \\ 0 & \text{otherwise} \end{cases}$$
Calculate the value of the constant c.

Sketch the graph of $f(x)$ for $0 \leqslant x \leqslant 1$.

State the mean cloud cover.

Determine, correct to 3 decimal places, the value x_0 which is such that $P(X > x_0) = 0.95$.

11.5 Obtaining f from F

Since F can be obtained by integrating f, f can be obtained by differentiating F. The value of f(b) is therefore the slope of F at the point where $x = b$.

Example 13

The random variable X has cdf given by:

$$F(x) = \begin{cases} 0 & x \leqslant 1 \\ \dfrac{(x-1)^3}{8} & 1 \leqslant x \leqslant 3 \\ 1 & x \geqslant 3 \end{cases}$$

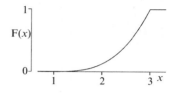

Find the form of the pdf of X.

———

Evidently f(x) is equal to 0 for $x < 1$ and for $x > 3$, since F(x) is unchanging in these regions. Writing F$'(x)$ for $\dfrac{dF(x)}{dx}$, for the interval $1 < x < 3$ we calculate:

$$\text{f}(x) = \text{F}'(x)$$

$$= \frac{3}{8}(x-1)^2$$

Hence:

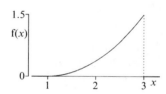

$$f(x) = \begin{cases} \dfrac{3}{8}(x-1)^2 & 1 < x < 3 \\ 0 & \text{otherwise} \end{cases}$$

Example 14

The continuous random variable X has cdf given by:

$$F(x) = \begin{cases} 0 & x \leqslant -1 \\ \alpha x + \alpha & -1 \leqslant x \leqslant 0 \\ 2\alpha x + \alpha & 0 \leqslant x \leqslant 1 \\ 3\alpha & x \geqslant 1 \end{cases}$$

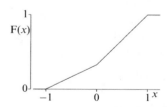

where α is a constant.
Determine (i) the value of α, (ii) the pdf of X.

———

We know that F(∞) = 1. Hence $3\alpha = 1$, implying that $\alpha = \frac{1}{3}$. Also it is clear that f(x) = 0 for $x < -1$ and for $x > 1$. Differentiating F(x) for each of the remaining intervals, and substituting for α, we get:

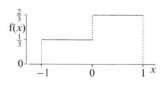

$$\text{f}(x) = \begin{cases} \frac{1}{3} & -1 < x < 0 \\ \frac{2}{3} & 0 < x < 1 \\ 0 & \text{otherwise} \end{cases}$$

Exercises 11d

In Questions **1–7** the pdf and cdf of the continuous random variable X are f and F respectively.

1 It is given that:

$$F(x) = \begin{cases} 0 & x \leqslant 1 \\ \dfrac{2(x-1)}{x} & 1 \leqslant x \leqslant 2 \\ 1 & x \geqslant 2 \end{cases}$$

Find (i) f(x), (ii) E(X).

2 It is given that:

$$F(x) = \begin{cases} 0 & x \leqslant 1 \\ a + bx^2 & 1 \leqslant x \leqslant 3 \\ 1 & x \geqslant 3 \end{cases}$$

Find the values of the constants (i) a and (ii) b.
Find (iii) E(X) and (iv) Var(X).

3 It is given that:

$$F(x) = \begin{cases} 0 & x \leqslant a \\ k(x-a)^2 & a \leqslant x \leqslant 2a \\ 1 & x \geqslant 2a \end{cases}$$

Find k in terms of a.
It is also given that $P(X > 2) = \frac{3}{4}$.
(i) Find the value of a.
(ii) Find f(x).
(iii) Find the median of X.

4 It is given that:

$$F(x) = \begin{cases} 0 & x \leqslant 4 \\ \frac{1}{4}(x-4) & 4 \leqslant x \leqslant 8 \\ 1 & x \geqslant 8 \end{cases}$$

(i) Find f(x) and sketch its graph.
Find also:
(ii) the median of X,
(iii) the lower and upper quartiles of X,
(iv) E(X).

5 It is given that:

$$F(x) = \begin{cases} 0 & x \leqslant -4 \\ \frac{1}{8}(x+4) & -4 \leqslant x \leqslant 0 \\ \frac{1}{2} & 0 \leqslant x \leqslant 4 \\ \frac{1}{8}x & 4 \leqslant x \leqslant 8 \\ 1 & x \geqslant 8 \end{cases}$$

Find (i) f(x) and sketch its graph.
Find also:
(ii) E(X),
(iii) $P[|X - E(X)| < 2]$,
(iv) Var(X).

6 It is given that:

$$F(x) = \begin{cases} 0 & x \leqslant 0 \\ \frac{1}{4}x & 0 \leqslant x \leqslant 2 \\ a + bx & 2 \leqslant x \leqslant 3 \\ 1 & x \geqslant 3 \end{cases}$$

Find:
(i) the constants a and b, (ii) f(x).
Sketch the graph of f.
(iii) Find the lower and upper quartiles of X.

7 It is given that:

$$F(x) = \begin{cases} 0 & x \leqslant 0 \\ \frac{1}{2}x^2 & 0 \leqslant x \leqslant 1 \\ a + bx^3 & 1 \leqslant x \leqslant 2 \\ 1 & x \geqslant 2 \end{cases}$$

Find:
(i) the constants a and b, (ii) f(x).
Sketch the graph of f.
Find:
(iii) the mode, (iv) the median,
(v) the mean of X.

11.6 The uniform (rectangular) distribution

We encountered a random variable having a uniform distribution in Example 4 of this chapter. Its characteristic is that, for the entire range of possible values of X (from a to b, say), f is constant:

$$f(x) = \begin{cases} \dfrac{1}{b-a} & a < x < b \\ 0 & \text{otherwise} \end{cases} \qquad (11.13)$$

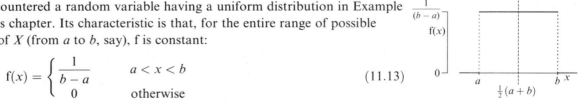

Between a and b the probability density is **uniform** and the resulting shape is **rectangular**. The rectangle has width $(b-a)$ and height $\dfrac{1}{b-a}$, so that its area (equal to *height* × *width*) is equal to 1, as required.

Since the probability density is symmetrical about the line $x = \frac{1}{2}(a+b)$, the mean, $E(X)$, and the median, m, are both equal to $\frac{1}{2}(a+b)$. The cumulative distribution function, F, is given by:

$$F(c) = P(X \leqslant c) = \int_a^c \frac{1}{b-a}\,dx$$

$$= \left[\frac{x}{b-a}\right]_a^c$$

$$= \frac{c-a}{b-a}$$

Formally, therefore, we have:

$$F(x) = \begin{cases} 0 & x \leqslant a \\ \dfrac{x-a}{b-a} & a \leqslant x \leqslant b \\ 1 & x \geqslant b \end{cases} \tag{11.14}$$

To find the variance of X, we use the transformation:

$$Y = \frac{X-a}{b-a}$$

which amounts to a translation followed by an enlargement. It follows that the distribution of Y is also uniform, but with a revised range. Since $X = a$ gives $Y = 0$ and $X = b$ gives $Y = 1$, the range for Y is from 0 to 1. The probability density function of Y, g say, is given by:

$$g(y) = \begin{cases} 1 & 0 < y < 1 \\ 0 & \text{otherwise} \end{cases}$$

We see that $E(Y) = \frac{1}{2}$, while $E(Y^2)$ is given by:

$$E(Y^2) = \int_0^1 y^2\,dy$$

$$= \left[\frac{y^3}{3}\right]_0^1$$

$$= \frac{1}{3}$$

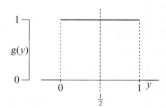

Hence:

$$\text{Var}(Y) = E(Y^2) - \{E(Y)\}^2 = \frac{1}{3} - \left(\frac{1}{2}\right)^2 = \frac{1}{12}$$

Since $X = a + (b-a)Y$

$$\text{Var}(X) = (b-a)^2\,\text{Var}(Y)$$

and hence, for a general uniform random variable, X:

$$\text{Var}(X) = \frac{(b-a)^2}{12}$$

Example 15

The distance between two points is to be measured correct to the nearest tenth of a kilometre.
Determine the mean and standard deviation of the associated round-off error. Four points A, B, C and D lie, in that order, on a straight line. The

lengths of the distances AB, BC and CD are each to be measured correct to the nearest tenth of a kilometre.

Determine the mean and standard deviation of the difference between the total of the three measured lengths and the true overall length.

————

Suppose the length of AB is given as 45.2 km. The true length of AB could be any value between 45.15 km and 45.25 km. The round-off error (in km), X, could therefore take any value between -0.05 and 0.05. The random variable X therefore has a uniform distribution with $b = 0.05$ and $a = -0.05$, so $\dfrac{1}{(b-a)} = \dfrac{1}{0.1} = 10$ and therefore:

$$f(x) = \begin{cases} 10 & -0.05 < x < 0.05 \\ 0 & \text{otherwise} \end{cases}$$

Evidently $E(X) = 0$. We need $E(X^2)$:

$$E(X^2) = \int_{-0.05}^{0.05} 10x^2 \, dx$$

$$= 10 \left[\frac{x^3}{3} \right]_{-0.05}^{0.05}$$

$$= 10 \left(\frac{0.000\,125}{3} - \frac{-0.000\,125}{3} \right)$$

$$= 0.000\,833\,33$$

Hence $\text{Var}(X) = 0.000\,833\,33$ and thus the standard deviation of the round-off error is $\sqrt{0.000\,833\,33} = 0.029$ (to 3 d.p.).

Denote the three round-off errors by X, Y and Z. These variables have independent uniform distributions each with mean 0 and variance $0.000\,833\,33$. Their sum therefore has mean 0 and variance $3 \times 0.000\,833\,33 = 0.0025$. The mean and standard deviation of the difference between the total of the three stated lengths and the true overall length are therefore 0 and $\sqrt{0.0025} = 0.05$.

Example 16

The radius of a circle, R cm, has a uniform distribution in the interval from 1 to 3.

Denoting the area of the circle by A cm^2, determine the pdf of A and the mean area of the circle.

————

The pdf of R is given by:

$$f(r) = \begin{cases} \frac{1}{2} & 1 < r < 3 \\ 0 & \text{otherwise} \end{cases}$$

For $1 \leqslant r \leqslant 3$, the cdf is given by:

$$F(r) = \tfrac{1}{2}(r - 1)$$

Since R takes values in the interval $(1,3)$, A takes values in the interval $(\pi, 9\pi)$. Let g be the required pdf of A, then g will be zero outside the interval $(\pi, 9\pi)$. In order to find g, we first find the cdf, G. Confining attention to values of A in the interval $(\pi, 9\pi)$, we use the fact that, since $A = \pi R^2$, it follows that $A \leqslant a$ is equivalent to $\pi R^2 \leqslant a$ and hence to $R \leqslant \sqrt{\dfrac{a}{\pi}}$. So:

$$G(a) = P(A \leqslant a)$$

$$= P\left(R \leqslant \sqrt{\frac{a}{\pi}}\right)$$

$$= F\left(\sqrt{\frac{a}{\pi}}\right)$$

$$= \frac{1}{2}\left(\sqrt{\frac{a}{\pi}} - 1\right)$$

$$= \frac{1}{2\sqrt{\pi}}\sqrt{a} - \frac{1}{2}$$

Differentiating G with respect to a we get, for $\pi < a < 9\pi$:

$$g(a) = \frac{1}{2\sqrt{\pi}}\frac{1}{2\sqrt{a}}$$

$$= \frac{1}{4\sqrt{a\pi}}$$

We will give two ways of obtaining $E(A)$. The first method uses the pdf of A in a straightforward fashion – the second method is much quicker but needs cunning!

Using $g(a)$ we have

$$E(A) = \int_{\pi}^{9\pi} a\frac{1}{4\sqrt{a\pi}}\,\mathrm{d}a$$

$$= \frac{1}{4\sqrt{\pi}}\int_{\pi}^{9\pi} a^{\frac{1}{2}}\,\mathrm{d}a$$

$$= \frac{1}{4\sqrt{\pi}}\left[\frac{2}{3}a^{\frac{3}{2}}\right]_{\pi}^{9\pi}$$

$$= \frac{1}{6\sqrt{\pi}}\left(9\pi\sqrt{9\pi} - \pi\sqrt{\pi}\right)$$

$$= \frac{\pi}{6}(27 - 1)$$

$$= \frac{13\pi}{3}$$

The second (cunning) method uses R. Since R is uniform in $(1, 3)$, we have $E(R) = 2$ and $\mathrm{Var}(R) = \frac{1}{12}(3 - 1)^2 = \frac{1}{3}$. Thus:

$$E(R^2) = \mathrm{Var}(R) + \{E(R)\}^2 = \frac{1}{3} + 2^2 = \frac{13}{3}$$

Thus $E(A) = E(\pi R^2) = \pi E(R^2) = \frac{13}{3}\pi$, as before.

The mean area of the circle is $\frac{13}{3}\pi$ cm^2. Note that this is not simply the area of the circle corresponding to the average radius (2 cm), since, in general, $E(R^2) \neq \{E(R)\}^2$.

Practical _____

> Any type of round-off error is likely to have a uniform distribution. An 'instantaneous' example is provided by a glance at a watch! On a given signal, all members of the class should record the number of seconds shown by their watches. The numbers recorded are likely to be observations from a continuous uniform distribution with range 0 to 60, though, since they are recorded as integers, this distribution is being approximated by its discrete counterpart.
> Summarise the class data using a stem-and-leaf diagram.
> Does the uniform distribution seem appropriate?
>   You can easily generate more observations from your own watch by recording the numbers of seconds that it shows at odd times during the day (i.e. every now and then, when you remember – this could be a way of livening up the lessons in which you are not studying Statistics!).

Exercises 11e _____

1 The continuous random variable X has a uniform distribution on the interval $0 < x < 2$. Find (i) the pdf of X, (ii) the cdf of X.

The random variable Y is defined by $Y = 2X$. Find (iii) $P(Y < y)$ and hence (iv) $g(y)$, where g is the pdf of Y.

Verify that $E(Y) = 2E(X)$.

2 The continuous random variable U is uniformly distributed on the interval $a < u < b$. Given that $E(U) = 4$ and $\text{Var}(U) = 3$, find (i) a and b, (ii) $P(U > 5)$.

3 The continuous random variable T is uniformly distributed on the interval $0 < t < 100$.
(i) Find $P(20 < T < 60)$.
(ii) Denoting the mean and standard deviation of T by μ and σ respectively, find $P(|T - \mu| < \sigma)$.

4 The continuous random variable S is uniformly distributed on the interval $c < s < d$. Given that $P(S < 3) = \frac{1}{4}$ and $P(S < 7) = \frac{3}{4}$, find c and d.

5 The continuous random variable Y is uniformly distributed on the interval $0 < y < 2$ and $X = 3Y + 4$. Show that X is uniformly distributed on an interval $a < x < b$, giving the values of a and b.

6 (a) A pointed arrow is thrown on to a table and the continuous random variable A is the angle (acute or obtuse and measured in degrees) between the direction of the arrow and due north, measured so that $0 < A < 180$. Find (i) $E(A)$, (ii) $\text{Var}(A)$.

(b) A pointed arrow is thrown on to a table and the continuous random variable B is the bearing (in degrees) of the direction of the arrow, measured so that $0 < B < 360$. Find (i) $E(B)$, (ii) $\text{Var}(B)$.

7 Many calculators and computers generate random numbers which are approximately uniformly distributed on the interval $0 < u < 1$. Let U be such a random variable.
(a) It is desired to find constants h and k such that $X = hU + k$ is uniformly distributed on the interval $a < x < b$. Find h and k in terms of a and b.
(b) It is desired to find constants r and s such that $rU + s$ is uniformly distributed with mean μ and standard deviation σ. Find r and s in terms of μ and σ.

(Strictly these random numbers have a discrete uniform distribution, since they only contain a fixed number of decimal places, say 9. As the difference between neighbouring numbers is 10^{-9}, the distribution may be taken to be continuous.)

8 Mrs Parent occasionally allows her daughter to borrow her car. When Mrs Parent leaves the car at home after driving it, the amount of petrol in the tank is uniformly distributed between 10 litres and 50 litres. When her daughter leaves the car at home after having borrowed it, the amount of petrol in the tank is uniformly distributed between 0 litres and 20 litres. (Mrs Parent is none too pleased!) Mrs Parent is the driver for 80% of the time and her daughter is the driver for the remaining 20% of the time.

Mr Parent checks the car at home, not knowing who drove it last.

Find the probability that there is less than 15 litres of petrol in the tank.

9 The continuous random variable Θ (capital theta) is uniformly distributed on the interval $0 < \theta < \frac{1}{2}\pi$.

Find:

(i) $E(\sin \Theta)$, (ii) $E(\cos \Theta)$, (iii) $E(\sin^2 \Theta)$, (iv) $E(\cos^2 \Theta)$.

Verify that:

$$E(\sin^2 \Theta) + E(\cos^2 \Theta) = 1$$

Verify also that $E(\sin \Theta) \neq \sin\{E(\Theta)\}$ and $E(\cos \Theta) \neq \cos\{E(\Theta)\}$.

10 When Luke throws a dart at a circular target of radius a cm the dart is certain to hit the target and all points in the circle are equally likely.

(i) Find the probability that a dart lands within x cm of the centre of the target.

(ii) The probability density function for the distribution of the distance of a dart from the centre is of the form

$$\phi(x) = \begin{cases} * & (0 \leqslant x \leqslant a) \\ 0 & \text{(otherwise)} \end{cases}$$

Show that the function of x that should replace the asterisk is $\dfrac{2x}{a^2}$.

(iii) Find the mean of the distribution.   [SMP]

11 X is a random variable having probability density function f, where

$$f(x) = \frac{1}{h}, \quad 0 < x < h,$$
$$f(x) = 0, \quad \text{otherwise.}$$

Given that $Y = X(h - X)$, find $E(Y)$.

Find also the probability that Y is greater than $\dfrac{3h^2}{16}$.   [ULSEB(P)]

11.7 The exponential distribution

When events occur at random points in time according to a Poisson process (see Chapter 10), the number of events in an interval of time has a Poisson distribution. Suppose that these events are occurring at a rate λ per unit time (where λ is some positive constant). Obviously the *mean* time between events will be $\dfrac{1}{\lambda}$, but the individual time intervals will vary about this mean. The distribution of these intervals is known as the **exponential distribution**. Later in this section we shall show that the pdf is given by:

$$f(x) = \begin{cases} \lambda e^{-\lambda x} & x > 0 \\ 0 & \text{otherwise} \end{cases} \tag{11.15}$$

For $b \geqslant 0$, the distribution function is given by:

$$F(b) = \int_0^b \lambda e^{-\lambda x} \, dx = \left[-e^{-\lambda x} \right]_0^b = \left[-e^{-\lambda x} \right]_0^b$$
$$= 1 - e^{-\lambda b}$$

so:

$$P(X < x) = 1 - e^{-\lambda x} \tag{11.16}$$
$$P(X > x) = e^{-\lambda x} \tag{11.17}$$
$$P(a < X < b) = e^{-\lambda a} - e^{-\lambda b} \tag{11.18}$$

Example 17

During the morning a busy office receives on average 20 telephone calls an hour. The office manager arrives at 10 minutes past 10 (he worked late the previous night!).
Determine the probability that there are no calls during the next 10 minutes.

A rate of 20 calls an hour implies a rate of $\frac{20}{60} = \frac{1}{3}$ per minute. There are no calls during the next 10 minutes if the time to the next call is at least 10 minutes. Denoting the time (in minutes) to the next call by X, we require $P(X > 10)$.

Since the calls occur at random points in time, X has an exponential distribution with $\lambda = \frac{1}{3}$. Now:

$$P(X > 10) = e^{-\frac{10}{3}} = 0.036 \text{ (to 3 d.p.)}$$

Shape of the exponential distribution

Like the geometric distribution, which is its discrete analogue, the exponential distribution has mode 0:

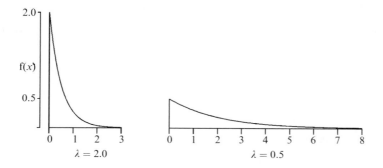

An unusual property of the exponential distribution is that we can throw away a lump of it (working from the left) and the remainder, when rescaled to an area of 1, will look just like the original.

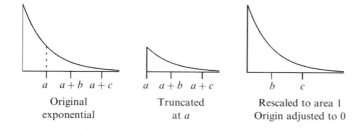

| Original | Truncated | Rescaled to area 1 |
| exponential | at a | Origin adjusted to 0 |

The result of this relation is that:

$$P[X > (a + b)|X > a] = P(X > b) \qquad (11.19)$$

This is known as the **lack of memory** property of a Poisson process. The probability of the next event not occurring in the next b units of time is independent of everything that has occurred up till now (time a):

The proof of the property is simple:

$$P(X > a + b | X > a) = \frac{P[(X > a + b) \cap (X > a)]}{P(X > a)}$$

$$= \frac{P(X > a + b)}{P(X > a)}$$

$$= \frac{e^{-\lambda(a+b)}}{e^{-\lambda a}}$$

$$= e^{-\lambda b}$$

$$= P(X > b)$$

An explanation of the result is as follows. Adding an extra event randomly to a sequence of n random events does not alter their randomness! All that happens is that the number of random events has increased by one. The general case was discussed in Section 10.7 (p. 250). In the present case the 'extra event' is 'our starting to look at the random sequence' – a peculiar event we agree!

Expectation and variance of the exponential distribution

The range of the exponential distribution is infinite and we require:

$$E(X) = \int_0^\infty x\lambda e^{-\lambda x}\, dx$$

To avoid problems with infinite integrals, we consider the interval $(0, G)$, where G is some Gigantic number. For $E(X)$ we need to determine the value of:

$$\int_0^G x\lambda e^{-\lambda x}\, dx$$

and then let G tend to ∞.

Now:

$$\int \lambda e^{-\lambda x}\, dx = -e^{-\lambda x}$$

or, equivalently:

$$\lambda e^{-\lambda x} = \frac{d}{dx}\left(-e^{-\lambda x}\right)$$

Integrating by parts therefore gives:

$$\int_0^G (x)\left(\lambda e^{-\lambda x}\right) dx = \left[(x)\left(-e^{-\lambda x}\right)\right]_0^G - \int_0^G (1)\left(-e^{-\lambda x}\right) dx$$

$$= \left\{G \times \left(e^{-\lambda G}\right)\right\} - \left\{0 \times (-1)\right\} + \int_0^G e^{-\lambda x}\, dx$$

As G increases so the first term on the right-hand side approaches zero (try it on your calculator!) and the second term is zero anyway. In the limit, therefore:

$$E(X) = \int_0^\infty e^{-\lambda x}\, dx$$

Now:

$$\int_0^\infty e^{-\lambda x}\, dx = \frac{1}{\lambda} \int_0^\infty \lambda e^{-\lambda x} dx$$

and $\lambda e^{-\lambda x}$ is the original pdf specified in Equation (11.15). Consequently:

$$\int_0^\infty \lambda e^{-\lambda x}\, dx = 1$$

and hence:

$$E(X) = \frac{1}{\lambda}$$

For $\mathrm{Var}(X)$ we need $E(X^2)$ which is the limit of $\int_0^G x^2 \lambda e^{-\lambda x}\, dx$. Integration by parts gives:

$$\int_0^G (x^2)(\lambda e^{-\lambda x})\, dx = \left[(x^2)(-e^{-\lambda x}) \right]_0^G - \int_0^G (2x)(-e^{-\lambda x})\, dx$$

$$= \left\{ G^2 \times (e^{-\lambda G}) \right\} - \left\{ 0 \times (-1) \right\} + \int_0^G 2x e^{-\lambda x}\, dx$$

Letting G increase towards ∞, the first term on the right-hand side approaches zero and so:

$$E(X^2) = 2 \int_0^\infty x e^{-\lambda x}\, dx$$

$$= \frac{2}{\lambda} \int_0^\infty x \lambda e^{-\lambda x}\, dx$$

$$= \frac{2}{\lambda} E(X)$$

$$= \frac{2}{\lambda^2}$$

Using $\mathrm{Var}(X) = E(X^2) - \{E(X)\}^2$, we therefore have that:

$$\mathrm{Var}(X) = \frac{1}{\lambda^2}$$

Connection with a Poisson process

Consider a sequence of independent events occurring at random points in time at a rate λ; in other words a Poisson process with parameter λ. We start examining this process at an arbitrary time $t = 0$ and denote the random variable 'the time to the first event' by X. Some possible alternative futures are shown in the diagram.

Now:

$$P(X > x) = P[\text{No events occur in the time interval } (0, x)]$$

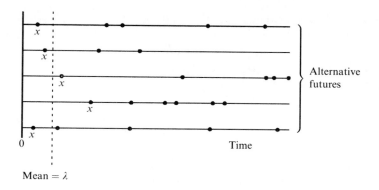

Mean $= \lambda$

To find this probability, we note that the mean number of events occurring in a time interval of length x is λx, and that the probability of obtaining the value 0 from a Poisson distribution with mean λx is:

$$\frac{(\lambda x)^0 e^{-\lambda x}}{0!} = e^{-\lambda x}$$

So:

$$P(X > x) = e^{-\lambda x}$$

and:

$$F(x) = 1 - P(X > x) = 1 - e^{-\lambda x}$$

Finally, differentiating with respect to x, we obtain the pdf of X:

$$f(x) = \lambda e^{-\lambda x}$$

We therefore see that in a Poisson process having events occurring at rate λ per unit time, the time to the first occurrence is an observation from an exponential distribution with mean $\dfrac{1}{\lambda}$.

Example 18

The random variable X is the lifetime, in years, of a particular type of bulb which is in constant use as part of an advertising display. The pdf of X is given by:

$$f(x) = \begin{cases} k e^{-2x} & x > 0 \\ 0 & \text{otherwise} \end{cases}$$

Determine the value of the constant k and the median lifetime of a bulb.

Two bulbs are chosen at random. One bulb is found to be 3 months old and the other to be 5 months old.
Determine the probability that both bulbs will still be working in 4 months' time.

$$F(b) = \int_0^b k e^{-2x} \, dx = k \left[-\frac{1}{2} e^{-2x} \right]_0^b$$

$$= \frac{k}{2} \left(1 - e^{-2b} \right)$$

This implies that $F(\infty) = \dfrac{k}{2}$. Since, for any distribution, $F(\infty) = 1$, it follows that $k = 2$.

We therefore have that $F(b) = 1 - e^{-2b}$. The median, m, is the solution of the equation $F(m) = 0.5$. Hence we need to solve:

$$e^{-2m} = 0.5$$

Taking natural logarithms of both sides we get:

$$-2m = \ln(0.5)$$

Hence:

$$m = -\tfrac{1}{2}\ln(0.5) = 0.347 \text{ (to 3 d.p.)}$$

The median lifetime of a bulb is 0.347 years (a little over 4 months).

Since the lifetimes of the bulbs have an exponential distribution, we can use the 'lack of memory' property and argue that the probability of a three-month-old bulb lasting for at least a further four months is the same as the probability that a new bulb will last for that long. We therefore require $P(X > \tfrac{1}{3})$, since X was measured in years. Now:

$$P(X > \tfrac{1}{3}) = 1 - F(\tfrac{1}{3}) = e^{-\tfrac{2}{3}}$$

The probability that both bulbs last this long is therefore:

$$(e^{-\tfrac{2}{3}})^2 = e^{-\tfrac{4}{3}} = 0.264 \text{ (to 3 d.p.)}$$

Example 19

A bargain-hunter discovers a large roll of material that is being sold at a greatly reduced price because it contains flaws. These flaws occur at random locations down the length of the roll. The mean length of cloth between successive flaws is 0.5 metres.
Determine the probability that the first 1.5 metres of the roll contain no flaws.
Determine the probability that the first flaw occurs at between 0.6 and 0.8 metres from the start of the roll.

_____

Random locations imply a Poisson process. The mean length between flaws of 0.5 metres implies a rate of occurrence (λ) of 2 per metre. The probability of no flaws in the first 1.5 metres may be obtained using either the Poisson or the exponential distributions. Using the latter with $\lambda = 2$ we have:

$$P(X > 1.5) = e^{-3} = 0.050 \text{ (to 3 d.p.)}$$

To determine the probability of the first flaw occurring between 0.6 and 0.8 metres from the start we need:

$$F(0.8) - F(0.6) = (1 - e^{-1.6}) - (1 - e^{-1.2}) = 0.099 \text{ (to 3 d.p.)}$$

Practical _____

> *This practical requires a Geiger counter and an accurate watch! Record the times of occurrence of the clicks of the meter and hence derive the 'inter-click' times. Represent the data using a histogram. You will need to use wider class ranges for the longer time intervals. Compute $\bar{x}$, the mean time between clicks.*

Compare the relative frequencies for your classes with the corresponding theoretical probabilities based on an exponential distribution having

$$\lambda = \frac{1}{\bar{x}}.$$

Project

This is another traffic project. Choose an 'event' to be something that occurs reasonably frequently (so that you don't have to wait for hours and so that you are not overrun by events!). Examples are 'a lorry', 'a red car', 'a woman driver'. It will be sensible to decide on your event after you have watched the traffic for a while and have identified something that occurs on average every 1 to 2 minutes.

Record the times of occurrence of your chosen events. Then represent the data using a histogram and follow the remaining instructions from the Geiger counter practical (above).

Does it appear that your chosen event occurs at random points in time or is there evidence of clustering (too many short time intervals) or regularity (too few short time intervals)?

Exercises 11f

1 The continuous random variable X has an exponential distribution with mean 3. Find
(i) the pdf of X, (ii) the cdf of X, (iii) $P(X < 3)$, (iv) $P(X > 4)$, (v) $P(3 < X < 4)$.

2 The continuous random variable X has an exponential distribution with mean μ. Given that $P(X < 1) = \frac{1}{4}$, find μ.

3 The continuous random variable T has an exponential distribution.
Given that $\text{Var}(T) = 4$, find (i) $P(T > 1)$, (ii) $P(T < 2)$.

4 A child is given a plastic toy. The time before he loses it has an exponential distribution. The probability that he loses it in the first 10 days is 0.35.
Find the probability that he has not lost it in the first 30 days.

5 The lifetime of a toaster element has an exponential distribution with mean 9 years. A particular element is still working after 4 years. Find the probability that it will still be working after 5 more years.

6 A motorist joins a motorway. The distance, X miles, that she travels before she sees a police car has an exponential distribution with mean 50 miles. This is also the distribution of the distances travelled between subsequent sightings. The distance that she travels between seeing the first and second police cars is Y miles.
Find (i) $P(X > 100)$, (ii) $P(Y > 50)$.

7 The lifetime of a new type of lightbulb has an exponential distribution with mean 4000 hours.
Determine the probability that:
(i) a randomly chosen bulb will last more than 3000 hours,
(ii) two randomly chosen bulbs will each last more than 3000 hours.

There are ten of the new bulbs in a house. Assuming independence, determine the probability that none of these bulbs will last for less than 1000 hours.

8 Faults occur at random locations along a nylon line. The average rate of occurrence is 2 per 100 m.
Determine the probability that:
(i) 200 m of the line includes exactly three faults,
(ii) the length of line between the second and third faults exceeds 100 m.

294 Understanding Statistics

Baron Augustin-Louis Cauchy (1789–1857) was a French mathematician who, after Euler, may well have been the second most prolific writer of mathematics ever. (Leonhard Euler, the 18th-century Swiss mathematician, is often regarded as the founder of modern pure mathematics, and his collected works are still being published!) Cauchy's work covered many branches of pure and applied mathematics and his name is linked to a variety of tests, formulae and equations. He was a keen royalist and was therefore excluded from his professorship for 18 years because he refused to take the oath of allegiance to the Republican Government.

11.8 The Cauchy distribution

This distribution has pdf:

$$f(x) = \frac{1}{\pi\{1 + (x-a)^2\}} \qquad -\infty < x < \infty$$

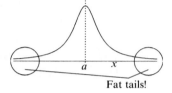

Fat tails!

While it is possible to think of situations in which observations from a Cauchy distribution occur naturally, the real importance of this distribution is as an AWFUL WARNING! The sketch of the distribution shows a symmetric curve that seems entirely unproblematical. However, read on!

By symmetry, the median is equal to a and it appears as though this will also be the mean of the distribution. However, the Cauchy distribution actually has *no mean!* A further problem is that $E(X - a)^2$ is infinite. The mathematics involves infinite integrals and is a bit messy, so, instead of proving these statements, we look at some sample data. The diagram below shows the changing value of the sample mean as a random sample increases in size.

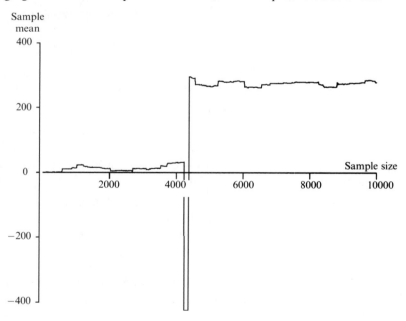

Very occasionally, incredibly large or incredibly small values of X occur. When they occur, they cause a huge and unpredictable jump in the size of the sample mean. There is no evidence of the relatively smooth progress towards some limiting value that we have come to expect.

The Cauchy distribution is the statistician's favourite counter-example to show that the 'obvious' is not always true!

11.9 Moments

Older books on Statistics often started with a chapter on moments, but this seems to have gone out of fashion. They are mentioned here for completeness and because of the title of the following section. For a given random variable X, the kth **central moment** is simply $E(X^k)$, so that $\mu = E(X)$ is the first central moment. The kth **moment about the mean** is $E[(X - \mu)^k]$ so that $\sigma^2 = E[(X - \mu)^2]$ is the second moment about the mean.

11.10 $E(e^{tX})$ – the moment generating function

The **moment generating function (mgf** for short) of a random variable X, provides an alternative procedure for calculating central moments ($E(X)$, $E(X^2)$, ...) which is sometimes much easier to use than the direct method of Section 11.4 (p. 275) and is also useful in establishing some standard results. The mgf of X is usually denoted by $M_X(t)$, or, when there is no possibility of confusion, by $M(t)$. It is defined by:

$$M(t) = E(e^{tX}) \tag{11.20}$$

where t is a variable that can be restricted to take values close to zero. The mgf can be used with either discrete or continuous random variables, though the pgf is usually preferred for the discrete case. In either case, using the series expansion for e^c:

$$e^c = 1 + \frac{c}{1!} + \frac{c^2}{2!} + \frac{c^3}{3!} + \cdots$$

together with the result:

$$E[g(X) + h(X)] = E[g(X)] + E[h(X)]$$

we can write:

$$M(t) = 1 + E\left(\frac{tX}{1!}\right) + E\left(\frac{t^2 X^2}{2!}\right) + E\left(\frac{t^3 X^3}{3!}\right) + \cdots$$

$$= 1 + \frac{t}{1!} E(X) + \frac{t^2}{2!} E(X^2) + \frac{t^3}{3!} E(X^3) + \cdots$$

We see that $E(X^k)$ **is the coefficient of** $\dfrac{t^k}{k!}$ **in the series expansion of** $M(t)$.

The Maclaurin expansion for $M(t)$ is:

$$M(0) + \frac{t}{1!} M'(0) + \frac{t^2}{2!} M''(0) + \frac{t^3}{3!} M'''(0) + \cdots$$

where $M'(t)$, $M''(t)$, etc are successive derivatives of $M(t)$ and each is being evaluated at $t = 0$. Comparing coefficients of $\dfrac{t^k}{k!}$ in the two formulae we see at once that:

$$M'(0) = E(X) \tag{11.21}$$

$$M''(0) = E(X^2) \tag{11.22}$$

so that:

$$\mathrm{Var}(X) = M''(0) - \{M'(0)\}^2 \tag{11.23}$$

Evidently putting $t = 0$ in the kth derivative of $M(t)$ will give us the value of $E(X^k)$.

Example 20

Use the moment generating function to find the mean and variance of the continuous random variable X which has an exponential distribution with parameter λ.

Since the pdf of X is given by $f(x) = \lambda e^{-\lambda x}$, for $x > 0$, the mgf is given by:

$$\mathbf{M}(t) = \mathrm{E}(e^{tX}) = \int_0^\infty e^{tx}\lambda e^{-\lambda x}\,dx$$

Now:

$$\int_0^G e^{tx}\lambda e^{-\lambda x}\,dx = \lambda \int_0^G e^{-(\lambda - t)x}\,dx$$

$$= \lambda\left[-\frac{e^{-(\lambda - t)x}}{\lambda - t}\right]_0^G$$

$$= \frac{\lambda}{\lambda - t}\left\{-e^{-(\lambda - t)G} + 1\right\}$$

As G increases so the exponential term approaches zero (provided t is small enough for $\lambda - t$ to be positive) and hence:

$$\mathbf{M}(t) = \frac{\lambda}{\lambda - t}$$

Differentiating we get:

$$\mathbf{M}'(t) = \frac{\lambda}{(\lambda - t)^2}$$

$$\mathbf{M}''(t) = \frac{2\lambda}{(\lambda - t)^3}$$

Finally, putting t equal to zero, we get $\mathbf{M}'(0) = \frac{1}{\lambda}$ and $\mathbf{M}''(0) = \frac{2}{\lambda^2}$, so that $\mathrm{E}(X) = \frac{1}{\lambda}$ and $\mathrm{Var}(X) = \frac{2}{\lambda^2} - \left(\frac{1}{\lambda}\right)^2 = \frac{1}{\lambda^2}$.

Alternatively, instead of differentiating $\mathbf{M}(t)$, we can use the binomial expansion, which is valid for $|t| < \lambda$:

$$\mathbf{M}(t) = \frac{\lambda}{\lambda - t} = \left(1 - \frac{t}{\lambda}\right)^{-1}$$

$$= 1 + \frac{t}{\lambda} + \left(\frac{t}{\lambda}\right)^2 + \ldots$$

$$= 1 + \frac{1}{\lambda}t + \left(\frac{2}{\lambda^2}\right)\frac{t^2}{2!} + \ldots$$

The coefficient of t is $\mathrm{E}(X)$, the coefficient of $\frac{t^2}{2!}$ is $\mathrm{E}(X^2)$, and so forth.

These are, of course, the same results as we obtained earlier – but the actual calculations using the mgf are very much simpler.

Mgf of the sum of independent random variables

Suppose U and V are two independent random variables, and suppose that the random variable W is defined by $W = U + V$. Let the moment generating functions of U, V and W be denoted by $M_U(t)$, $M_V(t)$ and $M_W(t)$, respectively. Then:

$$M_W(t) = E(e^{tW}) \qquad \text{by definition}$$
$$= E(e^{tU+tV})$$
$$= E(e^{tU} \times e^{tV})$$
$$= E(e^{tU}) \times E(e^{tV}) \qquad \text{since } U \text{ and } V \text{ are independent}$$
$$= M_U(t) \times M_V(t)$$

This simple result obviously extends to more than two independent random variables. It comes in particularly handy when dealing with independent identically distributed random variables. First of all we introduce the standard alternative notation for the exponential function, which overcomes the problem of small print. Instead of writing e^x, we write $\exp(x)$, so that, for example, $e^{tW} = \exp(tW)$, the mgf of W is $E\{\exp(tW)\}$ and $\exp(tU + tV) = \exp(tU) \times \exp(tV)$.

Let the random variable S be defined by:

$$S = \sum_{i=1}^{n} X_i$$

where $X_1, \ldots, X_n$ are independent and identically distributed random variables, each with mgf $M_X(t)$. The mgf of S is simply:

$$M_S(t) = E[\exp(tS)] = E[\exp(t\Sigma X_i)]$$
$$= E[\exp(tX_1) \times \exp(tX_2) \times \cdots \times \exp(tX_n)]$$
$$= E[\exp(tX_1)] \times E[\exp(tX_2)] \times \cdots E[\exp(tX_n)]$$
$$= \{M_X(t)\}^n$$

with the penultimate step relying on the independence of the random variables. From this expression, on differentiating both sides with respect to t, we obtain:

$$M_S'(t) = n\{M_X(t)\}^{n-1}M_X'(t)$$

Setting $t = 0$, we see that $M_S'(0) = nM_X'(0)$, since $M_X(0) = 1$, and hence we have shown once again that:

$$E(S) = E(\Sigma X_i) = nE(X)$$

One final useful result concerns the mgf of a linear function of a random variable. If $Y = aX + b$, where a and b are constants, then:

$$M_Y(t) = E[\exp(tY)] = E[\exp(atX + bt)]$$
$$= \exp(bt) \times E[\exp(atX)]$$
$$= e^{bt}M_X(at)$$

▼ ▼

Example 21

The random variable X has moment generating function given by:

$$M(t) = 1 + 2t + 3t^2 + 4t^3 + \cdots$$

Determine the mean and variance of X.

—————

The coefficient of t is 2, so $E(X) = 2$. Writing $3t^2$ as $\dfrac{6t^2}{2!}$ reveals that $E(X^2) = 6$ and hence $Var(X) = 6 - 2^2 = 2$.

Alternatively, differentiating, we get $M'(t) = 2 + 6t + \cdots$, so that $M'(0) = 2$, while $M''(t) = 6 + \cdots$ and $M''(0) = 6$.

Example 22

The continuous random variable Y has moment generating function $M_Y(t)$ given by:

$$M_Y(t) = 1 + a_1 t + a_2 t^2 + \cdots$$

(i) Give an expression for the variance of Y.

The continuous random variable Z, which is independent of Y, has moment generating function $M_Z(t)$ given by:

$$M_Z(t) = 1 + b_1 t + b_2 t^2 + \cdots$$

(ii) Obtain an expression for the moment generating function of the random variable W defined by $W = Y - Z$.

(iii) Hence show that $E(Y - Z) = E(Y) - E(Z)$ and that $Var(Y - Z) = Var(Y) + Var(Z)$.

(i) In the expression for $M_Y(t)$, the coefficients of t and t^2 are, by definition, the values of $E(Y)$ and $\dfrac{1}{2}E(Y^2)$, respectively. Hence:

$$Var(Y) = E(Y^2) - \{E(Y)\}^2 = 2a_2 - a_1^2$$

(ii) From the definition:

$$
\begin{aligned}
M_W(t) = E(e^{tW}) &= E(e^{t(Y-Z)}) \\
&= E(e^{tY}e^{-tZ}) \\
&= E(e^{tY}) \times E(e^{-tZ}) \\
&= (1 + a_1 t + a_2 t^2 + \cdots)\{1 + b_1 \times (-t) + b_2 \times (-t)^2 + \cdots\} \\
&= (1 + a_1 t + a_2 t^2 + \cdots)(1 - b_1 t + b_2 t^2 + \cdots) \\
&= 1 + (a_1 - b_1)t + (a_2 - a_1 b_1 + b_2)t^2 + \cdots
\end{aligned}
$$

(iii) $E(W)$, which is the coefficient of t, is given by $(a_1 - b_1)$, which is equal to $E(Y) - E(Z)$, as required.

The coefficient of t^2 is equal to $\frac{1}{2}E(W^2)$. The variance of W is therefore given by:

$$
\begin{aligned}
Var(W) &= E(W^2) - \{E(W)\}^2 \\
&= 2(a_2 - a_1 b_1 + b_2) - (a_1 - b_1)^2 \\
&= (2a_2 - a_1^2) + (2b_2 - b_1^2) \\
&= Var(Y) + Var(Z)
\end{aligned}
$$

as required.

Exercises 11g

1 A line through the point $(0,1)$ in the x–y plane makes a random angle Θ with the y-axis, where Θ has a uniform distribution on the interval $-\frac{1}{2}\pi < \theta < \frac{1}{2}\pi$. The line cuts the x-axis at the point with co-ordinates $(X, 0)$.

Show that:

$$P(X < x) = \frac{1}{2} + \frac{\tan^{-1}x}{\pi}$$

Deduce:

(i)   the pdf of X, and

(ii) that X has a Cauchy distribution.

(iii) Find the value of q such that:

$$P(-q < X < q) = \tfrac{1}{2}$$

2 Using the notation of Section 11.10 and writing $E(X) = \mu$, $\text{Var}(X) = \sigma^2$, show that:

$$M_S''(0) = n(n-1)\mu^2 + nE(X^2)$$

and deduce that $\text{Var}(S) = n\sigma^2$.

3 The random variable U has a continuous uniform distribution on the interval $0 < u < 1$. Show that:

$$M_U(t) = \frac{e^t - 1}{t}$$

Deduce the values of (i) $E(U)$, (ii) $\text{Var}(U)$, (iii) $E(U^3)$, (iv) $E(U^4)$.

Check your answers to (iii) and (iv) by direct evaluation.

4 The random variable D has a discrete uniform distribution given by:

$$P\left(D = \frac{r}{n}\right) = \frac{1}{n}, \quad r = 0, 1, 2, \ldots, n-1$$

Show that:

$$M_D(t) = \frac{\exp(t) - 1}{n\left\{\exp\left(\dfrac{t}{n}\right) - 1\right\}}$$

Suppose that n is large. Use the result that $\exp\left(\dfrac{t}{n}\right) - 1 \approx \dfrac{t}{n}$ to show that:

$$M_D(t) \approx \frac{e^t - 1}{t}$$

(Taken with the previous question, this result demonstrates that a continuous uniform distribution can be thought of as the limit of a discrete uniform distribution.)

5 The discrete random variable X has a Poisson distribution with mean λ.

Show that the mgf of X is given by
$$M_X(t) = \exp(\lambda e^t - \lambda).$$

6 The discrete random variable X has a geometric distribution with $P(X = r) = p(1-p)^{r-1}$, $r = 1, 2, 3, \ldots$

Show that:

$$M_X(t) = \frac{p}{e^{-t} - (1-p)}$$

Suppose that ε is small. Put $Y = X\varepsilon$ and $p = \lambda\varepsilon$. Use the result $\exp(-t\varepsilon) - 1 \approx -t\varepsilon$ to show that:

$$M_Y(t) \approx \frac{\lambda}{\lambda - t}$$

(This result demonstrates that an exponential distribution can be thought of as a continuous version of a geometric distribution.)

7 Using the notation of Section 11.10 and writing $E(X) = \mu$, $\text{Var}(X) = \sigma^2$, show that:

$$M_{X-\mu}(t) = 1 + \frac{1}{2}\sigma^2 t^2 + \cdots$$

Use the fact that:

$$\bar{X} - \mu = \frac{1}{n}\sum_{i=1}^{n}(X_i - \mu)$$

to show that:

$$\ln\{M_{\bar{X}-\mu}(t)\} = n\ln\left\{1 + \frac{1}{2}\sigma^2\left(\frac{t}{n}\right)^2 + \cdots\right\}$$

Use the result $\ln(1 + \varepsilon) \approx \varepsilon$ for small ε to deduce that, if n is large,

$$M_{\bar{X}-\mu} \approx \exp\left\{\frac{1}{2}\left(\frac{\sigma^2}{n}\right)t^2\right\}.$$

(We shall see in Section 12.14 that this is the mgf of a $N\left(0, \dfrac{\sigma^2}{n}\right)$ distribution, which establishes the 'Central Limit Theorem'.)

Chapter summary

◆ **Probability density function** (pdf): f(x)

$$f(x) \geqslant 0 \qquad \int_{-\infty}^{\infty} f(x)\,dx = 1$$

$$P(a < X < b) = \int_a^b f(x)\,dx$$

◆ **Cumulative distribution function** (cdf): F(x)

$$F(b) = P(X \leqslant b) = \int_{-\infty}^b f(x)\,dx \qquad f(x) = F'(x)$$

$$F(-\infty) = 0 \qquad P(a < X < b) = F(b) - F(a) \qquad F(\infty) = 1$$

- Median: the value of x for which $F(x) = 0.5$

◆ **Expectations**:

$$E(X) = \int_{-\infty}^{\infty} x\,f(x)\,dx \qquad E[g(X)] = \int_{-\infty}^{\infty} g(x)f(x)\,dx$$

$$E(X^2) = \int_{-\infty}^{\infty} x^2\,f(x)\,dx$$

$$Var(X) = E[(X - \mu)^2] = E(X^2) - E(X)^2$$

◆ **Uniform (rectangular) distribution**:

- $f(x) = \begin{cases} \dfrac{1}{b-a} & a < x < b \\ 0 & \text{otherwise} \end{cases}$

- $E(X) = \dfrac{a+b}{2} \qquad Var(X) = \dfrac{(b-a)^2}{12}$

◆ **Exponential distribution**:

- $f(x) = \begin{cases} \lambda e^{-\lambda x} & x > 0 \\ 0 & \text{otherwise} \end{cases}$

- $E(X) = \dfrac{1}{\lambda} \qquad Var(X) = \dfrac{1}{\lambda^2}$

◆ **Moment generating function (mgf)**:

$$M(t) = E(e^{tX})$$

$$M'(0) = E(X)$$

$$M''(0) = E(X^2)$$

$$M_{U+V}(t) = M_U(t) \times M_V(t), \text{ if } U \text{ and } V \text{ are independent}$$

Exercises 11h (Miscellaneous)

1 The probability density function f of a continuous random variable X is given by

$$f(x) = \begin{cases} kx(2-x) & 0 \leqslant x \leqslant 1, \\ 0 & \text{otherwise.} \end{cases}$$

Show that $k = \frac{3}{2}$ and calculate
(i)   the mean and variance of X,
(ii) $P(X < \frac{1}{2})$,
(iii) the probability that all of three independent observed values of X will be less than $\frac{1}{2}$,
(iv) $P(X > \frac{1}{4} | X < \frac{1}{2})$.   [WJEC]

2 The continuous random variable X has probability density function f given by

$$f(x) = \begin{cases} k(4-x^2) & \text{for } 0 \leqslant x \leqslant 2, \\ 0 & \text{otherwise,} \end{cases}$$

where k is a constant. Show that $k = \frac{3}{16}$ and find the values of $E(X)$ and $Var(X)$.

Find the cumulative distribution function of X, and verify by calculation that the median value of X is between 0.69 and 0.70.

Find also $P(0.69 < X < 0.70)$, giving your answer correct to one significant figure.   [UCLES]

3 The continuous random variable X has probability density function given by

$$f(x) = \begin{cases} \dfrac{k}{x} & \text{for } 1 \leqslant x \leqslant 9, \\ 0 & \text{otherwise} \end{cases}$$

where k is a constant. Giving your answers correct to three significant figures where appropriate,
(i)   find the value of k, and find also the median value of X,
(ii) find the mean and variance of X,
(iii) find the cumulative distribution function, F, of X, and sketch the graph of $y = F(x)$.   [UCLES]

4 The continuous random variable X has cumulative distribution function given by

$$F(x) = \begin{cases} 0, & \text{for } x < 0, \\ 2x - x^2, & \text{for } 0 \leqslant x \leqslant 1, \\ 1, & \text{for } x > 1. \end{cases}$$

(i)   Find $P(X > \frac{1}{2})$.
(ii) Find the value of q such that $P(X < q) = \frac{1}{4}$.
(iii) Find the probability density function of X, and sketch its graph.
(iv) Find $E(X)$ and $E(\sqrt{X})$.   [UCLES]

5 The continuous random variable U has a uniform distribution on $0 < u < 1$. The random variable X is defined as follows:

$$X = 2U \text{ when } U \leqslant \frac{3}{4},$$
$$X = 4U \text{ when } U > \frac{3}{4}.$$

(i)   Give a reason why X cannot take values between $\frac{3}{2}$ and 3, and write down the values of $P(0 < X \leqslant \frac{3}{2})$ and $P(3 < X < 4)$.
(ii) Sketch the complete graph of the probability density function of X.
(iii) Find the lower quartile q of X, i.e. the value of q such that $P(X < q) = \frac{1}{4}$.
(iv) Three independent observations are taken of X. Find the probability that they all exceed q.
(v)   Show that $E(X) = \frac{23}{16}$ and find $E(X^2)$.   [UCLES]

6 The total amount of fuel used by a road haulage firm in a month is a random variable X (thousands of gallons) which has the following probability density function.

$$f(x) = \begin{cases} cx & 0 < x < 1, \\ c(3-x) & 1 \leqslant x < 3, \\ 0 & \text{otherwise,} \end{cases}$$

(a) Find the value of c.
(b) Find the probability that the firm uses less than 900 gallons in a month.
(c) Find the probability that the firm uses between 900 and 1600 gallons a month.
(d) Given that the firm used over 900 gallons in a particular month, find the probability that over 2000 gallons were used during the month.
(e) The supplier of the fuel charges the firm £1.20 per gallon for the first 900 gallons supplied per month, £1.10 per gallon for the next 700 gallons and £1.00 per gallon for the remainder. Find the probability that the monthly cost exceeds £2250.   [AEB 90]

7 The amount of vegetables eaten by a family in a week is a random variable W kg. The probability density function is given by

$$f(w) = \begin{cases} \dfrac{20}{5^5} w^3 (5-w) & 0 \leqslant w \leqslant 5, \\ 0 & \text{otherwise.} \end{cases}$$

(continued)

(a) Find the cumulative distribution function of W.

(b) Find, to 3 decimal places, the probability that the family eats between 2kg and 4kg of vegetables in one week.

(c) Given that the mean of the distribution is $3\frac{1}{3}$, find, to 3 decimal places, the variance of W.

(d) Find the mode of the distribution.

(e) Verify that the amount, m, of vegetables which is such that the family is equally likely to eat more or less than m in any week is about 3.431 kg.

(f) Use the information above to comment on the skewness of the distribution.   [ULSEB]

8 The overall mark obtained by a ski-jumper is based on the distance, X metres, jumped in excess of 80 metres and marks for style, Y.

Assuming that X is a continuous random variable with probability density function

$$f(x) = \begin{cases} ax & 0 \leqslant x \leqslant 10, \\ 0 & \text{elsewhere,} \end{cases}$$

and that Y is a discrete random variable with probability function

$$g(y) = \begin{cases} by & y = 1, 2, 3, 4, 5, \\ 0 & \text{elsewhere,} \end{cases}$$

find a and b.

Evaluate

(a) the expected distance jumped in excess of 80 metres,

(b) the expected style marks obtained by the ski-jumper,

(c) the expected overall mark for the ski-jumper if the total mark, T, is given by $T = X^2 + \frac{1}{2} Y$.

Assuming X and Y are independent, determine the probability that the ski-jumper exceeds 85 metres and obtains more than 3 style marks for a particular jump.   [AEB 91]

12 The normal distribution

Thank God we're normal, normal, normal
Thank God we're normal
Yes, this is our finest shower!

The Entertainer, John Osborne

The normal distribution is the most important of all distributions because it describes the situation in which very large values are rather rare, very small values are rather rare, but middling values are rather common. Since this is a good description of lots of things the 'normal' distribution is indeed normal!

Here are some examples:

♦ Heights and weights (bean poles and Humpty Dumpties are not too common!)
♦ Times taken by students to run 100 m.
♦ The precise volumes of lager in 'pints' of lager at the local pub.

The distribution can also be applied as an approximation in the case of some discrete variables:

♦ Marks obtained by students on an A-level paper.
♦ The IQ scores of the population.

We will see later that the distribution can also be used as an approximation to the binomial and Poisson distributions.

Formally, a normal distribution is a unimodal symmetric continuous distribution having two parameters, μ (the mean) and σ^2 (the variance). Because of the symmetry, the mean is equal to both the mode and the median. As a shorthand we refer to a $N(\mu, \sigma^2)$ distribution – note that it is $N(\mu, \sigma^2)$ and *not* $N(\mu, \sigma)$.

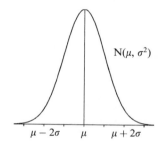

12.1 The standard normal distribution

Since for any distribution, changes in μ and σ can be regarded as changes of location and scale, all normal distributions can be related to a single reference distribution, the so-called **standard normal** distribution, which has mean 0 and variance 1. Traditionally the random variable with this distribution is denoted by Z. Hence:

$$Z \sim N(0, 1)$$

The pdf for Z is usually designated by ϕ (a lower-case Greek letter, pronounced 'phi'):

$$\phi(z) \propto e^{-\frac{1}{2}z^2} \qquad -\infty < z < \infty$$

The corresponding distribution function is denoted by Φ (the capital letter version of ϕ and also pronounced 'phi'):

$$\Phi(a) = P(Z \leqslant a) = P(Z < a) = \int_{-\infty}^{a} \phi(z) \; dz$$

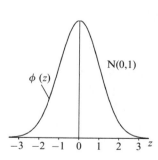

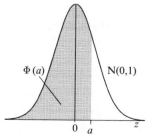

12.2 Tables of $\Phi(z)$

Although $\phi(z)$ looks simple, it cannot be integrated explicitly and so tables are required for $\Phi(z)$. The tables available vary in style quite considerably. You should make sure that you are familiar with the tables supplied for your particular examination.

Most tables give the values of either $\Phi(z)$ or $1 - \Phi(z)$, and generally do so only for non-negative values of z. The tables may refer to $\Phi(z)$ as $P(z)$ or to $1 - \Phi(z)$ as $Q(z)$, as shown in the diagrams.

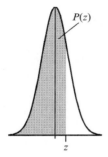 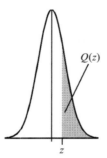

The tables given at the back of this book are tables of $\Phi(z)$ for $z \geqslant 0$. Values for negative values of z can be obtained by using the symmetry of the distribution:

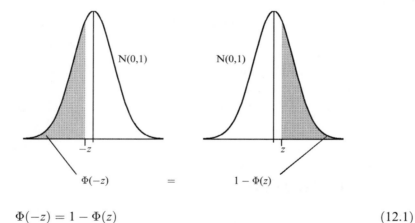

$$\Phi(-z) = 1 - \Phi(z) \qquad (12.1)$$

Here is an extract from the first column of the tables of values of $\Phi(z)$ given in the Appendix (p. 619):

z	$\Phi(z)$	z	$\Phi(z)$	z	$\Phi(z)$
0.0	0.5000	1.0	0.8413	2.0	0.9772
0.2	0.5793	1.2	0.8849	2.2	0.9861
0.4	0.6554	1.4	0.9192	2.4	0.9918
0.6	0.7257	1.6	0.9452	2.6	0.9953
0.8	0.7881	1.8	0.9641	2.8	0.9974

The first entry in the table, $\Phi(0) = 0.5000$ is one we already knew! All normal distributions are symmetric about their mean, and the standard normal distribution has mean 0.

Here are some examples of the use of the table:

$P(Z < 1.0) = 0.8413$

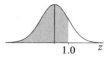

$P(Z < 2.0) = 0.9772$

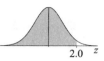

$P(Z > 1.6) = 1 - \Phi(1.6) = 1 - 0.9452 = 0.0548$

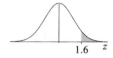

$P(Z > -0.6) = P(Z < 0.6) = \Phi(0.6) = 0.7257$

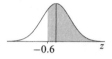

$P(Z < -2.4) = P(Z > 2.4) = 1 - \Phi(2.4) = 1 - 0.9918 = 0.0082$

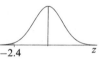

To find the probability associated with a range of values we use:
$$P(a < Z < b) = P(Z < b) - P(Z < a) = \Phi(b) - \Phi(a)$$

In terms of $Q(z)$ this becomes:
$$P(a < Z < b) = P(Z > a) - P(Z > b) = Q(a) - Q(b)$$

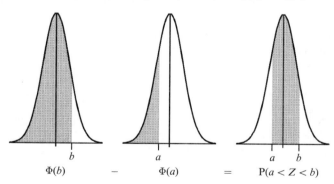

$$\Phi(b) \qquad - \qquad \Phi(a) \qquad = \qquad P(a < Z < b)$$

Notes

- With this type of calculation it is easy to muddle the plus and minus signs. A final answer that is negative or that is greater than 1 often occurs as a consequence of a mistaken sign.
- $\Phi(z) > 0.5$ for $z > 0$, whereas $\Phi(z) < 0.5$ for $z < 0$.
- As elsewhere in Statistics, a quick sketch is useful! If your sketch shows a large part of the area under the curve as shaded, then the probability ought to be large. If your answer is small then your calculations were wrong (or your sketch was mistaken!).
- In the calculations here and elsewhere in the book we will be using the tabulated probabilities, which are correct to 4 decimal places only. Each one may be in error by as much as 0.00005. Calculations involving sums or differences of k such quantities may therefore be in error by as much as $k \times 0.00005$. Many of our answers are therefore likely to be slightly in error in the final decimal place.

Example 1

$$P(1.0 < Z < 2.0) = \Phi(2.0) - \Phi(1.0)$$
$$= 0.9772 - 0.8413$$
$$= 0.1359$$

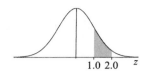

Using $Q(z)$, we would have:

$$P(1.0 < Z < 2.0) = Q(1.0) - Q(2.0) = 0.1587 - 0.0228 = 0.1359$$

Example 2

$$P(-1.0 < Z < 2.0) = \Phi(2.0) - \Phi(-1.0)$$
$$= 0.9772 - \{1 - \Phi(1.0)\}$$
$$= 0.9772 - 1 + 0.8413$$
$$= 0.8185$$

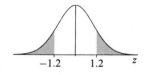

Using $Q(z)$, we would have:

$$P(-1.0 < Z < 2.0) = \{1 - Q(1.0)\} - Q(2.0)$$
$$= (1 - 0.1587) - 0.0228 = 0.8185$$

Every problem involving normal distributions can be solved using either of Φ or Q. This is the last example for which we will provide both solutions, since there is no essential difference between one solution and the other.

This problem also illustrates the limitations of 4-figure tables. The quantities 0.9772 and 0.8413, which are correct to four decimal places, would be 0.97725 and 0.84134 if expressed to five decimal places. With the extra accuracy the final result would be 0.81859 which rounds to 0.8186 and not to our stated answer of 0.8185. However, there is no need to worry too much about this! A reasonable tolerance is always built into marking schemes, and in real life one would be quoting only one or two significant figures – not four. We can't tell the difference between 82% and 80%!

Example 3

$$P(|Z| > 1.2) = P(Z > 1.2) + P(Z < -1.2)$$
$$= \Phi(-1.2) + \{1 - \Phi(1.2)\}$$
$$= \{1 - \Phi(1.2)\} + \{1 - \Phi(1.2)\}$$
$$= 2\{1 - \Phi(1.2)\}$$
$$= 2(1 - 0.8849) = 2 \times 0.1151 = 0.2302$$

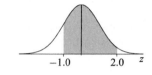

Example 4

The random variable Z has a standard normal distribution. Determine the value of a, where $P(Z < a) = 0.9953$.

From the table we see that $\Phi(2.6) = 0.9953$: hence $a = 2.6$.

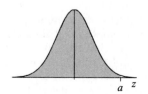

Example 5

The random variable Z has a standard normal distribution.
Determine the value of a, where $P(Z > a) = 0.2743$.

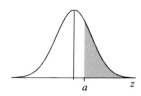

If $P(Z > a) = 0.2743$, then $P(Z < a) = 1 - 0.2743 = 0.7257$. Scanning through the tables we see that $P(Z < 0.6) = 0.7257$. Hence $a = 0.6$.

Example 6

The random variable Z has a standard normal distribution.
Determine the value of a, where $P(|Z| < a) = 0.5762$.

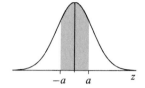

If $P(|Z| < a) = 0.5762$, then $P(0 < Z < a) = \frac{1}{2}(0.5762) = 0.2881$ and hence $P(Z < a) = 0.5 + 0.2881 = 0.7881$. The tables show that $\Phi(0.8) = 0.7881$. Hence $a = 0.8$.

Example 7

The random variable Z has a standard normal distribution.
Determine the value of a, where $P(a < Z < 1.6) = 0.7865$.

From the tables we know that $P(Z < 1.6) = 0.9452$.
Since:
$$P(a < Z < 1.6) = P(Z < 1.6) - P(Z < a)$$
it follows that:
$$\begin{aligned} P(Z < a) &= P(Z < 1.6) - P(a < Z < 1.6) \\ &= 0.9452 - 0.7865 \\ &= 0.1587 \end{aligned}$$

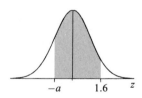

This is less than 0.5 and therefore corresponds to a negative value of z. We need to use Equation (12.1). Instead of finding the value of z corresponding to 0.1587, we find the value that corresponds to $1 - 0.1587 = 0.8413$. From the table we find that this is the value 1.
Since $\Phi(-z) = 1 - \Phi(z)$, we have therefore deduced that
$\Phi(-1) = 1 - 0.8413 = 0.1587$.
The required value of a is -1.

Exercises 12a

In Questions **1–5**, the random variable Z has a standard normal distribution, with mean zero and variance 1.
Use the table given in Section 12.2 (p. 304) to answer the following questions.

1 Find:
(i) $P(Z < 1.2)$, (ii) $P(Z > 1.8)$,
(iii) $P(Z < -1.4)$, (iv) $P(Z < -0.8)$.

2 Find:
(i) $P(2.2 < Z < 2.8)$, (ii) $P(-1.2 < Z < 0.4)$,
(iii) $P(-1.8 < Z < -0.2)$.

3 Find:
(i) $P(|Z| < 0.8)$, (ii) $P(|Z| > 1.6)$,
(iii) $P(0.6 < |Z| < 2.2)$.

4 Find a such that:
(i) $P(Z < a) = 0.9192$, (ii) $P(Z < a) = 0.3446$,
(iii) $P(Z > a) = 0.8849$, (iv) $P(Z > a) = 0.0047$,
(v) $P(1 < Z < a) = 0.1039$,
(vi) $P(a < Z < -0.8) = 0.1760$.

5 Find a such that:
(i) $P(|Z| < a) = 0.4514$,
(ii) $P(|Z| > a) = 0.1096$.

12.3 Probabilities for other normal distributions

Suppose $X \sim N(\mu, \sigma^2)$. It can be shown that, if we use the transformation:

$$Z = \frac{X - \mu}{\sigma}$$

then $Z \sim N(0, 1)$. This is equivalent to the change of location and scale referred to in Section 12.1. Thus:

$$P(X < x) = P(X - \mu < x - \mu) = P\left(\frac{X - \mu}{\sigma} < \frac{x - \mu}{\sigma}\right)$$
$$= P\left(Z < \frac{x - \mu}{\sigma}\right)$$
$$= \Phi\left(\frac{x - \mu}{\sigma}\right)$$

The relevant value for Z is simply the value of $\frac{X - \mu}{\sigma}$ when X is replaced by the value of interest.

As an example, if $X \sim N(8, 4)$ and we want $P(X < 10)$, then this is given by:

$$\Phi\left(\frac{10 - 8}{2}\right) = \Phi(1) = 0.8413$$

The link between the normal distribution for X and the standard normal distribution for Z is conveniently summarised by showing two scales.

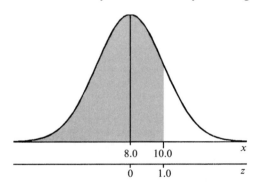

The value of zero for z always occurs under the mean μ for x. A value of k for z would occur under the value $\mu + k\sigma$ for x, where σ^2 is the variance of the distribution of X.

▼ ▼

Example 8

The random variable $X \sim N(3.4, 0.09)$.
Determine $P(X < 3.1)$.

————

Now $X < 3.1$ corresponds to $Z < \dfrac{3.1 - 3.4}{\sqrt{0.09}} = -1$, so that:

$$P(X < 3.1) = P(Z < -1)$$
$$= 1 - P(Z < 1)$$
$$= 1 - 0.8413$$
$$= 0.1587$$

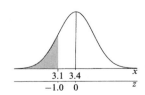

▲ ▲

Example 9

The random variable $X \sim N(21, 16)$.
Determine $P(17.8 < X < 29.0)$.

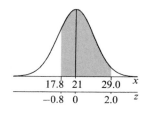

Now $X < 29.0$ corresponds to $Z < \dfrac{29.0 - 21}{\sqrt{16}} = 2$

and $X > 17.8$ corresponds to $Z > \dfrac{17.8 - 21}{\sqrt{16}} = -0.8$, so that:

$$
\begin{aligned}
P(17.8 < X < 29.0) &= P(-0.8 < Z < 2) \\
&= \Phi(2) - \Phi(-0.8) \\
&= \Phi(2) - \{1 - \Phi(0.8)\} \\
&= 0.9772 - 1 + 0.7881 \\
&= 0.7653
\end{aligned}
$$

Example 10

The random variable $X \sim N(50, 36)$.
Denoting $E(X)$ by μ, determine $P(|X - \mu| > 4.8)$.

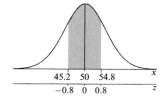

We are given that X has mean 50. Hence $\mu = 50$.

$$
\begin{aligned}
P(|X - \mu| > 4.8) &= P\left(\frac{|X - \mu|}{\sqrt{36}} > \frac{4.8}{\sqrt{36}}\right) \\
&= P(|Z| > 0.8) \\
&= 1 - P(|Z| < 0.8) \\
&= 1 - \{P(Z < 0.8) - P(Z < -0.8)\} \\
&= 1 - \{0.7881 - (1 - 0.7881)\} \\
&= 0.4238
\end{aligned}
$$

Project

This practical needs a sensitive pair of scales or balance. The aim is to study the variability in the masses of some packaged goods. A good choice would be packets of crisps – how variable are their masses? Does it seem as though a normal distribution is appropriate?

This project can be extended to compare different brands and flavours. Is there any evidence that some brands are more variable than others? Is there a difference between plain crisps and flavoured crisps? In each case represent your data using a histogram. For comparing flavours or types, box–whisker diagrams will be useful.

Some ways of formally testing for differences will be discussed later in Chapters 15 to 18.

Exercises 12b

Use the table given in Section 12.2 (p. 304) to answer the following questions.

1 Given that $X \sim N(12, 9)$, find:
(i) $P(X > 15)$, (ii) $P(X < 16.8)$,
(iii) $P(X < 8.4)$, (iv) $P(X > 9.6)$.

2 Given that $X \sim N(50, 100)$, find:
(i) $P(36 < X < 62)$, (ii) $P(40 < X < 50)$,
(iii) $P(56 < X < 70)$, (iv) $P(38 < X < 42)$.

3 Given that $X \sim N(1.6, 4)$, find:
(i) $P(X > 0)$, (ii) $P(X < -1.6)$,
(iii) $P(|X| < 2)$, (iv) $P(0 < X < 2)$.

4 Given that $X \sim N(-4, 25)$, find:
(i) $P(X > 0)$, (ii) $P(-5 < X < -2)$,
(iii) $P(-2 < X < 1)$, (iv) $P(|X| > 1)$.

5 IQ scores are normally distributed with mean 100 and standard deviation 15.

Determine the proportion of people with an IQ:
(a) below 118,
(b) above 112,
(c) below 94,
(d) above 73,
(e) between 100 and 112,
(f) between 73 and 118,
(g) between 73 and 94.

12.4 Finer detail in the tables of $\Phi(z)$

The table of $\Phi(z)$ in Section 12.2 was very abbreviated. We now give a (very short) section from the full table given in the Appendix (p. 619):

z	0	1	2	3	4	5	6	7	8	9	1	2	3	4	5	6	7	8	9
															ADD				
0.0	.5000	.5040	.5080	.5120	.5160	.5199	.5239	.5279	.5319	.5359	4	8	12	16	20	24	28	32	36
0.1	.5398	.5438	.5478	.5517	.5557	.5596	.5636	.5675	.5714	.5753	4	8	12	16	20	24	28	32	36
0.2	.5793	.5832	.5871	.5910	.5948	.5987	.6026	.6064	.6103	.6141	4	8	12	15	19	23	27	31	35
0.3	.6179	.6217	.6255	.6293	.6331	.6368	.6406	.6443	.6480	.6517	4	7	11	15	19	22	26	30	34

As an example of the use of this table, suppose we require $P(Z < 0.100)$. We look first at the row with 0.1 in the first column. Since the value in the second decimal place of 0.100 is a '0', we look in the column headed '0' and find the value '.5398' implying a probability of 0.5398.

If we require instead $P(Z < 0.140)$ then we look at the value in the row labelled 0.1, and the column headed '4', which gives 0.5557. Some further examples:

$$P(Z < 0.070) = 0.5279$$
$$P(Z < 0.260) = 0.6026$$

All the previous examples referred to cases where the z had a '0' in the third decimal place. For other cases we need to modify the value found using the values given in the right-hand section of the table.

For example, if we require $P(Z < 0.175)$ then we must first find $P(Z < 0.170)$. Looking in the main part of the table we find '.5675'. For the adjustment due to the third decimal place (which is 5) we look at the right-hand column headed '5'. The value in this column for the row labelled 0.1 is 20. This value '20' is a shorthand for 0.0020. The instruction at the top of this section is ADD. Therefore the required probability is $0.5675 + 0.0020 = 0.5695$. Here are some further examples:

$$P(Z < 0.246) = 0.5948 + 0.0023 = 0.5971$$
$$P(Z < 0.302) = 0.6179 + 0.0007 = 0.6186$$

The tables can also be used 'in reverse'. For example, to find the value of z which is such that $P(Z < z) = 0.6443$, we look through the table for the probability 0.6443 and observe that this corresponds to $z = 0.37$. A negative value for z would be signalled by a probability less than 0.5000. For example, if $P(Z < z) = 0.3783$, we look instead for the probability $1 - 0.3783 = 0.6217$, which corresponds to $z = 0.31$. The required z-value is therefore -0.31.

An example of a case where the given probability does not appear in the table is given by part (v) of Example 11.

Notes
- The tables report probabilities as, for example, '.5948'. This is done to save space in the table: the probability should be reported as '0.5948'.
- It is easy to get confused over the decimal places. One guide is that after adding the adjustment from the right-hand side of the table, the sum should not be larger than the next item in the body of the table. For example, if we had calculated $P(Z < 0.302)$ as $0.6179 + 0.0070 = 0.6249$, then we would know there was an error because this value is greater than $P(Z < 0.31) = 0.6217$.
- When the tables are of $Q(z)$, the adjustment factors given in the right-hand section will have to be *subtracted* from the value given in the body of the table.
- Some tables do not contain the fine detail provided by the right-hand side of our table. For these tables the user has to interpolate 'by hand'.
- Some tables use a type of shorthand to deal with probabilities that are very close to either 0 or 1. Thus:

$$.0^4 3 \rightarrow 0.000\,03$$

$$.9^3 7 \rightarrow 0.9997$$

Example 11

The random variable Z has a normal distribution with mean 0 and variance 1.
Determine (i) $P(Z < 0.27)$, (ii) $P(Z < 0.345)$, (iii) $P(Z > 0.004)$,
(iv) the value of a for which $P(Z < a) = 0.6217$,
(v) the value of b for which $P(Z < b) = 0.6000$.

───────

(i)   From the table, $\Phi(0.27) = 0.6064$.
(ii) From the main part of the table, $\Phi(0.34) = 0.6331$. From the fifth column of the additional part of the table we need to add '19' (i.e. 0.0019) and so the required probability is $0.6331 + 0.0019 = 0.6350$.
(iii) From the table $\Phi(0.004) = 0.5000$ with an addition of '16', so $P(Z < 0.004) = 0.5016$. Hence $P(Z > 0.004) = 1 - 0.5016 = 0.4984$.
(iv) Scanning through the main part of the table we find that $\Phi(0.31) = 0.6217$. Thus $a = 0.31$.
(v)   Scanning through the table we see that $\Phi(0.25) = 0.5987$, while $\Phi(0.26) = 0.6026$. The required value for a must lie between 0.25 and 0.26. If we use the supplementary part of the table we find that $\Phi(0.253) = 0.5999$ and $\Phi(0.254) = 0.6002$ so the value of a must be about 0.2533.

Example 12

The random variable X has a normal distribution with mean 12 and variance 5.
Determine $P(X < 13)$.

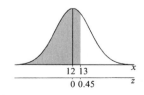

───────

$$P(X < 13) = P\left(\frac{X - 12}{\sqrt{5}} < \frac{13 - 12}{\sqrt{5}}\right) = P\left(Z < \frac{1}{\sqrt{5}}\right) = \Phi(0.4472)$$

$$= 0.673 \text{ (to 3 d.p.)}$$

Example 13

The entrance qualification for membership of a society is that the aspiring member should score highly on a particular IQ test. The scores obtained on the test have a normal distribution, with mean 100.0 and standard deviation 16. All those obtaining 124.0 or more automatically join the society.

Determine the median score of the members of the society.

———

The median score obtained by those taking the test is 100.0, since the distribution of scores is normal. However, it is the median of those who are admitted that is required. We need first to find the proportion of those taking the test that obtain scores of 124.0 or more and then divide this proportion in two. As usual, a diagram helps.

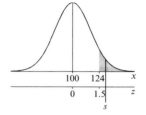

The proportion of those taking the test that are admitted to the society

is $1 - \Phi\left(\dfrac{124 - 100}{16}\right) = 1 - \Phi(1.5) = 0.0668$. We therefore require the

score that is exceeded by $\frac{1}{2}(0.0668) = 0.0334$ of the population.

Using tables of $\Phi(z)$, we need to find the value of z corresponding to $\Phi(z) = 1 - 0.0334 = 0.9666$. The required value is 1.833 and the required median score, s, is therefore the solution of:

$$\frac{s - 100}{16} = 1.833$$

The median score of the members of the society is therefore:
$s = 100 + 16 \times 1.833 \approx 129.3$.

Exercises 12c

Use the tables in the Appendix, or the tables that you will use in your examination, to answer the following questions.

1   Given that $Z \sim N(0, 1)$, find:
(i) P$(Z < 0.932)$, (ii) P$(Z > 1.235)$,
(iii) P$(Z < -1.414)$, (iv) P$(Z > -0.519)$.

2   Given that $X \sim N(0, 1)$, find:
(i) P$(X > 3.213)$, (ii) P$(X < 3.615)$,
(iii) P$(X < -2.841)$, (iv) P$(X > -2.818)$.

3   Given that $Y \sim N(3.7, 2.4)$, find:
(i) P$(Y > 4)$, (ii) P$(Y < 4.5)$,
(iii) P$(3.1 < Y < 4.2)$, (iv) P$(2.8 < Y < 3.5)$.

4   Given that $X \sim N(23, 12)$, find:
(i) P$(X < 25)$, (ii) P$(20 < X < 25)$,
(iii) P$(X > 27)$, (iv) P$(23 < X < 30)$.

5   Given that $X \sim N(3, 4)$, find:
(i) P$(3X < 7)$, (ii) P$(\frac{1}{2}X < 2)$,
(iii) P$(2X + 1 > 10)$, (iv) P$(3 - X < 2)$.

6   The mass of a small loaf of bread is normally distributed with mean 500 g and standard deviation 20 g.
Find the probability that a randomly chosen loaf has a mass:

(i) not exceeding 475 g,
(ii) not less than 495 g,
(iii) at most 510 g,
(iv) at least 515 g.

7   Chicken eggs have mean mass 60 g with standard deviation 15 g, and the distribution of their masses may be taken to be normal. Eggs of less than 45 g are classed as 'small'. The remainder are classed as either 'standard' or 'large'. It is desired that these two classifications should occur with approximately equal frequency.
Suggest the mass at which the division between standard and large should be made.

8 An aircraft dropping a bomb towards a straight railway track has an error in the impact point which is normally distributed with standard deviation 40 m. The aircraft can carry either 6 light bombs or 3 heavy bombs. A light bomb must fall within 15 m of the track in order to inflict damage. A heavy bomb must fall within 30 m of the track. Suppose that the aircraft is vertically above the track.

   Given that it is sufficient if one bomb damages the track, show that the probability that the track is undamaged if the plane carries light bombs is approximately 0.126.

   Determine the corresponding probability if the plane carries heavy bombs.

   Is the best strategy to carry the light bombs or to carry the heavy bombs?

   Determine the best strategy if the plane is mistakenly positioned above a point 20 m away from the track.

12.5 Tables of percentage points

In addition to the extensive table of probabilities corresponding to values of z, most books of tables also present the information in the reverse fashion. That is to say that, for a range of values of $\Phi(z)$, this second table gives the corresponding value of z. This table may be described as providing 'upper quantiles'.

An abbreviated version of the table given in the Appendix (p. 620) is given below:

$q(\%)$	z	$q(\%)$	z	$q(\%)$	z
50	0.000	5	1.645	0.5	2.576
40	0.253	4	1.751	0.1	3.090
30	0.524	3	1.881	0.01	3.719
25	0.674	2.5	1.960	0.0^21	4.265
20	0.842	2	2.054	0.0^31	4.753
10	1.282	1	2.326	0.0^41	5.199

Here $q\%$ is the upper tail probability corresponding to the given value of z. Thus:

> The value of a such that $P(Z > a) = 2\%$ is 2.054.
> The value of a such that $P(Z < a) = 70\%$ is 0.524.
> The value of a such that $P(Z > a) = 0.03$ is 1.881.
> The value of a such that $P(Z < a) = 0.95$ is 1.645
> The value of a such that $P(Z < a) = 0.1$ is -1.282.
> The value of a such that $P(Z < a) = 0.999\,990$ is 4.265.

Example 14

The random variable $Z \sim N(0, 1)$.
Determine the value of a which is such that $P(-a < Z < a) = 0.4$.

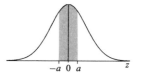

A normal distribution is symmetric about its mean. If a is such that $P(-a < Z < a) = 0.4$ then $P(0 < Z < a) = 0.2$. But $P(Z > 0) = 0.5$ and so combining these results we have $P(Z > a) = 0.5 - 0.2 = 0.3 = 30\%$. The required value for a is therefore 0.524.

Example 15

The random variable $X \sim N(19, 49)$.
Determine the value of a which is such that $P(X < a) = 0.90$.

$$P(X < a) = P\left(\frac{X - \mu}{\sigma} < \frac{a - \mu}{\sigma}\right)$$
$$= \Phi\left(\frac{a - \mu}{\sigma}\right)$$

We want to find the value of a such that $\Phi\left(\frac{a - 19}{7}\right) = 0.90$, which means

that there is 10% in the upper tail. But we know, from the table, that the upper 10% point of a standard normal distribution is 1.282. The inescapable conclusion is that:

$$\frac{a - 19}{7} = 1.282$$

which implies that $a = 19 + (7 \times 1.282) = 27.974$. Hence, to 1 d.p., $a = 28.0$.

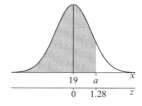

Example 16

A machine is supposed to cut up logs into pieces that are each 2 metres long. However, the machine is an old one, and while the pieces that it produces do have an *average* length of 2 metres, 10% of the pieces that it produces are less than 1.95 metres long.
Assuming that the lengths produced are normally distributed, determine the proportion of the pieces that are longer than 2.10 metres.

Let X be the random variable corresponding to the length of a log.
We have the information:

$$X \sim N(2.00, \sigma^2)$$
$$P(X < 1.95) = 0.10$$

We need to find $P(X > 2.10)$ and we must evidently first determine the value of σ. To do this we relate the known probability to the standard normal distribution.
  Now:

$$P(X < 1.95) = \Phi\left(\frac{1.95 - 2.00}{\sigma}\right)$$

and also:

$$0.10 = \Phi(-1.282)$$

using the information in the table of percentage points. Hence it must be the case that:

$$\Phi(-1.282) = \Phi\left(\frac{1.95 - 2.00}{\sigma}\right)$$

which implies that:

$$-1.282 = \frac{1.95 - 2.00}{\sigma}$$

Hence:

$$\sigma = \frac{-0.05}{-1.282}$$

$$= 0.0390$$

We now know that $X \sim \text{N}(2.00, 0.0390^2)$. In order to find $\text{P}(X > 2.10)$ we first find $\text{P}(X < 2.10)$ which equals:

$$\Phi\left(\frac{2.10 - 2.00}{0.039}\right) = \Phi(2.564)$$

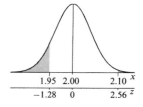

Using the tables in the Appendix, the corresponding probability is found to be 0.9948. The required probability is therefore $1 - 0.9948 = 0.0052$, in other words just over 0.5% of the pieces of wood have lengths greater than 2.10 metres.

Example 17

The random variable Y has a normal distribution with mean μ and variance σ^2.
Given that 10% of the values of Y exceed 17.24 and that 25% of the values of Y are less than 14.37, find the values of μ and σ.

───────

We know from the tables of percentage points that the upper 10% point of a standard normal distribution is 1.282 and the lower 25% point is -0.674. The values of μ and σ are therefore the solutions of:

$$\frac{17.24 - \mu}{\sigma} = 1.282$$

$$\frac{14.37 - \mu}{\sigma} = -0.674$$

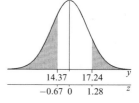

Multiplying through by σ, and subtracting, we get:

$$
\begin{array}{rcl}
17.24 - \mu & = & 1.282\sigma \\
14.37 - \mu & = & -0.674\sigma \\
\hline
2.87 & = & 1.956\sigma
\end{array}
$$

Thus $\sigma = 1.47$ and hence $\mu = 17.24 - 1.282\sigma = 15.4$.

Example 18

The random variable X has a normal distribution. It is known that $\text{P}(X > 9) = 0.9192$ and that $\text{P}(X < 11) = 0.7580$.
Determine $\text{P}(X > 10)$.

───────

We are not told the values of μ or σ, so must begin by finding these, using the two pieces of information that we do have. An initial sketch shows that μ must lie between 9 and 11, and, since 0.9192 is larger than 0.7580, the mean must be slightly above 10.

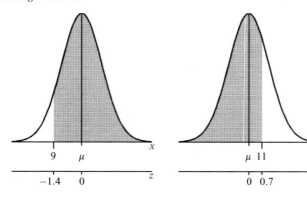

Now $1 - 0.9192 = 0.0808$, so we need to search the tables of $\Phi(z)$ to find the z-values corresponding to $\Phi(z) = 0.0808$ and $\Phi(z) = 0.7580$. These values are -1.400 and 0.700, respectively. We now solve the simultaneous equations:

$$\frac{9 - \mu}{\sigma} = -1.400$$

$$\frac{11 - \mu}{\sigma} = 0.700$$

Multiplying through by σ we get:

$$9 - \mu = -1.400\sigma$$
$$11 - \mu = 0.700\sigma$$

Subtracting one equation from the other we get $2 = 2.100\sigma$, so that $\sigma = \frac{20}{21}$ (approximately).

Substituting in either equation we get $\mu = \frac{31}{3}$.

We want $P(X > 10)$, so first find the corresponding value of z:

$$z = \frac{10 - \frac{31}{3}}{\frac{20}{21}} = -\frac{7}{20}$$

Hence:

$$P(X > 10) = 1 - \Phi(-\tfrac{7}{20}) = 1 - \{1 - \Phi(0.35)\} = 0.6368$$

The probability that X exceeds 10 is 0.637 (to 3 d.p.).

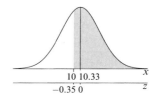

Exercises 12d

Use the tables of percentage points in Section 12.5 (p. 313), or the tables that you will use in your examination, to answer the following questions.

1 Given that $Z \sim N(0, 1)$, find a such that:
(i) $P(Z < a) = 0.97$, (ii) $P(Z > a) = 0.05$,
(iii) $P(Z < a) = 0.001$, (iv) $P(Z > a) = 0.99$,
(v) $P(Z > a) = 0.0001$, (vi) $P(Z < a) = 0.999$.
Give your answers to 3 decimal places.

2 Given that $X \sim N(20, 25)$, find a such that:
(i) $P(X < a) = 0.97$, (ii) $P(X > a) = 0.05$,
(iii) $P(X < a) = 0.001$, (iv) $P(X > a) = 0.99$,
(v) $P(X > a) = 0.0001$, (vi) $P(X < a) = 0.999$.
Give your answers to 3 decimal places.

3 Given that $X \sim N(\mu, 2.5)$ and that $P(X > 3.5) = 0.970$, find μ.

4 Given that $X \sim N(\mu, 0.5)$ and that $P(X < -1.2) = 0.050$, find μ.

5 Given that $X \sim N(32.4, \sigma^2)$ and that $P(X > 45.2) = 0.300$, find σ.

6 Given that $X \sim N(-7.21, \sigma^2)$ and that $P(X < 0) = 0.900$, find σ.

7 Given that $X \sim N(\mu, \sigma^2)$, $P(X > 0) = 0.800$ and $P(X < 5) = 0.700$, find μ and σ.

8 The mass M of a randomly chosen tin of baked beans is such that $M \sim N(420, 100)$.
Find, correct to 1 decimal place:
(i) the 20th percentile,
(ii) the 90th percentile.

9 The quantity of milk in a bottle is normally distributed with mean 1000 ml. Find the standard deviation given that the probability that there is less than 990 ml in a randomly chosen bottle is 5%.
What can you say about the standard deviation if the probability is to be less than 5% ?

10 A variety of hollyhock grows to great heights. Assuming a normal distribution of heights, find the mean and standard deviation, given that the 30th and 70th percentiles are 1.83 m and 2.31 m respectively.

11 Due to manufacturing variations, the length of string in a randomly chosen ball of string can be modelled by a normal distribution.
Find the mean and standard deviation given that 95% of balls of string have lengths exceeding 495 m, and 99% have lengths exceeding 490 m, giving your answers to two decimal places.

12 The length, in cm, of a brass cylinder has a normal distribution with mean μ and variance σ^2, both μ and σ^2 being unknown. Suppose that a large sample reveals that 10% of the cylinders are longer than 3.68 cm and that 3% are shorter than 3.52 cm. Find the values of μ and σ^2.

13 Nylon cords produced by a certain manufacturer are known to have breaking tensions that are normally distributed with mean 74 N and standard deviation 2.5 N. Find, correct to two decimal places:
(i) the probability that the breaking tension of a randomly chosen cord will be less than 75 N,
(ii) the value c given that 75% of such cords have breaking tensions less than c N.   [WJEC]

14 The random variable X is normally distributed with mean μ and variance σ^2. Given that $P(X > 58.37) = 0.02$ and $P(X < 40.85) = 0.01$, find μ and σ.   [ULEAC]

12.6 Using calculators

Some calculators have the ability to evaluate probabilities for normal random variables. They are able to provide values for three quantities:

$$P(z) = \Phi(z) = P(Z < z)$$
$$Q(z) = P(0 < Z < z)$$
$$R(z) = P(Z > z)$$

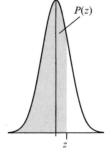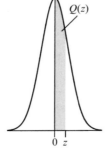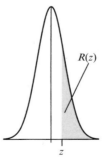

Notes
- The definition of Q is different to that used in some tables.
- Calculators also vary! Check what yours actually does.

12.7 Applications of the normal distribution

Without any doubt, normal distributions are the most important distributions in Statistics. The reasons are various:

- When *n* is large, a binomial distribution with parameters *n* and *p* can be well approximated by a normal distribution.
- When λ is large, a Poisson distribution with parameter λ can be well approximated by a normal distribution.
- Because of the **Central Limit Theorem** (discussed later), the distribution of the mean and of the total of a set of independent observations on a random variable are often well approximated by a normal distribution.
- Any linear combination of normal random variables will again have a normal distribution.
- Many naturally occurring variables have distributions closely resembling the normal. Examples are:
  - masses of '1 kg' bags of flour,
  - diameters of ball-bearings,
  - measurement errors.

In addition, the normal distribution often provides a good approximation for discrete random variables with many categories, such as the marks obtained by students on an A-level Statistics paper.

Note

- The distribution is called *normal* precisely because it occurs so frequently. At one time, if a set of data did not appear to be well approximated by a normal distribution it was thought that the data must be in error!

Practical ───────────────────────────────

    Depending on the measuring instruments available, each of the following is likely to give rise to a set of observations that are approximately normally distributed: heights, weights, maximum hand span (thumb to little finger outstretched), span (maximum distance between finger tips of left hand and finger tips of right hand, length of right foot (may be smelly!), circumference of wrist. Data for males should be kept apart from that for females, since, at most ages, the latter are, on average, distinctly smaller (i.e. there are two distinct populations).

    Represent your data using a histogram. Where is the peak of the histogram? Does the histogram appear symmetric?

Carl Friedrich Gauss (1777–1855) was a German mathematician and astronomer who has been described (with Archimedes and Newton) as one of the three greatest mathematicians of all time. One of his books, published in 1809, was entitled (after translation!) *The Theory of the Motion of Heavenly Bodies Moving about the Sun in Conic Sections*. As its title suggests, the book was principally concerned with the mathematics of planetary orbits. However, one section, towards the end, deals with the problem of reconciling measurement errors. It is here that Gauss proposed a series of properties that such errors should possess. From these properties, Gauss deduced the form of the normal distribution. The argument that Gauss used was slightly flawed, but his conclusion was correct! Engineers usually refer to the normal distribution as the **Gaussian distribution**.

12.8 General properties

Gauss proposed that if one makes a number of independent observations on the value of some quantity then:

1   A positive error of given magnitude should be as probable as a negative error of the same magnitude.
2   Large errors should be less likely than small errors.
3   The mean of the observations should be the most likely value of the quantity being measured.

The consequence of (1) is that the distribution is *symmetric* and therefore has mean equal to median. The consequence of (2) is that both are equal to the mode. The three propositions taken together led Gauss to deduce that a random error, X, was likely to have pdf:

$$f(x) = \frac{1}{\sigma\sqrt{2\pi}} \exp\left(-\frac{(x-\mu)^2}{2\sigma^2}\right) \qquad -\infty < x < \infty$$

which is, in fact, the pdf for a $N(\mu, \sigma^2)$ random variable.
(The use of 'exp' was explained in Section 11.10, p. 295.)
Here π and e have their usual values (3.14... and 2.78...) and the two parameters μ and σ^2 are the mean and the variance of the distribution.

Notes
◆ Although the nominal range for X is infinite, most values (about 99.7%!) of X fall in the interval:

$$\mu - 3\sigma, \mu + 3\sigma$$

while 95% fall in the interval:

$$\mu - 2\sigma, \mu + 2\sigma$$

which is sometimes referred to as the **2 sigma rule**.
◆ For a normal distribution, changes in the values of μ and σ can be regarded simply as changes of location and scale. The fundamental shape is unaltered, though its appearance will vary.

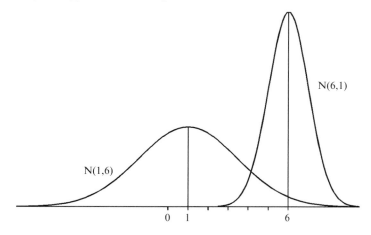

Calculator practice

If your calculator is able to evaluate integrals numerically then you will be able to satisfy yourself that the area under the normal curve is indeed equal to 1. Choose any values that you please for μ and σ. Try a variety of upper and lower limits for the integration and observe the way in which the area slowly approaches 1 as the limits become further apart.

12.9 Linear combinations of independent normal random variables

Normal random variables behave very nicely! We shall prove the following result later (in Section 12.14, p. 352):

If X and Y are two independent normally distributed random variables, and if a and b are constants, then $aX + bY$ also has a normal distribution.

The mean and variance of the resulting normal distribution follows from the results of Section 11.4 (p. 274):

$$E(aX + bY) = aE(X) + bE(Y)$$
$$\text{Var}(aX + bY) = a^2\text{Var}(X) + b^2\text{Var}(Y)$$

Example 19

The random variables X and Y are independent with $X \sim N(2, 3)$ and $Y \sim N(6, 4)$. The random variable W is defined by $W = X + Y$. Determine (i) $P(W > 8)$, (ii) $P(W > 10)$.

We begin by determining the mean and variance of W:

$$E(W) = E(X + Y) = E(X) + E(Y) = 2 + 6 = 8$$
$$\text{Var}(W) = \text{Var}(X + Y) = \text{Var}(X) + \text{Var}(Y) = 3 + 4 = 7$$

so $W \sim N(8, 7)$.

(i) Since W has a normal distribution with mean 8 we can write down immediately that $P(W > 8) = \frac{1}{2}$.

(ii) To find $P(W > 10)$ we need the z-value corresponding to $w = 10$:

$$z = \frac{10 - 8}{\sqrt{7}} = 0.756$$

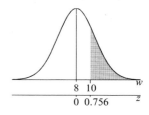

Hence:

$$P(W > 10) = 1 - \Phi(0.756) = 1 - 0.7752 = 0.225 \text{ (to 3 d.p.)}.$$

Example 20

The random variables X and Y are independent with $X \sim N(3, 1)$ and $Y \sim N(7, 5)$. The random variable W is defined by $W = Y - 2X$. Determine the probability that W takes a positive value.

We begin by determining the mean and variance of W:

$$E(W) = E(Y - 2X) = E(Y) - 2E(X) = 7 - (2 \times 3) = 1$$
$$\text{Var}(W) = \text{Var}(Y - 2X) = \text{Var}(Y) + (-2)^2 \, \text{Var}(X)$$
$$= 5 + (4 \times 1) = 9$$

so $W \sim N(1, 9)$.

To find $P(W > 0)$ we need the z-value corresponding to $w = 0$:

$$z = \frac{0 - 1}{\sqrt{9}} = -\frac{1}{3}$$

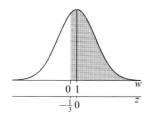

Hence:

$$P(W > 0) = 1 - \Phi(-\tfrac{1}{3}) = \Phi(\tfrac{1}{3}) = 0.6304$$

The probability that W takes a positive value is 0.630 (to 3 d.p.).

Example 21

The random variables X and Y are independent with $X \sim \mathrm{N}(16,4)$ and $Y \sim \mathrm{N}(8, 9)$.
Find (i) $\mathrm{P}(X - 2Y > 0)$, (ii) $\mathrm{P}(X + 2Y < 30)$.

(i)   Denote $X - 2Y$ by V. Then V has a normal distribution with mean:

$$\mathrm{E}(X) - 2\mathrm{E}(Y) = 16 - (2 \times 8) = 0$$

and variance:

$$\mathrm{Var}(X) + \{(-2)^2 \times \mathrm{Var}(Y)\} = 4 + (4 \times 9) = 40$$

We require $\mathrm{P}(V > 0)$. Since $\mathrm{E}(V) = 0$, the required probability is $\frac{1}{2}$.

(ii) Denote $X + 2Y$ by W. Then W has a normal distribution with mean:

$$\mathrm{E}(X) + 2\mathrm{E}(Y) = 16 + (2 \times 8) = 32$$

and variance:

$$\mathrm{Var}(X) + \{2^2 \times \mathrm{Var}(Y)\} = 4 + (4 \times 9) = 40$$

We require $\mathrm{P}(W < 30)$. The corresponding z-value is

$\dfrac{30 - 32}{\sqrt{40}} = -0.316$. The required probability is therefore

$\Phi(-0.316) = 0.376$ (to 3 d.p.).

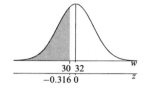

Example 22

The random variable $X \sim \mathrm{N}(16, 4)$. The independent random variables X_1 and X_2 have the same distribution as X. The random variable $\bar{X}$ is defined by $\bar{X} = \frac{1}{2}(X_1 + X_2)$.
Determine $\mathrm{P}(\bar{X} > 18)$.

The distribution of $\bar{X}$ is normal with mean $\frac{1}{2}(16 + 16) = 16$ and variance $(\frac{1}{4})(4 + 4) = 2$. The z-value of interest is therefore $\dfrac{18 - 16}{\sqrt{2}} = 1.414$ and hence the required probability is

$1 - \Phi(1.414) = 1 - 0.9213 = 0.0787 = 0.079$ (to 3 d.p.).

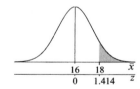

Example 23

The diameter in mm, X, of the circular mouth of a bottle has a normal distribution with mean 20 and standard deviation 0.1. The diameter in mm, Y, of the circular cross-section of a glass stopper, has a normal distribution with mean 19.7 and standard deviation 0.1.
Determine the probability that a randomly chosen stopper will fit in the mouth of a randomly chosen bottle.

We require $\mathrm{P}(X > Y)$, which looks like a rather difficult quantity to calculate. However, $\mathrm{P}(X > Y) = \mathrm{P}(X - Y > 0)$ and $X - Y$ is a linear combination of independent normal random variables and therefore has a normal distribution. Let $G = X - Y$, where G mm is the gap between the diameters of the stopper and the mouth. Now

$E(G) = E(X) - E(Y) = 20 - 19.7 = 0.3$
and $\text{Var}(G) = \text{Var}(X) + \text{Var}(Y) = 0.1^2 + 0.1^2 = 0.02$, so:

$$G \sim N(0.3, 0.02)$$

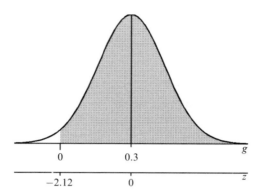

The z-value of interest is $\dfrac{0 - 0.3}{\sqrt{0.02}} = -2.121$, so:

$$P(G > 0) = 1 - P(G < 0) = 1 - \Phi(-2.121) = \Phi(2.121) = 0.9830$$

The probability that a randomly chosen stopper will fit inside the mouth of a randomly chosen bottle is 0.983 (to 3 d.p.).

Extension to more than two variables

This follows immediately. Suppose that W, X and Y are independent normal random variables, and that a, b and c are constants. Consider the random variable U defined by:

$$U = aW + bX + cY$$

and let $V = aW + bX$. From the previous result we know that V also has a normal distribution. Thus we can write:

$$U = V + cY$$

and, since the right-hand side is once again a linear combination of independent normal random variables, it follows that U has a normal distribution. The results of Section 11.4 (p. 274) give the mean and variance of U:

$$E(U) = aE(W) + bE(X) + cE(Y)$$
$$\text{Var}(U) = a^2\text{Var}(W) + b^2\text{Var}(X) + c^2\text{Var}(Y)$$

This argument can be extended indefinitely.

A linear combination of any number of independent normal random variables has a normal distribution.

Example 24

The independent normal random variables W, X and Y each have mean 0 and variance 1. The random variable V is defined by $V = W + 2X + 3Y$. Determine the probability that V is less than 4.

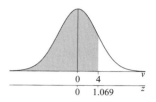

V has mean 0 and variance $1 + (2^2 \times 1) + (3^2 \times 1) = 14$. The z-value of interest is $\dfrac{4 - 0}{\sqrt{14}} = 1.069$. The required probability is $\Phi(1.069) = 0.857$ (to 3 d.p.).

Example 25

The independent random variables W, X and Y are such that $W \sim N(3, 5)$, $X \sim N(5, 5)$ and $Y \sim N(7, 5)$.
Determine the probability that the sum of W and X exceeds Y.

Let $V = W + X - Y$. Then V has a normal distribution with mean equal to $3 + 5 - 7 = 1$ and variance equal to $5 + 5 + \{(-1)^2 \times 5\} = 15$. We want $P(V > 0)$. The z-value of interest is therefore $\dfrac{0 - 1}{\sqrt{15}} = -0.258$. The required

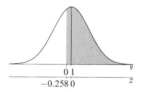

probability is $1 - \Phi(-0.258) = \Phi(0.258) = 0.602$ (to 3 d.p.).

Example 26

The normal random variable X has mean and variance equal to 1. Determine the probability that 16 independent values of X have a total in excess of 24.

Denote the random variables corresponding to the 16 values by $X_1, X_2, \ldots, X_{16}$. Let T be given by $T = X_1 + \cdots + X_{16}$. Then T has a normal distribution with mean and variance given by:

$$\mathrm{E}(T) = \mathrm{E}(X_1) + \cdots + \mathrm{E}(X_{16}) = 16$$
$$\mathrm{Var}(T) = \mathrm{Var}(X_1) + \cdots + \mathrm{Var}(X_{16}) = 16$$

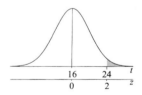

Hence $T \sim N(16, 16)$. The z-value of interest is $\dfrac{24 - 16}{\sqrt{16}} = 2$. The required probability is $1 - \Phi(2) = 0.0228 = 0.023$ (to 3 d.p.).

Example 27

The total mass (in g) of a packet of biscuits is made up of the mass (in g) of the packaging, Y, and the masses (in g) of the fifteen biscuits $X_1, \ldots, X_{15}$. The mass of a biscuit has a normal distribution, with mean 30 and standard deviation 1, while the mass of the packaging has a normal distribution with mean 5 and standard deviation 0.2.
Determine the probability that the total mass of a packet of biscuits lies between 450 and 460 g.

Let W denote the total mass. So:

$$W = Y + X_1 + \cdots + X_{15}$$

and hence:

$$E(W) = 5 + (15 \times 30) = 455$$

Assuming that the masses of the sixteen items are independent of one another, we also have:

$$\text{Var}(W) = 0.2^2 + (15 \times 1^2) = 15.04$$

so that $W \sim N(455, 15.04)$.

The z-values corresponding to the values of interest, 450 and 460, are

$$\frac{460 - 455}{\sqrt{15.04}} \text{ and } \frac{450 - 455}{\sqrt{15.04}}, \text{ respectively. These expressions simplify to } 1.289$$

and -1.289, respectively.

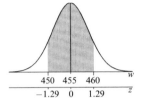

Hence:

$$\begin{aligned}
P(450 < W < 460) &= P(W < 460) - P(W < 450) \\
&= \Phi(1.289) - \Phi(-1.289) \\
&= 2\Phi(1.289) - 1 \\
&= (2 \times 0.9013) - 1 \\
&= 0.8026
\end{aligned}$$

The probability that the total mass of a packet of biscuits lies between 450 and 460 g is 0.803 (to 3 d.p.).

Exercises 12e

1   Given that $X \sim N(4, 9)$, $Y \sim N(7, 4)$, and that X and Y are independent, find:
    (i) $P(X + Y < 15)$, (ii) $P(Y - X > 1)$,
    (iii) $P(X > Y)$, (iv) $P(X > 10 - Y)$.

2   Given that $X \sim N(-1, 5)$, $Y \sim N(2, 7)$ and that X and Y are independent, find:
    (i) $P(3X + 2Y > 1)$, (ii) $P(Y < 2X + 6)$,
    (iii) $P(X + 2Y + 1 < 0)$, (iv) $P(Y > X)$.

3   Given that $X \sim N(-1, 4)$, $Y \sim N(1, 4)$, and that X and Y are independent, find:
    (i) $P(2X > 3Y)$, (ii) $P(X + Y > 0)$,
    (iii) $P(50X + 100Y > 300)$,
    (iv) $P(100Y - 50X > 200)$.

4   It is given that $X \sim N(7, 5)$ and that $X_1, X_2, \ldots, X_{10}$ are ten independent observations of X. Let $S = X_1 + X_2 + \cdots + X_{10}$. Find:
    (i) $P(10X_1 > 80)$, (ii) $P(S > 80)$,
    (iii) $P(10X_{10} < 50)$, (iv) $P(S < 50)$.

5   The mass of a biscuit is a normal variable with mean 50 g and standard deviation 4 g. A packet contains 10 randomly selected biscuits. The mass of the packing material is a normal random variable with mean 40 g and standard deviation 3 g and is independent of the masses of the biscuits.
    Find the probability that the total mass of a packet is less than 566 g.

6   A toy company manufactures plastic nuts whose internal diameters are normally distributed with mean 50 mm and standard deviation 4 mm. The company also manufactures plastic bolts whose external diameters are normally distributed with mean 48 mm and standard deviation 3 mm.
    By considering the difference between their diameters, determine the probability that a randomly chosen bolt can be inserted into a randomly chosen nut.

7 The amount of jam in a standard jar has a normal distribution with mean 340 g and standard deviation 10 g. The mass of the jar has a normal distribution with mean 150 g and standard deviation 8 g.
Find the probability that:
(i) a randomly chosen jar of jam has total mass exceeding 500 g,
(ii) a randomly chosen pack of 20 jars of jam has a total mass exceeding 10 000 g.

8 Three women and four men enter a lift. Assume that women have masses that are normally distributed with mean 60 kg and standard deviation 10 kg, and that men have masses that are normally distributed with mean 80 kg and standard deviation 15 kg.
Find the probability that the total mass of the seven people in the lift exceeds 550 kg.

9 The continuous random variable X is normally distributed with mean 212.6 and standard deviation 2. Calculate, correct to three decimal places, the probabilities that
(i) a randomly observed value of X will lie between 212 and 213,
(ii) the mean of four randomly observed observations of X will exceed 213.   [WJEC]

10 [In this question give three places of decimals in each answer.] The mass of tea in "Supacuppa" teabags has a normal distribution with mean 4.1 g and standard deviation 0.12 g. The mass of tea in "Bumpacuppa" teabags has a normal distribution with mean 5.2 g and standard deviation 0.15 g.
(i) Find the probability that a randomly chosen Supacuppa teabag contains more than 4.0 g of tea.
(ii) Find the probability that, of two randomly chosen Supacuppa teabags, one contains more than 4.0 g of tea and one contains less than 4.0 g of tea.
(iii) Find the probability that five randomly chosen Supacuppa teabags contain a total of more than 20.8 g of tea.
(iv) Find the probability that the total mass of tea in five randomly chosen Supacuppa teabags is more than the total mass of tea in four randomly chosen Bumpacuppa teabags.

[UCLES]

11 Monto sherry is sold in bottles of two sizes – standard and large. For each size, the content, in litres, of a randomly chosen bottle is normally distributed with mean and standard deviation as given in the table.

	Mean	Standard deviation
Standard bottle	0.760	0.008
Large bottle	1.010	0.009

(i) Show that the probability that a randomly chosen standard bottle contains less than 0.750 litres is 0.1056, correct to 4 places of decimals.
(ii) Find the probability that a box of 10 randomly chosen standard bottles contains at least three bottles whose content are each less than 0.750 litres. Give three significant figures in your answer.
(iii) Find the probability that there is more sherry in four randomly chosen standard bottles than in three randomly chosen large bottles.   [UCLES]

12 The weight of the contents of a randomly chosen packet of breakfast cereal A may be taken to have a normal distribution with mean 625 grams and standard deviation 15 grams. The weight of the packaging may be taken to have an independent normal distribution with mean 25 grams and standard deviation 3 grams. Find, giving three significant figures in your answers,
(i) the probability that a randomly chosen packet of A has a total weight exceeding 630 grams,
(ii) the probability that the total weight of the contents of four randomly chosen packets of A exceeds 2450 grams.
The weight of the contents of a randomly chosen packet of breakfast cereal B may be taken to have a normal distribution with mean 465 grams and standard deviation 10 grams. Find the probability that the contents of four randomly chosen packets of B weigh more than the contents of three randomly chosen packets of A.   [UCLES]

13 A fertiliser is manufactured in batches. The percentage of phosphate in each batch is a random variable which is normally distributed

(*continued*)

about a mean of 35.0 with standard deviation 2.4. Find the probability that a batch chosen at random will contain more than 30% of phosphate. Two batches are chosen at random.

(i) Find the probability that, of these two batches, one contains more than 30% of phosphate and the other contains less than 30% of phosphate.

(ii) These batches are thoroughly mixed together. Find the probability that the mixture contains less than 30% of phosphate. [O&C]

14 The random variable X is distributed $N(\mu_1, \sigma_1^2)$ and the random variable Y is distributed $N(\mu_2, \sigma_2^2)$. X and Y are independent variables. State the form of the distribution of $(X + Y)$ and of $(X - Y)$ and give the mean and variance for each distribution.

A factory makes both rods and copper tubes. The internal diameter, X cm, of a copper tube is distributed $N(2.2, 0.0009)$.

(a) Find, to 3 significant figures, the proportion of tubes with internal diameter less than 2.14 cm.
The diameter, Y cm, of a rod is distributed $N(2.15, 0.0004)$.

(b) Find, to 3 decimal places, the proportion of rods with diameter greater than 2.1 cm and less than 2.2 cm.

(c) A rod and a tube are chosen at random. Find, to 3 decimal places, the probability that the rod will not pass through the tube.

(d) A rod and a tube are chosen at random. A second rod and a second tube are chosen at random and then a third rod and a third tube are chosen at random. Find, to 3 decimal places, the probability that each of two rods will pass through the tube which was selected at the same time and the other will not. [ULSEB]

15 X and Y are continuous random variables having independent normal distributions. The means of X and Y are 10 and 12 respectively, and the standard deviations are 2 and 3 respectively. Find

(i) $P(Y < 10)$,
(ii) $P(Y < X)$,
(iii) $P(4X + 5Y > 90)$,
(iv) the value of x such that $P(X_1 + X_2 > x) = \frac{1}{4}$, where X_1 and X_2 are independent observations of X. [UCLES]

16 The mass, X g, of a grade A apple sold in a supermarket is a normally distributed random variable having mean 212 g and standard deviation 12 g. The mass, Y g, of a grade B apple sold on a market stall is a normally distributed random variable having mean 150 g and standard deviation 20 g. Find, to three decimal places,

(i) $P(X < 230)$,
(ii) the probability that a random sample of 9 grade B apples sold on the market stall has a total mass exceeding 1.5 kg,
(iii) $P(X - Y > 37)$. [NEAB]

17 Independently for each week, the number of miles travelled by a motorist per week is a normally distributed random variable having mean 235 and standard deviation 20.

(i) Calculate the probability that the motorist will travel between 200 and 240 miles in a week.

(ii) Determine the probability that the motorist will travel a distance of less than 1000 miles in a period of four successive weeks.

(iii) Find the probability that the difference between the distances travelled by the motorist in two successive weeks will be more than 30 miles. [NEAB]

18 The length, in cm, of a rectangular tile is a normal variable with mean 19.8 and standard deviation 0.1. The breadth, in cm, is an independent normal variable with mean 9.8 and standard deviation 0.1.

(i) Find the probability that the sum of the lengths of five randomly chosen tiles exceeds 99.4 cm.

(ii) Find the probability that the breadth of a randomly chosen tile is less than one half of the length.

(iii) S denotes the sum of the lengths of 50 randomly chosen tiles and T denotes the sum of the breadths of 90 randomly chosen tiles. Find the mean and variance of $S - T$. [UCLES]

19 A group of students are weighing lead weights said to have a nominal mass of 10 grams. They discover that the weights produced by manufacturer A have a mean mass of 9.82 grams and standard deviation 0.1 gram.

(a) Using the Normal distribution and these values as estimates of the population

(*continued*)

parameters, calculate the probability that a randomly chosen weight from manufacturer *A* has a mass of 10 grams or more.

(*b*) Similar weights from manufacturer *B* are known to have a mass which is Normally distributed with mean 10.05 grams and standard deviation 0.05 gram. Calculate the probability that a randomly chosen weight from manufacturer *A* has a mass which is greater than the mass of a randomly chosen weight from manufacturer *B*. You are expected to make clear the parameters of the distribution you use in answering this part of the question.   [UODLE(P)]

20 Mass-produced laminated-wood beams are constructed using five layers of wood. A study of the ends of individual layers reveals that the thickness of each of the two outside layers is normally distributed with mean 52 mm and standard deviation 3 mm. The thickness of the ends of each of the three middle layers is also normally distributed but with mean 31 mm and standard deviation 2 mm. In the assembly of the beams both the two outside layers and the three middle layers are selected at random. Find the mean thickness of the beam ends. Show that, correct to two decimal places, the standard deviation of the thickness of the beam ends is 5.48 mm. Determine the probability that the thickness of an end of a beam exceeds 200 mm. Each end of a beam is fitted into a slot in a steel plate. The widths of the slots are normally distributed with mean 206 mm and standard deviation 2 mm, independently of the thicknesses of the beam ends. Determine the probability that an end of a beam will fit into its slot. Hence, assuming that the thicknesses of the two ends of a beam are independent, determine the probability that both ends of a beam will fit into their slots. The mean width of the slots in the steel plates can be altered without affecting the standard deviation. Determine the mean slot width that will ensure that 95% of the beams fit into *both* their slots.   [JMB]

Pierre Simon Laplace (1749–1827) was eulogized by Poisson as being 'the Newton of France'. He had been elected to membership of the French Royal Academy of Sciences by the age of 24, and during his long life held a large number of influential positions including being a professor at the Ecole Militaire during the time that Napoleon was a student. His greatest interest was in celestial mechanics, which involves, amongst other things, the accurate determination of the positions of heavenly bodies. The paper in which Laplace derived the Central Limit Theorem was read to the Academy in 1810, and was a direct consequence of the work by Gauss in the previous year. In France the normal distribution is often referred to as the **Laplacean distribution**.

12.10 The Central Limit Theorem

An informal statement of this extremely important theorem is as follows:

Suppose $X_1, X_2, \ldots, X_n$ are n independent random variables, each having the *same* distribution.

Then, as n increases, the distributions of $X_1 + \cdots + X_n$ and of $\frac{X_1 + \cdots + X_n}{n}$ come increasingly to resemble normal distributions.

The importance of the **Central Limit Theorem** lies in the fact that:

- The common distribution of the *X*-variables is not stated – it can be almost *any* distribution.
- In most cases the resemblance to a normal distribution holds for remarkably small values of *n*.
- Totals and means are quantities of interest!

As an example, consider the following data which constitute part of a random sample of observations on a random variable having a continuous uniform distribution in the interval (0, 1).

Original observations	0.020		0.706		0.536		0.580
Means of pairs		0.363				0.558	
Means of fours				0.4605			
Original observations	0.290		0.302		0.776		0.014
Means of pairs		0.296				0.395	
Means of fours				0.3455			

The successive diagrams below show (i) the original distribution, (ii) a histogram of the first 50 randomly chosen single observations from that distribution, (iii) a histogram of the first 50 means of pairs of observations from the distribution, (iv) the same for groups of four observations, and finally, (v) the same for groups of eight observations. As the group size increases so the means become increasingly clustered in a symmetrical fashion about 0.5 (the mean of the original population).

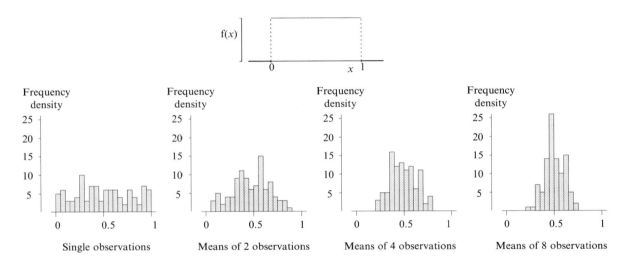

Single observations · Means of 2 observations · Means of 4 observations · Means of 8 observations

As a second example, consider successive averages of observations from a V-shaped distribution:

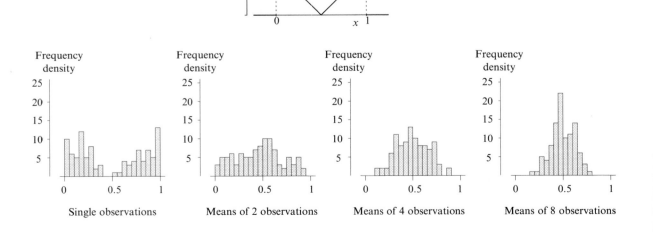

Single observations · Means of 2 observations · Means of 4 observations · Means of 8 observations

Here the original distribution has a trough in the middle, whereas the normal distribution has a peak, but once we start looking at averages of even as few as $n = 2$ observations, a peak starts to appear.

As a final example we look at a triangular distribution, which is quite heavily skewed. In this case the histograms of the means of two observations and four observations still appear skewed, but this skewness has almost vanished by the time we are working with means of eight observations.

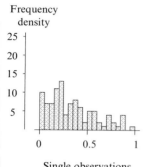

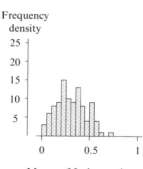

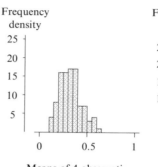

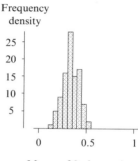

| Single observations | Means of 2 observations | Means of 4 observations | Means of 8 observations |

Note

♦ It is clear that the practical consequences of the Central Limit Theorem were understood well before the time of Laplace. A 16th-century German treatise on surveying instructs the surveyor to establish the length of a rood (a standard unit of length) in the following manner:

'Stand at the door of a church on a Sunday and bid 16 men to stop, tall ones and small ones, as they happen to pass out when the service is finished; then make them put their left feet one behind the other, and the length thus obtained shall be a right and lawful rood to measure and survey the land with, and the 16th part of it shall be a right and lawful foot.'

The distribution of the sample mean, $\bar{X}$

Denote the *i*th observation in a sample by x_i. Different samples would give different values for x_i (see Section 8.5, p.203). Thus x_i is an observation on a random variable that we will denote by X_i. Suppose that X_1 has mean μ and variance σ^2. The same will be true for each of $X_1, X_2, \ldots, X_n$, which are therefore identically distributed random variables, each with mean μ and variance σ^2. We define $\bar{X}$ by:

$$\bar{X} = \frac{1}{n}(X_1 + X_2 + \cdots + X_n)$$

and we showed in Section 8.5 that $\bar{X}$ has mean μ and variance $\dfrac{\sigma^2}{n}$. By the Central Limit Theorem, $\bar{X}$ has an approximate normal distribution for large n, and so:

$$\bar{X} \sim \mathrm{N}\left(\mu, \frac{\sigma^2}{n}\right)$$

Standardising, we get the equivalent result that:

$$\frac{\bar{X} - \mu}{\frac{\sigma}{\sqrt{n}}} \sim \mathrm{N}(0, 1)$$

Notes

♦ If the distribution of the individual X variables is normal, then, since $\bar{X}$ is then a linear combination of normally distributed random variables, the result is true even for small values of n.

♦ The equivalent result for a sum is that:

$$\Sigma X_i \sim \text{N}(n\mu, n\sigma^2)$$

Practical ────────────────────

This is a slightly tiring practical! Roll a die and record the value obtained. Roll the die again ... and again ... – about 200 rolls should suffice! Record each outcome and total groups of two and four as indicated below:

Singles	2	4	4	2	2	1	6	2
Sums of 2	6		6		3		8	
Sums of 4		12			11			
Singles	1	4	4	5	1	5	4	4
Sums of 2	5		9		6		8	
Sums of 4		14			14			

Draw up a frequency distribution for the original values and also for the two sets of sums. Illustrate the three distributions using bar charts. In each case, state the smallest and largest values that could possibly have occurred. Compare these extreme values with the ranges of values actually observed and comment on the results.

Computer project ────────────────────

The random numbers generated by a computer (or a calculator) may be thought of as independent observations from a uniform distribution on the interval (0, 1).

Write a program to examine the distribution of the means of k observations. A convenient way of keeping count of the values generated is as follows.

1 *Let XBAR be an array of size 100, and set all members of the array to 0. (The choice of 100 is arbitrary, but the array needs to be quite large in order to cope with the case where k is large.)*

Now, for each sample, repeat 2 to 5:

2 *Calculate the sample mean, which will have a value between 0 and 1.*

3 *Multiply the mean by 100 (the array size) to give a value between 0 and 100.*

4 *Calculate the integer part of this quantity. Call this M.*

5 *Add 1 to the Mth item in the array XBAR.*

This was the method used to produce the first of the three earlier examples. The result will be a set of counts for each of 100 classes, of width 0.01. As k increases it will be found that increasing numbers of these classes will have zero frequencies.

Computer project ────────────────────

To simulate from other than uniform distributions requires a little more work! Consider, for example, the third example (the triangular density function) for which the pdf was given by:

$$f(x) = \begin{cases} 2(1-x) & 0 < x < 1 \\ 0 & \text{otherwise} \end{cases}$$

This corresponds to the cdf:

$$F(x) = \begin{cases} 0 & x \leqslant 0 \\ x(2-x) & 0 \leqslant x \leqslant 1 \\ 1 & x \geqslant 1 \end{cases}$$

Suppose that u is a number in the range (0,1). If we solve the equation:

$$u = x(2-x)$$

which can be written as:

$$x^2 - 2x + u = 0$$

choosing the root in the interval (0,1), we get:

$$x = 1 - \sqrt{(1-u)}$$

We now know that, with this choice of value for x,

$$P(X \leqslant x) = P(U \leqslant u)$$

where U is a random variable having a uniform distribution with range (0,1).

Successive uniform random numbers u_1, u_2, ... can therefore be used to generate successive 'random observations' from the triangular distribution, if the latter are chosen to have the values $1 - \sqrt{(1-u_1)}$, $1 - \sqrt{(1-u_2)}$, and so on.

Adjust the program that you devised for the previous project and verify that this approach works!

Example 28

The continuous random variable X has mean 5 and variance 25. A random sample of 100 observations are taken on X.
Determine the probability that the sample mean exceeds 5.4.

The random variable $\bar{X}$, corresponding to the sample mean, has expectation 5 and variance $\frac{25}{100} = \frac{1}{4}$. By the Central Limit Theorem it has an approximate normal distribution.

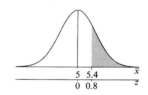

The z-value of interest is therefore $\dfrac{5.4 - 5}{\sqrt{\frac{1}{4}}} = 0.8$. Since $\Phi(0.8) = 0.788$ the required probability is 0.212.

Example 29

The discrete random variable X has probability distribution given by
$P(X = 0) = \frac{1}{4}$, $P(X = 1) = \frac{1}{2}$, $P(X = 2) = P(X = 3) = \frac{1}{8}$.
Determine an approximation to the probability that a random sample of 500 observations on X will have a total less than 520, giving the answer to the nearest percentage point.

---

x	0	1	2	3
P_x	$\frac{1}{4}$	$\frac{1}{2}$	$\frac{1}{8}$	$\frac{1}{8}$

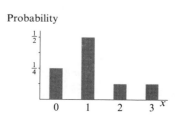

Probability

We begin by determining the mean and variance of X:

$$E(X) = \left(0 \times \frac{1}{4}\right) + \left(1 \times \frac{1}{2}\right) + \left(2 \times \frac{1}{8}\right) + \left(3 \times \frac{1}{8}\right) = \frac{9}{8}$$

and:

$$\mathrm{E}(X^2) = \left(0^2 \times \frac{1}{4}\right) + \left(1^2 \times \frac{1}{2}\right) + \left(2^2 \times \frac{1}{8}\right) + \left(3^2 \times \frac{1}{8}\right) = \frac{17}{8}$$

so that $\mathrm{Var}(X) = \frac{17}{8} - \frac{81}{64} = \frac{55}{64}$. Denoting the random variable corresponding to the total of the 500 observations by T, we have $\mathrm{E}(T) = (500 \times \frac{9}{8}) = 562.5$ and $\mathrm{Var}(T) = (500 \times \frac{55}{64}) = \frac{6875}{16}$.

By the Central Limit Theorem the distribution of T will be

approximately normal. The z-value of interest is $\dfrac{520 - 562.5}{\sqrt{\frac{6875}{16}}} = -2.050$.

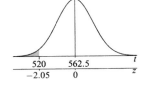

Hence the approximate probability is $1 - \Phi(2.050) = 0.0202$, which is approximately 2%.

(An improved approximation would use a continuity correction – see Section 12.11, p.336.)

Example 30

Bags of rice are marked as containing 1 kg of rice. In reality, the mean mass of rice per bags is 1.05 kg. The mass of rice varies from bag to bag, and has a standard deviation of 20 g.

Making a suitable assumption, estimate the proportion of bags that contain less than 1 kg of rice.

No mention has been made of a probability distribution for the mass of the contents. However, the total mass of a bag is the aggregate of the masses of thousands of individual grains of rice and we can reasonably assume that the mass of a bag, X kg, has a normal distribution.

The question mentions both kg and g as units of weight: we will work in kg, noting that $20\,\mathrm{g} = 0.02\,\mathrm{kg}$. Thus $X \sim \mathrm{N}(1.05, 0.02^2)$. The z-value corresponding to $x = 1$ is $\dfrac{1.00 - 1.05}{0.02}$, which equals -2.5. Hence:

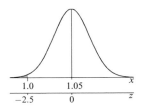

$$\mathrm{P}(X < 1.00) = \Phi(-2.5) = 0.006\,21$$

This value was obtained from the table given in the Appendix. The probability is reassuringly low – only about 1 in 160 bags actually have masses less than the stated value.

Example 31

A builder orders 200 planks of walnut and 50 planks of mahogany. The mean and standard deviation of the masses (in kg) of walnut planks are 15 and 1, respectively. The corresponding figures for the mahogany planks are 20 and 1.1, respectively.

Assuming that the planks delivered to the builder are random samples from the populations of planks, determine the probability that the wood delivered has a total mass of wood that is:

(i)   less than 4000 kg,

(ii) between 3980 kg and 4000 kg.

From the Central Limit Theorem, the total mass of the walnut planks has an approximate normal distribution. The same is true for the mahogany planks. Also, since linear combinations of independent normal random variables have a normal distribution, the combined mass, X, has an approximate normal distribution.

(i) If we denote the masses of the walnut planks by $W_1, \ldots, W_{200}$ and the masses of the mahogany planks by $M_1, \ldots, M_{50}$, then we can write:

$$X = W_1 + \cdots + W_{200} + M_1 + \cdots + M_{50}$$

Thus:

$$\begin{aligned}
\mathrm{E}(X) &= \mathrm{E}(W_1) + \cdots + \mathrm{E}(W_{200}) + \mathrm{E}(M_1) + \cdots + \mathrm{E}(M_{50}) \\
&= (200 \times 15) + (50 \times 20) \\
&= 3000 + 1000 \\
&= 4000
\end{aligned}$$

The probability that the wood delivered has a total mass of less than 4000 kg is therefore exactly $\frac{1}{2}$.

(ii) In a similar way, we see that:

$$\mathrm{Var}(X) = (200 \times 1^2) + (50 \times 1.1^2) = 200 + 60.5 = 260.5$$

The z-value corresponding to $x = 3980$ is therefore $\dfrac{3980 - 4000}{\sqrt{260.5}}$

which equals -1.239 (to 3 d.p.). Thus:

$$\begin{aligned}
\mathrm{P}(3980 < X < 4000) &= \Phi(0) - \Phi(-1.239) \\
&= 0.5 - (1 - 0.8924) = 0.3924
\end{aligned}$$

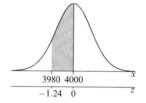

The probability that the total mass of wood delivered is between 3980 kg and 4000 kg is 0.392 (to 3 d.p.).

Exercises 12f

1 The random variable X has mean 15 and variance 25. The random variable $\bar{X}$ is the mean of a random sample of 70 observations on X.
State the approximate distribution of $\bar{X}$ and hence find, approximately, $\mathrm{P}(15 < \bar{X} < 16)$.

2 The random variable Y has mean 50 and standard deviation 20. The random variable $\bar{Y}$ is the mean of a random sample of n observations on Y.
Find, approximately, $\mathrm{P}(45 < \bar{Y} < 55)$ in the cases (i) $n = 50$, (ii) $n = 100$.

3 The random variable W has mean 20 and variance 72. The random variable S is defined to be the sum of 80 independent observations on W.

Find, approximately, (i) $\mathrm{P}(S > 1700)$, (ii) $\mathrm{P}(1400 < S < 1700)$.

4 Size 1 eggs have a uniform distribution of mass, in g, between 70 and 75. Size 2 eggs have a uniform distribution of mass, in g, between 65 and 70. Size 3 eggs have a uniform distribution of mass, in g, between 60 and 65. The variance of the mass, in g, of each size of egg is $\frac{25}{12}$. Find the probability that:
(i) the average mass, in g, of 120 randomly chosen Size 2 eggs lies between 67.2 and 67.8,
(ii) the average mass, in g, of a collection of 60 randomly chosen Size 1 eggs and 60 randomly chosen Size 3 eggs lies between 67.2 and 67.8.

5 The random variable X has a continuous distribution with pdf as shown in the diagram.

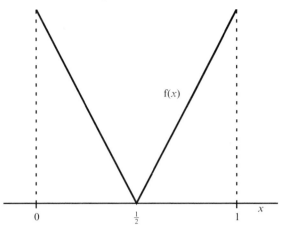

It is given that $E(X) = \frac{1}{2}$ and $Var(X) = \frac{1}{8}$. The random variable $\bar{X}$ is the mean of n independent observations of X.
Using the Central Limit Theorem, find the smallest value of n such that
$P(|\bar{X} - \frac{1}{2}| < \frac{1}{20}) \geqslant 0.95$.

6 The random variable Y has a continuous distribution with pdf as shown in the diagram.

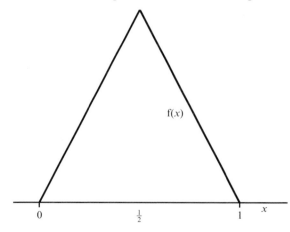

It is given that $E(Y) = \frac{1}{2}$ and $Var(Y) = \frac{1}{24}$. The random variable $\bar{Y}$ is the mean of n independent observations of Y.
Using the Central Limit Theorem, find the smallest value of n such that
$P(|\bar{Y} - \frac{1}{2}| < \frac{1}{20}) \geqslant 0.95$.

7 The random variables $X_1, X_2, \ldots, X_n$ denote n independent observations of the random variable X, where $X \sim N(\mu, \sigma^2)$. Writing
$\bar{X} = \dfrac{X_1 + X_2 + \cdots + X_n}{n}$, find $P(|\bar{X} - \mu| < \sigma)$ in
the following cases:
(i) $n = 2$, (ii) $n = 6$, (iii) $n = 10$.

8 A random sample of 16 observations is to be drawn from a Normal distribution having mean 11 and standard deviation 3. Let $\bar{X}$ denote the sample mean. Find, correct to three decimal places,
(i) the probability that $\bar{X}$ will have a value between 9.2 and 12.2,
(ii) the value of c for which $P(\bar{X} < c) = 0.03$.
[WJEC]

9 The mass of coffee in a randomly chosen jar sold by a certain company may be taken to have a normal distribution with mean 203 g and standard deviation 2.5 g.
(i) Find the probability that a randomly chosen jar will contain at least 200 g of coffee.
(ii) Find the mass m such that only 3% of jars contain more than m grams of coffee.
(iii) Find the probability that two randomly chosen jars will together contain between 400 g and 405 g of coffee.
(iv) The random variable $\bar{C}$ denotes the mean mass (in grams) of coffee per jar in a random sample of 20 jars. Find the value of a such that
$P(|\bar{C} - 203| < a) = 0.95$. [UCLES]

10 A continuous random variable X is uniformly distributed between 144 and 156. Find
$P(149.8 < X < 150.4)$.
$\bar{X}$ is the mean of 300 randomly observed values of X. Find an approximate value for
$P(149.8 < \bar{X} < 150.4)$. [JMB]

11 The random variable X is normally distributed with mean μ and variance σ^2.
(*a*) Write down the distribution of the sample mean $\bar{X}$ of a random sample of size n.
An efficiency expert wishes to determine the mean time taken to drill a fixed number of holes in a metal sheet.
(*b*) Determine how large a random sample is needed so that the expert can be 95% certain that the sample mean time will differ from the true mean time by less than 15 seconds. Assume that it is known from previous studies that $\sigma = 40$ seconds. [ULEAC]

12 Explain briefly how you acquired empirical evidence for the Central Limit theorem.
The weights of the trout at a trout farm are normally distributed with mean 1 kg and standard deviation 0.25 kg.

(a) Find, to 4 decimal places, the probability that a trout chosen at random will weigh more than 1.25 kg.

(b) Two trout are chosen independently and at random. If their total weight is denoted by X kg, find $E(X)$ and $Var(X)$.

(c) If $\bar{Y}$ kg represents the mean weight of a sample of 10 trout chosen at random at the farm, state the distribution of $\bar{Y}$. Evaluate the mean and the variance of this distribution.

Find, to 3 decimal places, the probability that the mean weight of a sample of 10 trout chosen at random will be less than 0.9 kg.   [ULSEB(P)]

13 The contents of bags of oats are normally distributed with mean 3.05 kg, standard deviation 0.08 kg.

(a) What proportion of bags contain less than 3.11 kg?

(b) What proportion of bags contain between 3.00 and 3.15 kg?

(c) What weight is exceeded by the contents of 99.9% of the bags?

(d) If 6 bags are selected at random what is the probability that the mean weight of the contents will be between 3.00 and 3.15 kg?

The weight of the bags, when empty, is normally distributed with mean 0.12 kg, standard deviation 0.02 kg. Full bags are packed into boxes each of which holds 6 bags.

(e) What is the distribution of the weight in a box, i.e. 6 bags together with their contents? Assume that the weight of all bags and contents are independent of each other.

(f) Within what limits will the weight in a box lie with probability 0.9?   [AEB 93]

14 The lengths of the petals of a particular variety of flower are approximately normally distributed with mean 32 mm and standard deviation 5 mm.

(i) Explain briefly why the assumption of a normal distribution for the lengths of the petals may be reasonable even though such a length cannot possibly be negative.

(ii) Calculate the proportions of all petals that have lengths

(a) greater than 34 mm,

(b) between 29 mm and 38 mm.

(iii) If the lengths of the petals were to be measured correct to the nearest millimetre, calculate the proportion of the petals whose measured lengths would be 35 mm or less.

(iv) Find the length, correct to the nearest mm, which is exceeded by 60% of all petals.

(v) Calculate the probability that the mean of the lengths of a random sample of ten petals will be greater than 33 mm.

(vi) Determine how many petals should be sampled to ensure that there is a probability of at least 0.95 that the sample mean length will be within 1 mm of 32 mm, the mean length of all petals.   [WJEC]

15 (a) The percentage of a metal extracted from a batch of raw material is normally distributed with mean 53.5% and standard deviation 2.5%.

Determine the probability that for any batch selected at random the percentage of metal extracted from that batch

(i) exceeds 58%,

(ii) lies between 50% and 60%.

Two batches are selected at random. Find the probability that both batches will have between 50% and 60% of metal extracted from them.

Find the minimum number of batches which need to be randomly sampled to ensure that the probability of the sample mean of the batches being within 1.5% of 53.5% is at least 0.9.

(b) In a blending process two liquids, A and B, are poured into the same randomly chosen empty container. The volumes of the containers are normally distributed with mean 9.2 cm³ and standard deviation 0.55 cm³.

The volume of liquid A poured in is normally distributed with mean 4 cm³ and standard deviation 0.32 cm³. Liquid B is added independently and its volume is normally distributed with mean 5 cm³ and standard deviation 0.45 cm³.

Determine the probability that the contents of the container do not overflow.   [AEB 92]

Abraham de Moivre (1667–1754) was a French Protestant who emigrated to London in 1688. By the age of 30 he had been elected a Fellow of the Royal Society as a consequence of his work in various branches of mathematics. His first book, the *Doctrine of Chances*, deals with various aspects of probability, and in its second edition (1738) he carried out the essential calculations that lead to the approximation of a binomial distribution with $p = \frac{1}{2}$ by a normal distribution with mean $\frac{1}{2}n$. It is said that, during his final illness, he noted that, each day, he needed a quarter of an hour's more sleep than on the preceding day. He was thus able to predict accurately the day of his death – when he needed 24 hours of sleep.

12.11 The normal approximation to a binomial distribution

Suppose $X \sim B(n, p)$. In Section 9.7 (p. 229) we noticed that we could write:

$$X = Y_1 + Y_2 + \cdots + Y_n$$

where the Y-variables had independent Bernoulli distributions, each with parameter p. By the Central Limit Theorem, the sum of independent identically distributed random variables has an approximate normal distribution. For large n, therefore, the binomial distribution must resemble a normal distribution. This is illustrated in the diagram, which shows three binomial distributions having the same value of p (0.3) but differing values of n.

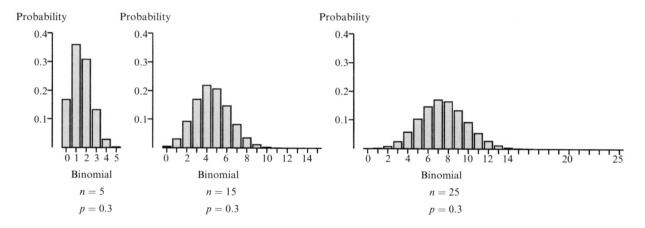

The limiting normal distribution must have the same mean and variance as its binomial counterpart and hence, if we denote the normal counterpart of X by W:

$$X \sim B(n, p) \rightarrow W \sim N(np, npq)$$

where $q = 1 - p$.

The normal distribution is continuous, with probabilities associated with all small intervals between $-\infty$ and ∞. The binomial is discrete, with 'chunks' of probability, like slices of a slab of butter, associated with each integer between 0 and n, inclusive. If the bars of binomial probability really were made of butter what would happen if we trod on them? They would spread out sideways – an equal amount on each side. This is precisely how we deal with the move from X to W – we imagine that the probability originally associated with the single point value x becomes identified with the interval $(x - \frac{1}{2}, x + \frac{1}{2})$.

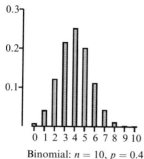

Binomial: $n = 10$, $p = 0.4$

'Point'
probabilities

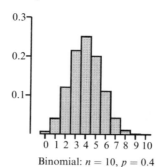

Binomial: $n = 10$, $p = 0.4$

'Continuous'
version

The normal approximation is:

$$P(X = x) \approx P(x - \tfrac{1}{2} < W < x + \tfrac{1}{2}) \tag{12.2}$$

where $W \sim N(np, npq)$. The adjustment by $\tfrac{1}{2}$ in each direction is referred to as using the **continuity correction**.

To calculate the approximate probability, we must transform to the standard normal distribution by writing:

$$Z = \frac{W - np}{\sqrt{npq}}$$

Hence:

$$P(X = x) \approx \Phi\left(\frac{(x + \tfrac{1}{2}) - np}{\sqrt{npq}}\right) - \Phi\left(\frac{(x - \tfrac{1}{2}) - np}{\sqrt{npq}}\right) \tag{12.3}$$

Example 32

Using the normal approximation, determine the probability that exactly 30 heads are obtained when a fair coin is tossed 64 times.

Here $n = 64$, $p = q = 0.5$, so:

$$P(X = 30) \approx \Phi\left(\frac{30.5 - 32}{4}\right) - \Phi\left(\frac{29.5 - 32}{4}\right)$$
$$= \Phi(-0.375) - \Phi(-0.625)$$
$$= 0.3538 - 0.2660$$
$$= 0.0878$$

(This value agrees, to four decimal places, with the exact value calculated using the binomial distribution.)

Inequalities

To see how the approximation works for inequalities we will look at $P(X \leqslant x)$:

$$P(X \leqslant x) = P(X = 0) + \cdots + P(X = x)$$
$$\approx P(-\tfrac{1}{2} < W < \tfrac{1}{2}) + \cdots + P(x - \tfrac{1}{2} < W < x + \tfrac{1}{2})$$
$$= P(-\tfrac{1}{2} < W < x + \tfrac{1}{2})$$
$$= \Phi\left(\frac{(x + \tfrac{1}{2}) - np}{\sqrt{npq}}\right) - \Phi\left(\frac{-\tfrac{1}{2} - np}{\sqrt{npq}}\right)$$

For sufficiently large n the second term will be close to zero and we then have:

$$P(X \leqslant x) \approx \Phi\left(\frac{(x + \frac{1}{2}) - np}{\sqrt{npq}}\right) \qquad (12.4)$$

Similarly

$$P(X \geqslant x) \approx 1 - \Phi\left(\frac{(x - \frac{1}{2}) - np}{\sqrt{npq}}\right) \qquad (12.5)$$

These formulae need adjustment if the inequalities are strict:

$$P(X < x) \approx \Phi\left(\frac{(x - \frac{1}{2}) - np}{\sqrt{npq}}\right) \qquad (12.6)$$

$$P(X > x) \approx 1 - \Phi\left(\frac{(x + \frac{1}{2}) - np}{\sqrt{npq}}\right) \qquad (12.7)$$

It is impossible to give precise rules for when the normal approximation should be used. All one can say is that the approximation is better for p near to $\frac{1}{2}$ and that it improves as n increases.

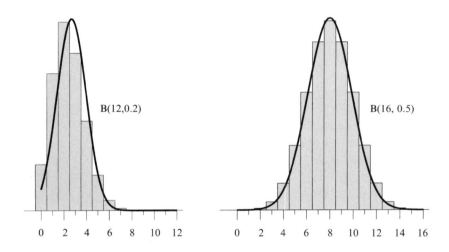

As the diagram shows, the approximation of the skewed B(12, 0.2) distribution is rather poor, particularly in the lower tail (where the normal approximation anticipates an appreciable proportion of negative outcomes!). By contrast, the approximation of the symmetric B(16, 0.5) distribution is quite impressive.

On the grounds that accurate working is better than inaccurate working (!) binomial probabilities should be calculated exactly if this is feasible in the time available.

Notes
- The approximation is most accurate when $p = \frac{1}{2}$ because this is the case when the binomial distribution is symmetric.
- A continuity correction is needed whenever a discrete distribution is being approximated by a continuous distribution.

Example 33

It is given that $X \sim B(25, 0.6)$. Using a normal approximation, determine $P(X \leqslant 16)$.

It is quite easy to calculate this probability using a computer or programmable calculator, but would be very tedious to calculate if each individual probability had to be calculated separately. The normal approximation is easy to calculate, using $np = 15$ and $npq = 6$:

$$P(X \leqslant 16) \approx \Phi\left(\frac{16.5 - 15}{\sqrt{6}}\right)$$

$$= \Phi(0.612)$$

$$= 0.7298$$

The probability that $X \leqslant 16$ is approximately 0.730 (to 3 d.p.).

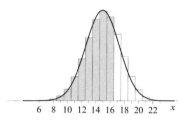

Example 34

A fair coin is tossed 1000 times. The number of heads obtained is denoted by X.
Using a normal approximation, determine the largest value of x for which $P(X \leqslant x) < 0.95$.

As n is extremely large, this is a case for the normal approximation. Since $X \sim B(1000, 0.5)$, the corresponding normal distribution has mean 500 and variance 250. Hence:

$$P(X \leqslant x) \approx \Phi\left(\frac{(x + \frac{1}{2}) - 500}{\sqrt{250}}\right)$$

From the table of percentage points, we have $\Phi(1.645) = 0.95$. The required value of x is therefore the largest integer for which:

$$\frac{(x + \frac{1}{2}) - 500}{\sqrt{250}} < 1.645$$

Rearranging, x is the largest integer less than:

$$499.5 + (1.645 \times \sqrt{250}) = 525.51$$

The largest value of x satisfying $P(X \leqslant x) < 0.95$ is 525 (and not 526, which exceeds 525.51).

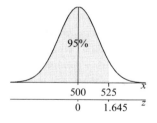

Francis Galton (1822–1911) was a first cousin of Charles Darwin, the author of *The Origin of Species*. Galton studied medicine at Cambridge, but, on coming into money, abandoned this career and spent the period 1850–52 exploring Africa, receiving the gold medal of the Royal Geographical Society in recognition of his achievements. In the 1860s he turned to meteorology and devised an early form of the weather maps used by the modern meteorologist. Subsequently, possibly inspired by Darwin's work, Galton turned to inheritance and the relationships between the characteristics of successive generations. His best known work was published in 1889 and entitled *Natural Inheritance*. He made great use of the normal distribution and illustrated it in a lecture to the Royal Institution in 1874 using a quincunx (see below).

Project _____

This is a construction project! The idea is to construct a quincunx, which is a simple arrangement of nails (or pins, or pegs) on a board in the manner shown in the diagram. The horizontal gaps between the nails in a row should be just sufficient that a marble (or a ball-bearing, or something similar) is just able to pass between them. The successive rows should be parallel, with the nails on one row being placed centrally with respect to the gaps on the previous row. The vertical distance between the rows should be just sufficient to permit the passage of the marble.

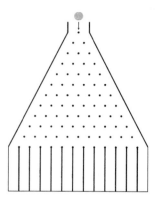

The board should be placed nearly vertical on level ground. A succession of marbles are placed at the start and allowed to fall. As a marble descends it will encounter a number of nails. If the board is not tilted and if it has been constructed accurately, then when the marble hits a nail it will be equally likely to divert to the right or to the left.

Each encounter constitutes a Bernoulli trial, with p equal to $\frac{1}{2}$. The total effect of n rows of nails is therefore a $B(n,\frac{1}{2})$ distribution. The vertical channels at the bottom of the board enable one to observe the counts of the various outcomes when a large number of balls have been allowed to drop and provide a visual bar chart of a sample from a $B(n, p)$ distribution. By starting the balls at a point below the top of the board the value of n can be reduced. If a very large quincunx (with a correspondingly large number of balls) is available then the resemblance of a $B(n, \frac{1}{2})$ distribution to a normal distribution will be apparent.

Other binomial distributions can be simulated by tilting the quincunx from the horizontal, so that there is a bias towards the left (low x) or right (high x).

We have found that a version using wooden dowelling as pegs and ping-pong balls in place of marbles works well. The cost of such a large-scale quincunx can easily be retrieved by using it to generate income at the school fete; 'suitable' prizes can be given for balls that land in the end channels – where 'suitable' means that a healthy profit may be expected!

Choosing between the normal and Poisson approximations to a binomial distribution

It is not possible to give any strict rules beyond the advice to calculate a probability exactly if this is feasible. However the following may help:

♦   The Poisson should not be used for values of *p* near 0.5.
♦   The normal should not be used when either *np* or $n(1 - p)$ is less than 5.

◆   If p is small (e.g. < 0.1) and the question requires, for example, $P(X < x)$ or $P(X > x)$, where x is small, then the *Poisson* approximation is probably best.

◆   If n is large (e.g. > 30) and the question requires, for example, $P(X > x)$ or $P(X < x)$, where x is not close to n or 0, then the *normal* approximation is probably best.

Example 35

Only 0.1% of 10p postage stamps display a particular error. Using a normal approximation, determine the probability that:

(i)   in a random sample of 1000 stamps, there are 4 or fewer that display the error,

(ii) in a random sample of 20 000 stamps, there are more than 10 that display the error.

———————

(i)   Let X be the number of stamps displaying the error. Now:

$$X \sim B(1000, 0.001)$$

Since $np = 1$, which is less than 5, the normal approximation cannot be used. The exact binomial calculation is just about possible, but the Poisson approximation is easy, and since n is so large and p is so small, the Poisson approximation is certain to be very accurate.

$$P(X \leqslant 4) \approx \frac{1^0 e^{-1}}{0!} + \frac{1^1 e^{-1}}{1!} + \frac{1^2 e^{-1}}{2!} + \frac{1^3 e^{-1}}{3!} + \frac{1^4 e^{-1}}{4!}$$

$$= \left(1 + 1 + \frac{1}{2} + \frac{1}{6} + \frac{1}{24}\right) e^{-1}$$

$$= \frac{65}{24} e^{-1}$$

$$= 0.996\,34$$

The exact binomial answer is 0.996 36. As expected, the difference is trivial. To three decimal places the probability is 0.996.

(ii) Let Y be the number of stamps displaying the error. Now:

$$Y \sim B(20\,000, 0.001)$$

It would be tedious to calculate $P(Y > 10)$ exactly, so we would like a simple approximation. We could use the Poisson, but this is nearly as tedious. The normal distribution, however, is easy to use and should not be too inaccurate since the product of n and p is reasonably large:

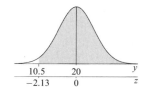

$$P(Y > 10) = 1 - P(Y \leqslant 10)$$

$$\approx 1 - \Phi\left(\frac{10.5 - 20}{\sqrt{19.98}}\right)$$

$$= 1 - \Phi(-2.125)$$

$$= \Phi(2.125)$$

$$= 0.9832$$

In fact the exact probability is 0.989 (to 3 d.p.).

Exercises 12g

1 It is given that $X \sim B(40, 0.3)$.
Use a suitable approximation to find $P(X > 15)$.

2 It is given that $Y \sim B(50, 0.7)$.
Use a suitable approximation to find $P(30 < Y < 40)$.

3 It is given that $X \sim B(100, 0.5)$.
Use a suitable approximation to find $P(|X - 50| > 10)$.

4 It is given that $W \sim B(81, 0.2)$.
Use a suitable approximation to find $P(W > 12)$.

5 A fair coin is tossed 300 times.
Using a suitable approximation, find the probability that the number of heads obtained (i) exceeds 160, (ii) is not greater than 135.

6 An artificial lake contains a large number of fish, of which $\frac{2}{5}$ are carp and $\frac{3}{5}$ are roach. A fisherman catches a random sample of five fish. Determine the probability that this contains 2 or more roach.
The fisherman was taking part in a competition. Altogether a random sample of 100 fish were caught. Determine the approximate probability that more than 70 roach were caught.

7 In Atlantis 45% of the population are Republicans and the remainder are Monarchists. Determine the (approximate) probability that a random sample of 50 Atlanteans contains a majority of Republicans. Determine also the probability that the sample contains as many Republicans as Monarchists.

8 Between 0800 and 0900, 40% of the traffic on a busy road travels from west to east, with the remainder going from east to west. A random sample of 250 cars is sampled from the cars passing during the stated time period. Using a continuity correction to improve the normal approximation, determine the approximate probability that more than 80 of the cars sampled were travelling from west to east.

9 A farmer believes that 60% of the apples that are brought into his grading shed are classed as grade A.
Assuming that he is right, use a suitable approximation to find the probability that, in a random sample of 100 apples, the proportion of Grade A apples is less than 65%.

10 A politician believes that 55% of his constituents agree with his views on environmental issues.
Assuming that he is right, use a suitable approximation to find the probability that in a random sample of 200 constituents, less than 50% agree with these views.

11 Whenever a particular gymnast performs a certain routine, the probability that she will do so faultlessly is 0.7. Find, to three decimal places, the probabilities that
(i) she will perform the routine faultlessly on four occasions out of six;
(ii) she will perform the routine faultlessly on at least 14 occasions out of 20.
Use a suitable approximate method to evaluate, to three decimal places, the probability that she will perform the routine faultlessly on more than 130 occasions out of 200.
Comment briefly on the assumption implied in the first sentence of this question.   [JMB]

12 State the conditions under which a normal approximation may be used as an approximation to the binomial distribution $B(n, p)$, giving the parameters of the normal approximation.
Give an example, from your projects where possible, of the use of this approximation.
In the production of compact discs at a certain factory, the proportion of faulty discs is known to be $\frac{1}{5}$. A random sample of 20 discs is taken from production. Using the tables provided, or otherwise, find the probability that the sample will contain at least 6 faulty discs.
Find the smallest value of n if there is a probability of at least 0.85 that a random sample of n discs will contain at least one faulty disc.
Each week the factory produces 2000 discs. Estimate, to 2 significant figures, the probability that there will be at most 349 faulty discs produced in one week.
It costs 60p to produce a disc. A faulty disc has to be discarded, while a non-faulty disc is sold for £9. Find the expected profit made by the factory per week.   [ULSEB]

13 (a) The random variable X follows a binomial distribution with parameters n and p.
  (i) Prove that $E[X] = np$.
  (ii) Show that

$$P(X = r) = \frac{(n - r + 1)p}{(1 - p)r} \times P(X = r - 1).$$

  Hence, given that X follows a binomial distribution with $E[X] = 5$ and $P(X = 4) = 1.75 P(X = 3)$, find n and p.

(b) A manufacturer of wine glasses sells them in presentation boxes of twelve. Records show that three in a hundred glasses are defective. Find the probability that a randomly chosen box of glasses contains
  (i) no defective glasses,
  (ii) at least two defective glasses.
  Find the probability that a consignment of 10 000 such glasses contains at most 250 defective glasses. [AEB 89]

14 Describe, briefly, the conditions under which the binomial distribution $\text{B}(n, p)$ may be approximated by
  (a) a normal distribution,
  (b) a Poisson distribution,
giving the parameters of each of the approximate distributions.
Among the blood cells of a certain animal species, the proportion of cells which are of type A is 0.37 and the proportion of cells that are of type B is 0.004. Find, to 3 decimal places, the probability that in a random sample of 8 blood cells at least 2 will be of type A.
Find, to 3 decimal places, an approximate value for the probability that
  (c) in a random sample of 200 blood cells the combined number of type A and type B cells is 81 or more,
  (d) there will be 4 or more cells of type B in a random sample of 300 blood cells. [ULSEB]

12.12 Normal approximation to a Poisson distribution

We saw in Chapter 10 that the shape of a Poisson distribution depends upon the value of its parameter λ. Although the distribution is always skewed, as λ increases this skewness becomes less visible and the distribution increasingly resembles the normal in appearance.

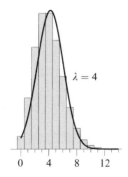
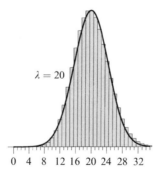

A Poisson random variable X with parameter λ has mean and variance both equal to λ. The approximating normal random variable, Y, therefore has a $\text{N}(\lambda, \lambda)$ distribution. As in the case of the approximation to a binomial distribution, a **continuity correction** is required:

$$P(X = x) \approx P(x - \tfrac{1}{2} < Y < x + \tfrac{1}{2})$$

After standardising to a $\text{N}(0,1)$ distribution we get:

$$P(X = x) \approx \Phi\left(\frac{(x + \tfrac{1}{2}) - \lambda}{\sqrt{\lambda}}\right) - \Phi\left(\frac{(x - \tfrac{1}{2}) - \lambda}{\sqrt{\lambda}}\right)$$

An example of the approximation for an inequality is:

$$P(X \leqslant x) \approx \Phi\left(\frac{(x + \frac{1}{2}) - \lambda}{\sqrt{\lambda}}\right)$$

Notes
- When possible, the Poisson probabilities should be calculated exactly. The approximation will usually be rather poor for $\lambda < 10$, but should be adequate for $\lambda \geqslant 25$.
- The approximation is worst for values in the tails of the distribution.
- Because the sum of independent Poisson random variables is another Poisson random variable, it is possible to think of any Poisson distribution as arising from a summation – which leads once again to the Central Limit Theorem being an explanation for the success of the normal approximation.

Example 36

The random variable X has a Poisson distribution with mean 9. Compare the exact values with those given by the normal approximation for the following probabilities:
(i) $P(X = 9)$, (ii) $P(X \leqslant 2)$.

(i) The exact probability is not difficult to determine:

$$P(X = 9) = \frac{9^9 e^{-9}}{9!} = 0.1318$$

The normal approximation is not at all bad since we are looking at a central point of the distribution. The relevant z-values are
$\frac{(9.5 - 9)}{\sqrt{9}} = 0.167$ and $\frac{(8.5 - 9)}{\sqrt{9}} = -0.167$, so:

$$
\begin{aligned}
P(X = 9) &\approx \Phi(0.167) - \Phi(-0.167) \\
&= 2\Phi(0.167) - 1 \\
&= (2 \times 0.5664) - 1 \\
&= 0.1328
\end{aligned}
$$

To three decimal places, the normal approximation gives 0.133 compared to the exact value of 0.132.

(ii) The exact probability is:

$$\frac{9^0 e^{-9}}{0!} + \frac{9^1 e^{-9}}{1!} + \frac{9^2 e^{-9}}{2!} = 0.0001 + 0.0011 + 0.0050 = 0.0062$$

The normal approximation is:

$$
\begin{aligned}
\Phi\left(\frac{2.5 - 9}{\sqrt{9}}\right) &= \Phi(-2.167) \\
&= 1 - \Phi(2.167) \\
&= 1 - 0.9849 \\
&= 0.0151
\end{aligned}
$$

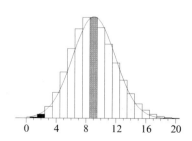

which is substantially in error, because we are looking at a tail of the distribution.

Example 37

The numbers of accidents per day on a given stretch of road are found to have a Poisson distribution with mean 1.4.
Determine the probability that more than 50 accidents occur during a 4-week period.

_____

The number of accidents, X, occurring during a 4-week period has a Poisson distribution with mean $(1.4 \times 28) = 39.2$. We require $P(X > 50)$. Without a computer or programmable calculator it is not feasible to calculate this probability exactly. However, since λ is quite large the normal approximation should be sufficiently accurate.

The relevant value of z is $\dfrac{50.5 - 39.2}{\sqrt{39.2}} = 1.805$. Hence:

$$\begin{aligned} P(X > 50) &= 1 - P(X \leqslant 50) \\ &\approx 1 - \Phi(1.805) \\ &= 1 - 0.9645 \\ &= 0.0355 \end{aligned}$$

On about 3.6% of 4-week periods more than 50 accidents will occur.

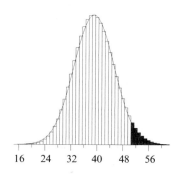

16   24   32   40   48   56

Exercises 12h _____

1 It is given that X has a Poisson distribution with mean 50.
Using a suitable approximation, find $P(X < 40)$.

2 It is given that Y has a Poisson distribution with mean 30.
Using a suitable approximation, find $P(Y > 20)$.

3 In deal planks the number of knots per metre has a Poisson distribution with mean 3.2.
Use a suitable approximation to find the probability that two 5 m lengths contain a total of at least 40 knots.

4 The number of faulty light bulbs returned to a shop in a week has a Poisson distribution with mean 0.7. Using a suitable approximation, find the probability that in a period of 50 weeks not more than 45 faulty bulbs are returned.

5 Large boulders deposited by glaciers are known as erratics. On average a square kilometre of glacial valley contains 25 erratics.
Using a normal approximation to a Poisson distribution, find the probability that a randomly chosen square kilometre contains between 15 and 35 (inclusive) erratics.

6 When pond water is examined under the microscope there are nasty bugs to be seen. The average concentration of these bugs is three per millilitre.
Determine the probability that a random sample of 5 millilitres of pond water contains exactly 14 bugs:
(a) using the Poisson distribution,
(b) using the normal approximation to the Poisson distribution.
Determine also the probability that the number of bugs to be found in a random sample of 200 millilitres of pond water is between 560 and 620 *inclusive*.

7 The number of cars arriving at a petrol station in a period of t minutes may be assumed to have a Poisson distribution with mean $\frac{7}{20}t$.
(i) Find, to three decimal places, the probability that fewer than 6 cars will arrive in a 10 minute period.
(ii) Find, to three decimal places, the probability that exactly 3 cars will arrive in a 5 minute period.
(iii) Find, to two decimal places, an approximate value for the probability that more than 24 cars will arrive in an hour.

[JMB]

8 Analysis of the scores in football matches in a local league suggests that the total number of goals scored in a randomly chosen match may be modelled by the Poisson distribution with parameter 2.7. The numbers of goals scored in different matches are independent of one another.

(i) Find the probability that a match will end with no goals scored.

(ii) Find the probability that 4 or more goals will be scored in a match.

One Saturday afternoon, 11 matches are played in the league.

(iii) State the expected number of matches in which no goals are scored.

(iv) Find the probability that there are goals scored in all 11 matches.

(v) State the distribution for the total number of goals scored in the 11 matches. Using a suitable approximating distribution, or otherwise, find the probability that more than 30 goals are scored in total.   [MEI]

9 (i) X, Y and Z are random variables having Poisson distributions with means a, b and $a + b$ respectively. X and Y are independent. Show that
$$P(X + Y = 2) = P(Z = 2).$$

(ii) The number of printing errors on any page of a certain book has a Poisson distribution with mean 0.4.

(1) Find the probability that the total number of errors in the first 10 pages is exactly 3.

(2) Find the probability that the total number of errors in the first 10 pages is more than 3.

(3) N is the smallest integer such that the probability of there being more than 2 errors in the first N pages is greater than 0.88. Verify that $N = 13$.

(4) The book has 250 pages. If there are more than 110 errors then the publishers will have the book corrected and reprinted. Use a suitable approximation to find the probability of this happening.   [O&C]

10 Independently for each year, the number of road accidents per year at a certain blackspot may be regarded as having a Poisson distribution with 2.3 as mean. Calculate, correct to three decimal places, the probabilities that

(i) more than two accidents will occur in a given year,

(ii) the first accident after 31st December 1989 will occur during 1992,

(iii) exactly two of the six years 1990 to 1995 inclusive will be free from accidents.

Given that the sum of independent Poisson variables is also a Poisson variable, use a suitable approximation to calculate, to three decimal places, the probability that there will be at least 30 accidents in a given ten-year period.   [JMB]

11 The number of flaws in a length of cloth, λ m long, produced on a certain machine, has a Poisson distribution with mean 0.04λ.

(i) Find, to three decimal places, the probability that a 10 m length of cloth has fewer than 2 flaws.

(ii) Find, to three decimal places, the probability that a 100 m length of cloth has more than 4 flaws.

(iii) Find, to two decimal places, an approximate value for the probability that a 1000 m length of cloth has at least 46 flaws.

(iv) Given that the cost of rectifying X flaws in a 1000 m length of cloth is X^2 pence, find the expected value of this cost.   [JMB]

12 Data files on computers have sizes measured in megabytes. When files are sent from one computer to another down a communications link, the number of errors has a Poisson distribution. On average, there is one error for every 10 megabytes of data.

(i) Find the probability that a 3 megabyte file is transmitted

(A) without error,

(B) with 2 or more errors.

(ii) Show that a file which has a 95% chance of being transmitted without error is a little over half a megabyte in size.

A commercial organisation transmits 1000 megabytes of data per day.

(iii) State how many errors per day they will incur on average.

Using a suitable approximating distribution, show that the number of errors on any randomly chosen day is virtually certain to be between 70 and 130.   [MEI]

12.13 Normal probability paper

How can we find out whether it is reasonable to assume that a set of data comes from a normal distribution? One simple informal method is to plot the data as a scatter diagram, using special graph paper in which one axis corresponds to the data values and the other axis corresponds to cumulative probabilities.

Commercial normal probability paper generally only shows the left-hand probability scale. We have added the right-hand scale (which is linear in z) so that you can see how **normal probability paper** is constructed.

Normal probability paper

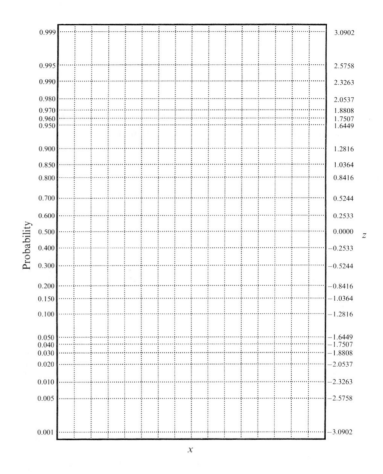

Suppose we have a sample of n observations. The first step is to arrange these observations in ascending order. We will number the observations so that:

$$x_1 \leqslant x_2 \leqslant \cdots \leqslant x_n$$

Assuming that $x_k < x_{k+1}$, the proportion of the *sample* that is less than or equal to x_k is evidently $\dfrac{k}{n}$, and the proportion that is less than or equal to x_n is 1. However, we don't believe that the latter is true for the population. To get around this problem we plot:

$$x_k \text{ against } \frac{k}{(n+1)}$$

If the data come from a normal distribution then the plot should be an approximately straight line. If the plot is clearly not straight then the population distribution is probably not normal.

Notes

- To estimate the mean from the graph draw a line (by eye) through the middle part of the data to fit the points as best you can. The value of x corresponding to the point where this line crosses the 0.5 line gives an estimate of the median, and hence of μ.
- The same line can be used to provide an estimate of the standard deviation, though this is often rather inaccurate. A sensible method is to read off the x-values corresponding to 0.84 and 0.16 – the difference between these values is approximately 2σ.

Example 38

The following data are the levels of cholesterol (in milligrams per millilitre) of a random sample of 20 youthful patients in a hospital.

210, 209, 212, 208, 217, 207, 210, 203, 208, 210, 210, 199, 215, 221, 213, 218, 202, 218, 200, 214

Plot these on normal probability paper and use your graph to estimate the population mean and standard deviation.

———

Note that some values are repeated and that all the values are given correct to the nearest milligram. The upper end-points are therefore 199.5, 200.5, ... A complete summary of the plotting positions is given in the following table.

Upper end-point	199.5	200.5	202.5	203.5	207.5	208.5	209.5	210.5
Proportion $\frac{k}{n+1}$	$\frac{1}{21}$	$\frac{2}{21}$	$\frac{3}{21}$	$\frac{4}{21}$	$\frac{5}{21}$	$\frac{7}{21}$	$\frac{8}{21}$	$\frac{12}{21}$

Upper end-point	212.5	213.5	214.5	215.5	217.5	218.5	221.5
Proportion $\frac{k}{n+1}$	$\frac{13}{21}$	$\frac{14}{21}$	$\frac{15}{21}$	$\frac{16}{21}$	$\frac{17}{21}$	$\frac{19}{21}$	$\frac{20}{21}$

When the results are plotted on the normal probability paper, the result is a distinctly wiggly line! However, there is no obvious sign of skewness and the assumption of a normal distribution may not be unreasonable. Reading off estimates of the values of x corresponding to the 0.16, 0.50 and 0.84 cumulative proportions, we get 203.5, 210.4 and 217.5. These give estimates of the population mean and standard deviation as 210.4 and 7.0, respectively. For comparison, the sample mean and standard deviation obtained from the data by direct calculation are 210.2 and 5.9.

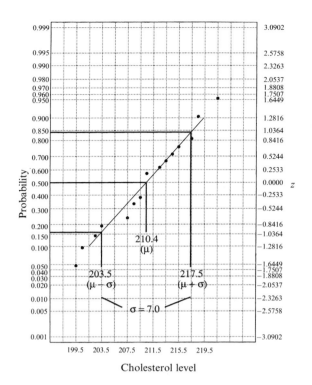

Cholesterol level

Example 39

The following data refer to the pH values of samples of river water downstream from an industrial centre.

> 5.5, 6.1, 5.7, 5.9, 4.9, 4.7, 4.6, 4.8, 5.6, 4.8, 4.4, 6.0, 5.3, 5.5, 4.7, 4.8,
> 4.9, 4.9, 4.7, 4.4, 4.2, 3.9, 3.8, 3.8, 3.9, 4.2, 4.2, 4.3, 4.2, 4.3, 4.4, 4.4,
> 4.5, 4.6, 4.7, 4.8, 4.4, 4.5, 4.4, 4.7, 5.1, 5.2, 4.7, 4.9, 5.8, 5.4, 6.0, 5.1

Plot the data on normal probability paper.
Do the data appear to come from a normal distribution? If they do, then obtain estimates of the mean and variance.

We begin by summarising the data using a stem-and-leaf diagram.

3	8, 8, 9, 9
4	2, 2, 2, 2, 3, 3, 4, 4, 4, 4, 4, 4
4	5, 5, 6, 6, 7, 7, 7, 7, 7, 7, 8, 8, 8, 8, 9, 9, 9, 9
5	1, 1, 2, 3, 4
5	5, 5, 6, 7, 8, 9
6	0, 0, 1

Key: 5|2 = 5.2

It looks as though the data are somewhat skew, with a missing tail of really low (acid) values. The fish in the river will be pleased!

When the data are plotted on normal probability paper (using the proportions $\frac{2}{49}, \frac{4}{49}, \frac{8}{49}, \frac{10}{49}, \ldots, \frac{48}{49}$) the result is not a straight line. The peak between 4.7 and 4.9 produces a noticeable distortion in the plot. Although we have plotted a line and deduced values for the mean and variance, we cannot expect these to be very accurate. In fact the sample mean is 4.80 and the sample standard deviation, σ_n, is equal to 0.6.

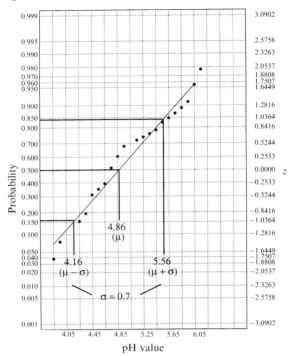

pH value

Example 40

The following table shows the numbers of marriages contracted in Australia between 1907 and 1914, arranged according to the age of the bridegroom! The ages shown are the upper limits of (for the most part) three-year age ranges.

Age	No.	Cum. Propn.	Age	No.	Cum. Propn.	Age	No.	Cum. Propn.
18	294	0.001	36	20 569	0.848	54	2190	0.982
21	10 995	0.037	39	14 281	0.895	57	1655	0.987
24	61 001	0.239	42	9320	0.926	60	1100	0.991
27	73 054	0.481	45	6236	0.947	63	810	0.994
30	56 501	0.669	48	4770	0.963	66	649	0.996
33	33 478	0.780	51	3620	0.975	91	1262	1.000

Plot the data on normal probability paper.
Do the data appear to come from a normal distribution? If they do, then obtain estimates of the mean and variance.

———

This time the diagram reveals without any doubt that the data are *not* normally distributed: the graph shows a sweeping curve which is a consequence of the very skewed distribution: most marriages involve bridegrooms aged between 20 and 40, but some are more than twice as old. It would not be sensible to attempt to estimate μ and σ from the graph.

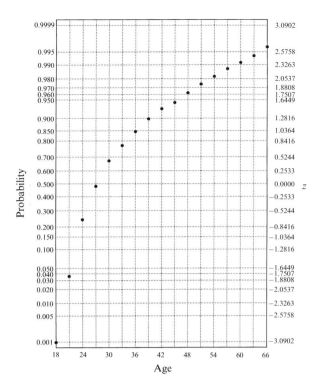

▲

Exercises 12i

1 The Table gives the relative blood cholesterol levels of a random sample of 275 workers in the chemical industry.

Cholesterol level	Frequency
140–159	10
160–179	26
180–199	34
200–219	45
220–239	70
240–259	42
260–279	25
280–299	15
300–319	8
	275

(a) Plot the data on arithmetic (normal) probability paper.

(b) The mean cholesterol level is 225 and the standard deviation is 37. Calculate the 10th and the 90th percentiles of the Normal distribution which has this mean and standard deviation. Hence plot the cumulative distribution function of the Normal distribution with mean 225 and standard deviation 37 on the same diagram as your plot of the data. Comment. [O&C]

2 The time taken to service and repair vehicles is recorded by a garage. A random sample of 400 of those times is summarised below.

Time (mins)	Number of vehicles
30–	16
50–	20
60–	26
70–	44
80–	62
90–	72
100–	46
110–	82
130–180	32

(a) Calculate estimates of
 (i) the mean and the standard deviation,
 (ii) the upper and lower quartiles,
  for these data.

(b) Plot these data on arithmetic probability paper.
 Comment on the shape of the distribution.

(*continued*)

(c) From your graph estimate the mean and the standard deviation.

Evaluate the percentage difference in this estimate of the mean compared to that from part (a).

For a normal distribution the value of :

$$\frac{\text{semi-interquartile range}}{\text{standard deviation}}$$

would be approximately 0.67.

(d) Evaluate the corresponding value for the above data using your results from (a). Comment on whether or not your value is consistent with your comment in (b).

[AEB 92]

3 A user of a certain gauge of steel wire suspects that its breaking strength, in newtons (N), is different from that specified by the manufacturer. Consequently the user tests the breaking strength, x N, of each of a random sample of nine lengths of wire and obtains the following *ordered* results.

72.2, 72.9, 73.4, 73.8, 74.1,
74.5, 74.8, 75.3, 75.9.

$$\left(\sum x = 666.9, \sum x^2 = 49\,428.25\right)$$

Using arithmetic (normal) probability paper, show that the manufacturer's specification that the breaking strength of the wire is normally distributed is reasonable. [AEB 93(P)]

4 The table gives the weights in kg, recorded to the nearest kg and then grouped as shown, of a random sample of 1000 children of a certain age in a certain area; the table also gives the cumulative frequencies:-

Wt (kg)	23–25	26–28	29–31	32–34	35–40
Freq.	3	13	52	178	484
Cum. freq.	3	16	68	246	730

Wt (kg)	41–43	44–46	47–49	50–52
Freq.	166	80	22	2
Cum. freq.	896	976	998	1000

(i) Draw a histogram to illustrate these data.
(ii) Plot the data on arithmetic (normal) probability paper, and hence discuss whether it appears reasonable to assume an underlying normal distribution for these weights. On the assumption that such a distribution is appropriate, estimate from your plot the median, the lower and upper quartiles, the mean and the standard deviation.

[O&C]

12.14 Mgf of a normal distribution

In order to keep the algebra from being too messy, we begin with the special case of the standard normal distribution, which, for a random variable Z, has probability density function:

$$\phi(z) = \frac{1}{\sqrt{2\pi}} \exp\left(-\tfrac{1}{2}z^2\right)$$

(See Section 11.10, p. 295 for an explanation of 'exp'.)

The mgf is given by:

$$E(e^{tZ}) = \frac{1}{\sqrt{2\pi}} \int_{-\infty}^{\infty} \exp(tz) \times \exp(-\tfrac{1}{2}z^2)\,dz$$

$$= \frac{1}{\sqrt{2\pi}} \int_{-\infty}^{\infty} \exp[-\tfrac{1}{2}(z^2 - 2tz)]\,dz$$

$$= \frac{1}{\sqrt{2\pi}} \int_{-\infty}^{\infty} \exp\{\tfrac{1}{2}t^2 - \tfrac{1}{2}(z^2 - 2tz + t^2)\}\,dz$$

$$= \exp(\tfrac{1}{2}t^2) \times \frac{1}{\sqrt{2\pi}} \int_{-\infty}^{\infty} \exp\{-\tfrac{1}{2}(z - t)^2\}\,dz$$

Note the clever completion of the square by the addition and subtraction of the $\frac{1}{2}t^2$ term in the exponential. The second piece of cunning is the recognition that:

$$\frac{1}{\sqrt{2\pi}}\exp\{-\tfrac{1}{2}(z-t)^2\}$$

is simply the pdf of a normal random variable with mean t and variance 1. Since any pdf integrates to 1:

$$E(e^{tZ}) = \exp(\tfrac{1}{2}t^2) \times 1$$

and hence a standard normal random variable has moment generating function:

$$\exp(\tfrac{1}{2}t^2)$$

The mgf for the general normal distribution (derived below) is:

$$M_X(t) = \exp(\mu t + \tfrac{1}{2}\sigma^2 t^2)$$

Notes

- Suppose $X \sim N(\mu, \sigma^2)$. Then X is related to Z by $X = \mu + \sigma Z$. Hence:

$$\begin{aligned}
M_X(t) = E[\exp(tX)] &= E[\exp(t\mu + t\sigma Z)] \\
&= \exp(t\mu) \times E[\exp\{(\sigma t)Z\}] \\
&= \exp(\mu t) \times \exp\{\tfrac{1}{2}(\sigma t)^2\} \\
&= \exp(\mu t + \tfrac{1}{2}\sigma^2 t^2)
\end{aligned}$$

- Suppose that X has a normal distribution with mean μ_x and variance σ_x^2, and that Y has a normal distribution with mean μ_y and variance σ_y^2. Suppose also that X and Y are independent. Let $W = X + Y$. Then:

$$\begin{aligned}
M_W(t) = E[\exp(tW)] &= E[\exp(tX + tY)] \\
&= E[\exp(tX)] \times E[\exp(tY)] \\
&= M_X(t) \times M_Y(t) \\
&= \exp(\mu_x t + \tfrac{1}{2}\sigma_x^2 t^2) \times \exp(\mu_y t + \tfrac{1}{2}\sigma_y^2 t^2) \\
&= \exp\{(\mu_x + \mu_y)t + \tfrac{1}{2}(\sigma_x^2 + \sigma_y^2)t^2\}
\end{aligned}$$

which is the moment generating function of a $N(\mu_x + \mu_y, \sigma_x^2 + \sigma_y^2)$ distribution.
This shows that the sum of two independent normal random variables has a normal distribution.
This is the basis of the result, stated earlier, that any linear combination of independent normal random variables has a normal distribution.

Chapter summary

♦ **Notation**: $N(\mu, \sigma^2)$ denotes a random variable having a normal distribution with mean μ and variance σ^2.
  - The standard normal random variable, Z, has a $N(0, 1)$ distribution.
  - $P(Z < z)$ is denoted by $\Phi(z)$.
  - $\Phi(-z) = 1 - \Phi(z)$.
  - If $X \sim N(\mu, \sigma^2)$ then $P(X < x) = \Phi\left(\dfrac{x - \mu}{\sigma}\right)$.

♦ The mean of a set of n independent identically distributed random variables has a distribution that increasingly resembles the normal distribution as n increases (the **Central Limit Theorem**). The same is true for the sum.

♦ **A linear combination of independent normal random variables** has a normal distribution.

♦ **Approximation to binomial distribution**:
  If $X \sim B(n, p)$ with $np > 5$ and $n(1 - p) > 5$ then:

  $$P(X = x) \approx P(x - \tfrac{1}{2} < W < x + \tfrac{1}{2})$$

  where $W \sim N(np, npq)$ and $q = 1 - p$.

♦ **Approximation to Poisson distribution**:
  If X has a Poisson distribution with parameter λ, and if $\lambda > 25$ then:

  $$P(X = x) \approx P(x - \tfrac{1}{2} < W < x + \tfrac{1}{2})$$

  where $W \sim N(\lambda, \lambda)$.

♦ **Mgf of normal distribution**:

  $$M_X(t) = \exp(\mu t + \tfrac{1}{2}\sigma^2 t^2)$$

Exercises 12j (Miscellaneous)

1 'Disks R Us' purchases a consignment of 50 000 grade A computer disks and 25 000 grade B disks. The probability that a grade A disk contains at least one bad sector is 0.0001, whereas, for a grade B disk this probability is 0.0005. Determine the approximate probability that the consignment contains between 15 and 20 (inclusive) disks having bad sectors.
Grade A disks come in two colours: black and red. Assuming that the colours of the disks in the consignment were independent of one another, and given that the proportion of red disks in the population is 40%, determine the probability that fewer than 20 200 of the grade A disks were red.

2 In a large shipment of peaches, 10% are bad. In most cases the peaches have gone bad because of bruising. However, 10% of the bad peaches can be attributed to the presence of an insect called a 'peach-borer'.
(a) Determine the probability that a random sample of ten peaches contains precisely two bad peaches.
(b) Using a Poisson approximation, determine the approximate probability that a random sample of 100 peaches contains more than eight bad peaches.
(c) Using a Poisson approximation, determine the approximate probability that a random
(*continued*)

sample of 100 peaches contains no more than one that has gone bad because of a peach-borer.

(d) Using a normal approximation, determine the approximate probability that a random sample of 1000 peaches contains more than eight peaches that have gone bad because of peach-borers.

3 A newspaperman sells papers at random points in time: the number of papers sold in an hour is an observation from a Poisson distribution with mean 50.
  (a) Determine the probability that he sells more than 180 in a 3-hour period.
  (b) If I have just purchased a paper from him, what is the probability that it will be at least 2 minutes until he sells another?
  (c) Given that it is already 5 minutes since his last sale, what is the probability that it will be at least 2 more minutes before his next sale?

4 The random variables X_1 and X_2 are both normally distributed such that $X_1 \sim N(\mu_1, \sigma_1^2)$ and $X_2 \sim N(\mu_2, \sigma_2^2)$. Given that $\mu_1 < \mu_2$ and $\sigma_1^2 < \sigma_2^2$, sketch both distributions on the same diagram.

State the '2σ rule' for a normal random variable. Explain how you used, or could have used, a normal distribution in a project.

The weights of vegetable marrows supplied to retailers by a wholesaler have a normal distribution with mean 1.5 kg and standard deviation 0.6 kg. The wholesaler supplies 3 sizes of marrow:

  Size 1, under 0.9 kg,
  Size 2, from 0.9 kg to 2.4 kg,
  Size 3, over 2.4 kg.

Find, to 3 decimal places, the proportions of marrows in the three sizes.

Find, in kg to one decimal place, the weight exceeded on average by 5 marrows in every 200 supplied.

The prices of the marrows are 16p for Size 1, 40p for Size 2 and 60p for Size 3. Calculate the expected total cost of 100 marrows chosen at random from those supplied. [ULSEB]

5 Give **two** reasons why the normal distribution is so important in statistics.

In a study of the dimensions of fibre glass particles produced by manufacturing process A, the particle diameters, in micrometers (μm), were established as being normally distributed with a mean of 1.52 μm and a standard deviation of 0.44 μm. A filter with porosity 0.80 will remove all particles with diameters greater than 0.80 μm. Calculate the proportion of particles removed by such a filter. Determine the maximum porosity of a filter that will remove at least 99% of the particles. An analysis of fibre glass particles from process B revealed that 28.1% had diameters greater than 2.60 μm and that 10.2% had diameters less than 1.30 μm. Assuming these diameters to be normally distributed, determine their mean and standard deviation.

In fact the two manufacturing processes, A and B, are in the same building. The number of fibre glass particles produced by B is three times the number produced by A. Determine the proportion of particles in the building not removed by a filter with a porosity of 0.80. [JMB]

6 Describe the main features of a normal distribution. Give an example of an experiment that you would expect to produce a random variable that is normally distributed.

A small firm has three machines producing ball bearings. The diameters of bearings produced by each machine are normally distributed. The firm rejects as undersize all bearings with diameter less than 9.490 mm, and rejects as oversize all bearings with diameter greater than 9.520 mm. Bearings produced on Machine I have diameters with mean 9.506 mm and standard deviation 0.006 mm. Calculate
  (i) the percentage of bearings produced on this machine that are rejected as undersize,
  (ii) the percentage of bearings produced on this machine that are considered acceptable.

Machine II produces bearings with a mean diameter of 9.504 mm, of which 2.28% are rejected as oversize. Calculate the standard deviation of the bearings produced on Machine II. Of the bearings produced on Machine III, 0.5% are rejected as undersize and 4.35% are rejected as oversize. Calculate the mean and the standard deviation of the bearing diameters from Machine III. [JMB]

7 The lifetimes of Surecell batteries are normally distributed with mean 200 hours and standard deviation 25 hours.

(*continued*)

Calculate, to 3 decimal places, the probability that a battery chosen at random will

(*a*) last longer than 230 hours,

(*b*) have a lifetime between 190 and 210 hours. The manufacturers of Surecell batteries wish to offer a guaranteed life T hours on their batteries. They conduct an experiment on batteries they use in the factory. It is found that 9.85% of batteries have a lifetime less than T hours.

(*c*) Calculate, to 2 decimal places, the value of T. Batteries are removed from the production line and placed in groups of 6.

(*d*) Calculate, to 3 decimal places, the probability that, in a group of 6 batteries, exactly 2 batteries will have a lifetime less than T hours.

Batteries are packed in boxes of 6.

(*e*) Use your answer to (*d*) to calculate, to 3 decimal places, the probability that, in a batch of ten boxes, exactly three boxes will contain 2 batteries with lifetimes less than T.

[ULSEB]

8 The weights of pieces of home made fudge are normally distributed with mean 34 g and standard deviation 5 g.

(a) What is the probability that a piece selected at random weighs more than 40 g?

(b) For some purposes it is necessary to grade the pieces as small, medium or large. It is decided to grade all pieces weighing over 40 g as large and to grade the heavier half of the remainder as medium. The rest will be graded as small. What is the upper limit of the small grade?

(c) A bag contains 15 pieces of fudge chosen at random. What is the distribution of the total weight of fudge in a bag?

What is the probability that the total weight is between 490 g and 540 g?

(d) What is the probability that the total weight of fudge in a bag containing 15 pieces exceeds that in another bag containing 16 pieces?   [AEB 91]

9 Part of an assembly requires the fitting of a cylinder through a circular hole in a metal plate. It is known that the diameters of the cylinders, D_c, are distributed with mean 24.96 mm and standard deviation 0.04 mm and the diameters of the holes, D_h, are distributed with mean 25.00 mm and standard deviation 0.03 mm.

(a) Find the mean and standard deviation of the difference, $D_h - D_c$, between the diameters of randomly chosen components.

(b) Assuming that both distributions are normal and the components are chosen at random, find the percentage of cases for which the cylinder will not fit the hole.

(c) A plate is chosen at random and then a cylinder is chosen randomly. This is done 4 *more* times. Find the probability that in 3 of the cases the cylinder will not fit its plate.

(d) The percentage of cases for which the cylinder will not fit its plate is to be fixed at 5%. If the standard deviation remains unchanged, determine the increase in mean diameter of hole needed to meet this requirement.

(e) The maximum tolerance – the amount by which D_h exceeds D_c – is set at 0.16 mm. Using the increased mean diameter of the hole, find the percentage of assemblies that do not satisfy this tolerance.   [AEB 90]

10 In a survey into working practices, the distance walked each day by a postal worker delivering mail in a residential district was recorded. For each weekday (Monday to Friday) the distribution had mean 12 km and standard deviation 0.9 km. For Saturdays the distribution had mean 10 km and standard deviation 0.5 km. No mail was delivered on Sundays. The distances walked on different days may be assumed to be independent of each other and normally distributed.

(i) Find the probability that, on a randomly chosen Saturday, the postal worker walked between 8.5 km and 11 km.

(ii) Find the probability that, in a randomly chosen week, the postal worker walked further on the Saturday than on the Friday.

(iii) Find the probability that, in a randomly chosen week, the mean daily distance walked by the postal worker for the six-day period was less than 11 km.   [UCLES]

11 A machine makes metal rods. A rod is oversize if its diameter exceeds 1.05 cm. It is found from experience that 1% of the rods produced by the machine are oversize. The diameters of the rods are normally distributed with mean 1.00 cm and standard deviation σ cm. Find the value of σ, giving 3 decimal places in your answer.

(*continued*)

Two hundred rods are chosen at random. Using a suitable approximation, find the probability that four or more of the rods are oversize, giving 3 decimal places in your answer.

Another machine makes metal rings. The internal diameters of the rings are normally distributed with mean $(1.00 + 2\sigma)$ and standard deviation 2σ cm, where σ has the value found in the first paragraph. Find the probability that a randomly chosen ring can be threaded on a randomly chosen rod, giving 3 decimal places in your answer.   [UCLES]

12 The mass of flour in bags produced by a particular supplier is normally distributed with mean μ grammes and standard deviation 7.5 grammes, where the actual value of μ may be set accurately by the supplier. Any bag containing less than 500 grammes of flour is said to be underweight. The trading standards inspector takes a sample of n bags of flour at random from the bags packed by the supplier. The supplier will be prosecuted if the mean mass of flour in the n bags is less than 500 grammes.

(i) Given that $\mu = 505$ and $n = 10$, find the probability that the supplier will be prosecuted.

(ii) The supplier wishes to ensure that, when $n = 10$, the probability of being prosecuted is not greater than 0.001. Calculate, to one decimal place, the least value at which μ should be set.

(iii) The inspector wishes to ensure that, if the supplier's mean setting produces 80% of bags underweight, the chance that he will escape prosecution is less than one in a thousand. Determine the least value of n that the inspector can use in taking his sample.   [JMB]

13 A food packaging company produces tins of baked beans. The empty cans have weights which are normally distributed with mean 40 grams and standard deviation 3.5 grams. The weights of the contents are independent of the weights of the cans and are normally distributed with mean 450 grams and standard deviation 12 grams. Find, to two decimal places, the probability that the total weight of a randomly chosen can of beans is greater than 500 grams.

Show that, in approximately 91% of the cans of beans, the weight of the contents is more than ten times the weight of the empty can. It is decided to change the procedure for packing beans into cans. The weights of the empty cans have the same distribution as before. The cans are filled with beans, so that the total weight is independent of the weight of the can and is a normal variable with mean 490 grams and standard deviation 12.5 grams. Calculate the mean and (to two decimal places) the standard deviation of the weights of the contents of the cans, and explain briefly the significance to a consumer of the change on the packing system.   [JMB]

14 An octahedral die has eight faces numbered from 1 to 8. The random variable X is the score obtained when the die is thrown. The bias of the die is such that

$$P(X = r) = c \text{ for } r = 1, 2, 3, 4, 5$$
$$P(X = r) = d \text{ for } r = 6, 7, 8$$
$$P(X < 6) = P(X \geqslant 6).$$

(i) Find the values of c and d, show that $E(X) = 5$ and find the variance of X.

(ii) The die is thrown twice. Calculate the probability that the sum of the two scores is 10.

(iii) The random variable Y is the sum of the scores when this die is thrown 48 times. Find the mean and variance of Y. Assuming that Y has a normal distribution with this mean and variance, find the probability that Y lies between 220 and 260 inclusive.   [O&C]

15 A class of 35 third-form pupils conduct a physics experiment in which each measures the time for one complete swing of a pendulum. The experiment is repeated until each pupil has six measurements. The mean time for a complete swing was 1.015 seconds and the standard deviation of the times was 0.045 second. Using the Normal distribution and these values as estimates of the population parameters calculate:

(a) the probability that a recorded time is less than 1.1 seconds;

(b) the number of recordings of a time less than 1.0 seconds;

(c) the number of recorded times that are more than two standard deviations away from the mean time.   [UODLE(P)]

16 State the conditions under which a normal distribution can be used as an approximation to (*a*) a binomial distribution, (*b*) a Poisson distribution.

Typesetting errors on a galley proof occur randomly and at an average rate of 100 per galley proof. Assuming these errors to be Poisson distributed, calculate, to 3 decimal places, the probability that less than 110 errors, but not less than 80 errors, occur on a given galley proof.

It is known that on average 5% of the errors are not subsequently corrected. Calculate, to 3 decimal places, the probability that, in a random sample of 400 errors, more than 28 will not be corrected.   [ULSEB]

17 On average a Coastguard station receives one distress call every two days. A "bad" week is a week in which 5 or more distress calls are received. Show that the probability that a week is a bad week is 0.275, correct to three significant figures.

Find the probability that, in 8 randomly chosen weeks, at least 2 are bad weeks.

Find the probability that, in 80 randomly chosen weeks, at least 30 are bad weeks.   [UCLES]

18 The operational lifetimes of brand A electric light bulbs are normally distributed with mean 1512 hours.

(i) Given that 20% of these bulbs have lifetimes exceeding 1600 hours calculate the standard deviation of the lifetimes, giving your answer correct to the nearest hour.

(ii) Use a distributional approximation to calculate the probability that in a random sample of 120 such bulbs at least 30 of them will have operational lifetimes exceeding 1600 hours; give your answer correct to three decimal places.

The operational lifetime of a brand B electric light bulb is X thousand hours, where X is a continuous random variable having probability density function f, where

$$f(x) = \begin{cases} 2(2 - x), & \text{for } 1 \leqslant x \leqslant 2, \\ 0, & \text{otherwise.} \end{cases}$$

(iii) Determine which of brand A and brand B bulbs

(*a*) has the larger mean lifetime,

(*b*) has lifetimes with larger standard deviation.   [WJEC]

19 [In this question give three places of decimals in each answer.] When a telephone call is made in the country of Japonica, the probability of getting the intended number is 0.95.

(i) Ten independent calls are made. Find the probability of getting eight or more of the intended numbers. Find also the conditional probability of getting all ten intended numbers given that at least eight of the intended numbers are obtained.

(ii) Three hundred independent calls are made. Find the probability of failing to get the intended number on at least ten but not more than twenty of the calls.

(iii) Four hundred independent calls are made. For each call the probability of getting "number unobtainable" is 0.004. Find the probability of getting "number unobtainable" fewer than three times.   [UCLES]

20 (*a*) Records of accidents at a factory over a period of 100 working days showed that there were 53 days on which no accident occurred, 30 days on which only one accident occurred, 12 days on which two accidents occurred, and 5 days on which 3 or 4 accidents occurred. The mean number of accidents per day over the 100-day period was 0.7.

(i) Show that the number of days on which 3 accidents occurred was four.

(ii) Calculate the standard deviation of the number of accidents per day over the period.

(*b*) The number of accidents that will occur in one day at another factory has a Poisson distribution with mean 0.7.

(i) Find the probability that on a randomly chosen day there will be 2, 3 or 4 accidents at the factory.

(ii) Find the probability that, on a randomly chosen day, there will be no accident at the factory. Use a distributional approximation to find the probability that in a period of 60 days there will be 35 days on which no accident occurs.   [WJEC]

21 State the conditions under which the binomial distribution B(n, p) can be approximated by

(*a*) a Poisson distribution,

(*b*) a normal distribution.

(*continued*)

State the parameters of each of the approximate distributions.

Over a long period it is found that in music examinations 25% of the candidates fail, while 9% of the candidates obtain a distinction.

(c) Calculate, to 3 decimal places, the probability that in a random group of 8 candidates, exactly 2 will fail the examination.

Estimate, giving your answer to 3 decimal places, the probability that

(d) in a random group of 100 candidates, at most 20 will fail the examination,

(e) in a random group of 40 candidates, more than 2 will obtain a distinction.   [ULSEB]

22 A machine produces sheets of glass of thickness X mm, where $X \sim N(\mu, \sigma^2)$. Over a long period of time it has been found that 2% of the sheets are less than 2 mm thick and 5% of the sheets are more than 3 mm thick. Find, to 2 decimal places, the values of μ and σ.

In the manufacture of a certain type of windscreen two of these sheets are taken at random and put together to form a sheet of double thickness. Find, to 2 decimal places, the mean and the standard deviation of the thickness, in mm, of these sheets of double thickness. Find, to 2 decimal places, the probability that the thickness of a sheet of double thickness will lie between 4 mm and 6 mm.

Given that a sheet of double thickness is between 4 mm and 6 mm thick, find, to 2 decimal places, the probability that the two single sheets from which it is made are between 2 mm and 3 mm thick.   [ULSEB]

23 A computer is used to add up a series of numbers. Each addition introduces an error which may be regarded as a random variable, X, which has the rectangular distribution on the interval $[-a, a]$, where a is, of course, very small. Find, in terms of a,

(i)   f(x), the probability density function for X,

(ii) E(X) and Var(X).

Now suppose that the total number of additions performed is n, and that successive errors are independent. Let the total error be Y.

(iii) Write down the values of E(Y) and Var(Y) in terms of a. State the approximate form which the distribution of Y will take.

(iv) For the case $a = 0.5 \times 10^{-10}$ and $n = 1000$ state the greatest possible value of the total error, Y. Find also the value k such that P($Y > k$) = 0.01. Comment on these two results.   [MEI(P)]

24 Machine A delivers sand into bags whose contents have a quoted weight of 25 kg. The weight of sand delivered by the machine is known to be normally distributed with a mean of 25.05 kg and a standard deviation of 0.25 kg. Determine the probability that the weight of sand in a bag exceeds 25 kg.

Calculate the weight of sand exceeded by exactly 95% of the bags.

A builder purchases 40 bags of sand. Determine the probability that the *mean* weight per bag of sand in these 40 bags exceeds 25 kg. The weight of sand in bags filled by machine B may be regarded as a normally distributed random variable with a mean of 25.05 kg and a standard deviation of 0.15 kg. The builder's next purchase of sand is made up of n bags from machine A and the rest from machine B. Specify the distribution of the *total* weight of sand in

(i)   the n bags filled by machine A,

(ii) the $(40 - n)$ bags filled by machine B,

(iii) the builder's 40 bags.

Calculate the largest possible value of n such that the probability of the total weight of sand in the 40 bags exceeding 1000 kg is at least 0.95.   [JMB]

25 The thicknesses of steel plates made by a certain process are normally distributed with mean 30 cm and standard deviation 0.8 cm.

(i)   Find the probability that a plate is more than 29.5 cm thick.

(ii) What thickness is exceeded by 85% of the plates?

(iii) A specification for the plates is that their thicknesses should be between 28.6 cm and 30.8 cm. Find the probability that a plate will meet this specification.

(iv) For another usage, the plates have to meet a different specification. It is calculated that the probability of a plate meeting this specification is 0.9. Find the probability that, of 10 randomly chosen plates, this specification is met by

(1) 8 or more plates,

(2) exactly 7 plates.

(*continued*)

(v) It is now required that the probability of a plate being less than 31 cm thick must be 0.95. If the mean thickness of the plates cannot be changed, what value of the standard deviation is required?   [O&C]

26 A component has a length which is normally distributed with a mean of 15 cm and a standard deviation of 0.05 cm. An acceptable component is one whose length is between 14.92 cm and 15.08 cm inclusive. The cost of production is 50p per component. An acceptable component can be sold for £1. Undersized components can be sold for scrap at 10p each, and oversized components can be corrected at an additional cost of 20p each and then sold as acceptable. Find the expected profit per 1000 components.

Of these components with acceptable length, the company estimates that 6 in every 1000 are defective in some other way.

(*a*) If X represents the number of defective items in a sample, state the distribution associated with X.

A customer is considering buying some of these components, but will only place an order if there is less than 5% risk that a sample of 150 components contains more than 3 defectives.

(*b*) Use the Poisson approximation to decide whether or not this customer is likely to place an order.

(*c*) State why a Poisson approximation is appropriate in this situation.   [ULSEB]

13 Sampling and simulation

'Data! data! data!', he cried impatiently. 'I can't make bricks without clay.'

The Adventures of Sherlock Holmes, Sir Arthur Conan Doyle

13.1 The purpose of sampling

Some might say that without sampling there would be no Statistics! This is because:

by careful study of a relatively small amount of data
(the *sample*)
we draw conclusions about a very much larger set of data
(the *population*),
without actually studying the whole of the larger set.

For the sample to be useful:

♦ *it must not be biased*
If we are interested in the distribution of the shoe sizes of army officers, then our sample should not be restricted to tall officers.

♦ *the sample must be taken from the correct population*
If we are interested in the characteristics of army officers we should not be studying submariners!

13.2 Methods for sampling a population

The simple random sample

Most sampling methods endeavour to give every member of the population the same probability of being included in the sample. If each member of the sample is selected by the equivalent of drawing lots, then the sample selected is described as being a **simple random sample**.

One procedure for drawing lots is the following:

1 Make a list of all N members of the population.
2 Assign each member of the population a different number.
3 For each member of the population place a correspondingly numbered ball in a bag.
4 Draw n balls from the bag, without replacement. The balls should be chosen at random.
5 The numbers on the balls identify the chosen members of the population.

This is like the procedure used in deciding the draw for the Cup competition in football. The drawing of the balls from the bag is sometimes televised.

An automated version would use the computer to simulate the drawing of the balls from the bag. The principles of simulation are discussed in the following sections of this chapter.

The principal difficulty with the above procedure is the first step: the creation of a list of all N members of the population. This list is known as the **sampling frame**. In many cases there will be no such central list. For example, suppose it was desired to test the effect of a new cattle feed on a random sample of British cows. Each individual farm may have a list of its own cows (Daisy, Buttercup,...), but the Government keeps no central list.

For the country as a whole there is not even a 100% accurate list of people (because of births, deaths, immigration and emigration).

Because of the straightforward nature of the simple random sample, most analyses (and almost all exam questions) assume that this kind of sample has been used to obtain the data. The necessary adjustments that may be required when dealing with other methods of sampling are well beyond the scope of this book. However, the nature of these other methods of sampling needs to be discussed.

Cluster sampling

Even if there were a 100% accurate list of the population, simple random sampling of the entire British population would almost certainly not be performed because of the expense. It is easy to imagine the groans emitted by the pollsters on drawing a ball from the bag corresponding to an inhabitant of Land's End, or the Shetland Isles. The intrepid interviewer would be a much travelled individual!

To avoid this problem, populations that are geographically scattered are usually divided into conveniently sized regions. A possible procedure is then:

1   Choose a region at random.
2   Choose individuals at random from that region.

The consequences of this procedure are that instead of a random scatter of selected individuals there are scattered **clusters** of individuals. The selection probabilities for the various regions are not equal, but are adjusted to be in proportion to the number of individuals that the regions contain. If the ith region contains N_i individuals, then the chance that it is selected is chosen to be $\dfrac{N_i}{N}$, where $N = \Sigma N_i$.

The size of the chosen region is usually sufficiently small that a single interviewer can perform all the interviews in that region without incurring huge travel costs. In practice, because of the sparse population and the difficulties of travel in the highlands and islands of Scotland, studies of the British population are usually confined to the region south of the Caledonian Canal.

Stratified sampling

Most populations contain identifiable **strata**, which are distinctive non-overlapping subsets of the population. For example, for human populations, useful strata might be 'males' and 'females', or 'receiving education', 'working' and 'retired', or combinations such as 'retired female'. From census data we might know the proportions of the population falling into these different categories. With stratified sampling, we ensure that these proportions are reproduced by the sample. Suppose, for example, that the age distribution of the adult population in a particular district is as given in the table below.

Aged under 40	Aged between 40 and 60	Aged over 60
38%	40%	22%

A simple random sample of 200 adults would be unlikely to reproduce these figures *exactly*. If we were very unfortunate, over half the individuals in the sample might be aged under 40. If the sample were concerned with people's taste in music, then, by chance, the simple random sample might provide a misleading view of the population.

A **stratified sample** is made up of separate simple random samples for each of the strata. In the present case, we would choose a simple random sample of 76 adults aged under 40, a separate simple random sample of 80 adults aged between 40 and 60, and a separate simple random sample of 44 adults aged over 60.

Stratified samples exactly reproduce the characteristics of the strata and this almost always increases the accuracy of subsequent estimates of population parameters. Their slight disadvantage is that they are a little more difficult to organise.

Systematic sampling

Both cluster sampling and stratified sampling subdivide the population into components. In both cases the final stage consists of selecting a random sample from a portion of the population. One possible method of doing the final selection is by simple random sampling. An alternative is to use **systematic sampling**, which is described below for the case of a sample of n individuals to be drawn from a listed population of N individuals.

1 Choose one individual at random.
2 Choose every kth individual thereafter, returning to the beginning of the list when the end is reached. The value of k is not crucial, but should be chosen beforehand. A popular choice is a convenient value close to $\frac{N}{n}$.
   The use of this wide spacing guards against the list consisting of clusters of similar individuals.

For example, suppose we wish to choose six individuals from a list of 250. A convenient value for k might be 40. Suppose that the first individual selected is number 138. The remainder would be numbers 178, 218, 8, 48 and 88.

If the list has been ordered by some relevant characteristic (e.g. age, or years of service), then, with $k \approx \frac{N}{n}$, this procedure produces a spread of values for the characteristic – a type of informal stratification.

Quota sampling

This is the method often used for street interviews. The interviewer is given a series of targets. For example, he or she might be instructed to interview equal numbers of men and women, of whom one-quarter should be aged over 60 and one-third should be in low-paid jobs. The instructions would be more detailed than these, with the idea being that each interviewer will select a representative cross-section of the population. It is easy to see that an interviewer might have some difficulty in completing his or her **quota** – as night falls the search for an elderly red-bearded giant might still be on!

The results of quota sampling must always be viewed with a little suspicion, since the interviewees are not chosen at random.

Self-selection

However bad quota sampling may be, it is wonderful by comparison with self-selection! The latter is exemplified by radio or television 'phone-ins' where listeners or viewers record their 'vote'. The views of the apathetic majority are seriously under-represented (though maybe they don't have any to represent!).

A national survey

To illustrate the methods of sampling discussed in the previous section, we now give a brief outline of the method of selection for the households included in the BHPS (*British Household Panel Study*), which is an annual study conducted by staff at the University of Essex.

The *sampling frame* for the BHPS was the Postcode Address File, a (computer-based) master list of all the 1.5 million postcodes in Britain. Each individual postcode (e.g. CO5 8JU) is a member of a so-called Postcode Sector (e.g. CO5 8). There are around 9000 postcode sectors, each of which identifies a *cluster* of about 2500 households.

A simplified version of the stages involved in the selection of the households is as follows.

1 *Selection of sectors*
   Sector selection was accomplished by using *systematic* sampling of a cleverly reordered list. The reordering process was as follows:

   (a) The 9000 postcode sectors were subdivided into 18 geographical regions.

   (b) Within each region the postcode sectors were arranged in an ordered list. The first sector in the list was the postcode sector having the highest proportion of professional heads of households, with the last in the list being that sector with the lowest proportion. These proportions were determined using data from the 1981 Census.

   (c) Each regional group was now split into a 'high' half and a 'low' half, to form two subgroups.

   (d) Each subgroup was reordered by using descending order of percentage of pensioners as a criterion and was again split into two.

   (e) The reordering process was repeated once more, using another social characteristic, to give a total of 144 subgroups, each separately ordered.

   The systematic selection of sectors from the reordered list will result in the selected sectors being spread across the country and across the characteristics used for the reordering. The effect is similar to that of *stratification*.

2 *Selection of households*
   Simple! These were a *simple random sample* of about 35 households chosen from each selected sector. In all, about 8000 households were selected.

It can be seen that a large survey is likely to combine several different types of sampling procedure. A school survey would obviously be nothing like as complicated, but it is still true that care will be needed to avoid bias.

13.3 Random numbers

Suppose that we have a box containing ten balls labelled 0–9 and otherwise identical. A ball is drawn at random from the box, its number is noted and the ball is then returned to the box. A sequence of draws of this nature will result in a sequence of digits, occurring in an unpredictable fashion, but nevertheless having the property that, in a sequence of $10N$ digits, the expected number of occurrences for each digit would be N. An example of the start of such a sequence is shown below:

```
85113   47660   38795   86932   04334
60952   44952   45981   54876   87666
44303   61914   54504   18774   29845
50836   38781   50084   98521   78069
19190   50125   54011   39418   12020
```

Note that these **random numbers** may be read as individual digits (8, 5, 1, 1, ...) or as pairs of digits (85, 11, 34, 76, ...) or in whatever is a convenient manner. In each case they are in effect observations from a discrete uniform distribution.

The numbers may also be interpreted as observations from a *continuous* uniform distribution by the simple expedient of introducing some decimal points:

.85113   .47660   .38795   .86932   .04334

In effect these are random observations from a uniform distribution with range 0 to 1, truncated to an accuracy of 5 decimal places.

Pseudo-random numbers

Suppose we require a sample of 1000 random digits. A thousand draws of balls from a box would be feasible but very tedious. Instead therefore, we make use of computer-generated **pseudo-random numbers**. These are numbers which are generated by a mathematical formula. They have the following properties:

♦ someone who did not know where they came from would be unable to deduce that they had not been generated using a ten-sided die, but

♦ the computer could generate exactly the same sequence time after time if this was required.

In practice the description 'pseudo-' is usually dropped and these numbers are also described as **random numbers**.

Tables of random numbers

Although it is easy to produce pseudo-random numbers using the computer, and many calculators also produce numbers of this type, it is often convenient to be able to refer to a printed list of random numbers. Many books of tables and text-books (including this one!) therefore have such a list (see Appendix, p. 628).

To use such a list, it is sufficient that the starting point, and the direction (up, down, left, right, diagonal, etc.) should be decided upon before the numbers in the table can be seen. This is to guard against biased selection of an 'interesting' number with which to start.

Computer project ────────────────────────────────

Many calculators claim to generate a sequence of random numbers. These are usually pseudo-random numbers that are approximately uniformly distributed in the interval (0,1). Similarly, virtually all computer languages have a readily available function (often called RAN or RAND) to generate uniform numbers in this interval.
Investigate the commands required by your calculator/computer.
Do the numbers really seem unpredictable?

13.4 Simulation

Have you ever been inside a flight simulator? The inside is made to look just like the inside of a real aeroplane, with the 'view' from the window being a film that has been taken from a real aeroplane. When the machine starts the simulator is moved up and down and from side to side in a way that matches the film. The result is that anyone in the simulator obtains the illusion of flying. The machine **simulates** reality.

We can easily use the computer to simulate the outcomes of situations described by probability distributions. Using the random numbers listed in Section 13.3, and working along the first row (85113 47660 38795 . . .), we can simulate the rolling of a fair six-sided die. One method is to treat each digit separately, with the digits 7, 8, 9 and 0 being ignored. The first 10 computer 'rolls' are therefore:

    5, 1, 1, 3, 4,   6, 6, 3, 5, 6

Using the random numbers, we can 'roll' other sorts of dice! We do not need to worry about whether such dice actually exist. Here are the first ten simulated 'rolls' of an unbiased die with 64 sides labelled 1 to 64:

    11, 34, 60, 38, 58,   32, 4, 33, 46, 9

In this case the digits were taken in pairs and discarded if they did not correspond to a feasible outcome.

The last two examples both demonstrated cases in which all the outcomes were equally likely. We can easily extend the method to cases of unequally likely events. Suppose that the discrete random variable X has probability distribution given by:

$$P(X = r) = \begin{cases} \dfrac{r}{10} & r = 1,2,3,4 \\ 0 & \text{otherwise} \end{cases}$$

The following scheme will have the desired effect:

Observed random digit	Simulated value of X
1	1
2 or 3	2
4, 5 or 6	3
7, 8, 9 or 0	4

Since each random digit is equally likely, we see that the probability of, for example, one of the digits 4, 5 or 6 occurring is $\frac{3}{10}$, which is exactly the same as the probability that X takes the value 3. The first ten simulated observations on X (using the first 10 digits of the first row of the random digits given earlier) are therefore:

    4, 3, 1, 1, 2   3, 4, 3, 3, 4

Example 1

Use the following sequence of random digits:

    37297048762551119802394671925524436693773331927334592884

to generate a sequence of observations from a binomial distribution with $n = 7$ and $p = 0.1$.

There are two possible approaches. One is to subdivide the digits into groups of size 7, and to nominate one digit (1, say) to be a 'success'. This leads to the following sequence of results:

3729704 8762551 1198023 9467192 5524436 6937733 3192733 4592884
   0   1   2   1   0   0   1   0

The simulated binomial outcomes are the sequence 0, 1, 2, 1, 0, 0, 1 and 0.

The alternative is to use a table of cumulative probabilities for the required distribution. A reasonable approximation would be provided by the tables in the Appendix, but it is quite straightforward to calculate the individual binomial terms precisely using a calculator or computer and to accumulate them as follows:

r	$P(X \leqslant r)$	Range of random numbers
0	0.478 2969	0000001 to 4782969
1	0.850 3056	4782970 to 8503056
2	0.974 3085	8503057 to 9743085
3	0.997 2720	9743086 to 9972720
4	0.999 8235	9972721 to 9998235
5	0.999 9936	9998236 to 9999936
6	0.999 9999	9999937 to 9999999
7	1.000 0000	0000000 only

Grouping the random digits in eights we get:

37297048 76255111 98023946 71925524 43669377 33319273 34592884
   0   1   3   1   0   0   0

The simulated binomial outcomes are the sequence 0, 1, 3, 1, 0, 0 and 0.

Simulating non-numerical occurrences

As an example, suppose that the traffic lights for a set of roadworks are arranged so that they show Red for 0.45 of the time, Green for 0.45 of the time, and Amber for 0.1 of the time. Treating the random digits in pairs, the following scheme will work:

Range of random numbers	Traffic light colour
01 to 45	Red
46 to 90	Green
91 to 00	Amber

The first ten colours generated are therefore:

    Green, Red, Red, Green, Green,   Red, Green, Green, Green, Red

Computer project ————————————————————————————

> *Write a program to simulate the rolling of a fair six-sided die. Check that it works by arranging for the program to keep count of the frequencies of the six possible outcomes.*

Exercises 13a ————————————————————————————

1 Use the following sequence of random digits:

    05141   56911   63712   65643
    88081   50168   09126   79697
    17144   21103   76294

to generate a sequence of simulated observations from a geometric distribution with (i) $p = \frac{1}{5}$, (ii) $p = \frac{1}{3}$.
Explain your method clearly.

2 Use the following sequence of random digits:

    70748   13945   01576   13577
    64729   04681   39907   61398
    30339   58094   81367

to generate a sequence of simulated observations from a binomial distribution with (i) $n = 5$, $p = \frac{1}{5}$, (ii) $n = 6$, $p = \frac{1}{6}$.
Explain your method clearly.

3 Use the following sequence of random digits:

33696   77300   48644   57598
10243   82181   13321   11121
30206   43239   66612

to generate a sequence of simulated observations from a uniform distribution taking values 0, 1, 2, 3, 4, 5, 6. Explain your method clearly.

4 Use the following sequence of random digits:

42291   38319   14447   55642
13954   15553   62205   74508
24051   39207   01418

to generate a sequence of simulated observations from tossing an unbiased coin. Explain your method clearly.

5 Use the following sequence of random digits:

19913   79213   09786   51646
24845   59514   48193   40973
43734   22163   96771

to generate a sequence of simulated observations of a random variable X, with probability distribution given by $P(X = 0) = \frac{1}{4}$, $P(X = 2) = \frac{1}{4}$, $P(X = 5) = \frac{1}{2}$. Explain your method clearly.

6 Use the following sequence of random digits:

54426   94706   37483   00974
11046   88690   80331   41220
80713   60890   07142

to generate a sequence of simulated observations of a hand of three cards dealt from a pack of 52 playing cards. Explain your method clearly.

7 In each of the following cases describe fully a method by which you could obtain a random sample of 40 students from your school or college that is (i) stratified by sex, (ii) stratified by year group.

8 Describe fully a method by which you could obtain a random sample of 30 students from your school or college that is systematic, with a spread of age.

9 Describe fully a method by which you could obtain a random sample of 20 students who travel by public transport to your school or college.

10 Comment on the following methods of obtaining a sample from your school or college, suggesting improvements where possible.

(i)   Circulate a questionnaire, asking for replies, and take the first twenty who reply.

(ii) Take the third person from the alphabetical list of each class.

(iii) Stand at one of the entrances and choose every tenth person who enters.

11 Give a brief explanation of and an example of the use of
(*a*) a census,
(*b*) a sample survey.   [ULEAC]

12 Write brief notes on
(*a*) simple random sampling,
(*b*) quota sampling.
Your notes should include a description of each method, and an advantage and a disadvantage associated with it.   [ULEAC]

13 The following table gives results obtained in the General Household Survey of 1984 relating to adult male cigarette smokers in Great Britain.

Age group	16–19	20–24	25–34	35–49
Sample size	672	789	1572	2116
% smokers	29	40	40	39

Age group	50–59	60 *or over*	*Total*
Sample size	1170	2098	8417
% smokers	39	30	36

(i) Suggest a reason why different sample sizes were taken from the various age groups.

(ii) Show the calculation that gave the overall percentage of male adults as being 36 (the lower entry under *Total* in the table).
                                               [WJEC]

Simulating a continuous random variable

This requires an extension of the approach used in Example 1. Suppose the continuous random variable X has probability density function $f(x)$ and cumulative distribution function $F(x)$, and suppose the random variable U has a uniform distribution in (0,1), so that, for $0 \leqslant u \leqslant 1$:

$$P(U \leqslant u) = u$$

Both u and $F(x)$ have values in (0,1), and any particular value of $F(x)$ corresponds to a unique value of x. Hence, if we let $u = F(x)$, there is a one-to-one correspondence between a value u and a value of x.

As an example, suppose that X has an exponential distribution with mean $\frac{1}{\lambda}$:

$$f(x) = \begin{cases} \lambda e^{-\lambda x} & x > 0 \\ 0 & \text{otherwise} \end{cases}$$

The cumulative distribution function is (from Section 11.7, p. 287):

$$F(x) = 1 - e^{-\lambda x}$$

Setting $F(x) = u$, we get:

$$e^{-\lambda x} = 1 - u$$

Taking natural logarithms for each side, we get:

$$-\lambda x = \ln(1 - u)$$

and hence:

$$x = -\frac{1}{\lambda}\ln(1 - u) \tag{13.1}$$

Given a value for u, we can use Equation (13.1) to obtain a corresponding value for x. If the value of u is taken from a continuous uniform distribution on (0, 1), then, because of the derivation of Equation (13.1), x will be a (simulated) observation from an exponential distribution with mean $\frac{1}{\lambda}$. As an example, we show the calculations required for the simulation of observations from an exponential distribution with mean 3 ($\lambda = \frac{1}{3}$) using the random numbers from the first row of the list in Section 13.3.

Random number, u	$\ln(1-u)$	Exponential observation, x
0.851 13	−1.9047	5.71
0.476 60	−0.6474	1.94
0.387 95	−0.4909	1.47
0.869 32	−2.0350	6.11
0.043 34	−0.0443	0.13

Example 2

Customers enter a small bank at random points in time, with a mean arrival rate of 20 per hour. There is only one bank employee available to serve the customers. The service times for the customers are observations from an exponential distribution with mean 2.5 minutes. Simulate the activity that takes place during the first 15 minutes that the bank is open.

The phrase 'at random points in time' is a coded message saying 'Poisson process'. The time intervals between events in a Poisson process are observations from an exponential distribution (in this case with mean 3 minutes).

Using the same random numbers as previously, the first few inter-arrival times are the values given in the table above, namely 5.71, 1.94, 1.47, 6.11 and 0.13. Here times are given (in decimal form) in minutes. The times of arrival of the first four customers are therefore 5.71, 7.65 ($= 5.71 + 1.94$), 9.12 and 15.23. In this particular simulation, only 3 customers arrive in the first 15 minutes.

To simulate the service times for the first three customers, we again use random digits. For convenience, we take these from the second row of the original list in Section 13.3 (0.609 52, 0.449 52 and 0.459 81). Transformed into observations from an exponential distribution with mean 2.5 these become 2.35, 1.49 and 1.54.

Combining the two sets of information, the details of the first 15 minutes are as follows.

Time	Event	Server	Queue length
0.00		Idle	0
5.71	First customer arrives	Busy	1
7.65	Second customer arrives	Busy	2
8.06	First customer leaves	Busy	1
9.12	Third customer arrives	Busy	2
9.55	Second customer leaves	Busy	1
11.09	Third customer leaves	Idle	0

In summary, during this 15 minutes, the maximum queue length is 2, the total time spent by customers waiting for someone else to be served is about 50 seconds (0.84 minutes) and the server is idle for a total of about 9 minutes and 37 seconds (9.62 minutes).

This very short simulation is sufficient to give a flavour of this type of pragmatic approach to the difficult problems that may occur in real life. Usually a simulation would be repeated several thousand times (with different random numbers!) in order to get accurate estimates of the features of interest, and to get an idea of the potential variability (how likely is it that customers will be queueing out of the door?!).

Computer project

Write a computer subroutine to generate observations from an exponential distribution with mean λ. Use the subroutine to generate 100 8-hour days at the bank described in the previous example. For each day calculate the following:

1 *The total time (in person-minutes) spent by customers in waiting to be served.*
2 *The mean queue length and the maximum queue length.*
3 *The total time that the server is idle.*

Examine the overall characteristics of the bank over the 100 days.
Is there much variation from day to day?
What was the longest queue length ever observed?
What was the longest time that an individual spent waiting to be served?
What would happen if the mean service time was a little longer than the mean inter-arrival time? Try varying the service and arrival rates and watch what happens. Advanced programmers might like to introduce a second server into the simulation and see the effect on queue build-up.

Chapter summary

♦ **Sampling frame**: The list of population members.

♦ **Simple random sampling**: Selection of individuals directly from the sampling frame with equal probability of selection for each individual.

♦ **Cluster sampling**: Selection from randomly chosen groups of neighbouring individuals.

♦ **Stratified sampling**: Division of the sampling frame into non-overlapping subsets called **strata**, with proportionate simple random sampling from each subset.

♦ **Systematic sampling**: Sampling at regular intervals from an ordered sampling frame.

♦ **Quota sampling**: Non-random sampling of targeted types of individuals.

♦ **Pseudo-random numbers**: Numbers, created using a mathematical formula, that appear indistinguishable from genuinely random numbers.

♦ **Simulation**: The use of pseudo-random numbers, usually with a computer program, to mimic mathematically a real-life process.

Exercises 13b (Miscellaneous)

1 Use the following sequence of random digits:

43858 49773 92862 29151
74528 60679 81884 54998
47875 35550 96911

to generate a sequence of simulated observations of a random variable having
(i) a standard normal distribution,
(ii) a N(5,4) distribution.

2 Use the following sequence of random digits:

13128 84217 53904 03236
75703 29180 42901 10958
95051 79393 20277

to generate a sequence of simulated observations of a random variable having a uniform distribution on $2 < x < 6$.

3 Use the following sequence of random digits:

50771 83084 38473 65522
55547 55364 01833 13683
54911 92009 50977

to generate a sequence of simulated observations of a random variable having a Cauchy distribution with cdf given by:

$$F(x) = \frac{1}{2} + \frac{1}{\pi} \, \tan^{-1}(x)$$

4 A (drunken) fly is initially at the point O on a long straight wire. Each second it moves one unit to the right, or one unit to the left, with equal probability. Use the following sequence of random digits:

51438 88520 20306 29627
40366 88470 88655 35464
92365 57758 45195

to generate a simulation of the movements of the fly. (This is a 'one-dimensional random walk'.)

5 A grasshopper is initially at the point O on horizontal ground. Every 10 seconds it jumps and it always lands at unit distance from its

(contnued)

point of takeoff. The direction is random, having a bearing Θ which has a uniform distribution on $0 < \theta < 2\pi$.
Use the following sequence of random digits:

10836 55233 12274 10206
36474 30607 26884 92952
07282 17817 47458

to generate a sequence of simulated observations of Θ, and hence a simulation of the movements of the grasshopper. (This is a 'two-dimensional random walk'.)
Explain your method fully.

6 An alarm system is monitored each minute. There are three possible states for the alarm system: ready, green (alert) and red (alert). If it is ready at time n, then, at time $n + 1$, the probability that it is ready is 0.9 and the probability that it is green is 0.1. If it is green at time n, then at time $n + 1$, the probability that it is ready is 0.3, the probability that it is green is 0.5 and the probability that it is red is 0.2. If it is red at time n, then, at time $n + 1$, the probability that it is red is 0.6 and the probability that it is green is 0.4.
Use the following sequence of random digits:

55582 67995 81685 52260
63779 48251 65440 08428
37108 82711 90483

to simulate the states of the system as time varies. Explain your method fully.

7 A college of 3000 students has students registered in four Departments, Arts, Science, Education and Crafts. The Principal wishes to take a sample from the student population to gain information about likely student response to a rearrangement of the college timetable so as to hold lectures on Wednesday, previously reserved for sports. What sampling method would you advise the Principal to use? Give reasons to justify your choice.   [ULEAC]

8 (*a*) Explain briefly
    (i) why it is often desirable to take samples,
    (ii) what you understand by a sampling frame.
   (*b*) State two circumstances when you would consider using
    (i) clustering,
    (ii) stratification,
    when sampling from a population.

(*c*) Give two advantages and two disadvantages associated with quota sampling.   [ULEAC]

9 Define what is meant by *simple random sampling*.

A simple random sample of size 20 is required from a population of size 100. The members of the population are labelled 00, 01, 02, ..., 98, 99. Draw such a sample, using the pairs of random digits given in the first column of Table 27 of the *New Cambridge Statistical Tables [see note at the end of this question]*.

Define what is meant by *systematic sampling*. Draw a systematic sample of size 20 from the population specified in the previous paragraph, making clear how you have done so.

Define what is meant by *stratified sampling*. Discuss the circumstances when stratified rather than simple random sampling might usefully be used, and the advantages that might hope to be gained.
[**Note**. *Candidates who use statistics tables other than "New Cambridge" may take suitable random sampling numbers from such tables, but MUST clearly indicate the tables that have been used. It is NOT permitted, in this question, to take random sampling numbers from a calculator.*]   [O&C]
[**Readers may use the table in Section 13.3.**]

10 Explain briefly the difference between a census and a sample survey. Give an example to indicate the practical use of each method.

A school held an evening disco which was attended by 500 pupils. The disco organisers were keen to assess the success of the evening. Having decided to obtain information from those attending the disco, they were undecided whether to use a census or a sample survey. Which method would you recommend them to use? Give one advantage and one disadvantage associated with your recommendation.
    [ULEAC]

11 (a) Give one advantage and one disadvantage of using
    (i) a census,
    (ii) a survey.
   (b) It is decided to take a sample of 100 from a population consisting of 5000 elements. Explain how you would obtain a simple random sample without replacement from this population.   [ULEAC(P)]

12 Listed below are the daily numbers of pupils, x, absent from school during a period of 30 successive school days.

8 12 11 9 7 6
9 5 10 10 11 7
10 3 9 9 8 9
8 5 10 9 6 9
3 10 8 5 4 10

(a) Calculate the mean and variance of this population. (Use $\sum x^2 = 2088$.)

(b) Use the **first row** of the table of random numbers in the booklet of mathematical formulae to take a random sample of size 6 **without** replacement.
[**Readers may use the table in Section 13.3.**]

(c) Use the **second** row of the same table to take a random sample of size 6 **with** replacement.

(d) For each of your samples, find the sample mean and the standard error of the mean, giving your answers to 2 decimal places.

(When sampling with replacement from a population of size N, the variance of the sample mean is $\dfrac{N-n}{N-1} \cdot \dfrac{\sigma^2}{n}$.)

(e) State, giving a reason, which of these two methods is the better one for estimating the population mean. [ULEAC]

14 Point and interval estimation

Oh! let us never, never doubt what nobody is sure about!

More Beasts for Worse Children, Hilaire Belloc

The first part of this book (Chapters 1–4) concentrated on the collection, portrayal and summary of raw data. The second part (Chapters 5–12) presented the mathematics of probability and probability distributions. This chapter marks the start of the final part in which we begin to draw conclusions about a population on the evidence of sample data. This process is known as **statistical inference**.

14.1 Point estimates

A **point estimate** is a numerical value, calculated from a set of data, which is used as an estimate of an unknown parameter in a population. The random variable corresponding to an estimate is known as the **estimator**. The most familiar examples of point estimates are:

the sample mean, $\bar{x}$ used as an estimate of the population mean, μ

the sample proportion, $\dfrac{r}{n}$ used as an estimate of the population proportion, p

$s^2 = \dfrac{\Sigma(x_i - \bar{x})^2}{n - 1}$ used as an estimate of the population variance, σ^2

These three estimates are very natural ones and they also have a desirable property: the expected value of the estimator is exactly equal to the corresponding population parameter. For example, $E(\bar{X}) = \mu$. Estimates for which this is true are said to be **unbiased**. We discuss unbiasedness and other properties in Section 14.8 (p. 394).

14.2 Confidence intervals

We hope that our point estimate will be close to the true population value, but we cannot know for certain how close it is. We can, however, say that it is likely to lie in some interval $(\bar{x} - \delta, \ \bar{x} + \delta)$, where the quantity δ is chosen in a sensible way. If we use a procedure that, with a specified probability, creates an interval that includes the true population value, then the interval is known as a **confidence interval**.

14.3 Confidence interval for a population mean

There are four cases that we need to consider:

1 A sample (large or small) is taken from a normally distributed population, with known variance.
2 A large sample is taken from a population with known variance.
3 A large sample is taken from a population with unknown variance.
4 A sample (large or small) is taken from a normally distributed population with unknown variance (in Section 14.7, p.390).

Note

♦ The distinction between 'large' and 'small' sample sizes is arbitrary, but, typically, 'large' is taken in this context to mean 30 or more observations.

Normal distribution with known variance

A sample of n observations is taken from a $N(\mu, \sigma^2)$ distribution. We denote the random variable corresponding to the sample mean by $\bar{X}$. Since $\bar{X}$ is a linear combination of independent normal random variables, it too has a normal distribution. From Section 8.5 (p. 204) we know that $\bar{X}$ has expectation μ and variance $\dfrac{\sigma^2}{n}$. Hence:

$$\bar{X} \sim N\left(\mu, \frac{\sigma^2}{n}\right)$$

Supposing, for the moment, that μ was known, we could work with the random variable Z, given by:

$$Z = \frac{\bar{X} - \mu}{\dfrac{\sigma}{\sqrt{n}}}$$

where the quantity $\dfrac{\sigma}{\sqrt{n}}$ is often called the **standard error** of the mean.

Since the distribution of Z is known to be $N(0,1)$ we find, by looking at the table of percentage points for a standard normal distribution, that:

$$P(Z > 1.96) = 0.025$$

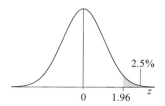

from which it follows that:

$$P(Z < -1.96) = 0.025$$

and hence that:

$$P(|Z| < 1.96) = 0.95$$

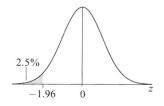

Substituting for Z, this implies that:

$$P\left(\frac{|\bar{X} - \mu|}{\dfrac{\sigma}{\sqrt{n}}} < 1.96\right) = 0.95$$

Multiplying the inequality through by $\dfrac{\sigma}{\sqrt{n}}$, this statement becomes:

$$P\left(|\bar{X} - \mu| < 1.96 \frac{\sigma}{\sqrt{n}}\right) = 0.95$$

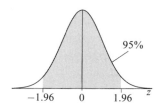

In words, this states that the probability that the distance between μ and $\bar{X}$ is less than $1.96 \dfrac{\sigma}{\sqrt{n}}$ is 0.95. We can conveniently rewrite this as:

$$P\left(\bar{X} - 1.96 \frac{\sigma}{\sqrt{n}} < \mu < \bar{X} + 1.96 \frac{\sigma}{\sqrt{n}}\right) = 0.95$$

Note that, despite its present appearance, this is still a probability statement concerning the random variable $\bar{X}$. It is *not* a probability statement about μ which is a constant (albeit an unknown constant).

Suppose we now collect our n observations on X and compute the sample mean, $\bar{x}$. The interval:

$$\left(\bar{x} - 1.96 \frac{\sigma}{\sqrt{n}}, \quad \bar{x} + 1.96 \frac{\sigma}{\sqrt{n}} \right) \tag{14.1}$$

is called a **95% symmetric confidence interval** for μ. Often the adjective symmetric is omitted and we just write **95% confidence interval**. The two limiting values that define the interval are known as the **95% confidence limits**.

As the diagram shows, different samples will lead to different values of $\bar{x}$ and hence to different 95% confidence intervals: on average, 95% will include the true population value.

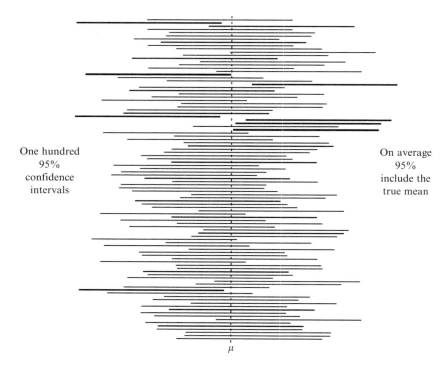

One hundred
95%
confidence
intervals

On average
95%
include the
true mean

μ

If we wish to be more confident that our interval includes the true value of μ, all we need do is to replace 1.96 by a larger value. This will make the intervals wider! If we wish to have a smaller interval, then we must either take a larger sample or be less confident that the interval includes μ.

The most common percentage points used in the construction of symmetric confidence intervals based on the normal distribution are given in the table below.

Degree of confidence	90%	95%	98%	99%
Percentage point	1.645	1.960	2.326	2.576

Example 1

A machine cuts metal tubing into pieces. It is known that the lengths of the pieces have a normal distribution with standard deviation 4 mm. After the machine has undergone a routine overhaul, a random sample of 25 pieces is found to have a mean length of 146 cm. Assuming the overhaul has not affected the variance of the tube lengths, determine a 99% symmetric confidence interval for the population mean length.

Working in centimetres, the confidence interval is:

$$\left(146 - 2.576 \times \frac{4}{\sqrt{25}}, \ 146 + 2.576 \times \frac{4}{\sqrt{25}}\right)$$

which simplifies to (143.94, 148.06). This particular interval either does or does not include the true population mean length – we cannot say which is true! What we *can* say is that 99% of the intervals constructed in this way will include the true population mean length.

Unknown population distribution, known population variance, large sample

By the Central Limit Theorem (Section 12.10, p. 327), the distribution of $\bar{X}$ will be approximately normal:

$$\bar{X} \sim N\left(\mu, \frac{\sigma^2}{n}\right)$$

This case is therefore equivalent to the last case and nothing further need be added.

Unknown population distribution, unknown population variance, large sample

Once again, from the Central Limit Theorem, we can assume that the distribution of $\bar{X}$ is approximately normal. In place of the unknown population variance, σ^2, we use s^2, the unbiased estimate of the population variance (as an approximation). If the sample size is reasonably large (say 30 or more) then the approximation should not be bad. The 95% confidence interval for μ becomes:

$$\left(\bar{x} - 1.96\frac{s}{\sqrt{n}}, \ \bar{x} + 1.96\frac{s}{\sqrt{n}}\right) \tag{14.2}$$

Note

- ♦ A more accurate procedure using the *t*-distribution is discussed in Section 14.7 (p. 390).

Example 2

A random sample of 64 sweets is selected. The sweets are found to have a mean mass of 0.932 g, and the value of s is 0.100 g.
Determine an approximate 99% confidence interval for the population mean mass.

The confidence interval will be approximate, since the population variance is unknown and we will use $s(= 0.100)$ in place of σ. The percentage point for a 99% symmetric confidence interval is 2.576, and so the interval becomes:

$$\left(0.932 - 2.576 \times \frac{0.100}{\sqrt{64}}, \ 0.932 + 2.576 \times \frac{0.100}{\sqrt{64}}\right)$$

which simplifies to:

(0.900, 0.964)

giving the 99% confidence limits correct to three decimal places.

Example 3

A random sample of 100 men is measured and they are found to have heights (x cm) summarised by $\Sigma x = 17\,280$ and $\Sigma x^2 = 2\,995\,400$. Determine an unbiased estimate of the population variance. Determine also an approximate 98% symmetric confidence interval for the population mean. Give your answers correct to one decimal place.

The unbiased estimate of the population variance is s^2, given by

$$s^2 = \frac{1}{99}\left\{2\,995\,400 - \frac{17\,280^2}{100}\right\} = 95.11$$

which is 95.1 to one decimal place.

The corresponding approximate 98% symmetric confidence interval is:

$$\left(172.8 - 2.326\sqrt{\frac{95.11}{100}},\ 172.8 + 2.326\sqrt{\frac{95.11}{100}}\right)$$

which simplifies to:

$$(170.5,\ 175.1)$$

Example 4

Stingy Stephen takes a random sample of 20 observations from a population with unknown mean μ and unknown variance σ^2. His sample has a mean of 16.2 and an unbiased estimate of the population variance equal to 27.34. Independently, Gorgeous Gertie takes a random sample of 16 observations from the same population. Her sample has a mean of 18.0 and an unbiased estimate of the population variance equal to 35.40. Combining their results to give a single sample, obtain an approximate 95% confidence interval for the population mean, giving the confidence limits correct to two decimal places.

In order to obtain the combined mean and combined variance we need to find the overall sum of the 36 observations and also the overall sum of squares (see Section 2.23, p. 72).

Obtaining the overall sum is easy:

$$\Sigma x = (20 \times 16.2) + (16 \times 18.0) = 324.0 + 288.0 = 612.0$$

The combined mean is therefore $\frac{1}{36} \times 612.0 = 17.0$.

Since:

$$s^2 = \frac{1}{n-1}\left\{\Sigma x^2 - \frac{(\Sigma x)^2}{n}\right\}$$

simple algebraic manipulation gives:

$$\Sigma x^2 = (n-1)s^2 + \frac{(\Sigma x)^2}{n}$$

The combined sum of squares for our data is therefore:

$$\Sigma x^2 = \left\{ (19 \times 27.34) + \frac{324^2}{20} \right\} + \left\{ (15 \times 35.40) + \frac{288^2}{16} \right\}$$

$$= 519.46 + 5248.8 + 531.00 + 5184.0$$

$$= 11\,483.26$$

The unbiased estimate of the population variance for the combined sample of 36 observations is:

$$\frac{1}{35} \left(11\,483.26 - \frac{612^2}{36} \right) = \frac{1079.26}{35} = 30.836$$

An approximate 95% confidence interval for the population mean is therefore given by:

$$\left(17.0 - 1.96 \sqrt{\frac{30.836}{36}}, \ 17.0 + 1.96 \sqrt{\frac{30.836}{36}} \right)$$

which simplifies to:

$$(15.19, \ 18.81)$$

Poisson distribution, large mean

When the mean of a Poisson distribution is large, the normal distribution provides a reasonable approximation (see Section 12.12, p. 343). Since the variance of a Poisson distribution is equal to its mean, there is no need to estimate its value from the data. We simply use the value of the sample mean as its estimate. In this case, therefore, the 95% confidence interval for the population mean is given (approximately) by:

$$\left(\bar{x} - 1.96 \sqrt{\frac{\bar{x}}{n}}, \ \bar{x} + 1.96 \sqrt{\frac{\bar{x}}{n}} \right) \tag{14.3}$$

Example 5

An environmentalist takes a random sample of water from a river. She discovers that her 100 ml sample contains 64 organisms of a particular (undesirable!) type. Give a 99% confidence interval for the mean number of these organisms in a litre of this river water.

———

We must first obtain a confidence interval for a water sample of the size obtained. We can then scale this to the required size. The 99% confidence interval for 100 ml is:

$$(64 - 2.576\sqrt{64}, \ 64 + 2.576\sqrt{64})$$

since in this case $n = 1$. This interval simplifies to (43.4, 84.6).
The required confidence interval for a litre of the river water is therefore (434, 846).

Exercises 14a

1 The random variable X has a normal distribution with mean μ and variance 9. A random sample of 10 observations of X has mean 8.2.
Find:

(i) a 95% symmetric confidence interval for μ,

(ii) a 99% symmetric confidence interval for μ.

2 The random variable Y has a normal distribution with mean μ and unknown variance. A random sample of 200 observations of Y gives $\sum y_i = 541.2$, $\sum y_i^2 = 1831.42$.
Find:

(i) a 90% symmetric confidence interval for μ,

(ii) a 98% symmetric confidence interval for μ.

3 The random variable W has a distribution with mean μ and unknown variance. A random sample of 150 observations of W gives $\sum w_i = 1601$, $\sum w_i^2 = 18\,048$.
Giving your answers to two decimal places, find:

(i) a 90% symmetric confidence interval for μ,

(ii) a 95% symmetric confidence interval for μ.

4 The number of telephone calls arriving at a school was monitored on 10 randomly chosen days. The total number of calls was 1053. Assuming a Poisson distribution, find a 95% symmetric confidence interval for the mean number of calls per day.

5 A field of area $7000\,\text{m}^2$ is sown with grass seed. Fifteen non-overlapping squares, each of side $0.1\,\text{m}$ are chosen at random and the number of seeds falling on each square is counted. The results are summarised by $\sum x = 2874$.
Assuming a Poisson distribution, find a 90% symmetric confidence interval for:

(i) the mean number of seeds per square metre,

(ii) the number of seeds on the whole field.

6 The weights of 4-month-old pigs are known to be normally distributed with standard deviation $4\,\text{kg}$. A new diet is suggested and a sample of 25 pigs given this new diet have an average weight of $30.42\,\text{kg}$.
Determine a 99% confidence interval for the mean weight of 4-month-old pigs that are fed this diet.

7 The result X of a stress test is known to be a normally distributed random variable with mean μ and standard deviation 1.3. It is required to have a 95% symmetric confidence interval for μ with total width less than 2. Find the least number of tests that should be carried out to achieve this.   [ULSEB(P)]

8 The frequency table below summarises the lengths of time in minutes that it took to service an aeroplane between flights on 24 occasions chosen at random.

Time (centre of interval)	55	60	65	70	75
Frequency	2	5	8	6	3

(i) Find the mean and standard deviation of these data.

(ii) Assuming that this sample comes from a normally distributed population with the same standard deviation as you have found in (i), find symmetric 98% confidence limits for the population mean.   [O&C]

9 Packets of soap powder are filled by a machine. The weights of powder (to the nearest gram) in 32 packets chosen at random are summarised below.

Weight	999	1000	1001	1002	1003	1004
Packets	1	7	12	8	3	1

Find

(i) the amount by which the mean exceeds $1000\,\text{g}$

(ii) the standard deviation

(iii) the standard error of the mean.

Assuming that this sample comes from a normally distributed population, find, correct to the nearest $0.1\,\text{g}$, 99.8% symmetrical confidence limits for the population mean.   [O&C]

10 A plant produces steel sheets whose weights are known to be normally distributed with a standard deviation of $2.4\,\text{kg}$. A random sample of 36 sheets had a mean weight of $31.4\,\text{kg}$. Find 99% confidence limits for the population mean.   [ULEAC]

11 A random sample of 80 electrical elements produced by a manufacturer have resistances $x_1, x_2, \ldots, x_{80}$ ohms, where $\sum x_i = 790$, and $\sum x_i^2 = 7821$.

(continued)

(i) Calculate unbiased estimates of the mean and the variance of the resistances of the elements produced by the manufacturer.

(ii) Use a normal distribution to calculate approximate 98% confidence limits for the mean resistance of the elements produced by the manufacturer.   [WJEC]

12 Every week a boy buys a packet of his favourite sweets. Each packet carries the statement: "Average contents 150 sweets". Suspecting that this is not the case, the boy decides to count the number of sweets, x, in each of the 52 packets bought during a given year, and finds that $\sum x = 7540$ and $\sum x^2 = 1\,104\,775$.
Calculate

(i) an unbiased estimate of the mean number, μ, of sweets in a packet,

(ii) an unbiased estimate of the variance of the number of sweets in a packet,

(iii) an approximate symmetrical 95% confidence interval for μ.   [JMB(P)]

13 A machine is regulated to dispense liquid into cartons in such a way that the amount of liquid dispensed on each occasion is normally distributed with a standard deviation of 20 ml. Find 99% confidence limits for the mean amount of liquid dispensed if a random sample of 40 cartons had an average content of 266 ml.   [ULEAC]

14 Describe the work you did to obtain empirical evidence to demonstrate the Central Limit Theorem. State the parameters of your distributions.

$\bar{X}$ is the mean of a large random sample of size n_1 from a population with mean μ_1 and variance σ_1^2.

$\bar{Y}$ is the mean of a large random sample of size n_2 from a population with mean μ_2 and variance σ_2^2.

State the form of the sampling distribution of $(\bar{Y} - \bar{X})$, giving its mean and variance.

Buildrite and Constructall are two building firms. The amount, X thousand pounds, paid to Buildrite by each of 100 randomly chosen customers is summarised as follows:
$$\sum x = 160, \quad \sum x^2 = 265$$
Find approximate 95% symmetrical confidence limits for the amount paid per customer to Buildrite.

The amount paid to Constructall by each customer was Y thousand pounds. Based on a random sample of 200 customers, unbiased estimates of the mean and variance of Y were 1.8 and 0.3216 respectively. Find, to the nearest pound, approximate 90% confidence limits for the value by which the mean amount paid per customer to Constructall exceeds that paid to Buildrite.   [ULSEB]

14.4 Confidence interval for a population proportion

Suppose that a random sample of n observations is taken from a population in which the proportion of successes is p and the proportion of failures is $q\ (= 1 - p)$. Suppose the number of successes in the sample is denoted by r (an observation on the random variable R). The observed proportion of sucesses is $\frac{r}{n}$, which is denoted by $\hat{p}$, so that $\hat{p} = \frac{r}{n}$ with the corresponding random variable, $\hat{P}$, being given by $\hat{P} = \frac{R}{n}$.

The random variable R has a binomial distribution with parameters n and p and therefore $E(R) = np$ and $Var(R) = npq$. Hence:
$$E(\hat{P}) = E\left(\frac{R}{n}\right) = \frac{1}{n}E(R) = \frac{1}{n}np = p$$
which shows that $\hat{P}$ is an unbiased estimator of p. Its variance is given by:
$$Var(\hat{P}) = Var\left(\frac{R}{n}\right) = \left(\frac{1}{n}\right)^2 Var(R) = \left(\frac{1}{n}\right)^2 npq = \frac{pq}{n}$$

We restrict attention to cases where n is sufficiently large that the normal approximation to the binomial distribution may be used, so that then:

$$\hat{P} \sim N\left(p, \frac{pq}{n}\right)$$

The standardised variable, Z, given by:

$$Z = \frac{\hat{P} - p}{\sqrt{\dfrac{pq}{n}}}$$

will have an approximate $N(0,1)$ distribution.

As before, we note that:

$$P(|Z| < 1.96) = 0.95$$

Substituting for Z, this implies that:

$$P\left(\frac{|\hat{P} - p|}{\sqrt{\dfrac{pq}{n}}} < 1.96\right) = 0.95$$

Multiplying the inequality through by $\sqrt{\dfrac{pq}{n}}$, and rearranging, we obtain:

$$P\left\{\left(\hat{P} - 1.96\sqrt{\frac{pq}{n}}\right) < p < \left(\hat{P} + 1.96\sqrt{\frac{pq}{n}}\right)\right\} = 0.95$$

Replacing $\hat{P}$ by its observed value $\hat{p}$, and replacing the unknown pq by the approximation $\hat{p}\hat{q}$, where $\hat{q} = 1 - \hat{p}$, an approximate 95% symmetric confidence interval for p is given by:

$$\left(\hat{p} - 1.96\sqrt{\frac{\hat{p}\hat{q}}{n}}, \ \hat{p} + 1.96\sqrt{\frac{\hat{p}\hat{q}}{n}}\right) \tag{14.4}$$

Notes

♦ There are three approximations involved in the production of this confidence interval:

1 the normal approximation to the binomial,
2 the replacement of pq by $\hat{p}\hat{q}$,
3 the omission of a continuity correction for the normal approximation.

All three can be avoided by working directly with the binomial distribution. This procedure requires a computer, but tables of exact confidence intervals do exist.

♦ Some books suggest working with:

$$\frac{(\hat{p} - p)^2}{\dfrac{p(1 - p)}{n}} < 1.96^2$$

and solving the resulting quadratic inequality for p to get more accurate approximations to the confidence limits. In practice the improvement is usually very slight.

♦ The uncertainty in the value of pq is rarely a problem, since for p in $(0.3, 0.7)$ the value of pq only varies between 0.21 and 0.25.

Example 6

An importer has ordered a large consignment of tomatoes. When it arrives he examines a randomly chosen sample of 50 boxes and finds that 12 contain at least one bad tomato. Assuming that these boxes may be regarded as being a random sample from the boxes in the consignment, obtain an approximate 99% confidence interval for the proportion of boxes containing at least one bad tomato, giving your confidence limits correct to three decimal places.

We have $\hat{p} = 0.24$, $\hat{q} = 0.76$. The percentage point is 2.576. The confidence interval is therefore:

$$\left(0.24 - 2.576 \sqrt{\frac{0.24 \times 0.76}{50}}, \ 0.24 + 2.576 \sqrt{\frac{0.24 \times 0.76}{50}} \right)$$

which simplifies to:

$$(0.084, \ 0.396)$$

or from about 8% to 40%.

Example 7

It is known that p, the proportion of voters supporting the Conservative party, is (at the time of writing!) about 40%. A market research organisation intends to interview a random sample of n voters, and wishes to ensure that the probability is about 0.90 that its sample estimate of the proportion of Conservative voters lies within two percentage points of the population percentage.
What size sample (to the nearest hundred) should the organisation take? Assume that all voters interviewed do reveal which party they support!

The requirement implies that the 90% confidence interval for p should take the form:

$$(\hat{p} - 0.02, \ \hat{p} + 0.02)$$

We therefore choose n so that:

$$1.645 \sqrt{\frac{\hat{p}\hat{q}}{n}} \approx 0.02$$

Rearranging, and taking $\hat{p}$ to be 0.4, we get:

$$n \approx \frac{1.645^2 \times 0.4 \times 0.6}{0.02^2} = 1623.6$$

so that a sample size of about 1600 people should be satisfactory.

Practical ———————————————————————————————

> This practical answers a question first posed by the Comte de Buffon in
> 1777. If a needle of length l is dropped randomly (i.e. without looking!)
> on to a grid of equi-spaced parallel lines (separated by a distance d) then,
> the Comte enquired, what is the probability that the needle crosses a line?
>
> For the case $l < d$ the Comte showed that the answer is $\dfrac{2l}{\pi d}$. Using a
> matchstick rather than a needle, and using a grid in which d is chosen to
> be about $\dfrac{4l}{3}$, perform Buffon's experiment 100 times. Show that, with this
> number of tosses and this choice of d, a result of r successes (the crossing
> of a line), corresponds to an estimate of π as $\dfrac{150}{r}$.
>
> Obtain a 99% symmetric confidence interval for the proportion crossing
> a line and deduce a 99% confidence interval for π.
>
> Compare your results with those of the rest of the class. Obtain a
> narrower confidence interval based on the pooled information from the
> entire class.

Exercises 14b ———————————————————————————————

1 A random sample of 75 two-year-old rockets
   was tested and it was found that 67 fired
   successfully.
   Find a 95% confidence interval for the
   proportion of two-year-old rockets that would
   fire successfully.

2 A coin which is possibly biased is thrown 400
   times. The number of heads obtained is 217.
   Find a 90% confidence interval for the
   probability of obtaining a head.

3 A random sample of 120 library books is taken
   as they are borrowed. They are classified as
   fiction or non-fiction, and hardback or
   paperback. 88 books are found to be fiction,
   and, of these, 74 are paperback.
   Find a 90% confidence interval for:

   (i) the proportion of books borrowed that are
        fiction,

   (ii) the proportion of fiction books borrowed
        that are paperback.

4 A pilot survey reveals that about 1% of the
   population have a particular physical
   characteristic.
   Approximately how large would the main
   survey have to be in order to be 99% confident
   of obtaining an estimate of this proportion that
   is correct to within 0.1%?
   Comment on your answer.

5 A random sample of 1000 voters are
   interviewed, of whom 349 state that they
   support the Conservative party.
   Determine a 98% symmetric confidence
   interval for the proportion of Conservative
   supporters in the population.

6 Shivering on a traffic island in December, I
   study the number plates of the cars that pass
   by. Of the first 250 cars that pass me, 36 have
   K registrations.
   Assuming that these cars can be regarded as
   forming a random sample of the cars in the
   country, determine a 95% symmetric
   confidence interval for the proportion of cars in
   the country that have a K registration.

7 A market researcher performs a survey in order
   to determine the popularity of SUDZ washing
   powder in the Manchester area. He visits every
   house on a large housing estate in Manchester
   and asks the question: "Do you use SUDZ
   washing powder?" Of 235 people questioned,
   75 answered "Yes". Treating the sample as
   being random, calculate a symmetric 95%
   confidence interval for the proportion of
   households in the Manchester area which use
   SUDZ.
   Comment on the assumption of randomness
   and also on the question posed.   [JMB]

14.5 One-sided confidence intervals

In Section 14.3 (p. 374) when we introduced the idea of a confidence interval we blithely assumed that small values and large values were equally of interest. Writing Z as a random variable having a N(0,1) distribution, our 95% confidence intervals were based on the probability statement:

$$P(-1.96 < Z < 1.96) = 0.95$$

which led to, for example:

$$\left(\bar{x} - 1.96 \frac{\sigma}{\sqrt{n}},\; \bar{x} + 1.96 \frac{\sigma}{\sqrt{n}} \right)$$

We called this interval a symmetric confidence interval for μ, but we could also have called it a **95% two-sided confidence interval** for μ, since equal attention was paid to both tails of the distribution.

Suppose instead we consider a one-sided probability statement, such as:

$$P(-1.645 < Z) = 0.95$$

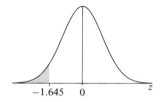

If we now substitute for Z using, for example:

$$Z = \frac{\bar{X} - \mu}{\dfrac{\sigma}{\sqrt{n}}}$$

then, with a simple rearrangement, we arrive at:

$$P\left(\bar{X} + 1.645 \frac{\sigma}{\sqrt{n}} > \mu \right) = 0.95$$

Replacing the random variable $\bar{X}$ by the sample value $\bar{x}$, we obtain the following **95% one-sided confidence interval** for μ:

$$\left(-\infty,\; \bar{x} + 1.645 \frac{\sigma}{\sqrt{n}} \right)$$

Alternatively, using the opposite tail, we would get:

$$\left(\bar{x} - 1.645 \frac{\sigma}{\sqrt{n}},\; \infty \right)$$

With a large sample and an unknown variance, approximate one-sided confidence intervals are obtained by replacing the population standard deviation σ by its sample counterpart s.

Equivalent arguments lead to 95% one-sided confidence intervals for a Poisson mean:

$$\left(0,\; \bar{x} + 1.645 \sqrt{\frac{\bar{x}}{n}} \right) \text{ and } \left(\bar{x} - 1.645 \sqrt{\frac{\bar{x}}{n}},\; \infty \right)$$

and for binomial proportions:

$$\left(0,\; \hat{p} + 1.645 \sqrt{\frac{\hat{p}\hat{q}}{n}} \right) \text{ and } \left(\hat{p} - 1.645 \sqrt{\frac{\hat{p}\hat{q}}{n}},\; 1 \right)$$

The only noticeable difference with these latter situations is the restriction to values in $(0, \infty)$ for the Poisson mean and to $(0, 1)$ for the binomial proportion.

Example 8

A vet is called to examine a large herd of cattle. Out of a random sample of 60 cows, the vet finds that eight show signs of a particular disease. Find a 99% one-sided confidence interval of the form $(0, \theta)$ for the proportion of cows in the herd that show signs of the disease.

———

The estimate of the population proportion is $\hat{p} = \frac{2}{15}$. The upper 1% point of a standard normal distribution is 2.326. Hence the required interval is:

$$\left(0, \ \frac{2}{15} + 2.326 \sqrt{ \frac{2}{15} \times \frac{13}{15} \times \frac{1}{60} } \right)$$

which simplifies to:

(0, 0.235)

or from 0 to 24%.

Exercises 14c ———

1 The random variable X has a normal distribution with mean μ and variance 16. A random sample of 10 observations of X has mean 8.2.
Find a 95% confidence interval for μ:

(i) of the form (θ_1, ∞),

(ii) of the form $(-\infty, \theta_2)$.

2 The random variable Y has a normal distribution with mean μ and unknown variance. A random sample of 500 observations of Y gives $\sum y = 811.2$, $\sum y^2 = 2741.24$.
Find a 90% confidence interval for μ:

(i) of the form (θ_1, ∞),

(ii) of the form $(-\infty, \theta_2)$.

3 A random sample of 90 one-year-old fireworks was tested. It was found that 72 went off successfully.
Find a 95% confidence interval of the form $(\theta, 1)$ for the proportion of one-year-old fireworks that go off satisfactorily.

4 A die which is possibly biased is thrown 500 times. The number of '1's obtained is 53.
Find a 90% confidence interval of the form $(0, \theta)$ for the probability of obtaining a '1'.

5 The number of customers entering a shop was monitored on 8 randomly chosen Saturdays. The total number was 4781. Assuming a Poisson distribution, find a 90% confidence interval, of the form $(0, \theta)$, for the mean number of customers per day.

6 A field of area $8500 \, \mathrm{m}^2$ is sown with clover seed. Ten squares, each of side 0.1 m are chosen at random and the number of seeds falling on each square is counted. The results are summarised by $\sum x = 1216$.
Assuming a Poisson distribution, find a 95% confidence interval, of the form (θ, ∞), for:

(i) the mean number of seeds per square metre,

(ii) the number of seeds on the whole field.

William Sealy Gossett (1876–1937) studied chemistry at Oxford University and, in 1899, joined the staff of Arthur Guinness Son & Co. Ltd. as a 'brewer'. One of his early tasks was to investigate the relationship between the quality of the final product and the quality of the raw materials (such as barley and hops). The difficulty with this task was the expense and time involved in obtaining an observation, so large samples were not available. Gossett correctly mistrusted the existing theory and, in a paper published in 1908, entitled *The Probable Error of a Mean*, he conjectured the form of the *t*-distribution (see below) relevant for small samples. Guinness company policy at the time meant that Gossett was obliged to publish under a pseudonym and, being naturally modest, he chose the pen-name 'Student'. The distribution with which he is associated is still occasionally referred to as 'Student's *t*-distribution'.

14.6 The *t*-distribution

The crucial statistic in the construction of a confidence interval for the mean of a normal distribution is Z, given by:

$$Z = \frac{\bar{X} - \mu}{\frac{\sigma}{\sqrt{n}}}$$

When σ was unknown, and n was large, we replaced σ by s and continued to use the normal distribution. However, this was only an approximation.

The random variable T, defined by:

$$T = \frac{\bar{X} - \mu}{\frac{S}{\sqrt{n}}}$$

involves *two* random variables: $\bar{X}$ in the numerator and S (the random variable corresponding to s) in the denominator. The values of T vary from sample to sample not only because of variations in $\bar{X}$ (as in the case of Z) but also because of variations in S.

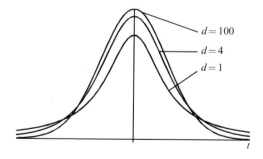

The distribution of T is a member of a family of distributions known as *t*-distributions. All *t*-distributions are symmetric about zero and have a single parameter, d, which is a positive integer. This parameter is known as the number of **degrees of freedom** of the distribution. As a shorthand we replace the phrase 'a *t*-distribution with d degrees of freedom' by 'a t_d-distribution'. It can be shown that T has a t_{n-1}-distribution.

As d increases so the corresponding t_d-distribution increasingly resembles the limiting standard normal distribution (which corresponds to $d = \infty$). When d is 30 or more, the differences between the t_d-distribution and the normal distribution are very slight – which explains why the normal distribution could continue to be used for cases where n was large.

Notes

♦ The result 'T has a t_{n-1}-distribution' requires that $X_1, \ldots, X_n$ have independent and identical normal distributions.

♦ The t_1-distribution is the Cauchy distribution (see Section 11.8, p. 294).

♦ The phrase 'degrees of freedom' is used because of a link between the t-distribution and the chi-squared distribution (which is introduced in Chapter 18).

Tables of the *t*-distribution

Since the use of the t-distribution is largely confined to situations involving pre-specified tail probabilities, the tables concentrate on giving the percentage points (which depend on d, the number of degrees of freedom) for a limited number of cases. There is some variation in the way in which the tables are set out and you should make sure that you are familiar with the tables available to you.

Here is a brief extract from the table given in the Appendix (p. 621) which gives values of t such that $P(T < t) = p\%$, where T, has a t_d distribution:

d	75	90	95	97.5	99	99.5	99.75	99.9	99.95
					$p(\%)$				
1	1.000	3.078	6.314	12.71	31.82	63.66	127.3	318.3	636.6
2	0.816	1.886	2.920	4.303	6.965	9.925	14.09	22.33	31.60
3	0.765	1.638	2.353	3.182	4.541	5.841	7.453	10.21	12.92
.	.	.	.	.	.	.	.	.	.
.	.	.	.	.	.	.	.	.	.
.	.	.	.	.	.	.	.	.	.
∞	0.674	1.282	1.645	1.960	2.326	2.576	2.807	3.090	3.291

Note

♦ A problem with tables of the t-distribution is finding the correct column. You may find it helpful to begin by locating the corresponding normal percentage point, which will be given in the final row of the table.

Example 9

The random variable T has a t-distribution with 3 degrees of freedom. Determine the values of t for which:

(i) $P(T < t) = 0.999$,

(ii) $P(T < t) = 0.25$,

(iii) $P(|T| > t) = 0.05$,

(iv) $P(|T| < t) = 0.98$.

(i)   From the column headed '99.9' we find $t = 10.21$.

(ii) From the column headed '75' we find $P(T < 0.765) = 0.75$, which implies that $P(T > 0.765) = 0.25$. By symmetry (see the sketch), $P(T < -0.765) = 0.25$, and hence the required value of t is -0.765.

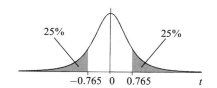

(iii) A combined probability of 0.05 in the two tails implies 0.025 in the upper tail, and hence $p = 0.975$. The required value of t is 3.182.

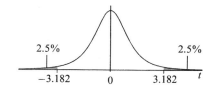

(iv) We see that the required value for t is the value corresponding to an upper-tail probability of 0.01, and hence to a p-value of 0.99. The required value of t is therefore 4.541.

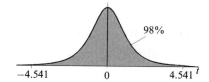

Example 10

The random variable T has a t_2-distribution.
Determine (i) $P(T > -2.920)$, (ii) $P(-4.303 < T < 14.09)$.

(i)   By symmetry, $P(T > -2.920) = P(T < 2.920)$. The required probability is therefore 0.95.

(ii) We begin by rewriting the required probability in terms of tail probabilities:

$$P(-4.303 < T < 14.09) = P(T < 14.09) - P(T < -4.303)$$

Now:

$$P(T < -4.303) = P(T > 4.303)$$
$$= 1 - P(T < 4.303)$$

and so:

$$P(-4.303 < T < 14.09) = P(T < 14.09) - 1 + P(T < 4.303)$$
$$= 0.9975 - 1 + 0.9750$$
$$= 0.9725$$

Exercises 14d

1 Given that T has a t-distribution with 10 degrees of freedom, find:
   (i) $P(T < -2.764)$,
   (ii) $P(T > 1.812)$,
   (iii) $P(-2.228 < T < 0.700)$.

2 Given that T has a t_5-distribution, find:
   (i) $P(T < 2.571)$,
   (ii) $P(T > -4.032)$,
   (iii) $P(|T| < 2.015)$.

3 Given that T has a t-distribution with 25 degrees of freedom, find:
   (i) $P(|T| < 2.060)$,
   (ii) $P(|T| > 1.316)$,
   (iii) $P(1.316 < |T| < 2.060)$.

4 The random variable T has a t-distribution with d degrees of freedom.
   Find the value of t in each of the following cases.
   (i) $d = 8$, $P(T < t) = 0.95$,
   (ii) $d = 15$, $P(T > t) = 0.01$,
   (iii) $d = 5$, $P(|T| < t) = 0.995$,
   (iv) $d = 12$, $P(T < t) = 0.01$,
   (v) $d = 4$, $P(T > t) = 0.975$,
   (vi) $d = 10$, $P(|T| > t) = 0.10$.

14.7 Confidence interval for a population mean using the t-distribution

A small sample has been taken from a normal distribution with unknown variance. In this case, the random variable T, given by:

$$T = \frac{\bar{X} - \mu}{\dfrac{S}{\sqrt{n}}}$$

has a t_{n-1}-distribution.

Suppose that the relevant percentage point of this distribution is c. Then an argument exactly parallel to that used for the case of a normal distribution with σ known would lead to:

$$P\left(\bar{X} - c\frac{S}{\sqrt{n}} < \mu < \bar{X} + c\frac{S}{\sqrt{n}}\right) = 0.95$$

The symmetric confidence interval for μ therefore becomes:

$$\left(\bar{x} - c\frac{S}{\sqrt{n}}, \ \bar{x} + c\frac{S}{\sqrt{n}}\right) \tag{14.5}$$

▼ ─────────────────────────────────────── ▲

Example 11

A random sample of 16 sweets is chosen from a sack of sweets and the mass, x g, of each sweet is determined. The measurements are summarised by $\Sigma x = 13.3$ and $\Sigma x^2 = 15.13$.
Assuming that the masses have a normal distribution, determine a 99% symmetric confidence interval for the population mean, giving the confidence limits correct to three decimal places.

─────────

Since the sample size is less than 30, and the population has a normal distribution, it is appropriate to base a confidence interval on a t-distribution – in this case the t_{15}-distribution. From the tables in the Appendix the percentage point for $d = 15$ and $p = 0.995$ is 2.947.

The unbiased estimate of the population variance, s^2, is given by:

$$s^2 = \frac{1}{15}\left\{15.13 - \frac{13.3^2}{16}\right\} = 0.271\,625$$

so that $s = 0.521\,18$ and $\bar{x} = \dfrac{13.3}{16} = 0.831\,25$. The 99% symmetric confidence interval is therefore:

$$\left(0.831\,25 - 2.947 \times \frac{0.521\,18}{\sqrt{16}},\ 0.831\,25 + 2.947 \times \frac{0.521\,18}{\sqrt{16}}\right)$$

which simplifies to:

$$(0.447,\ 1.215)$$

A 99% symmetric confidence interval for the population mean is from 0.447 g to 1.215 g.

Note that the intermediate working has been carried out using far more than just the three decimal places required for the answer. Premature rounding of intermediate calculations is liable to adversely affect final accuracy.

Example 12

Ten students independently performed an experiment to estimate the value of π. Their results were:

$$3.12,\ 3.16,\ 2.94,\ 3.33,\ 3.00,\ 3.11,\ 3.50,\ 2.81,\ 3.02,\ 3.10$$

(i) Calculate the sample mean and the value of s^2.

(ii) Stating any necessary assumption that you make, calculate a 95% symmetric confidence interval for π based on these data, giving the confidence limits correct to two decimal places.

(iii) Estimate the minimum number of results that would be needed if it is required that the width of the resulting 95% symmetric confidence interval should be at most 0.02.

—————

(i) The data are summarised by $\Sigma x = 31.09$ and $\Sigma x^2 = 97.0011$, giving $\bar{x} = 3.109$ and $s^2 = 0.038\,032$.

(ii) We have to assume that the underlying distribution is normal with mean π. The percentage point of the t_9-distribution is 2.262, leading to the symmetric confidence interval:

$$\left(3.109 - 2.262\sqrt{\frac{0.038\,032}{10}},\ 3.109 + 2.262\sqrt{\frac{0.038\,032}{10}}\right)$$

which simplifies to give a 95% symmetric confidence interval for π as (2.97, 3.25).

(iii) The width of a symmetric confidence interval is:

$$2c \, \frac{s}{\sqrt{n}}$$

where c is the percentage point. With 10 observations the width was 0.28, which is much greater than the desired 0.02. Far more observations will be required to achieve the desired accuracy. Since n will be large, the value of c will be that for the limiting normal distribution, in other words 1.96.

We do not know what value will be obtained for s^2, so our best guess is the value provided by the present sample, namely 0.038 032. To find the required value for n we must solve the equation:

$$2 \times 1.96 \times \sqrt{\frac{0.038\,032}{n}} = 0.02$$

The solution is:

$$n = \frac{2^2 \times 1.96^2 \times 0.038\,032}{0.02^2} = 1461.04$$

and hence the required number of observations (rounding up to the next integer) is estimated as being 1462.

▲ _____ ▲

Project _____

> *How many words are there in this book? Choose a number of pages at random and count (or estimate) the number of words on each chosen page. Assume that these numbers may be regarded as arising from a (discretised) normal distribution.*
>
> *Calculate their variance and hence obtain a 95% symmetric confidence interval for the number of words in the book.*

Exercises 14e _____

1 The random variable X has a normal distribution with mean μ. A random sample of 10 observations of X is taken and gives $\sum x_i = 83.3$, $\sum x_i^2 = 721.41$.
Find:
(i) a 95% confidence interval for μ,
(ii) a 99% confidence interval for μ.

2 The quantity of milk in a bottle may be assumed to have a normal distribution.
A random sample of 16 bottles was taken and the quantity of milk was measured, with the following results, in ml.

   1005, 1003, 998, 1001, 1002, 999, 1000, 1001, 1007, 1003, 1010, 1001, 1003, 1002, 1005, 995

Find a 99% confidence interval for the mean quantity of milk in a bottle, giving your answers to 2 decimal places.

3 A random sample of 12 hollyhock plants, grown from the seeds in a particular packet, was taken, and the height of each plant was measured, in m. The results are summarised by $\sum x_i = 28.43$, $\sum x_i^2 = 88.4704$.
Making a suitable assumption about the distribution of heights, which should be stated, find a 90% confidence interval for the mean height of hollyhock plants grown from that packet.

4 A lorry is transporting a large number of red apples. As it passes over a bump in the road 10 apples fall off its back and are collected by a passing boy. The masses (in g) of the fallen apples are summarised by $\sum(x-100)=23.7$, $\sum(x-100)^2=1374.86$.

Treating the fallen apples as being a random sample, determine a symmetric 99% confidence interval for the mean mass of a red apple, stating any assumptions that you have made.

[UCLES(P)]

5 The total costs (in £) of the telephone calls from an office during six randomly chosen weeks of the year are given below.

   113.20, 87.60, 109.40,
   131.20, 201.10, 142.90

Regarding these values as being independent observations from a normal distribution, obtain a symmetric 99% confidence interval for the mean weekly cost of telephone calls made from the office.

[UCLES(P)]

6 The speed at which a baseball is thrown is measured (in km h^{-1}) at the instant that it leaves the pitcher's hand. The results for 10 randomly chosen throws on a cool day are summarised by $\sum(x_i-128)=7.9$, $\sum(x_i-128)^2=338.4$, where x_i is the speed of throw i.

Assuming that these results are observations from a normal distribution, obtain unbiased estimates of the mean and variance of this distribution, and obtain a symmetric 99% confidence interval for the mean. [UCLES(P)]

7 A customer obtained a trial supply of wire from a manufacturer and measured the breaking strength, y N, of each of a random sample of 12 lengths of wire, obtaining the results shown below.

   80.2   83.5   76.2   79.2   88.7   90.2
   93.4   75.1   87.2   83.4   82.6   81.2

   $(\sum y=1000.9, \sum y^2=83\,826.27)$

Use the sample data to obtain a symmetric 99% confidence interval for the mean breaking strength of lengths of wire from the manufacturer. State any distributional assumptions you have made in obtaining your confidence interval.

Explain carefully the meaning of 99% *confidence* as applied to an interval in this context. [JMB(P)]

8 The random variable X has a normal distribution with mean μ. A random sample of 10 observations of X is taken and gives $\sum x_i=83.3$, $\sum x_i^2=721.41$.

Find a 95% confidence interval for μ

(i) of the form (θ_1,∞),
(ii) of the form $(-\infty,\theta_2)$.

9 The quantity of milk in a bottle may be assumed to have a normal distribution. A random sample of 16 bottles was taken and the quantity of milk was measured, with the following results, in ml.

   1005, 1003, 998, 1001, 1002, 999,
   1000, 1001, 1007, 1003, 1010,
   1001, 1003, 1002, 1005, 995

Find a 99% confidence interval, of the form (θ,∞), for the mean quantity of milk in a bottle.

10 A random sample of 12 hollyhock plants, grown from the seeds in a particular packet, was taken, and the height of each plant was measured, in m. The results are summarised by $\sum x_i=28.43$, $\sum x_i^2=88.4704$.

Making a suitable assumption about the distribution of heights, which should be stated, find a 90% confidence interval, of the form (θ,∞), for the mean height of hollyhock plants grown from that packet.

11 In a classroom experiment to estimate the mean height, μ cm, of seventeen-year-old boys, the heights, x cm, of 10 such pupils were obtained. The data were summarised by $\sum x=1727$, $\sum x^2=298\,834$.

(i) Find the mean and variance of the data, and use them to find the symmetrical 95% confidence interval for μ. State clearly but briefly the two important assumptions which you need to make.

A large experiment is planned using the heights of 150 seventeen-year-old boys.

(ii) What effect will the use of a larger sample have on the width of the confidence interval for μ? Identify two distinct mathematical reasons for this effect.

(iii) To what extent are the assumptions made in (i) still necessary with the larger sample size? [MEI]

14.8 Desirable properties of an estimator

Recall that the word estimator is used to describe the random variable corresponding to an estimate. The estimator has a distribution, whereas the estimate has a specific value.

Suppose that θ is some population parameter (e.g. μ) whose value we wish to estimate. Let U be an estimator of this parameter. Desirable properties are:

1 $E(U) = \theta$.
   If this is true, then U is said to be **unbiased**.

2 $\text{Var}(U)$ should be as small as possible.
   If U and V are two unbiased estimators of θ with:

$$\text{Var}(U) < \text{Var}(V)$$

   then we naturally prefer U, because it seems likely that U will be closer than V to θ. In this case U is said to be more **efficient** than V.

Notes
   ◆ If $E(U) = \theta + b$, with $b \neq 0$, then U is said to be **biased**. Sometimes the **bias**, b, is a function of the sample size, n, and often it may reduce to zero as the sample size increases to infinity. In this case the estimator is described as being **asymptotically unbiased**.

   ◆ If U is unbiased (or asymptotically unbiased), and if $\text{Var}(U)$ reduces to zero as the sample size increases, then U is said to be **consistent**.

▼—————————————————————————▼

Example 13

A random sample of 2 observations (X_1, X_2) is to be taken from a population with unknown mean μ and variance σ^2. Three estimators for μ have been suggested. These are U_1, U_2 and U_3, defined by:

$$U_1 = X_1 \qquad U_2 = \frac{X_1 + X_2}{2} \qquad U_3 = 2X_1 - X_2$$

Show that all three estimators are unbiased and determine which is the most efficient, and which is the least efficient.

———————

Since $E(U_1) = E(X_1) = \mu$, U_1 is an unbiased estimator of μ, with variance σ^2.
   For U_2 we have:

$$E(U_2) = \frac{1}{2}E(X_1 + X_2)$$

$$= \frac{1}{2}\{E(X_1) + E(X_2)\}$$

$$= \frac{1}{2}(\mu + \mu) = \mu$$

showing that U_2 (which is the sample mean) is unbiased. It has variance given by:

$$\text{Var}(U_2) = \text{Var}\left\{\frac{1}{2}(X_1 + X_2)\right\}$$

$$= \left(\frac{1}{2}\right)^2 \{\text{Var}(X_1) + \text{Var}(X_2)\} \quad \text{since } X_1 \text{ and } X_2 \text{ are independent}$$

$$= \frac{1}{4}(\sigma^2 + \sigma^2)$$

$$= \frac{1}{2}\sigma^2$$

which is less than $\text{Var}(U_1)$ and implies that U_2 is more efficient than U_1.

Discovering that U_2, which uses information from both observations, is more efficient than U_1, which uses one observation only, should not be a surprise. However, how does U_3 fare?

$$E(U_3) = E(2X_1 - X_2)$$

$$= 2E(X_1) - E(X_2)$$

$$= 2\mu - \mu$$

$$= \mu$$

confirming that all three estimators are unbiased. Finally we calculate:

$$\text{Var}(U_3) = \text{Var}(2X_1 - X_2)$$

$$= 2^2\text{Var}(X_1) + (-1)^2\text{Var}(X_2) \quad \text{since } X_1 \text{ and } X_2 \text{ are independent}$$

$$= 4\sigma^2 + \sigma^2$$

$$= 5\sigma^2$$

The variance of U_3 is ten times that of U_2 and five times that of U_1. The most efficient estimator is U_2 and the least efficient is U_3.

Example 14

A random sample of n observations is taken from a distribution with mean μ and variance σ^2. Show that the sample mean $\bar{X}$ is a consistent estimator of μ.

An independent random sample of $(n + 1)$ observations is taken from the same distribution. The sample mean is denoted by $\bar{X}'$.
Show that $\bar{X}'$ is more efficient than $\bar{X}$ as an estimator of μ.

We denote the n individual observations by $X_1, \ldots, X_n$ and begin by calculating $E(\bar{X})$:

$$E(\bar{X}) = E\left(\frac{1}{n}X_1 + \cdots + \frac{1}{n}X_n\right)$$

$$= \frac{1}{n}\mu + \cdots + \frac{1}{n}\mu = \mu$$

Thus $\bar{X}$ is an unbiased estimator of μ. To show that it is consistent we also need to show that its variance approaches zero as the sample size increases. Now:

$$\text{Var}(\bar{X}) = \text{Var}\left(\frac{1}{n}X_1 + \cdots + \frac{1}{n}X_n\right)$$

$$= \left(\frac{1}{n}\right)^2 \sigma^2 + \cdots + \left(\frac{1}{n}\right)^2 \sigma^2 \qquad \text{since the observations are independent}$$

$$= \frac{\sigma^2}{n}$$

As n increases, so $\frac{\sigma^2}{n}$ approaches zero as required. We have therefore shown that $\bar{X}$ is a consistent estimator of μ.

For the second sample, replacing n by $n+1$, we obtain $\text{E}(\bar{X}') = \mu$, and $\text{Var}(\bar{X}') = \frac{\sigma^2}{n+1}$. Both $\bar{X}$ and $\bar{X}'$ are unbiased estimators of μ. Since $\bar{X}'$ has the smaller variance it is the more efficient.

Example 15

A random sample of n observations is taken from a distribution with mean μ and variance σ^2. The random variable S^2 is defined by:

$$S^2 = \frac{1}{n-1} \sum_{i=1}^{n} (X_i - \bar{X})^2$$

where X_i and $\bar{X}$ are defined in the usual way.
Show that S^2 is an unbiased estimator of σ^2.

One proof of this result was given in Section 8.6 (p. 206). Here is another, in which we commence by manipulating the summation:

$$\Sigma(X_i - \bar{X})^2 = \Sigma(X_i^2 - 2\bar{X}X_i + \bar{X}^2)$$

$$= \Sigma X_i^2 - 2\bar{X}\Sigma X_i + n\bar{X}^2$$

$$= \Sigma X_i^2 - n\bar{X}^2$$

since $\Sigma X_i = n\bar{X}$.
Now:

$$\text{Var}(\bar{X}) = \text{E}(\bar{X}^2) - \{\text{E}(\bar{X})\}^2$$

so that:

$$\text{E}(\bar{X}^2) = \text{Var}(\bar{X}) + \{\text{E}(\bar{X})\}^2$$

Since $\text{E}(\bar{X}) = \mu$ and $\text{Var}(\bar{X}) = \frac{\sigma^2}{n}$, we get:

$$\text{E}(\bar{X}^2) = \frac{\sigma^2}{n} + \mu^2$$

The corresponding result for a single observation, X_i, is:

$$\text{E}(X_i^2) = \sigma^2 + \mu^2$$

Combining these results we have:

$$E[\Sigma(X_i - \bar{X})^2] = E(\Sigma X_i^2 - n\bar{X}^2)$$

$$= \{\Sigma E(X_i^2) - nE(\bar{X}^2)\}$$

$$= \left\{\Sigma(\sigma^2 + \mu^2) - n\left(\frac{\sigma^2}{n} + \mu^2\right)\right\}$$

$$= (n\sigma^2 + n\mu^2 - \sigma^2 - n\mu^2)$$

$$= (n-1)\sigma^2$$

Thus S^2 is an unbiased estimator of the population variance, σ^2.

Example 16

A random sample of two observations, X_1 and X_2, is taken from a distribution with mean μ and variance σ^2.
Determine the most efficient unbiased estimator of μ of the form $aX_1 + bX_2$, where a and b are constants whose values are to be determined.

Let $Y = aX_1 + bX_2$. For the estimator to be unbiased, we require $E(Y) = \mu$. Now, since:

$$E(Y) = E(aX_1 + bX_2) = a\mu + b\mu = (a+b)\mu$$

it is evident that a and b must satisfy:

$$a + b = 1$$

Also:

$$Var(Y) = a^2 Var(X_1) + b^2 Var(X_2) \qquad \text{since } X_1 \text{ and } X_2 \text{ are independent}$$

$$= (a^2 + b^2)\sigma^2$$

$$= \{a^2 + (1-a)^2\}\sigma^2 \qquad \text{using } a + b = 1$$

$$= (2a^2 - 2a + 1)\sigma^2$$

To maximise efficiency, the variance must be as small as possible. By completing the square and writing:

$$2a^2 - 2a + 1 = 2\left(a - \frac{1}{2}\right)^2 + \frac{1}{2}$$

we see that the minimum occurs when $a = \frac{1}{2}$. Since $a + b = 1$, this implies that b is also equal to $\frac{1}{2}$.

Note

◆ We have shown that, of all possible linear combinations, the unbiased estimator of μ that has the minimum variance is, in fact, $\bar{X}$. This result extends to samples of any size.

Exercises 14f

1 A random variable has a distribution with mean μ and variance σ^2. Two independent samples of observations are taken. The first sample has size m, and the observations are denoted by $X_1, X_2, \ldots, X_m$, with $\bar{X} = \dfrac{1}{m}\displaystyle\sum_{i=1}^{m} X_i$.

The second sample has size n, and the observations are denoted by $Y_1, Y_2, \ldots, Y_n$, with $\bar{Y} = \dfrac{1}{n}\displaystyle\sum_{j=1}^{n} Y_j$.

Show that:

$$\frac{m\bar{X} + n\bar{Y}}{m+n} \quad \text{and} \quad \frac{\sum(X_i - \bar{X})^2 + \sum(Y_j - \bar{Y})^2}{m+n-2}$$

are unbiased estimators of μ and σ^2 respectively.

2 When a biased cubical die is rolled, the probability that a six will be obtained is an unknown constant p. The die is rolled 40 times and the number, X, of sixes obtained is recorded. The number, Y, of sixes obtained when the die is rolled a further 60 times is also recorded. Show that

$$T_1 = \frac{3X + 2Y}{240} \quad \text{and} \quad T_2 = \frac{X + Y}{100}$$

are both unbiased estimators for p. Find, in terms of p, the standard errors of T_1 and T_2, and state, with a reason, which of these two estimates you consider the better. [JMB]

3 A population consists of the four numbers 0,2,4,4. Calculate the mean μ and variance σ^2 of this population.

A random sample of two numbers is to be drawn *without replacement* from the above population. Let $\bar{X}$ denote the mean and V the variance of this sample.

(i) Find the four distinct values which $\bar{X}$ can take, and calculate the probability of each.

(ii) Find all the possible values which V may take, and calculate the probability of each.

Verify that $\bar{X}$ is an unbiased estimator of μ, but that V is not an unbiased estimator of σ^2. [JMB]

Chapter summary

♦ **Properties of point estimators**
  • **Unbiasedness**
    An estimator U is an unbiased estimator of a population parameter θ if $E(U) = \theta$.
  • **Efficiency**
    If U and V are two unbiased estimators of θ, with $\text{Var}(U) < \text{Var}(V)$ then U is more efficient than V.
  • **Consistency**
    If U is an unbiased estimator of θ, and if $\text{Var}(U)$ approaches 0 as the sample size increases, then U is a consistent estimator of θ.

(continued)

♦ **Confidence intervals for a population mean**
For a sample of size n, denote the appropriate percentage point from a normal distribution by c_N, and its counterpart from a t-distribution with $n-1$ degrees of freedom by c_t.

Condition	Confidence interval	Notes
X normal, σ^2 known	$\left(\bar{x} - c_N \dfrac{\sigma}{\sqrt{n}},\ \bar{x} + c_N \dfrac{\sigma}{\sqrt{n}} \right)$	Exact
σ^2 known, n large	$\left(\bar{x} - c_N \dfrac{\sigma}{\sqrt{n}},\ \bar{x} + c_N \dfrac{\sigma}{\sqrt{n}} \right)$	Approximate
σ^2 unknown, n large	$\left(\bar{x} - c_N \dfrac{s}{\sqrt{n}},\ \bar{x} + c_N \dfrac{s}{\sqrt{n}} \right)$	Approximate
X normal, σ^2 unknown	$\left(\bar{x} - c_t \dfrac{s}{\sqrt{n}},\ \bar{x} + c_t \dfrac{s}{\sqrt{n}} \right)$	Exact

♦ **Confidence interval for a population proportion**
Write $\hat{p} = \dfrac{r}{n}$ and $\hat{q} = \dfrac{n-r}{n}$, where r is the number of successes in a large sample of size n. An approximate confidence interval is provided by:

$$\left(\hat{p} - c_N \sqrt{\frac{\hat{p}\hat{q}}{n}},\ \hat{p} + c_N \sqrt{\frac{\hat{p}\hat{q}}{n}} \right)$$

♦ **One-sided confidence intervals**
These are typified by:

$$\left(-\infty,\ \bar{x} + 1.645 \frac{\sigma}{\sqrt{n}} \right)$$

Exercises 14g (Miscellaneous) _____

1 Foresters are interested in estimating the number of beech trees in a large wood which has an area of 60 hectares. Ten widely separated square sites are selected for examination and are found to contain a total of 16 beech trees. The total area of these sites is 2 hectares.

(a) Give a point estimate of the total number of beech trees in the wood.

(b) Assuming that the beech trees grow at random places within the wood, use a normal approximation to a Poisson distribution to obtain a symmetric 95% confidence interval for the mean number of beech trees in a randomly chosen 2-hectare region.

Hence obtain a symmetric 95% confidence interval for the number of trees in the wood.

2 A soft-drink machine is regulated so that the amount it delivers per cup is approximately normally distributed with standard deviation 1.2 ml. The amounts delivered, in ml, to 5 cups were:

212.6, 210.4, 211.5, 209.8, 210.7

(i) Calculate an estimate of the mean amount that is delivered per cup by the machine.

(ii) Calculate 90% confidence limits for the mean amount that is delivered per cup by the machine.   [WJEC]

3 The random variable X is distributed normally with mean μ and variance σ^2.
Write down the distribution of the sample mean $\bar{X}$ of a random sample of size n.

Records from a dental practice showed that during 1991 the number of minutes per visit spent in the dentist's chair can be taken to be normally distributed with mean 14.5 minutes and standard deviation 2.9 minutes.

(*a*) Calculate an interval within which 90% of the times spent in the dentist's chair will lie.

In 1992 it was assumed that the standard deviation remained unchanged, and the distribution can be assumed to be normal. A random sample of 16 consultations gave the following times in minutes.

13.2	18.7	14.9	12.1
11.6	17.2	10.6	9.4
14.6	12.9	11.2	13.5
12.9	11.8	14.1	12.5

(*b*) For 1992, calculate a 95% confidence interval for the mean length of visit to the dentist.   [ULEAC(P)]

4 A piece of apparatus used by a chemist to determine the weight of impurity in a chemical is known to give readings that are approximately normally distributed with a standard deviation of 3.2 mg per 100 g of chemical.

(*a*) In order to estimate the amount of impurity in a certain batch of the chemical, the chemist takes 12 samples, each of 100 g, from the batch and measures the amount of impurity in each sample. The results, obtained in mg/100 g are as follows:

7.6	3.4	13.7	8.6	5.3	6.4
11.6	8.9	7.8	4.2	7.1	8.4

(i) Find 95% central confidence limits for the mean weight of impurity present in a 100 g unit from the batch.

(ii) The chemist calculated a 95% confidence interval for the mean weight of impurity of 100 g units from the batch. The interval was of the form $-\infty \leqslant \text{mean} \leqslant \alpha$. Find the value of α. Suggest why the chemist might prefer to use the value α rather than the limits in (i).

(iii) Calculate an interval within which approximately 90% of the measured weights of impurity of 100 g units from the batch will lie.

(*b*) Estimate how many samples of 100 g the scientist should take in order to be 95% confident that an estimate of the mean weight of impurity per 100 g is within 1.5 mg of the true value.

(*c*) Over a period of months the chemist found that of 150 samples of the chemicals, 18 yielded a level of impurity which was unacceptable.
Calculate an approximate 95% confidence interval for the proportion of samples having an unacceptable level of impurity.   [AEB 91]

5 Sugar is produced and bagged by a large company. In a random sample of 80 bags, 18 were found to be bags on which the printing was not clear. Calculate an approximate 95% confidence interval for the proportion of bags with unclear printing.

Explain what you understand by a 95% confidence interval in this context.

The sugar produced is classified as granulated or castor and the masses of the bags of both types are known to be normally distributed. The mean of the masses of bags of granulated sugar is 1022.51 g and the standard deviation for both types of sugar is 8.21 g.

Calculate an interval within which 90% of the masses of bags of granulated sugar will lie.

A sample of 10 bags of castor sugar had masses, measured to the nearest gram, as follows.

1062	1008	1027	1031	1011
1007	1072	1036	1029	1041

Find a 99% confidence interval for the mean mass of bags of castor sugar.

To produce a bag of castor sugar of mass x g costs, in pence,

$$(32 + 0.023x)$$

and it is sold for 65p.

If the company produces 10,000 bags of castor sugar per day, derive a 99% confidence interval for its daily profit from castor sugar.   [AEB 92]

6 (i) Explain briefly, referring to your projects where possible, what you understand by a 90% confidence interval.

A normal population has variance 25. Find the size of the smallest sample which could be taken from the population so that the symmetrical 90% confidence interval for the mean has width less than 3 units.

(ii) Rainfall records in a certain town show that it rains on average 2 days in every 5. Taking Monday as the first day of the week, find, to 3 significant figures, the probability that, in a given week,

    (a) the first 3 days will be without rain and on the remaining days there will be rain,

    (b) rain will fall on exactly 4 days in the week,

    (c) Friday will be the first day on which it rains.

Find, to 3 decimal places, the probability that there will be rain in that town on exactly 160 days in a given year of 365 days. [ULSEB]

7 People attending a particular theatre during a week of performances were asked to complete a questionnaire. One of the questions asked the person to indicate his/her age-group. A random sample of 400 of the completed questionnaires produced the following grouped frequency distribution of the ages of the respondents.

Age	Under 25	25–39	40–49
Number of people	28	75	80

Age	50–59	60 or over	Total
Number of people	150	67	400

(a) Estimate the proportion of the people who attended the theatre who were under 30 years old.

(b) Estimate the median and the semi-interquartile range of the ages of the people who attended the theatre, giving each answer to the nearest month.

(c) State why it is not possible to obtain a reliable estimate of the mean age from the information given in the table.

(d) Give a reason why the median would be preferred to the mean as a representative value of the average age of the people who attended the theatre.

(e) Calculate approximate 95% confidence limits for the proportion of the people who attended the theatre who were 50 years old or more. [WJEC]

8 A random sample of ten quartz watches of a particular make were tested for accuracy over a period of four weeks.
The times, in seconds, gained by the ten watches were: $-3, +7, +2, +6, +8, +2, -3, +6, +11, +8$.

(i) Calculate unbiased estimates of the mean, μ, and the variance, σ^2, of the times gained over a period of four weeks.

(ii) Stating any assumptions you make, find a 95% confidence interval for the mean time gained by such watches over a period of four weeks.

(iii) Another random sample of ten of the watches was taken and the times, in seconds, they gained over a period of four weeks gave unbiased estimates of μ and σ^2 equal to 3.8 and 23.4, respectively.
Use the combined set of twenty observations to determine a 95% confidence interval for the mean time gained by such watches over a period of four weeks. [WJEC(P)]

15 Hypothesis tests

It is a good morning exercise for a research scientist to discard a pet hypothesis every day before breakfast. It keeps him young

On Aggression, Konrad Lorenz

Having nothing better to do, you decide to weigh some tins of extraordinarily cheap RIPOFF baked beans. To your amazement, the twelve tins have an average mass that is 10 g less than the mass stated on the tins. Should you report the manufacturers to the authorities? Still thinking about this, you visit the local casino to play Roulette. There are 37 numbers on the wheel and you keep betting on number 1. Should you be surprised that on 25 successive occasions you lose? Perhaps the wheel is biased? You go home in despair to eat baked beans (and learn about hypothesis tests – also known as **significance tests**).

15.1 The null and alternative hypotheses

The first stage of any hypothesis test is to write down the two hypotheses. Usually the **null hypothesis** specifies *a particular value* for some population parameter whereas the **alternative hypothesis** specifies *a range of values*. Here are some examples:

Parameter	Null hypothesis	Alternative hypothesis
Mean, μ	$\mu = 435$	$\mu \neq 435$
Proportion, p	$p = \frac{1}{37}$	$p \neq \frac{1}{37}$
Mean, μ	$\mu = 435$	$\mu < 435$
Proportion, p	$p = \frac{1}{37}$	$p < \frac{1}{37}$

To save writing out 'null hypothesis' and 'alternative hypothesis' lots of times, we denote the hypotheses by H_0 and H_1, respectively. Thus the first pair of hypotheses in the table above would become:

$$H_0: \mu = 435 \qquad H_1: \mu \neq 435$$

The first part of this chapter concentrates on cases where the sample size, n, is large. The situations considered are the following:

Unknown parameter	Sample statistic	Random variable	Condition	Distribution
μ	$\bar{x}$	$\bar{X}$	Normal distribution, σ^2 known	$\dfrac{\bar{X} - \mu}{\frac{\sigma}{\sqrt{n}}} \sim N(0, 1)$ (exact)
μ	$\bar{x}$	$\bar{X}$	Any distribution, σ^2 known, n large	$\dfrac{\bar{X} - \mu}{\frac{\sigma}{\sqrt{n}}} \sim N(0, 1)$
μ	$\bar{x}$	$\bar{X}$	Any distribution, σ^2 unknown, n large	$\dfrac{\bar{X} - \mu}{\frac{s}{\sqrt{n}}} \sim N(0, 1)$
λ	x	X	Poisson distribution, λ large	$\dfrac{X - \lambda}{\sqrt{\lambda}} \sim N(0, 1)$
p	$\hat{p}$	$\hat{P}$	n large	$\dfrac{\hat{P} - p}{\sqrt{\frac{p(1 - p)}{n}}} \sim N(0, 1)$

In each case the null hypothesis specifies a value for the unknown parameter. Using this value we can determine the probabilities of events of interest (such as the sample mean being greater than 450, or the sample proportion being less than 0.01). This enables us to develop rules for deciding whether or not to accept the null hypothesis.

Example 1

Ten independent observations are to be taken from a $N(\mu, 40)$ distribution. The hypotheses are $H_0: \mu = 20$, $H_1: \mu > 20$. The following procedure has been proposed:
'Reject H_0 (and accept H_1) if $\bar{X} > 23.29$; accept H_0 otherwise'.
Assuming H_0, determine the probabilities of accepting and rejecting H_0 when using this procedure.

———

Assuming that $\mu = 20$ the distribution of $\bar{X}$ is $N(20, \frac{40}{10})$ so that:

$$\frac{\bar{X} - 20}{2} \sim N(0, 1)$$

Thus:

$$P(\bar{X} > 23.29) = P\left(Z > \frac{23.29 - 20}{2}\right) = P(Z > 1.645)$$

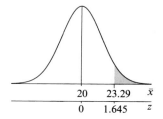

where $Z \sim N(0, 1)$. This tail probability is 5%. Hence, assuming H_0, the probabilities of accepting and rejecting H_0, when using the procedure, are 95% and 5%, respectively.

Note

◆ In English law the prisoner in the dock is considered to be innocent until 'proven' guilty. In the same way, the null hypothesis is accepted until the evidence suggests that, compared to the alternative hypothesis, it is implausible.

15.2 Critical regions and significance levels

The set of values that leads to the rejection of H_0 in favour of H_1 is called the **rejection region** or the **critical region**. The set of values that leads to the acceptance of H_0 is referred to as – wait for it! – the **acceptance region**.

When the population parameter has the value specified by H_0, the probability that H_0 is nevertheless rejected in favour of H_1 is called the **significance level**. Changing the significance level changes the size of the critical region. In Example 1, the significance level was 5% and the critical region was values of $\bar{x}$ greater than 23.29. In this context 23.29 would be described as the **critical value**.

Hypothesis tests in which H_1 involves either a '>' sign (as in Example 1) or a '<' sign are called **one-tailed tests**. The critical regions in these cases involve values in the corresponding tail of the distribution specified by H_0.

Hypothesis tests in which H_1 involves a '≠' sign are called **two-tailed tests**. In these cases the 'critical region' actually consists of two regions – one in each tail of the distribution specified by H_0.

Three examples of critical regions (with probabilities shown shaded) are illustrated below for the case of a single observation on a random variable X having a normal distribution with variance 1. As usual, it is convenient to work with Z, where $Z = \dfrac{X - \mu}{\sigma}$, and so both x and z scales are shown.

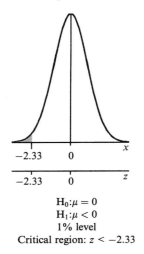

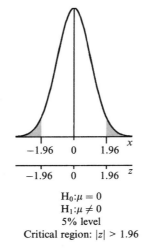

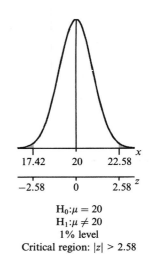

$H_0: \mu = 0$
$H_1: \mu < 0$
1% level
Critical region: $z < -2.33$

$H_0: \mu = 0$
$H_1: \mu \neq 0$
5% level
Critical region: $|z| > 1.96$

$H_0: \mu = 20$
$H_1: \mu \neq 20$
1% level
Critical region: $|z| > 2.58$

In the examples above, the critical values of z are given, in addition to those of x. In the context of hypothesis tests, z is often referred to as the **test statistic**. If the value of z falls in the critical (rejection) region then the result is said to be '**significant**'. If the significance level were $\alpha\%$, then the result would be described as being '**significant at the $\alpha\%$ level**'.

Notes

◆ The most commonly chosen significance levels are 5%, 1% and 0.1%. Note, however, that professional statisticians regard 'significance at the 5% level' as being no more than an indicator that further sampling should take place.
◆ A result that is significant at the $\alpha\%$ level is also significant at the $\beta\%$ level, for all $\beta > \alpha$.
◆ Smaller significance levels result in smaller rejection regions.

15.3 The general test procedure

Following the determination of the underlying probability distribution, the full test procedure is as follows:

1 Write down H_0 and H_1.
2 Determine the appropriate test statistic and the distribution of the corresponding random variable (using the parameter value specified by H_0).
3 Determine the significance level.
4 Determine the acceptance and rejection regions.

   Now collect the data

5 Calculate the value of the test statistic.
6 Determine the outcome of the test.

It is important to decide upon the critical (rejection) region *before* looking at the actual data so as not to be accidentally biased. We might otherwise have carefully selected our region so as to get a 'significant' result! This would be cheating!

15.4 Test for mean, known variance, normal distribution or large sample

Evidence concerning the value of the population mean is provided by the sample mean. If the population variance is known to be σ^2, and the null hypothesis specifies a mean μ, then, by the Central Limit Theorem (see Section 12.10, p. 327), for a large sample of size n, the distribution of $\bar{X}$ is approximately:

$$N\left(\mu, \frac{\sigma^2}{n}\right)$$

If the individual observations are themselves normally distributed, then this result is exact and n need not be large.

▼ ▼

Example 2

A random sample of 36 observations is to be taken from a distribution with variance 100. In the past the distribution has had a mean of 83.0, but it is believed that recently the mean may have changed.

(i) Using a 5% significance level, determine an appropriate test of the null hypothesis, H_0, that the mean is 83.0.

When the sample is actually taken it is found to have a mean of 86.2. Does this provide significant evidence against H_0?

(ii) Suppose it is known that, if the population mean has changed, then it can only have increased.

How would this knowledge affect the conclusions?

———

(i) We will go through the test procedure one stage at a time.

  1 *Write down H_0 and H_1*

  There is no suggestion in the initial question that any change can only be in one direction. The test is therefore two-tailed:

  H_0: $\mu = 83$

  H_1: $\mu \neq 83$

  2 *Determine the appropriate test statistic and the distribution of the corresponding random variable (using the parameter value specified by H_0).*

  The sample size is sufficiently large for us to assume that the distribution of $\bar{X}$ is approximately normal. Since $\sigma^2 = 100$ and $n = 36$, the appropriate test statistic is:

  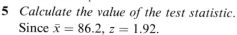

  $$z = \frac{\bar{x} - 83.0}{\sqrt{\dfrac{100}{36}}}$$

  Assuming H_0, z is an observation from a standard normal distribution.

  3 *Determine the significance level.*

  The question specifies 5%.

  4 *Determine the acceptance and rejection regions.*

  The test is two-tailed. Since $P(Z > 1.96) = 0.025$, and $P(Z < -1.96) = 0.025$, an appropriate procedure is to accept H_0 if z lies in the interval $(-1.96, 1.96)$ and otherwise to reject H_0 in favour of H_1.

  5 *Calculate the value of the test statistic.*

  Since $\bar{x} = 86.2$, $z = 1.92$.

  6 *Determine the outcome of the test.*

  Since z lies in the interval $(-1.96, 1.96)$, we accept H_0. In other words there is no significant evidence, at the 5% level, that the mean has changed from its previous value of 83.0. Note that this does *not* imply that the mean is unchanged; simply that the mean of our particular sample did not happen to fall in the rejection region.

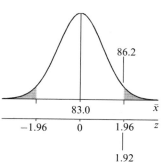

(ii) If it is known that the population mean cannot have decreased then we will only be persuaded to reject H_0 if $\bar{x}$ is unusually large. The test is now one-tailed with H_1: $\mu > 83.0$. Since $P(Z > 1.645) = 0.05$, an appropriate procedure is now to reject H_0 in favour of H_1 if z is greater than 1.645.

Since 1.92 *is* greater than 1.645, we reject the null hypothesis and accept the alternative hypothesis. In other words, we now have significant evidence, at the 5% level, that the population mean has increased from its previous value.

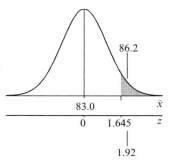

Exercises 15a

1 Jars of honey are filled by a machine. It has been found that the quantity of honey in a jar has mean 460.3 g, with standard deviation 3.2 g. It is believed that the machine controls have been altered in such a way that, although the standard deviation is unaltered, the mean quantity may have changed. A random sample of 60 jars is taken and the mean quantity of honey per jar is found to be 461.2 g. State suitable null and alternative hypotheses, and carry out a test using a 5% level of significance.

2 Observations of the time taken to test an electrical circuit board show that it has mean 5.82 minutes with standard deviation 0.63 minutes. As a result of the introduction of an incentive scheme, it is believed that the inspectors may be carrying out the test more quickly. It is found that, for a random sample of 150 tests, the mean time taken is 5.68 minutes.
State suitable null and alternative hypotheses. Assuming that the population variance remains unchanged, carry out a test at the 5% significance level.

3 A lightbulb manufacturer has established that the life of a bulb has mean 95.2 days with standard deviation 10.4 days. Following a change in the manufacturing process which is intended to increase the life of a bulb, a random sample of 96 bulbs has mean life 96.6 days.
State suitable hypotheses.
Assuming that the population standard deviation is unchanged, test whether there is significant evidence, at the 1% level, of an increase in life.

4 The length of string in the balls of string made by a particular manufacturer has mean μ m and variance 27.4 m². The manufacturer claims that $\mu = 300$. A random sample of 100 balls of string is taken and the sample mean is found to be 299.2 m. Test whether this provides significant evidence, at the 3% level, that the manufacturer's claim overstates the value of μ.   [UCLES(P)]

5 Climbing rope produced by a manufacturer is known to be such that one-metre lengths have breaking strengths that are normally distributed with mean 170.2 kg and standard deviation 10.5 kg.
A new component material is added to the ropes being produced. The manufacturer believes that this will increase the mean breaking strength without changing the standard deviation. A random sample of 50 one-metre lengths of the new rope is found to have a mean breaking strength of 172.4 kg. Perform a significance test at the 5% level to decide whether this result provides sufficient evidence to confirm that the mean breaking strength is increased. State clearly the null and alternative hypotheses which you are using.   [ULSEB(P)]

6 The distance driven by a long distance lorry driver in a week is a normally distributed variable having mean 1130 km and standard deviation 106 km.
New driving regulations are introduced and, in the first 20 weeks after their introduction, he drives a total of 21900 km. Assuming that the standard deviation of the weekly distances he drives is unchanged, test, at the 10% level of significance, whether his mean weekly driving distance has been reduced. State clearly your null and alternative hypotheses.   [ULSEB(P)]

7 In a large population of chickens, the distribution of the mass of a chicken has mean μ kg and standard deviation σ kg. A random sample of 100 chickens is taken from the population. The mean mass for the sample is $\bar{X}$ kg. State the approximate distribution of $\bar{X}$, giving its mean and standard deviation.
The sample values are summarised by $\Sigma x = 189.1$, where x kg is the mass of a chicken. Given that, in fact, $\sigma = 0.71$, test, at the 1% level of significance, the null hypothesis $\mu = 1.75$ against the alternative hypothesis $\mu > 1.75$, stating whether you are using a one-tail or a two-tail test and stating your conclusion clearly.
[UCLES(P)]

8 A normal distribution has unknown mean μ and known variance σ^2. A random sample of n observations from the distribution has sample mean $\bar{x}$. The null hypothesis $\mu = \mu_0$ is being tested. Find, in terms of μ_0, σ and n, the set of values of $\bar{x}$ for which $\mu = \mu_0$ is rejected in favour of $\mu \neq \mu_0$ at the 1% level of significance. Find also, in terms of $\bar{x}$, σ and n, the set of values of μ_0 for which the hypothesis $\mu = \mu_0$ is rejected in favour of $\mu < \mu_0$ at the 5% level of significance.
[UCLES(P)]

9 A fruit grower uses a machine to sort apples into various grades. Grade C apples have weights uniformly distributed in the interval 100 to 110 grams. Find the variance of the weight of a grade C apple.
Ten randomly chosen grade C apples are packed in a bag. Using the central limit theorem, find an approximate value for the probability that the weight of the ten apples in the bag exceeds 1030 grams.
The grower suspects that the machine is not working correctly and that the mean weight, μ grams, of a grade C apple may be less than 105 grams. Devise a test, at the 10% level of significance, based on the weight of the apples in five randomly chosen bags, each containing ten apples, of the null hypothesis $\mu = 105$, with alternative hypothesis $\mu < 105$.
[UCLES]

10 Every day Wombles collect litter from Wimbledon Common. They take it home, weigh it (in Womblegrams) and record the daily total. The recorded daily totals for a randomly chosen week during the last year were

173, 149, 181, 151, 178, 185, 194.

Assuming that these figures are independent observations from the population distribution of daily totals, obtain an unbiased estimate of the population mean and show that the unbiased estimate of the population variance is 289.
A Scottish relation, MacWomble, claims that they will find more litter if they have porridge for breakfast. During the first week that they have porridge they collect a daily average of 180.0 Womblegrams of litter. Assuming a normal distribution, with variance 289, test whether this week's daily average is significantly greater, at the 5% level, than that of the week whose daily results are given in the first paragraph.
[UCLES]

11 'Kruncho' biscuits have weights that are normally distributed with mean 10 g and standard deviation 1.5 g. If the biscuits are sold in packets of 16, what distribution do the weights of randomly chosen packets follow? Following maintenance adjustments to the moulding equipment (that are not thought to affect the standard deviation of the biscuit weights) an inspector finds that the average weight of a random sample of 25 packets is 156.9 g. Examine whether there is significant evidence that the adjustments have affected the mean weight of a biscuit.
If the inspector were to weigh a sample of 100 packets, determine over what range of average weights he should conclude that the adjustments have had a significant effect.
[SMP]

12 The variables $X_1, X_2, \ldots, X_{12}$ are independent with common probability density
$$f(x) = \begin{cases} 1 & \text{for } 0 \leqslant x \leqslant 1, \\ 0 & \text{otherwise.} \end{cases}$$
Give the mean and variance of X_1 and deduce the mean and variance of the variable $Y = X_1 + X_2 + \cdots + X_{12}$. What is the approximate distribution of Y?
As a check on the random number generator of a microcomputer the following sample of ten values of Y was obtained:

4.85, 5.11, 8.06, 4.20, 6.04,
4.82, 6.28, 5.68, 5.49, 5.58.

Use a test based on the normal distribution to determine whether the mean of these values differs significantly from the expected value.
[SMP]

13 It is known that lengths of steel wire of a certain gauge have breaking strengths that are normally distributed with standard deviation 8.5 newtons (N). A customer suspected that the mean breaking strength of such lengths was lower than the mean of 80 N specified by the manufacturer. Consequently the customer tested the breaking strength, x N, of each of a random sample of 8 lengths of wire and obtained the following results.

$$80.7, 80.2, 68.2, 73.1, 70.4, 87.1, 62.2, 73.3.$$

Carry out an appropriate test to decide, at the 5% significance level, whether the customer has sufficient evidence to justify the suspicion that the wire was not up to specification.   [JMB(P)]

15.5 Identifying the two hypotheses

It is often easy to identify a question on hypothesis tests, because the word 'test' appears in the question! The significance level is also usually stated. However, it can sometimes be difficult to identify the two hypotheses.

The null hypothesis

This states that a parameter has some precise value:

1 The value that occurred in the past.
2 The value claimed by some person.
3 The (target) value that is supposed to occur.

Sometimes the null hypothesis may not appear to refer to a precise value:

The mean breaking strengths of types of climbing rope have never exceeded 200 kg, and have sometimes been considerably less. It is claimed that a new rope brought on to the market has a breaking strength in excess of this figure. A random sample of 12 pieces of the new rope are tested . . .

Here it appears that the hypotheses are:

$H_0: \mu \leqslant 200$ kg
$H_1: \mu > 200$ kg

In order to see how to proceed, consider two specific null hypotheses such as $H_0': \mu = 200$ kg and $H_0'': \mu = 190$ kg. Suppose we use H_0' and suppose that the outcome of the test is that H_0' is rejected in favour of H_1. Can we say what would have happened if we had used H_0''? The answer is that it too would have been rejected – if the mean of the sample values is so large that $\mu = 200$ kg is not accepted, then $(\bar{x} - 200)$ must be unacceptably large. Since $(\bar{x} - 190) > (\bar{x} - 200)$ this too must be unacceptably large. The same argument would apply for any value of μ less than 200 kg. Hence we can cover all the cases where μ is less than 200 kg by using:

$H_0: \mu = 200$ kg
$H_1: \mu > 200$ kg

The alternative hypothesis

The alternative hypothesis involves the use of one of the signs $>$, $<$ or $\neq$. A decision has to be made as to which is appropriate. Generally, exam questions attempt to signal which sign is to be used by means of suitable phrases:

$\neq$ 'change', 'different', 'affected'
$>$ or $<$ 'less than', 'better', 'increased', 'overweight'

In real life the choice is not usually so clear cut! Suppose, for example, that we have a situation such as the following:

> The mean breaking strength of a type of climbing rope is 200 kg. Scientists make an adjustment to the method of construction which, they claim, will result in an increase in the breaking strength. A random sample of 12 pieces of the new rope are tested ...

This appears very straightforward. We would use the hypotheses:

H_0: $\mu = 200$ kg

H_1: $\mu > 200$ kg

Suppose now that the 12 pieces of new rope have the following breaking strengths:

187, 196, 193, 187, 194, 193, 197, 194, 191, 195, 194, 199

We evidently do not reject H_0 in favour of H_1 – but would we really want to accept H_0? The new rope appears to have a mean breaking strength of about 193 or 194, and not 200. Some statisticians argue that, because of this type of situation, one-tailed tests should never be used. However, in the context of exam questions they certainly *can* be used.

15.6 Test for mean, large sample, variance unknown

The unbiased estimate of the population variance is given by:

$$s^2 = \frac{1}{n-1}\left\{\Sigma x^2 - \frac{(\Sigma x)^2}{n}\right\}$$

If the sample size is large then this should be a reasonably accurate estimate of σ^2. For such large samples the Central Limit Theorem will also apply and hence, assuming the population mean is μ as specified by H_0, the distribution of $\bar{X}$ is approximately:

$$N\left(\mu, \frac{s^2}{n}\right)$$

The approximation improves as n increases, but should not be used for cases where $n < 30$.

▼   ▼

Example 3

In an experiment on people's perception, a class of 100 students were given a piece of paper which was blank except for a line 120 mm long. The students were asked to judge by eye the centre point of the line, and to mark it. The students then measured the distance, x, between the left-hand end of the line and their mark. Working with $y = x - 60$, the results are summarised by $\Sigma y = -143.5$, $\Sigma y^2 = 1204.00$.

Determine whether there is significant evidence, at the 1% level, of any overall bias in the students' perception of the centre of the lines.

————————

It is simplest to work with $Y = X - 60$.

1   *Write down H_0 and H_1.*

The test is two-tailed since there is no suggestion in the question that any bias will necessarily be to the left. With the mean of Y denoted by μ, the hypotheses are therefore:

H_0: $\mu = 0$

H_1: $\mu \neq 0$

2 *Determine the appropriate test statistic and the distribution of the corresponding random variable (using the parameter value specified by H_0).*
Since σ^2 is unknown, but n is large (100), we use:

$$s^2 = \frac{1204.00 - \dfrac{(-143.5)^2}{100}}{99} = 10.0816$$

The test statistic is therefore:

$$z = \frac{\bar{y} - 0}{\sqrt{\dfrac{10.0816}{100}}}$$

which, assuming H_0, will be an observation from an approximate standard normal distribution.

3 *Determine the significance level.*
The question specifies 1%.

4 *Determine the acceptance and rejection regions.*
The test is two-tailed. Since $P(Z > 2.576) = 0.005$, and $P(Z < -2.576) = 0.005$, an appropriate procedure is to accept H_0 if z lies in the interval $(-2.576, 2.576)$ and otherwise to reject H_0 in favour of H_1.

5 *Calculate the value of the test statistic.*
Since $\bar{y} = -1.435$, $z = -4.52$.

6 *Determine the outcome of the test.*
Since $z < -2.576$, we reject H_0 and accept H_1. There is significant evidence, at the 1% level, that the students' results are biased. Indeed, the result would also have been deemed significant at the 0.001% level!

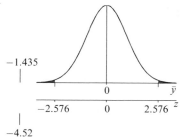

Practical ———————————————————————

Why not try out the experiment in Example 3 in class? In order to have a sufficiently large sample size it may be necessary for everyone to divide two lines. A good idea is to make the line 5 inches long and then to measure to the marked 'centre' point in millimetres. Changing the unit of measurement helps to avoid 'accidental' cheating in which the recorded answer is miraculously correct!

If time allows, there is much scope for experimentation. For example, does the length of the line affect accuracy?
Does the angle of inclination of the line make a difference?

Exercises 15b ———————————————————————

1 The mean IQ score is adjusted to be 100 for each age group of the population. A random sample of 3-year-old children is given vitamin supplements for five years. At the end of the period the 180 children have mean IQ score 102.4. The value of s^2 is 219.4.
Test whether there is significant evidence at the 1% level to support the theory that vitamin supplements increase IQ scores.

2 An inspector wishes to determine whether eggs sold as Size 1 have mean weight 70.0 g. She weighs a sample of 200 eggs and her results are summarised by $\Sigma x = 13\,824$, $\Sigma x^2 = 957\,320$, where x is the weight of an egg in grams.
Test whether there is significant evidence, at the 1% level, that the mean weight is not 70.0 g.

3 Rumour has it that the average length of a leading article in the 'Daily Intellectual' is 960 words. As part of a project, a student counts the number of words in each of 55 randomly chosen leading articles from the paper. His results give $\Sigma x = 51\,452$, $\Sigma x^2 = 49\,146\,729$. Test, at the 10% significance level, the truth of the rumour.

4 A teacher notes the time that she takes to drive to school. She finds that, over a long period, the mean time is 24.5 minutes. After a new bypass is opened, she notes the time on 72 randomly chosen journeys to school. Her results are summarised by $\Sigma(x - 20) = 215$, $\Sigma(x - 20)^2 = 3234$, where x minutes is the time for a journey.
Using a 5% significance level, test whether the journey now takes less time.

5 A supermarket manager investigated the lengths of time that customers spent shopping in the store. The time, x minutes, spent by each

of a random sample of 150 customers was measured and it was found that $\Sigma x = 2871$, $\Sigma x^2 = 60\,029$. Test, at the 5% level of significance, the hypothesis that the mean time spent shopping by customers is 20 minutes, against the alternative that it is less than this.
[UCLES(P)]

6 An electronic device is advertised as being able to retain information stored in it "for 70 to 90 hours" after power has been switched off. In experiments carried out to test this claim, the retention time in hours, X, was measured on 250 occasions, and the data obtained is summarised by $\Sigma(x - 76) = 638$ and $\Sigma(x - 76)^2 = 26\,132$. The population mean and variance of X are denoted by μ and σ^2 respectively.
(i) Show that, correct to one decimal place, an unbiased estimate of σ^2 is 97.5.
(ii) Test the hypothesis that $\mu = 80$ against the alternative hypothesis that $\mu < 80$, using a 5% significance level.
[UCLES(P)]

15.7 Test for large Poisson mean

If X has a Poisson distribution with a large mean, λ, then the distribution of X is well approximated by:

$$N(\lambda, \lambda)$$

providing a continuity correction is used (see Section 12.12, p. 343).

There is no need to consider the sample size. If there are n observations from a Poisson distribution with hypothesised mean λ, then their sum may be considered as a single observation from a Poisson distribution with mean $n\lambda$. This is because the sum of independent Poisson random variables is a Poisson random variable (see Section 10.7 p. 250).

▼ ▼

Example 4

In a particular river a certain micro-organism occurs at an average rate of 10 per millilitre. A random sample of 0.5 litres of water is taken from a nearby stream and is found to contain 3478 micro-organisms. Does this provide significant evidence, at the 5% level, of a difference in the incidence of the micro-organisms between the stream and the river?

———

1 *Write down H_0 and H_1.*
If the incidence in the stream were the same as that in the river, then 0.5 litres (i.e. 500 millilitres) of stream water would contain an average of $10 \times 500 = 5000$ micro-organisms. The question refers to a 'difference' and there is no implication that a low count was

anticipated when sampling the stream; we can take the alternative hypothesis to be a two-sided one.

$$H_0: \lambda = 5000$$

$$H_1: \lambda \neq 5000$$

2 *Determine the appropriate test statistic and the distribution of the corresponding random variable (using the parameter value specified by H_0).*

We assume that the micro-organisms are randomly distributed in the stream water, so that a Poisson distribution is appropriate. The single count, x, is therefore an observation from a Poisson distribution with mean 5000. The test statistic is therefore:

$$z = \frac{x - 5000}{\sqrt{5000}}$$

When the population mean is indeed 5000, z will be an observation from an approximate standard normal distribution.

The approximation is improved by introducing a continuity correction which would reduce the magnitude of the numerator in the expression for z by 0.5.

3 *Determine the significance level.*

The question prescribes a significance level of 5%.

4 *Determine the acceptance and rejection regions.*

The test is two-tailed. Since $P(Z > 1.645) = 0.025$, and $P(Z < -1.645) = 0.025$, an appropriate procedure is to accept H_0 if z lies in the interval $(-1.645, 1.645)$ and otherwise to reject H_0 in favour of H_1.

5 *Calculate the value of the test statistic.*

Using $x = 3478$ and introducing the continuity correction we calculate z:

$$z = \frac{(3478 + 0.5) - 5000}{\sqrt{5000}} = -21.52$$

6 *Determine the outcome of the test.*

Since -21.5 is considerably(!) less than -1.645, we need have no hesitation in rejecting H_0 at the 5% significance level. There can be no real doubt that there is a difference in the incidence of micro-organisms between the stream and the river.

Exercises 15c

1 Rolls of plastic sheeting from a given manufacturer have been established to have minor faults at an average rate of 0.32 per metre. A 100-metre roll is obtained from a second manufacturer and is found to have 27 minor faults.

Is there significant evidence, at the 10% level, that the second manufacturer's plastic sheeting has fewer faults per metre than the first?

2 A traffic survey shows that, between 9 a.m. and 10 a.m., cars pass a particular census point at an average rate of 4.5 per minute. After the opening of a supermarket in the vicinity, the total number of cars passing the census point, between 9 a.m. and 10 a.m. on 5 days, is found to be 1258.

Test, at the 1% level of significance, whether there is evidence of a change in the rate at which cars pass the census point.

3 A rail company claims that 4.3% of its trains are late. A Passenger Association believes this to be an under-estimate, and carries out a check on a random sample of 500 trains, finding that 30 trains are late.

Test the Association's belief, using a 5% significance level.

4 At a small telephone exchange, the number of calls arriving in a period of t minutes has a Poisson distribution with mean λt, where λ is an unknown constant. Use a 5 per cent significance level to test the null hypothesis $H_0: \lambda = 1$ against the alternative hypothesis $H_1: \lambda > 1$, when 74 calls arrive in 1 hour.

[JMB(P)]

15.8 Test for proportion, large sample size

With a sample of size n that contains r 'successes', evidence concerning the population success probability, p, is provided by the sample proportion $\hat{p}$, defined by $\hat{p} = \dfrac{r}{n}$, with the corresponding random variable being denoted by $\hat{P}$. When n is large, the normal approximation to the binomial distribution is valid. Writing $q = 1 - p$, the distribution of $\hat{P}$ is approximately:

$$N\left(p, \frac{pq}{n}\right)$$

(see Section 14.4, p. 381). The approximation is improved by the use of a continuity correction.

Example 5

A golf professional sells wooden tees. The type that he usually sells are very brittle, and 25% break on the first occasion that they are used. The golfers are not very pleased about this, so the golf professional buys a batch of 'Longlast' tees (which are supposed to last longer!). The professional chooses a random sample of 100 of these tees and tries them out. Only 18 break on the first occasion that they are used.

Does this provide significant evidence, at the 1% level, that the proportion of 'Longlast' tees that break on the first occasion they are used is less than 25%?

In this case a 'success' is a breakage!

1 *Write down H_0 and H_1.*

$H_0: p = 0.25$
$H_1: p < 0.25$

2 *Determine the appropriate test statistic and the distribution of the corresponding random variable (using the parameter value specified by H_0).*

Since $n = 100$ and H_0 specifies that $p = 0.25$ and $q = 0.75$, the test statistic is:

$$z = \frac{\hat{p} - 0.25}{\sqrt{\dfrac{0.25 \times 0.75}{100}}}.$$

which is an observation from an approximate standard normal distribution.

The approximation is improved by reducing the absolute magnitude of the numerator of z by $\dfrac{1}{2n}$ (which is the continuity correction).

3   *Determine the significance level.*
   The question prescribes a significance level of 1%.

4   *Determine the acceptance and rejection regions.*
   The test is one-tailed. Since $P(Z > 2.326) = 0.01$,
   $P(Z < -2.326) = 0.01$, and an appropriate (approximate) test procedure is therefore to accept H_0 if $z > -2.326$ and otherwise to reject H_0 in favour of H_1.

5   *Calculate the value of the test statistic.*
   Since $\hat{p} = \frac{18}{100}$, the value of $\hat{p} - p$ is $0.18 - 0.25 = -0.07$. The continuity correction, $\dfrac{1}{2n}$, is equal to 0.005, and hence z is given by:

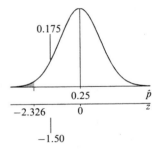

$$z = \frac{-0.065}{\sqrt{\dfrac{0.25 \times 0.75}{100}}} = -1.50$$

6   *Determine the outcome of the test.*
   Since -1.50 is greater than -2.326 it lies in the acceptance region. Therefore, at the (approximate) 1% level, the sample result does not give significant evidence that the proportion of 'Longlast' tees that break when first used is less than 25%.

▲ ── ▲

Exercises 15d ──

1 A coin is thrown 500 times and 267 heads are obtained.
   Test whether the coin is unbiased, using a 10% significance level.

2 A seed company sells pansy seeds in mixed packets and claims that at least 20% of the resulting plants will have red flowers. A packet of seeds is sown by a gardener who finds that only 11 out of 82 plants have red flowers.
   Test the seed company's claim, using a 2.5% significance level.

3 The 'Daily Intellectual' claims that 60% of its readers are car owners. In a random sample of 312 readers there are 208 car owners.
   Test whether there is significant evidence, at the 2% level, to support the claim.

4 A survey in a university library reveals that 12% of returned books are overdue. After a big increase in fines, a random sample of 80 returned books reveals that only 6 are overdue.
   Test, at the 10% level of significance, whether the proportion of overdue books has decreased.

5 When a "Thumbnail" drawing pin is dropped on to the floor, the probability that it lands "point up" is p. A teacher drops a "Thumbnail" drawing pin 900 times and observes that it lands "point up" 315 times.
   Test, at the 1% level, the hypothesis $p = 0.4$ against the alternative $p < 0.4$.

[UCLES(P)]

6 In a public opinion poll, 1000 randomly chosen electors were asked whether they would vote for the "Purple Party" at the next election and 357 replied "Yes". The leader of the "Purple Party" believes that the true proportion is 0.4. Test, at the 8% level, whether he is overestimating his support. [UCLES(P)]

7 The owner of a large apple orchard states that 10% of the apples on the trees in his orchard have been attacked by birds. A random sample of 2500 apples is picked and 274 apples are found to have been attacked by birds. Test, at the 8% significance level, whether there is significant evidence that the owner has understated the proportion of the apples on the trees in his orchard that have been attacked. State your hypotheses clearly. [UCLES(P)]

8 A drug company tested a new pain-relieving drug on a random sample of 100 headache suffers. Of these, 75% said that their headache was relieved by the drug. With the currently marketed drug, 65% of users say that their headache is relieved by it. Test, at the 4% level, whether the new drug will have a greater proportion of satisfied users. [UCLES(P)]

9 A questionnaire was sent to a large number of people, asking for their opinions about a proposal to alter an examination syllabus. Of the 180 replies received, 134 were in favour of the proposal. Assuming that the people replying were a random sample from the population, test, at the 5% level, the hypothesis that the population proportion in favour of the proposal is 0.7 against the alternative that it is more than 0.7. [UCLES(P)]

10 A schoolmaster wishes to estimate how many of the 36 pupils in his class smoke at least one cigarette every day. Since they may not answer truthfully if he asks this question directly, each pupil is asked to carry out the following procedure:
Toss a coin, concealing the outcome from all but yourself. If you obtain a head then answer "yes". If you obtain a tail then answer "yes" only if you smoke at least one cigarette every day, otherwise answer "no".
Using this procedure, the pupils may be assumed to answer truthfully, and the number who answer "yes" is 24.
(i) Given that the probability of a head is $\frac{1}{2}$, estimate the proportion of pupils in the class who smoke at least one cigarette every day.
(ii) Using a normal distribution, test, at the 4% significance level, the null hypothesis that there is no pupil in the class who smokes at least one cigarette every day, stating clearly your alternative hypothesis. [UCLES]

15.9 Test for mean, small sample, variance unknown

When the sample size is small, we have to use the *t*-distribution rather than the normal distribution (see Sections 14.6 and 14.7, pp. 387–93). The test statistic, *t*, is defined by:

$$t = \frac{\bar{x} - \mu}{\frac{s}{\sqrt{n}}}$$

where μ is the population mean specified by the null hypothesis. The distribution of the corresponding random variable, T, is t_{n-1}.

Notes
- The distribution of T is only exactly a *t*-distribution if the population is normal.
- Strictly, the *t*-distribution should be used in preference to the normal whenever s is used in place of σ, and not only when n is small.

Example 6

Bottles of wine are supposed to contain 75 cl of wine. An inspector takes a random sample of six bottles of wine and determines the volumes of their contents, correct to the nearest half millilitre. Her results are:

  747.0, 751.5, 752.0, 747.5, 748.0, 748.0

Determine whether these results provide significant evidence, at the 5% level, that the population mean is less than 75 cl.

──────────

It is simplest to work in millilitres. The target quantity, 75 cl, is $\frac{75}{100}$ of a litre, which is the same as 750 millilitres.

1 *Write down H_0 and H_1.*
   The test is one-tailed. The hypotheses are:

   H_0: $\mu = 750$
   H_1: $\mu < 750$

2 *Determine the appropriate test statistic and the distribution of the corresponding random variable (using the parameter value specified by H_0).*
   Since σ^2 is unknown, we must calculate s^2. The numbers are simpler if we work with y, given by $y = x - 750$. This transformation does not alter the variability of the observations, which become:

   $-3.0, 1.5, 2.0, -2.5, -2.0, -2.0$

   These are summarised by $\Sigma y = -6.0$ and $\Sigma y^2 = 29.50$, so that:

   $$s^2 = \frac{1}{5}\left\{29.50 - \frac{(-6.0)^2}{6}\right\} = 4.70$$

   The test statistic is therefore:

   $$t = \frac{\bar{x} - 750}{\sqrt{\dfrac{4.70}{6}}}$$

   which, assuming H_0, is an observation from a t_5-distribution.

3 *Determine the significance level.*
   The question specifies 5%.

4 *Determine the acceptance and rejection regions.*
   The test is one-tailed. The upper 5% point of a t_5-distribution is 2.015, and hence, by symmetry, the lower 5% point is -2.015. An appropriate procedure is therefore to accept H_0 if t is greater than -2.015 and otherwise to reject H_0 in favour of H_1.

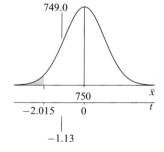

5 *Calculate the value of the test statistic.*
   Since $\bar{x} = 749.0$, $t = -1.13$.

6 *Determine the outcome of the test.*
   Since $t > -2.015$, we accept H_0: there is no significant evidence, at the 5% level, that the population mean is less than 75 cl.

Exercises 15e

1 A manufacturer claims that the mean lifetime of the light bulbs he produces is at least 1200 hours. A random sample of 10 bulbs is taken and the lifetimes are observed. The results are summarised by $\Sigma(x - 1000) = 1890.0$ and $\Sigma(x - 1000)^2 = 362\,050.2$, where x is measured in hours. Assuming the lifetimes of the bulbs to be normally distributed, and using a 5% significance level, test whether there are grounds to dispute this claim.   [UCLES(P)]

2 A national company owns a chain of laboratories at which routine chemical tests are carried out. In order to ensure the accuracy of the analyses a sample with a known (but undisclosed) sulphur content of 10.57 grams/litre is sent to each laboratory for testing by its senior analyst. The results from 10 such laboratories are shown below.

Lab	1	2	3	4
Result (x)	10.37	10.49	10.51	10.39

Lab	5	6	7	8
Result (x)	10.56	10.56	10.70	10.46

Lab	9	10
Result (x)	10.44	10.59

These results are summarised by
$\Sigma(x - 10.5) = 0.07$,
$\Sigma(x - 10.5)^2 = 0.0897$.
Assuming that the errors of analysis are normally distributed, test, at the 5% significance level, whether there is any indication of an overall bias in these results.   [UCLES(P)]

3 A sample of eight containers is selected at random from a large batch. The containers have powder contents with masses x g,
1998.5, 2000.4, 1999.9, 2005.8,
2011.5, 2007.6, 2001.3, 2002.4,
which are summarised by
$\Sigma(x - 2000) = 27.4$ and
$\Sigma(x - 2000)^2 = 233.52$.
Assuming a normal distribution for the masses of the contents, show that there is significant evidence, at the 5% level, that the mean mass of the contents of the containers in this batch is greater than 2000 g.   [UCLES(P)]

4 After a nuclear accident, government scientists measured radiation levels at 20 randomly chosen sites in a small area. The measuring instrument used is calibrated so as to measure the ratio of present radiation to the previous known average radiation in that small area. The measurements are summarised by $\Sigma x_i = 22.8, \Sigma x_i^2 = 27.55$. Making suitable assumptions, test, at the 5% level, the hypothesis that there has been no increase in the radiation level.   [UCLES(P)]

5 A random sample of size n is taken from a normal distribution with mean μ and variance σ^2. The statistic z is defined by $z = \dfrac{\bar{x} - \mu}{\frac{\sigma}{\sqrt{n}}}$, where $\bar{x}$ is the sample mean. A statistician who wants to use the statistic z to test a hypothesis about μ does not know the population variance σ^2 and so replaces σ in the statistic z by an estimate of σ.
(i)   State what estimate of σ the statistician should use.
(ii) Name the distribution of z and the distribution of the statistic which results from z when σ is replaced by the estimate of σ.
(iii) Sketch these two distributions on the same diagram.   [UCLES(P)]

6 A marmalade manufacturer produces thousands of jars of marmalade each week. The mass of marmalade in a jar is an observation from a normal distribution having mean 455 g and standard deviation 0.8 g. Determine the probability that a randomly chosen jar contains less than 454 g.
Following a slight adjustment to the filling machine, a random sample of 10 jars is found to contain the following masses (in g) of marmalade:
454.8, 453.8, 455.0, 454.4, 455.4,
454.4, 454.4, 455.0, 455.0, 453.6.
(i) Assuming that the variance of the distribution is unaltered by the adjustment, test, at the 5% significance level, the hypothesis that there has been no change in the mean of the distribution.
(ii) Assuming that the variance of the distribution may have altered, obtain an unbiased estimate of the new variance and, using this estimate, test, at the 5% significance level, the hypothesis that there has been no change in the mean of the distribution.   [UCLES(P)]

7 A plant breeder A claims that a new variety of fruit bush he has produced gives a higher yield than the variety it will replace. A random sample of ten bushes of the new variety is grown and the yields of the bushes recorded. The old variety had an average yield of 5.2 kg per bush. The yields of the sample of new bushes are summarised by $\Sigma x_i = 61.0, \Sigma x_i^2 = 383.96$. It may be assumed that each item of data is an independent observation from a normal distribution. Test, at the 5% level, breeder A's claim.

[UCLES(P)]

8 You are given a random sample of n observations from a normal distribution and you are required to test the null hypothesis that the normal distribution has mean μ_0. How would you decide whether to use a normal distribution or a t-distribution when carrying out the test?

An owner of a small distillery uses an old machine for dispensing whisky into bottles. The mean volume of whisky dispensed into a bottle can be altered using a setting dial on the machine, but for any setting the volume of whisky dispensed is a normal variable with a standard deviation of 0.85 cl. Bottles of whisky have nominal contents of 75 cl. When the machine is set so that it dispenses a mean volume of 76.5 cl calculate the percentage of bottles that have contents below the nominal volume of 75 cl.

Regulations stipulate that 99% of bottles must contain at least the nominal volume of whisky. If the distillery owner were to adjust the setting dial to comply with the regulations, calculate the lowest mean volume that could be set. Instead of adjusting the setting dial on his old machine the owner decides to invest in a new machine. To check the calibration of the setting dial on the new machine he sets the dial to dispense a mean volume of 76 cl. He then measures the volume, x cl, of whisky in each of a random sample of 18 bottles, obtaining the following values for x.

76.1	75.7	76.8	76.2	75.8	77.1
76.2	75.4	76.9	75.7	76.1	76.7
76.5	75.8	76.3	76.1	75.8	76.4

$$(\Sigma x = 1371.6 \qquad \Sigma x^2 = 104\,519.62)$$

Assuming that, for any given setting, the volumes dispensed by this machine are normally distributed, test, at the 5% significance level, the hypothesis that the calibration on the dial is correct when set to dispense a mean volume of 76 cl.   [JMB]

15.10 The *p*-value approach

Stages 3–6 of our test procedure are:

3 Determine the significance level.
4 Determine the acceptance and rejection regions.
5 Calculate the value of the test statistic.
6 Determine the outcome of the test.

An alternative approach is:

3 Calculate the value of the test statistic.
4 Determine the corresponding tail probability (the *p*-value), using one or two tails depending on the form of H_1.
5 Compare with possible significance levels.
6 Determine the outcome of the test.

An advantage of this approach is that we can highlight an extreme value of z by noting a result as being significant (say) at the 0.001% level. Some results are more significant than others and this approach makes this explicit.

The main disadvantage is the potential for bias through choosing a significance level in order to make the observed result 'significant'.

Another disadvantage arises in connection with the small-sample cases considered later in Sections 16.4 and 16.5 (pp. 436–42). Since Professors of Statistics disagree over how to report such tail probabilities, we will stick with our original procedure!

15.11   Hypothesis tests and confidence intervals

There is a simple rule that usually works in the case of a two-sided alternative hypothesis:

> If a $c\%$ symmetric confidence interval excludes the population value of interest, then the null hypothesis that the population parameter takes this value will be rejected at the $100(1 - c)\%$ level.

For example, if the symmetric 95% confidence interval for a population mean, μ, is (83.0, 85.1), then the null hypothesis that $\mu = 85.2$ will be rejected at the 5% level since the interval excludes 85.2. Indeed *any* hypothesised value for μ that is greater than 85.1, or is less than 83.0 will be rejected at the 5% level. Conversely, the hypothesis that μ takes *any* specific value in the range (83.0, 85.1) will be accepted at the 5% level.

Example 7

A machine cuts wood to form stakes, which are supposed to be 2 metres long. A random sample of 40 stakes is taken, the stakes are accurately measured, and their lengths (x cm) are summarised (using a coding method with reference value 200 cm) by:

$$\Sigma(x - 200) = 41.56, \qquad \Sigma(x - 200)^2 = 107.4673$$

Determine a 95% confidence interval for the mean stake length.
Test, at the 5% significance level, the null hypothesis that the population mean is 2 metres against the alternative that this is not the case.

The unbiased estimate of the population variance is given by:

$$s^2 = \frac{1}{39}\left(107.4673 - \frac{(41.56)^2}{40}\right) = 1.6484$$

The sample size is sufficiently large that the distribution of the sample mean can be taken to be normal (by the Central Limit Theorem). There will be little loss of accuracy in treating $\dfrac{\bar{X} - \mu}{\frac{s}{\sqrt{n}}}$ as having a N(0,1) distribution. The 95% confidence interval is therefore:

$$200 + \frac{41.56}{40} \pm 1.96\sqrt{\frac{1.6484}{40}} = (200.64, 201.44)$$

Since the interval excludes 200.0, the hypothesis that the mean is 2 metres is rejected, at the 5% significance level, in favour of the alternative that this is not the case.

Notes

♦ The rule does not work perfectly in the case of a binomial proportion because the variance used in the calculation of the confidence interval $\left(\dfrac{\hat{p}\hat{q}}{n}\right)$ will usually be slightly different from that used in the context of a hypothesis test $\left(\dfrac{pq}{n}\right)$.

♦ In the case of one-sided alternative hypotheses the link is with the corresponding one-sided confidence interval. Suppose, for example, that H_0 states that $\mu = 15$ with H_1 being that $\mu > 15$. Unusually large values of $\bar{x}$ will lead to rejection of H_0. Since H_1 is concerned with the upper tail, the relevant confidence interval includes all that tail. If, for example, the interval has the form $(15.2, \infty)$, which excludes the hypothesised 15, then H_0 is rejected.

Exercises 15f

1 Jars of honey are filled by a machine. It has been found that the quantity of honey in a jar has mean 460.3 g, with standard deviation 3.2 g. It is believed that the machine controls have been altered in such a way that, although the standard deviation is unaltered, the mean quantity may have changed. A random sample of 60 jars is taken and the mean quantity of honey per jar is found to be 461.2 g. State suitable null and alternative hypotheses, and carry out a test using a 5% level of significance:
  (a) using the *p*-value method,
  (b) by finding an appropriate confidence interval.

2 Observations of the time taken to test an electrical circuit board show that it has mean 5.82 minutes with standard deviation 0.63 minutes. As a result of the introduction of an incentive scheme, it is believed that the inspectors may be carrying out the test more quickly. It is found that, for a random sample of 150 tests, the mean time taken is 5.68 minutes.
State suitable null and alternative hypotheses. Assuming that the population variance remains unchanged, carry out a test at the 5% significance level:
  (a) using the *p*-value method,
  (b) by finding an appropriate confidence interval.

3 A lightbulb manufacturer has established that the life of a bulb has mean 95.2 days with standard deviation 10.4 days. Following a change in the manufacturing process which is intended to increase the life of a bulb, a random sample of 96 bulbs has mean life 96.6 days.

State suitable hypotheses.
Assuming that the population standard deviation is unchanged, test whether there is significant evidence, at the 1% level, of an increase in life:
  (a) using the *p*-value method,
  (b) by finding an appropriate confidence interval.

4 The mean IQ score is adjusted to be 100 for each age group of the population. A random sample of 3-year-old children is given vitamin supplements for five years. At the end of the period the 180 children have mean IQ score 102.4. The sample variance is 219.4.
Test whether there is significant evidence at the 1% level to support the theory that vitamin supplements increase IQ scores:
  (a) using the *p*-value method,
  (b) by finding an appropriate confidence interval.

5 An inspector wishes to determine whether eggs sold as Size 1 have mean weight 70.0 g. She weighs a sample of 200 eggs and her results are summarised by $\Sigma x = 13\,824$, $\Sigma x^2 = 957\,320$, where *x* is the weight of an egg in grams.
Test whether there is significant evidence, at the 1% level, that the mean weight is not 70.0 g:
  (a) using the *p*-value method,
  (b) by finding an appropriate confidence interval.

6 Rumour has it that the average length of a leading article in the 'Daily Intellectual' is 960 words. As part of a project, a student counts the number of words in each of 55 randomly chosen leading articles from the paper. His results give $\Sigma x = 51\,452$, $\Sigma x^2 = 49\,146\,729$.
Test, at the 10% significance level, the truth of the rumour:

(*continued*)

(a) using the *p*-value method,

(b) by finding an appropriate confidence interval.

7 Explain what you understand by the term 'Central Limit Theorem', illustrating your answer with reference to any experiment you may have conducted.

In 1988 a meteorologist recorded the length of time (hours) the sun shone at her work station for each of the 31 days during December. She then calculated the mean daily figure for that month. Her data can be summarised as

$$\sum x_i = 44.48, \qquad \sum x_i^2 = 83.5008,$$

where $i = 1$ to 31 and x_i represents the daily sunshine in hours.

(*a*) Write down a point estimate for the mean daily hours of sunshine. Calculate an unbiased estimate for the variance of the daily sunshine. Hence find the 'standard error' of the mean.

(*b*) Calculate a 95% confidence interval for the expected hours of sunshine for a day in December. In December 1989, the sun shone for a total of 62.62 hours. Is this sufficient evidence to suggest that there was a change in the average daily sunshine? Justify your response. [UODLE]

8 A politician, speaking to a journalist, claims that school leavers in his constituency have, on average, 6 GCSEs. The journalist checks the claim by interviewing a random sample of 100 school leavers. The data he obtains are summarised below; x denotes the number of GCSEs per person.

$$n = 100 \quad \sum x = 431 \quad \sum x^2 = 2578$$

(i) Obtain the mean and standard deviation of the data.

(ii) Construct a 95% confidence interval for the mean number of GCSEs per person.

(iii) Explain, without further calculation, whether or not the politician's claim is consistent with the journalist's findings.

(iv) By considering again the mean and standard deviation of the data as calculated in (i), explain why the number of GCSEs per person seems unlikely to be Normally distributed. Show in a sketch a possible shape for the distribution.

(v) Explain whether lack of Normality does or does not invalidate the confidence interval found in (ii). [MEI]

9 A credit card company is interested in estimating the proportion of card holders who, at some time, have carried a non-zero balance at the end of a month and so have incurred interest charges. A random sample of 400 credit card holders reveals that 168 have, at some time, incurred interest charges. Calculate an approximate 99% confidence interval for the proportion of all credit card holders who have, at some time, incurred interest charges. Hence comment on the claim that this proportion is 0.5. [AEB(P) 90]

10 Packets of baking powder have a nominal weight of 200 g. The distribution of weights is normal and the standard deviation is 7 g. Average quantity legislation states that, if the nominal weight is 200 g

(i) the average weight must be at least 200 g,

(ii) not more than 2.5% of packages may weigh less than 191 g,

(iii) not more than 1 in 1000 packages may weigh less than 182 g.

A random sample of 30 packages had the following weights

218,	207,	214,	189,	211,	206,
203,	217,	183,	186,	219,	213,
207,	214,	203,	204,	195,	197,
213,	212,	188,	221,	217,	184,
186,	216,	198,	211,	216,	200

(a) Calculate a 95% confidence interval for the mean weight.

(b) Find the proportion of packets in the sample weighing less than 191 g and use your result to calculate an approximate 95% confidence interval for the proportion of all packets weighing less than 191 g.

(c) Assuming that the mean is at the lower limit of the interval calculated in (a), what proportion of packets would weigh less than 182 g?

(d) Discuss the suitability of the packets from the point of view of the average quantity system. A simple adjustment will change the mean weight of future packages. Changing the standard deviation is possible, but very expensive. Without carrying out any further calculations, discuss any adjustments you might recommend. [AEB 90]

11 A food processor produces large quantities of jars of jam. In each batch, the gross weight of a jar is known to be normally distributed with standard deviation 7.5 g. (The gross weight is the weight of the jar plus the weight of the jam.) The gross weights, in grams, of a random sample from a particular batch were:

> 514, 485, 501, 486, 502,
> 496, 509, 491, 497, 501,
> 506, 486, 498, 490, 484,
> 494, 501, 506, 490, 487,
> 507, 496, 505, 498, 499.

(a) Estimate the proportion of this batch with gross weight over 500 g. Calculate an approximate 95% confidence interval for this proportion.

(b) Calculate a 90% confidence interval for the mean gross weight of this batch.

The weight of an empty jar is known to be normally distributed with mean 40 g and standard deviation 4.5 g. It is independent of the weight of the jam.

(c) (i) What is the standard deviation of the weight of the jam in a batch of jars?

(ii) Assuming that the mean gross weight is at the upper limit of the confidence interval calculated in (b), calculate limits within which 99% of the weights of the contents would lie.

(d) The jars are claimed to contain 454 g of jam. Comment on this claim as it relates to this batch of jars.

[AEB 94]

Chapter summary

♦   Two hypotheses:
   ● **Null hypothesis**, H_0: Parameter has usual value or target value.
   ● **Alternative hypothesis**, H_1: Parameter has changed, or has increased, or has decreased.

♦   **Critical region or rejection region**: range of values of the test statistic for which H_0 is rejected in favour of H_1. Remainder is acceptance region.

♦   **Significance level**: P(test statistic falls in rejection region given the situation described by H_0).

♦   **The test procedure:**

   1 Write down H_0 and H_1.
   2 Determine the appropriate test statistic and the distribution of the corresponding random variable (using the parameter value specified by H_0).

   3 Determine the significance level.
   4 Determine the acceptance and rejection regions.

   Now collect the data

   5 Calculate the value of the test statistic.
   6 Determine the outcome of the test.

◆   **Test statistic:**

●   **For a mean**, assuming a normal distribution or n large:

$$\frac{\bar{X} - \mu}{\frac{\sigma}{\sqrt{n}}} \sim N(0, 1)$$

Replace σ by s if σ is unknown and n is large.

●   **For a mean**, assuming a normal distribution with σ unknown and n not large:

$$\frac{\bar{X} - \mu}{\frac{\sigma}{\sqrt{n}}} \sim t_{n-1}$$

●   **For a proportion**:

$$\frac{\hat{p} - p}{\sqrt{\frac{p(1 - p)}{n}}} \sim N(0, 1)$$

This requires large n. The accuracy is improved by reducing the absolute value of the numerator by the continuity correction $\frac{1}{2n}$.

●   **For a large Poisson mean**, λ, and a single observation:

$$\frac{X - \lambda}{\sqrt{\lambda}} \sim N(0, 1)$$

16 Hypothesis tests: errors and other problems

To err is human; to forgive, divine

An Essay on Criticism, Alexander Pope

16.1 Type I and Type II errors

The statistician's life is not a happy one! When conducting hypothesis tests there are two types of error that may occur, which are summarised in the table below.

		Our decision	
		We accept H_0	We reject H_0
Reality	H_0 correct	Correct!	TYPE I ERROR
	H_0 incorrect	TYPE II ERROR	Correct!

As the table shows, a **Type I error** is made if a correct null hypothesis is rejected. The probability of this error is under our control since:

$$P(\text{Type I error}) = \text{significance level} \qquad (16.1)$$

Calculation of the probability of a **Type II error** is not so straightforward, since the probability depends on the extent to which H_0 is false. If H_0 is only slightly incorrect then we may not notice that it is wrong and the probability of a Type II error will be large. On the other hand, if H_0 is nothing like correct then the probability of a Type II error will be low.

In a more positive frame of mind, rather than asking about the probability of making an error, we can ask how good a test is at detecting a false null hypothesis. This is known as the **power** of a test. Formally:

$$\text{power} = 1 - P(\text{Type II error}) \qquad (16.2)$$

The general procedure
This closely follows that for the construction of hypothesis tests:

1 Write down the two hypotheses, for example, $H_0: \mu = \mu_0$ and $H_1: \mu > \mu_0$.
2 Determine the appropriate test statistic and the distribution of the corresponding random variable (using the parameter value specified by H_0).
3 Determine the significance level. This is P(Type I error).
4 Determine the acceptance and rejection regions.

Consider now the case when the value of the parameter is not that specified by H_0

5 Determine the distribution of the random variable corresponding to the test statistic given $\mu = \mu_1$, say.
6 Calculate the probability of an outcome falling in the acceptance region (given $\mu = \mu_1$). This is P(Type II error) for the case where $\mu = \mu_1$.

Notes
- As usual, it is sensible to avoid premature rounding during intermediate calculations.
- When calculating the probability of a Type II error for a test concerning a proportion, remember that a change in the value of p will change the value of the quantity pq that occurs in the variance of the test statistic.

▼   ▼

Example 1

A machine is supposed to fill bags with 38 kg of sand. It is known that the quantities in the bags vary and have a standard deviation of 0.5 kg. When a new employee starts using the machine it is standard practice to determine the masses of a random sample of 20 bags taken from the first batch produced by the employee in order to verify that the mean of the machine has been set correctly.

Determine an appropriate test procedure, given that it is desired that the probability of a Type I error should be 4%.

Suppose that an employee has set the machine so that it fills bags with an average of μ kg.

Determine the probability of a Type II error in the cases $\mu = 38.1$ and $\mu = 38.4$.

_____

1 *Write down H_0 and H_1.*

The test is two-tailed with the hypotheses being:

H_0: $\mu = 38.0$

H_1: $\mu \neq 38.0$

2 *Determine the appropriate test statistic and the distribution of the corresponding random variable (using the parameter value specified by H_0).*

The appropriate test is one that uses the sample mean, $\bar{x}$. Assuming H_0, by the Central Limit Theorem the distribution of $\bar{X}$ is approximately:

$$N\left(38.0, \frac{(0.5)^2}{20}\right)$$

so that the appropriate test statistic is:

$$z = \frac{\bar{x} - 38.0}{\sqrt{\dfrac{0.250}{20}}}$$

3 *Determine the significance level. This is P(Type I error).*

This is given as 4% (which makes a change from the more usual 5%).

4 *Determine the acceptance and rejection regions.*

Tables show that the upper 2% point of a standard normal random variable is 2.054. The acceptance region (in terms of z) is therefore $(-2.054, 2.054)$. In order to calculate the probability of a Type II error we will need this as an interval for $\bar{x}$:

$$\left(38.0 - 2.054\sqrt{\frac{0.250}{20}}, \; 38.0 + 2.054\sqrt{\frac{0.250}{20}}\right)$$

which simplifies to (37.7704, 38.2296). This is the acceptance region for $\bar{x}$. Values of $\bar{x}$ outside this interval are in the rejection region.

Consider now the case $\mu = 38.1$

5   *Determine the distribution of the random variable corresponding to the test statistic.*

The distribution of $\bar{X}$ is:

$$N\left(38.1, \frac{0.250}{20}\right)$$

6   *Calculate the probability of an outcome falling in the acceptance region. This is P(Type II error).*

We require $P(37.7704 < \bar{X} < 38.2296)$, given that $\bar{X} \sim N\left(38.1, \frac{0.250}{20}\right)$.

Now:

$$P(\bar{X} < 38.2296) = P\left(Z < \frac{38.2296 - 38.1}{\sqrt{\frac{0.250}{20}}}\right) = P(Z < 1.159)$$

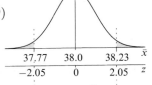

where $Z \sim N(0, 1)$. Similarly:

$$P(\bar{X} < 37.7704) = P\left(Z < \frac{37.7704 - 38.1}{\sqrt{\frac{0.250}{20}}}\right) = P(Z < -2.948)$$

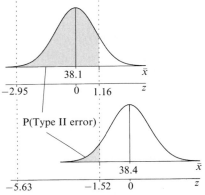

Thus:

$$P(37.7704 < \bar{X} < 38.2296) = \Phi(1.159) - \Phi(-2.948)$$
$$= 0.8767 - (1 - 0.9984)$$
$$= 0.8751$$

When the mean is only slightly too large (38.1, rather than 38.0) the probability of a Type II error is high, being 0.875 (to three decimal places).

Consider now the case $\mu = 38.4$. We need to calculate:

$$\frac{38.2296 - 38.4}{\sqrt{\frac{0.250}{20}}} = -1.524$$

and:

$$\frac{37.7704 - 38.4}{\sqrt{\frac{0.250}{20}}} = -5.631$$

so that now:

$$P(37.7704 < \bar{X} < 38.2296) = \Phi(-1.524) - \Phi(-5.631)$$
$$= (1 - 0.9362) - 0$$
$$= 0.0638$$

When a more substantial shift in the mean is considered, the probability of a Type II error is considerably reduced. When $\mu = 38.4$ the value is 0.064 (to three decimal places).

Example 2

A coin, believed to be fair, is tossed 100 times. The hypothesis that the coin is fair will be accepted if the number of heads obtained lies between 40 and 60, inclusive.
Determine the probability of a Type I error.
Determine also the probability of a Type II error for the case where the probability of a head is 0.6.
State the power of the test in this case.

———

In this question the acceptance region is given, but it is still useful to follow through the general procedure. We will let X be the number of heads obtained and let p be the probability of a head.

1 *Write down H_0 and H_1.*
   The hypotheses are:
   H_0: $p = 0.5$
   H_1: $p \neq 0.5$.

2 *Determine the appropriate test statistic and the distribution of the corresponding random variable (using the parameter value specified by H_0).*
   The test statistic is the number of heads obtained.

3, 4 *Determine the acceptance and rejection regions and the significance level*
   We are given that the acceptance region is $40 \leqslant X \leqslant 60$. We require
   P(Type I error) which is therefore equal to $1 - P(40 \leqslant X \leqslant 60)$.
   Assuming H_0, $X \sim B(100, 0.5)$. Since the number of trials is large we can use the normal approximation, together with continuity corrections, to determine the required probability. The distribution of X is approximated by:

   $$N(50, 100 \times 0.5 \times 0.5) = N(50, 25)$$

   Hence, using continuity corrections:

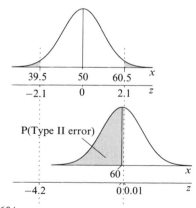

   $$P(\text{Type I error}) \approx 1 - \left\{ \Phi\left(\frac{60.5 - 50}{5}\right) - \Phi\left(\frac{39.5 - 50}{5}\right) \right\}$$

   $$= 1 - \{\Phi(2.1) - \Phi(-2.1)\}$$

   $$= 1 - \{0.9821 - (1 - 0.9821)\}$$

   $$= 0.0358$$

The probability of a Type I error (the significance level) is about 3.6%.

Consider now the case $p = 0.6$.

5 *Determine the distribution of the random variable corresponding to the test statistic.*

   The normal approximation becomes

   $$N(60, 100 \times 0.6 \times 0.4) = N(60, 24)$$

   Note that the variance has slightly altered as a consequence of the change in the value of p.

6   *Calculate the probability of an outcome falling in the acceptance region.*
This is P(Type II error).

The probability of a type II error is given by:

$$P(40 \leqslant X \leqslant 60) \approx \Phi\left(\frac{60.5 - 60}{\sqrt{24}}\right) - \Phi\left(\frac{39.5 - 60}{\sqrt{24}}\right)$$

$$= \Phi(0.102) - \Phi(-4.185)$$

$$= 0.5406 - 0$$

$$= 0.5406$$

The probability of a Type II error is about 54.1%.

The power of the test is $1 - P(\text{Type II error}) = 1 - 0.5406 = 0.460$ (to 3 decimal places).

Example 3

The random variable X has a binomial distribution with unknown p.
Devise a test of H_0: $p = 0.4$ against H_1: $p < 0.4$.
   The test is to be based on 240 observations, and should have a significance level of about 1%.
Determine the approximate probability of a Type II error for your test in the case where $p = 0.3$.

1   *Write down H_0 and H_1.*

H_0: $p = 0.4$
H_1: $p < 0.4$

2   *Determine the appropriate test statistic and the distribution of the corresponding random variable (using the parameter value specified by H_0).*
Assuming H_0, the distribution of X is approximated by:

$N(240 \times 0.4,\ 240 \times 0.4 \times 0.6) = N(96, 57.6)$

3   *Determine the significance level. This is P(Type I error).*
The significance level is to be about 1%. It may not be *exactly* 1% because the binomial distribution is a discrete distribution with probability that comes in 'chunks'.

4   *Determine the acceptance and rejection regions.*
For a standard normal random variable Z, $P(Z < -2.326) = 0.01$. The appropriate critical value for X is therefore:

$96 - (2.326\sqrt{57.6}) = 78.35$

However, the only possible values for X are integers. The resulting procedure is:

Accept H_0 unless the observed value of X is 78 or less, in which case H_0 should be rejected in favour of H_1.

Suppose now that $p = 0.3$

5 *Determine the distribution of the random variable corresponding to the test statistic.*

Using the normal approximation the distribution of X is approximately:

$$N(240 \times 0.3, \ 240 \times 0.3 \times 0.7) = N(72, 50.4)$$

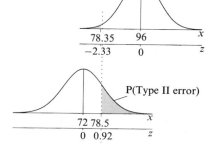

6 *Calculate the probability of an outcome falling in the acceptance region. This is P(Type II error).*

The probability of a Type II error is the probability of observing a value greater than 78. Using a continuity correction, we proceed as follows:

$$P(X > 78) \approx 1 - \Phi\left(\frac{78.5 - 72}{\sqrt{50.4}}\right)$$

$$= 1 - \Phi(0.916)$$

$$= 1 - 0.8201$$

$$= 0.1799$$

The probability of a Type II error, when the value of p is 0.3, is 0.180 (to three decimal places).

▲ ▲

Exercises 16a

1 It is given that $X \sim N(\mu, 16)$. It is desired to test the null hypothesis $\mu = 12$ against the alternative hypothesis $\mu > 12$, with the probability of a Type I error being 1%. A random sample of 15 observations of X is taken and the sample mean $\bar{X}$ is taken to be the test statistic.

  (i) Find the acceptance and rejection regions.
 (ii) For the case $\mu = 15$, find the probability of a Type II error and the power of the test.

2 It is given that $Y \sim N(\mu, 25)$. It is desired to test the null hypothesis $\mu = 20$ against the alternative hypothesis $\mu < 20$, with the probability of a Type I error being 5%. A random sample of 100 observations of Y is taken and the sample mean $\bar{Y}$ is taken to be the test statistic.

  (i) Find the acceptance and rejection regions.
 (ii) For the case $\mu = 19$, find the probability of a Type II error and the power of the test.

3 The temperature of an item taken from a freezer cabinet is $X°C$. X may be taken to be a normal variable with mean μ and standard deviation 1.8. A random sample of 11 items is taken from the cabinet, and the mean $\bar{X}$ of their temperatures is to be used as test statistic. It is desired to test the null hypothesis $\mu = -5.5$ against the alternative hypothesis $\mu \neq -5.5$, with the probability of a Type I error equal to 0.10.

  (i) Find the acceptance region.
 (ii) For the case $\mu = -7.0$, find the probability of a Type II error and the power of the test.

4 The random variable X is normally distributed with mean μ and standard deviation 11. The null hypothesis $\mu = 52$ is to be tested against the alternative hypothesis $\mu > 52$ using a 5 per cent significance level. The mean $\bar{X}$ of a random sample of 150 observations of X is to be used as the test statistic.

  (i) Find the range of values of the test statistic which lie in the critical region.
 (ii) When $\mu = 54$ calculate, to two decimal places, the probability of a type-2 error and the power of the test. [JMB]

5 *[Use a t-distribution in this question.]*
The haemoglobin levels (in g/100 l) of a random sample of ten elderly male cancer patients are as follows:

13.6, 11.9, 13.4, 12.4, 12.4,
13.2, 12.7, 15.7, 14.8, 12.0

Extensive evidence suggests that for healthy elderly males, the mean haemoglobin level is 13.0. The null hypothesis, H_0, is therefore that the mean haemoglobin level of the population of elderly male cancer patients is 13.0.

(a) State what is meant by a *Type I error*.
Suppose that the alternative hypothesis, H_1, is that the mean haemoglobin level is not 13.0. Use a 5% significance level to test the null hypothesis.

(b) State what is meant by a *Type II error*.
Show that, if the true mean haemoglobin level is 15.2, then the power of the test used in (a) is approximately 0.98.

(c) Explain how the test used in (a) would be modified if the alternative hypothesis had been that the mean haemoglobin level is less than 13.0.
What would the outcome of the modified test have been?

(d) Explain carefully what is meant by the phrase *confidence interval*.

Obtain a symmetric 99% confidence interval for the mean haemoglobin level of patients of the type sampled.

6 A bag contains a very large number of marbles, identical except for their colour. Of these, an unknown proportion p are red. It is required to test the null hypothesis

$$H_0: p = 0.3$$

against the alternative hypothesis

$$H_1: p < 0.3$$

In order to perform the test, a random sample of 100 marbles is taken and the number X of red marbles noted. The distribution of X is to be approximated by a normal distribution.

(i) If the significance level is 10%, determine whether the null hypothesis should be accepted in the case when $X = 25$.

(ii) If the significance level of the test is to be as close as possible to 5%, find the critical region in the form $0 \leqslant X \leqslant a$, where a is an integer.

(iii) Calculate the power of the test in the case when the critical region is $0 \leqslant X \leqslant 24$ and $p = 0.2$.   [JMB]

16.2 Comparing precise hypotheses

Up to now the alternative hypothesis, H_1, has been very vague! However, there will often be situations in which a precise alternative will seem appropriate.

For example, suppose we get consignments of some product from two sources. Source A is the cheaper, but produces an average of 5% defective items compared to the average of 2% of defectives from source B. We have a consignment from an unknown origin. The natural hypotheses are $p = 0.05$ and $p = 0.02$ (though it is not clear which is the null hypothesis!). The criteria for choosing between specific hypotheses such as these are beyond the scope of this book. However, once these criteria have been stipulated, calculations of Type I and Type II errors can proceed as usual.

As the next example shows, precise alternatives may specify different distributions, rather than different parameter values.

Example 4

The continuous random variable X only takes values in the interval $(0,1)$. The null hypothesis, H_0, is that X has pdf $f_0(x)$ given by:

$$f_0(x) = 2(1 - x) \qquad 0 < x < 1$$

The alternative hypothesis, H_1, states that the pdf of X is $f_1(x)$ given by:

$$f_1(x) = 2x \qquad 0 < x < 1$$

A single observation is to be taken on X and the suggested test procedure is to accept H_0 unless the observation is greater than a constant c. The value of c is to be chosen so as to minimise the value of P, where:

$$P = P(\text{Type I error}) + P(\text{Type II error})$$

Determine the values of c and P.

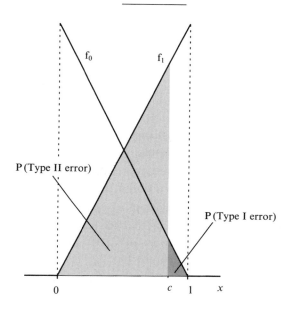

Accept H_0 if $x < c$

To answer this question we need first to determine expressions for the Type I and Type II errors:

$$P(\text{Type I error}) = P(X > c \,|\, \text{pdf is } f_0)$$

$$= \int_c^1 2(1 - x)\,dx$$

$$= \left[-(1 - x)^2 \right]_c^1$$

$$= (1 - c)^2$$

$$P(\text{Type II error}) = P(X < c \,|\, \text{pdf is } f_1)$$

$$= \int_0^c 2x\,dx$$

$$= \left[x^2 \right]_0^c$$

$$= c^2$$

The sum of the two errors, P, is given by:

$$P = (1 - 2c + c^2) + c^2 = 2c^2 - 2c + 1$$

To find the minimum of P we can complete the square as follows:

$$P = 2\left(c - \frac{1}{2}\right)^2 + \frac{1}{2}$$

The minimum value of P is $\frac{1}{2}$ which occurs when $c = \frac{1}{2}$.

Exercises 16b

1 It is desired to test between the two following cumulative distribution functions for the random variable X.

$$H_0: F(x) = \begin{cases} 0 & x \leqslant 0 \\ 1 - e^{-x} & x \geqslant 0 \end{cases}$$

$$H_1: F(x) = \begin{cases} 0 & x \leqslant 0 \\ 1 - \dfrac{1}{1 + x} & x \geqslant 0 \end{cases}$$

Find the medians of the two distributions, and find also their density functions.

A single observation of X is made and H_0 is accepted if and only if $X < c$. The significance level of the test is 5%.

Find the value of c and find also the power of the test.

2 The random variable X is hypothesised to have one of the following two distributions.

$$H_0: f(x) = \begin{cases} \dfrac{x}{25} & 0 < x < 5 \\ \dfrac{10 - x}{25} & 5 < x < 10 \\ 0 & \text{otherwise} \end{cases}$$

$$H_1: f(x) = \begin{cases} \dfrac{x}{50} & 0 < x < 10 \\ 0 & \text{otherwise} \end{cases}$$

Sketch the graphs of the two density functions.

The test procedure, based on a single observation of X, is to accept H_0 if and only if $X < c$.

(i) Find c if the significance level of the test is to be 5%, and find the power of the test in this case.

(ii) Find c if the probability of a Type I error is to be equal to the probability of a Type II error and state the common value of these probabilities.

3 Two hypotheses are under consideration for the random variable X.

$$H_0: f(x) = \begin{cases} \dfrac{1}{4} & 0 < x < 4 \\ 0 & \text{otherwise} \end{cases}$$

$$H_1: f(x) = \begin{cases} \dfrac{4 - x}{8} & 0 < x < 4 \\ 0 & \text{otherwise} \end{cases}$$

Sketch the graphs of the two density functions.

The test procedure, based on a single observation of X, is to accept H_0 if and only if $X > c$.

(i) Find c if the significance level of the test is to be 10%, and find the power of the test in this case.

(ii) Find c if the probability of a Type I error is to be twice the probability of a Type II error and state the power of the test in this case.

4 It is desired to test whether the random variable T has an exponential distribution, with mean 2, or a half-Cauchy distribution, with median 10. The corresponding density functions are:

$$H_0: f(t) = \begin{cases} \frac{1}{2} e^{-\frac{1}{2}t} & t > 0 \\ 0 & \text{otherwise} \end{cases}$$

$$H_1: f(t) = \begin{cases} \dfrac{20}{\pi(100 + t^2)} & t > 0 \\ 0 & \text{otherwise} \end{cases}$$

A single observation of T is made and H_0 is accepted if $T < 6$.

Find the significance level and the power of the test.

5 A sample of size n is taken from a normal distribution with variance 1 and with unknown mean μ. The null hypothesis is that $\mu = 0$ while the alternative hypothesis is that $\mu > 0$.
Using a significance test at the 5% level, find, in terms of n, the range of values of the sample mean that leads to rejection of the null hypothesis.

Suppose that, when using the test given above, the probability of rejecting the null hypothesis is required to be at least 0.99 when the true value of μ is 1.
Find the value of n.

16.3 The power curve

We know that the probability of a Type II error, and therefore the power, are dependent upon the parameter value. We therefore refer to the **power function**. A plot of the power function (on the y-axis) against the parameter value (on the x-axis) is known as the **power curve**.

Example 5

A random variable X has a normal distribution with mean μ and variance 1.
Devise a test of the null hypothesis H_0: $\mu = 20$ against the alternative hypothesis H_1: $\mu \neq 20$, using a 5% significance level and a random sample of 25 observations.
By calculating the power for a few values of μ, sketch the power curve for your test.

1 *Write down H_0 and H_1.*

$$H_0: \mu = 20$$
$$H_1: \mu \neq 20$$

2 *Determine the appropriate test statistic and the distribution of the corresponding random variable (using the parameter value specified by H_0).*
Assuming H_0, the distribution of $\bar{X}$ is:

$$N\left(20, \frac{1}{25}\right)$$

3 *Determine the significance level. This is P(Type I error).*
The significance level is 5%.

4 *Determine the acceptance and rejection regions.*
For a standard normal random variable Z, $P(Z > 1.960) = 0.025$.
Since $\sqrt{\dfrac{1}{25}} = 0.2$, an appropriate procedure is therefore to accept H_0 only if the value of $\bar{X}$ lies in the range:

$$(20 - 1.96 \times 0.2, \ 20 + 1.96 \times 0.2)$$

which simplifies to $(19.608, 20.392)$.

We must now consider a sequence of alternative values for μ.

5 *Determine the distribution of the random variable corresponding to the test statistic and hence determine the power.*

(a) Suppose, for example, that $\mu = 20.2$.

$$\begin{aligned} \text{Power} &= 1 - P(\bar{X} \text{ is in acceptance region}) \\ &= 1 - P(19.608 < \bar{X} < 20.392) \\ &= 1 - \left\{ \Phi\left(\frac{20.392 - 20.2}{0.2}\right) - \Phi\left(\frac{19.608 - 20.2}{0.2}\right) \right\} \\ &= 1 - \{\Phi(0.960) - \Phi(-2.960)\} \\ &= 1 - (0.8315 - 0.0015) \\ &= 0.170 \end{aligned}$$

(b) By symmetry, if the actual value of μ had been 19.8, the power would have been the same.

(c) Suppose next that $\mu = 20.4$. This time:

$$\begin{aligned} \text{Power} &= 1 - \left\{ \Phi\left(\frac{20.392 - 20.4}{0.2}\right) - \Phi\left(\frac{19.608 - 20.4}{0.2}\right) \right\} \\ &= 1 - \{\Phi(-0.040) - \Phi(-3.960)\} \\ &\approx \Phi(0.04) \\ &= 0.516 \end{aligned}$$

(d) Changing the value of μ by 0.2 changes the argument of the Φ function by 1.0. Thus, for $\mu = 20.6$ the power will be approximately $\Phi(1.04) = 0.8508$, for $\mu = 20.8$ the power will be approximately $\Phi(2.04) = 0.9793$, and so on.

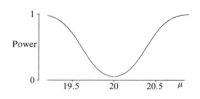

(e) Corresponding values apply for $\mu = 19.4$ and $\mu = 19.2$. Since power cannot exceed 1.0, the power curve therefore has the form shown.

Note

◆ At the point on the power curve where the parameter has the value specified by H_0 we have left a gap (which should really be invisible!). This is because power applies only to cases where H_0 is incorrect.

Example 6

The random variable X has a uniform distribution with mean μ, and probability density function given by:

$$f(x) = \begin{cases} 0.5 & \mu - 1 < x < \mu + 1 \\ 0 & \text{otherwise} \end{cases}$$

The hypotheses of interest are H_0: $\mu = 5$ and H_1: $\mu > 5$. The suggested test procedure, based on a single observation on X, is to reject H_0 in favour of H_1 only if the observation is greater than 5.9.
Determine the significance level and sketch the power curve.

To see what is happening, a sketch is useful. From the sketch we can easily see that the significance level is

$$P(X > 5.9 \mid \mu = 5) = (6 - 5.9) \times 0.5 = 0.05$$

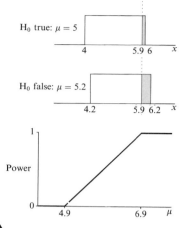

Suppose now that the actual value of μ is not 5, but 5.2. This amounts to a slide of the rectangle representing f(x) by 0.2 to the right. This increases $P(X$ falls in the rejection region) by $0.2 \times 0.5 = 0.1$. Similarly, if $\mu = 6$, $P(X > 5.9) = 0.55$, while for values of μ less than 4.9 there is no chance of rejecting H_0.

The power 'curve' is illustrated right. Note that, because the test is one-tailed, the power function is not symmetrical about the value specified by the null hypothesis.

Exercises 16c

1 Underground cables are placed in pipes which are liable to corrode, with pits forming on their surface. After a year's burial, standard pipes have pits with a mean depth of 0.0042 inches and a standard deviation of 0.0003 inches. To test a new coating for the pipes, ten newly coated pipes are to be buried for a year. A *two-tailed* 5% significance test will then be conducted to see if there is evidence of any change in the mean pit depth.

Determine the power of this test for the cases where the new population mean is 0.0043, 0.0044, 0.0045 and 0.0046.

Sketch the entire power curve.

The actual results obtained were 0.0039, 0.0041, 0.0038, 0.0044, 0.0040, 0.0036, 0.0034, 0.0046, 0.0035 and 0.0036.

Do these values provide significant evidence of a change in mean?

2 *[Use a normal approximation in this question.]*

A factory produces components of which a proportion p prove faulty. The factory claims that only 1% are faulty. A potential buyer examines a random sample of 1000 components and finds that 12 are faulty.

Does this provide significant evidence, at the 5% level, that the true proportion of defective components is in excess of 1%?

Determine the smallest number of defectives that would lead to rejection of the null hypothesis.

Determine the probability that future samples of size 1000 would lead to rejection of the claim made by the factory if the true proportion was (i) 1.25%, (ii) 1.5%, (iii) 2%.

Sketch the power curve.

3 The random variable X has a rectangular distribution with mean 0 and range denoted by $2r$. The probability distribution is given by:

$$f(x) = \begin{cases} \dfrac{1}{2r} & -r < x < r \\ 0 & \text{otherwise} \end{cases}$$

The hypotheses are $H_0: r = a$ and $H_1: r > a$. A significance test at the $100\alpha\%$ level is desired. The procedure is based on a single observation of X and is to reject H_0 if and only if $|X| > c$. Find c in terms of α and a.

Find, in terms of r and c, the power of the test. Sketch the power curve.

4 The random variable X has mean denoted by μ, and has a triangular distribution given by:

$$f(x) = \begin{cases} \dfrac{2(3\mu - x)}{9\mu^2} & 0 < x < 3\mu \\ 0 & \text{otherwise} \end{cases}$$

The hypotheses are $H_0: \mu = k$ and $H_1: \mu > k$. A significance test at the $100\alpha\%$ level is desired. The procedure is based on a single observation of X and is to reject H_0 if and only if $X > c$.

Find c in terms of α and k.

Find, in terms of μ and c, the power of the test. Sketch the power curve.

5 The random variable T has an exponential distribution with mean μ. The hypotheses are H_0: $\mu = k$ and H_1: $\mu > k$. A significance test at the $100\alpha\%$ level is desired. The procedure is based on a single observation of T and is to reject H_0 if and only if $T > c$.
Find c in terms of α and k.

Find, in terms of μ and c, the power of the test. Sketch the power curve.

6 Repeat the preceding question with H_1: $\mu < k$ and test $T < c$.

7 The random variable Y has distribution with mean μ, and is given by:

$$f(y) = \frac{1}{2}\, e^{-|y-\mu|} \qquad -\infty < y < \infty$$

The hypotheses are H_0: $\mu = 0$ and H_1: $\mu \neq 0$.
The test procedure is to accept H_0 if $|Y| < c$.
Find, in terms of c, the significance level of the test.

Show that the power $P(\mu)$ of the test is given by:

$$P(\mu) = \begin{cases} e^{-c}\left(\dfrac{e^{\mu} + e^{-\mu}}{2}\right) & 0 < |\mu| < c \\[3mm] 1 - e^{-|\mu|}\left(\dfrac{e^{c} - e^{-c}}{2}\right) & |\mu| \geqslant c \end{cases}$$

Sketch the power curve.

16.4 Hypothesis tests for a proportion based on a small sample

The difficulties associated with this type of hypothesis test are best illustrated with an example. Consider the following problem.

The standard treatment for a particular disease is successful on only 40% of occasions. A new treatment is introduced that is supposed to be better. Initially the treatment is given to just ten patients: the treatment is successful eight times.

Does this provide significant evidence, at the 5% level, that the new treatment is significantly better than the standard treatment?

This clearly requires a hypothesis test, so let us follow through our standard procedure.

1 *Write down H_0 and H_1*
We will ignore the fact that the results of the new treatment might have appeared worse than the standard treatment, since there appears to be a strong expectation that the new treatment is at least as good as the old. So the test is one-tailed and the hypotheses are:

$$H_0: p = 0.4$$
$$H_1: p > 0.4$$

2 *Determine the appropriate test statistic and the distribution of the corresponding random variable (using the parameter value specified by H_0).*
For a test of a hypothesis about p it is natural to use $\hat{p}$. However, we cannot use a normal approximation since n is small, so it is simpler to work directly with the number of 'successes', x. Assuming H_0, the distribution of the corresponding random variable, X, is:

$$B(10, 0.4)$$

3 *Determine the significance level.*
The question specifies 5%.

4 *Determine the acceptance and rejection regions.*

The test is one-tailed: only if we see a large value of x are we going to reject H_0 in favour of H_1. We therefore need to look at the probabilities of large values of x occurring under the conditions specified by H_0. Either from tables of the $B(10, 0.4)$ distribution (Appendix, p. 617), or by direct calculation (Section 9.5, p. 225), we have the following:

r	$P(X = r)$	$P(X \geqslant r)$
10	0.0001	0.0001
9	0.0016	0.0017
8	0.0106	0.0123
7	0.0425	0.0548
6	0.1115	0.1662

There is a problem here – probability comes in chunks! Suppose that we decide on the following strategy.

Reject H_0 in favour of H_1 if $X \geqslant 7$, accept H_0 otherwise.

The significance level was defined as the probability of an observation falling into the rejection region by chance under the conditions specified by H_0. The rejection region is the set of values 7, 8, 9 and 10 and the significance level is therefore 5.48% and not 5%. A significance level of exactly 5% is not obtainable. In this case we refer to '5%' as the **nominal significance level**, with '5.48%' being the **actual significance level** corresponding to the rejection region selected.

A conservative approach would be to define the critical region so that the actual significance level was not greater than the nominal significance level. In the present example this would imply using the following strategy.

Reject H_0 in favour of H_1 if $X \geqslant 8$, accept H_0 otherwise.

This strategy has an actual significance level of 1.23%.

This approach is adopted in some textbooks (and, implicitly, by some syllabuses), but we prefer an 'average' approach. This defines the rejection region so as to obtain an actual significance level as close to the nominal level as possible, without the restriction that it may not exceed it. Applying this strategy in lots of cases should lead to an average significance level close to the nominal value. In the present case this approach means working with an actual significance level of 5.48%.

Whichever approach is adopted, it is good practice to report the actual significance level, rather than the nominal significance level, when reporting the final decision.

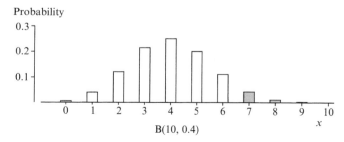

B(10, 0.4)

5 *Calculate the value of the test statistic.*

The observed value of X was 8.

6 *Determine the outcome of the test.*
Since 8 lies in the rejection region {7, 8, 9, 10} we can state that there is significant evidence, at an actual level of 5.48%, that the new treatment is successful in more than 40% of cases.

Notes
- The distinction between the nominal and actual significance levels is glossed over in some texts (and in some syllabuses!). We recommend that, where possible, the actual significance level is reported.

- Sometimes the possible significance levels are quite remote from the nominal level. For a nominal 5% test, the achievable significance levels may be 3% and 7%. If the outcome of the test is unaffected by the level chosen, then this does not matter. If the outcome is significant at the 7% level, but not at the 3% level, then the two results should be reported – the results might reasonably be described as inconclusive, if really the 5% level is required.

- The problem caused by discreteness also affects two-tailed tests. For a two-tailed 5% hypothesis test for a continuous distribution, it is simple to assign 2.5% to each tail. However, this is unlikely to be true for a discrete distribution, and we recommend applying the 'average' approach to each tail separately.

▼ ▼

Example 7

According to a genetic theory, $\frac{1}{4}$ of a certain group of plants should have red flowers. A random sample of 12 plants is examined. Six have red flowers. Does this provide significant evidence, at a nominal 5% level, to reject the hypothesis?

———

1 *Write down H_0 and H_1.*
The test is two-tailed. The hypotheses are:

$$H_0: p = 0.25$$
$$H_1: p \neq 0.25$$

2 *Determine the appropriate test statistic and the distribution of the corresponding random variable (using the parameter value specified by H_0).*
We use X, the number of plants with red flowers as our test statistic. Assuming H_0, the distribution of X is:

$$B(12, 0.25)$$

3 *Determine the significance level.*
The question specifies 5%.

4 *Determine the acceptance and rejection regions.*
The test is two-tailed, so we need to consider the cumulative probabilities associated with each tail of the B(12, 0.25) distribution:

	Upper tail			Lower tail	
r	$P(X = r)$	$P(X \geqslant r)$	r	$P(X = r)$	$P(X \leqslant r)$
12	0.0000	0.0000	0	0.0317	0.0317
11	0.0000	0.0000	1	0.1267	0.1584
10	0.0000	0.0000			
9	0.0004	0.0004			
8	0.0024	0.0028			
7	0.0115	0.0143			
6	0.0401	0.0544			

In the upper tail the nearest that we can get to 2.5% is 1.43%. In the lower tail the nearest that we can get is 3.17%. We therefore propose the decision rule:

> Reject H_0 in favour of H_1 if the observed value of X is either 0 or at least 7. Otherwise accept H_0.

The actual significance level is $1.43\% + 3.17\% = 4.60\%$.

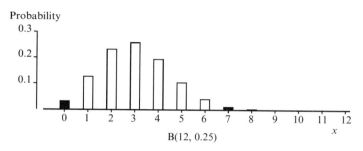

B(12, 0.25)

5 *Calculate the value of the test statistic.*
The observed value of X was 6.

6 *Determine the outcome of the test.*
Since 6 lies in the acceptance region $\{1, 2, 3, 4, 5, 6\}$ we accept the hypothesis H_0: $p = 0.25$, using an actual significance level of 4.60%.

16.5 Hypothesis tests for a Poisson mean based on a small sample

The problems here are essentially the same as those of the previous section. However, since Poisson distributions have infinite range, it is always sensible to work upwards from the outcome zero. The following example illustrates the procedure.

Example 8

A company uses a large number of floppy disks. At random intervals in time, disks develop faults: on average 0.4% of black disks fail per month. The company also has blue disks. During a randomly chosen nine-month period a random sample of 100 blue disks develop a total of 7 faults. Is there significant evidence, at the 5% level, that the failure rate of the blue disks is not 0.4% per month?

1 *Write down H_0 and H_1.*
The test is two-tailed. For convenience we will work with X, the total number of faults in 100 disks during a nine-month period. This has a Poisson distribution since faults occur at random intervals in time. The hypotheses are therefore:

> H_0: Rate $= 0.4\%$
> H_1: Rate $\neq 0.4\%$

2 *Determine the appropriate test statistic and the distribution of the corresponding random variable (using the parameter value specified by H_0).*
Assuming H_0, the distribution of X is Poisson with mean
$0.004 \times 100 \times 9 = 3.6$

3 *Determine the significance level.*
The question specifies 5%.

4 *Determine the acceptance and rejection regions.*
The test is two-tailed, so we need to consider the cumulative probabilities associated with each tail of the distribution, using either tables or direct calculation:

r	$P(X = r)$	$P(X \leqslant r)$	$P(X \geqslant r)$
0	0.0273	0.0273	1.0000
1	0.0984	0.1257	0.9727
2	0.1771		0.8743
3	0.2125		0.6973
4	0.1912		0.4848
5	0.1377		0.2936
6	0.0826		0.1559
7	0.0425		0.0733
8	0.0191		0.0308
9	0.0076		0.0117

In the lower tail the nearest that we can get to 2.5% is 2.73%. In the upper tail the nearest that we can get is 3.08%. We therefore propose the decision rule:

Reject H_0 in favour of H_1 if the observed value of X is either 0 or at least 8. Otherwise accept H_0.

The actual significance level is $2.73\% + 3.08\% = 5.81\%$.

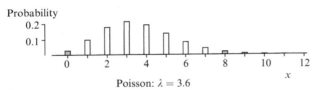
Poisson: $\lambda = 3.6$

5 *Calculate the value of the test statistic.*
The observed value of X was 7.

6 *Determine the outcome of the test.*
Since 7 lies in the acceptance region {1, 2, 3, 4, 5, 6, 7} we accept the null hypothesis (that the rate is 0.4%) using a significance level of 5.81%. The findings do not provide significant evidence that the blue disks have a different failure rate to the black disks.

Exercises 16d

1 A die is suspected of being biased towards the score of 6. It is thrown 10 times and the number n of sixes is observed.

State suitable null and alternative hypotheses and determine the acceptance and rejection regions for a test at a nominal significance level of 10%.

State the actual significance level of the test.

What are the conclusions in the cases
(i) $n = 3$, (ii) $n = 5$?

2 The proportion of £1 coins that bear the motto *Nemo me impune lacessit* is denoted by p. A random sample of 9 coins is taken and the number n that bear the motto is observed. It is desired to test the null hypothesis $p = 0.40$ against the alternative hypothesis $p \neq 0.40$ at a nominal significance level of 10%.

Determine the appropriate acceptance region and the corresponding actual significance level.

3 Two methods are proposed to test whether a coin is biased.

  (a) In the first method, the coin will be tossed 10 times and it will be considered biased if at least 8 heads or at least 8 tails are obtained.

   (i) Show that the probability of making a type-1 error is approximately 0.11.

   (ii) Determine, to two decimal places, the probability of making a type-2 error when the probability of obtaining a head on each toss is actually 0.6.

  (b) In the second method, the coin will be tossed 100 times and it will be considered biased if at least 60 heads or at least 60 tails are obtained. Calculate, to two decimal places, an approximate value of the significance level of the test.   [JMB]

4 Explain the terms *critical region, significance level* and *power* in the context of hypothesis testing.

National publicity was given to a university microbiologist's claim that 30% of pre-cooked chicken portions sold in supermarkets are contaminated with listeria. A large supermarket chain arranged to test, in its own laboratory, a random sample of 20 chicken portions from its supplier. Although it believed that the microbiologist was overstating the problem, the supermarket chain decided that it would contest the microbiologist's claim only if fewer than 3 of the chicken portions tested proved to be contaminated with listeria.

Considering this as a hypothesis testing problem, state suitable null and alternative hypotheses.

State the critical region and determine the significance level of the test.

Determine the power of the test if 15% of the supermarket chain's chicken portions are contaminated with listeria.

Subsequently the chain commissioned an independent laboratory to carry out tests on a random sample of 120 chicken portions, of which 22 proved to be contaminated.

Using a 1% significance level, conduct a test to decide whether the chain has sufficient evidence to conclude that the listeria contamination affects less than 30% of its chicken portions.

[JMB]

5 A certain production process is said to be out of control when the proportion p of its output which is defective exceeds 5%. A test is required to decide between the hypotheses: $H_0 : p = 0.05$ and $H_1 : p > 0.05$. The test suggested is to take a random sample of 20 items and reject H_0 if more than 2 items are defective. Calculate

  (i) the significance level of this test,

  (ii) the power of the test when 10% of the output is defective.

Without carrying out any further calculations, state briefly why the answers to (i) and (ii) should cause some concern.

Suggest a modification which could be made to the test in order that both the significance level and the power might be improved.   [JMB]

6 Dr Zed believes he possesses psychic powers. He claims that, when shown the back of a normal playing card, he has a better than 25% chance of predicting the suit of the card. A statistician is invited to investigate this claim. She asks Dr Zed to make independent predictions of the suits of 20 cards. The number of times Dr Zed predicts correctly is denoted by X and p is the probability that any given prediction is correct. To decide between the hypotheses:

$$H_0 : p = 0.25 \quad \text{and} \quad H_1 : p > 0.25$$

the statistician decides to accept H_0 if $X \leqslant r$, where r is the least integer for which the probability of a type I error is less than 5%. Using appropriate tables, or otherwise, find the value of r.

Dr Zed claims that the true value of p is 0.6. Show that, if this claim is true, the probability that a type II error will be made using the above test is approximately 6%.   [JMB]

7 Describe the roles of the null and alternative hypotheses in a test of significance. Explain how to decide whether the use of a one-tail or a two-tail test is appropriate.

Over a long period it has been found that the ratio of females to males attending classical ballet performances is 13 females to 7 males.

(a) On the afternoon of a football cup match, a random sample of 20 people attending a classical ballet performance is found to contain 4 males. Carry out a significance test to determine whether or not the proportion of males attending is lower than usual. State clearly your null and alternative hypotheses, and use a 10% significance level.

(b) At a contemporary ballet performance, a random sample of 100 people attending is found to contain 44 males. Set up null and alternative hypotheses and test whether the mean number of males attending contemporary ballet performances is different from that associated with classical ballet performances. Use a normal approximation and a 5% level of significance.   [ULSEB]

8 Explain what is meant by the *null hypothesis* and the *alternative hypothesis* in significance testing.

Explain the difference between a *one-sided* and a *two-sided* test, briefly describing how you would decide which one to use.

The annual number of accidents at a certain road junction could be modelled by a Poisson distribution with mean 8.5. New road markings are introduced at the junction which may reduce the accident rate, and which are not expected to increase it. In order to test their effectiveness it is planned to use the number of accidents at the junction in the following year. You may assume that, after the change in road markings, a Poisson model is still appropriate.

State suitable null and alternative hypotheses for the test.

If the new road markings are to be considered effective when fewer than 5 accidents occur in the year, determine the significance level of the test.

Find the power of the test if the new road markings have the effect of reducing the mean annual number of accidents at the junction to 3.

Suppose a test were designed so that its power would be approximately 0.6 when the new markings had the effect of reducing the mean annual number of accidents to 5. Obtain the significance level of the test, giving a clear justification for your answer.   [JMB]

9 The standard treatment for a particular ailment cures 75% of the patients treated. A new treatment, aimed at improving on this cure rate, was applied to 20 patients and cured 18 of them. To test whether the new treatment increases the cure rate when applied to a very large number of patients a statistician formulated appropriate null and alternative hypotheses. On carrying out the appropriate statistical test the statistician calculated the p-value to be 0.091. Write down the null and alternative hypotheses and verify that the p-value obtained by the statistician is correct.

State, giving the reason for your choice, which of the following recommendations you would make from the results obtained using the new treatment;

(*a*) adopt the new treatment,

(*b*) dismiss the new treatment as being no better than the standard treatment,

(*c*) conduct further trials using the new treatment before making a decision on whether or not to adopt the new treatment.   [WJEC]

10 Past records have indicated that the number of accidents along a particular roadway has a Poisson distribution, the average number of accidents per month being 10. During a month when additional warning signs had been placed along the roadway the number of accidents was 5. The Road Safety Officer, who had received some statistical training, decided to test whether the mean number of accidents per month had been reduced.

Having set up appropriate null and alternative hypotheses, the Officer calculated the p-value of the null hypothesis to be approximately 0.07. Show how this p-value was calculated and state the form of conclusion that the Officer can make from the calculated p-value.   [WJEC]

16.6 Type II errors in small samples

These are no different to type II errors in large samples! However, their probabilities can be particularly easy to 'calculate' since the values can often be taken directly from the tables of cumulative probabilities. The following example illustrates the procedure that will work for both Poisson means and binomial proportions.

Example 9

Construct a test, having a significance level of about 5% and a sample size of 7, for contrasting the hypotheses H_0: $p = 0.35$ and H_1: $p < 0.35$. Determine and sketch the corresponding power curve.

This is a question about lower tail probabilities of binomial distributions with $n = 7$. Consulting the table of cumulative probabilities in the Appendix, we see the following.

n	r						p				
		0.05	0.10	0.15	0.20	0.25	0.30	0.35	0.40	045	0.50
7	0	.6983	.4783	.3206	.2097	.1335	.0824	.0490	.0280	.0152	.0078
	1	.9556	.8503	.7166	.5767	.4449	.3294	.2338	.1586	.1024	.0625

From the column corresponding to $p = 0.35$ we see that $P(X \leqslant 0) = 0.049$ which is about 5%, while $P(X \leqslant 1) = 0.2338$ which is much too large. The required test procedure is therefore:

> Accept H_0 unless the observed value is 0 in which case H_0 should be rejected in favour of H_1.

The power of a test is the probability that it rejects H_0 when H_0 is false. In this case that means the probability of an observed value of 0, which is conveniently given for a range of values of p by the table entries.

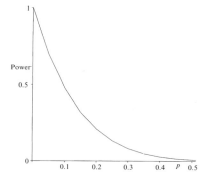

16.7 Quality control

Important applications of hypothesis testing occur in an industrial context. In this section we give a brief outline of two major areas in which testing takes place.

Process control

Imagine going into a factory in which baked beans are being put into tins. The baked beans will arrive from one direction and the tins from another: a complicated process will then get the right amount of beans into the tins (in a hygienic way and without spilling beans all over the floor!).

There are all sorts of regulations governing food, but one important set of rules governs the allowable variation in the mass of the contents. Suppose that something goes slightly wrong with the machinery and it starts underfilling tins. If this is noticed by the consumers, they will complain to the shops, who will complain to the canning firm and the factory manager will be out of a job.

To try and avoid problems like these the factory manager will keep a constant check on the filling process. It would be impractical to weigh the contents of every tin of beans because that would mean reopening all the tins! Instead, therefore, small numbers of tins might be examined at regular intervals (say every half hour). This idea of frequent small samples is known as **process control**.

The most common method of monitoring performance is by means of a **control chart**, which is simply a plot of the value of the sample statistic (for example, the sample mean) on the y-axis against 'sample number' on the x-axis. Here 'sample number' simply identifies the sample – the first sample each day might be numbered 1.

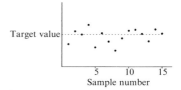

Typically each sample consists of just a few observations, with the usefulness of the procedure being captured by the appearance of the diagram. If, for example, the sequence of means appeared to 'wander' up and down during the day, then this might indicate some dependence of the manufacturing process on temperature or on how wide-awake the workers feel! If the sequence of means appeared to 'drift' upwards or downwards, then this might suggest that some part of the machinery was possibly 'shaking loose'. Alternatively, if there appeared to be a sharp break in the series, then this might suggest a corresponding sharp break in the machinery!

Suppose that the process is working correctly, with a mean μ and variance σ^2. The sample mean will then be an observation from (at least approximately) a $N\left(\mu, \dfrac{\sigma^2}{n}\right)$ distribution, where n is the sample size. By chance, some sample means may be considerably greater than μ, with about 1 in 40 being than greater than $\mu + \dfrac{2\sigma}{\sqrt{n}}$. Since there may be many means being calculated during a week's production, there will be quite a number that fall outside the interval $\left(\mu - \dfrac{2\sigma}{\sqrt{n}}, \mu + \dfrac{2\sigma}{\sqrt{n}}\right)$ simply due to chance. The usual procedure, when σ is known, is therefore something like this:

Take action if:

1   A sample mean falls outside the **action lines**, which are drawn at $\mu - \dfrac{3\sigma}{\sqrt{n}}$ and $\mu + \dfrac{3\sigma}{\sqrt{n}}$, or

2   Two successive sample means fall outside the same **warning line** $\left(\text{the warning lines drawn at } \mu - \dfrac{2\sigma}{\sqrt{n}} \text{ and } \mu + \dfrac{2\sigma}{\sqrt{n}}\right)$, or

3   Seven successive sample means fall on the same side of the target value, μ.

Examples of all these situations are illustrated in the diagram. Hollow circles represent occasions after which no action is taken, with filled circles representing occasions where action is taken. Vertical lines indicate the start of new sampling sequences.

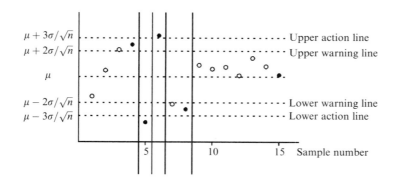

The rules control for sudden shifts in the population mean (rules 1 and 2) as well as a gentle drift in the population mean (rule 3).

Notes
  ◆ Control charts are also used to investigate changes in process variability (by plotting range or standard deviation against sample number), or variations in quality as measured by the proportions of defectives in a batch. A **batch** is a collection of items that were all produced under the same conditions (i.e. by the same workers and with no adjustments to machines).
  ◆ Control charts are also known as **Shewhart charts**.

Control charts in practice: (i) The mean chart

In practice neither the population mean (μ) nor the population standard deviation (σ) will be known so both must be estimated. The usual procedure is as follows.

1 Take a number of samples in the usual way.
2 For each sample, calculate the sample mean, $\bar{x}$, and the range, R.
3 Calculate the average of the sample means, $\bar{\bar{x}}$ ('x bar bar'), and the average of the sample ranges, $\bar{R}$ ('R bar').
4 Calculate action limits using $\bar{\bar{x}} \pm A_2 \bar{R}$, where the value of A_2 (which is the internationally agreed standard notation!) depends upon the sample size. Calculate warning limits using $\bar{\bar{x}} \pm \frac{2}{3} A_2 \bar{R}$. A full table of values of A_2 is given in the Appendix. A shortened version is given later in this section.

Notes
  ◆ The range is preferred to the standard deviation because it is easier to calculate. Increased simplicity also explains why the multipliers 2 and 3 are used in place of 1.96 and 3.09.

  ◆ The quantity $A_2 \bar{R}$ is being used in place of $\dfrac{3\sigma}{\sqrt{n}}$. It follows that an estimate of σ is $\dfrac{A_2 \bar{R} \sqrt{n}}{3}$.

Control charts in practice: (ii) The range chart

The samples taken to monitor possible fluctuations in the mean also provide information, in the form of the successive sample ranges, about any changes in the standard deviation.

A plot of R against sample number constitutes a range chart. Unusually large ranges may indicate a machine breakage, while an unusually small range may indicate some form of equipment malfunction.

Standard practice is to arrange the mean chart and the range chart one above the other on a single piece of paper. The critical (action) values corresponding to unusually small or large ranges are obtained by multiplying $\bar{R}$ by the tabulated values D_3 and D_4 (using, once again, an agreed notation).

An abbreviated table of the critical values for control charts is given here. A fuller table (including cases where $D_3 > 0$) is given in the Appendix.

Factors for control chart action lines			
Sample size	Factor for mean	Factors for range	
n	A_2	D_3	D_4
2	1.880	0	3.267
3	1.023	0	2.575
4	0.729	0	2.282
5	0.577	0	2.115

Notes

- ◆ There are in fact a number of different control schemes. In some schemes there will be warning limits for the range chart.
- ◆ With modern computer-controlled machines more sophisticated monitoring is possible without human intervention.

▼ ── ▼

Example 10

Random samples of four bags are taken every half hour from the output of a machine filling 0.5 kg bags of flour. The bags are accurately weighed and their masses (less 0.5 kg) are given (in tenths of a gram) in the table below.

					Sample				
1	2	3	4	5	6	7	8	9	10
−2	−9	9	−1	9	9	8	2	−1	1
2	7	9	2	−9	−4	4	1	0	3
−2	0	−4	7	1	−7	−9	−9	−6	−1
9	9	4	3	1	−8	0	2	−9	0

Assuming that the standard deviation is unaltered, obtain an estimate of its value (in tenths of a gram). Draw up mean and range charts and plot the given observations. Is there evidence that the machine has gone out of control?

──────────

We begin by calculating the values of $\bar{x}$ and R:

					Sample					
	1	2	3	4	5	6	7	8	9	10
$\bar{x}$	1.75	1.75	4.50	3.75	0.50	−2.50	0.75	−1.00	−4.00	0.75
R	11	18	13	8	18	17	17	11	9	4

From these we can obtain the average values: $\bar{\bar{x}} = 0.625$ and $\bar{R} = 12.6$.

Since $n = 4$, the table above gives $A_2 = 0.729$ and hence σ is estimated as:

$$\frac{A_2 \bar{R} \sqrt{n}}{3} = \frac{0.729 \times 12.6 \times \sqrt{4}}{3} = 6.12$$

For the mean chart the action lines are at
$0.625 \pm (0.729 \times 12.6)$, i.e. at -8.56 and 9.81
and the warning lines are at
$0.625 \pm \frac{2}{3}(0.729 \times 12.6)$, i.e. at -5.50 and 6.75.
All measurements are given in tenths of a gram.

All ten sample means lie well within the warning lines, so the output mean appears to be in control.

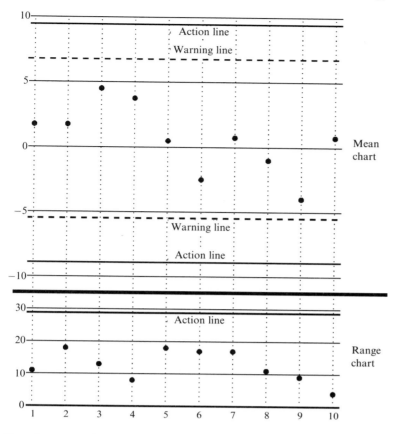

For the range chart the action lines are at 0 and $2.282 \times 12.6 = 28.8$, so there is no evidence that the process has gone out of control in any way.

Acceptance sampling

Acceptance sampling is typically concerned with the proportion of 'defectives' in a batch of production. In the context of tins of baked beans a 'defective' might be a damaged tin. It is often the case that there are a small proportion of defectives and the customer usually agrees to accept a small proportion of defective items (since the cost of checking every item would make the items too expensive). However, if the production process goes out of control the proportion of defectives may become unacceptably high. In order to guard against this, a number of sampling procedures have been devised. The simplest procedure is the **single sample**:

1 Take a random sample of n items from the batch.
2 If there are fewer than r defectives in the sample, accept the batch without further inspection. Otherwise examine every item in the batch to check for further defectives.

The single sample procedure is just like the test procedures discussed earlier, with the null hypothesis being that the proportion of defectives is as required, and the (one-sided) alternative hypothesis being that the proportion is higher than the target value.

Because this type of acceptance sampling is a crucial part of the production process, there are both international and British **standards** that give advice on the values of n and r. Larger batches require larger sample sizes because of the more expensive consequences of halting the production process.

An interesting variant on the single sample is the so-called **double sample**. The typical procedure is as follows:

1   Take a random sample of n items from the batch.
2   (a) If there are fewer than r defectives in the sample, accept the batch without further inspection.
     (b) If the number of defectives is between r and s, inclusive, take a second sample of size n from the same batch.
     (c) If the number of defectives is greater than s, reject the batch, examining every item individually to check for further defectives.
3   If a second sample has been taken, determine the total number of defectives in the two samples. If this is less than t then accept the batch without further inspection. Otherwise, reject the batch, examining every item individually and retaining all non-defectives.

Once again we note that there are both international and British standards governing the values of n, r, s and t.

▼   ▼

Example 11

For a batch size of 2000, hopefully containing no more than 1% of items that are defective, the standard double sampling scheme is as follows.
    Sample 80 items, chosen at random from the batch. If the sample contains no more than one defective item then accept the batch. If the sample contains four or more defectives, then reject the batch. If the sample contains two or three defective items, then take a second random sample, also of 80 items. If the two samples combined contain five or more defectives, then reject the batch; otherwise, accept the batch.
Determine the probability of accepting the batch when the proportion of defectives is indeed equal to 1%.

───────────

Let P_k denote the probability of obtaining k defectives in a sample of size 80. Using the binomial distribution, we have:

$$P_k = \binom{80}{k}(0.01)^k(0.99)^{80-k}$$

which is approximated by the Poisson probability:

$$\frac{0.8^k}{k!}e^{-0.8}$$

The required probability is:

$$(P_0 + P_1) + \{P_2(P_0 + P_1 + P_2) + P_3(P_0 + P_1)\}$$

where the first bracketed term represents the probability of accepting the batch directly and the second bracketed term gives the probability of acceptance after the second sample. Direct calculation using the binomial, which is rather tedious, gives the result 0.9774.
Using the Poisson approximation the expression simplifies to:

$$1.8e^{-0.8} + \{(0.32 \times 2.12) + (0.256 \times 0.6)\}e^{-1.6}$$

which is equal to 0.9768 – a good approximation to the true value.
    When the proportion of defectives is acceptably low, the recommended double sampling scheme has a low probability (about 2%) of rejecting a batch.

▲   ▲

The OC curve

This is shorthand for **operating characteristic** curve. The OC curve is really just the complement of the power curve: it is a plot of the probability of accepting a batch (the Type II error) against the proportion of defectives in the batch (or the batch population mean).

In the quality control context, the probabilities of the Type I and Type II errors are also given fancy names! The probability of a Type I error is called the **producer's risk** and the probability of a Type II error is called the **consumer's risk**.

To be more specific, suppose a machine makes goods that are either effective or defective – a binomial situation. Let the proportion of defectives be denoted by p. Following consultations, the producer and consumer agree that if $p \leqslant p_1$ the batch will be entirely satisfactory to the consumer, whereas if $p \geqslant p_2$ then the consumer would be most unhappy (which includes going very red in the face and leaping up and down!). The value p_1 is usually called the **acceptable quality level** or AQL.

International standards prescribe sampling procedures where the sample size is determined by the size of batch being sampled. Suppose that $p = p_1$; by chance the sampled data may lead to rejection of the batch even though it was acceptable to the consumer – the probability of this (significant result) is the producer's risk. Alternatively, suppose that $p = p_2$; the sampled data may nevertheless lead to acceptance of the batch (because of a result that was not significant) – the probability of this is the consumer's risk.

Example 12

A certain type of goods is produced in batches of size 1000. The AQL is 0.4% and the prescribed sample procedure is:

> Take a sample of size 125 and reject the batch if the sample contains two or more defectives.

Determine the producer's risk and also the consumer's risk in the case where the true proportion of defectives is 2%.

Let p be the probability of an item being defective, and let $q = 1 - p$. The probability of accepting the batch is:

$$\binom{125}{0} p^0 q^{125} + \binom{125}{1} p^1 q^{124}$$

When $p = 0.004$ this is equal to 0.9101 while, when $p = 0.02$, this falls to 0.2842. The producer's risk is therefore 9% (the probability of a Type I error) while, when the proportion of defectives in the batch is 2%, the consumer's risk (the probability of a Type II error) is 28.4%.

In this case n is large and the p values are small, so that a Poisson approximation should work well. The formula becomes:

$$e^{-125p} + \frac{(125p)^1}{1!} e^{-125p}$$

which simplifies to $(1 + 125p)e^{-125p}$. This approximation gives the approximate values 0.9098 and 0.2873, which are indeed close to the exact binomial values.

Exercises 16e

1 (a) A hotel group buys large quantities of towels for use by guests. When a batch is received a sample of 25 towels is subjected to a test of water absorption. If no more than one towel fails the test the batch is accepted. If two or three towels fail, a further sample of 25 towels is tested. The batch is then accepted if a total of no more than three (out of 50) fail the test. Otherwise it is rejected.

If a batch of towels, containing 7% which would fail the test, is submitted what is

  (i) the probability of its being rejected,
  (ii) the expected number of towels inspected?

 (b) In another test the towels are checked for visual defects. If the defects are distributed at random with a mean of 2 defects per towel, how many defects would be exceeded (on a particular towel) with a probability of just over 5%?

 (c) In a final check the lengths of 25 towels are measured and the batch rejected if the mean length is less than a specified value, k. What should be the value of k to give a probability of 0.99 of accepting a batch with mean length 106 cm and standard deviation 6 mm?

 (d) For towels from a particular supplier the probabilities of a batch failing these tests are p_1, p_2 and p_3 respectively. Write down an expression for the probability of the batch passing all three tests stating any assumptions you have needed to make.

[AEB 89]

2 A quality control inspector has been sent a large batch of items for testing. In deciding whether or not to accept the batch he is considering using one of the two following schemes:

Scheme A: Take a random sample of 100 items from the batch and accept the whole batch if two or fewer defectives are found.

Scheme B: Take a random sample of 50 items from the batch. Accept the batch if no defectives are found and reject the batch if two or more defectives are found. Otherwise take a further random sample of 100 items and accept the batch if this new sample contains two or fewer defectives.

Suppose the batch contains an unknown small proportion p of defectives. Use a suitable Poisson approximation to find the probabilities of the batch being accepted under each of the above schemes.

Show that, whatever the (small) value of p, the expected total sample size under *Scheme B* is less than that under *Scheme A*.

[You may find it helpful to consider the maximum value of xe^{-x} for $x \geqslant 0$.]   [SMP]

3 A pottery produces batches of 5000 mugs decorated with forenames. The pottery is considering adopting one of the two following batch sampling methods for controlling the quality of finished mugs.

Method I *(single sampling plan)*
Select 20 mugs at random from the batch and accept the batch if there are fewer than three defective mugs, otherwise reject the batch.

Method II *(double sampling plan)*
Select 10 mugs at random from the batch and accept the batch if there are no defective mugs, reject the batch if there are three or more defective mugs, otherwise select another 10 mugs at random. If the total number of defective mugs in the combined sample of 20 is fewer than three, the batch is accepted, otherwise it is rejected.

Explain why a binomial distribution is appropriate as a model for the number of defective mugs in a sample.

The proportion of defective mugs in a batch is p.

Copy and complete the following table of probabilities that a batch of mugs is *accepted*.

	p			
	0.01	0.05	0.10	0.20
Method I		0.925		
Method II	0.999		0.702	0.240

For Method II, the *expected sample size* per batch is given by

10(1+probability of selecting second sample).

Use this information, together with your table of probabilities, to decide which of the two sampling plans is to be preferred. Justify your choice.   [JMB]

4 A manufacturing process produces large batches of light bulbs. These are then packed in boxes of five. One hundred boxes are chosen at random and the bulbs are tested to find defectives. The results are summarised in the table below.

Number of defectives in a box	0	1	2	3	4	5
Number of boxes	76	10	7	4	1	2

Show that the mean number of defectives per box is 0.5. Supposing that the number of defectives per box follows a binomial distribution with its mean equal to 0.5, calculate the number of boxes with r defectives ($r = 0, 1, 2, 3, 4, 5$) which you would expect in a sample of one hundred boxes.

Comment briefly on the adequacy of the fit of this binomial model in relation to the sample above.

In another method for checking for defectives, light bulbs are chosen at random from a large batch and tested one by one until the first defective is found. If 10% of the batch are defective, find the smallest number m such that the probability of obtaining a defective on or before the mth choice is greater than 0.95.

At the end of each day a random sample of n bulbs is taken from the day's batch and the batch is rejected if more than 13% of this sample are defective. Calculate the value of n if there is a 0.05 probability of rejecting a batch of which in fact 10% are defective. [Assume that the size of the batch and of n are so large that the normal approximation to the binomial distribution is appropriate, and continuity corrections may be neglected.] [O&C]

5 (a) An acceptance sampling scheme consists of taking a sample of 25 from a large batch of components and rejecting the batch if 3 or more defectives are found.

What is the probability of accepting batches containing 2%, 4%, 6%, 10%, 15% and 20% defective?

Use your results to draw an operating characteristic.

From the operating characteristic, estimate

(i) the probability of accepting a batch containing 11% defective,

(ii) the proportion defective in a batch that has a probability of 0.6 of being rejected.

(b) An alternative plan requires a sample of 40 to be taken from the batch and the batch to be rejected if 4 or more defectives are found. Verify that both plans have a similar probability of rejecting batches containing 4% defective and comment on the advantages and disadvantages of the second plan compared to the first.

(c) If more than one out of any eight successive batches from a particular supplier are rejected, a more stringent form of inspection is introduced. What is the probability of more than one of the next eight batches being rejected if the first plan is used and all batches contain 4% defective?

(d) The more stringent inspection requires samples of 100 from each batch. The original form of inspection is reinstated if a sample contains no defectives. What is the proportion defective in the batch if the probability of no defectives in a sample of 100 is 0.5? [AEB 91]

Chapter summary

♦ **Type I error**:
This is the rejection of H_0 and acceptance of H_1 when H_0 is true. *The probability of a Type I error is the significance level.*

♦ **Type II error**:
This is the acceptance of H_0 rather than H_1 when H_0 is false. *The probability of a Type II error is a function of the parameter value.*

♦ **Power**:
This is given by:

$$\text{power} = 1 - P(\text{Type II error}).$$

A plot of power against parameter value is called **the power curve**.

♦ **Nominal and actual significance levels**:
With a discrete distribution, a particular standard level (such as 5%) may not be attained exactly. Such a level is therefore referred to as being a 'nominal' level, with the level attained being called the 'actual' level.

17 Two samples and paired samples

Between two evils, I always pick the one I never tried before

Mae West in *Klondike Annie*

Here are some examples of the types of question with which this chapter is concerned.

◆ Is it true that, on a motorway, male drivers drive faster than female drivers?
  ● We need to collect independent random samples of the speeds of male drivers and of female drivers. For each sex we calculate the mean speed. We then *compare the means* of the two samples.

◆ Is it true that male drivers are more likely to be prosecuted for speeding?
  ● We need to collect independent random samples of male drivers and of female drivers. For each sex we calculate the proportion prosecuted for speeding. We then *compare the proportions* in the two samples.

◆ Is it true that, for right-handed people, the span of their right hand is greater than the span of their left hand?
  ● This time each randomly chosen individual, who may range in size from Tarzan to Tom Thumb, provides two observations – one for the right-hand data set and one for the left. The variation in the sizes of the individuals does not matter – what matters is the difference *within* each of these pairs of observations.

17.1 Comparison of two means

Suppose we wish to test the hypothesis that two populations have the same mean. We need to go through the same six stages as in the case of a single sample:

1 Write down H_0 and H_1.
2 Determine the appropriate test statistic and, assuming H_0, the distribution of the corresponding random variable.
3 Determine the significance level.
4 Determine the acceptance and rejection regions.

   Now collect the data.

5 Calculate the value of the test statistic.
6 Determine the outcome of the test.

The form of the hypothesis test for comparing the means of two populations depends on whether or not the population variances are known. There are two reasonably simple cases:

◆ The population variances are known.
◆ Although the population variances are unknown, they can be assumed to have the same values as each other.

17.2 Comparison of two means – known population variances

The random variable X has unknown mean μ_x and known variance σ_x^2. The independent random variable Y has unknown mean μ_y and known variance σ_y^2. The null hypothesis is:

$$H_0: \mu_x = \mu_y, \text{ or equivalently, } \mu_x - \mu_y = 0$$

The alternative hypothesis may be two-sided:

$$H_1: \mu_x \neq \mu_y$$

or one-sided, e.g.

$$H_1: \mu_x > \mu_y$$

Since the hypotheses concern the population means, the test statistic will involve the sample means $\bar{x}$ and $\bar{y}$. Suppose that the samples have sizes n_x and n_y. Then, if X and Y have normal distributions, the same will be true for $\bar{X}$ and $\bar{Y}$:

$$\bar{X} \sim N\left(\mu_x, \frac{\sigma_x^2}{n_x}\right)$$

$$\bar{Y} \sim N\left(\mu_y, \frac{\sigma_y^2}{n_y}\right)$$

These results also hold, approximately, for large samples from other distributions, because of the Central Limit Theorem. In both cases $\bar{X} - \bar{Y}$ has a normal distribution with mean $\mu_x - \mu_y$ and variance $\frac{\sigma_x^2}{n_x} + \frac{\sigma_y^2}{n_y}$. Hence:

$$\frac{(\bar{X} - \bar{Y}) - (\mu_x - \mu_y)}{\sqrt{\frac{\sigma_x^2}{n_x} + \frac{\sigma_y^2}{n_y}}} \sim N(0, 1)$$

Assuming H_0, so that $\mu_x - \mu_y = 0$, we can calculate:

$$z = \frac{\bar{x} - \bar{y}}{\sqrt{\frac{\sigma_x^2}{n_x} + \frac{\sigma_y^2}{n_y}}}$$

and then proceed as usual.

Confidence interval for the common mean

According to H_0, $\mu_x = \mu_y$. We denote their common value by μ. All the n_x observations on X (i.e. $x_1, x_2, \ldots, x_{n_x}$) and all the n_y observations on Y (i.e. $y_1, y_2, \ldots, y_{n_y}$) therefore come from populations having the same mean. A natural **pooled estimate of the population mean**, $\hat{\mu}$, is therefore given by:

$$\hat{\mu} = \frac{\sum_{i=1}^{n_x} x_i + \sum_{j=1}^{n_y} y_j}{n_x + n_y} = \frac{n_x \bar{x} + n_y \bar{y}}{n_x + n_y} \tag{17.1}$$

The distribution of the corresponding random variable is:

$$N\left(\mu, \frac{n_x \sigma_x^2 + n_y \sigma_y^2}{(n_x + n_y)^2}\right)$$

(see the note below).

Following the usual arguments, the corresponding 95% confidence interval for μ is:

$$\left(\hat{\mu} - 1.96\frac{\sqrt{n_x\sigma_x^2 + n_y\sigma_y^2}}{(n_x + n_y)}, \quad \hat{\mu} + 1.96\frac{\sqrt{n_x\sigma_x^2 + n_y\sigma_y^2}}{(n_x + n_y)} \right)$$

If $\sigma_x^2 = \sigma_y^2 (= \sigma^2$, say) then, writing $n_x + n_y$ as n, the confidence interval simplifies to become:

$$\left(\hat{\mu} - 1.96\frac{\sigma}{\sqrt{n}}, \quad \hat{\mu} + 1.96\frac{\sigma}{\sqrt{n}} \right)$$

Note

♦ Since $\bar{X}$ has variance $\dfrac{\sigma_x^2}{n_x}$, the quantity $\dfrac{n_x\bar{X}}{n_x + n_y}$ has variance:

$$\left\{ \frac{n_x}{(n_x + n_y)} \right\}^2 \times \frac{\sigma_x^2}{n_x} = \frac{n_x\sigma_x^2}{(n_x + n_y)^2}$$

Combining this expression with the corresponding expression for the variance of $\dfrac{n_y\bar{Y}}{(n_x + n_y)}$, results in the expression for the variance given above.

Example 1

Suppose that random samples from two independent normal populations give the following results:

Sample 1: $n_x = 100$, $\quad \bar{x} = 46.0$
Sample 2: $n_y = 120$, $\quad \bar{y} = 47.0$

Suppose that the specified significance level is 5%, that the population variances are known to be $\sigma_x^2 = 16.0$ and $\sigma_y^2 = 24.0$, and that the hypotheses being compared are:

H_0: $\mu_x = \mu_y$
H_1: $\mu_x \neq \mu_y$

Show that H_0 may be accepted, and obtain a symmetric 99% confidence interval for the common population mean.

Assuming H_0, the test statistic z, given by:

$$z = \frac{\bar{x} - \bar{y}}{\sqrt{\dfrac{\sigma_x^2}{n_x} + \dfrac{\sigma_y^2}{n_y}}}$$

is an observation from a N(0, 1) distribution. Since the alternative hypothesis is two-sided, and the significance level is 5%, we will accept H_0 only if $-1.96 < z < 1.96$.
The observed value of z is:

$$\frac{46.0 - 47.0}{\sqrt{\dfrac{16.0}{100} + \dfrac{24.0}{120}}} = \frac{-1.0}{\sqrt{0.16 + 0.20}} = -1.67$$

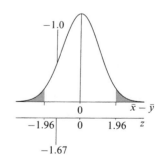

We therefore accept, at the 5% significance level, the hypothesis that there is no difference between the means of the two populations.
The pooled estimate of the common population mean is:

$$\frac{n_x\bar{x} + n_y\bar{y}}{n_x + n_y} = \frac{(100 \times 46.0) + (120 \times 47.0)}{220} = 46.545$$

The variance of the corresponding random variable is:

$$\frac{n_x\sigma_x^2 + n_y\sigma_y^2}{(n_x + n_y)^2} = \frac{(100 \times 16.0) + (120 \times 24.0)}{220^2} = 0.092\,56$$

and thus the 99% confidence interval for μ is given by:

$$(46.545 - 2.576 \times \sqrt{0.092\,56}, \quad 46.545 + 2.576 \times \sqrt{0.092\,56})$$

which simplifies to (45.76, 47.33).

Example 2

The standard deviation of the scores obtained on a particular test of mathematical ability is known to be 15. A school experiments with a new method of teaching which is supposed to increase general quantitative awareness. A group of 99 students are randomly assigned to one of two classes. The 50 students in the first class are given the new method of teaching, whereas the 49 students in the second class are taught in the standard way.

At the end of the year, the two classes are given the same test of mathematical ability. The mean for the first class is 116.0, whereas that for the second class is 113.1.

Does this provide significant evidence, at the 5% level, that the new method leads to a higher mean performance? State carefully any assumptions made.

We begin at the end! The assumptions that we need to make are the following:

1 The students allocated to the two classes originally had the same average ability (otherwise any apparent differences may be because there were better students in one class than the other).
2 The same degree of effort was put into the two types of teaching. (In practice this is usually *not* the case. A new method usually requires extra effort by all concerned and it may be this effort rather than the method itself that affects the results.)
3 The two classes are assumed to be effectively random samples from the population of national students. If this is not the case, then the results of the experiment cannot be generalised to the wider population.

Assumptions like these may seem pedantic and are often not spelt out. However, if they are *not* all true then the school's experiment is useless as a guide to future results.

Given that the assumptions *are* valid, we can proceed as usual. We denote the mean of the (hypothetical) population of students taught by the new method by μ_x, with the mean for students taught in the normal way being denoted by μ_y. The hypotheses are therefore:

$H_0: \mu_x = \mu_y$

$H_1: \mu_x > \mu_y$

The sample sizes are sufficiently large that we can assume that the sample means are observations from normal distributions. Assuming H_0, the test statistic z, given by:

$$z = \frac{\bar{x} - \bar{y}}{\sqrt{\dfrac{\sigma_x^2}{n_x} + \dfrac{\sigma_y^2}{n_y}}}$$

is therefore an observation from a N(0, 1) distribution. Since the alternative hypothesis is one-sided, and the significance level is 5%, we will accept H_0 only if $z > 1.645$.

The observed value of z is:

$$\frac{116.0 - 113.1}{\sqrt{\frac{15^2}{49} + \frac{15^2}{50}}} = \frac{2.9}{\sqrt{9.092}} = 0.962$$

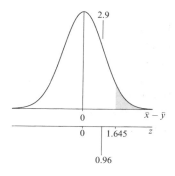

This value is considerably smaller than 1.645 and we can therefore confidently accept H_0: there is no significant evidence that the mean score using the new teaching method is greater than that using the traditional method. Subject to the assumptions made previously, the school need not switch to the new method.

Exercises 17a

1 The mean of a random sample of 10 observations from a population distributed as $N(\mu_1, 25)$ is 97.3. The mean of a random sample of 15 observations taken from a population distributed as $N(\mu_2, 36)$ is 101.2. Test, at the 5% level, (i) whether $\mu_1 < \mu_2$, (ii) whether $\mu_1 \neq \mu_2$.

2 A random sample of 85 observations is taken from a population with standard deviation 10.2, and the sample mean is 31.2. A random sample of 72 observations is taken from a second population with standard deviation 15.8, and the sample mean is 35.5. Test, at the 1% level, whether the second population has a greater mean then the first.

3 Two wine producers A and B have identical machines that fill bottles of wine. For A the quantity of wine put into a bottle is $(k_A + X)$ cl, where k_A is a constant and X is a normal random variable with mean 0 and standard deviation 0.180. For B the quantity of wine put into a bottle is $(k_B + Y)$ cl, where k_B is a constant and Y has the same distribution as X.

A retailer buys 8 bottles from A and measures the contents in cl. He finds the sample mean is 75.22 cl. He also buys 10 bottles from B and finds that the mean content is 74.91 cl.

Is there significant evidence, at the 5% level, that, on average, bottles from A contain more than bottles from B?

4 A machine assesses the life of a ball-point pen, by measuring the length of a continuous line drawn using the pen. A random sample of 80 pens of brand A have a total writing length of 96.84 km. A random sample of 75 pens of brand B have a total writing length of 93.75 km.

Assuming that the standard deviation of the writing length of a single pen is 0.15 km for both brands, test at the 5% level, whether the writing lengths of the two brands differ significantly.

Francis Ysidro Edgeworth 1845–1926 was an Irishman who obtained degrees in Classics from Trinity College, Dublin and Oxford University. On leaving Oxford he studied commercial law, becoming a barrister in 1877.

At the same time as his law studies, Edgeworth educated himself in mathematics and in 1880 he became lecturer in logic at King's College, London. Subsequently he turned his attention to Probability and Statistics, and in 1885 he delivered a paper entitled 'Methods of Statistics' to the Royal Statistical Society.

Edgeworth's early statistical work was concerned with formulating two-sample tests of means (though not with the structure shown in this chapter). His major contributions to Statistics were in the areas of correlation and regression, which will be encountered in Chapter 20.

17.3 Comparison of two means – common unknown population variance

In practice, situations in which the variance is known, but the mean is in doubt, are rare. Usually, if the mean is unknown, then so is the variance. Unfortunately the completely general situation in which the means and variances of two populations are all free to take any value leads to mathematical difficulties, so we now consider the restricted case in which *the unknown variances of the two populations may be assumed to be equal.*
Consider the hypotheses:

$$H_0: \mu_x = \mu_y$$
$$H_1: \mu_x > \mu_y$$

We are assuming that $\sigma_x^2 = \sigma_y^2 \ (= \sigma^2$, say). The samples have n_x and n_y observations, with sample means $\bar{x}$ and $\bar{y}$, respectively. If the two populations have the same mean then the pooled estimate $\hat{\mu}$ is given by:

$$\hat{\mu} = \frac{\sum_{i=1}^{n_x} x_i + \sum_{j=1}^{n_y} y_j}{n_x + n_y} = \frac{n_x \bar{x} + n_y \bar{y}}{n_x + n_y}$$

Assuming H_0, $\bar{X} - \bar{Y}$ has mean 0. Now $\text{Var}(\bar{X} - \bar{Y}) = \dfrac{\sigma^2}{n_x} + \dfrac{\sigma^2}{n_y}$, but σ^2 is unknown, so an estimate will be needed before progress can be made.

Information about variability is given by the squared deviations of the observations from their mean. The unbiased estimate of σ^2 based on the sample of x-values is given by s_x^2, where:

$$s_x^2 = \frac{1}{n_x - 1} \sum (x_i - \bar{x})^2$$

It follows that:

$$\sum (x_i - \bar{x})^2 \text{ is an unbiased estimate of } (n_x - 1)\sigma^2$$

Similarly, using:

$$s_y^2 = \frac{1}{n_y - 1} \sum (y_j - \bar{y})^2$$

we can state that:

$$\sum (y_j - \bar{y})^2 \text{ is an unbiased estimate of } (n_y - 1)\sigma^2$$

Adding these quantities together we find that:

$$\sum (x_i - \bar{x})^2 + \sum (y_j - \bar{y})^2 \text{ is an unbiased estimate of } (n_x + n_y - 2)\sigma^2$$

The so-called **pooled estimate of the common variance** is s^2, which is defined by:

$$s^2 = \frac{\sum (x_i - \bar{x})^2 + \sum (y_j - \bar{y})^2}{n_x + n_y - 2} \tag{17.2}$$

and is an unbiased estimate of σ^2. The estimate s^2 is best calculated as:

$$s^2 = \frac{\left\{ \sum x_i^2 - \dfrac{1}{n_x} \left(\sum x_i \right)^2 \right\} + \left\{ \sum y_j^2 - \dfrac{1}{n_y} \left(\sum y_j \right)^2 \right\}}{n_x + n_y - 2} \tag{17.3}$$

Equivalent expressions that make use of the quantities calculated by statistical calculators are:

$$s^2 = \frac{(n_x - 1)s_x^2 + (n_y - 1)s_y^2}{n_x + n_y - 2} \tag{17.4}$$

and:

$$s^2 = \frac{n_x \sigma_{n_x}^2 + n_y \sigma_{n_y}^2}{n_x + n_y - 2} \tag{17.5}$$

where $\sigma_{n_x}^2$ and $\sigma_{n_y}^2$ are the two sample variances.

What happens next depends on the sizes of the samples.

Large sample sizes

If the sample sizes are large then, because of the Central Limit Theorem, the distributions of $\bar{X}$ and $\bar{Y}$ will be approximately normal, so that, assuming H_0:

$$\frac{\bar{X} - \bar{Y}}{\sqrt{\sigma^2 \left(\dfrac{1}{n_x} + \dfrac{1}{n_y} \right)}} \sim N(0, 1)$$

If the sample sizes are very large then s^2, the pooled estimate of the common variance, will be an excellent approximation to the unknown σ^2. A natural test statistic is therefore z, given by:

$$z = \frac{\bar{x} - \bar{y}}{\sqrt{s^2 \left(\dfrac{1}{n_x} + \dfrac{1}{n_y} \right)}} \tag{17.6}$$

Assuming H_0, z may be considered to be an observation from a $N(0, 1)$ distribution and a test procedure can be constructed in the usual way.

Note

◆ If the sample sizes are very large, but the assumption that $\sigma_x^2 = \sigma_y^2$ cannot be made, then, assuming H_0, it is reasonable to base a test procedure on the test statistic:

$$z = \frac{\bar{x} - \bar{y}}{\sqrt{\dfrac{s_x^2}{n_x} + \dfrac{s_y^2}{n_y}}}$$

as in Section 17.2 (p. 454).

Since the sample sizes are large and this is only an approximation, either s_x^2 and s_y^2 or $\sigma_{n_x}^2$ and $\sigma_{n_y}^2$ can be used in this formula.

▼   ▼

Example 3

The marks obtained in a statistics paper by a random sample of 200 male students have $\bar{x} = 54.6$ and $s^2 = 101.3$. On the same paper, an independent random sample of 150 female students had a mean mark of 57.1, with $s^2 = 92.4$. Assuming a common population variance, obtain the pooled estimate of this variance, and test, at the 1% significance level, whether there is significant evidence of a difference in the two population means.

Denoting the mark of a randomly chosen male by X, and the mark of a randomly chosen female by Y, the two hypotheses are:

$H_0: \mu_x = \mu_y$

$H_1: \mu_x \neq \mu_y$

The test statistic is:

$$z = \frac{\bar{x} - \bar{y}}{\sqrt{s^2\left(\dfrac{1}{n_x} + \dfrac{1}{n_y}\right)}}$$

where s^2, the pooled estimate of the common variance, is given by:

$$s^2 = \frac{(n_x - 1)s_x^2 + (n_y - 1)s_y^2}{n_x + n_y - 2}$$

$$= \frac{(199 \times 101.3) + (149 \times 92.4)}{348}$$

$$= 97.49$$

Hence:

$$z = \frac{54.6 - 57.1}{\sqrt{97.49 \times \left(\frac{1}{200} + \frac{1}{150}\right)}}$$

$$= \frac{-2.5}{\sqrt{1.1374}}$$

$$= -2.344$$

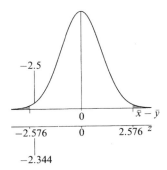

In this case the sample sizes are so large that a normal approximation can be used. The two-tailed 1% point of the standard normal distribution is 2.576. The test procedure is therefore to accept H_0, at the 1% significance level, if the value of z falls in the interval $(-2.576, 2.576)$.

Since the observed value, -2.344, does fall in this range, the hypothesis that male and female students have the same mean score can be accepted.

Example 4

A market inspector randomly samples the produce on two market stalls. A random sample of 80 apples from Rufus Russett's stall had masses (in g) having a sample mean of 74.2 and a sample variance of 24.21. An independent random sample of 100 of the apples sold by Granny Smith had a sample mean of 68.8 and a sample variance of 43.23.

Assuming a common population variance, obtain the pooled estimate of this variance and test, at the 0.1% significance level, whether there is significant evidence that the population of apples sold by Granny Smith has a lower mean mass than that of the population of apples on Rufus Russett's stall.

Denoting the mass (in g) of an apple from Rufus Russett's stall by X and the mass of an apple sold by Granny Smith by Y, the two hypotheses are:

$$H_0: \mu_x = \mu_y$$
$$H_1: \mu_x > \mu_y$$

The test statistic is:

$$z = \frac{\bar{x} - \bar{y}}{\sqrt{s^2\left(\dfrac{1}{n_x} + \dfrac{1}{n_y}\right)}}$$

where s^2, the pooled estimate of the common variance, is given by:

$$s^2 = \frac{n_x \sigma_{n_x}^2 + n_y \sigma_{n_y}^2}{n_x + n_y - 2}$$

$$= \frac{(80 \times 24.21) + 100 \times 43.23)}{178}$$

$$= 35.167$$

Hence:

$$z = \frac{74.2 - 68.8}{\sqrt{35.167 \times \left(\frac{1}{80} + \frac{1}{100}\right)}}$$

$$= \frac{5.4}{0.8895}$$

$$= 6.07$$

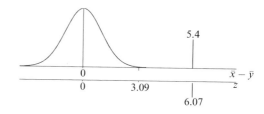

In this case the sample sizes are so large that a normal approximation can be used. The one-tailed 0.1% point of the standard normal distribution is 3.090. The test procedure is therefore to accept H_0 at the 0.1% significance level, if the value of z is less than 3.090.

Since the observed value, 6.07, is much greater than 3.09, we can confidently reject the null hypothesis in favour of the alternative hypothesis that the apples on Rufus Russett's stall have a greater population mean mass than that of the population of apples sold by Granny Smith.

Project

Are the cars in the local station car park newer than those in the supermarket car park? This could be the case if the 'bread-winner' uses the newest car to drive to the station (and thence to work), while the bread-winner's partner takes an older car to do the shopping.

Assume that all cars with this year's registration letter are 0 years old, that all cars with last year's letter are 1 year old, and so forth. Choose random samples of 50 cars in each situation, record the ages of the cars and perform an appropriate two-sample test to determine whether there is significant evidence to reject the hypothesis that the populations have a common mean.

Exercises 17b

1 A supermarket suspects that the average weight of Grade A melons from supplier X is less than that for Grade A melons from supplier Y. Two random samples are taken and weighed. For 82 melons from X, the results, in kg, are summarised by $\sum x = 58.65$, $\sum x^2 = 51.6460$. For 78 melons from Y, the results are summarised by $\sum y = 61.23$, $\sum y^2 = 55.3425$. Is there evidence at the 5% level to support the supermarket's suspicion:
   (i) assuming the population variances are equal,
   (ii) without assuming this?

2 In a traffic census drivers are asked the distance, in miles, of their current journey. The figures for a random sample of 120 drivers, between 8 and 9 am, are summarised by $\sum x = 1873$, $\sum x^2 = 56\,285$. The figures for a random sample of 94 drivers, between 1 and 2 pm, are summarised by $\sum y = 1711$, $\sum y^2 = 89\,894$.
Without assuming a common variance, test, at the 10% level, whether the mean distance reported by the 8 to 9 am drivers is less than the mean distance reported by the 1 to 2 pm drivers.

3 Acorns are sown in seed compost A and, after three years, the resulting 105 oak trees have mean height 0.641 m, with the corresponding value of s^2 being 0.0453 m^2. Acorns are also sown in seed compost B and grown in similar circumstances. After three years the 97 trees have mean height 0.578 m, with the corresponding value of s^2 being 0.0712 m^2. Test whether there is significant evidence, at the 5% level, that taller trees are produced in seed compost A:
(i) without assuming that the population variances are equal,
(ii) assuming that the population variances are equal.

4 A consumers' association tests car tyres by running them on a machine until their tread depth reaches a prescribed minimum. 150 tyres of brand A were tested and the equivalent distance, measured in thousands of kilometres, was measured with summary results $\sum(x-30) = 974$, $\sum(x-30)^2 = 10\,051$. The corresponding results for 120 tyres of brand B were $\sum(y-30) = 587$, $\sum(y-30)^2 = 10\,473$. Assuming a common variance, test whether there is significant evidence, at the 5% level, of a difference in the two mean distances.

Small sample sizes

If the sample sizes are not large, then the normal distribution is no longer a reasonable approximation to the distribution of the test statistic. In order to progress we must not only assume a common variance, but also we must:

assume that X and Y have normal distributions.

With this assumption it can be shown that the test statistic t, given by:

$$t = \frac{\bar{x} - \bar{y}}{\sqrt{s^2\left(\frac{1}{n_x} + \frac{1}{n_y}\right)}} \qquad (17.7)$$

has a t-distribution, with $n_x + n_y - 2$ degrees of freedom. This is the statistic labelled z in the case of large sample sizes. With the exception of the resulting change in the percentage points (which are found from the t-tables), the test procedure is unchanged.

▼ ▼

Example 5

I have two alternative routes to work. The times taken on the 8 randomly chosen occasions that I use route 1 are summarised by $\sum x = 182$ and $\sum x^2 = 4202$, while the times taken on the 12 randomly chosen occasions that I take route 2 are summarised by $\sum y = 238$ and $\sum y^2 = 5108$, with time being measured in minutes.

Assuming that the times taken on either route have normal distributions, with a common variance, determine whether there is significant evidence, at the 5% level, of a difference in the mean times taken on the two routes.

The pooled estimate of the common variance s^2 is given by:

$$s^2 = \frac{\left\{\sum x^2 - \frac{1}{n_x}(\sum x)^2\right\} + \left\{\sum y^2 - \frac{1}{n_y}(\sum y)^2\right\}}{n_x + n_y - 2}$$

$$= \frac{\left(4202 - \frac{1}{8} \times 182^2\right) + \left(5108 - \frac{1}{12} \times 238^2\right)}{18}$$

$$= 24.95$$

The two hypotheses are:

H_0: $\mu_x = \mu_y$
H_1: $\mu_x \neq \mu_y$

The test statistic is:

$$t = \frac{\bar{x} - \bar{y}}{\sqrt{s^2\left(\frac{1}{n_x} + \frac{1}{n_y}\right)}}$$

$$= \frac{\frac{182}{8} - \frac{238}{12}}{\sqrt{24.95 \times \left(\frac{1}{8} + \frac{1}{12}\right)}}$$

$$= \frac{2.917}{\sqrt{5.20}}$$

$$= 1.279$$

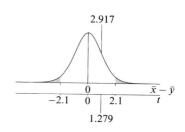

Assuming H_0, t is an observation from a t-distribution with 18 degrees of freedom. The two-tailed 5% point of a t_{18}-distribution is 2.101, so that H_0 will be accepted if t falls in the interval $(-2.101, 2.101)$.

The actual value of t is 1.279. There is no reason to reject the null hypothesis that the mean times taken using the two routes are equal.

Exercises 17c

1 The quantities of beer in a random sample of 7 'pints', bought at 'The Sensible Statistician', are measured in litres, and the results are summarised by $\sum x = 4.15$, $\sum x^2 = 2.4638$. The results for a random sample of 5 'pints' from 'The Mad Mathematician' are summarised by $\sum y = 2.79$, $\sum y^2 = 1.5585$.
Assuming the population variances are equal, find a pooled estimate of the common variance. Test, at the 5% level, whether there is more beer in a 'pint' from the first pub than the second.

2 A random sample of 10 yellow grapefruit is weighed and the average weight is found to be 201.4 g. The value of an unbiased estimate for the population variance is 234.1 g². The corresponding figures for a random sample of 8 pink grapefruit are 221.8 g and 281.9 g². Determine, using a 1% level of significance, whether there is a difference in the mean weights of the two kinds of grapefruit.

3 Explain the different circumstances in which a one-tailed or a two-tailed test of significance of a sample mean should be used.
The suppliers of a particular brand of jam claim that their pots of standard size have a mean mass greater than 346 g. A random sample of 8 pots from a particular delivery yielded the following masses, in grams.

342 354 348 344 349 350 347 345

Stating any necessary assumptions, perform a test, at the 10% significance level, to determine whether these data support the manufacturer's claim.
From the next delivery of jam, a random sample of 10 pots yielded the following masses, in grams.

340 341 350 348 342 350 346
344 347 342

Perform a test to determine whether there is evidence, at the 10% significance level, that the mean mass has changed from that of the previous delivery. State any further assumptions required.   [UCLES]

4 State the conditions under which a two-sample t-test may validly be applied.
The drug sodium aurothiomalate is sometimes used as a treatment for rheumatoid arthritis. Twenty patients were treated with the drug and, of these, 12 suffered an adverse reaction while 8 did not. The ages, in years, of the two groups were as follows.

Adverse reaction	53 29 53 67 54 57
	51 68 38 44 63 53
No adverse reaction	44 51 64 33
	39 37 41 72

(continued)

(i) Assuming that the conditions for validity are satisfied, carry out a two-sample *t*-test, at the 5% significance level, to test whether the data provide evidence that there is a difference between the average ages of patients who suffer an adverse reaction and those who do not.

Giving a brief reason, state whether you consider that the assumption that the conditions are satisfied is reasonable in this case.

(ii) Calculate an approximate 95% confidence interval for the proportion of patients, treated with the drug, that suffer an adverse reaction.   [UCLES]

5   A lorry is transporting a large number of red apples. As it passes over a bump in the road 10 apples fall off its back and are collected by a passing boy. The masses (in g) of the fallen apples are summarised by $\sum (x - 100) = 23.7$, $\sum (x - 100)^2 = 1374.86$.

On its return down the road the lorry is carrying a large number of green apples. When it passes over the bump 15 green apples fall off its back and are collected by the boy. The masses (in g) of these apples are summarised by $\sum (y - 110) = -73.2$, $\sum (y - 110)^2 = 2114.33$.

Treating the fallen apples as being random samples from those on the lorry, and assuming that the distribution of the masses of green apples has the same variance as that for red apples, obtain a pooled estimate of the common variance.

Test, at the 10% significance level, the hypothesis that the two distributions have the same mean.   [UCLES(P)]

6   A random sample of size *n* is drawn from a normal population with mean μ and variance σ^2. State, as fully as you can, the distribution of the sample mean.

In a supermarket the daily sales of packets of 'Crispo' cereal have mean 124.5 and variance 129.96. Following a television advertisement the mean daily sales over a period of 12 days increased to 132.5. Stating any necessary assumptions, test at the 1% significance level whether the mean number of packets sold per day has increased.

It later transpired that the mean of 124.5 and the unbiased estimate of variance, 129.96, were based on a sample of 12 days. The corresponding variance of the sample with mean 132.5 was 112.36. Determine whether there is any change in your previous conclusion.   [UCLES]

17.4 Confidence intervals for the difference between two means

Known variances

When σ_x^2 and σ_y^2 are known, we can obtain a confidence interval for the difference between the two population means by using an argument that by now has become rather familiar. We know that:

$$\frac{(\bar{X} - \bar{Y}) - (\mu_x - \mu_y)}{\sqrt{\dfrac{\sigma_x^2}{n_x} + \dfrac{\sigma_y^2}{n_y}}} \sim \mathrm{N}(0, 1)$$

and hence:

$$\mathrm{P}\left[-1.96 < \frac{(\bar{X} - \bar{Y}) - (\mu_x - \mu_y)}{\sqrt{\dfrac{\sigma_x^2}{n_x} + \dfrac{\sigma_y^2}{n_y}}} < 1.96\right] = 0.95$$

which, on rearrangement, gives:

$$\mathrm{P}\left[(\bar{X} - \bar{Y}) - 1.96\sqrt{\frac{\sigma_x^2}{n_x} + \frac{\sigma_y^2}{n_y}} < \mu_x - \mu_y < (\bar{X} - \bar{Y}) + 1.96\sqrt{\frac{\sigma_x^2}{n_x} + \frac{\sigma_y^2}{n_y}}\right] = 0.95$$

A 95% confidence interval for the difference is therefore given by:

$$\left(\bar{x} - \bar{y} - 1.96\sqrt{\frac{\sigma_x^2}{n_x} + \frac{\sigma_y^2}{n_y}}, \quad \bar{x} - \bar{y} + 1.96\sqrt{\frac{\sigma_x^2}{n_x} + \frac{\sigma_y^2}{n_y}} \right)$$

Note

◆ The arguments given above apply exactly when X and Y have normal distributions, and approximately (because of the Central Limit Theorem) in other cases in which the sample sizes are large.

Unknown common variance

In this case we use s^2, the pooled estimate of the common variance, together with the appropriate t-distribution:

$$\frac{(\bar{X} - \bar{Y}) - (\mu_x - \mu_y)}{\sqrt{s^2\left(\frac{1}{n_x} + \frac{1}{n_y}\right)}} \sim t_{n_x+n_y-2}$$

The confidence interval becomes:

$$\left(\bar{x} - \bar{y} - cs\sqrt{\frac{1}{n_x} + \frac{1}{n_y}}, \quad \bar{x} - \bar{y} + cs\sqrt{\frac{1}{n_x} + \frac{1}{n_y}} \right)$$

where c is the relevant percentage point from the t-distribution with $(n_x + n_y - 2)$ degrees of freedom.

Note

◆ The above argument only applies exactly if X and Y have normal distributions.

Large samples

If the samples are large, but the variances are not known and cannot be assumed to be equal to one another, then only approximate confidence intervals can be calculated. In such cases it is irrelevant whether you use the sample variances or the unbiased estimates of the population variances. Using the latter, an approximate 95% confidence interval would be given by:

$$\left(\bar{x} - \bar{y} - 1.96\sqrt{\frac{s_x^2}{n_x} + \frac{s_y^2}{n_y}}, \quad \bar{x} - \bar{y} + 1.96\sqrt{\frac{s_x^2}{n_x} + \frac{s_y^2}{n_y}} \right)$$

▼   ▼

Example 6

A headmaster is interested in whether there is any difference in the performance of the large numbers of students taught by Mrs White and Mr Green. He chooses students at random and gives them a statistics test. The 10 students taught by Mrs White have marks summarised by $\sum x = 612$, $\sum x^2 = 40\,104$, while the 8 students taught by Mr Green have marks summarised by $\sum y = 444$, $\sum y^2 = 27\,460$.

Determine a 95% confidence interval for the difference in the mean marks of students taught by the two teachers.

Is there significant evidence, at the 5% level, of a difference in the means?

─────────────

In this case the two populations consist of all students taught by the teachers. We assume that the marks obtained by the two groups of students have the same population variance. This is estimated by s^2, given by:

$$s^2 = \frac{(40\,104 - \frac{1}{10} \times 612^2) + (27\,460 - \frac{1}{8} \times 444^2)}{10 + 8 - 2} = 341.725$$

The two-tailed 5% point of a t-distribution having 16 degrees of freedom is 2.120. The means of the two groups are 61.2 and 55.5, so the 95% confidence interval for the difference between the population means is:

$$\left(5.7 - 2.120\sqrt{341.725\left(\frac{1}{10} + \frac{1}{8}\right)},\ 5.7 + 2.120\sqrt{341.725\left(\frac{1}{10} + \frac{1}{8}\right)}\right)$$

which simplifies to $(-12.9,\ 24.3)$.

The individual marks are clearly very variable and the headmaster's samples are therefore far too small to be able to measure the difference (if any) between the population means with any useful accuracy. Since the confidence interval comfortably includes 0, we can conclude, as in Section 15.11, that there is no significant evidence of a difference in the average marks obtained by the two groups of students.

Practical

How long does it take to thread a needle? Are females better at this task than males? Are left-handers better than right-handers? Are those wearing glasses better than those not wearing glasses?

The answers to these questions can quickly be discovered by timing all the people in your class! Record the times taken to the nearest second, obtain the class estimate of the common variance and perform a two-sample test for a difference between the means.

Decide beforehand what significance level to use and also whether you feel that a one-tailed or a two-tailed test is appropriate.

Exercises 17d

1 The mean of a random sample of 10 observations from a population distributed as $N(\mu_1, 25)$ is 97.3. The mean of a random sample of 15 observations taken from a population distributed as $N(\mu_2, 36)$ is 101.2. Find a 95% symmetric confidence interval for $\mu_2 - \mu_1$.

2 A random sample of 85 observations is taken from a population with standard deviation 10.2, and the sample mean is 31.2. A random sample of 72 observations is taken from a second population with standard deviation 15.8, and the sample mean is 35.5. Find a 90% symmetric confidence interval for the mean of the second population minus the mean of the first population.

3 Two wine producers A and B have identical machines that fill bottles of wine. For A the quantity of wine put into a bottle is $(k_A + X)$ cl, where k_A is a constant and X is a normal random variable with mean 0 and standard deviation 0.180. For B the quantity of wine put into a bottle is $(k_B + Y)$ cl, where k_B is a constant and Y has the same distribution as X. A retailer buys 8 bottles from A and measures the contents in cl. He finds the sample mean is 75.22 cl. He also buys 10 bottles from B and finds that the mean content is 74.91 cl. Find a 95% symmetric confidence interval for $k_A - k_B$.

4 A machine assesses the life of a ball-point pen, by measuring the length of a continuous line drawn using the pen. A random sample of 80 pens of brand A have a total writing length of 96.84 km. A random sample of 75 pens of brand B have a total writing length of 93.75 km.

Assuming that the standard deviation of the writing length of a single pen is 0.15 km, for both brands, find a 90% symmetric confidence interval for the mean writing life of brand A minus the mean writing life of brand B.

5 In a traffic census drivers are asked the distance, in miles, of their current journey. The figures for a random sample of 120 drivers, between 8 and 9 am, are summarised by $\sum x = 1873$, $\sum x^2 = 56\,285$. The figures for a random sample of 94 drivers, between 1 and 2 pm, are summarised by $\sum y = 1711$, $\sum y^2 = 89\,894$.

Without assuming a common variance, find a 95% symmetric confidence interval for the mean distance reported by the 1 to 2 pm drivers minus the mean distance reported by the 8 to 9 am drivers.

6 The quantities of beer in a random sample of 7 'pints', bought at 'The Sensible Statistician', are measured in litres, and the results are summarised by $\sum x = 4.15$, $\sum x^2 = 2.4638$. The results for a random sample of 5 'pints' from 'The Mad Mathematician' are summarised by $\sum y = 2.79$, $\sum y^2 = 1.5585$.

Assuming the population variances are equal, find a pooled estimate of the common variance.

Find a 95% symmetric confidence interval for the mean quantity in a 'pint' from 'The Sensible Statistician' minus the mean quantity in a 'pint' from 'The Mad Mathematician'.

7 A random sample of 10 yellow grapefruit is weighed and the average weight is found to be 201.4 g. The value of an unbiased estimate for the population variance is 234.1 g^2. The corresponding figures for a random sample of 8 pink grapefruit are 221.8 g and 281.9 g^2.

Find a 95% symmetric confidence interval for the mean weight of a pink grapefruit minus the mean weight of a yellow grapefruit.

8 A random sample of size 25 taken from a normal population with mean μ_1 and standard deviation 5 has a mean of 80. Independently, a random sample of size 36 taken from a normal population with mean μ_2 and standard deviation 3 has a mean of 75.

(*a*) Find a 95% confidence interval for μ_1.

(*b*) Find a 90% confidence interval for $\mu_1 - \mu_2$.

[ULEAC]

9 A random sample of 100 batteries produced by company A had a mean lifetime of 3.2 years and a standard deviation of 0.3 years, whilst a random sample of 150 batteries produced by company B had a mean lifetime of 2.9 years and a standard deviation of 0.5 years.

Explaining the basis for your calculations, find approximate 90% confidence limits for $\mu_A - \mu_B$, where μ_A and μ_B are the population mean lifetimes of batteries produced by the two companies. [ULEAC]

10 A study was made to assess the differences in salaries of engineers working in industry and those working in colleges. A random sample of 100 engineers from industry had a mean salary of £27 500, whilst a random sample of 100 college engineers had a mean salary of £21 000. The standard error of the difference between these two means was found to be £540. Find an approximate 95% confidence interval for the difference between the population mean salaries. [ULEAC]

11 The athletics coach of the Road Runners club monitored the number of miles, x, to the nearest mile, run in training by a random sample of 75 of the club members during a particular week.

The results are summarised as follows:

$$\Sigma x = 1876, \Sigma x^2 = 50186.$$

(*a*) Find approximate 90% confidence limits for the mean number of miles run that week by club members.

During the same week, unbiased estimates of the mean and variance of the number of miles run by members of the Veteran athletics club, based on a random sample of 35 members, were 22.20 miles and 26.12 miles^2 respectively.

(*b*) Find approximate 90% confidence limits for the number of miles by which the mean distance run by the Road Runners exceeds that run by the Veterans. [ULEAC]

12 For each person in a random sample of 100 persons following diet A for one month, the weight loss, x lb, was measured. The following results were calculated:

$$\Sigma x = 702, \Sigma x^2 = 6214.$$

For each person in a random sample of 100 persons following diet B for one month, the weight loss, y lb, was measured. The following results were calculated:

$$\Sigma y = 690, \Sigma y^2 = 5791.$$

(i)   Calculate, to two decimal places, unbiased estimates for the means μ_A, μ_B and the variances of the weight losses of persons following the diets A and B for one month.

(ii) Calculate an approximate 90 per cent symmetric confidence interval for μ_A.

(iii) Calculate an approximate 95 per cent symmetric confidence interval for $\mu_A - \mu_B$. State, with a reason, whether or not the confidence interval discredits the assertion that there is no difference between μ_A and μ_B. [JMB]

13 In a survey of school children, the variable measured was weight in kilograms. A random sample of 100 children had a mean weight of 37.2 kg and standard deviation of 3.6 kg.

(a) Find an unbiased estimate of the variance of the population from which this sample was taken.

(b) Find a 90% confidence interval for the mean weight.

A second sample of 100 children was taken and the mean weight was found to be 39.4 kg. The standard error of the difference in mean weights for the two samples was found to be 0.55 kg.

(c) Find a 95% confidence interval for the difference between the two means.

(d) Explain the importance of the fact that zero does not lie in this interval. [ULEAC]

17.5 Comparison of two proportions

We denote the proportions of successes in two populations by p_1 and p_2. The method used for comparing the hypotheses:

$$H_0: p_1 = p_2$$
$$H_1: p_1 \neq p_2$$

depends upon the sample sizes. In this chapter we confine our attention to the case where the sample sizes, n_1 and n_2, are sufficiently large that the binomial distributions involved can be well approximated by normal distributions.

Suppose that the two samples contain respectively r_1 and r_2 successes. The proportions of successes in the two samples are $\hat{p}_1$ and $\hat{p}_2$, where $\hat{p}_1 = \dfrac{r_1}{n_1}$ and $\hat{p}_2 = \dfrac{r_2}{n_2}$. Writing $q_1 = 1 - p_1$ and $q_2 = 1 - p_2$, the normal approximations state that the distributions of the random variables corresponding to $\hat{p}_1$ and $\hat{p}_2$ are:

$$N\left(p_1, \frac{p_1 q_1}{n_1}\right) \text{ and } N\left(p_2, \frac{p_2 q_2}{n_2}\right)$$

An alternative way of writing H_0 is as:

$$H_0: p_1 - p_2 = 0$$

from which it is clear that an appropriate test will be based on the value of $\hat{p}_1 - \hat{p}_2$.

Suppose that H_0 is true and that the common value of p_1 and p_2 is p. Then the distribution of $\hat{p}_1 - \hat{p}_2$ is (approximately):

$$N\left(0, \frac{pq}{n_1} + \frac{pq}{n_2}\right)$$

where $q = 1 - p$.

Standardising, we see that a test of H_0 would be provided by calculating:

$$z = \frac{\hat{p}_1 - \hat{p}_2}{\sqrt{pq\left(\dfrac{1}{n_1} + \dfrac{1}{n_2}\right)}}$$

However, there is a serious problem with the above! The expression for the variance involves the *unknown* common population proportion. Ever resourceful (and bearing in mind that the distribution is in any case only *approximately* normal), we replace the unknown pq with $\hat{p}\hat{q}$, where:

$$\hat{p} = \frac{r_1 + r_2}{n_1 + n_2} = \frac{n_1\hat{p}_1 + n_2\hat{p}_2}{n_1 + n_2}$$

is the **pooled estimate of the common proportion**, and $\hat{q} = 1 - \hat{p}$.

The test statistic used with large samples is therefore:

$$z = \frac{\hat{p}_1 - \hat{p}_2}{\sqrt{\hat{p}\hat{q}\left(\dfrac{1}{n_1} + \dfrac{1}{n_2}\right)}}$$

If the hypothesis of a common proportion of 'successes' is acceptable, then we can amalgamate the two samples and calculate a confidence interval for the common proportion in the usual way, using $\hat{p}$ and the combined sample size $(n_1 + n_2)$.

Notes

- ◆ Although the formula for z involves an estimate of variance, the t-distribution is not appropriate.
- ◆ Since continuous approximations are being made to discrete distributions, a continuity correction should be used. However, this makes the formula somewhat complicated and is usually omitted. Paradoxically, when using the equivalent chi-squared test (Section 18.5, p. 500) the correction is routinely used.
- ◆ Part of the reason why we can be quite happy substituting $\hat{p}\hat{q}$ for pq is that over a wide range of values of p the value of pq alters only very slowly, so that the precise value of p is of little importance. For example:

p	0.3	0.4	0.5	0.6	0.7
pq	0.21	0.24	0.25	0.24	0.21

Example 7

In 1989, across the European Community, the characteristics of a randomly chosen sample of trees were recorded. A total of 13 468 trees were described as growing at an altitude of less than 250 m. Of these, 11 879 were slightly defoliated, having lost only a few leaves. Of the 11 594 trees growing at heights of between 250 m and 500 m, 10 345 were slightly defoliated.

Show that, at the 1% significance level, the hypothesis that the proportion of slightly defoliated trees is the same for each group is acceptable. Determine a symmetric 95% confidence interval for the common proportion.

———

In this question a 'success' is a slightly defoliated tree. The hypotheses under comparison are:

$$H_0: p_1 = p_2$$
$$H_1: p_1 \neq p_2$$

With very large numbers such as those in this question, it is particularly advisable to avoid premature rounding of intermediate calculations. Intermediate values are therefore shown to an unusual degree of precision.

470 Understanding Statistics

The sample proportions of slightly defoliated trees are

$$\hat{p}_1 = \frac{11\,879}{13\,468} = 0.882\,017 \text{ and } \hat{p}_2 = \frac{10\,345}{11\,594} = 0.892\,272, \text{ while the pooled}$$

estimate of the common proportion is $\hat{p} = \frac{22\,224}{25\,062} = 0.886\,761$.

The test statistic is:

$$z = \frac{0.882\,017 - 0.892\,272}{\sqrt{0.886\,761 \times 0.113\,239 \times \left(\frac{1}{13\,468} + \frac{1}{11\,594}\right)}}$$

$$= \frac{-0.010\,255}{0.004\,015} = -2.554$$

The two-tailed 1% point of a standard normal distribution is 2.576, which is just larger than 2.554, and thus the null hypothesis can be accepted.

A symmetric 95% confidence interval for the overall proportion of slightly defoliated trees is:

$$\left((0.886\,761 - 1.96\sqrt{\frac{0.886\,761 \times 0.113\,239}{25\,062}}, \right.$$

$$\left. 0.886\,761 + 1.96\sqrt{\frac{0.886\,761 \times 0.113\,239}{25\,062}} \right)$$

which simplifies to (0.8828, 0.8907).

Example 8

In the United States a **double blind** experiment was conducted to investigate whether a daily dose of aspirin would alter the risk of having a heart attack. A 'double blind' experiment is one in which neither the person administering the drug, nor the patient, knows what drug is being administered. This eliminates possible biases.

In the aspirin experiment, the doctors were also the patients! Each doctor was sent a large box of pills and was instructed to take one pill a day. All the pills looked identical, but the pills in half of the boxes were aspirins, whereas the pills in the other boxes contained no drugs whatsoever. A dummy medicine of this type is called a **placebo**.

The outcome of the trial was as follows. Of 11 037 doctors given aspirin, 104 had heart attacks during the period of the trial, whereas of 11 034 doctors given the placebo, 189 had heart attacks.
Investigate the significance of this result.

In this question a 'success' is a heart attack! The hypotheses being compared are:

$$H_0: p_1 = p_2$$
$$H_1: p_1 \neq p_2$$

Once again we need to preserve high accuracy in intermediate calculations.

Assuming H_0, the pooled estimate of the common proportion is

$$\hat{p} = \frac{104 + 189}{11\,037 + 11\,034} = 0.013\,275. \text{ The sample proportions are}$$

$\frac{104}{11\,037} = 0.009\,423$ for aspirin and $\frac{189}{11\,034} = 0.017\,129$ for the placebo, so

the test statistic is:

$$z = \frac{0.009\,423 - 0.017\,129}{\sqrt{0.013\,275 \times 0.986\,725\left(\frac{1}{11\,037} + \frac{1}{11\,034}\right)}} = \frac{-0.007\,706}{0.001\,541} = -5.00$$

Given the 'life or death' nature of the data, we would wish to be very clear about the significance of the outcome and it is therefore natural to choose an extreme significance level, such as 0.1%, which implies acceptance of H_0 if z falls in the interval $(-3.29, 3.29)$. The observed z-value is -5, which does not fall in this interval.

We can therefore reject H_0 in favour of H_1: it appears that a regular dose of aspirin significantly alters the risk of a heart attack. (Despite this result, it is also true that aspirin is harmful to some people. You should *not* take aspirin regularly therefore unless it is prescribed by your doctor.)

▲————————————————————————————▲

Project ————————————————————————————

> *Here is another project involving car registration plates. The general question is whether the proportion of new cars in one location differs from that in another location. Here are some pairs of possible locations:*
> 1   *Outside detached houses and outside semi-detached houses.*
> 2   *In the station car park and in a supermarket car park.*
> 3   *Travelling at 0800 ('business' cars?) and travelling at 1100 ('shopping' cars?).*

Exercises 17e ————————————————————————————

1   A gardener sows two varieties of runner bean seeds. 152 seeds of Variety A yield 82 healthy plants. 100 seeds of Variety B yield 59 healthy plants.
   Using a 10% level, test whether there is a difference between the proportions of healthy plants that can be expected from the two varieties.

2   A random sample of 600 men with supposedly bad eating habits were observed for a period of 15 years. During this time 34 died from heart disease. Another random sample of 600 men had a supposedly healthy diet for the same period, and the number who died from heart disease was 14.
   Test, at the 1% level, whether the proportion of deaths amongst those on the supposedly healthy diet is significantly lower than the proportion amongst those with bad eating habits.

3   A region has a large population, 60% of whom have surnames beginning with a letter between A and M in the alphabet. In a random sample of 400 people from a town in the region it is found that 260 have surnames beginning with a letter between A and M. Test whether this result indicates any significant difference between the town and the region. Give full details of your test, stating any assumptions made and the hypotheses under test.
   In a sample of 300 from another town in the region, 200 have surnames beginning with a letter between A and M. Test whether the results from the two towns indicate a significant difference between them. Give full details of your test.   [UCLES]

4   In a random sample of 200 people from the electorate in a given constituency, 43% reply that they would vote for political party A. Determine 95% confidence limits for the proportion of the electorate who would reply in the same way.
   A sample of 100 people is taken from another constituency and 35% reply that they would vote for party A. Test whether there is any significant difference between the two constituencies. Give full details of your test, stating any assumptions and the hypotheses under test.   [UCLES]

17.6 Paired samples

Before dealing with the testing of paired samples, we give some idea of the reasons for wanting to carry out such tests.

Experimental design

It is widely thought that people's reaction times are shorter in the morning, and generally increase as the day goes on. One way of testing reaction times uses a buzzer and a light. The light is programmed to flash at random intervals and the experimental subject has to press the buzzer as soon after as possible. A linked computer records the delay between the two actions. Some people (particularly sportsmen and sportswomen) have amazingly fast reactions.

How might we test this idea in the context of a school? Here is a suggestion:

Experiment 1

Two random samples of 40 students are selected from the school register. One of these samples, chosen at random, uses the apparatus during the first period of the day, while the second sample uses the apparatus during the last period of the day. The means of the two samples are then compared.

This will require a standard two-sample comparison of means, assuming a common variance and using the test described in Section 17.3 (p. 458). There is nothing actually *wrong* with the procedure, but we could be misled. Suppose that all the bookworms were in the first sample and all the athletes were in the second – we might well conclude that reaction times improve with time of day!

There is a second, more subtle, problem with Experiment 1. We have said that reaction times may vary greatly from student to student. These variations between students may be much greater than any changes in individual students over the time of day. The latter may then pass unnoticed.

Here is an improved suggestion:

Experiment 2

A random sample of 40 students is selected from the school register. Each student in the sample is tested in the first period of the day, and again in the final period of the day. The differences in reaction times between the two periods for each student are calculated. The mean difference is compared with zero.

The problems with Experiment 1 have vanished. The variability *between* students plays no part. All that matters is the variability of the changes *within* each student's readings.

Experiment 2 is a simple example of a paired-sample test.

Notes
- The two-sample test (Experiment 1) uses a pooled estimate of variance. The variance being estimated is $\sigma_s^2 + \sigma_e^2$, where σ_s^2 represents the variability of the students themselves and σ_e^2 represents the usual random errors.
  The paired-sample test (Experiment 2) uses the estimated variance of the differences. Here the student variability cancels out, leaving the quantity being estimated as $2\sigma_e^2$. The factor 2 occurs because 2 measurements are involved. The paired-sample test is therefore more efficient and preferable whenever (as is usually the case) σ_s^2 noticeably exceeds σ_e^2.
- A paired-sample test is a very simple example of the application of **experimental design** in Statistics. Before we continue with the analysis of a paired-sample test, here is an example of a more complicated design:

Experiment 3

The school is divided into two populations: athletes and non-athletes. These sub-populations are themselves divided into two parts: older students and younger students. From each of the resulting four groups of students a random sample of 5 boys and 5 girls is chosen. Testing then proceeds as in Experiment 2.

This experiment also uses the paired sample idea, and the values to be analysed will be the differences between reaction times early and late in the day. However, with this experiment we can also examine differences between athletes and non-athletes, between boys and girls, and between older and younger students. Without taking any more readings than in the earlier experiments, we can answer questions about four separate possible effects all at once. Experimental design is a powerful idea!

Distinguishing between the paired-sample and two-sample cases

The rules are simple:

- If the two samples are of unequal size then they are not paired.
- Two samples of equal size are paired *only* if we can be *certain* that each observation from the second sample is associated with a corresponding observation from the first sample.

The paired-sample comparison of means

Writing μ_d for the mean of the distribution of differences between the paired values, the hypothesis becomes:

$$H_0: \mu_d = 0$$

with a one-sided or two-sided alternative as appropriate.

We have a single set of n pairs of values and are interested in the *differences* $d_1, d_2, \ldots, d_n$, which, assuming H_0, are a random sample from a population with mean 0. An unbiased estimate of the unknown variance of this population, σ_d^2, is provided by s_d^2, defined by:

$$s_d^2 = \frac{\sum d_i^2 - \frac{1}{n}\left(\sum d_i\right)^2}{n - 1}$$

Although the data arise from two sets of measurements, by working with differences we have effectively created a single sample situation, so that the methods of Chapters 14 and 15 apply. For example, if the differences can be assumed to have a normal distribution, or if n is sufficiently large that a normal approximation can be used, then a 95% confidence interval for μ_d is provided by:

$$\left(\bar{d} - 1.96\sqrt{\frac{s_d^2}{n}}, \quad \bar{d} + 1.96\sqrt{\frac{s_d^2}{n}}\right)$$

where:

$$\bar{d} = \frac{\sum d_i}{n}$$

Alternatively, if the differences can be presumed to have a normal distribution, but n is small, then a t_{n-1}-distribution can be used.

Example 9

Suppose that Experiment 2 on the reaction times is carried out, with the following results (in units of 0.001 seconds).

Subject	1	2	3	4	5	6	7	8	9	10	11	12	13	14
First period	23	50	31	44	92	70	33	44	58	39	44	42	60	61
Last period	29	71	50	50	68	52	55	38	53	61	66	82	59	68
Difference	6	21	19	6	−24	−18	22	−6	−5	22	22	40	−1	7

Subject	15	16	17	18	19	20	21	22	23	24	25	26	27	28
First period	77	33	31	22	25	82	44	69	38	55	29	70	88	81
Last period	63	40	49	51	35	68	68	68	45	57	43	75	99	62
Difference	−14	7	18	29	10	−14	24	−1	7	2	14	5	11	−19

Subject	29	30	31	32	33	34	35	36	37	38	39	40
First period	43	55	61	69	29	31	27	73	81	48	59	61
Last period	50	61	60	73	45	40	29	64	84	58	57	92
Difference	7	6	−1	4	16	9	2	−9	3	10	−2	31

Analyse these data to determine whether there is significant evidence, at the 1% level, of an increase in reaction times,

- (i) efficiently,   using a paired-sample test,
- (ii) inefficiently,   using a two-sample test.

Comment on the results.

(i) *Efficient analysis, using the paired-sample test*

The summary statistics are $\sum d_i = 266, \sum d_i^2 = 9574$ so that $\bar{d} = 6.650$ and $s_d^2 = 200.1308$. The hypotheses being compared are:

$$H_0: \mu_d = 0, \; H_1: \mu_d > 0$$

The test statistic is z, given by:

$$z = \frac{\bar{d}}{\sqrt{\dfrac{s_d^2}{n}}} = \frac{6.650}{2.237} = 2.97$$

This value greatly exceeds the upper 1% point of the standard normal distribution (2.326). The hypothesis of no change in mean reaction time can confidently be rejected in favour of the alternative hypothesis that, by the final period, reaction times have increased. The analysis is efficient because it uses the knowledge that two observations have been generated by each individual.

Paired samples (small variance)

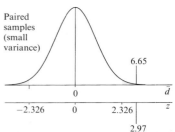

It would be more accurate to compare the observed value, 2.97, with the upper 1% point of a *t*-distribution with 39 degrees of freedom (2.43). The conclusion is unaltered in this case.

(ii) *Inefficient analysis, using the two-sample test*
The summary statistics are $\sum x_i = 2072$, $\sum x_i^2 = 122\,528$, $\sum y_j = 2338$,
$\sum y_j^2 = 146\,300$. Thus $\bar{x} = 51.80$, $\bar{y} = 58.45$ and the difference in the
means is 6.65, as before.

Assuming that the two sets of data arise from populations

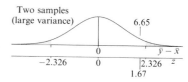

that have common variance, we use the pooled estimate of the
common variance, s^2, given by:

$$s^2 = \frac{\left(122\,528 - \frac{1}{40} \times 2072^2\right) + \left(146\,300 - \frac{1}{40} \times 2338^2\right)}{40 + 40 - 2} = 318.491$$

The test statistic, z, is given by:

$$z = \frac{\bar{y} - \bar{x}}{\sqrt{s^2\left(\dfrac{1}{n_x} + \dfrac{1}{n_y}\right)}} = \frac{6.650}{\sqrt{318.491\left(\frac{1}{40} + \frac{1}{40}\right)}} = 1.67$$

This value does not exceed 2.326, so with this test we would fail to
reject the null hypothesis.

The conclusion is unaltered by using the more accurate
t-distribution with 78 degrees of freedom.

Comment

In the two-sample test, the large variations in the individual reaction times,
which range from 0.022 seconds to 0.092 seconds during the first period,
have obscured the relatively small mean increase (0.006 65 seconds)
between the two periods.

Practical

*Does being right-handed imply that your right hand is more flexible? In
order to find out, use a ruler to measure R, the span of your right hand
from outstretched thumb tip to the end of your little finger. Now find L,
the corresponding distance for your left hand. Calculate d, which is equal
to R − L for right-handers and L − R for left-handers.*

*The null hypothesis is that there is no difference between hand spans
($\mu_d = 0$) and the alternative is that the favoured hand has the larger span
($\mu_d > 0$).*

Is it true for you?

How about for the class in general?

Perform a paired-sample test.

Exercises 17f

1 An experiment to discover the movement of
antibiotics in a certain variety of broad bean
plants was made by treating 10 cut shoots and
10 rooted plants for 18 hours with a solution
containing 200 micrograms per millilitre of
chloramphericol. After treatment, the
concentrations of chloramphericol in a part of
the treated plants were as given below:

Cut shoots	55	61	57	60	52
Rooted plants	53	50	43	46	35

Cut shoots	65	48	58	68	63
Rooted plants	48	39	44	56	51

(continued)

Find a 95% confidence interval for the difference in the mean levels of chloramphericol in the two populations:

(i) assuming that the samples are independent, but come from normal distributions having the same variance,

(ii) assuming that the samples were paired.

2 (i) A group of seven sunflower plants were given fertiliser during April and a second group (of six plants) were given fertiliser during May. The yields (total seed weight in g) of the remainder are summarised below.

| April | 203, 342, 199, 286, 313, 301, 277 |
| May | 177, 276, 231, 299, 218, 188 |

Assuming that the two sets of observations can be regarded as arising from normal distributions having the same variance, σ^2:

(a) obtain the pooled two-sample unbiased estimate of σ^2,

(b) perform a two-sample t-test, at the 5% level, of the hypothesis that there is no difference in the means of the underlying distributions.

(ii) In a second experiment, twelve pairs of plants were positioned close to each other in various different locations in a large greenhouse. One plant in each pair was given fertiliser in April and the other in May. The yields are given below.

Pair	1	2	3	4	5	6
April	344	307	339	256	398	267
May	315	289	317	277	363	258

Pair	7	8	9	10	11	12
April	256	407	335	381	300	388
May	283	385	269	355	275	363

(a) Explain briefly why this is a better experiment than that described earlier in the question.

(b) Assuming normality, use a t-test, at the 5% level, to ascertain whether there is a significant difference between the population means.

3 An area of land was sampled at the same ten points on a day in April after prolonged rain in 1989 and in 1990. The percentage of water in each sample was calculated, giving the following results.

Sample point	1	2	3	4	5
% water in 1989	20	15	26	19	19
% water in 1990	19	24	21	29	23

Sample point	6	7	8	9	10
% water in 1989	17	24	29	19	23
% water in 1990	28	30	21	32	26

It was believed that the water content of the area, at the time sampled, was greater in 1990 than in 1989. Determine if the data supports this belief, at the 5% level of significance, using a t-test. State the assumptions about the data which are necessary for the validity of the t-test used.

[UCLES(P)]

4 Nine swimmers were timed over a 100 metre distance, first in an outdoor pool and then in an indoor pool. Their times, in seconds, given in the alphabetical order of the swimmers' names were, in the outdoor pool,

63.1 65.0 65.1 62.0 67.1 65.2 64.3 68.2 65.9,

and in the same order, in the indoor pool,

62.6 64.2 64.7 62.0 66.8 65.2 63.7 67.4 65.7.

An amateur statistician suggests performing a two-sample t-test to test whether there is a significant difference between the times taken in the indoor and outdoor pools. Carry out this suggestion, stating your null and alternative hypotheses.

Carry out a more appropriate test and state why you consider it to be more appropriate.

[UCLES(P)]

5 State conditions under which it is valid to use the two-sample (i.e. unpaired) t-test to test for a difference between the means of two populations. Ten hospital patients, selected at random, were given a drug A and their reaction times, in milliseconds, to a certain stimulus were measured with the following results.

303 289 291 288 293 280 285 297 283 298

Ten other patients were given drug B and their reaction times to the same stimulus were:

295 294 278 291 284 282 275 293 272 283

Assuming that the necessary conditions apply, show that the data do not provide evidence at the 5% significance level of a difference between mean reaction times after the administration of the two drugs.

(continued)

In fact the two sets of figures had been obtained from the same ten patients in the same order. Carry out a more appropriate *t*-test, using a 5% significance level.   [UCLES]

6 A national company owns a chain of laboratories at which routine chemical tests are carried out. A junior analyst is employed at one of these laboratories. In order to check his accuracy the senior analyst retested 6 samples chosen at random from amongst those analysed by this junior. The results were as follows.

Sample	1	2	3
Senior	8.51	14.70	3.59
Junior	8.62	13.97	4.07

Sample	4	5	6
Senior	7.63	5.55	4.57
Junior	7.97	5.83	4.62

Test, at the 5% significance level, whether there is a disparity between the sets of results.   [UCLES(P)]

7 The ability to withstand pain is known to vary from individual to individual. In a standard test a tiny electric shock is applied to the finger until a tingling sensation is felt. When this test was applied to a random sample of ten adults, the times recorded, in seconds, before they experienced a tingling sensation were

4.2, 4.5, 3.9, 4.4, 3.9, 4.5, 3.7, 4.8, 3.9, 4.2.

Making suitable assumptions, which should be stated, test, at the 5% level, the hypothesis that the average time before an adult would experience a tingling sensation is 4.0 seconds. It is believed that physical exercise increases the time until a tingling sensation is felt. The same ten adults were retested after performing a prescribed set of exercises. With the adults in the same order, the results, in seconds, were

4.4, 4.9, 4.0, 4.9, 3.8, 5.1, 3.6, 5.5, 3.8, 4.5.

Test this belief, at the 5% level, choosing an appropriate *t*-test and justifying your choice.   [UCLES]

Chapter summary

♦ **The comparison of two means:**
$H_0: \mu_x = \mu_y$, $H_1: \mu_x \neq \mu_y$ or $H_1: \mu_x > \mu_y$, or $H_1: \mu_x < \mu_y$

● **Variances known**

$$z = \frac{\bar{x} - \bar{y}}{\sqrt{\frac{\sigma_x^2}{n_x} + \frac{\sigma_y^2}{n_y}}}$$

If the distributions are normal, or n_x and n_y are large, then, assuming H_0, z is an observation from a N(0,1) distribution.

● **Unknown common variance**

$$z = \frac{\bar{x} - \bar{y}}{\sqrt{s^2\left(\frac{1}{n_x} + \frac{1}{n_y}\right)}}$$

(continued)

where the pooled estimate of the common variance is:

$$s^2 = \frac{\left\{\sum x_i^2 - \frac{1}{n_x}\left(\sum x_i\right)^2\right\} + \left\{\sum y_j^2 - \frac{1}{n_y}\left(\sum y_j\right)^2\right\}}{n_x + n_y - 2}$$

$$= \frac{(n_x - 1)s_x^2 + (n_y - 1)s_y^2}{n_x + n_y - 2}$$

$$= \frac{n_x \sigma_{n_x}^2 + n_y \sigma_{n_y}^2}{n_x + n_y - 2}$$

$$= \frac{\sum(x_i - \bar{x})^2 + \sum(y_j - \bar{y})^2}{n_x + n_y - 2}$$

If the observations come from normal distributions, then, assuming H_0, z is an observation from a $t_{n_x+n_y-2}$-distribution. If $(n_x + n_y)$ is large (say 30 or more), then z is an observation from an approximate $N(0,1)$ distribution.

- **Pooled estimate of common mean**

$$\hat{\mu} = \frac{\sum x_i + \sum y_j}{n_x + n_y} = \frac{n_x \bar{x} + n_y \bar{y}}{n_x + n_y}$$

- **Paired samples**

$$z = \frac{\bar{d}}{\sqrt{\frac{s_d^2}{n}}}$$

where:

$$s_d^2 = \frac{\sum d_i^2 - \frac{1}{n}\left(\sum d_i\right)^2}{n - 1}$$

If the observations come from normal distributions, then, assuming H_0, z is an observation from a t_{n-1}-distribution. If n is large (say 30 or more), then z is an observation from an approximate $N(0,1)$ distribution.

- **The comparison of two proportions:**
  H_0: $p_1 = p_2$, H_1: $p_1 \neq p_2$ or H_1: $p_1 > p_2$ or H_1: $p_1 < p_2$

$$z = \frac{\hat{p}_1 - \hat{p}_2}{\sqrt{\hat{p}\hat{q}\left(\frac{1}{n_1} + \frac{1}{n_2}\right)}}$$

where the pooled estimate of the common proportion is:

$$\hat{p} = \frac{n_1\hat{p}_1 + n_2\hat{p}_2}{n_1 + n_2}$$

and $\hat{q} = 1 - \hat{p}$.

If n_1 and n_2 are large, then assuming H_0, z is an observation from an approximate $N(0,1)$ distribution.

18 Goodness of fit

To observations which ourselves we make,
We grow more partial for th' observer's sake

Moral Essays, Alexander Pope

Previous chapters have assumed that a particular type of distribution is appropriate for the data given and have focused on estimating and testing hypotheses about the parameter(s) of this distribution. In this chapter the focus switches to the distribution itself, and we ask the question 'Does the data support the assumption that a particular type of distribution is appropriate?'.

Suppose, for example, that we roll an apparently normal six-sided die 60 times and obtain the following observed frequencies:

Outcome	1	2	3	4	5	6
Observed frequency	4	7	16	8	8	17

In this sample (of possible results from rolling the die) there seems to be a rather large number of 3s and 6s. Is this die fair, or is it biased? With a fair die the probability of each outcome is $\frac{1}{6}$. With 60 tosses the expected frequencies would each be $60 \times \frac{1}{6} = 10$:

Outcome	1	2	3	4	5	6
Expected frequency	10	10	10	10	10	10

The question of interest is whether the observed frequencies (O) and the expected frequencies (E) are reasonably close or unreasonably different. We add the differences ($O - E$) to the table:

Observed frequency, O	4	7	16	8	8	17
Expected frequency, E	10	10	10	10	10	10
Difference, $O - E$	−6	−3	6	−2	−2	7

The larger the magnitude of the differences (i.e. ignoring the sign), the more the observed data differs from that expected according to our model (that the die was fair).

Suppose we now roll a second die 660 times, and obtain the following results:

Observed frequency, O	104	107	116	108	108	117
Expected frequency, E	110	110	110	110	110	110
Difference, $O - E$	−6	−3	6	−2	−2	7

This time the observed and expected frequencies seem remarkably close, yet the $O - E$ values are the same as before: it is not simply the size of $O - E$ that matters, but also its size relative to the expected frequency, $\dfrac{O - E}{E}$.

Combining the ideas that both 'difference' and 'relative size' matter might suggest using the product $(O - E) \times \dfrac{O - E}{E}$, so that the goodness of fit for outcome i is measured using $\dfrac{(O_i - E_i)^2}{E_i}$. The smaller this quantity is, the better the fit. An aggregate measure of **goodness of fit** of the model is therefore provided by X^2, defined by:

$$X^2 = \sum_{i=1}^{m} \frac{(O_i - E_i)^2}{E_i} \tag{18.1}$$

where m is the number of different outcomes (6, in the case of a die). Significantly large values of X^2 suggest lack of fit.

Different samples (e.g. different sets of 60 rolls of the die) will give different sets of observed frequencies and hence different values for X^2. Thus X^2 has a probability distribution! Karl Pearson (see below) showed that when the probabilities of the various outcomes are correctly specified by the null hypothesis, X^2 is (approximately) an observation from a so-called chi-squared distribution.

> The X^2 goodness-of-fit test was first proposed in 1900 by Karl Pearson (1857-1936). Pearson, a Yorkshireman, graduated in Mathematics from Cambridge and spent most of his working life at University College, London. Originally appointed as Professor of Applied Mathematics and Mechanics in 1884, in 1890 he added the title of Gresham lecturer in Geometry. It was not until 1893 that Pearson started publishing articles on Statistics. By that time he already had 100 publications to his name (including a number on German history and folklore!). His initial statistical work included two volumes entitled *The Chances of Death and other Studies in Evolution*, and much of his subsequent work on statistical theory had a similar focus. In 1911 he was appointed Professor of Eugenics (the study of human evolution), a post he held until 1933.

18.1 The chi-squared distribution

'Chi' is the Greek letter χ, pronounced 'kye'. The chi-squared distribution is continuous and has a positive integer parameter d which determines its shape.

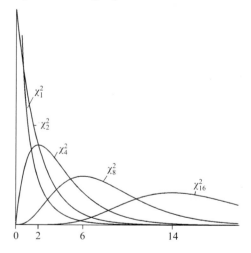

As its name implies, χ^2 cannot take a negative value. The parameter d is known as the **degrees of freedom** of the distribution and we refer to a 'chi-squared distribution with d degrees of freedom'. For simplicity, we write this as:

$$\chi_d^2$$

Properties of the chi-squared distribution

The following properties are included here only for completeness. They are important in later work but have no direct bearing on tests of goodness of fit.

♦   A χ_d^2 distribution has mean d and variance $2d$.

♦   A χ_d^2 distribution has mode at $d - 2$ for $d \geqslant 2$. This is useful when doing a quick sketch.

♦ If Z has a N(0,1) distribution, then Z^2 has a χ_1^2 distribution.
♦ If U and V are independent random variables having χ_u^2 and χ_v^2 distributions, respectively, then their sum $U + V$ has a χ_{u+v}^2 distribution.
♦ The χ_2^2 distribution is an exponential distribution with mean 2.

Tables of the chi-squared distribution

The usual layout consists of a few selected percentage points giving the columns of the table, with rows referring to different values for d. Here is an extract from the table given in the Appendix (p. 622) at the end of this book:

d	\multicolumn{6}{c}{$p(\%)$}
	90	95	97.5	99.0	99.5	99.9
1	2.706	3.841	5.024	6.635	7.879	10.83
2	4.605	5.991	7.378	9.210	10.60	13.82
3	6.251	7.815	9.348	11.34	12.84	16.27
4	7.779	9.488	11.14	13.28	14.86	18.47
5	9.236	11.07	12.83	15.09	16.75	20.52

If X has a χ_d^2 distribution, then a tabulated value x is such that $P(X < x) = p\%$. Thus $P(\chi_1^2 < 2.706) = 0.900$, $P(\chi_5^2 > 20.52) = 0.001$ and the upper 1% percentage point of a χ_3^2 distribution is 11.34.

Exercises 18a

1 Find: (i) $P(\chi_4^2 > 11.14)$, (ii) $P(\chi_5^2 < 11.07)$, (iii) $P(\chi_3^2 > 12.84)$, (iv) $P(\chi_1^2 < 6.635)$, (v) $P(\chi_1^2 > 1.96^2)$.

2 Find:
(i) $P(7.779 < \chi_4^2 < 13.28)$,
(ii) $P(11.07 < \chi_5^2 < 16.75)$,
(iii) $P(7.378 < \chi_2^2 < 9.210)$.

3 Find c such that:
(i) $P(\chi_4^2 > c) = 0.005$, (ii) $P(\chi_5^2 > c) = 0.025$,
(iii) $P(\chi_1^2 > c) = 0.100$, (iv) $P(\chi_1^2 < c) = 0.995$,
(v) $P(\chi_3^2 < c) = 0.975$.

4 Verify that the upper percentage points of χ_1^2, given in the table above, are (except for rounding errors) the squares of the corresponding two-tail percentage points of N(0, 1).

5 By finding the cumulative distribution function of an exponential distribution with mean 2, verify the entries in the above table for the case $d = 2$.

18.2 Goodness of fit to prescribed probabilities

The goodness-of-fit statistic proposed earlier was X^2, where:

$$X^2 = \sum_{i=1}^{m} \frac{(O_i - E_i)^2}{E_i}$$

Here O_i and E_i are, respectively, observed and expected frequencies and m is the number of categories being compared. H_0 specifies the probabilities of the various categories, and the expected frequencies are the product of the sample size and these probabilities. The alternative hypothesis is that H_0 is incorrect. Assuming H_0, X^2 is *approximately* an observation from a chi-squared distribution with $m - 1$ degrees of freedom (χ_{m-1}^2).

We now continue with the calculations of X^2 for the 60 throws of the die with which we started the chapter, using the null hypothesis that the die is fair, and (for variety!) a 2.5% significance level.

O_i	E_i	$O_i - E_i$	$\dfrac{(O_i - E_i)^2}{E_i}$
4	10	−6	3.6
7	10	−3	0.9
16	10	6	3.6
8	10	−2	0.4
8	10	−2	0.4
17	10	7	4.9
Total 60	60	0	13.8

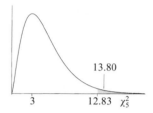

In this case $m = 6$ and the relevant χ^2 distribution therefore has 5 degrees of freedom. The upper 2.5% point of a χ_5^2 distribution is 12.83, which is less than the value of X^2 (13.8): there is therefore significant evidence, at the 2.5% level, that the die is biased.

Notes

- The O_i are observed frequencies and are always whole numbers (not percentages).
- The E_i will not usually be whole numbers.
- The total of the $O_i - E_i$ column should always be zero (barring possible round-off errors).
- Round-off errors can accumulate, so it is advisable to use more decimal places than usual in X^2 calculations.
- In all X^2 goodness-of-fit tests the null hypothesis, H_0, is that the results are a random sample from the supposed distribution; the alternative hypothesis, H_1, simply says that H_0 is incorrect.
- The test is often called the **chi-squared test** (or, by lovers of Greek, the χ^2 test!). Similarly, X^2 is often called χ^2.

Example 1

According to a genetic theory, when sweet peas having red flowers are crossed with sweet peas having blue flowers, the next generation of sweet peas have red, blue and purple flowers in the proportions $\frac{1}{4}$, $\frac{1}{4}$ and $\frac{1}{2}$, respectively. The outcomes in an actual experiment are as follows: 84 with red flowers, 92 with blue flowers and 157 with purple flowers.
Using a 5% significance level, determine whether these results support the theory.

The null hypothesis is that the three proportions are $\frac{1}{4}$, $\frac{1}{4}$ and $\frac{1}{2}$, with the alternative being simply that the null hypothesis is false. We set out the calculations as before:

Type	O_i	E_i	$O_i - E_i$	$\dfrac{(O_i - E_i)^2}{E_i}$
red	84	83.25	0.75	0.007
blue	92	83.25	8.75	0.920
purple	157	166.50	−9.50	0.542
Total	333	333	0	1.469

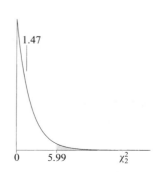

With 3 categories, $m = 3$, there are 2 degrees of freedom and the critical value is the upper 5% point of a χ_2^2 distribution (5.991). Since X^2 (1.469) is less than this value, the results are consistent with the theory.

Example 2

Four coins are tossed 100 times. The numbers of heads obtained on each toss were recorded and are summarised below.

Number of heads	0	1	2	3	4
Frequency	5	23	39	19	14

Using a 2.5% significance level, test the hypothesis that all four coins are unbiased.

___

The null hypothesis is that the probability of a head is $\frac{1}{2}$ for each of the four coins. As usual with these tests, the alternative hypothesis states simply that the null hypothesis is untrue.

If the null hypothesis is true, then the number of heads obtained when the four coins are tossed is an observation from a binomial distribution with $n = 4$ and $p = 0.5$. The probability of obtaining r heads, P_r, is given by:

$$P_r = \binom{4}{r} \left(\frac{1}{2}\right)^r \left(\frac{1}{2}\right)^{4-r} = \frac{1}{16}\binom{4}{r}$$

with the corresponding expected frequency being equal to $100P_r$. Thus, for $r = 0$ we get $P_0 = 0.0625$ and the corresponding expected frequency is $100 \times 0.0625 = 6.25$. The probabilities of the various outcomes, together with the expected frequencies and the stages in the calculation of the X^2 test are summarised in the table below.

No. of heads	O_i	Probability	E_i	$O_i - E_i$	$\dfrac{(O_i - E_i)^2}{E_i}$
0	5	0.0625	6.25	−1.25	0.25
1	23	0.2500	25.00	−2.00	0.16
2	39	0.3750	37.50	1.50	0.06
3	19	0.2500	25.00	−6.00	1.44
4	14	0.0625	6.25	7.75	9.61
Total	100	1.0000	100.00	0	11.52

There are 5 categories and hence 4 degrees of freedom. The upper 2.5% point of a χ_4^2 distribution is 11.14. Since 11.52 is greater than this critical value, there is significant evidence that the four coins are not unbiased. Examination of the table shows that the principal contribution to X^2 arises from the final row of the table: there were too many occasions when all four coins showed a head.

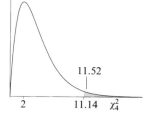

Example 3

A university is subdivided into faculties of Arts, Engineering, Humanities, Law and Science. The numbers of students in these faculties are 1300, 800, 1100, 500 and 1400, respectively.

A questionnaire concerning library usage is sent to all the students. The first 300 replies opened contained 101 from Arts students, 30 from Engineering students and 69, 17 and 83 from students of the other three faculties, respectively.

Using a 0.1% significance level, test whether the replies appear to be providing an unbiased representation of the students in the university.

___

The null hypothesis is that each student is equally likely to reply. The alternative hypothesis is that the sample of replies is in some way biased.

The university contains a total of 5100 students. The probability that a randomly chosen student belongs to the Arts faculty is therefore $\frac{1300}{5100}$. The expected number of Arts students under the hypothesis that the sample of replies is unbiased would therefore be $\frac{13}{51} \times 300 = 76.47$. The remaining expected frequencies can be calculated in the same way. The results are summarised below:

Faculty	O_i	Probability	E_i	$O_i - E_i$	$\dfrac{(O_i - E_i)^2}{E_i}$
Arts	101	$\frac{13}{51}$	76.471	24.529	7.868
Engineering	30	$\frac{8}{51}$	47.059	−17.059	6.184
Humanities	69	$\frac{11}{51}$	64.706	4.294	0.285
Law	17	$\frac{5}{51}$	29.412	−12.412	5.238
Science	83	$\frac{14}{51}$	82.353	0.647	0.005
Total	300	1.000	300.000	0	19.580

Since there are 5 faculties the relevant χ^2 distribution has 4 degrees of freedom. The upper 0.1% point of a χ_4^2 distribution is 18.47, which is less than the observed 19.58. There is therefore significant evidence, at the 0.1% level, that the sample is not a fair cross-section of the university. Studying the table it is apparent that the sample contains a higher than expected proportion of Arts students, while the proportions of Engineering and Law students are unduly low.

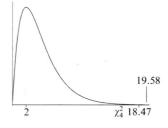

Example 4

Let S denote the sum of 12 uniform random numbers in $(0, 1)$. It is known that, if the random number generator is working correctly, the distribution of S will be approximately normal, with mean 6 and variance 1.

The distribution of 500 values of S is summarised in the table below:

$s < 4$	$4 \leqslant s < 5$	$5 \leqslant s < 6$	$6 \leqslant s < 7$	$7 \leqslant s < 8$	$8 \leqslant s$
10	75	163	174	66	12

Is there evidence, at the 5% significance level, that the random number generator is working incorrectly?

The null hypothesis is that $S \sim N(6,1)$, with the alternative being that this is not the case.

If $S \sim N(6,1)$ then $P(S < 4) = P\left(Z < \dfrac{4 - 6}{\sqrt{1}}\right) = \Phi(-2) = 0.0228$.

For $P(4 \leqslant S < 5)$ we need to calculate $P(S < 5) - P(S < 4)$. This is $\Phi(-1) - \Phi(-2) = 0.1587 - 0.0228 = 0.1359$. We obtain probabilities for the remaining categories in the same way. The calculations are summarised as follows:

Interval	O_i	Probability	E_i	$O_i - E_i$	$\dfrac{(O_i - E_i)^2}{E_i}$
$s < 4$	10	0.0228	11.40	−1.40	0.172
$4 \leqslant s < 5$	75	0.1359	67.95	7.05	0.731
$5 \leqslant s < 6$	163	0.3413	170.65	−7.65	0.343
$6 \leqslant s < 7$	174	0.3413	170.65	3.35	0.066
$7 \leqslant s < 8$	66	0.1359	67.95	−1.95	0.056
$s \geqslant 8$	12	0.0228	11.40	0.60	0.032
Total	500	1.0000	500.00	0	1.400

On this occasion there are 6 categories and hence 5 degrees of freedom. The value of X^2 (1.400) is much less than the upper 5% point of the χ_5^2 distribution (11.07) and hence there is no significant evidence (at the 5% level) that the random number generator is working incorrectly.

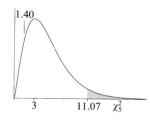

Computer project _____

The repetitive nature of these calculations suggests the use of a spreadsheet in which successive columns hold the observed frequencies, the expected frequencies, their differences and the contributions to X^2. Use a spreadsheet to reproduce the working of the previous example.

Practical _____

Roll a die 30 times, recording the results. Now perform a goodness-of-fit test to determine whether there is significant evidence, at the 10% level, of any bias.

Assuming that your die is fair, what proportion of the time would a sample of 30 rolls give rise to a result that was 'significant at the 10% level'?

What proportion of individuals in your class obtained 'significant' results?

Project _____

In Statistics we frequently use phrases such as 'randomly chosen' or 'at random'. The object of this project is to determine whether people can really choose things 'at random'. If they can't, then tables of random numbers are needed!

Write the letters A B C D E in a horizontal line on a sheet of paper. Then ask people to 'choose a letter at random'. Record their choice. After you have recorded the choices of at least 25 people, test the hypothesis that all letters are chosen with equal probability.

Combine your results with those of the other members of your class. Most research suggests that people are biased towards the ends of lists, and particularly towards the left of a horizontal list. You might like to repeat the experiment with a vertical list.

Exercises 18b

1 A random sample of 80 observations on X is summarised below:

x	0	1	2	$\geqslant 3$
Frequency	14	30	17	19

Test, at the 5% significance level, the hypothesis that X has a Poisson distribution with mean equal to 1.6.

2 The null hypothesis is that the discrete random variable X has probability distribution given by:

$$P(X = x) = \begin{cases} \dfrac{x}{10} & (x = 1, 2, 3, 4) \\ 0 & \text{otherwise} \end{cases}$$

A random sample of 70 observations of X is summarised in the table below.

Value of X	1	2	3	4
Frequency	4	20	18	28

Use a chi-squared test to determine whether there is evidence, at the 1% significance level, to reject the null hypothesis.

3 The population of a country consists of three ethnic groups, A, B and C, which make up 30%, 60% and 10% of the population, respectively. A sample of 80 people includes 21 from group A, 57 from group B and 2 from group C.
Is there significant evidence, at the 5% level, that the sample is biased?

4 If a slot machine is working correctly then the symbol in its left window should be equally likely to be a lemon, a banana, a cherry or a bar. In 100 trials the machine shows 23 lemons, 16 bananas, 34 cherries and 27 bars.
Does this provide significant evidence, at the 5% level, that the machine is not working correctly?

5 If a cross-pollination experiment has been performed correctly then $\frac{1}{8}$ of the subsequent plants should have red flowers, $\frac{3}{8}$ should have pink flowers and the remainder should have white flowers. A random sample of 80 plants is found to comprise 4 with red flowers, 40 with pink flowers and 36 with white flowers.
Does this provide significant evidence, at the 0.1% level, that the experiment has been performed incorrectly?

6 A very large population consists of equal numbers of males and females. Forty per cent of each sex have black hair. A sample of 100 members of this population consists of 23 black-haired females, 18 black-haired males, 36 other females and 23 other males.
Determine whether there is significant evidence, at the 5% level, that the sample has not been chosen at random with respect to hair colour.

7 The random variable X takes all real values between 0 and 4 and has probability density function $\dfrac{5}{4(1 + x)^2}$.
Show that $P(X < 1)$ is 0.625 and find $P(X > 2)$. The delay in a computer between asking for a particular result and its appearance on the screen was measured on forty occasions chosen at random. The results are summarised below.

Delay (minutes)	< 1	$1 - 2$	> 2
Frequency	26	4	10

Use χ^2 with three classes and a 5% significance level to test the hypothesis that this delay has the same probability distribution as X.   [O&C(P)]

8 The continuous random variable X has a normal distribution with mean 254.0 and standard deviation 2.4. The numbers x_1, x_2, x_3 and x_4, where $x_1 < x_2 < x_3 < x_4$, divide the range of X ($-\infty$ to $+\infty$) into five intervals such that the probability of X lying in any one of the intervals is $\frac{1}{5}$. Show that $x_1 = 251.98$, correct to 2 decimal places, and find the values of x_2, x_3 and x_4.

A random sample of 40 observations of X is taken. Write down the expected frequencies for each of the five intervals found above.

Bottles of a certain brand of hair shampoo are marked as containing 250 ml. A random sample of 40 bottles was tested and the volumes of their contents measured in ml. The results, in numerical order, were as follows:

248.1 249.0 249.1 250.3 250.8 251.4 252.2
252.2 252.4 252.4 252.5 252.5 252.7 252.9
253.1 253.4 253.5 253.5 253.6 253.8 253.9
253.9 254.1 254.3 254.3 254.4 254.5 254.5
254.6 254.8 254.9 255.0 255.3 255.4 255.8
256.2 257.0 257.2 258.3 259.4

(continued)

(i) Carry out a suitable test, at the 5% significance level, to determine whether the sample supports the manufacturer's claim that the bottles contain amounts which are normally distributed with mean 254.0 ml and standard deviation 2.4 ml. State your conclusions clearly.

(ii) The sample mean is 253.68 ml. Assuming that the population mean and standard deviation are as claimed in (i), calculate the probability that a random sample of size 40 will yield a sample mean of at least 253.68 ml.   [UCLES]

18.3 Small expected frequencies

The distribution of X^2 is discrete – the χ^2 distribution is continuous and is simply a convenient approximation which becomes less accurate as the expected frequencies become smaller. The rule often stated for deciding whether the approximation may be used is:

'All expected frequencies must be equal to at least 5'.

If the original categories chosen lead to expected frequencies less than 5, then it will be necessary to combine categories together. This combination may be done on any sensible grounds, but should be done *without reference to the observed frequencies* so as to avoid biasing the results. With numerical data it is natural to combine adjacent categories: for example, we might replace the three categories '7', '8' and '9' by the single category '7–9'.

Note

◆ The rule given ('at least 5') errs on the safe side. Many researchers happily permit a small proportion of expected frequencies to be less than 5.

Example 5

A test of a random number generator is provided by studying the lengths of 'runs' of digits. The probability of a run of length k (i.e. that a randomly chosen digit is followed by exactly $k-1$ similar digits) is $0.9 \times 0.1^{k-1}$. This is a geometric distribution (see Section 7.4, p. 174).

A sequence of supposedly random numbers are generated, and the following results are obtained:

Length of run	1	2	3	4	5	6 or more
Frequency	8083	825	75	9	1	0

Use a 10% significance level to decide whether these results suggest that there is anything wrong with the random number generator.

The null hypothesis specifies that a run of length k has probability
$0.9 \times 0.1^{k-1}$, with the alternative hypothesis stating that the null hypothesis
is incorrect.

Run length	O_i	Probability	E_i	$O_i - E_i$	$\dfrac{(O_i - E_i)^2}{E_i}$
1	8083	0.900	8093.700	−10.700	0.014
2	825	0.090	809.370	15.630	0.302
3	75	0.009	80.937	−5.937	0.435
4 ⎫ 5 ⎬ 4+ 6+ ⎭	9 ⎫ 1 ⎬ 10 0 ⎭	} 0.001	8.094 ⎫ 0.809 ⎬ 8.993 0.090 ⎭	1.007	0.113
Total	8993	1.000	8993.000	0	0.864

The expected frequencies are shown in the table. The frequency for
run lengths of 6 or more is obtained by subtraction. We then find that
the last two expected frequencies are very small and we combine these
with the previous category to form a category '4+'.

After combining these categories m becomes 4 and we use the χ_3^2
distribution. The upper 10% point of this distribution is 6.251 which
greatly exceeds the observed value (0.864). There is no significant evidence
for rejecting the null hypothesis that the random number generator is
working correctly.

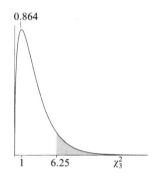

Exercises 18c

1 A random sample of 50 observations on the
discrete random variable X is summarised
below:

x	0	1	2	3
Frequency	12	16	20	2

Test, at the 1% significance level, the
hypothesis that $X \sim B(3, 0.45)$.

2 A random sample of 100 observations on the
discrete random variable X is summarised
below:

x	0	1	2	3	4
Frequency	18	36	36	8	2

Test, at the 1% significance level, the
hypothesis that $X \sim B(4, 0.3)$.

3 A random sample of 80 observations on the
discrete random variable X is summarised below:

x	0	1	2	3	$\geqslant 4$
Frequency	24	30	17	5	4

Test, at the 5% significance level, the
hypothesis that X has a Poisson distribution
with mean equal to 1.2.

4 A random sample of 150 observations on the
continuous variable X is summarised below:

x	<5	5–	10–	15–
Frequency	2	6	24	51

x	20–	25–	30–	$\geqslant 35$
Frequency	35	28	3	1

Test, at the 1% significance level, the
hypothesis that $X \sim N(20, 36)$.

5 The null hypothesis is that the random variable X has a binomial distribution with $n = 6$ and $p = \frac{1}{3}$. A random sample of 60 observations gave the following results:

x	0	1	2	3	4	5	6
Frequency	8	16	18	15	3	0	0

Test the null hypothesis using a 5% significance level.

6 Timothy believes that his lucky coin is fair. To test his belief he tosses the coin and counts the numbers of tails between successive heads. If the coin is fair then $P(r \text{ tails}) = (\frac{1}{2})^{r+1}$ for $r = 0, 1, 2, \ldots$

Timothy's results are as follows:

r	0	1	2	3	4	$\geqslant 5$
Frequency	25	18	9	6	2	0

Determine whether there is significant evidence, at the 5% level, that the coin is biased.

7 A certain plant population has flowers of various colours whose proportions are given in the table below. A gardener has 80 of the plants, as specified in the table.

Colour	red	white	pink	orange	yellow
Population	0.4	0.3	0.2	0.05	0.05
Plants	24	28	24	3	1

Test, at the 5% level, the hypothesis that the gardener's plants may be regarded as a random sample from the population.

8 A botanist conjectures that the specimens of a particular plant are growing at random locations in a bog, with mean rate 4 plants per square metre. To test the conjecture she divides a randomly chosen region of the bog into non-overlapping regions of area $0.5\,\text{m}^2$ and counts the numbers of plants in each region. The results were as follows:

No. of plants	0	1	2	3	4	$\geqslant 5$
No. of regions	4	14	9	7	6	0

Test the conjecture using a 5% significance level.

9 A random number program is claimed to produce random integers in the range

  00000 to 99999

inclusive. One way of testing this supposed randomness is to let X be the total number of 3s *and* 7s in a five-digit number that has been produced. For example, for 02037 the value of X is 2, and for 30703 it is 3. If the program is working properly, what would you predict the probability distribution of X to be?

A test run produces the following results:

Value of X	0	1	2	3	4	5
Frequency	1001	1302	624	175	23	0

Test the consistency of this sample with the predicted distribution of X, stating the basis of your calculations. [SMP]

10 A random variable X has a normal distribution with mean 35 and variance 100. The first table below shows the probability that the value of a single reading, x, lies in some particular interval. Copy and complete this table.

x	less than 10	10–	20–	30–
Probability	0.0062			0.3830

x	40–	50–	60 and above
Probability			

The second table shows the frequency distribution of the times, in seconds, required by 200 ten-year-old children to tie both their shoe laces.

Time	less than 10	10–	20–	30–
Frequency	8	11	40	59

Time	40–	50–	60 and above
Frequency	66	10	6

Perform a χ^2 goodness-of-fit test to show that there is evidence to suggest that the times taken by ten-year-old children to tie both their shoe laces do not follow a normal distribution with mean 35 seconds and standard deviation 10 seconds. [JMB(P)]

11 The heights (x) of one hundred police officers recruited to a police force in a particular year are summarised in the table below.

Height (cm)	Frequency
$x < 175$	2
$175 \leqslant x < 177$	15
$177 \leqslant x < 179$	29
$179 \leqslant x < 181$	25
$181 \leqslant x < 183$	12
$183 \leqslant x < 185$	10
$185 \leqslant x$	7

The population of police officers has a mean height of 180 cm with a standard deviation of 3 cm. Test the hypothesis that the distribution of heights is normal.

12 Pseudo-random numbers are generated, usually by computer, using mathematical algorithms. The numbers generated are supposed to mimic the properties of genuine (unpredictable) random numbers, but are called pseudo-random because the computer will usually generate the same sequence each time the computer is switched on.

Various tests for randomness are applied to a set of supposed pseudo-random numbers, for example:

(i) A set of digits should contain each of $0, 1, \ldots, 9$ with approximately equal frequency.

(ii) The number, K, of non-zero digits occurring between successive occurrences of a multiple of 3 should be an observation from a geometric distribution:

$$P(K = k) = \frac{2^k}{3^{k+1}}, \text{ for } k = 0, 1, \ldots$$

Using the X^2 goodness-of-fit test, perform these two tests of randomness on the following set of 80 digits (working along the rows).

6	5	1	9	1		2	1	4	8	6
8	9	7	2	9		8	7	9	8	9
0	7	3	7	4		9	0	3	4	5
4	5	1	0	7		2	9	5	0	5
7	0	3	9	5		3	1	9	9	7
1	2	3	2	0		9	9	5	6	1
9	3	4	5	5		3	4	9	5	6
4	4	6	7	1		2	7	0	2	6

18.4 Goodness of fit to prescribed distribution type

We now turn to cases where the null hypothesis states that the data 'has a particular named distribution', but does *not* specify all the parameters of the distribution. A typical example would be the hypothesis:

H$_0$: the masses of a certain brand of digestive biscuit have a normal distribution.

The hypothesis does not specify *which* normal distribution, so we choose the most plausible one – this is the one having the same mean and variance as the observed data. Because of this deliberate matching we are imposing constraints on the expected frequencies. The value of X^2 will be a little smaller because of the better fit and this alters the value of d, the degrees of freedom of the approximating χ^2 distribution. The general rule is:

$$d = m - 1 - k \tag{18.2}$$

where m is the number of different outcomes (after amalgamations to eliminate small expected frequencies) and k is the number of parameters estimated from the data. In the cases just considered in Sections 18.2 and 18.3, k was equal to 0.

Example 6

Eggs are packed in cartons of six. On arrival at a supermarket each pack is checked to make sure that no eggs are broken. Fred, the egg-checker, attempts to relieve his boredom by recording the numbers of broken eggs in a pack. After examining 5000 packs his results look like this:

Number of broken eggs	0	1	2	3	4	5	6
Number of packs	4704	273	22	0	0	1	0

Test, at the 0.1% level, whether these results are consistent with the null hypothesis that egg breakages are independent of one another, with each of the six eggs in a pack being equally likely to break.

According to the null hypothesis, all six eggs are equally likely to break and the breakages are independent of one another. If this is the case, then the number of broken eggs in a pack is an observation from a binomial distribution with $n = 6$. Fred examined a total of 30 000 eggs, of which 322 were found to be broken. The sample estimate of p is therefore $\frac{322}{30\,000}$. The X^2 calculations now proceed as usual, with, in this case, the last five categories being combined.

Broken		O_i	Probability	E_i	$O_i - E_i$	$\dfrac{(O_i - E_i)^2}{E_i}$
0		4704	0.937 30	4686.518	17.482	0.065
1		273	0.061 02	305.086	−32.086	3.374
2 3 4 2+ 5 6	22 0 0 23 1 0		0.001 68	8.396	14.604	25.402
Total		5000	1.000 00	5000.000	0	28.841

After combining the categories, $m = 3$. One parameter (p) was estimated from the data and consequently $d = 3 - 1 - 1 = 1$. The observed value of X^2 greatly exceeds the upper 0.1% point of a χ_1^2 distribution (10.83) so we can confidently reject the null hypothesis.

Examining the data we see that there were too many packs containing two (or more) broken eggs. It seems likely that egg breakages are not independent events, but are caused by packs being dropped and other accidents.

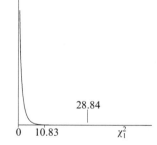

Example 7

A typist produces a 50-page typescript and gives it to its author for checking. She notes the numbers of errors on each page. The results are summarised below:

Number of errors	0	1	2	3	4	5	6	7	8	9	10+
Number of pages	2	5	16	11	6	3	1	2	3	1	0

Test, at the 5% significance level, the null hypothesis that the errors are randomly distributed through the typescript.

The null hypothesis states that the errors are distributed at random. If this is true then the counts should be observations from a Poisson distribution.

In order to test the hypothesis we need first to estimate the mean of the Poisson distribution. In total there are $(0 \times 2) + (1 \times 5) + \ldots = 162$ errors and hence the mean number per page is 3.24. The probability of a page containing r errors is therefore estimated as being P_r, where:

$$P_r = \frac{(3.24)^r \, e^{-3.24}}{r!}$$

and the corresponding expected frequency is $50P_r$. The calculations are set out below.

Number of errors	O_i	Probability	E_i	$O_i - E_i$	$\dfrac{(O_i - E_i)^2}{E_i}$
0 } 1 }	2 } 5 } 7	0.0392 } 0.1269 } 0.1661	1.9582 } 6.3446 } 8.3027	−1.3027	0.204
2	16	0.2056	10.2782	5.7218	3.185
3	11	0.2220	11.1004	−0.1004	0.001
4	6	0.1798	8.9913	−2.9913	0.995
5	3	0.1165	5.8264	−2.8264	1.371
6 } 7 } 8 } 6+ 9 } 10+ }	1 } 2 } 3 } 7 1 } 0 }	0.1100	5.5009	1.4991	0.409
Total	50	1.0000	50.0000	0	6.165

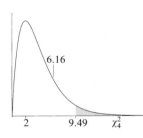

Notice that, in this case, it was necessary to combine categories in both tails of the distribution. After these combinations, $m = 6$, and hence $d = 6 - 1 - 1 = 4$, since one parameter was estimated from the data. The observed value of X^2 (6.165) does not exceed the upper 5% point of a χ_4^2 distribution (9.488), and we can therefore accept the hypothesis that the typist's errors occurred at random.

Example 8

The following table gives the distribution of systolic blood pressure (in mm of mercury) for a random sample of 250 men aged between 30 and 39.

Blood pressure (b)	$80 < b \leqslant 100$	$100 < b \leqslant 110$	$110 < b \leqslant 120$	$120 < b \leqslant 130$
Number of men	3	12	52	74

Blood pressure (b)	$130 < b \leqslant 140$	$140 < b \leqslant 150$	$150 < b \leqslant 160$	$160 < b \leqslant 180$
Number of men	67	26	12	4

Determine whether there is significant evidence, at the 5% level, to reject the null hypothesis that blood pressure has a normal distribution.

To test the null hypothesis we need estimates of the population mean and variance. For this we use the methods of Chapter 2, using class mid-points and coded values. Treating b as a continuous variable and denoting a class mid-point by x, we shall use the coded values given by $y = \dfrac{(x - 130)}{5}$. The basic calculations are shown in the table below.

Blood pressure, b	Mid-point, x	Coded value, y	Frequency, f	fy	fy^2
$80 < b \leqslant 100$	90	-8	3	-24	192
$100 < b \leqslant 110$	105	-5	12	-60	300
$110 < b \leqslant 120$	115	-3	52	-156	468
$120 < b \leqslant 130$	125	-1	74	-74	74
$130 < b \leqslant 140$	135	1	67	67	67
$140 < b \leqslant 150$	145	3	26	78	234
$150 < b \leqslant 160$	155	5	12	60	300
$160 < b \leqslant 180$	170	8	4	32	256
Total			250	-77	1891

The sample mean and the value of s_y^2 for the coded values are given by:

$$\bar{y} = \frac{-77}{250} = -0.308$$

$$s_y^2 = \frac{1}{249}\left\{1891 - \frac{(-77)^2}{250}\right\} = 7.4991$$

Hence the sample mean and the value of s_x^2 for the grouped data are:

$$\bar{x} = 130 + 5\bar{y} = 128.46$$

$$s_x^2 = 5^2 s_y^2 = 187.4783$$

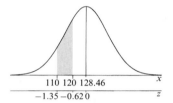

110 120 128.46 x

-1.35 -0.62 0 z

We now calculate the expected frequencies from a normal distribution having precisely these values for its mean and variance. For example, for the class $110 < b \leqslant 120$, the theoretical probability is:

$$\Phi\left(\frac{120 - 128.46}{\sqrt{187.4783}}\right) - \Phi\left(\frac{110 - 128.46}{\sqrt{187.4783}}\right) = \Phi(-0.618) - \Phi(-1.348)$$

$$= 0.2683 - 0.0888 = 0.1795$$

The corresponding expected frequency is therefore $250 \times 0.1795 = 44.875$. The remaining calculations are summarised in the table below, where u denotes the upper class boundaries.

u	$z = \dfrac{u - \bar{x}}{s_x}$	$\Phi(z)$	Class probability	E_i
80	-3.539	0.0002	0.0002	0.050
100	-2.079	0.0188	0.0186	4.650
110	-1.348	0.0888	0.0700	17.500
120	-0.618	0.2683	0.1795	44.875
130	0.112	0.5446	0.2763	69.075
140	0.843	0.8003	0.2557	63.925
150	1.573	0.9422	0.1419	35.475
160	2.303	0.9894	0.0472	11.800
180	3.764	0.9999	0.0105	2.625
∞	∞	1.0000	0.0001	0.025

We need to combine categories at both ends of the distribution. Using the rule that no expected frequencies should be less than 5, the number of categories is reduced to 6. We can now carry out the X^2 calculations:

Category	O_i	E_i	$O_i - E_i$	$\dfrac{(O_i - E_i)^2}{E_i}$
$b \leqslant 110$	15	22.200	−7.200	2.335
$110 < b \leqslant 120$	52	44.875	7.125	1.131
$120 < b \leqslant 130$	74	69.075	4.925	0.351
$130 < b \leqslant 140$	67	63.925	3.075	0.148
$140 < b \leqslant 150$	26	35.475	−9.475	2.531
$150 < b$	16	14.450	1.550	0.166
Total	250	250	0	6.662

There are $6 - 1 - 2 = 3$ degrees of freedom, since both the mean and the variance have been estimated from the data. The value of X^2 (6.662) does not exceed the upper 5% point of a χ_3^2 distribution (7.815) and we therefore accept the null hypothesis that the observations of systolic blood pressure have a normal distribution.

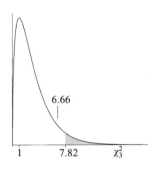

Exercises 18d

1 It is hypothesised that the random variable X has a binomial distribution with $n = 4$. The table below summarises 200 observations on X.

x	0	1	2	3	4
Frequency	46	77	69	7	1

Test the hypothesis at the 1% level.

2 It is hypothesised that the random variable X has a binomial distribution with $n = 4$. The table below summarises 100 observations on X.

x	0	1	2	3	4
Frequency	53	37	7	3	0

Test the hypothesis at the 5% level.

3 It is hypothesised that the random variable X has a Poisson distribution. The table below summarises 80 observations on X.

x	0	1	2	3	4	$\geqslant 5$
Frequency	23	41	10	5	1	0

Test the hypothesis at the 5% level.

4 It is hypothesised that the random variable X has a Poisson distribution. The table below summarises 200 observations on X.

x	0	1	2	3	$\geqslant 4$
Frequency	125	57	11	7	0

Test the hypothesis at the 5% level.

5 Sam regards a day as being 'good' if the telephone does not ring that day! The numbers of good days during the 50 weeks of a year (discounting the Xmas period) were as follows:

No. of days	0	1	2	3	$\geqslant 4$
No. of weeks	25	16	8	1	0

Perform a test, at the 5% level, of the hypothesis that the number of good days in a week has a binomial distribution.

6 The following data represent the lengths (in mm to the nearest 0.1 mm) of cuckoo eggs found in nests of meadow pipits:

19.5	21.6	23.0	18.9	20.0
22.6	19.1	22.4	18.9	23.2
19.7	21.0	19.4	21.5	23.8
19.6	19.9	21.4	20.7	19.7
21.0	20.3	21.7	19.2	19.5
22.1	20.8	20.1	19.2	24.7
22.3	20.2	20.5	19.9	19.4
21.0	25.6	21.3	18.8	22.1

The sum and sum of squares of these 40 observations are 835.6 and 17 561.16, respectively.

(a) Display these data on a histogram using the group intervals:

18.0–18.9, 19.0–19.9, 20.0–20.9,
21.0–21.9, 22.0–22.9, 23.0–25.9

(b) Using the sample mean and the value of s^2 as estimates of the population mean and variance, and without combining any categories, carry out a goodness-of-fit test for the null hypothesis that the data are normally distributed.

7 A gardener sows 4 seeds in each of 100 plant pots. The number of pots in which r of the 4 seeds germinate is given in the table below.

Number of seeds germinating	0	1	2	3	4
Number of pots	21	27	27	21	4

Estimate the probability p of an individual seed germinating. Fit a binomial distribution and test for goodness of fit at the 1% level.

[UCLES(P)]

8 A research study into complications arising from knee surgery collected data on 15 operations from each of 40 surgeons. Patients who had other serious health problems were excluded from the study. The following table summarises the number of each surgeon's patients who suffered from complications, together with the expected number when a binomial distribution is fitted to the data:

Number of patients (out of 15) with complications	0	1	2	3
observed no. of surgeons, O	2	9	8	7
expected no. of surgeons, E	1.41	5.28	9.24	10.01

Number of patients (out of 15) with complications	4	5	6	7 or more
observed no. of surgeons, O	3	6	3	2
expected no. of surgeons, E	7.50	4.12	1.72	0.72

Test, at the 5% significance level, whether the binomial distribution is an adequate model for the data.

State, giving a reason, whether your conclusion supports the theory that the probability of a knee operation leading to complications is independent of the surgeon who carries it out.

[AEB(P) 91]

9 A botanist examining the distribution of daisies in a field counts the number of daisies in 200 randomly chosen, non-overlapping, small areas each one metre square. The results are given in the table below.

Number of daisies per square metre (x)	0	1	2	3	4
Number of squares with x daisies per square metre	37	49	49	32	15

Number of daisies per square metre (x)	5	6	7	8 or more
Number of squares with x daisies per square metre	12	5	1	0

(i) Find $\bar{x}$, the mean of this frequency distribution.

(ii) Test, at the 5% level, the hypothesis that the distribution of the daisies is Poisson with mean $\bar{x}$. [UCLES(P)]

10 Explain how to calculate the number of degrees of freedom associated with a χ^2-test for goodness of fit.

A study of the distribution of lichens found on stone walls in Derbyshire was carried out. As part of the study, 100 randomly chosen sections of wall, all of the same width and all of the same height, were examined. The number, X, of lichens in each section was recorded and the results are summarised in the table.

Number of lichens (x)	0	1	2	3	4	5	6	7
Number of sections	8	21	32	15	12	6	4	2

(i) Calculate the mean and variance of this sample, and state why the results might suggest that X has a Poisson distribution.

(ii) Perform a test, at the 5% significance level, to determine whether the data could be a sample from a Poisson distribution. [UCLES]

11 A shop that repairs television sets keeps a record of the number of sets brought in for repair each day. The numbers brought in during a random sample of 40 days were as follows.

```
4 0 0 0 2 1 1 0 0 0
0 1 1 0 3 0 0 0 1 0
4 0 0 0 0 0 2 0 1 0
0 0 0 1 1 1 0 2 0 0
```

Test, at the 5% significance level, the hypothesis that these numbers are observations from a Poisson distribution. [UCLES(P)]

12 Describe briefly how the number of degrees of freedom is calculated in a χ^2 goodness-of-fit test.

The following set of grouped data from 100 observations has mean 1.03. The data are thought to come from a normal distribution with variance 1 but unknown mean. Using an appropriate χ^2-distribution, test this hypothesis at the 1% significance level.

Lower value of grouping interval	$-\infty$	-2.0	-1.5	-1.0	-0.5
Number of observations	0	1	0	6	10

Lower value of grouping interval	0.0	0.5	1.0	1.5	2.0
Number of observations	12	15	23	16	13

Lower value of grouping interval	2.5	3.0	3.5
Number of observations	3	1	0

[UCLES]

13 A weaving mill sells lengths of cloth with a nominal length of 70 m. The customer measured 100 lengths and obtained the following frequency distribution:

Length (m)	Frequency
61–67	1
67–69	16
69–71	26
71–73	19
73–75	20
75–81	18

(a) Use a χ^2 test at the 5% significance level to show that the normal model is not an adequate model for the data.

(b) The contract provides for the mill to pay compensation to the customer for any lengths less than 67 m supplied. Comment on the distribution of the lengths of cloth in the light of this further information. [AEB 89]

18.5 Contingency tables

Often data are collected on several variables at a time. For example, a questionnaire will usually contain more than one question! A table that gives the frequencies for two or more variables simultaneously is called a **contingency table**. We first met such a table in Section 1.23 (p. 31). Here is another example which shows information on voting:

	Conservative	Liberal Democrat	Labour	
Male	313	124	391	828
Female	344	158	388	890
	657	282	779	1718

Sample data of this type are collected in order to answer interesting questions about the behaviour of the population, such as 'Are there differences in the way that males and females vote?'. If there are differences then the variables vote and gender are said to be **associated**, whereas if there are no differences then the variables are said to be **independent**.

The null hypothesis is that the variables are independent. If this is true then, in the population, the proportion of Conservative supporters who are male will be equal to the proportion of Liberal Democrat supporters who are male and to the proportion of Labour supporters who are male. Furthermore, the proportion of males who support the Conservative party will be equal to the proportion of females who support the Conservative party, and so on.

If the null hypothesis of independence is true then the best estimate of the population proportion voting for the Conservatives is $\frac{657}{1718}$ ($= 0.3824$). The expected number of males voting Conservative would therefore be:

$$828 \times \tfrac{657}{1718} = 316.64$$

and the number of females would be:

$$890 \times \tfrac{657}{1718} = 340.36$$

These expected values are easily calculated using the formula:

$$\frac{\text{row total} \times \text{column total}}{\text{grand total}} \qquad\qquad (18.3)$$

The value of X^2 is calculated as before:

	Observed frequencies		
	Conservative	Liberal Democrat	Labour
Male	313	124	391
Female	344	158	388

	Expected frequencies		
	Conservative	Liberal Democrat	Labour
Male	316.64	135.91	375.44
Female	340.36	146.09	403.56

$$X^2 = \frac{(313 - 316.64)^2}{316.64} + \cdots + \frac{(388 - 403.56)^2}{403.56} = 3.34$$

To complete the test we compare the value of X^2 with the upper tail of the relevant χ^2 distribution. Here there is a simple rule:

For a contingency table with r rows and c columns,

$$d = (r - 1)(c - 1) \qquad\qquad (18.4)$$

To see the reason for this look again at the expected frequencies:

316.64	135.91	?	828
?	?	?	890
657	282	779	1718

After calculating the $(r-1)(c-1)\ (=2)$ expected frequencies that are shown, the remainder are not 'free' – their values are fixed by the need for them to sum to the known row and column totals.

If we decide to use a 5% significance level, then for the voting data we conclude that the voting patterns of males and females do not differ significantly, since the observed value (3.34) does not exceed the upper 5% point of a χ_2^2 distribution (5.991).

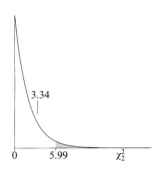

Example 9

The following data refer to visits to patients in a mental hospital.

| | Length of stay in hospital | | | |
	2 to 20 years	10 to 20 years	More than 20 years	
Visited regularly	43	16	3	62
Visited sometimes	6	11	10	27
Never visited	9	18	16	43
	58	45	29	132

Verify that the association between length of stay and the frequency with which a patient is visited is significant at the 0.1% level.

The null hypothesis states that length of stay and frequency of visit are independent of one another, with the alternative being that the two are associated.

Since there are 3 rows and 3 columns, the relevant χ^2 distribution has $(3-1)\times(3-1)=4$ degrees of freedom. If the variables were independent, then the expected frequencies would be given by the $\dfrac{\text{row total}\times\text{column total}}{\text{grand total}}$ formula. Thus the expected frequency for

regular visits to recent patients is $\dfrac{62\times 58}{132}=27.24.$

The complete set of expected frequencies, together with the calculations of X^2 are shown in the table below.

O_i	E_i	$O_i - E_i$	$\dfrac{(O_i - E_i)^2}{E_i}$
43	27.24	15.76	9.11
16	21.14	−5.14	1.25
3	13.62	−10.62	8.28
6	11.86	−5.86	2.90
11	9.20	1.80	0.35
10	5.93	4.07	2.79
9	18.89	−9.89	5.18
18	14.66	3.34	0.76
16	9.45	6.55	4.55
132	132	0	35.17

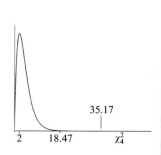

Since the value of X^2 greatly exceeds the upper 0.1% point of a χ_4^2 distribution (18.47), it is clear that there is a strong association between the classifying variables.

It is often useful to set out the individual cell contributions to X^2 in a table like that of the data, since this helps to show up any pattern in the lack of fit.

Contributions to X^2		
9.11	1.25	8.28
2.90	0.35	2.79
5.18	0.76	4.55

For this data set, focusing on the cells with large observed frequencies, the major lack of fit arises from the corner cells – for example, the NW and SE corners have much larger frequencies than would have been expected if there had been no association. These cells correspond to the regular visiting of recently admitted patients and the infrequent visits to the very long-stay patients.

Project

Does the age distribution of the cars travelling along a road vary with the time of day?

Use the registration letter as an indicator of age. Choose a reasonably busy road and draw up a tally chart of registration letters for the first 100 cars to pass. Repeat at a different time of day.

Pool the rarer (older) registration letters together, and assemble all the results in a contingency table looking something like this:

	G or earlier	H	J	K	L	M or N
Morning	13	15	18	17	21	16
Midday	19	21	22	18	12	8

Test the hypothesis that the age distribution is independent of the time of day, using the X^2 goodness-of-fit test.

Frank Yates (1902–94) was the son of a Manchester seed merchant. He was educated at Clifton College, Bristol and Cambridge University, where he obtained a first class degree in Mathematics. After brief periods as a schoolmaster and a mathematics adviser, he joined the statistics staff at Rothamsted (the agricultural research institute in Hertfordshire). He was made a Fellow of the Royal Society in 1948. Although he retired in 1967, he maintained his links with Rothamsted and published his last paper in 1990 at the age of 88. He is best known for the correction introduced below.

The Yates correction

In the special case of a 2×2 table, where the approximating χ^2 distribution has just 1 degree of freedom, the approximation is improved by making the small adjustment suggested by Frank Yates in 1926. Denoting the 4 cells of the table with the suffices 1, 2, 3 and 4, in place of X^2 we use X_c^2 defined by:

$$X_c^2 = \sum_{i=1}^{4} \frac{(|O_i - E_i| - 0.5)^2}{E_i}$$

There are a number of ways of simplifying this expression. For each of the 4 cells the difference between the expected and observed frequencies has the same magnitude. So we can write:

$$X_c^2 = (|O_1 - E_1| - 0.5)^2 \left(\frac{1}{E_1} + \cdots + \frac{1}{E_4} \right)$$

Alternatively, if we denote the four cells by a, b, c and d, with marginal totals m, n, r and s, and with grand total $N(= m + n = r + s)$, as illustrated below:

a	b	m
c	d	n
r	s	N

then the Yates-corrected goodness-of-fit statistic is:

$$X_c^2 = \frac{N(|ad - bc| - \frac{1}{2}N)^2}{mnrs}$$

Notes

- A 2×2 table provides one way of summarising the numbers of successes (a and c, say) and the numbers of failures (b and d) in samples of sizes m and n. In Section 17.5 (p. 469) we presented a comparison of the proportions $\frac{a}{m}$ and $\frac{c}{n}$ using an approximating normal distribution. Squaring the z-value derived there would lead to a test directly comparable to the *uncorrected* X^2 statistic. The equivalence occurs because the square of a $N(0,1)$ variable has a χ_1^2 distribution (see Section 18.1, p. 481).
- The correction suggested by Yates is a disguised form of the continuity correction used when approximating a binomial distribution by a normal distribution. Seventy years after its introduction it continues to be a bone of contention. We recommend its use.

Example 10

The following data come from a study concerning a possible cure for the common cold. A random sample of 279 French skiers was divided into two groups. Members of both groups took a pill each day. The pills taken by one group contained the possible cure, whereas the identical looking pills taken by the other group contained only sugar.

Of the 139 skiers taking the possible cure, 17 caught a cold. Of the 140 skiers taking the sugar pill, 31 caught a cold.

Does this provide significant evidence, at the 5% level, that the possible cure has worked?

The null hypothesis is that the outcome (cold or not) is independent of the treatment (sugar or cure), with the alternative being that there is an association.

As mentioned in the preceding notes, the question could also be treated as a comparison of proportions. A summary of the observed and expected frequencies is given below:

	Observed				Expected		
	Cold	No cold			Cold	No cold	
Sugar	31	109	140	Sugar	24.09	115.91	140
Cure	17	122	139	Cure	23.91	115.09	139
	48	231	279		48	231	279

Since there are just 2 rows and 2 columns, the Yates correction is needed. We could calculate X_c^2 using:

$$X_c^2 = \frac{(|31 - 24.09| - 0.5)^2}{24.09} + \cdots + \frac{(|122 - 115.09| - 0.5)^2}{115.09} = 4.14$$

or using:

$$X_c^2 = \frac{279(|(31 \times 122) - (109 \times 17)| - \frac{1}{2} \times 279)^2}{48 \times 231 \times 140 \times 139} = 4.14$$

Since 4.14 is a little greater than the upper 5% point (3.84) of a χ_1^2 distribution, there does appear to be significant evidence that the cure has worked!

(But remember that results that are 'significant' at the 5% level occur in 5% of samples even where there is independence!)

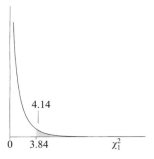

Exercises 18e

1 A survey of the effectiveness of three hospitals in treating a particular illness revealed the following results.

	Complete recovery	Partial recovery	Died
A	37	23	7
Hospital B	52	44	12
C	22	30	13

Do the data reveal significant evidence, at the 5% level, of differences in the effectiveness of the hospitals?

2 A survey obtains the following information concerning the party supported by a voter and the greatest academic attainment of the voter.

	Con.	Lab.	Lib.
GCSE	62	111	38
A-level	57	53	25
Degree	24	15	20

Is there significant evidence, at the 1% level, of an association between greatest academic attainment and party supported?

3 A survey of the traffic passing along a particular road concentrates on the age of the car and the sex of the driver. The results are as follows.

	New car	Old car
Male	117	63
Female	52	48

Using a 5% significance level, test whether there is an association between the sex of the driver and the age of the car.

4 A random sample of individuals is broken down by age and sex:

Age	Male	Female
16–17	3	2
18–24	15	14
25–60	147	177
61–75	24	30
76–	4	12

Test whether there are significant differences, at the 5% level, between the age distributions of the males and females.

5   A Danish survey investigated attitudes towards early retirement. Of 317 people in bad health, 276 were in favour of early retirement. Of 258 in moderate health, 232 were in favour. Of 86 in excellent health, 73 were in favour. All those not in favour were against the idea.

   (i)   Test the hypothesis that attitudes towards early retirement are independent of health.
   (ii) Repeat the test ignoring the data on those who were in moderate health.
         State your conclusions.

6   The contingency table below shows the results of random samples of pupils in three schools A, B and C on taking a certain examination. Use a χ^2 test to determine, at the 5% level of significance, whether there is any evidence of association between the school and the pass-rate in the examination. State the null hypothesis.

   Give your conclusion, explaining clearly what it means.

	A	B	C
Pass	25	20	15
Fail	10	15	15

[O&C]

7   A market research organisation interviewed a random sample of 120 users of launderettes in London and found that 37 preferred brand X washing powder, 66 preferred brand Y, and the remainder preferred brand Z. A similar survey was carried out in Birmingham. In this survey, of 80 people interviewed, 19 preferred brand X, 40 preferred brand Y and the remainder preferred brand Z. Test whether these results provide significant evidence, at the 5% level, of different preferences in the two cities.   [UCLES(P)]

8   A university department recorded the A-level grade in a particular subject and the class of degree obtained by 120 students in a given year. The data are summarised in the following table.

		Class of degree		
		I	II	III
A level grade	A	10	9	11
	B	4	24	8
	C	4	22	10
	D or E	2	5	11

Test, at the 1% level, the hypothesis that the class of degree is independent of A-level grade.   [UCLES(P)]

9   In 1988 the number of new cases of insulin-dependent diabetes in children under the age of 15 years was 1495. The table below breaks down this figure according to age and sex.

Age (yrs)	0–4	5–9	10–14	Total
Boys	205	248	328	781
Girls	182	251	281	714
Total	387	499	609	1495

Perform a suitable test, at the 5% significance level, to determine whether age and sex are independent factors.   [UCLES(P)]

10   The political inclinations of a random sample of students from two subject areas are given in the table below, which shows the number of students from each area supporting each political party.

	Labour	Alliance	Conservative
Arts	27	18	20
Science	23	25	12

Test, at the 2.5% level, the hypothesis that political inclination is independent of subject area.   [UCLES(P)]

11   A hospital employs a number of visiting surgeons to undertake particular operations. If complications occur during or after the operation the patient has to be transferred to a larger hospital nearby where the required back up facilities are available.

   A hospital administrator, worried by the effects of this on costs, examines the records of three surgeons. Surgeon A had 6 out of her last 47 patients transferred, surgeon B 4 out of his last 72 patients and surgeon C 14 of his last 41. Form the data into a 2×3 contingency table and test, at the 5% significance level, whether the proportion transferred is independent of the surgeon.

   The administrator decides to offer as many operations as possible to surgeon B. Explain why and suggest what further information you would need before deciding whether the administrator's decision was based on valid evidence.   [AEB(P) 91]

12 A research worker studying the ages of adults and the number of credit cards they possess obtained the results shown in the table.

Number of cards possessed	$\leqslant 3$	> 3
Age < 30	74	20
Age $\geqslant 30$	50	35

Use the χ^2 statistic and a significance test at the 5% level to decide whether or not there is an association between age and number of credit cards possessed.

[ULSEB(P)]

18.6 The dispersion test

This is a special goodness-of-fit test for testing the hypothesis that a sample of observations has been taken from a population having a Poisson distribution with unknown mean. It is a more powerful alternative to the X^2 test.

If a random variable does have a Poisson distribution then it will have mean equal to variance. If the ratio of the sample variance to the sample mean is very different from 1 then this would provide evidence against the Poisson hypothesis.

The test statistic, often known as **the index of dispersion**, is:

$$I = \frac{(n-1)s^2}{\bar{x}}$$

where n is the sample size (and *not* the number of classes).

If the data come from a Poisson distribution, then I is an observation from a χ^2_{n-1} distribution. The test is usually two-tailed, with unusually small or unusually large values leading to rejection of the null hypothesis of a Poisson distribution.

With n individual observations I is calculated using:

$$I = \frac{\sum (x_i - \bar{x})^2}{\bar{x}} = \frac{n \sum x_i^2 - (\sum x_i)^2}{\sum x_i}$$

If the data are summarised in a frequency table in which the value x_j occurs with frequency f_j, and $n = \Sigma f_j$, the formula is:

$$I = \frac{n \sum f_j x_j^2 - (\sum f_j x_j)^2}{\sum f_j x_j}$$

Example 11

We return to the data of Example 7 which referred to a typist's errors. It is repeated here for convenience:

Number of errors	0	1	2	3	4	5	6	7	8	9	10+
Number of pages	2	5	16	11	6	3	1	2	3	1	0

Carry out a two-tailed dispersion test, at the 5% level, to determine whether to accept the null hypothesis that the errors are randomly distributed throughout the manuscript.

There are 50 observations and, for a two-tailed 5% test, we therefore need the upper and lower 2.5% points of the χ^2_{49} distribution. By interpolation in the tables of critical values given in the Appendix, these are found to be approximately 32 and 71. Hence, if the value of I lies in the interval $(32,71)$, we will accept the Poisson hypothesis.

We now calculate the sum and sum of squares of the 50 observations:

$$\sum f_j x_j = 162 \qquad \sum f_j x_j^2 = 746$$

The value of I is given by:

$$I = \frac{50 \times 746 - 162^2}{162} = 68.25$$

Since 68.25 lies inside the interval we accept the null hypothesis. This agrees with our previous conclusion.

Exercises 18f

1 A botanist examining the distribution of daisies in a field counts the number of daisies in 200 randomly chosen, non-overlapping, small areas each one metre square. The results are given in the table below.

Number of daisies per square metre (x)	0	1	2	3	4
Number of squares with x daisies per square metre	37	49	49	32	15

Number of daisies per square metre (x)	5	6	7	8 or more
Number of squares with x daisies per square metre	12	5	1	0

Use a dispersion test, at the 5% level, to test the hypothesis that the distribution of the daisies is Poisson.   [UCLES(A)]

2 A study of the distribution of lichens found on stone walls in Derbyshire was carried out. As part of the study, 100 randomly chosen sections of wall, all of the same width and all of the same height, were examined. The number, X, of lichens in each section was recorded and the results are summarised in the table.

Number of lichens (x)	0	1	2	3	4	5	6	7
Number of sections		8	21	32	15	12	6	4	2

Perform a dispersion test, at the 5% significance level, to determine whether the data could be a sample from a Poisson distribution. [UCLES(A)]

3 A shop that repairs television sets keeps a record of the number of sets brought in for repair each day. The numbers brought in during a random sample of 40 days were as follows.

```
4 0 0 0 2 1 1 0 0 0
0 1 1 0 3 0 0 0 1 0
4 0 0 0 0 0 2 0 1 0
0 0 0 1 1 1 0 2 0 0
```

Test, at the 5% significance level, the hypothesis that these numbers are observations from a Poisson distribution, using a dispersion test.   [UCLES(A)]

4 Gulmon and Mooney investigated the distribution of the shrub *Atriplex hymenelytra* on a portion of the floor of Death Valley in California. Using quadrats of area $25\,\text{m}^2$, they obtained the frequencies given in the table below.

No. of plants	0	1	2	3	4	5
No. of quadrats	20	22	16	3	2	1

Test whether these figures provide evidence of significant departures from randomness:

(i) using the X^2 test,

(ii) using the dispersion test.

Estimate the probability of a randomly chosen area of $100\,\text{m}^2$ containing exactly one shrub.

Determine the maximum value of k for which P(area of $100\,\text{m}^2$ contains more than k shrubs) exceeds 0.5.

18.7 Comparing distribution functions

A problem with data from continuous distributions is that in order to use the X^2 test we need to group the data. Using different groups with the same data we could reach different conclusions.

An alternative is to compare the theoretical cumulative distribution function, $F(x)$, with its sample approximation. A large difference between these quantities would suggest that the theoretical distribution was incorrect. Two tests that use this type of approach are the **Kolmogorov–Smirnov** and **Cramer–Von Mises** tests. Special tables are needed for these tests and they are mentioned here only for completeness.

Chapter summary

♦ **The X^2 test of goodness of fit:**

$$X^2 = \sum_{i=1}^{m} \frac{(O_i - E_i)^2}{E_i}$$

- The O_i values are non-negative integers.
- $\Sigma(O_i - E_i) = 0$.
- Providing all E_i-values are reasonably large (say, $\geqslant 5$), X^2 has an approximate χ^2 distribution.
- Small expected frequencies should be eliminated by combining categories.
- The χ_d^2 distribution has d degrees of freedom. When there are m categories (after combination), d is given by:

  1 $d = m - 1$, for the case of probabilities prescribed by H_0,
  2 $d = m - 1 - k$, for the case where k parameters are estimated from the data,

The null and alternative hypotheses are:

  H_0: the results are a random sample from the supposed distribution
  H_1: H_0 is incorrect

♦ **Independence in an $r \times c$ contingency table:**
For a table having (after combination) r rows and c columns:
- X^2 has approximately a χ^2 distribution with $(r-1)(c-1)$ degrees of freedom.
- For 2×2 tables, the Yates correction is needed:

$$X_c^2 = \sum_{i=1}^{m} \frac{(|O_i - E_i| - 0.5)^2}{E_i}$$

The null and alternative hypotheses are:

  H_0: the variables are independent
  H_1: the variables are associated

♦   **The dispersion test:**

$$I = \frac{n \sum x_i^2 - (\sum x_i)^2}{\sum x_i}$$

I is only appropriate for testing

   H_0: the results are a random sample from a Poisson distribution
against

   H_1: H_0 is incorrect.

Assuming H_0, I has a χ^2_{n-1} distribution.

Exercises 18g (Miscellaneous)

1 A factory makes a certain type of sweet in eight different colours, with equal numbers being made of each colour.

(a) A tube of these sweets, purchased at random, has the following numbers of the eight colours:

   7, 4, 5, 3, 6, 1, 3, 11

Test the hypothesis that the sweets in the tube constitute a random sample of the output of the factory.

(b) The quality control staff are interested in the behaviour of the storage hopper used to fill the tubes. The hopper is continually revolved in an attempt to obtain an even mixture of colours. In order to test the mixing process, random samples of five sweets are taken at five-minute intervals. The numbers obtained are summarised below.

No. of blue sweets	0	1	2	3	4	5
No. of groups	22	12	5	1	0	0

Determine whether there is significant evidence to reject the null hypothesis that the blue sweets have been randomly distributed, with the probability of a randomly chosen sweet being blue being $\frac{1}{8}$.

(c) One hundred tubes of the sweets are examined to see whether any of the sweets are grossly misshapen. The numbers found are summarised in the table below.

No. misshapen	0	1	2	3	4	$\geqslant 5$
No. of tubes	22	36	22	15	5	0

Use a dispersion test to test the hypothesis that these frequencies have a Poisson distribution.

2 A meteorologist conjectures that, at a certain location, the rainfall (x mm) on June 30th may be regarded as an observation from the exponential distribution with probability density function given by

$$f(x) = \lambda e^{-\lambda x}, \ 0 \leqslant x < \infty.$$

Show that $E(X) = 1/\lambda$.

It is known that during the 25-year period 1936 to 1960 a total of 260 mm of rain fell on June 30th. Use this information to provide an estimate of λ. Hence show that the probability that more than 20 mm of rain will fall at this location on June 30th, 1989, is 0.146, correct to three significant figures.

The individual rainfall measurements on June 30th for the period 1961 to 1985 are summarised in the table below.

Rainfall (x mm)	Number of days
$x \leqslant 4$	10
$4 < x \leqslant 9$	5
$9 < x \leqslant 16$	6
$x > 16$	4

Using your estimate of λ obtained above, test the conjecture of the meteorologist, using a 5% significance level and showing your working clearly.   [UCLES]

3 At the top of a table of Random Sampling Numbers it is stated that each digit in the table is an independent sample from a population in which each of the digits 0 to 9 is equally likely. A count of 40 such digits yielded the following frequencies.

Digit	0	1	2	3	4	5	6	7	8	9
Frequency	4	4	3	3	2	4	3	5	10	2

Give a reason why it would not be valid to apply a χ^2-test using the individual frequencies given above.

A larger sample of 80 digits yielded the following frequencies.

Digit	0	1	2	3	4	5	6	7	8	9
Frequency	5	9	5	6	6	8	9	10	16	6

Using this larger sample, perform a χ^2-test, at the 10% significance level, of the hypothesis that each digit is equally likely.

The complete table was read in 1000 consecutive pairs and the number of doubles (i.e. 00, 11, 22, ..., 99) was found to be 77. Using a 2.5% significance level, determine whether there are significantly fewer doubles than would be expected.   [UCLES]

4 A game contains 20 pieces, each of which has probability 0.08 of being defective.

(a) Suggest a suitable distribution to model the number of defective pieces in a game.

Let X represent the number of defective pieces in a game.

(b) Copy and complete the following probability distribution.

x	0	1	2	3
$P(X=x)$	0.1887			0.1414

x	4	5	6 or more
$P(X=x)$	0.0523	0.0145	

(c) Estimate, giving your answer to 3 decimal places, the probability that a consignment of 10 000 such *pieces* contains

(i) at most 750 defective pieces,

(ii) between 750 and 850 (inclusive) defective pieces.

A random sample of 1000 *games* was checked for defective pieces and the following table produced to summarise the number of defective *pieces* in each game.

Number of defective pieces	0	1	2	3
Number of games	194	344	266	137

Number of defective pieces	4	5	6 or more
Number of games	46	10	3

(d) Use a χ^2 test to test, at the 5% level, whether or not the observed results are consistent with those expected under the model specified in (a).   [ULSEB]

5 (i) The discrete random variable X has the geometric distribution given by:
$$P(X=r) = p(1-p)^{r-1}, \quad r = 1, 2, 3, \ldots$$
Find $E(X)$.

(ii) A sample of 200 values of the discrete random variable Y is summarised by:

Y	1	2	3	4
Frequency	140	35	17	3

Y	5	6	7 or more
Frequency	3	2	0

Find the sample mean $\bar{y}$, of these 200 values. It is believed that Y has a geometric distribution with parameter p. Using $1/\bar{y}$ as an estimate of p complete the table of expected frequencies given below.

Y	1	2	3	4
Frequency				

Y	5	6	7 or more
Frequency	1.6	0.5	0.3

Test, at the 10% significance level, whether the sample supports this belief.   [UCLES]

6 It is thought that there is an association between the colour of a person's eyes and the reaction of the person's skin to ultraviolet light. In order to investigate this each of a random sample of 120 people was subjected to a

(continued)

standard dose of ultraviolet light. The degree of their reaction was noted, "−" indicating no reaction, "+" indicating a slight reaction and "++" indicating a strong reaction. The results are shown in the table below.

Eye colour

		Blue	Grey or Green	Brown
Reaction	−	7	8	18
	+	29	10	16
	++	21	9	2

(i)   Perform an appropriate test at the 5% significance level, stating your null and alternative hypotheses. Find the least value of k given in the mathematical tables for which the null hypothesis can be rejected at the $k\%$ significance level.

(ii) Estimate the percentage of people in the population from which the sample was drawn who would not suffer a reaction to ultraviolet light, and calculate an approximate 95% symmetric confidence interval for this percentage.   [UCLES]

7 (a) Each object in a large collection belongs to one of k classes. The null hypothesis is that a randomly chosen object is equally likely to belong to any of the classes. A random sample of N objects is taken. The observed frequency in class $i\,(i = 1, 2, \ldots, k)$ is denoted by f_i and the corresponding expected frequency by e_i. The value of the χ^2 goodness-of-fit statistic is given by

$$X^2 = \sum_{i=1}^{k} \frac{(f_i - e_i)^2}{e_i}$$

Show that X^2 may be written as

$$X^2 = \left(\frac{k}{N} \sum_{i=1}^{k} f_i^2 \right) - N$$

A certain type of button is made in nine colours. A random collection of 137 buttons is found to contain the following numbers of the different colours:

7, 10, 13, 14, 15, 17, 17, 19, 25.

Test, at the 5% significance level, the hypothesis that a randomly chosen button is equally likely to have any one of the colours.

(b) The numbers of buttons lost off their clothes by 100 schoolboys during 1989 are summarised in the table below.

Form

		6th	5th	4th
Colour	White	24	9	27
	Grey	26	13	31
	Black	2	5	8
	Other colours	4	3	8

Test, at the 5% significance level, the hypothesis that the colour of a lost button is independent of the schoolboy's form.   [UCLES]

8 (a) If Y has a χ^2 distribution with v degrees of freedom it can be shown that, for large v,

(i)   the distribution of Y is approximately normal with mean v and variance $2v$,

(ii) the distribution of $\sqrt{2Y} - \sqrt{2v - 1}$ is approximately the standard normal distribution.

For the case $v = 55$ use each of the above approximations to obtain estimates for $P(Y > 75)$.

(b) The accident rates, taken over a twelve-month period, for the workers in a particular company, classified by age, are given in the following table.

Age (years)	18–25	26–40	41–50
At least 1 accident	112	156	75
No accidents	175	267	179
Totals	287	423	254

Age (years)	over 50	Total
At least 1 accident	77	420
No accidents	228	849
Totals	305	1269

Show that the data provide evidence at the 0.1% significance level that age and accident rate are not independent.

Comment on the relation between age and accident rate.   [UCLES]

19 Handling variances

Variety's the very spice of life,
That gives it all its flavour

The Garden, William Cowper

Chapters 14 to 17 presented estimates and tests for means and for proportions, but not for variances. The reason is that the mathematics required for dealing with variances is more difficult. In this chapter we give no more than a rough sketch of the results.

19.1 Confidence intervals for σ^2

Suppose that we have a sample of n independent observations, $x_1, x_2, \ldots, x_n$, with sample mean $\bar{x}$. The unbiased estimate of the population variance, σ^2, is s^2, defined by:

$$s^2 = \frac{1}{n-1} \sum_{i=1}^{n} (x_i - \bar{x})^2$$

Different samples from the same population will contain different observations, with different sample means and different values of s^2. We can therefore regard s^2 as an observation on a random variable that we will denote by S^2.

Denoting the random variable corresponding to the ith observation by X_i, and defining the random variable $\bar{X}$ by:

$$\bar{X} = \frac{1}{n}(X_1 + X_2 + \cdots + X_n)$$

we therefore define the random variable S^2 by:

$$S^2 = \frac{1}{n-1} \sum_{i=1}^{n} (X_i - \bar{X})^2$$

To get an idea of the distribution of S^2, we will look at a single term involved in the summation, $(X_1 - \bar{X})^2$, say, and we begin by determining the mean and variance of $(X_1 - \bar{X})$. Denoting the population mean by μ, we have:

$$\mathrm{E}(X_1) = \mathrm{E}(\bar{X}) = \mu$$

so that:

$$\mathrm{E}(X_1 - \bar{X}) = 0$$

Now $X_1 - \bar{X}$ is a linear combination of independent X-variables:

$$X_1 - \bar{X} = X_1 - \frac{1}{n}(X_1 + X_2 + \cdots + X_n)$$

$$= \left(1 - \frac{1}{n}\right)X_1 - \frac{1}{n}(X_2 + \cdots + X_n)$$

and so:

$$\text{Var}(X_1 - \bar{X}) = \left(1 - \frac{1}{n}\right)^2 \text{Var}(X_1)$$

$$+ \left(-\frac{1}{n}\right)^2 \{\text{Var}(X_2) + \cdots + \text{Var}(X_n)\}$$

$$= \left(\frac{n-1}{n}\right)^2 \sigma^2 + \left(-\frac{1}{n}\right)^2 (n-1)\sigma^2$$

$$= \frac{n-1}{n}\sigma^2$$

In order to make further progress, for the remainder of this chapter:

we assume that the individual observations have come from a normal distribution.

Since linear combinations of normally distributed random variables have a normal distribution, we have:

$$(X_1 - \bar{X}) \sim \text{N}\left(0, \frac{n-1}{n}\sigma^2\right)$$

and hence:

$$\frac{(X_1 - \bar{X})}{\sqrt{\frac{n-1}{n}\sigma^2}} \sim \text{N}(0, 1)$$

In Section 18.1, when introducing the chi-squared distribution, we noted that if Z has a N(0,1) distribution, then Z^2 has a χ_1^2 distribution. Hence:

$$\frac{(X_1 - \bar{X})^2}{\frac{n-1}{n}\sigma^2} \sim \chi_1^2$$

The above argument would hold for each component of S^2:

$$\frac{n}{\sigma^2}\frac{(X_i - \bar{X})^2}{(n-1)} \sim \chi_1^2$$

Since the sum of independent χ^2 random variables has a χ^2 distribution, if the separate components of the summation in the definition of S^2 were independent, we would have:

$$\frac{n}{\sigma^2}S^2 \sim \chi_n^2$$

However, the components are not independent, since $\Sigma(X_i - \bar{X}) = 0$. A slight adjustment is required, and we now quote the result that:

$$\frac{(n-1)}{\sigma^2}S^2 \sim \chi_{n-1}^2 \qquad\qquad (19.1)$$

Suppose we denote the lower and upper 2.5% points of a χ_{n-1}^2 distribution by L and U. Then we have:

$$P\left[L < \frac{(n-1)S^2}{\sigma^2} < U\right] = 0.95$$

Taking reciprocals, we need to reverse the inequalities:

$$P\left[\frac{1}{U} < \frac{\sigma^2}{(n-1)S^2} < \frac{1}{L}\right] = 0.95$$

giving:

$$P\left[\frac{(n-1)S^2}{U} < \sigma^2 < \frac{(n-1)S^2}{L}\right] = 0.95$$

and so the 95% confidence interval for σ^2 is:

$$\left(\frac{(n-1)s^2}{U}, \quad \frac{(n-1)s^2}{L}\right) \tag{19.2}$$

In calculating the interval we can use the fact that:

$$(n-1)s^2 = \sum_{i=1}^{n}(x_i - \bar{x})^2 = \sum_{i=1}^{n}x_i^2 - \frac{1}{n}\left(\sum_{i=1}^{n}x_i\right)^2$$

and, since $(n-1)s^2 = n\sigma_n^2$, the interval can also be written as:

$$\left(\frac{n\sigma_n^2}{U}, \frac{n\sigma_n^2}{L}\right)$$

Example 1

A machine fills containers with orange juice. A random sample of 10 containers was examined in a laboratory, and the amounts of orange juice in each container were determined correct to the nearest 0.05 ml. The results (after subtracting 500 ml from each) were as follows:

   3.45, 7.80, 1.10, 6.45, 5.85, 4.40, 3.45, 5.50, 2.95, 7.75

Obtain an unbiased estimate of the population variance.
Assuming that the observations have a normal distribution, obtain a 95% symmetric confidence interval for the population variance, explaining the sense in which the confidence interval is symmetric.

We begin by calculating the sum, and the sum of squares, of the observations: $\Sigma x = 48.70$, $\Sigma x^2 = 280.055$. Now:

$$(n-1)s^2 = n\sigma_n^2 = \Sigma x^2 - \frac{1}{n}(\Sigma x)^2$$

$$= 280.055 - \tfrac{1}{10} \times 48.70^2$$

$$= 42.886$$

An unbiased estimate of the population variance is therefore $\frac{1}{9} \times 42.886 = 4.77$ (to two decimal places).

In order to obtain a 95% confidence interval we need to find the lower and upper 2.5% points of the relevant χ^2 distribution. In this case, with $n = 10$, this is the χ_9^2 distribution, for which the percentage points are $L = 2.70$ and $U = 19.02$. The 95% confidence interval is therefore:

$$\left(\frac{42.886}{19.02}, \frac{42.886}{2.70}\right)$$

which simplifies to (2.25, 15.88).

The confidence interval is symmetric in the sense that it uses the central 95% of the χ^2 distribution, with 2.5% being excluded in each tail.

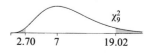

Exercises 19a

1 For a random sample of 12 observations from a normal population the value of the unbiased estimate of the population variance is given by $s^2 = 3.56$.
   Determine a 99% confidence interval for the population variance.

2 A random sample of 20 observations on Y, a normally distributed random variable, is summarised by $\Sigma y = 211.4$, $\Sigma y^2 = 3487.88$.
   Determine a 95% confidence interval for the population variance.

3 The following sample was randomly chosen from a population of normally distributed values: 1.92, 3.16, -0.62, -0.30, 2.78, 0.76, 1.30, 3.52, 4.12, 5.20
   (i)   Determine the sample mean and an unbiased estimate of the population variance.
   (ii) Determine a symmetric 95% confidence interval for the population mean.
   (iii) It is hypothesised that the population variance is equal to 4.0.
         By constructing a two-sided 95% confidence interval for σ^2, test this hypothesis at the 5% significance level.
   (iv) Assuming that the population variance is indeed 4.0, determine a symmetric 95% confidence interval for the population mean.

4 A random sample of 21 observations is taken from a normal distribution with unknown variance σ^2. Let T be defined by $T = \Sigma(x_i - \bar{x})^2$, where x_i is the value of the ith observation and $\bar{x}$ is the sample mean.
   Determine, in terms of σ^2, an expression for the value t which is such that $P(T > t) = 0.05$.

Hence obtain a one-sided 95% confidence interval for σ^2 for the case where the observed value of T is 17.3.

5 A random sample of 16 observations from a normal distribution has sum equal to 83.4 and sum of squares equal to 639.44.
   Determine a 99% symmetric confidence interval for σ^2, the variance of the distribution.
   Determine also the value v_0 which is such that $0 < \sigma^2 < v_0$ is a one-sided 95% confidence interval for σ^2.

6 A class of students perform the following experiment: each student is blindfolded and given a stop watch. They are asked to start the watch and then stop it after a time that they estimate to be 60 seconds. The actual times taken (in secs) are shown in the table below.

Time	51	55	56	57	58	59
Frequency	1	1	3	6	3	4

Time	60	61	62	63	64	66
Frequency	4	4	2	1	1	1

   (i)   Obtain unbiased estimates of the population mean and variance.
   (ii) Assuming a normal distribution, obtain a 95% confidence interval for the population variance.
   (iii) Suppose the population variance is equal to the upper confidence limit. Using a normal distribution (rather than a t-distribution) determine a 95% confidence interval for the population mean.
   (iv) Repeat (iii) supposing the population variance to be equal to the lower confidence limit.

19.2 Testing a hypothesis about σ^2

As usual, hypothesis tests may be either one-tailed or two-tailed. In either case, the null hypothesis assumes a particular value, σ_0^2 say, for the population variance. If the assumed value is correct, then $\dfrac{(n-1)s^2}{\sigma_0^2}$ will be an observation from a χ_{n-1}^2 distribution. The critical values separating the acceptance and rejection regions are the relevant percentage points of the χ_{n-1}^2 distribution.

Example 2

A garage claims that its standard service for a particular type of car should take on average 75 minutes, with a standard deviation of 7 minutes. A garage inspector does not believe that the garage is as consistent as they claim. The inspector times a random sample of 15 services and finds that they have a sample standard deviation of 9.34 minutes.
Does this provide significant evidence, at the 5% level, to refute the garage's claim?

Here we have:

$$H_0: \sigma^2 = 49$$
$$H_1: \sigma^2 > 49$$

The test is one-tailed, so there is a single critical value separating the acceptance and rejection regions. The significance level is 5% and the sample size is 15, so, assuming a normal distribution for the service times, the critical value is 23.68, the upper 5% point of a χ^2_{14} distribution. The test procedure is therefore to reject H_0 in favour of H_1 if the observed value of $\dfrac{(n-1)s^2}{\sigma_0^2}$ exceeds 23.68, and to accept H_0 otherwise.

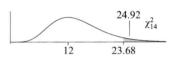

The observed value of $\dfrac{(n-1)s^2}{\sigma_0^2}$ is $\dfrac{14 \times 9.34^2}{49} = 24.92$. This is greater than the critical value, and we conclude that there is significant evidence, at the 5% level, to refute the garage's claim.

Example 3

Prior to overhaul, a machine is cutting planks of wood with a standard deviation of 4.7 mm. After the overhaul, a random sample of 30 planks is examined and their lengths are found to give $s = 3.84$ mm.
Does this provide significant evidence, at the 5% significance level, of a change in population standard deviation? A normal distribution may be assumed.

Here we have:

$$H_0: \sigma^2 = 4.7^2 = 22.09$$
$$H_1: \sigma^2 \neq 22.09$$

The test is two-tailed, so there are two critical values separating the acceptance and rejection regions. The significance level is 5% and the sample size is 30, so the critical values are 16.05 and 45.72, the lower and upper 2.5% points of a χ^2_{29} distribution. The test procedure is therefore to accept H_0 if the observed value of $\dfrac{(n-1)s^2}{\sigma_0^2}$ lies between 16.05 and 45.72, and otherwise to reject H_0 in favour of H_1.

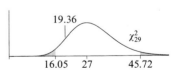

The observed value of $\dfrac{(n-1)s^2}{\sigma_0^2}$ is $\dfrac{29 \times 3.84^2}{22.09} = 19.36$. This value falls inside the acceptance region and there is therefore no significant evidence, at the 5% level, to suggest that the machine's accuracy has altered.

Exercises 19b _____

1 A sample of 25 observations is taken from a normal distribution with variance σ^2. The unbiased estimate of the population variance is found to be 14.63.

Test, at the 5% level, the hypothesis that $\sigma^2 = 10$ against the alternative that $\sigma^2 > 10$.

2 A sample of 15 observations is taken from a normal distribution with variance σ^2. The unbiased estimate of the population variance is found to be 74.26.

Test, at the 2.5% level, the hypothesis that $\sigma^2 = 100$ against the alternative that $\sigma^2 < 100$.

3 A sample of 10 observations is taken from a normal distribution with variance σ^2. The unbiased estimate of the population variance is found to be 0.074.

Test, at the 1% level, the hypothesis that $\sigma^2 = 0.3$ against the alternative that $\sigma^2 \neq 0.3$.

4 The following observations are taken from a normal distribution that is believed to have unit variance.

> 16.2, 14.4, 17.9, 11.6,
> 18.3, 15.5, 17.2, 16.6

Determine whether there is significant evidence, at the 5% level, that the population variance is not equal to 1.

5 A machine saws logs of woods into lengths that are supposed to have a standard deviation of 3 cm. If the machine goes wrong then the standard deviation increases. A random sample of 10 logs have lengths, in cm, as follows:

> 997, 1004, 1009, 999, 1006,
> 1014, 998, 999, 1001, 1000.

Assuming a normal distribution, determine whether there is significant evidence, at the 1% level, that the machine has gone wrong.

Sir Ronald Aylmer Fisher (1890–1962) was educated at Harrow and at Cambridge, where he studied mathematics. His initial interest in Statistics developed because of his interest in genetics, and he pursued both subjects for the rest of his life. His first paper (in 1912) introduced the method of maximum likelihood, his second the mathematical derivation of the *t*-distribution (Chapter 14), his third the distribution of the correlation coefficient (Chapter 20). Following World War I (which saw Fisher employed as a teacher because his dreadful eyesight prevented him from fighting), he joined the agricultural research station at Rothamsted. During his time there he virtually invented the subject of experimental design and the methods of analysis of variance (Chapter 21) which involve the comparison of variances and led to his derivation of the *F*-distribution (see below). In 1933 he became Professor of Eugenics at University College, London and subsequently, in 1943, Professor of Genetics at Cambridge. He became a Fellow of the Royal Society in 1929 and was knighted in 1952 – the first statistical knight!

19.3 The *F*-distribution

We end this chapter by looking at a method for comparing the variances of two populations. In order to do this we need yet another distribution. Formally, if the random variables U and V are independent, with:

$$U \sim \chi_u^2 \quad \text{and} \quad V \sim \chi_v^2$$

then:

$$\frac{U}{u} \div \frac{V}{v} \sim F_{u,v} \tag{19.3}$$

We describe the ratio as having 'an *F*-distribution with *u* and *v* degrees of freedom'.

All *F*-distributions have range $(0, \infty)$. They have mean equal to $\dfrac{v}{(v-2)}$, for $v > 2$. The precise shape of an *F*-distribution depends upon the values of *u* and *v*: when both are large the general shape is that illustrated.

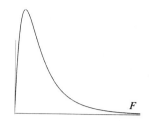

Note

- If $\dfrac{U}{u} \div \dfrac{V}{v} \sim F_{u,v}$ then $\dfrac{V}{v} \div \dfrac{U}{u} \sim F_{v,u}$

Tables of the *F*-distribution

With two parameters, these are both more extensive and more limited than any previous tables that we have met. A separate table is used for each significance level (typically, the upper 5%, 2.5% and 1% levels), with the columns of the table referring to the first of the degrees of freedom and the rows of the table referring to the second. Here is a short extract from the more detailed tables given in the Appendix (p. 623).

Upper 5% points

v	\multicolumn{12}{c}{u}
	1	2	3	4	5	6	7	8	12	24	40	∞
1	161.4	199.5	215.7	224.6	230.2	234.0	236.8	238.9	243.9	249.1	251.1	254.3
2	18.51	19.00	19.16	19.25	19.30	19.33	19.35	19.37	19.41	19.45	19.47	19.50
3	10.13	9.55	9.28	9.12	9.01	8.94	8.89	8.85	8.74	8.64	8.59	8.53
4	7.71	6.94	6.59	6.39	6.26	6.16	6.09	6.04	5.91	5.77	5.72	5.63
20	4.35	3.49	3.10	2.87	2.71	2.60	2.51	2.45	2.08	2.28	1.99	1.84
40	4.08	3.23	2.84	2.61	2.45	2.34	2.25	2.18	2.00	1.79	1.69	1.51
∞	3.84	3.00	2.60	2.37	2.21	2.10	2.01	1.94	1.75	1.52	1.39	1.00

The table refers to the upper 5% points only. However, the table can also be used to obtain the lower 5% points, because of the relation:

$$P(F_{u,v} < c) = P\left(F_{v,u} > \frac{1}{c}\right) \qquad (19.4)$$

Here are some examples of lower and upper 5% points:

Upper 5% point of $F_{3,1}$	215.7
Upper 5% point of $F_{1,3}$	10.13
Upper 5% point of $F_{12,20}$	2.28
Lower 5% point of $F_{4,3}$	$\dfrac{1}{6.59} = 0.152$
Lower 5% point of $F_{3,4}$	$\dfrac{1}{9.12} = 0.110$
Upper 5% point of $F_{3,4}$	6.59

Note that, when taking a reciprocal to obtain a lower 5% point, the degrees of freedom are interchanged.

Exercises 19c

1 Find the upper 5% point of an $F_{a,b}$-distribution for each of the following cases:
(i) $a = 2, b = 4$, (ii) $a = 4, b = 2$,
(iii) $a = 12, b = 40$, (iv) $a = 40, b = 12$.

2 Find the lower 1% point of an $F_{a,b}$-distribution for each of the following cases:
(i) $a = 5, b = 7$, (ii) $a = 4, b = 6$,
(iii) $a = 12, b = 40$, (iv) $a = 10, b = 8$.

3 Find the lower and upper 2.5% points of an $F_{a,b}$-distribution for each of the following cases:
(i) $a = 4, b = 4$, (ii) $a = 4, b = 8$,
(iii) $a = 2, b = 12$, (iv) $a = 5, b = 6$.

4 If the random variable A has a t_d-distribution, then A^2 has an $F_{1,d}$-distribution. It follows that $P(|A| > a) = P(A^2 > a^2)$.
Verify that this is the case by reference to the tables.

19.4 Comparison of two variances

We can now consider hypotheses such as:

$$H_0: \sigma_x^2 = \sigma_y^2$$
$$H_1: \sigma_x^2 \neq \sigma_y^2$$

where σ_x^2 and σ_y^2 are the variances of two different populations. This test would be a natural preamble to a comparison of two means as in Section 17.3 (p. 458).

Suppose we have two independent random samples of sizes n_x and n_y, with unbiased estimates of the respective population variances being denoted by s_x^2 and s_y^2. The corresponding random variables are S_x^2 and S_y^2. If the distributions sampled are normal, with common variance σ^2, then:

$$\frac{n_x - 1}{\sigma^2} S_x^2 \sim \chi_{n_x-1}^2 \quad \text{and} \quad \frac{n_y - 1}{\sigma^2} S_y^2 \sim \chi_{n_y-1}^2$$

From the definition of an F-distribution, taking the scaled ratio of random variables with χ^2 distributions, we get:

$$\frac{\dfrac{n_x - 1}{\sigma^2} S_x^2}{n_x - 1} \div \frac{\dfrac{n_y - 1}{\sigma^2} S_y^2}{n_y - 1} \sim F_{n_x-1, n_y-1}$$

which simplifies greatly to give:

$$\frac{S_x^2}{S_y^2} \sim F_{n_x-1, n_y-1} \tag{19.5}$$

while also:

$$\frac{S_y^2}{S_x^2} \sim F_{n_y-1, n_x-1} \tag{19.6}$$

Since the tables of percentage points of the F-distribution refer only to the upper tails in which the values are greater than 1, if $s_x^2 > s_y^2$ then we use Equation (19.5) and otherwise we use Equation (19.6). Because of this deliberate concentration on the upper tails, we have to remember to halve the tail probability when consulting the tables. For example, for a two-tailed 5% test, in which $s_x^2 > s_y^2$, we use the one-tail 2.5% percentage point of F_{n_x-1, n_y-1}.

Note

♦ If the null hypothesis is accepted, then a pooled estimate of the common variance is given by s^2, where:

$$s^2 = \frac{(n_x - 1)s_x^2 + (n_y - 1)s_y^2}{n_x + n_y - 2}$$

as noted in Section 17.3.

Example 4

An experiment was performed to investigate the effects of two alternative fertilisers on the growth of spinach plants. The plants were grown in controlled conditions, with 12 randomly selected plants being given fertiliser A, and a further 12 being given fertiliser B. Before the end of the experiment some plants were attacked by a fungus, and they were removed from the experiment. The final masses (x g) are summarised in the table below:

Fertiliser	Sample size	Σx	Σx^2
A	11	1098	175 644
B	10	1083	145 350

Show that, at the 5% significance level, the hypothesis that the two populations have equal variances is accepted.
Obtain an estimate of the common variance and determine a symmetric 99% confidence interval for the common value.

For fertilisers A and B, the unbiased estimates of the population variance are, respectively, $s_A^2 = 6604.36$ and $s_B^2 = 3117.90$. Since $s_A^2 > s_B^2$, we calculate the ratio $\dfrac{s_A^2}{s_B^2}$, which is equal to 2.12. Implicitly the test is two-tailed and we therefore compare the value 2.12 with the upper 2.5% (not 5%) point of an $F_{10,9}$-distribution. This is 3.96. Since the ratio is less than this, there is no need to reject the null hypothesis that the variances are the same.

The pooled estimate of the variance is given by:

$$s^2 = \frac{1}{19}\left\{\left(175\,644 - \frac{1}{11} \times 1098^2\right) + \left(145\,350 - \frac{1}{10} \times 1083^2\right)\right\} = 4952.88$$

For the symmetric 99% confidence interval, we need the upper and lower 0.5% points of the χ_{19}^2 distribution, which are respectively 38.58 and 6.844. The confidence interval is therefore:

$$\left(\frac{19 \times 4952.88}{38.58}, \frac{19 \times 4952.88}{6.844}\right)$$

which simplifies to (2440, 13 750).

19.5 Confidence interval for a variance ratio

Suppose that the variances σ_x^2 and σ_y^2 are not the same. We can repeat the arguments of the previous section, again assuming normal distributions, and see where they lead. We now have:

$$\frac{n_x - 1}{\sigma_x^2} S_x^2 \sim \chi_{n_x-1}^2 \qquad \text{and} \qquad \frac{n_y - 1}{\sigma_y^2} S_y^2 \sim \chi_{n_y-1}^2$$

From the definition of an F-distribution, taking the scaled ratio of random variables with χ^2 distributions, we get:

$$\frac{\dfrac{n_x - 1}{\sigma_x^2} S_x^2}{n_x - 1} \div \frac{\dfrac{n_y - 1}{\sigma_y^2} S_y^2}{n_y - 1} \sim F_{n_x-1,\,n_y-1}$$

which simplifies considerably to give:

$$\frac{S_x^2}{S_y^2} \frac{\sigma_y^2}{\sigma_x^2} \sim F_{n_x-1,\,n_y-1} \tag{19.7}$$

For a symmetric 95% confidence interval, we need the upper and lower 2.5% points, which we denote by U and L. So U is the upper 2.5% point of an $F_{n_x-1,\,n_y-1}$-distribution, and L is the reciprocal of the upper 2.5% point of an $F_{n_y-1,\,n_x-1}$-distribution. We can then write:

$$P\left(L < \frac{S_x^2}{S_y^2} \frac{\sigma_y^2}{\sigma_x^2} < U \right) = 0.95$$

Rearranging, we have:

$$P\left(\frac{S_y^2}{S_x^2} L < \frac{\sigma_y^2}{\sigma_x^2} < \frac{S_y^2}{S_x^2} U \right) = 0.95$$

Consequently, using the observed s_x^2 and s_y^2, the 95% confidence interval for the ratio $\dfrac{\sigma_y^2}{\sigma_x^2}$ is:

$$\left(\frac{s_y^2}{s_x^2} L, \quad \frac{s_y^2}{s_x^2} U \right)$$

Example 5

Obtain a 95% confidence interval for the ratio $\dfrac{\sigma_B^2}{\sigma_A^2}$, where σ_A^2 is the variance associated with the use of fertiliser A to treat the spinach plants of Example 4, and σ_B^2 is the variance associated with fertiliser B. The sample sizes were 11 and 10 and the corresponding values of s^2 were 6604.36 and 3117.90, respectively.

The required critical values are not given explicitly in the tables. Using linear interpolation the approximate values are $U = 3.98$ and $L = \dfrac{1}{3.78} = 0.265$. The ratio $\dfrac{s_B^2}{s_A^2} = 0.4721$, and hence the interval is $(0.4721 \times 0.265,\ 0.4721 \times 3.98)$ which simplifies to $(0.125, 1.88)$.

Chapter summary

♦ For observations from a normal distribution, the **distribution of the unbiased estimator, S^2, of the population variance**, σ^2, is given by

$$\frac{(n-1)}{\sigma^2} S^2 \sim \chi^2_{n-1}$$

♦ A **confidence interval for the population variance** is

$$\left(\frac{(n-1)s^2}{U}, \frac{(n-1)s^2}{L} \right), \text{ where } U \text{ and } L \text{ are upper and lower}$$

percentage points of a χ^2_{n-1} distribution.

♦ The **F-distribution:** if U and V are independent, with $U \sim \chi^2_u$ and

$V \sim \chi^2_v$ then $\dfrac{U}{u} \div \dfrac{V}{v} \sim F_{u,v}$.

♦ The **distribution of the ratio of the unbiased estimators of the**

population variances is given by $\dfrac{S^2_x}{S^2_y} \sim F_{n_x-1,n_y-1}$

♦ A **confidence interval for the ratio of population variances**, $\dfrac{\sigma^2_y}{\sigma^2_x}$, is

$$\left(\frac{s^2_y}{s^2_x} L, \frac{s^2_y}{s^2_x} U \right), \text{ where } U \text{ is the upper percentage point of an}$$

F_{n_x-1,n_y-1}-distribution and L is the reciprocal of the upper percentage point of an F_{n_y-1,n_x-1}-distribution.

Exercises 19d (Miscellaneous)

1 Independent random samples are taken from two normal populations. One sample of 8 observations gives 14.22 as the unbiased estimate of the population variance, while the other sample, with 6 observations gives 19.11 as the unbiased estimate of the population variance.

Test, at the 5% significance level, the hypothesis that the two populations have the same variance against the alternative that this is not the case.

2 Bill, an experienced barman, fills 16 glasses. The quantities of beer in the glasses give a value of 1.2 centilitres for s, where s^2 is the unbiased estimate of the population variance. Sid, a novice barman, fills 9 glasses. The quantities of beer in these give a value of 2.2 centilitres for s.

Assuming normal distributions, test, at the 5% level, the hypothesis that the two barmen are 'equally variable' in the amounts that they put in the glasses, against the alternative that Sid is more variable.

3 A sample of 8 observations is taken on a random variable having a normal distribution with unknown mean μ and unknown variance σ^2. Write down an expression (in terms of s, where s^2 is the unbiased estimate of the population variance) for the width, W, of a 90% symmetric confidence interval for μ.

By considering the distribution of s^2, show that, if the true value of σ^2 is 1.0, then $P(W > 2.49) \approx 0.001$.

Determine the corresponding probability when $\sigma^2 = 2.0$.

4 A sample of 13 observations, $(x_1, x_2, \ldots, x_{13})$, from a normal distribution with unknown mean μ_x and unknown variance σ^2_x is summarised by $\sum x_i = 207.4$, $\sum x^2_i = 4996.42$. An independent sample of 9 observations, $(y_1, y_2, \ldots, y_9)$, from a normal distribution with unknown mean μ_y and unknown variance σ^2_y is summarised by $\sum y_j = 76.2$, $\sum y^2_j = 1793.66$.

(continued)

(i) Determine a 95% confidence interval for
$$\frac{\sigma_x^2}{\sigma_y^2}.$$

(ii) Assuming that the two populations have the same variance, obtain a 95% confidence interval for the common value.

5 In an experiment to compare the yields of two varieties of barley, there were 10 observations on variety A and 12 observations on variety B. After subtracting 30 from each observation, the resulting figures were summarised as follows:

A: $\sum x = 50, \quad \sum x^2 = 323$
B: $\sum x = 36, \quad \sum x^2 = 167$

Verify that the hypothesis that observations on the two varieties are equally variable is accepted at the 5% significance level.
Test the hypothesis that there is no difference in yield between the two varieties against the alternative hypothesis that there is a difference. State and justify any assumptions that you make, and state your conclusions clearly.

6 The cellulose contents of the leaves of a certain plant are determined for random samples of leaves taken from two different locations. The results are shown below.

Location 1
15.4, 13.9, 15.1, 14.8, 14.4, 14.8, 15.0, 13.9, 15.4, 14.6, 14.8
Location 2
13.8, 14.4, 13.0, 15.3, 14.7, 14.3, 14.1, 12.9, 14.9

Let the population variances for the two locations be denoted by σ_1^2 and σ_2^2, respectively, and assume that the distributions of cellulose content are normal.

(a) Obtain unbiased estimates of σ_1^2 and σ_2^2.
(b) Determine a two-sided 95% confidence interval for σ_1^2.
(c) Determine a two-sided 95% confidence interval for $\dfrac{\sigma_1^2}{\sigma_2^2}$.
(d) Test the hypothesis that $\sigma_1^2 = \sigma_2^2$.
(e) Find the value L which is such that $P(L < \sigma_2^2) = 0.01$.

7 The resistances (in ohms) of a sample from a batch of resistors were

2314 2456 2389 2361 2360 2332 2402

(a) Calculate a 95% confidence interval for the standard deviation of the batch. Past experience suggests that the standard deviation, σ, is 35 ohms.
(b) Calculate a 95% confidence interval for the mean resistance of the batch
(i) assuming $\sigma = 35$,
(ii) making no assumption about the standard deviation.
(c) Compare the merits of the confidence intervals calculated in (b).   [AEB 91]

8 Catering size packets of lentils have a nominal net weight of 6000 g. A machine for filling the packets is believed to operate with a constant standard deviation, but with a mean which may alter during the day. Samples were taken at three different times during the day and the net weights, (g), measured with the following results;

Sample 1	6106	6044	6085	6072
Sample 2	5931	6001	6010	5950
Sample 3	6271	6340	6311	6353

(a) Use an F test to test whether samples 1 and 3 could have come from populations with the same variance. Use a 5% significance level.
(b) Calculate a 90% confidence interval for the standard deviation of the population from which sample 2 was drawn.
(c) Regarding all 12 observations as a sample from the same population estimate its standard deviation.
(d) Discuss to what extent your results confirm the belief that the mean may vary but the standard deviation, σ, remains constant. Suggest (but do not carry out) a suitable method of estimating σ.   [AEB 89]

20 Regression and correlation

A poor relation – is the most irrelevant thing in nature

Poor Relations, Charles Lamb

Previous chapters have concentrated principally on developing methods and models for a *single* random variable, but many data sets provide information about *several* variables, with the question of interest being whether or not there are connections between these variables. In this chapter we concentrate on methods suitable for use with two quantitative variables, usually denoted by x and y, where the values of x and y are observed in pairs.

Often all the data are collected at more or less the same time, for example:

x	y
Take-off speed of ski-jumper	Distance jumped
No. of red blood cells in sample of blood	No. of white blood cells in sample
Hand span	Foot length
Size of house	Value of house
Depth of soil sample from lake bottom	Amount of water content in sample

However, sometimes data are collected later on one variable than on the other variable, though the link (the same individual, same plot of land, same family, etc) is clear:

x	y
Mark in mock exam	Mark in real exam (three months later)
Amount of fertiliser	Amount of growth
Height of father	Height of son when 18

In some of the above cases, while the left-hand variable, x, may affect the right-hand variable, y, the reverse cannot be true – these are cases in which regression methods are particularly suitable. In other cases (such as the counts of blood cells) both variables are influenced by some unmeasured variable (e.g. the condition of the patient) and their mutual association may be best described using the so-called correlation coefficient. Correlation is discussed later in this chapter.

As an example of a case where regression methods are appropriate, consider the following. The petrol consumption of a car depends upon the speed at which it is driven. Cars are driven around a test circuit at a variety of (approximately) constant speeds and the fuel consumptions are monitored. The results are summarised in the table below:

mph	35	35	35	35	40	40	40	40
mpg	48.4	47.6	47.8	46.2	45.8	45.6	45.0	44.9

mph	45	45	45	45	50	50	50	50
mpg	43.0	42.8	42.7	42.2	39.9	40.3	38.9	39.6

Suggest the likely average fuel consumption of a car travelling at a constant 42 mph.

A sensible first step, whenever possible, is to plot the data on a **scatter diagram** which is simply a plot of all the pairs of values of x and y. For interest, we will also plot the means of each of the four groups (using horizontal bars).

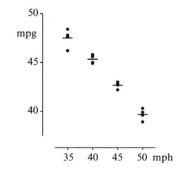

The group means lie approximately on a straight line. If we knew the equation of this line then we could easily arrive at an average mpg for a car travelling at a constant 42 mph.

In this chapter we shall concentrate on *linear* relationships and therefore, before continuing, we review the mathematical description of a straight line.

Note

 ◆ Plotting the group means provides a useful pictorial summary, but it is the individual values that are used in subsequent calculations.

20.1 The equation of a straight line

Denoting the measured quantities by x and y, the equation of a general straight line is often written in Pure Mathematics as:

$$y = mx + c$$

Perversely, the standard notation in Statistics is different (though, of course, the line is just as straight as before!). In Statistics we use:

$$y = a + bx$$

where a is a constant known as the **intercept** and b is a constant known as the **slope** or **gradient**.

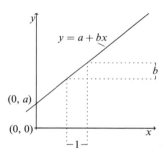

If we increase x to $(x + 1)$, the value of y changes from $a + bx$ to $a + b(x + 1)$, which is a change of b; thus b measures the amount of change in y for unit change in x. The quantity a represents the value of y when x is equal to zero, and therefore prescribes where the line crosses the y-axis.

Determining the equation

Suppose that we have drawn a line on a scatter diagram. How do we determine the equation of that line? The answer is quite simple. We first determine the co-ordinates of two points lying on the line. These can be any points, though it is a good idea to choose points near the edges of the diagram. Denote the points by (x_1, y_1) and (x_2, y_2). Then:

$$y_1 = a + bx_1$$
$$y_2 = a + bx_2$$

Subtracting the two left-hand sides and the two right-hand sides, we get:

$$y_1 - y_2 = b(x_1 - x_2)$$

and hence:

$$b = \frac{y_1 - y_2}{x_1 - x_2} \qquad\qquad (20.1)$$

To find the value of a it is easiest to substitute our value for b into one of the original equations:

$$a = (y_1 - bx_1) = (y_2 - bx_2)$$

Notice that the equation of the line is a function of the *x* and *y* values rather than of positions on the graph paper. The latter are affected by the choice of scale. A good choice is usually one that makes the graph 'fill' the space available!

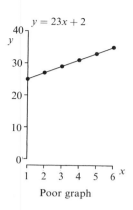

Poor graph

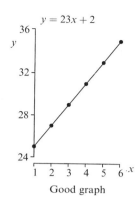

Good graph

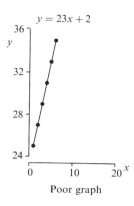

Poor graph

Example 1

A line goes through the points (1,15) and (10,33). The point (5, *y*) also lies on the line.
Determine the value of *y*.

We know that:

$$15 = a + b$$
$$33 = a + 10b$$

Hence $b = \dfrac{33 - 15}{10 - 1} = 2$ and hence $a = 15 - 2 = 13$. The line is therefore:

$$y = 13 + 2x$$

and the value of *y* corresponding to $x = 5$ is $13 + (2 \times 5) = 23$.

Calculator practice

If you have a graphical calculator then you can adjust the scales on the x-axis and the y-axis.
Experiment to see how changes of scale result in different appearances of the diagram.

20.2 The estimated regression line

Suppose that we have *n* points on a scatter diagram: (x_1, y_1), (x_2, y_2), ..., (x_n, y_n), and the line summarising the relation between *x* and *y* has equation $y = a + bx$.

Denoting the means of the entire sample of *x*-values and *y*-values by $\bar{x}$ and $\bar{y}$, respectively, the 'centre' of the scatter diagram could be represented by the point $(\bar{x}, \bar{y})$. It would therefore seem sensible for the line to pass through this point. The condition for this to occur is that *a* and *b* satisfy:

$$\bar{y} = a + b\bar{x}$$

so that:

$$a = \bar{y} - b\bar{x} = \frac{1}{n}(\Sigma y_i - b\Sigma x_i) \qquad (20.2)$$

The second form is often more convenient. To fix the line we need a value for
b. We will show later that a good choice is the least-squares estimate:

$$b = \frac{S_{xy}}{S_{xx}} \qquad (20.3)$$

where the quantities S_{xy} and S_{xx} (and, for completeness, the quantity S_{yy}) are
given by:

$$S_{xy} = \Sigma x_i y_i - \frac{\Sigma x_i \Sigma y_i}{n} \qquad (20.4)$$

$$S_{xx} = \Sigma x_i^2 - \frac{(\Sigma x_i)^2}{n} \qquad (20.5)$$

$$S_{yy} = \Sigma y_i^2 - \frac{(\Sigma y_i)^2}{n} \qquad (20.6)$$

Using the above values for the least-squares estimates a and b, the resulting
line is described as the estimated **regression line of y on x**, and b is called the
estimated **regression coefficient**.

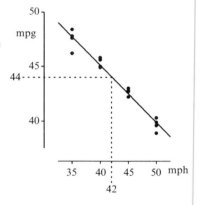

For the car data (with y denoting mpg and x denoting mph) we have
$n = 16$, $\Sigma x_i = 680$, $\Sigma x_i^2 = 29\,400$, $\Sigma y_i = 700.7$, $\Sigma y_i^2 = 30\,828.05$,
$\Sigma x_i y_i = 29\,518.5$. These sums and sums of squares are often referred to as the
summary statistics. We now calculate S_{xy} and S_{xx}:

$$S_{xy} = 29\,518.5 - \left(\frac{680 \times 700.7}{16}\right) = -261.25$$

$$S_{xx} = 29\,400 - \frac{680^2}{16} = 500.00$$

Hence, the least-squares estimates b and a are given by:

$$b = \frac{S_{xy}}{S_{xx}} = \frac{-261.25}{500.00} = -0.5225$$

$$a = \bar{y} - b\bar{x} = \tfrac{1}{16}\{700.7 - (-0.5225 \times 680)\} = 66.0$$

The estimated regression line is therefore $y = 66.0 - 0.5225x$ which is
illustrated on the scatter diagram.
The predicted average mpg for a car travelling at a steady 42 mph is therefore
$66.0 - 0.5225 \times 42 = 44.055$: approximately 44 mpg.

Notes

 Since $y = a + bx$ and $\bar{y} = a + b\bar{x}$, the regression line may written as:

$$(y - \bar{y}) = b(x - \bar{x})$$

◆ The formulae for a and b are for the estimated regression line of y on x. In
general, this regression line will not be the same as the regression line of x on y.
We discuss this at length in Section 20.10 (p. 541). However, both estimated
regression lines *pass through the point* $(\bar{x}, \bar{y})$.

◆ $\dfrac{S_{yy}}{n}$ is the sample variance for the y-values and $\dfrac{S_{xx}}{n}$ is the sample variance of the
x-values. The quantity $\dfrac{S_{xy}}{n}$ is called the **sample covariance**.

◆ The quantities S_{xx}, S_{xy} and S_{yy} can be written in a number of different ways.
The forms given in equations (20.4) to (20.6) are best for calculation, but other

forms will be useful in exploring the theory later in the chapter. We will now use the result that:

$$\Sigma(x_i - \bar{x}) = \Sigma x_i - n\bar{x} = \Sigma x_i - n\frac{\Sigma x_i}{n} = 0 \qquad (20.7)$$

to show that:

$$\Sigma\{(x_i - \bar{x})(y_i - \bar{y})\} = S_{xy}$$

We start by expanding the product inside the expansion:

$$\Sigma\{(x_i - \bar{x})(y_i - \bar{y})\} = \Sigma\{(x_i - \bar{x})y_i - (x_i - \bar{x})\bar{y}\}$$
$$= \Sigma\{(x_i - \bar{x})y_i\} - \bar{y}\Sigma(x_i - \bar{x})$$
$$= \Sigma\{(x_i - \bar{x})y_i\}$$

This follows because the second term is zero, as noted above. Continuing the expansion we get:

$$\Sigma\{(x_i - \bar{x})y_i\} = \Sigma x_i y_i - \bar{x}\Sigma y_i$$
$$= \Sigma x_i y_i - \frac{(\Sigma x_i)(\Sigma y_i)}{n}$$
$$= S_{xy}$$

Hence:

$$S_{xy} = \Sigma\{(x_i - \bar{x})y_i\} = \Sigma\{(x_i - \bar{x})(y_i - \bar{y})\}$$

Similarly:

$$\left.\begin{array}{l} S_{xx} = \Sigma\{(x_i - \bar{x})x_i\} = \Sigma(x_i - \bar{x})^2 \\[2mm] S_{yy} = \Sigma\{(y_i - \bar{y})y_i\} = \Sigma(y_i - \bar{y})^2 \end{array}\right\} \qquad (20.8)$$

Example 2

A factory uses steam to keep its radiators hot. Records are kept of y, the monthly consumption of steam for heating purposes (measured in lbs) and of x, the average monthly temperature (in degrees C). The results are as follows:

x	1.8	−1.3	−0.9	14.9	16.3	21.8	23.6	24.8
y	11.0	11.1	12.5	8.4	9.3	8.7	6.4	8.5

x	21.5	14.2	8.0	−1.7	−2.2	3.9	8.2	9.2
y	7.8	9.1	8.2	12.2	11.9	9.6	10.9	9.6

Estimate the regression line of y on x for these values.

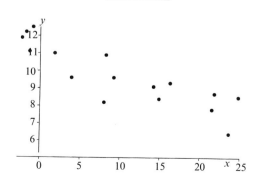

We begin by plotting the data. From the scatter diagram there does indeed appear to be a roughly linear relation between the variables, with the y-values decreasing as the x-values increase.

The summary statistics for the data are $n = 16$, $\Sigma x_i = 162.1$, $\Sigma x_i^2 = 3043.39$, $\Sigma y_i = 155.2$, $\Sigma y_i^2 = 1550.88$ and $\Sigma x_i y_i = 1353.11$, from which we get:

$$S_{xx} = 3043.39 - \frac{162.1^2}{16} = 1401.114$$

while $S_{yy} = 45.440$ and $S_{xy} = -219.260$. Hence:

$$b = \frac{-219.26}{1401.114} = -0.156\,49$$

$$a = \tfrac{1}{16}\{155.2 - (-0.156\,49 \times 162.1)\} = 11.2854$$

The estimated regression line of y on x is therefore approximately:

$$y = 11.3 - 0.156x$$

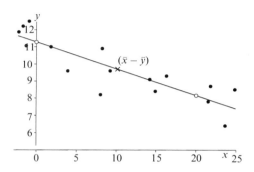

In order to check that we have not made any major arithmetic errors we now plot the fitted line on the scatter diagram. Since a line can be specified by two points that lie on it, we choose two x-values and calculate the corresponding y-values. For $x = 0$, we get $y = 11.3$, while for $x = 20$ we get $y = 11.2854 - (0.156\,49 \times 20) = 8.2$. Drawing the line through these points, $(0, 11.3)$ and $(20, 8.2)$, we see that the line does seem to provide a rough description of the data.

___

Note

- In choosing values for x for plotting two points on the estimated regression line, choose values near to the minimum and maximum values of x. The values should also be easy to use – so the value 0, as used in the example above, is ideal. Your line should also pass through $(\bar{x}, \bar{y})$, which is $(10.1, 9.7)$ for the example above.

Calculator practice _____

> *If you have a graphical calculator then you can plot the data as a scatter diagram and superimpose your estimated regression line. If you have done the calculations correctly then the line will go through the 'centre' of the data – specifically it will go through $(\bar{x}, \bar{y})$. If it misses then this doesn't mean that the method has 'gone wrong' – it means that you have gone wrong!*

Computer project _____

> *The representative nature of the regression calculations is terribly dull! this is just the sort of thing that a computer spreadsheet was designed to tackle. Draw up a spreadsheet in which successive columns give x, y, xy and x^2 values. Advanced spreadsheet designers might like to go further and calculate a and b and then have two further columns showing $a + bx$ (the estimated values) and $y - a - bx$ (the residuals, discussed later). Spreadsheets will also provide convenient diagrams.*

Exercises 20a

1 Eight pairs of observations on the variables x, y are given below.

x	1.2	0.5	0.8	0.1	2.3	1.1	1.8	2.2
y	8.1	4.3	7.1	3.5	12.8	8.4	9.9	11.4

(i) Plot a scatter diagram.
(ii) Calculate Σx, Σx^2, Σy, Σy^2, Σxy.
(iii) Find $\bar{x}$, $\bar{y}$, S_{xx}, S_{yy}, S_{xy}.
(iv) Find the values of a and b for the regression line $y = a + bx$.
(v) Mark on your diagram the point $(\bar{x}, \bar{y})$.
(vi) Use the answer to (iv) to estimate the value of y when $x = 0.2$, and mark the corresponding point on your diagram.
(vii) Draw the regression line on your diagram.

2 Six pairs of observations on the variables g, h are given below.

g	55.7	10.4	67.1	91.2	30.8	72.1
h	21.2	45.9	88.3	11.4	75.4	21.4

(i) Plot a scatter diagram, with values of g on the horizontal axis.
(ii) Calculate Σg, Σg^2, Σh, Σh^2, Σgh.
(iii) Find $\bar{g}$, $\bar{h}$.
(iv) Find the equation of the regression line of h on g.
(v) Mark on your diagram the point $(\bar{g}, \bar{h})$.
(vi) Draw the regression line on your diagram.

3 Five pairs of observations on the variables w, z are given below.

w	357.2	284.3	435.8
z	0.0149	0.0375	−0.0172

w	571.9	101.2
z	−0.0345	0.0651

(i) Plot a scatter diagram, with values of z on the horizontal axis.
(ii) Calculate Σw, Σw^2, Σz, Σz^2, Σwz.
(iii) Find $\bar{w}$, $\bar{z}$.
(iv) Find the equation of the regression line of w on z.
(v) Mark on your diagram the point $(\bar{w}, \bar{z})$.
(vi) Draw the regression line on your diagram.
(vii) Estimate the value of w when $z = 0.0572$.

4 For twelve consecutive months a factory manager recorded the number of items produced by the factory and the total cost of their production. The following table summarises the manager's data.

Number of items (x) thousands	18	36	45	22	69	72
Production Cost (y) £1000	37	54	63	42	84	91

Number of items (x) thousands	13	33	59	79	10	53
Production Cost (y) £1000	33	49	79	98	32	71

(a) Draw a scatter diagram for the data.
(b) Give a reason to support the use of the regression line $(y - \bar{y}) = b(x - \bar{x})$ as a suitable model for the data.
(c) Giving the values of $\bar{x}$, $\bar{y}$ and b to 3 decimal places, obtain the regression equation for y on x in the above form. (You may use $\Sigma x^2 = 27\,963$, $\Sigma xy = 37\,249$.)
(d) Rewrite the equation in the form $y = a + bx$ giving a to three significant figures.
(e) Give a practical interpretation of the values of a and b.
(f) The selling price of each item produced is £1.60. Find the level of output at which total income and estimated total costs are equal. Give a brief interpretation of this value. [ULSEB]

5 For the regression equation $y = a + bx$, the normal equations which give the estimates of a and b are

$$\Sigma y = na + b\Sigma x$$
$$\Sigma xy = a\Sigma x + b\Sigma x^2.$$

Show that the regression equation can be expressed in the form

$$(y - \bar{y}) = b(x - \bar{x}).$$

State the implication this form has for plotting the regression line.

For a period of three years a company monitors the number of units of output produced per quarter and the total cost of producing the units. The table below [overleaf] shows their results.

(continued)

Units of Output (*x*) (1000's)	14	29	55	74	11	23
Total Cost (*y*) (£1000)	35	50	73	93	31	42

Units of Output (*x*) (1000's)	47	69	18	36	61	79
Total Cost (*y*) (£1000)	65	86	38	54	81	96

(Use $\Sigma x^2 = 28740$, $\Sigma xy = 38286$)

(a) Draw a scatter diagram of these data.

(b) Calculate the equation of the regression line of *y* on *x* and draw this line on your scatter diagram.

The selling price of each unit of output is £1.60.

(c) Use your graph to estimate the level of output at which the total income and total costs are equal.

(d) Give a brief interpretation of this value.

[AEB 91]

20.3 Why 'regression'?

We met Francis Galton earlier, in Chapter 12. Galton's early work on inheritance included a study of the heights of successive generations of people. Some of his data are summarised below and are shown in the diagram, together with a dotted line corresponding to equality of the heights of the two generations.

Mean height of parents (inches)	72.5 70.5 68.5 66.5 64.5
Mean height of their adult children (inches)	72.2 69.5 68.2 67.2 65.8

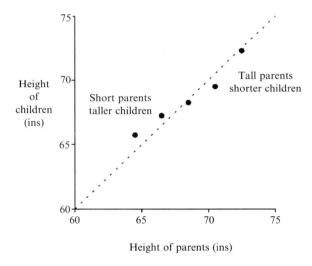

Height of parents (ins)

Galton plotted the data and noticed two things:

1 On average, the heights of adult children of tall parents were greater than the heights of adult children of short parents: the averages appeared to be (more or less) *linearly* related.

2 On average, the adult children of tall parents are shorter than their parents, (72.2 < 72.5, etc), whereas the adult children of short parents are taller than their parents, (65.8 > 64.5, etc): the values *regress* towards the mean.

These findings led Galton (in a talk entitled 'Regression towards mediocrity in hereditary stature' given in 1885 to the British Association for the Advancement of Science) to refer to his summary line drawn through the data as being a **regression line**, and this name is now used to describe quite general relationships.

20.4 The method of least squares

We want to draw a straight line on a scatter diagram so that it seems to fit the data as well as possible. The diagrams show two examples of a bad choice of line.

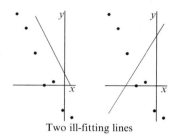

Two ill-fitting lines

In one case the line has the correct slope, but the wrong intercept and therefore 'misses' the data completely. In the other case the line goes through the data – but at a very strange angle.

The method of least squares, on the other hand, produces a line that is certain to appear satisfactory since it goes through the centre of the data and does so at a sensible angle. Note that, although the line 'goes through' the data, very few (if any) of the points actually lie on the line, which is simply a convenient way of summarising any apparent connection between the variables.

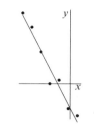

The least squares line

Suppose that an observation is made at (x_i, y_i). The discrepancy between y_i and the value given by the estimated regression line is called the **residual** and is denoted by r_i. Thus:

$$r_i = y_i - (a + bx_i) \qquad (20.9)$$

If the values of a and b are well chosen, then all of the residuals $r_1, r_2, \ldots, r_n$ will be small in magnitude. Here n is the number of pairs of (x, y) values. Some of the residuals will be negative (corresponding to points lying below the line), so it is mathematically convenient to work with their squares. A line that fits the data well will give a relatively small value for Σr_i^2. The **least-squares regression line** is the line that actually minimises this quantity and it is sometimes called the **line of best fit**.

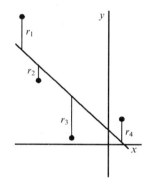

The values of a and b that minimise Σr_i^2 are the ones given by Equations (20.2) and (20.3). The sum of the residuals is given by:

$$\Sigma r_i = \Sigma y_i - \Sigma(a + bx_i)$$
$$= \Sigma y_i - na - b\Sigma x_i$$

The least-squares estimate a, given by Equation (20.2), is equal to $\bar{y} - b\bar{x}$. Substituting for a we therefore get:

$$\Sigma r_i = \Sigma y_i - n(\bar{y} - b\bar{x}) - b\Sigma x_i$$
$$= (\Sigma y_i - n\bar{y}) - b(\Sigma x_i - n\bar{x})$$

Since both bracketed terms are equal to zero – as in Equation (20.7) – we have:

$$\Sigma r_i = 0 \qquad (20.10)$$

This shows that, for the least-squares line, the residual discrepancies cancel out – which seems sensible.

Derivation of the estimates
The residual sum of squares, which is to be minimised, is given by:

$$\Sigma r_i^2 = \Sigma\{y_i - (a + bx_i)\}^2 = \Sigma(y_i - a - bx_i)^2$$

To find the 'least squares' values for a and b, we begin with a useful rearrangement:

$$\Sigma r_i^2 = \Sigma(y_i - a - bx_i)^2$$
$$= \Sigma\{(y_i - bx_i) - a\}^2$$
$$= \Sigma(y_i - bx_i)^2 - 2a\Sigma(y_i - bx_i) + na^2$$

We now 'complete the square':

$$\Sigma r_i^2 = n\left\{a - \frac{\Sigma(y_i - bx_i)}{n}\right\}^2 + \left(1 - \frac{1}{n}\right)\Sigma(y_i - bx_i)^2$$

The first term on the right-hand side is non-negative and the second term does not involve a. We can therefore minimise the right-hand side by reducing the first term to zero by letting:

$$a = \frac{\Sigma(y_i - bx_i)}{n}$$

This is equivalent to:

$$a = \frac{\Sigma y_i}{n} - b\frac{\Sigma x_i}{n} = \bar{y} - b\bar{x}$$

as in Equation (20.2).

Substituting this value for a and rearranging, we get:

$$\Sigma r_i^2 = \Sigma(y_i - a - bx_i)^2$$
$$= \Sigma\{(y_i - \bar{y}) - b(x_i - \bar{x})\}^2$$
$$= \Sigma(y_i - \bar{y})^2 - 2b\Sigma\{(x_i - \bar{x})(y_i - \bar{y})\} + b^2\Sigma(x_i - \bar{x})^2$$
$$= S_{yy} - 2bS_{xy} + b^2 S_{xx}$$

Once again we complete the square:

$$\Sigma r_i^2 = S_{xx}\left(b - \frac{S_{xy}}{S_{xx}}\right)^2 + S_{yy} - \frac{S_{xy}^2}{S_{xx}}$$

The first term on the right-hand side is non-negative and the second term does not involve b. We can therefore minimise the right-hand side by reducing the first term to zero by letting:

$$b = \frac{S_{xy}}{S_{xx}}$$

as in Equation (20.3).

Denoting the resulting minimum sum of squared residuals by D, we have:

$$D = S_{yy} - \frac{S_{xy}^2}{S_{xx}} \tag{20.11}$$

The modern name for D is the **deviance**, while you may also find it called the **residual sum of squares**.

Calculator practice

> *Many calculators have in-built routines for calculating least-squares regression lines. Often the values of the separate sums (e.g. Σx_i, $\Sigma x_i y_i$, Σx_i^2) are stored in memories that can be accessed by the user. Such calculators usually provide the values of a and b as well as other quantities discussed later in the chapter. If you have this type of calculator, practise using it to do regression calculations.*

Exercises 20b

1 Ten pairs of observations on the variables x, y are given below.

y	2.2	3.2	6.8	7.3	-1.3
x	1.2	0.5	0.0	-0.8	2.8

y	-0.8	1.7	9.5	12.3	1.7
x	3.4	1.7	-1.7	-4.2	1.1

(i)   Plot a scatter diagram.
(ii) Calculate Σx, Σx^2, Σy, Σy^2, Σxy.
(iii) Find $\bar{x}$, $\bar{y}$, S_{xx}, S_{yy}, S_{xy}.
(iv) Find the values of a and b for the regression line $y = a + bx$.
(v)   Draw the regression line on your diagram and mark also the point $(\bar{x}, \bar{y})$.
(vi) Find the value of D, and relate it to your diagram.
(vii) Find the sample variances for x and y.

2 Seven pairs of observations on the variables u, v are given below.

u	1.0	2.2	2.8	3.1	3.8	4.5	5.1
v	3.1	12.5	12.4	7.6	9.3	14.7	13.4

(i)   Plot a scatter diagram, with values of u on the horizontal axis.
(ii) Calculate Σu, Σu^2, Σv, Σv^2, Σuv.
(iii) Find $\bar{u}$, $\bar{v}$, S_{uu}, S_{vv}, S_{uv}.
(iv) Find the equation of the regression line of v on u.
(v)   Mark on your diagram the point $(\bar{u}, \bar{v})$.
(vi) Estimate the value of v when $u = 4.8$, and mark the corresponding point on your diagram.

(vii) Draw the regression line on your diagram.
(viii) Find the residual sum of squares and relate it to your diagram.
(ix)   Find the sample variances for u and v.

3 Draw a diagram showing a few non-collinear points and their regression line of y on x. Mark on your diagram the distances, the sum of whose squares is minimised by the regression line.
Five shells are fired from a gun standing on level ground. After each shell is fired, the angle of elevation of the gun is increased by 1 degree. The rth shell hits the ground at a distance of y_r km from the gun. Given that

$$\Sigma y_r = 9.6, \quad \Sigma r y_r = 29,$$

find the equation of the regression line of y on r in the form $y = \alpha + \beta r$.
A shell is fired at an elevation 2.4 degrees above the initial elevation. Estimate the distance from the gun at which this shell hits the ground.   [O&C]

4 Draw a sketch to illustrate the distances, the sum of the squares of which is minimised when fitting a regression line of y on x to given data.
The weights y of a certain substance that dissolved in a litre of water in five experiments at various temperatures x are summarised by $\Sigma x = 200$, $\Sigma y = 182$, $\Sigma x^2 = 9850$, $\Sigma xy = 8390$.
Find the regression line of y on x in the form $y = \alpha + \beta x$.
Estimate the weight that will dissolve in a litre of water at temperature $x = 90$.   [O&C]

20.5 Dependent random variable Y

In the previous section no mention was made of probability, probability distributions or random variables. The method of least squares was simply a procedure for determining sensible values for a and b, the parameters of a line that partially summarised the relation between y and x. We now introduce randomness!

   Suppose that the x-values are fixed, but the y-values are affected by sampling variation. Here are some examples:

x	y
Number of bricks in pile	Weight of pile of bricks
The price of a commodity	The number sold
The capacity of a car engine	The average mpg

The first example is a particularly obvious one. We could have lots of piles, each containing precisely 10 bricks. But each pile would have a different weight because of the variations in the weights of individual bricks. Thus while x is a fixed quantity, y is an observation on a random variable Y.

In these examples, Y is a **dependent variable** having an unpredictable value, whereas x is a so-called **independent variable** whose value has been fixed. If there is an underlying linear relationship, then this relationship connects x not to an individual y-value, but to the mean of the y-values *for that particular value of* x. The mean of Y, given the particular value x, is denoted by:

$$\mathrm{E}(Y|x)$$

which is more formally called the **conditional expectation** of Y. The **linear regression model** is:

$$\mathrm{E}(Y|x) = \alpha + \beta x$$

Here α and β are the (unknown) population values. (As usual, we use Greek letters for population parameters.) The least-squares estimates of α and β are a and b respectively, where, as before:

$$b = \frac{S_{xy}}{S_{xx}}$$

$$a = \bar{y} - b\bar{x}$$

Estimating a future *y*-value

Suppose that future observations are to be taken with $x = x_0$. Because Y is a random variable, the y-values will not be precisely predictable, but we can estimate their average value using the regression line with the least-squares estimates for a and b. The estimated value of $\mathrm{E}(Y|x_0)$ is:

$$a + bx_0 = (\bar{y} - b\bar{x}) + bx_0 = \bar{y} + \frac{S_{xy}}{S_{xx}}(x_0 - \bar{x})$$

Example 3

The data below refer to a chain of shops. The figures reported are the numbers of sales staff (x) and the average daily takings in thousands of pounds (y) for a random sample of shops. Use the least-squares line to estimate the average daily takings of a shop with 21 staff.

x	17	39	32	17	25	43	25	32	48	10	48	42	36	30	19
y	7	17	10	5	7	15	11	13	19	3	17	15	14	12	8

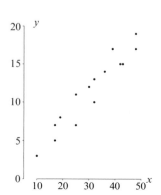

The summary statistics are $n = 15$, $\Sigma x_i = 463$, $\Sigma x_i^2 = 16\,275$, $\Sigma y_i = 173$, $\Sigma y_i^2 = 2315$, $\Sigma x_i y_i = 6102$. Hence $S_{xy} = 762.066\,67$ and $S_{xx} = 1983.733\,33$, so that the least-squares estimates are $b = 0.384\,16$ and $a = \frac{1}{15}\{173 - (0.384\,16 \times 463)\} = -0.324\,34$. Using these estimates of a and b, the estimated average y-value with $x = 21$ is $-0.324\,34 + (21 \times 0.384\,16) = 7.743$. The average daily takings are estimated to be a little under £7750.

Practical _____

Feet are smelly things, so it would be nice not to have to measure them! However, steel yourselves, as this practical involves just that. Use a tape measure to measure (in mm) the circumference of your right wrist and the length of your right foot. (Not so bad after all, since its your own foot!) Pool these results with others so as to obtain around 20 observations. Plot these on a scatter diagram.

For the benefit of future would-be foot measurers, determine the regression line of foot length on wrist circumference.

Exercises 20c _____

1 In the following six pairs of observations, the values of x are exact, but the values of y are liable to error.

x	1	2	3	4	5	6
y	4.61	18.13	32.35	45.29	48.61	72.13

(i) Plot a scatter diagram.
(ii) Calculate the equation of the estimated regression line of y on x.
(iii) Estimate the mean value of y corresponding to $x = 3.5$.

2 In the following twelve pairs of observations the values of u are exact but the values of v are liable to error.

u	5	5	5	6	6	6
v	0.71	0.63	0.46	0.56	0.82	0.71

u	7	7	7	8	8	8
v	1.03	0.99	0.98	1.05	1.42	1.07

(i) Plot a scatter diagram.
(ii) Calculate the equation of the estimated regression line of v on u.
(iii) Estimate the mean value of v when $u = 9$.

3 When a car is driven under specified conditions of load, tyre pressure and surrounding temperature, the temperature, $T°C$, generated in the shoulder of the tyre varies with the speed, V km h^{-1}, according to the linear model $T = a + bV$, where a and b are constants. Measurements of T were made at eight different values of V with the following results.

v	20	30	40	50	60	70	80	90
t	45	52	64	66	91	86	98	104

$(\Sigma v = 440, \Sigma v^2 = 28\,400, \Sigma t = 606,$
$\Sigma t^2 = 49\,278, \Sigma vt = 37\,000.)$

(i) Show these data on a scatter diagram.
(ii) Calculate the equation of the estimated regression line of T on V.
(iii) Estimate the expected value of T when $V = 60$.
(iv) It is given that, for each value of V, the measured value of T contains a random error which is normally distributed with zero mean and variance 16. Calculate the probability that, when $V = 60$, the measured value of T exceeds 91. Comment on the result. [UCLES(P)]

4 Six metal plates were immersed in a weak acid solution for various lengths of time. Their percentage losses in weight were then measured. The results are shown in the table below.

Time in solution (x hours)	150	200	200
Weight loss ($y\%$)	0.8	1.4	1.2

Time in solution (x hours)	300	450	500
Weight loss ($y\%$)	1.7	2.6	2.5

(i) Illustrate these data by drawing a graph.
(ii) Find the regression line of y on x in the form $y = \alpha + \beta x$.
(iii) Draw the line you have calculated on your graph.
(iv) Estimate the percentage weight loss of a plate immersed for 30 days.
(v) Mark on your graph the distances the sum of whose squares is minimised by the regression line. [O&C(P)]

5 An anemometer is used to estimate wind speed by observing the rotational speed of its vanes. This speed is converted to wind speed by means of an equation obtained from calibrating the instrument in a wind tunnel. In this calibration the wind speed is fixed precisely and the resulting anemometer speed is noted. For a particular anemometer this process produced the following set of data.

Actual wind speed (m/s) s	1.0	1.1	1.2	1.3	1.4
Anemometer (revs/min) r	30	38	48	58	68

Actual wind speed (m/s) s	1.5	1.6	1.7	1.8	1.9
Anemometer (revs/min) r	80	92	106	120	134

$[\Sigma s = 14.5, \Sigma s^2 = 21.85, \Sigma r = 774,$
$\Sigma r^2 = 71\,092, \Sigma rs = 1218.0.]$

(i) Obtain the equation of a suitable least squares regression line to summarise these results.

(ii) If the actual wind speed is 1.65 m/s, use the equation of the regression line to estimate the rotational speed of the anemometer.

(iii) Demonstrate, using the above regression line as an example, that it is unwise to extrapolate beyond the range of the data.
[UCLES(P)]

20.6 Transformations, extrapolation and outliers

Not all relationships are linear! However, there are quite a few non-linear relations which can be turned into the linear form. Here are some examples:

$y = ax^b$	Take logarithms	$\log(y) = \log(a) + b\log(x)$
$y = ae^{bx}$	Take natural logarithms	$\ln(y) = \ln(a) + bx$
$y = (a + bx)^k$	Take kth root	$\sqrt[k]{y} = a + bx$

In each case we can find a way of transforming the relationship into a linear one so that the formulae derived previously can be used.

For many relations no transformation is needed because, over the restricted range of the data, the relation does appear to be linear. As an example, consider the following fictitious data:

x Amount of fertiliser per m^2	y Yield of tomatoes per plant
10 g	1.4 kg
20 g	1.6 kg
30 g	1.8 kg

In this tiny data set there is an exact linear relation between the yield and the amount of fertiliser, namely $y = 0.02x + 1.2$. How we use that relation will vary with the situation. Here are some examples:

1 We can reasonably guess that, for example, if we had applied 25 g of fertiliser then we would have got a yield of about 1.7 kg. This is a sensible guess, because 25 g is a value similar to those in the original data.

2 We can expect that 35 g of fertiliser would give a yield of about 1.9 kg. This is reasonable because the original data involved a range from 10 g to 30 g of fertiliser and 35 g is only a relatively small increase beyond the end of that range.

3 We can expect that 60 g of fertiliser might lead to a yield in excess of 2 kg, as predicted by the formula. However, this is little more than a guess, since the original range of investigation (10 g to 30 g) is very different from the 60 g that we are now considering.

4 If we use 600 g of fertiliser then the formula predicts over 13 kg of tomatoes. This is obviously nonsense! In practice the yield would probably be zero because the poor plants would be smothered in fertiliser!

Our linear relation cannot possibly hold for *all* values of the variables, however well it appears to describe the relation in the given data.

The above shows that the least-squares regression line is *not* a substitute for common sense! Care is required, since thoughtless extrapolation can lead to stupid statements. If a cricketer were to make successive scores of 1, 10 and 100, we would be unwise to predict a score of 1000 for his next effort!

The third topic in this section is 'outliers'. An **outlier** is an observation that is very different from the rest of the data. Such a value should be treated with suspicion! The calculations for regression (and, later, for correlation) involve quantities such as $(x_i - \bar{x})$ and $(y_i - \bar{y})$. If one (or both) of these is large – because observation i is an outlier – then observation i is likely to dictate the values of the parameter estimates. The following diagram illustrates a case in which the precise location of a single outlier essentially dictates the equation of the regression line.

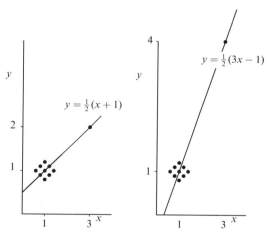

Nine of the points have the same values in both cases. These values show no special relation between Y and x. The slope of the regression line is decided by the location of the outlying point.

Example 4

The following observations on x and Y have been reported.

x	132	246	188	343	512	442	377	413	421	334
y	116	188	136	215	300	266	239	253	180	218

Plot the data using a scatter diagram and verify that there is an outlier. Supposing that this is due to a typographical error, suggest a correction.

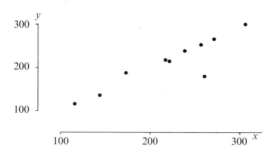

The diagram shows that all the observations except one lie close to a straight line. The exception is the point recorded as (421, 180). The most common typographical error is to interchange neighbouring digits. It is possible that the point should have been recorded as (241, 180).

20.7 Confidence interval for the population regression coefficient β

We assume that all the n observations on Y, whatever the value of x, come from independent normal distributions with variance σ^2. Information about this variability is provided by the deviance (the residual sum of squares), D, given by:

$$D = S_{yy} - \frac{S_{xy}^2}{S_{xx}}$$

When allowance has been made for the estimation of the values of α and β, there are effectively $(n-2)$ independent pieces of information available for the estimation of σ^2. A consequence is that an unbiased estimate of σ^2 is provided by $\dfrac{D}{n-2}$.

The estimated value b of the population regression coefficient β is obtained from the sampled values of Y. Different samples would give different estimated values. We can therefore think of our value, b, as being an observation on a random variable B with some distribution to be determined.

We can write S_{xy}, given by Equation (20.4), in the form:

$$S_{xy} = \Sigma x_i y_i - \bar{x}\Sigma y_i = \Sigma\{(x_i - \bar{x})y_i\}$$

Hence we can write the estimate b, given by Equation (20.3), in the form:

$$b = \frac{x_1 - \bar{x}}{S_{xx}}y_1 + \frac{x_2 - \bar{x}}{S_{xx}}y_2 + \cdots + \frac{x_n - \bar{x}}{S_{xx}}y_n$$

which shows that the estimate b is a linear combination of the y-values. Similarly the estimator B can be written:

$$B = \frac{x_1 - \bar{x}}{S_{xx}}Y_1 + \frac{x_2 - \bar{x}}{S_{xx}}Y_2 + \cdots + \frac{x_n - \bar{x}}{S_{xx}}Y_n$$

where Y_i is the 'random variable Y given that $x = x_i$'. This shows that B is a linear combination of $Y_1, Y_2, \ldots, Y_n$.

Assuming that the independent random variables $Y_1, Y_2, \ldots, Y_n$ are normally distributed, it follows that the estimator B also has a normal distribution (see Section 12.9, p. 320). With the further assumption that the distributions of the random variables have common variance σ^2, we shall show that b is an observation on a random variable having a $N\left(\beta, \dfrac{\sigma^2}{S_{xx}}\right)$ distribution.

Using an estimate of σ^2, because the true value is unknown, leads, as usual, to a move from a normal distribution to a t-distribution. In this case the estimate of σ^2 is $\dfrac{D}{n-2}$, which is associated with $(n-2)$ degrees of freedom. Denoting the upper 2.5% point of a t_{n-2}-distribution by c_t, a symmetric 95% confidence interval for β is given by:

$$\left(b - c_t\sqrt{\frac{D}{(n-2)S_{xx}}}\, , \; b + c_t\sqrt{\frac{D}{(n-2)S_{xx}}}\right)$$

Notes
- The usual sample variance, $\dfrac{S_{yy}}{n}$, is not an appropriate estimate of σ^2 since some of the variation in the y-values may be explained by the varying x-values.
- Traditionally the distinction between the population regression coefficient β and the sample regression coefficient b, which is an unbiased estimate of β, is glossed over. In some books and in some syllabuses you will see reference to a 'confidence interval for b'.

Mean and variance of the estimator of β

According to the model, $E(Y_i) = \alpha + \beta x_i$. The expected value of the estimator of B is:

$$E\left(\frac{1}{S_{xx}}\Sigma\{(x_i - \bar{x})Y_i\}\right) = \frac{1}{S_{xx}}\Sigma\{(x_i - \bar{x})E(Y_i)\}$$

$$= \frac{1}{S_{xx}}\Sigma\{(x_i - \bar{x})(\alpha + \beta x_i)\}$$

$$= \frac{\beta}{S_{xx}}\Sigma\{(x_i - \bar{x})x_i\} \qquad \text{since } \Sigma(x_i - \bar{x}) = 0$$

$$= \beta\frac{S_{xx}}{S_{xx}} = \beta \qquad \text{using Equation (20.8)}$$

Thus b is an unbiased estimate of β.

The variance, V, associated with the estimate is given by:

$$V = \left(\frac{x_1 - \bar{x}}{S_{xx}}\right)^2\sigma^2 + \left(\frac{x_2 - \bar{x}}{S_{xx}}\right)^2\sigma^2 + \cdots + \left(\frac{x_n - \bar{x}}{S_{xx}}\right)^2\sigma^2$$

$$= \frac{(x_1 - \bar{x})^2 + (x_2 - \bar{x})^2 + \cdots + (x_n - \bar{x})^2}{S_{xx}^2}\sigma^2$$

$$= \frac{S_{xx}}{S_{xx}^2}\sigma^2$$

$$= \frac{\sigma^2}{S_{xx}}$$

Significance test for the regression coefficient

This can be performed in the usual way. Alternatively, the result of such a test can be deduced by studying the corresponding confidence interval. For example, if the population value specified by the null hypothesis falls inside the confidence interval, then the hypothesis will be accepted at the corresponding level.

Example 5

A company has a fleet of similar cars of different ages. Examination of the company records reveals that the cost of replacement parts for the older cars is generally greater than that for the newer cars. A random sample of the records are reported below.

Age (years), x	1	1	2	2	2	3	3	3	4	4	4	5
Cost (£), y	163	382	478	466	549	495	723	681	619	1049	1033	890

Determine the least-squares estimates of the parameters of the regression line, $y = \alpha + \beta x$. Assuming that each y-value is an observation from a normal distribution with variance σ^2, obtain an estimate of σ^2 and obtain a 95% symmetric confidence interval for the value of β.

   Determine the outcomes of testing the null hypothesis $\beta = \beta_0$ against the alternative $\beta \neq \beta_0$, using a 5% test for (i) $\beta_0 = 0$ and (ii) $\beta_0 = 200$.

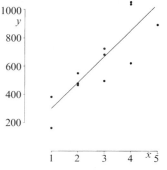

The summary statistics are $n = 12$, $\Sigma x_i = 34$, $\Sigma x_i^2 = 114$, $\Sigma y_i = 7528$, $\Sigma y_i^2 = 5\,493\,800$, $\Sigma x_i y_i = 24\,482$, leading to $S_{xx} = 17.6667$, $S_{yy} = 771\,234.6667$ and $S_{xy} = 3152.6667$. Note that the calculations retain many significant figures in order to avoid errors due to premature rounding.

The least-squares estimates are:

$$b = \frac{3152.6667}{17.6667} = 178.4528$$

$$a = \tfrac{1}{12}\{7528 - (178.4528 \times 34)\} = 121.7170$$

The least-squares line is approximately:

$$y = 122 + 178x$$

The deviance, D, is given by:

$$D = S_{yy} - \frac{S_{xy}^2}{S_{xx}} = 771\,234.6667 - \frac{(3152.6667)^2}{17.6667} = 208\,632.4$$

so that σ^2 is estimated as $\dfrac{208\,632.4}{12 - 2} \approx 20\,863$. The upper 2.5% point of a t_{10}-distribution is 2.23 and so the 95% confidence interval for β is:

$$178.4528 \pm \left(2.23 \times \sqrt{\frac{20\,863.24}{17.6667}} \right) = 178.4528 \pm 76.6335$$

The 95% confidence interval is $(102, 255)$.

   Since the interval excludes 0, the hypothesis that the slope is 0 is rejected, at the 5% level. On the other hand, since 200 falls within the interval, the hypothesis that the slope is 200 is accepted at the 5% level.

Exercises 20d

1 A large field of maize was divided into 6 plots of equal area and each plot was treated with a different concentration of fertilizer. The yield of maize from each plot is shown in the table.

Concentration (oz m^{-2})	0	1	2	3	4	5
Yield (tonnes)		15	22	31	40	48	54

Draw a scatter diagram for these data.
Obtain the equation of the regression line for yield on concentration, giving the values of the coefficients to 2 decimal places.
Use the equation of the regression line to obtain a value for the yield when the concentration applied is 3 oz m^{-2}. State precisely what is being estimated by this value. State any reservations you would have about making an estimate from the regression equation of the expected yield per plot if 7 oz m^{-2} of fertilizer are applied.   [ULSEB(P)]

2 A straight line regression equation is fitted by the least squares method to the n points $(x_r, y_r), r = 1, 2, \ldots, n$. For the regression equation $y = ax + b$, show in a sketch the distances whose sum of squares is minimised, and mark clearly which axis records the dependent variable and which axis records the independent (controlled) variable.
In a chemical reaction it is known that the amount, A grams, of a certain compound produced is a linear function of the temperature $T°$ C. Eight trial runs of this reaction are performed, two at each of four different temperatures. The observed values of A are subject to error. The results are shown in the table.

T	10	15	20	25
A	10	15	18	16
	12	12	16	20

Draw a scatter diagram for these data.
Calculate $\bar{A}$ and $\bar{T}$.
Obtain the equation of the regression line of A on T giving the coefficients to 2 decimal places. Draw this line on your scatter diagram.
Use the regression equation to obtain an estimate of the mean value of A when $T = 20$, and explain why this estimate is preferable to averaging the two observed values of A when $T = 20$.

Estimate the mean increase in A for a one degree increase in temperature.
State any reservations you would have about estimating the mean value of A when $T = 0$.   [ULSEB]

3 A shop sells home computers. The numbers of computers sold in each of five successive years are given in the table below.

Year (x)	1	2	3	4	5
Sales (y)	10	30	70	140	210
ln$y$		2.303	3.401	4.248	4.942	5.347

$\Sigma \ln y = 20.241, \Sigma(\ln y)^2 = 87.931,$
$\Sigma x \ln y = 68.353$

(i) Assuming that the sales, y, and the year, x, are related by the equation $y = ab^x$, find the least squares regression line of ln$y$ on x, and hence estimate the constants a and b.
(ii) The shop manager uses this relationship to predict the sales in the following (i.e. sixth) year. Find the predicted sales and comment on this prediction.
(iii) Give a symmetric two-sided 95% confidence interval for the slope of your regression line.   [UCLES]

4 It is believed that the probability of a randomly chosen pregnant woman giving birth to a Down's syndrome child is related to the woman's age x, in years, by the relation $p = ab^x$, $25 \leqslant x \leqslant 45$, where a and b are constants. The table gives observed values of p for 5 different values of x.

x	25	30	35	40	45
y	0.000 67	0.001 25	0.003 33	0.010 00	0.033 30

(i) By plotting ln$p$ against x show that the relation gives a reasonable model.
(ii) If ln$p = \alpha + \beta x$ is the regression equation of ln$p$ on x find the least squares estimates of α and β. Plot the estimated regression line on the graph plotted in (i).
(iii) Estimate the expected number of children with Down's syndrome that will be born to 5000 randomly chosen pregnant women of age 32.   [UCLES]

5 The average density of blackbirds (in pairs per thousand hectares) over very large areas of farmland and of woodland are shown, for the years 1976 to 1982, in the table below.

Year	1976	1977	1978	1979
Farmland density (f)	83	94	91	86
Woodland density (w)	313	342	366	350

Year	1980	1981	1982
Farmland density (f)	102	113	98
Woodland density (w)	376	438	400

$\Sigma f = 667, \Sigma f^2 = 64\,179, \Sigma w = 2585,$

$\Sigma w^2 = 964\,609, \Sigma fw = 248\,579.$

Counting blackbirds in woodland is easier than counting them in farmland. It is desired in future to determine only woodland density and hence to estimate farmland density.

(i) Treating the years as providing independent pairs of observations, use the given data to estimate the appropriate linear regression equation relating the farmland and woodland densities.

(ii) Given that the 1983 woodland density is 410, estimate the average farmland density for that year.

(iii) Give a symmetric two-sided 90% confidence interval for the slope of your regression line.   [UCLES]

20.8 Distinguishing x and Y

Here are two pairs of examples of x and Y-variables. In each case the x-variable has a non-random value set by the person carrying out the investigation, while the Y-variable has an unpredictable (random) value.

x	Y
Length of chemical reaction (min)	Amount of compound produced (g)
Amount of chemical compound (g)	Time taken to produce this amount (min)
An interval of time (h)	Number of cars passing during this interval
Number of cars passing junction	Time taken for these cars to pass (h)

To decide which variable is x and which is Y evidently requires some knowledge of how and why the data were collected. Actually this is generally true – we should always know why we are doing what we are doing!

20.9 Deducing x from a Y-value

Suppose x has a non-random value, as previously, *but we do not know what that value is*. If we have the resulting Y-value and the estimated regression line of Y on x, then this is not a problem! We simply use the line 'backwards':

$$x = \frac{y - a}{b}$$

Example 6

An experiment is conducted to determine the effects of varying amounts of fertiliser (x, measured in grams per square metre) on the average crop of potatoes (y, measured in kg per plant). The results are as follows:

x	3.2	1.5	0.8	0.0	2.5	8.0	4.0	*
y	2.4	2.2	2.3	2.0	2.3	3.6	3.0	2.8

One x-value, indicated by a * has been mislaid.
Estimate the value of the missing value, using a least-squares procedure.

It appears that specified quantities of fertiliser were used. The y-values are the random averages that arose from the selection of plants that happened to be treated. This is a case where the natural regression line is that of y on x.
    The summary statistics are $\Sigma x = 20$, $\Sigma y = 17.8$ (omitting the 8th value), $\Sigma x^2 = 99.38$, $\Sigma y^2 = 47.14$ and $\Sigma xy = 59.37$, leading to $S_{xx} = 42.2371$, $S_{yy} = 1.8771$ and $S_{xy} = 8.5129$. The estimated line is:

$$y = 1.967 + (0.201\,55)x$$

The estimated value of x corresponding to $y = 2.8$ is given by:

$$y = \frac{2.8 - 1.967}{0.201\,55} = 4.133$$

To 1 decimal place, the estimated value is 4.1 grams per square metre.

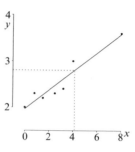

20.10 Two regression lines

Sometimes both X and Y are random variables. For example, if we measure the diameters and masses of a random sample of apples taken from the stock in a supermarket, then neither the diameters nor the masses are known in advance and neither can be regarded as 'causing' the other. In such cases there are two regression models to consider:

$$E(Y|x) = \alpha + \beta x \qquad \text{and} \qquad E(X|y) = \gamma + \delta y$$

These models are referred to as the 'Y on X model' and the 'X on Y model'.

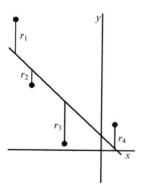 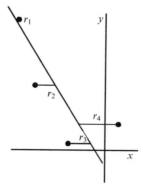

The corresponding estimated regression lines are:

$$y = a + bx \qquad \text{and} \qquad x = c + dy$$

The least squares estimates of γ and δ are denoted by c and d, respectively. The least-squares procedure minimises the deviations in the y-direction in one case and the deviations in the x-direction in the other case, as shown in the figure.

To see that the two regression models do not usually give rise to the same line, consider the following data on the heights (in inches, x) and weights (in lbs, y) of a random sample of 49 women. The sample was collected in the 1950s.

Height	65	62	62	60	65	63	67	62	64	58
Weight	188	178	168	164	164	158	157	154	153	147

Height	63	64	67	61	63	65	65	58	63	63
Weight	144	148	145	139	141	142	138	134	135	134

Height	64	64	66	58	59	62	65	64	66	59
Weight	133	135	136	130	127	129	126	129	123	123

Height	62	65	64	66	59	60	62	62	63	66
Weight	122	123	121	123	115	115	116	114	117	118

Height	61	63	62	65	66	60	62	58	62
Weight	111	110	108	112	111	105	103	96	98

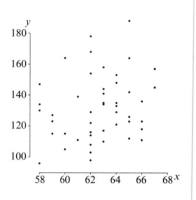

The summary statistics are $n = 49$, $\Sigma x_i = 3075$, $\Sigma x_i^2 = 193\,273$, $\Sigma y_i = 6460$, $\Sigma y_i^2 = 872\,160$ and $\Sigma x_i y_i = 405\,889$, giving $S_{xx} = 301.0612$, $S_{yy} = 20\,494.6939$, and $S_{xy} = 491.0408$. The estimates of the parameters of the regression line of Y on X are given by:

$$b = \frac{S_{xy}}{S_{xx}} = \frac{491.0408}{301.0612} = 1.631$$

$$a = \bar{y} - b\bar{x} = \frac{6460 - (1.631 \times 3075)}{49} = 29.48$$

In order to estimate the regression line of X on Y we interchange the x's and y's in the usual formulae, getting:

$$d = \frac{S_{xy}}{S_{yy}} \qquad\qquad (20.12)$$

$$c = \bar{x} - d\bar{y} \qquad\qquad (20.13)$$

Thus:

$$d = \frac{491.0408}{20\,494.6939} = 0.023\,96$$

$$c = \frac{3075 - (0.023\,96 \times 6460)}{49} = 59.60$$

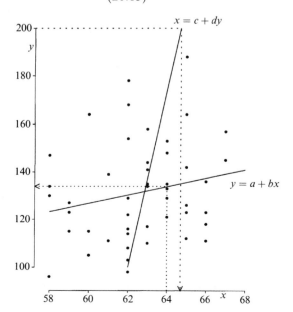

The first of these lines is used for estimating the expected value of Y for a given value of x. So we estimate that women with heights of 64 inches will have a mean weight of:

$$29.48 + (1.631 \times 64) \approx 134 \text{ lb}$$

The second line is used for estimating the mean height of women of known weight. For example, the mean height of women weighing 200 lbs would be predicted as:

$$59.60 + (0.023\,96 \times 200) \approx 64.4 \text{ in}$$

We can use the same predicted values for individuals, but we should not expect these individuals to have precisely the predicted values, since we know that individuals vary in shape from beanstalks to barrels!

Notes

◆ The two regression lines both pass through the point $(\bar{x}, \bar{y})$ which is therefore the point of intersection.

◆ To predict a value of x when, for the given data, the x-values are fixed (as opposed to being observations on a random variable) then it is appropriate to use the regression line of y on x 'in reverse' as in Section 20.9, rather than using the regression line of x on y. When in doubt you should specify *why* you are using the method that you choose.

Practical _____

Here is an obvious 'anatomical' practical. Collect the heights and weights of about 30 people. Plot the data on a scatter diagram. If the data refer to both males and females, then use different symbols for the two sexes and note whether there appear to be differences between the sexes.

For the data referring to your own sex, determine the regression lines of height on weight and weight on height.

Use these lines to estimate the average height of someone of your weight and the average weight of someone of your height.

Project _____

How do people 'see' scatter diagrams? If the previous scatter diagram, or one showing a more pronounced relation, were presented to 13-year-olds with the instruction 'Using a ruler, draw a line through these points so as to show the relationship between x and y as clearly as possible', what line would they draw? Would it approximate the regression line of Y on x, or the regression line of X on y, or would it lie half-way between these? Would an 18-year-old have a different preference?

Are people who have learnt about regression more likely to approximate the regression line of Y on x than the regression line of X on y?

Exercises 20e _____

1   Explain what is meant by the term "least squares" in the context of regression lines. Illustrate, with the aid of diagrams, the quantities being minimised in the cases of (*a*) the regression line of Y on X, (*b*) the regression line of X on Y.

Delegates who travelled by car to a statistics conference were asked to report d, the distance they travelled (in miles) and t, the time taken (in minutes). A random sample of the values reported is given in the table overleaf.   (*continued*)

Distance (miles) d	113	14	98	130	75
Time (minutes) t	130	25	180	148	100

Distance (miles) d	120	143	55	127
Time (minutes) t	120	196	48	165

$[\Sigma d = 875, \Sigma t = 1112, \Sigma d^2 = 99\,097,$
$\Sigma t^2 = 164\,174, \Sigma dt = 125\,443.]$

(i)   Fit a regression line of T on D, giving your result in the form $t = a + bd$.

(ii) Interpret the coefficient b in the context of this question.

(iii) Obtain an estimate of the average time taken by delegates who travelled 100 miles by car.

(iv) Obtain a 95% confidence interval for b.

[UCLES]

2 Explain, with the aid of a diagram, why the two least squares regression lines associated with a bivariate sample are, in general, different. State the circumstances under which the two regression lines coincide.

Flowers of a certain species are dissected. The number of carpels, X, and the number of stamens, Y, in each flower are counted. For a random sample of 15 flowers the numbers found are shown in the table.

No. of carpels (x)	19	21	24	25	26	29	30	31
No. of stamens (y)	41	70	32	34	35	72	57	64

No. of carpels (x)	31	33	34	36	36	37	38
No. of stamens (y)	67	74	69	76	74	75	70

These data are summarised by

$$\Sigma x = 450, \ \Sigma x^2 = 13\,992, \ \Sigma y = 910,$$
$$\Sigma y^2 = 59\,018, \ \Sigma xy = 28\,259.$$

(i)   Calculate the equations of the estimated regression lines of Y on X and of X on Y.

(ii) Calculate an estimate for the number of carpels in a flower with 50 stamens.

(iii) Calculate an estimate for the variance of the regression coefficient of X on Y.

[UCLES]

3 Explain, with the aid of diagrams, the difference in the definitions of the least squares regression line of Y on X and the least squares regression line of X on Y.

The following table shows the marks (x) obtained in a Christmas examination and the marks (y) obtained in the following summer examination by a group of nine students.

Student	A	B	C	D	E
Christmas (x)	57	35	56	57	66
Summer (y)	66	51	63	34	47

Student	F	G	H	I
Christmas (x)	79	81	84	52
Summer (y)	70	84	84	53

It is given that $\Sigma x = 567, \Sigma y = 552,$
$\Sigma xy = 36\,261, \Sigma x^2 = 37\,777, \Sigma y^2 = 36\,112.$

(i)   Find the equation of the estimated least squares regression line of Y on X.

(ii) A tenth student obtained a mark of 70 in the Christmas examination but was absent from the summer examination. Estimate the mark that this student would have obtained in the summer examinations.

(iii) An eleventh student took only the summer examination and obtained a mark of 55. Estimate the mark that this student would have obtained in the Christmas examination.

[UCLES]

4 The data in the following table show the length and breadth in mm of a group of skulls discovered during an excavation.

Length (x)	165	170	172	176	178
Breadth (y)	139	141	147	147	149

Length (x)	179	182	184	186	190
Breadth (y)	149	159	145	155	152

(You may assume that $\Sigma x^2 = 318\,086,$
$\Sigma y^2 = 220\,257, \Sigma xy = 264\,582$)

(a) Calculate the regression lines of length on breadth and breadth on length.

(b) Plot these data on a scatter diagram and draw both your regression lines on your diagram.

(c) State, in symbolic form, the point of intersection of your two lines.

(d) Using in each case the appropriate regression line, predict the breadth of a skull of length 185mm and the length of a skull of breadth 155mm.

(e) Under what circumstances would your two lines be co-incident?

[AEB 90]

5 (a) Given that the estimated regression line of Y on X has equation $y = \bar{y} + b(x - \bar{x})$, where $b = S_{xy}/S_{xx}$, write down corresponding formulae for the estimated regression line of X on Y and for its regression coefficient b^*.

   (b) Following a leak of radioactivity from a nuclear power station an index of exposure to radioactivity was calculated for each of 7 geographical areas close to the power station. In the subsequent 5 years the incidence of death due to cancer (measured in deaths per 100,000 person-years) was recorded. The data were as follows:

Area	1	2	3	4
Index (x)	7.6	23.2	3.2	16.6
Deaths (y)	62	75	51	72

Area	5	6	7
Index (x)	5.2	6.8	5.0
Deaths (y)	39	43	55

$[\Sigma x = 67.6,\ \Sigma x^2 = 980.08,\ \Sigma y = 397,$
$\Sigma y^2 = 23\,649,\ \Sigma xy = 4339.8.]$

   (i) Find the estimated regression line of Y on X.

   (ii) In another geographical area close to the power station the index of exposure was 6.0. Use the estimated regression line to predict the incidence, in this area, of death due to cancer (in deaths per 100 000 person-years).

   (iii) Estimate the incidence of death due to cancer (in deaths per 100 000 person-

years) that there would have been if there had been no leak from the power station (i.e. if the index of exposure to radioactivity were zero).   [UCLES(P)]

6 With the aid of suitable diagrams, describe the difference between the estimated regression line of Y on X and that of X on Y.
   The age of many types of tree can be determined by counting the growth rings in a cross-section of the trunk once it has been felled. A random sample of 10 oak trees in a timber yard was taken and the age and average girth of each trunk ascertained. The data are as follows:

Age (x years)	20	23	30	38	39
Girth (y units)	26	30	45	48	46

Age (x years)	45	45	48	55	71
Girth (y units)	60	64	68	70	92

$[\Sigma(x - 40) = 14,\quad \Sigma(x - 40)^2 = 2094,$
$\Sigma(x - 40)(y - 60) = 2604,$
$\Sigma(y - 60) = -51,\quad \Sigma(y - 60)^2 = 3825.]$

   (i) Obtain the equation of the estimated regression line of girth on age.

   (ii) Estimate the average girth of the trunk from a 35-year old oak tree.

   (iii) Calculate a 95% confidence interval for the slope of the regression line in (i).
                                                        [UCLES]

20.11 Correlation

The scatter diagram of heights and weights revealed a great deal of scatter! In the earlier examples the relation between y and x seemed quite well defined whereas with the height/weight data the relation was distinctly fuzzy. One effect of this imprecision was that the regression line of height on weight and the regression line of weight on height were very different from one another. We now introduce a measure of fuzziness that Galton called the index of co-relation. This is now called the **correlation coefficient**. It is most appropriate for the case where both X and Y are random variables.

We shall see that the coefficient can take any value from -1 to 1. In cases where increasing values of one variable, x, are associated with generally increasing values of the other variable, y, the variables are said to display **positive correlation**.

Positive correlation

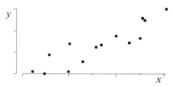

In this case the estimated regression line of Y on X, and the line of X on Y, have positive gradients.

The opposite situation (**negative correlation**) is where increasing values of one variable are associated with generally decreasing values of the other variable.

In this case both the estimated regression lines have negative gradients.

Negative correlation

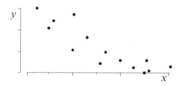

The intermediate case where the correlation coefficient is zero corresponds to cases where the two estimated regression lines are parallel to the axes. Some examples of data having zero correlation are shown in the diagrams.

Examples of zero correlation

The last of these diagrams serves to emphasise that the correlation coefficient is a statistic that is specifically concerned with *linear* relationships: zero correlation does not necessarily imply that X and Y are unrelated.

20.12 The product–moment correlation coefficient

We might expect that students who get high marks in Physics would tend to be the ones who get high marks in Mathematics, and *vice versa*. This seems clear enough – but what constitutes a high mark? If most students get 3 out of 20 on a test, but one student gets 9 out of 20, then this is a (relatively) high mark. So what we really mean is 'above the average of the values in the sample'. If, for a random sample of students, the individual marks in Mathematics are denoted by $x_1, x_2, \ldots, x_n$, with a mean $\bar{x}$, then we are concerned with the values of $x_1 - \bar{x}$, $x_2 - \bar{x}$, etc.

Suppose we denote the corresponding Physics marks by $y_1, y_2, \ldots, y_n$, with mean $\bar{y}$. The suggestion is that if $x_i - \bar{x}$ is positive (corresponding to a 'high' mark in Mathematics by the ith student) then $y_i - \bar{y}$ is also likely to be positive.

Of course, if high marks go together, then so do low marks, so we anticipate that a negative value for $x_i - \bar{x}$ will usually correspond to a negative value for $y_i - \bar{y}$.

To see how this works with actual numbers, here are some:

Student	Maths	Physics			
i	x_i	y_i	$x_i - \bar{x}$	$y_i - \bar{y}$	$(x_i - \bar{x})(y_i - \bar{y})$
1	65	60	9	−5	−45
2	45	60	−11	−5	55
3	40	55	−16	−10	160
4	55	70	−1	5	−5
5	60	80	4	15	60
6	50	40	−6	−25	150
7	80	85	24	20	480
8	30	50	−26	−15	390
9	70	70	14	5	70
10	65	80	9	15	135

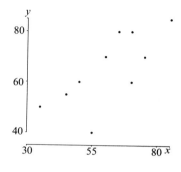

The mean Mathematics mark is 56. Of the five students who get less than this, four also get less than the average mark (65) in Physics. There is a corresponding result for the remaining five students. The consequence is that most of the $(x_i - \bar{x})(y_i - \bar{y})$ products are positive, as is their sum S_{xy}. The quantity S_{xy} is affected by changes in the scale. For example, if we marked out of 1000 instead of out of 100, the x-values and y-values would each be multiplied by 10 and the S_{xy} values would be multiplied by 100. For this reason Galton suggested calculating a quantity, r, that is *unaffected by scale changes*. This quantity is now called the (sample) **product–moment correlation coefficient** and is calculated using:

$$r = \frac{S_{xy}}{\sqrt{S_{xx}S_{yy}}} \tag{20.14}$$

We see that the value of S_{xy} determines whether the correlation coefficient is negative, zero or positive. We show in a note below that the values −1 and 1 correspond to cases where the points are collinear with, respectively, negative and positive slopes. Values between −1 and 0 correspond to cases where both regression lines have a negative slope, but the points are not collinear; while values of r between 0 and 1 correspond to cases where both the regression lines have positive slope.

Correlation = −1   Correlation = −0.8   Correlation = 1

The value of r is unaffected by working with linearly coded values, where for example, each value of x is replaced by x', where $x' = hx + k$ and h and k are constants. Such adjustments can greatly simplify the calculations.

Notes

- r is sometimes simply called 'the correlation'.
- r is also known as Pearson's correlation coefficient (after Karl Pearson – see Chapter 18).
- Considering the definitions of S_{xx}, S_{xy} and S_{yy}, we can write:

$$\text{correlation} = \frac{\text{sample covariance}}{\sqrt{(\text{sample variance of } x) \times (\text{sample variance of } y)}}$$

- If the points are perfectly collinear then $y_i = a + bx_i$ for all i. This means that:

$$S_{xy} = \Sigma\{(x_i - \bar{x})y_i\} = \Sigma(x_i - \bar{x})a + \Sigma\{(x_i - \bar{x})bx_i\} = bS_{xx}$$

using $\Sigma(x_i - \bar{x}) = 0$ and Equation (20.8). Similarly:

$$S_{yy} = \Sigma\{(y_i - \bar{y})y_i\} = \Sigma(y_i - \bar{y})a + \Sigma\{(y_i - \bar{y})bx_i\} = bS_{xy}$$
$$= b^2 S_{xx}$$

using the previous result. Substituting into the fraction $\dfrac{S_{xy}}{\sqrt{S_{xx}S_{yy}}}$ we see that

$r = \dfrac{b}{\sqrt{b^2}}$, which is equal to 1 or -1 depending upon whether b is positive or negative.

Hence *collinear points imply $r = \pm 1$*. We show later that the converse is also true.

- The value of r is unaffected by changes in the units of measurements.

Example 7

Determine the product–moment correlation coefficient for the exam marks data given earlier (i) using the raw data, and (ii) using the coded values x_i', and y_i' given by $x_i' = \dfrac{x_i - 50}{5}$, $y_i' = \dfrac{y_i - 50}{5}$.

(i) We set out the calculations in a table, as follows.

Student i	Maths x_i	Physics y_i	$x_i y_i$	x_i^2	y_i^2
1	65	60	3900	4225	3600
2	45	60	2700	2025	3600
3	40	55	2200	1600	3025
4	55	70	3850	3025	4900
5	60	80	4800	3600	6400
6	50	40	2000	2500	1600
7	80	85	6800	6400	7225
8	30	50	1500	900	2500
9	70	70	4900	4900	4900
10	65	80	5200	4225	6400
Total	560	650	37 850	33 400	44 150

We now calculate S_{xy}, S_{xx} and S_{yy}:

$$S_{xy} = \Sigma x_i y_i - \frac{\Sigma x_i \Sigma y_i}{n} = 37\,850 - \frac{560 \times 650}{10} = 1450$$

$$S_{xx} = \Sigma x_i^2 - \frac{(\Sigma x_i)^2}{n} = 33\,400 - \frac{560^2}{10} = 2040$$

$$S_{yy} = \Sigma y_i^2 - \frac{(\Sigma y_i)^2}{n} = 44\,150 - \frac{650^2}{10} = 1900$$

Hence:

$$r = \frac{1450}{\sqrt{2040 \times 1900}} = 0.7365$$

The two sets of marks display a strong positive correlation.

(ii) We now repeat the calculations using the coded values. It will be seen that the numbers become much simpler, while the value of r is unchanged.

Student i	Maths x_i	Physics y_i	x_i'	y_i'	$x_i' y_i'$	$x_i'^2$	$y_i'^2$
1	65	60	3	2	6	9	4
2	45	60	-1	2	-2	1	4
3	40	55	-2	1	-2	4	1
4	55	70	1	4	4	1	16
5	60	80	2	6	12	4	36
6	50	40	0	-2	0	0	4
7	80	85	6	7	42	36	49
8	30	50	-4	0	0	16	0
9	70	70	4	4	16	16	16
10	65	80	3	6	18	9	36
Total	560	650	12	30	94	96	166

We now calculate $S_{x'y'}$, $S_{x'x'}$ and $S_{y'y'}$:

$$S_{x'y'} = \Sigma x_i' y_i' - \frac{\Sigma x_i' \Sigma y_i'}{n} = 94 - \frac{12 \times 30}{10} = 58$$

$$S_{x'x'} = \Sigma x_i'^2 - \frac{(\Sigma x_i')^2}{n} = 96 - \frac{12^2}{10} = 81.6$$

$$S_{y'y'} = \Sigma y_i'^2 - \frac{(\Sigma y_i')^2}{n} = 166 - \frac{30^2}{10} = 76$$

Hence:

$$r = \frac{58}{\sqrt{81.6 \times 76}} = 0.7365$$

as before.

▲————————————————————————————————▲

Calculator practice ————————————————————————————

Many calculators have in-built routines for calculating r. If you have a calculator of this type then make sure that you know how to find the value of r. You could start with the exam mark data.

Computer project ————————————————————————————

Have you still got your regression spreadsheet? Why not extend it still further to enable the estimation of both regression lines and the calculation of r?

Exercises 20f

1 The yield (per hectare) of a crop, c, is believed to depend on the May rainfall, m. For nine regions records are kept of the average values of c and m, and these are recorded below.

c	8.3	10.1	15.2	6.4	11.8
m	14.7	10.4	18.8	13.1	14.9

c	12.2	13.4	11.9	9.9
m	13.8	16.8	11.8	12.2

$$[\Sigma c = 99.2, \ \Sigma m = 126.5, \ \Sigma c^2 = 1150.16,$$
$$\Sigma m^2 = 1832.07, \ \Sigma mc = 1427.15.]$$

(i)   Find the equation of the appropriate regression line.
(ii) Find r, the linear (product moment) correlation coefficient between c and m.
(iii) In a tenth region the average May rainfall was 14.6. Estimate the average yield of the crop for that region, giving your answer correct to one decimal place.   [UCLES(P)]

2 Given that the gradient of the least squares regression line of Y on X is b_1, and the gradient of the least squares regression line of X on Y is $1/b_2$, prove that $b_1 b_2 = r^2$, where r is the linear (product moment) correlation coefficient.
Hence show that if $r^2 = 1$ these two regression lines are identical.
The yield of a particular crop on a farm is thought to depend principally on the amount of rainfall in the growing season. The values of the yield Y, in tons per acre, and the rainfall X, in centimetres, for seven successive years are given in the table below.

x	12.3	13.7	14.5	11.2	13.2	14.1	12.0
y	6.25	8.02	8.42	5.27	7.21	8.71	5.68

$$[\Sigma xy = 654.006, \ \Sigma x = 91, \ \Sigma x^2 = 1191.72,$$
$$\Sigma y = 49.56, \ \Sigma y^2 = 362.1628]$$

(i)   Find the linear (product moment) correlation coefficient between X and Y.
(ii) Find the equation of the least squares regression line of Y on X and also that of X on Y.
(iii) Given that the rainfall in the growing season of a subsequent year was 14.0 cm, estimate the yield in that year.

(iv) Given that the yield in a subsequent year was 8.08 tons per acre, estimate the rainfall in the growing season of that year.   [UCLES]

3 The following data show the IQ and the score in an English test of a sample of 10 pupils taken from a mixed ability class.
The English test was marked out of 50 and the range of IQ values for the class was 80 to 140.

Pupil	A	B	C	D	E
IQ (x)	110	107	127	100	132
English score (y)	26	31	37	20	35

Pupil	F	G	H	I	J
IQ (x)	130	98	109	114	124
English score (y)	34	23	38	31	36

(a) Estimate the product-moment correlation coefficient for the class.
(b) What does this coefficient measure?
                                        [AEB (P) 90]

4 Explain, briefly, your understanding of the term "correlation".
Describe how you used, or could have used, correlation in a project or in classwork.
Twelve students sat two Biology tests, one theoretical and one practical. Their marks are shown in the table.

Marks in theoretical test (T)	5	9	7	11	20	4
Marks in practical test (P)	6	8	9	13	20	9

Marks in theoretical test (T)	6	17	12	10	15	16
Marks in practical test (P)	8	17	14	8	17	18

(continued)

(*a*) Draw a scatter diagram to represent these data.

(*b*) Find, to 3 decimal places:
 (i) the value of the sum of products, S_{TP},
 (ii) the product-moment correlation coefficient.

(*c*) Using evidence from (*a*) and (*b*) explain why a straight line regression model is appropriate for these data.

Another student was absent from the practical test but scored 14 marks in the theoretical test.

(*d*) Find the equation of the appropriate regression line and use it to estimate a mark in the practical test for this student. [ULSEB]

5 Draw a diagram to illustrate the lengths whose sum of squares is minimised in the least squares method for finding the regression line of y on x. State which is the independent and which is the dependent variable.

State, giving your reason, whether or not the equation of this line can be used to estimate the value of x for a given value of y.

The length (L mm) and width (W mm) of each of 20 individuals of a single species of fossil are measured. A summary of the results is:

$$\Sigma L = 400.20, \Sigma W = 176.00, \Sigma LW = 3700.20,$$
$$\Sigma L^2 = 8151.32, \Sigma W^2 = 1780.52.$$

(*a*) Obtain the product-moment correlation coefficient between the length and the width of these fossils. Without performing a significance test interpret your result.

(*b*) Obtain an equation of the line of regression from which it is possible to estimate the length of a fossil of the same species whose width is known, giving the values of the coefficients to 2 decimal places.

(*c*) From your equation find the average increase or decrease in length per 1 mm increase in width of these fossils. [ULSEB]

Correlation and regression

The same quantities (S_{xx}, S_{yy} and S_{xy}) feature in the discussion of both correlation and regression, since both are concerned with relationships between variables. Indeed, we can write:

$$r^2 = \frac{S_{xy}^2}{S_{xx}S_{yy}} = \frac{S_{xy}}{S_{xx}}\frac{S_{xy}}{S_{yy}} = bd$$

where b and d are the estimated regression coefficients of the two regression lines, and the quantity r^2 is known as the **coefficient of determination**.

We can link r^2 directly to the deviance, D:

$$D = S_{yy} - \frac{S_{xy}^2}{S_{xx}} = S_{yy}\left(1 - \frac{S_{xy}^2}{S_{xx}S_{yy}}\right) = S_{yy}(1 - r^2)$$

It is now easy to see that if $r = \pm 1$ then $D = 0$, which implies a perfect fit and hence implies that the data points are collinear.

The population product–moment correlation coefficient, ρ

The population product–moment correlation coefficient is usually denoted by the Greek letter ρ (written as rho and pronounced 'roe'). Since population characteristics are simply sample characteristics taken to the extreme, the definition of ρ is:

$$\rho = \frac{\text{population covariance}}{\sqrt{(\text{population variance of } X) \times (\text{population variance of } Y)}}$$

The **population covariance**, Cov(X, Y), is given by:

$$\text{Cov}(X, Y) = \text{E}(XY) - \text{E}(X)\text{E}(Y)$$

If X and Y are independent random variables, then $E(XY) = E(X)E(Y)$ and hence $\text{Cov}(X, Y) = 0$ and therefore $\rho = 0$.

In terms of expectations, ρ is calculated using:

$$\rho = \frac{E(XY) - E(X)E(Y)}{\sqrt{\{E(X^2) - E(X)^2\}\{E(Y^2) - E(Y)^2\}}}$$

If (as in the case of independent random variables) the covariance of X and Y is 0, then $\rho = 0$, and X and Y are said to be **uncorrelated**.

Note

◆ If variables X and Y are independent then they are uncorrelated. However, the fact that X and Y are uncorrelated does *not* imply that X and Y are independent. For example, consider a population in which $(-1, 1)$, $(0, 0)$ and $(1, 1)$ each has a probability $\frac{1}{3}$. The variables are not independent, since $Y = X^2$ and also $P(X = 1 \text{ and } Y = 1) = \frac{1}{3}$, but $P(X = 1) \times P(Y = 1) = \frac{1}{3} \times \frac{2}{3} = \frac{2}{9}$. The variables are, however, uncorrelated, since:

$E(XY) = \frac{1}{3}(-1) \times (1) + \frac{1}{3}(0) \times (0) + \frac{1}{3}(1) \times (1) = 0$

$E(X) = \frac{1}{3}(-1) + \frac{1}{3}(0) + \frac{1}{3}(1) = 0$

$E(Y) = \frac{1}{3}(1) + \frac{1}{3}(0) + \frac{1}{3}(1) = \frac{2}{3}$

so $\text{Cov}(X, Y) = E(XY) - E(X)E(Y) = 0$ and $\rho = 0$.

Testing the significance of r

If $\rho \neq 0$, then X and Y are related and this is likely to be interesting! We therefore concentrate on the hypotheses $H_0: \rho = 0$ and $H_1: \rho \neq 0$. We reject H_0, in favour of H_1, if $|r|$ is unusually far from zero.

The critical values of r depend on the distributions of X and Y. We only consider the case of random samples from normal distributions. An extract from the table of critical values for $|r|$ given in the Appendix (p. 625) is reproduced below.

n	5%	1%	n	5%	1%	n	5%	1%	n	5%	1%
4	.950	.990	7	.754	.874	10	.632	.765	13	.553	.684
5	.878	.959	8	.707	.834	11	.602	.735	14	.532	.661
6	.811	.917	9	.666	.798	12	.576	.708	15	.514	.641

The table is easy to use. For example, suppose that $n = 10$ and that $r = 0.7365$ (the values obtained in Example 7). The 5% critical value is 0.632, while the 1% critical value is 0.765. The observed value is therefore significant at the 5% level but not at the 1% level: there is some evidence to reject the hypothesis that X and Y are uncorrelated.

Notes

◆ The table can be used for a one-tailed test at the 2.5% or 0.5% levels.
◆ For larger values of n than those given in the table in the Appendix, use the result that, assuming H_0, $r\sqrt{\dfrac{n-2}{1-r^2}}$ is an observation from a t_{n-2}-distribution.

Example 8

The following data refer to the average temperature (in degrees Fahrenheit) and the average butterfat content for a group of cows (expressed as a percentage of the milk).

Temp.	64	65	65	64	61	55	39	41	46	59
Butterfat	4.65	4.58	4.67	4.60	4.83	4.55	5.14	4.71	4.69	4.65

Temp.	56	56	62	37	37	45	57	58	60	55
Butterfat	4.36	4.82	4.65	4.66	4.95	4.60	4.68	4.65	4.60	4.46

Assuming normal distributions, determine whether there is significant evidence, at the 1% level, of any correlation between the two variables.

———

The null hypothesis is that the two variables are uncorrelated with the alternative being that there is a correlation. The diagram does suggest that the variables are weakly negatively correlated. Since the correlation will not be affected by linear coding, we begin by making the numbers easier to handle by subtracting 50 from each of the temperatures (denoting the result by x). We also subtract 4.5 from the butterfat percentages and multiply the result by 100 to give simple y-values. The resulting values are:

x	14	15	15	14	11	5	−11	−9	−4	9
y	15	8	17	10	33	5	64	21	19	15

x	6	6	12	−13	−13	−5	7	8	10	5
y	−14	32	15	16	45	10	18	15	10	−4

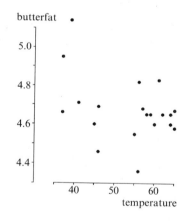

The resulting summary statistics are $n = 20$, $\Sigma x_i = 82$, $\Sigma x_i^2 = 2104$, $\Sigma y_i = 350$, $\Sigma y_i^2 = 11\,406$, $\Sigma x_i y_i = 50$. Hence $S_{xx} = 1767.8$, $S_{yy} = 5281$, and $S_{xy} = -1385$, and thus $r = \dfrac{-1385}{\sqrt{1767.8 \times 5281}} = -0.453$.

The sample size, 20, exceeds those tabulated so we calculate

$$r\sqrt{\frac{n-2}{1-r^2}} = -0.453 \times \sqrt{\frac{18}{1-(-0.453)^2}} = -2.158.$$

The lower 1% point of a t_{18}-distribution is -2.55, which is less than the observed -2.158, and we therefore conclude that there is no significant evidence, at the 1% level of any correlation between the variables.

Exercises 20g

1 The moisture content, M, of core samples of mud from a lake is measured as a percentage. It is believed that M is related to the depth d metres at which the core is collected. To test this surmise, samples were taken at a variety of depths, giving the results shown in the table opposite.

d_i	0	5	10	15
m_i	90	82	56	42

d_i	20	25	30	35
m_i	30	21	21	18

Some summary statistics are:

$$\sum m_i = 360, \quad \sum m_i^2 = 21\,830,$$

$$\sum d_i = 140, \quad \sum d_i^2 = 3500,$$

$$\sum d_i m_i = 3985$$

Calculate r, the product–moment correlation coefficient for these data.
State, briefly, the implications of the value that you obtain.

2 A technician monitoring water purity believes that there is a relationship between the hardness of the water and its alkalinity. Over a period of 10 days, she recorded the data (in mg/l) in the table:

Alkalinity (mg/l)	33.8	29.1	22.8	26.2
Hardness (mg/l)	51.0	45.0	41.3	46.0

Alkalinity (mg/l)	31.8	31.9	29.4	26.1
Hardness (mg/l)	48.0	50.0	46.3	45.0

Alkalinity (mg/l)	28.0	27.2
Hardness (mg/l)	45.3	43.0

(a) Plot the data on graph paper with 'Alkalinity' on the horizontal axis. Mark the mean point.

(b) Given that the product moment correlation coefficient for these data is 0.913, what conclusion is it reasonable to draw? You are expected to refer to appropriate tables to support your reply.

(c) The technician decides to calculate the equation for the least squares regression line of **hardness on alkalinity**. Show that this line has gradient 0.821 and find its equation.

(d) Estimate the hardness of water which has a measured alkalinity of 30mg/l.
Explain **briefly** why the technician would need to do further statistical work to be able to predict the alkalinity of water with a measured hardness of, say, 50 mg/l. **(You are not expected to carry out the work.)**
[UODLE]

3 A hospital doctor was interested in the percentage of a certain drug absorbed by patients. She obtained the following data on 10 patients taking the drug on two separate days.

	Percentage of drug absorbed				
Patient	1	2	3	4	5
Day 1, x	35.5	16.6	13.6	42.5	28.5
Day 2, y	27.6	15.1	12.9	34.1	35.5
Patient	6	7	8	9	10
Day 1, x	30.3	8.7	21.5	16.4	32.3
Day 2, y	32.5	84.3	21.5	11.1	36.4

$$\Sigma x = 245.9, \quad \Sigma y = 311, \quad \Sigma x^2 = 7107.55,$$
$$\Sigma y^2 = 13652.4, \quad \Sigma xy = 7405.07.$$

(a) Draw a scatter diagram of the data.

(b) Calculate the product moment correlation coefficient r. Assuming that the data come from a bivariate normal distribution, test, at the 5% significance level, that it has correlation coefficient $\rho = 0$.

(c) After examining the scatter diagram the doctor found that one of the points was surprising. Further checking revealed that this point was the result of abnormal circumstances. The value of r for the remaining 9 points is 0.863. State which point has been omitted.

(d) It has been said that the only valid use of the product moment correlation coefficient is to measure the strength of the relationship between two variables which are already known to be approximately linearly related. Give an example, including a rough sketch of a scatter diagram, where to ignore this advice would lead to a seriously misleading conclusion.
[AEB (P) 89]

Chapter summary

Define S_{xx}, S_{yy} and S_{xy} by, for example,

$$S_{xy} = \Sigma\{(x_i - \bar{x})(y_i - \bar{y})\} = \Sigma x_i y_i - \frac{\Sigma x_i \Sigma y_i}{n}$$

◆ The least-squares line of best fit which is **the estimated regression line of y on x**, is $y = a + bx$, where:

$$b = \frac{S_{xy}}{S_{xx}}$$

$$a = \bar{y} - b\bar{x}$$

This choice for a and b minimises the sum of the squared residuals: $\Sigma\{y_i - (a + bx_i)\}^2$. The resulting minimum is the **deviance**, or **residual sum of squares**, D, where:

$$D = S_{yy} - \frac{S_{xy}^2}{S_{xx}}$$

◆ A **confidence interval for the population regression coefficient**, β, is

$$b \pm c_t\sqrt{\frac{D}{(n-2)S_{xx}}},$$ where c_t is the relevant value from a

t_{n-2}-distribution.

◆ **The product–moment correlation coefficient, r:**

$$r = \frac{S_{xy}}{\sqrt{S_{xx}S_{yy}}}$$

If $r = \pm 1$ then the points $(x_1, y_1), \ldots, (x_n, y_n)$ are collinear.

Exercises 20h (Miscellaneous)

1 The acidity/alkalinity of a liquid is measured by its pH value (pure water has a pH value of 7 and lower values indicate acidity). The data in the following table refer to measurements of the pH values of samples of water collected from lakes in the vicinity of a Canadian copper smelting plant. It is believed that debris and dust from the smelter will be carried through the atmosphere and will contaminate the neighbourhood. The following data shows d, the distance (in km) of a lake from the smelter, and a, the pH value.

d	3.9	6.5	13.5	41.9	47.7
a	3.40	3.20	4.20	5.19	4.41

d	52.3	61.3	75.5	90.3
a	6.75	7.01	6.40	4.75

(a) Which is the dependent variable?
(b) Plot the data using a scatter diagram. Comment on any interesting features.

(c) Determine the value of the correlation coefficient r, defined by:

$$r = \frac{n\Sigma xy - \Sigma x \Sigma y}{\sqrt{\{n\Sigma x^2 - (\Sigma x)^2\}\{n\Sigma y^2 - (\Sigma y)^2\}}}$$

Explain carefully how this value should be interpreted.
(d) Fit an appropriate regression line of the form

$$E(Y|x) = \alpha + \beta x$$

giving the estimated values of α and β to the accuracy that you feel is appropriate.
(e) Explain carefully what the values of α and β mean in the context of the question.
(f) Where you feel that it is appropriate, give the estimated pH values for lakes at distances of 0 km, 50 km, 100 km and 200 km from the smelter. If you do not feel that it is appropriate to provide an estimated pH value, then explain the basis for your decision.

2 Suppose that it is known that the relationship between two variables x and Y is linear, with the random variable Y having expectation 0 when $x = 0$. The appropriate relation is therefore:

$$E(Y|x) = \beta x$$

Determine an expression for the least-squares estimator of β. Illustrate the procedure using a scatter diagram and showing in what sense the estimator is 'least squares'.

The data in the following table represent the average *per capita* income, I, and expenditure, E in the USA (measured in 1982 dollars).

	1960	1965	1970	1975	1980	1985
I	6036	7027	8134	8944	9722	10 622
E	5561	6326	7275	7926	8783	9830

(a) Represent the data on a scatter diagram.
(b) Propose a model for the data, explaining the reason(s) for your choice.
(c) Estimate the parameter(s) of your model.
(d) Calculate the predicted expenditure values for the six years under study.
(e) Explain what is meant by '1982 dollars'. What is the implication of the rise in the values over time?

3 The random variables X, Y and Z are independent of one another and each has mean 0 and variance σ^2. The variables A, B and C are defined by the relations:

$$A = X + Y, \quad B = X + Z, \quad C = X - Z$$

Using the algebra of expectations, determine the correlations of (i) A and B, (ii) B and C. In each case, state, with a reason, whether this necessarily implies that the two variables are independent.

4 Ten batches of raw material are selected. For each batch, the purity, x, before processing and the yield, Y, after processing, are measured. Here x has negligible error but Y is a random variable, observed as y. The results are as follows:

x	4.0	5.6	2.4	5.0	3.3
y	9.2	10.3	8.9	10.1	9.3

x	5.7	2.6	5.3	4.5	4.8
y	10.2	9.3	10.8	9.9	9.3

The model proposed is $E(Y_i|x_i) = \alpha_0 + \beta(x_i - \bar{x})$. Draw a rough scatter diagram and verify that the model appears plausible.
Determine the least-squares estimates of α_0 and β and explain the real-life interpretation of these values.

5 In a laboratory-based investigation of the photosynthesis rate (Y) of the plant *Larrea tridenta* and its relation to the level of irradiance (X), the following results were obtained:

x	1.9	2.9	2.9	4.0	4.0	5.5
y	13	34	40	34	39	73

x	5.5	5.5	8.0	8.0	8.0
y	52	65	98	82	84

The data are summarised by $\Sigma x = 56.2$, $\Sigma x^2 = 335.18$, $\Sigma y = 614$, $\Sigma y^2 = 41\,244$, $\Sigma xy = 3688.3$

(i) Plot a scatter diagram.
(ii) Calculate the equation of the estimated regression line of Y on X and draw this line on the scatter diagram.
(iii) Comment on the adequacy of the linear model.
(iv) Estimate the photosynthesis rate when the level of irradiance is 5.2.
(v) Obtain symmetric 95% confidence limits for the gradient of the regression line.   [UCLES]

6 Sketch a scatter diagram illustrating a bivariate sample having a linear (product moment) correlation coefficient of -1.
The resting metabolic rate (RMR), in kcal/24h, and body weight, in kg, of 12 randomly selected women are shown in the table below.

Body weight (x)	43.1	48.1	49.9	50.8
RMR (y)	870	1372	1079	1146

Body weight (x)	51.8	52.2	52.6	53.5
RMR (y)	1115	1132	1161	1172

Body weight (x)	57.6	61.4	62.3	64.9
RMR (y)	1325	1351	1402	1365

(*continued*)

The data are summarised by
$$\Sigma(x - 50) = 48.2, \quad \Sigma(y - 1200) = 90,$$
$$\Sigma(x - 50)^2 = 639.98, \quad \Sigma(y - 1200)^2 = 276\,650,$$
$$\Sigma(x - 50)(y - 1200) = 9031.6.$$

(i) Plot a scatter diagram for the above data and state what it indicates about the correlation between resting metabolic rate and body weight.

(ii) Calculate the linear (product moment) correlation coefficient for the sample.
                                    [UCLES(P)]

7 The radiation intensity I at time t, from a radioactive source, is given by the formula $I = I_0 e^{-kt}$, where I_0 and k are constants. Show that the relation between $\ln I$ and t is linear.
The following data were obtained from a particular source. The values of t maybe considered to be exact, while the values of I are subject to experimental error.

t	0.2	0.4	0.6	0.8	1.0
I	3.22	1.63	0.89	0.41	0.36

You are given that, correct to 5 decimal places,
$$\Sigma \ln I = -0.371\,82, \quad \Sigma(\ln I)^2 = 3.458\,46,$$
$$\Sigma t \ln I = 1.375\,54.$$

(i) Find the equation of the estimated regression line of $\ln I$ on t and hence give estimates for I_0 and k.

(ii) Calculate the radiation intensity that would be expected at time $t = 0.5$.

(iii) Calculate the linear (product moment) correlation coefficient between t and $\ln I$.

(iv) Explain why it is reasonable to use the regression equation obtained in (i) to estimate the value of t when $I = 1.5$. Obtain this value.   [UCLES]

8 (a) State, with a reason, the effect on the value of the product-moment correlation coefficient between two variables x and y of
  (i) changing the units of x,
  (ii) changing the origin of y.

(b) The following data relate to the percentage scores on a physical fitness test, the heights (in centimetres), the weights (in kilograms) and the ages (in years) of ten junior school pupils.

Pupil	Score (s)	Height (h)	Weight (w)	Age (a)
1	58	130	41.8	8
2	60	120	38.6	9
3	59	154	54.1	11
4	72	140	38.6	9
5	62	145	44.9	10
6	54	153	52.4	10
7	81	139	30.2	8
8	62	148	41.4	9
9	86	150	38.4	10
10	94	160	32.1	11

Plot a scatter diagram of weight and score. Given that
$$u = (w - 30)/0.1, \quad v = s - 50,$$
and that
$$\Sigma u = 1125, \quad \Sigma u^2 = 179671,$$
$$\Sigma v = 188, \quad \Sigma v^2 = 5226, \quad \Sigma uv = 13927,$$
calculate the value of the correlation coefficient, r_{uv}, between u and v.
State the value of r_{ws}.
Explain how your graph gives an indication that your value is correct.
A regression line is to be fitted between s and one of the other three variables in order to predict pupils' scores in the physical fitness test. Given that $r_{hs} = 0.357$ and $r_{as} = 0.188$, which of the three variables would you choose? Give a reason for your choice.   [JMB]

9 At the start of a certain card game a player determines, from the cards held, the value of a random variable P, which the player uses to estimate what the score at the end of the game will be. The random variable S denotes the actual score at the end of the game. The table below shows the values of S and P obtained in a random sample of twenty games. (For example, there were two games where the value of S was 10; in one of these the value of P was 26 and in the other it was 28.)

s	p
11	31
10	26, 28
9	24, 27
8	20, 25, 23, 23
7	23, 21, 18
6	18, 22, 25
5	13, 16
4	11, 13
3	10

(continued)

It is given that $\Sigma p = 417$, $\Sigma p^2 = 9351$.
(i)   Plot all the data on a scatter diagram.
(ii) Obtain an estimate of the value of the product moment correlation coefficient between S and P.
(iii) Obtain an estimate of the least squares regression line of S on P.
(iv) Give a symmetric 95% confidence interval for the slope of this regression line. [UCLES]

10 The scores obtained by 10 sumo wrestlers in two sumo wrestling competitions are given in the table below.

Wrestler	1	2	3	4	5
First competition score (x)	10	10	13	6	9
Second competition score (y)	12	9	11	8	8

Wrestler	6	7	8	9	10
First competition score (x)	7	6	9	5	6
Second competition score (y)	9	8	11	8	5

These results are summarised by $\Sigma x = 81$, $\Sigma x^2 = 713$, $\Sigma y = 89$, $\Sigma y^2 = 829$, $\Sigma xy = 753$.

(i)   Show these data clearly on a scatter diagram.
(ii) Obtain the value of the linear (product moment) correlation coefficient between x and y.
(iii) Obtain the least squares estimates of the values of the parameters α and β in the regression line $y = \alpha + \beta x$.
(iv) Assuming that it is valid to use the t-distribution, obtain an approximate 95% confidence interval for β. Hence test the hypothesis $\beta = 1$ at the 5% significance level. [UCLES]

11 The moisture content, M, in grams of water per 100 grams of dried solids, of core samples of mud from an estuary was measured at depth D metres. The results are shown in the table.

Depth (D)	0	5	10	15
Moisture content (M)	90	82	56	42

Depth (D)	20	25	30	35
Moisture content (M)	30	21	21	18

(a) On graph paper, draw a scatter diagram for these data.
(b) Obtain, to 3 decimal places, the product-moment correlation coefficient. Without performing a significance test, interpret the meaning of your result.
(c) Find the equation of the regression line of M on D, giving the coefficients to 2 decimal places.
(d) Find, to 2 decimal places, the minimum sum of squares of the residuals and explain using words and a diagram what this number represents.
(e) From your equation estimate, to 2 decimal places, the decrease in M when D increases by 1. [ULSEB]

12 The yield of a batch process in the chemical industry is known to be approximately linearly related to the temperature, at least over a limited range of temperatures. Two measurements of the yield are made at each of eight temperatures, within this range, with the following results:

Temperature (°C) x	180	190	200	210
Yield (tonnes) y	136.2 136.9	147.5 145.1	153.0 155.9	161.7 167.8

Temperature (°C) x	220	230	240	250
Yield (tonnes) y	176.6 164.4	194.2 183.0	194.3 175.5	196.5 219.3

$\Sigma x = 172$ $\Sigma x^2 = 374\,000$

(a) Plot the data on a scatter diagram.
(b) For each temperature, calculate the mean of the two yields. Calculate the equation of the regression line of this mean yield on temperature. Draw the regression line on your scatter diagram.
(c) Predict, from the regression line, the yield of a batch at each of the following temperatures: (i) 175, (ii) 185, (iii) 300. Discuss the amount of uncertainty in each of your three predictions.
(d) In order to improve predictions of the mean yield at various temperatures in the range 180 to 250 it is decided to take a further eight measurements of yield.

(continued)

Recommend, giving a reason, the temperatures at which these measurements should be carried out.   [AEB 91]

13 The values of x and y from a bivariate sample give a product–moment coefficient of r. Show that the line of regression of y on x may be written in the form:

$$\frac{y - \bar{y}}{d_y} = r\frac{x - \bar{x}}{d_x}$$

and the line of regression of x on y may be written in the form:

$$\frac{x - \bar{x}}{d_x} = r\frac{y - \bar{y}}{d_y}$$

where d_x and d_y are the sample standard deviations of x and y respectively.

The values of x and y are transformed to values of u and v by:

$$u = \frac{x - \bar{x}}{d_x}, \quad v = \frac{y - \bar{y}}{d_y}$$

Show that:

(i)   the sample values of u, and the sample values of v, have mean 0 and standard deviation 1,

(ii)   the value of the product–moment coefficient of the values of u and v is r,

(iii)   the line of regression of v on u is $v = ru$ and the line of regression of u on v is $u = rv$,

(iv)   on a (u, v) scatter diagram the line $v = u$ bisects the angle between the two regression lines in (iii).

21 Experimental design and the analysis of variance (ANOVA)

Deep in unfathomable mines
Of never failing skill,
He treasures up his bright designs,
And works his sovereign will

Olney Hymns, 35, William Cowper.

In this chapter we extend the ideas of experimental design and the comparison of populations introduced in Chapter 17. We use the results concerning variances that were given in Chapter 19 in a context that develops from the regression model introduced in Section 20.5 (p. 532).

21.1 The comparison of more than two means

Suppose we have samples from k populations having the same variance σ^2 but with possibly different means $\mu_1, \mu_2, \ldots, \mu_k$. The hypotheses of interest are these:

$\quad$ H_0: $\mu_1 = \mu_2 = \cdots = \mu_k$
$\quad$ H_1: Not all the means are equal.

Here is a typical problem. Adam, a market gardener, has three greenhouses containing miniature tomatoes. In each greenhouse, Adam chooses four tomato plants at random and keeps records of the masses of the tomatoes produced by each plant. The combined masses (in kg) for each plant are summarised in the following table.

Greenhouse 1	1.42	1.64	1.81	1.53
Greenhouse 2	1.85	1.74	2.23	2.18
Greenhouse 3	1.75	1.31	1.95	1.59

A quick plot of these data suggests that there may be a difference between the greenhouses.

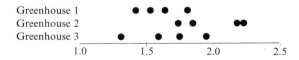

The natural inclination is to compare pairs of means using two-sample tests. To see why this won't work, consider the case $k = 4$, with H_0 being true, so that $\mu_1 = \mu_2 = \mu_3 = \mu_4$. Suppose that we use a 5% significance level, then the probability that we accept that $\mu_1 = \mu_2$ is 95% and, independently, the probability that we accept that $\mu_3 = \mu_4$ is 95%. The probability that we accept both is $0.95^2 \approx 0.90$. We still need to do more testing to compare, say, μ_1 with μ_3, so the probability of accepting that all four means are equal must be less than 0.90. Thus the overall significance level is at least 10% and not the desired 5%.

We consider instead the (apparently irrelevant) question of how we are going to estimate σ^2. If H_0 is correct, then all 12 observations come from the same distribution with unknown mean, μ, and unknown variance σ^2. The estimate of μ is the overall sample mean $\bar{y}$ given by:

$$\bar{y} = \frac{1.42 + 1.64 + \cdots + 1.95 + 1.59}{12}$$

$$= \frac{21.00}{12} = 1.75$$

The sum of the squares of the 12 observations is $1.42^2 + \cdots + 1.59^2 = 37.6076$, so that the unbiased estimate of σ^2, which we will denote by s_{Tot}^2, is given by:

$$s_{Tot}^2 = \frac{1}{11} \left(37.6076 - \frac{21.00^2}{12} \right) = 0.0780$$

Suppose, instead, that we do *not* assume that H_0 is correct. We now have three independent samples from populations with (possibly) different means. These are estimated by the sample means $\bar{y}_1 = \frac{1}{4}(6.40) = 1.60$, $\bar{y}_2 = \frac{1}{4}(8.00) = 2.00$ and $\bar{y}_3 = \frac{1}{4}(6.60) = 1.65$. Each sample provides an unbiased estimate of σ^2:

$$s_1^2 = \frac{1}{3} \left(10.3230 - \frac{6.40^2}{4} \right) = \frac{0.0830}{3} = 0.0277$$

$$s_2^2 = \frac{1}{3} \left(16.1754 - \frac{8.00^2}{4} \right) = \frac{0.1754}{3} = 0.0585$$

$$s_3^2 = \frac{1}{3} \left(11.1092 - \frac{6.60^2}{4} \right) = \frac{0.2192}{3} = 0.0731$$

Each estimate (given here to 3 sf) conveys some information about the value of σ^2. To *pool* that information we extend the argument of Section 17.3 (p. 458) and calculate an estimate, which we denote by s_W^2, as follows:

$$s_W^2 = \frac{0.0830 + 0.1754 + 0.2192}{3 + 3 + 3} = 0.0531$$

The quantity s_W^2 measures the variability *within* each sample, and is often called the **within samples estimate** of σ^2. Its value is unaffected by the truth, or otherwise, of H_0. To see this, suppose that the data had looked like this:

Greenhouse 1	2.42	2.64	2.81	2.53
Greenhouse 2	3.85	3.74	4.23	4.18
Greenhouse 3	1.75	1.31	1.95	1.59

We have added 1 to each observation in the first sample and 2 to each observation in the second sample. These additions have not made the separate samples more variable and s_W^2 is unaffected. However H_0 now looks decidedly unlikely!

Although s_W^2 is unaltered by the changes, s_{Tot}^2 is greatly inflated – the revised value is 1.060 (compared to $s_W^2 = 0.0531$). The reason is that, in addition to the within sample variability summarised by s_W^2, s_{Tot}^2 also contains a component due to the variations between the sample means and the overall mean – which is what we want to find out about!

One way of calculating this component is by using the weighted sum of squares of the differences between the individual sample means and the mean of the pooled sample, a quantity that will be zero only if all the sample means are equal. We shall see later that this weighted sum of squares is equal to $(k - 1)s_B^2$, where k is the number of samples and s_B^2 is called the **between samples estimate** of σ^2. We will show that a convenient way of calculating the value of s_B^2 is to use the formula:

$$s_B^2 = \frac{1}{k-1} \{(n-1)s_{Tot}^2 - (n-k)s_W^2\} \tag{21.1}$$

where n is the total number of observations.

In the present case, for the original data, we get:

$$s_B^2 = \frac{1}{2}\{(11 \times 0.0780) - (9 \times 0.0531)\} = 0.190$$

which is an unbiased estimate of σ^2 *only* if H_0 is true. If H_0 is false then, on average, s_B^2 will be greater than σ^2. A useful test statistic is the value of the ratio $\dfrac{s_B^2}{s_W^2}$; if this is unreasonably large then we conclude that s_B^2 has been inflated by differences between the means.

If we assume that all the observations come from normal distributions, then the answer to the question of how large is 'unreasonably large' is provided by the F-distribution introduced in Section 19.3 (p. 514). The test procedure consists of comparing $\dfrac{s_B^2}{s_W^2}$ with the upper tail critical value for an $F_{(k-1),(n-k)}$-distribution. If the critical value is exceeded then the null hypothesis of equality of the means is rejected and the best estimates of the individual population means are the corresponding sample values.

For the original tomatoes data we had $s_B^2 = 0.190$ and $s_W^2 = 0.0531$ so that $\dfrac{s_B^2}{s_W^2} = 3.58$. The upper tail 5% point of $F_{2,9}$ is 4.26. Since $3.58 < 4.26$, H_0 is not rejected.

Note

♦ The calculations frequently involve the subtraction of a large value from another, very similar, large value. In order for the difference to be calculated accurately it is sensible to carry through calculations using more significant figures than would ordinarily be used. For example, in calculating s_1^2, we used 10.3230 and not 10.3. The difference appears trivial, but would have resulted in a value of s_1^2 of 0.02 instead of 0.0277 – a serious discrepancy!

The problem can often be reduced by removing a constant from all the observations so as to reduce their average magnitude.

21.2 One-way ANOVA

Let n_i denote the number of observations in sample i, and let the jth observation in sample i be denoted by y_{ij}. Suppose there are k samples, containing a total of n observations, so that:

$$\sum_{i=1}^{k}(n_i - 1) = n - k$$

The quantities s_W^2 and s_{Tot}^2 are given by:

$$s_W^2 = \frac{1}{(n-k)}\sum_{i=1}^{k}\left(\sum_{j=1}^{n_i}y_{ij}^2 - \frac{y_{i+}^2}{n_i}\right) \tag{21.2}$$

and:

$$s_{Tot}^2 = \frac{1}{(n-1)}\left(\sum_{i=1}^{k}\sum_{j=1}^{n_i}y_{ij}^2 - \frac{y_{++}^2}{n}\right) \tag{21.3}$$

where:

$$y_{i+} = \sum_{j=1}^{n_i}y_{ij}, \qquad y_{i+}^2 = (y_{i+})^2 \quad \text{and} \quad y_{++} = \sum_{i=1}^{k}\sum_{j=1}^{n_i}y_{ij}, \qquad y_{++}^2 = (y_{++})^2$$

Equations (21.2) and (21.3) are used for calculating s_W^2 and s_{Tot}^2, but it is easier to see what is happening if we rewrite the expressions in terms of the overall mean, $\bar{y}$, and the individual sample means, $\bar{y}_1, \ldots, \bar{y}_k$. Thus:

$$(n-k)s_W^2 = \sum_{i=1}^{k}\sum_{j=1}^{n_i}(y_{ij} - \bar{y}_i)^2 \tag{21.4}$$

while:

$$(n-1)s_{Tot}^2 = \sum_{i=1}^{k}\sum_{j=1}^{n_i}(y_{ij} - \bar{y})^2 \tag{21.5}$$

We can rewrite the second expression by using the cunning device of introducing the sample means into the expression:

$$(n-1)s_{Tot}^2 = \sum_{i=1}^{k}\sum_{j=1}^{n_i}\{(y_{ij} - \bar{y}_i) + (\bar{y}_i - \bar{y})\}^2$$

$$= \sum_{i=1}^{k}\left\{\sum_{j=1}^{n_i}(y_{ij} - \bar{y}_i)^2 + 2(\bar{y}_i - \bar{y})\sum_{j=1}^{n_i}(y_{ij} - \bar{y}_i) + n_i(\bar{y}_i - \bar{y})^2\right\}$$

For each sample:

$$\sum_{j=1}^{n_i}(y_{ij} - \bar{y}_i) = 0$$

and so:

$$(n-1)s_{Tot}^2 = (n-k)s_W^2 + \sum_{i=1}^{k}n_i(\bar{y}_i - \bar{y})^2 \tag{21.6}$$

The second term on the right-hand side of Equation (21.6) is the weighted sum of squares referred to in the previous section. It quantifies the differences between the sample means and the overall mean. Since $(n-k)s_W^2$ has expectation $(n-k)\sigma^2$, and, when H_0 is true, $(n-1)s_{Tot}^2$ has expectation $(n-1)\sigma^2$, it follows that when H_0 is true $\sum_{i=1}^{k}n_i(\bar{y}_i - \bar{y})^2$ has expectation $(k-1)\sigma^2$. Therefore we can write:

$$(k-1)s_B^2 = \sum_{i=1}^{k}n_i(\bar{y}_i - \bar{y})^2 \tag{21.7}$$

so that, as implied by Equation (21.1):

$$(n-1)s_{Tot}^2 = (n-k)s_W^2 + (k-1)s_B^2 \tag{21.8}$$

Equations (21.5), (21.4) and (21.7) show that each of $(n-1)s_{Tot}^2$, $(n-k)s_W^2$ and $(k-1)s_B^2$ is a **sum of squares**. A convenient summary of the calculations is provided by an **analysis of variance table** or **ANOVA** table, as shown below.

Source of variation	Degrees of freedom	Sum of squares	Mean square	F-ratio
Between samples	$k - 1$	(by subtraction)	s_B^2	$\dfrac{s_B^2}{s_W^2}$
Within samples	$n - k$	$\sum\limits_{i=1}^{k} \left(\sum\limits_{j=1}^{n_i} y_{ij}^2 - \dfrac{y_{i+}^2}{n_i} \right)$	s_W^2	
Total	$n - 1$	$\sum\limits_{i=1}^{k} \sum\limits_{j=1}^{n_i} y_{ij}^2 - \dfrac{y_{++}^2}{n}$		

The entries in the column headed 'Mean square' consist of the sum of squares divided by the degrees of freedom.

To recapitulate, the within samples mean square s_W^2 is always an unbiased estimate of σ^2. The between samples mean square, s_B^2, is an unbiased estimate of σ^2 only if H_0 is true. The F-ratio is the ratio of these two quantities.

▼   ▼

Example 1

To test the lifetimes of batteries, 12 toy drummers are fitted with new batteries of one of three types: Amazing, Superlong and Endurance. The lengths of time (in hours) that the drummers continue to drum are summarised in the table below.

Amazing	4.7, 5.1, 5.2
Superlong	4.8, 5.1, 5.4, 5.4
Endurance	5.1, 5.2, 5.2, 5.4, 5.6

Determine whether there is significant evidence, at the 5% level, of a difference between the mean lifetimes of the three types of batteries. Summarise the findings in an ANOVA table.

We begin with a quick look at a plot of the data (which suggests that any differences are rather slight). We assume that the observations have normal distributions with a common variance. To calculate the various quantities required for the F-test, it is convenient to set out the information in a table, as shown.

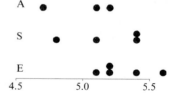

	Observations	n_i	y_{i+}	$\sum\limits_{j=1}^{n_i} y_{ij}^2$	$\sum\limits_{j=1}^{n_i} y_{ij}^2 - \dfrac{y_{i+}^2}{n_i}$
A	4.7, 5.1, 5.2	3	15.0	75.14	0.1400
S	4.8, 5.1, 5.4, 5.4	4	20.7	107.37	0.2475
E	5.1, 5.2, 5.2, 5.4, 5.6	5	26.5	140.61	0.1600
Totals		12	62.2	323.12	0.5475

From the table we have $9s_W^2 = 0.5475$ and

$$11s_{Tot}^2 = \left(323.12 - \frac{62.2^2}{12}\right) = 0.7167,$$ so that, by subtraction, $2s_B^2 = 0.1692$.

The test statistic therefore has value:

$$\frac{\frac{1}{2}(0.1692)}{\frac{1}{9}(0.5475)} = 1.39$$

This does not exceed the upper 5% point of an $F_{2,9}$-distribution (which is 4.26), so we accept the null hypothesis that the three types of battery have the same mean lifetime (estimated, by $\bar{y}$, to be about 5.2 hours).

The previous results are summarised in the following ANOVA table.

Source of variation	Degrees of freedom	Sum of squares	Mean square	F-ratio
Between samples	2	0.1692	0.0846	1.39
Within samples	9	0.5475	0.0608	
Total	11	0.7167		

Exercises 21a

1 Three fertilisers A, B and C were each applied to seven plots of strawberries (21 plots in all). The total crops (in kg) yielded by these plots over the entire season are given in the following table.

A	15	28	18	32	29	18	15
B	48	39	36	52	30	44	47
C	31	29	22	38	40	24	33

Analyse these data, summarising the results in an ANOVA table and stating your conclusions.

2 The following data refer to the per cent porosity of bricks fired at four temperatures.

2600 °F	2675 °F	2750 °F	2825 °F
18.8	16.7	15.7	14.4
14.7	14.3	13.1	11.9
14.9	12.5	11.5	9.8
14.2	12.6	11.7	11.3
15.5	13.7	11.3	11.3

Analyse these data using a one-way ANOVA table to determine whether the average porosity of bricks is affected by firing temperature.

3 The following data relate to measurements of the weights (in grams) of a number of copper wires, each of length 1 metre. A number of random samples were taken from the output of each of six machines (A–F). Carry out an analysis of variance for these data and test the hypothesis that there are no differences between the average densities of the outputs from the six machines.

			Machine			
A	B	C	D	E	F
1.31	1.31	1.29	1.33	1.32	1.32
1.30	1.33	1.34	1.29	1.28	1.31
1.34	1.35	1.32	1.30	1.34	1.33
1.30	1.36	1.35	1.29	1.30	1.31
	1.31		1.32	1.30	
			1.29		

21.3 Models

Much of Statistics is concerned with building simple models of situations. Early examples that we met were the binomial and Poisson models. For continuous variables we have used models based on the normal distribution.

1 For a sample from a single population, we could write simply:

$$E(Y_j) = \mu \qquad j = 1, \ldots, n$$

In Chapters 14 and 15 we were interested in estimating μ and testing hypotheses about its value.

2 When comparing two populations, denoting the observations by $y_{11}, \ldots, y_{1n_1}$ and $y_{21}, \ldots, y_{2n_2}$, we could write:

$$E(Y_{ij}) = \mu_i \qquad i = 1, 2 \qquad j = 1, \ldots, n_i$$

In Chapter 17 we were interested in testing the hypothesis $\mu_1 = \mu_2$.

3 The situation described in Sections 21.1 and 21.2 is a straightforward extension to several populations:

$$E(Y_{ij}) = \mu_i \qquad i = 1, \ldots, k \qquad j = 1, \ldots, n_i$$

The null hypothesis is $\mu_1 = \cdots = \mu_k$.

4 Sometimes the populations will be simply related to one another through another variable X. A simple example occurs when the means of the populations are linear functions of X:

$$\mu_i = a + bx_i$$

This corresponds to the regression model of Chapter 20:

$$E(Y_i) = a + bx_i$$

21.4 Randomised blocks

Let us return to Adam, our market gardener. Since we last met him, Adam has been reading up about tomatoes and Statistics and has discovered three things:

1 There is more than one variety of tomato.
2 There is more than one way of growing tomato plants.
3 When in doubt, randomise.

Adam is keen to compare four varieties of tomato (T_1, T_2, T_3, T_4) and four types of grow-bags (B_1, B_2, B_3, B_4). Adam has three plants of each type and the grow-bags are large enough to accommodate up to four plants. How should Adam plan his experiment?

Adam's first idea was to choose a variety at random, and a bag at random, and plant the three plants in the bag. If he had done that then he might have got a result like this:

T_1	T_1	T_1
1.94	1.89	1.99

Bag 1

T_4	T_4	T_4
1.42	1.53	1.46

Bag 2

T_2	T_2	T_2
2.10	1.97	2.03

Bag 3

T_3	T_3	T_3
1.85	1.76	1.84

Bag 4

At first glance, the results appear clear cut. Easily the best variety is T_2, and the best bag is B_3. However, are these good results due to the bag, or to the variety, or to both? Might T_1 have done even better if it had been grown in bag B_3? Might bag B_1 have seemed the best if only it had had variety T_2 growing in it? The differences between bags and the differences between the varieties are said to be **confounded** – we cannot separate one from the other.

While worrying about this problem, Adam's wife (Eve – of course!) removes one bag (B_4) to grow her begonias in. Left with just three bags, Adam sees the solution to his dilemma: grow one of each variety in each of the three remaining grow-bags. In this way, if a variety is particularly good it will be compared with every other variety under the same conditions. Conversely, if a bag is particularly good then it will benefit each variety equally.

There is still need for care: the allocation of the plants to bags must be done at random, and the positions of the plants should be chosen at random. The resulting arrangement is three **randomised blocks**. The results might look like this:

T_1	T_3	T_4	T_2
1.89	1.74	1.62	1.97

T_2	T_1	T_3	T_4
1.85	1.76	1.80	1.89

T_3	T_1	T_2	T_4
1.81	1.91	2.10	1.71

Bag 1   Bag 2   Bag 3

We will analyse these data in the next section.

21.5 Two-way ANOVA

Adam's tomato problem was an example of a situation in which the mean yield is affected not by a single x-variable, nor by a single 'treatment' (τ), but by two factors that are conventionally referred to as 'treatments' (the varieties) and 'blocks' (the grow-bags). An appropriate model must incorporate both these effects.

Let y_{ij} denote the yield for the ith treatment in the jth block. Then a simple additive model is:

$$E(Y_{ij}) = \mu + \tau_i + \beta_j \qquad (21.9)$$

with, conventionally, $\Sigma\tau_i = 0$ and $\Sigma\beta_j = 0$.

To see what to do next, consider the arrays of observed values and of expected values for the case of three blocks and four treatments:

	Observed values			
	Block 1	Block 2	Block 3	Total
Treatment 1	y_{11}	y_{12}	y_{13}	y_{1+}
Treatment 2	y_{21}	y_{22}	y_{23}	y_{2+}
Treatment 3	y_{31}	y_{32}	y_{33}	y_{3+}
Treatment 4	y_{41}	y_{42}	y_{43}	y_{4+}
Total	y_{+1}	y_{+2}	y_{+3}	y_{++}

	Expected values			
	Block 1	Block 2	Block 3	Total
Treatment 1	$\mu + \tau_1 + \beta_1$	$\mu + \tau_1 + \beta_2$	$\mu + \tau_1 + \beta_3$	$3(\mu + \tau_1)$
Treatment 2	$\mu + \tau_2 + \beta_1$	$\mu + \tau_2 + \beta_2$	$\mu + \tau_2 + \beta_3$	$3(\mu + \tau_2)$
Treatment 3	$\mu + \tau_3 + \beta_1$	$\mu + \tau_3 + \beta_2$	$\mu + \tau_3 + \beta_3$	$3(\mu + \tau_3)$
Treatment 4	$\mu + \tau_4 + \beta_1$	$\mu + \tau_4 + \beta_2$	$\mu + \tau_4 + \beta_3$	$3(\mu + \tau_4)$
Total	$4(\mu + \beta_1)$	$4(\mu + \beta_2)$	$4(\mu + \beta_3)$	12μ

The table of expected values shows that the block effects 'cancel out', so that differences between y_{1+}, y_{2+}, y_{3+} and y_{4+} can be attributed to a mixture of random variation and genuine differences between the treatments. By analogy with Equation (21.7), the sum of squares due to the differences between treatments is SS_T, given by:

$$SS_T = \sum_{i=1}^{4}\left\{3\left(\frac{y_{i+}}{3} - \frac{y_{++}}{12}\right)^2\right\} = \sum_{i=1}^{4}\frac{y_{i+}^2}{3} - \frac{y_{++}^2}{12}$$

Generalising to the case of b blocks and t treatments, and rearranging in a form suitable for calculations, this gives:

$$SS_T = \sum_{i=1}^{t}\frac{y_{i+}^2}{b} - \frac{y_{++}^2}{bt} \tag{21.10}$$

Interchanging the roles of blocks and treatments, we have the sum of squares for blocks, SS_B, being given by:

$$SS_B = \sum_{j=1}^{b}\frac{y_{+j}^2}{t} - \frac{y_{++}^2}{bt} \tag{21.11}$$

The quantities SS_T and SS_B are associated with, respectively, $(t-1)$ and $(b-1)$ degrees of freedom. If there were no differences due to blocks or treatments, σ^2 would have been estimated using the total sum of squares calculated from all bt observations, namely:

$$SS_{Tot} = \sum_{i=1}^{t}\sum_{j=1}^{b}y_{ij}^2 - \frac{y_{++}^2}{bt} \tag{21.12}$$

The quantity $\frac{y_{++}^2}{bt}$, which occurs in each of these formulae, is often called the **correction factor**. In doing the calculations it is sensible to calculate this first, storing it in a memory of the calculator if one is available.

After subtracting SS_B and SS_T from the total sum of squares, the remaining variation can only be attributable to chance and therefore forms the basis of an estimate of σ^2. We can summarise these findings in an ANOVA table.

Source of variation	Degrees of freedom	Sum of squares	Mean square	F-ratio
Blocks	$b-1$	SS_B	$\dfrac{SS_B}{(b-1)}$	$\dfrac{(t-1)SS_B}{D}$
Treatments	$t-1$	SS_T	$\dfrac{SS_T}{(t-1)}$	$\dfrac{(b-1)SS_T}{D}$
Residual	$(b-1)(t-1)$	D (by subtraction)	$\dfrac{D}{(b-1)(t-1)}$	
Total	$bt-1$	SS_{Tot}		

The F-tests require comparisons with the upper tail critical values of $F_{(b-1),(b-1)(t-1)}$ and $F_{(t-1),(b-1)(t-1)}$-distributions. If there are significant differences, then the block or treatment means are estimated by the corresponding sample means.

Notes

- Sometimes 'blocks' may be a second type of 'treatment'. For example, Adam might vary both the varieties of tomatoes and the amount of a fertiliser. In such a case there would not be blocks *per se* and the analysis would be better described as a **two-way analysis of variance**.
- It is possible to subdivide the treatments sum of squares into $t-1$ separate components, each associated with a single degree of freedom. Specific comparisons between the treatments can then be made. The procedures for doing this are quite simple – but this book is already sufficiently large!

Example 2

Analyse Adam's randomised block data to decide whether, at the 5% level, there are significant differences between (i) the varieties, (ii) the grow-bags.

———————

Arranging the data as a 4×3 table, and calculating the totals we get:

	T_1	T_2	T_3	T_4	Total (y_{+j})
B_1	1.89	1.97	1.74	1.62	7.22
B_2	1.76	1.85	1.80	1.89	7.30
B_3	1.91	2.10	1.81	1.71	7.53
Total (y_{i+})	5.56	5.92	5.35	5.22	22.05

In addition we need the following extra statistics:

$$\sum_{i=1}^{4}\sum_{j=1}^{3} y_{ij}^2 = 40.6955 \qquad \sum_{i=1}^{4} y_{i+}^2 = 121.8309 \qquad \sum_{j=1}^{3} y_{+j}^2 = 162.1193$$

The correction factor is $\frac{22.05^2}{12} = 40.516\,875$, using many significant figures to achieve a reasonable number of significant figures in the result.

The total sum of squares is therefore:

$$40.6955 - 40.516\,875 = 0.178\,625$$

while SS_T and SS_B are given by:

$$SS_T = \frac{121.8309}{3} - 40.516\,875 = 0.093\,425$$

$$SS_B = \frac{162.1193}{4} - 40.516\,875 = 0.012\,950$$

The resulting ANOVA table is:

Source of variation	Degrees of freedom	Sum of squares	Mean square	F-ratio
Differences between grow-bags	2	0.012 950	0.006 475	0.54
Differences between varieties	3	0.093 425	0.031 142	2.59
Residual	6	0.072 250	0.012 042	
Total	11	0.178 625		

The F-ratios have been given to the same accuracy as the tabulated values. At the 5% level there are no significant differences between either the grow-bags ($0.54 < 5.14$) or the varieties ($2.59 < 4.76$).

Using coded values

The calculations are often made much simpler by reducing all the data by a constant. At the same time we can often conveniently get rid of the decimal places by multiplying by a power of 10. Although the magnitudes of the resulting sums of squares will be changed, the values of the F-ratios will be unaltered.

Example 3

Use the coding $z = 100(y - 1.80)$ to analyse Adam's data.

The revised table is

	T_1	T_2	T_3	T_4	Total (z_{+j})
B_1	9	17	-6	-18	2
B_2	-4	5	0	9	10
B_3	11	30	1	-9	33
Total (z_{i+})	16	52	-5	-18	45

The extra statistics are now:

$$\sum_{i=1}^{4}\sum_{j=1}^{3} z_{ij}^2 = 1955 \qquad \sum_{i=1}^{4} z_{i+}^2 = 3309 \qquad \sum_{j=1}^{3} z_{+j}^2 = 1193$$

and the correction factor becomes $\dfrac{45^2}{12} = 168.75$.

The total sum of squares is therefore:

$$1955 - 168.75 = 1786.25$$

while SS_T and SS_B are given by:

$$SS_T = \frac{3309}{3} - 168.75 = 934.25$$

$$SS_B = \frac{1193}{4} - 168.75 = 129.50$$

The resulting ANOVA table is:

Source of variation	Degrees of freedom	Sum of squares	Mean square	F-ratio
Differences between grow-bags	2	129.50	64.75	0.54
Differences between varieties	3	934.25	311.42	2.59
Residual	6	722.50	120.42	
Total	11	1786.25		

The sums of squares are scaled versions of their predecessors and, as anticipated, the F-ratios are unaltered.

Exercises 21b

1 An experiment was conducted to compare the consequences of different methods of pruning on the resulting apple crop. Three similar apple trees were chosen and each was allocated to a different method of pruning. The resulting crops (in kg) during a four-year period are given in the table.

Is there significant evidence, at the 5% level, of a difference in the masses of the crops from the three trees?

Year	Method A	B	C
1	78	19	53
2	88	17	33
3	57	18	62
4	67	15	50

2 An experiment was conducted to investigate the effects of three different diets on the growth of pigs. There were five different litters of pigs used in the experiment. From each litter, one pig, chosen at random, was assigned to each of the three diets. The weight gains (in kg) over a fortnight's period are summarised in the table below. Determine whether there are significant differences, at the 5% level (i) between the litters, (ii) between the diets.

Litter	Diet A	Diet B	Diet C
1	15.8	14.7	12.7
2	15.4	16.1	14.8
3	15.0	13.1	12.5
4	14.2	15.0	10.7
5	15.2	16.7	15.9

3 In order to investigate whether different fats are absorbed in different amounts by a doughnut mix during cooking, batches of six doughnuts were cooked on five different days in each of the four fats. The results are grams of fat absorbed.

Day	Fat			
	1	2	3	4
1	164	172	177	178
2	177	197	184	196
3	168	167	187	177
4	156	161	169	181
5	172	180	179	184

Analyse the data, summarising your results in an ANOVA table.

21.6 Latin squares

Since last year Adam has been speaking to Noah, an expert on water. Water, says Noah, can make a difference to one's view of life. Adam thinks it might make a difference to his tomatoes. Adam also decides to try four quite different soil types, so, this year he grows his tomatoes in large flower pots, so that he can try the different soils and the different watering systems. For Christmas, Eve gave Adam a book on Latin squares, so his experiment looks like this:

	Soil 1	Soil 2	Soil 3	Soil 4
Water system 1	T_1	T_2	T_3	T_4
Water system 2	T_2	T_1	T_4	T_3
Water system 3	T_3	T_4	T_1	T_2
Water system 4	T_4	T_3	T_2	T_1

You can easily see the special features of this experiment. Each tomato type occurs once in each row and once in each column. This careful planning enables us to distinguish between the three potential causes of differences between the yields (the differences between types of tomatoes, the differences between soils and the differences between watering systems.)

If we denote the tomato types by A, B, C and D instead of T_1 to T_4 then, with rows and columns denoting the watering systems and soil types as before, Adam's design is summarised by a square of (roman script) letters, therefore known as a **Latin square**:

```
A B C D
B A D C
C D A B
D C B A
```

For this design we need parameters for tomato types (e.g. τ_i), rows (e.g. ρ_j) and columns (e.g. κ_k). For the general case where there are m 'treatments' corresponding to the letters in the Latin square, and denoting a typical observed value by y_{ijk}, the model for an $m \times m$ table is:

$$E(Y_{ijk}) = \mu + \tau_i + \rho_j + \kappa_k \qquad (i,j,k = 1,\ldots,m)$$

Note that, because of the design, there are only m^2 observations – so not all combinations of i, j and k occur.

The sums of squares associated with the three sources of variation are simple extensions of those given by Equations (21.10) and (21.11):

$$SS_T = \sum_{i=1}^{m} \frac{y_{i++}^2}{m} - \frac{y_{+++}^2}{m^2} \tag{21.13}$$

$$SS_R = \sum_{j=1}^{m} \frac{y_{+j+}^2}{m} - \frac{y_{+++}^2}{m^2} \tag{21.14}$$

$$SS_C = \sum_{k=1}^{m} \frac{y_{++k}^2}{m} - \frac{y_{+++}^2}{m^2} \tag{21.15}$$

The correction factor, $\frac{y_{+++}^2}{m^2}$ is also needed in the calculation of the total sum of squares which is given by:

$$SS_{Tot} = \sum_{i=1}^{m} \sum_{j=1}^{m} y_{ijk}^2 - \frac{y_{+++}^2}{m^2}$$

Note that, because of the cunning design, we could sum the observations over any two of the three subscripts.

Each of the SS_T, SS_R, SS_C sums of squares is associated with $m - 1$ degrees of freedom, while there are $m^2 - 1$ degrees of freedom in total. The corresponding ANOVA looks like this:

Source of variation	Degrees of freedom	Sum of squares	Mean square	F-ratio
Treatments	$m - 1$	SS_T	$\dfrac{SS_T}{m - 1}$	$\dfrac{(m - 2)SS_T}{D}$
Rows	$m - 1$	SS_R	$\dfrac{SS_R}{m - 1}$	$\dfrac{(m - 2)SS_R}{D}$
Columns	$m - 1$	SS_C	$\dfrac{SS_C}{m - 1}$	$\dfrac{(m - 2)SS_C}{D}$
Residual	$(m - 1)(m - 2)$	D (by subn.)	$\dfrac{D}{(m - 1)(m - 2)}$	
Total	$m^2 - 1$	SS_{Tot}		

The F-tests involve comparison with the upper tail critical values of an F-distribution with $(m - 1)$ and $(m - 1)(m - 2)$ degrees of freedom.

Example 4

It has been suggested that the brand of petrol is important in the calculation of average fuel consumption. To test this idea, three fuels, A, B and C were tested. The experiment used three cars: a Rover, a Citroën and a Ford. Each car was driven at a steady speed (either 30 mph, 35 mph, or 40 mph) around an oval track. The cars were filled with 5 gallons of fuel and were driven until the fuel ran out. The distance travelled was recorded to the nearest 0.1 mile.

The experimental results were as follows:

Car	Fuel	Speed	Distance
Rover	A	40	195.1
Citroën	B	30	217.4
Ford	C	50	186.8
Ford	A	30	228.4
Citroën	C	40	222.1
Citroën	A	50	172.4
Ford	B	40	231.1
Rover	C	30	199.3
Rover	B	50	169.9

Analyse these data, assuming the observations come from normal distributions, and state whether there is significant evidence that the fuel consumption is dependent upon the brand of petrol.

———

We begin by rearranging the data as a Latin square, using cars as rows and speeds as columns:

	30	40	50
Citroën	B (217.4)	C (222.1)	A (172.4)
Ford	A (228.4)	B (231.1)	C (186.8)
Rover	C (199.3)	A (195.1)	B (169.9)

The row (car) totals are 611.9, 646.3 and 564.3; the column (speed) totals are 645.1, 648.3 and 529.1; the treatment (petrol) totals are 595.9, 618.4 and 608.2. The overall total is 1822.5,

so that the correction factor is $\dfrac{1822.5^2}{9} = 369\,056.25$.

The overall sum of squared observations is $373\,431.45$ and therefore the total sum of squares in the ANOVA table is $373\,431.45 - 369\,056.25 = 4375.20$.

The remaining calculations are summarised in the ANOVA table.

Source of variation	Degrees of freedom	Sum of squares	Mean square	F-ratio
Petrols	2	84.62	42.31	1.0
Cars	2	1130.35	565.18	13.3
Speeds	2	3074.99	1537.50	36.1
Residual	2	85.24	42.62	
Total	8	4375.20		

The upper 5% point of an $F_{2,2}$-distribution is 19.0. The results confirm that petrol consumption varies with speed. There is some evidence of a difference between cars, which we would expect. This evidence is not significant at the 5% level, though we could reasonably expect that a more extensive experiment would have found the differences to be significant. The result of interest is the small non-significant F-ratio for differences between petrols.

Note

◆ An extension of the Latin square to include a fourth factor results in the so-called Graeco-Latin square, in which a further set of treatments are indicated by Greek letters. Here is the simplest example:

Aα   Bβ   Cγ
Bγ   Cα   Aβ
Cβ   Aγ   Bα

Exercises 21c

1 Four different doses of insulin, A, B, C and D, were tested on rabbits and compared in terms of the subsequent sugar contents in the rabbit's blood. Individual rabbits differ widely in their blood-sugar levels. The effect of an insulin injection wears off in a few days, so enabling further doses to be tested on the same rabbit in successive weeks. The results, in mg of glucose per 100 cc of blood, were as follows.

Week	Rabbit			
	I	II	III	IV
1	B:47	A:90	C:79	D:50
2	D:46	C:74	B:63	A:68
3	A:61	B:62	D:58	C:65
4	C:76	D:61	A:87	B:60

Summarise the results in an ANOVA table and test whether there are significant differences between the doses at the 1% level.

2 In a field trial it is required to compare the effects of five different pesticides on the growth of turnips. The field is divided into a lattice of 25 plots and the treatments (pesticides) are allocated using a randomly chosen Latin square design. The results (using coded values) are as follows.

1	2	3	4	5
A: 111	B: 80	C: 173	D: 64	E: 120
B: 263	C: 147	E: 62	A: 105	D: 57
E: 71	D: 71	B: 202	C: 203	A: 145
C: 52	A: 31	D: 122	E: 17	B: 123
D: 123	E: 65	A: 267	B: 201	C: 88

Do there appear to be significant differences between treatments?

3 The abrasion resistance of rubber is altered by treating the rubber with a chlorinating agent. Four different machines (1, 2, 3, 4) are available so that it is possible to study the effects of four different chlorinating agents (A, B, C, D). Four different samples of rubber (a, b, c, d) were available. The results are given below in the form (Rubber, Machine, Agent: Resistance).

(a,1,A: 16.6), (a,2,B: 16.2), (a,3,D: 15.8), (a,4,C: 16.0), (b,1,B: 17.2), (b,2,C: 16.8), (b,3,A: 17.6), (b,4,D: 16.5), (c,1,C: 17.4), (c,2,D: 17.6), (c,3,B: 17.8), (c,4,A: 18.4), (d,1,D: 18.2), (d,2,A: 20.2), (d,3,C: 19.6), (d,4,B: 20.4)

Determine whether there are significant differences in abrasion resistance due to the chlorinating agents.

21.7 Replication

In the last two sections there has been (at most) a single observation corresponding to each combination of effects. If it is possible, then it is a good idea to repeat the entire experiment (the technical word is **replicate**).

If two observations are taken under the same conditions, then these so-called **replications** will differ only through experimental error. The greenhouse experiment of Section 21.1 (p. 560) is a typical example. If there are r replications of a particular set of conditions then these provide $(r - 1)$ pieces of information (degrees of freedom) concerning the size of σ^2, the error variance.

One advantage of replication is that it gives us an opportunity to assess whether the model is reasonable. When we do our sums there will be some degrees of freedom 'left over' – these can be used to test for lack of fit.

We return once again to Adam and his tomatoes. Adam now knows that there are no differences between the grow-bags, but he is not convinced that his greenhouses do not differ. Once again he compares four types of tomatoes, and his results now look like this:

T_2	T_3	T_4	T_2
1.86	1.82	1.64	2.03
T_3	T_4	T_1	T_1
1.76	1.82	1.81	2.03

Greenhouse 1

T_4	T_1	T_3	T_4
1.83	1.75	2.13	1.79
T_3	T_2	T_2	T_1
1.69	1.86	2.33	1.86

Greenhouse 2

T_3	T_1	T_2	T_3
1.87	1.82	1.82	1.94
T_1	T_4	T_2	T_4
1.72	1.88	2.02	1.76

Greenhouse 3

Rearranging the data into a manageable array, we see that we have two replications of a randomised block experiment with three 'blocks' (the greenhouses) and four 'treatments' (the types of tomato):

	Type of tomato				
	1	2	3	4	Total
Greenhouse 1	1.81, 2.03	1.86, 2.03	1.82, 1.76	1.64, 1.82	14.77
Greenhouse 2	1.86, 1.75	1.86, 2.33	2.13, 1.69	1.83, 1.79	15.24
Greenhouse 3	1.82, 1.72	1.82, 2.02	1.87, 1.94	1.88, 1.76	14.83
Total	10.99	11.92	11.21	10.72	44.84

The sum of the squares of the 24 observations is 84.3022. The correction factor is $\dfrac{44.84^2}{24} = 83.7761$, and hence SS_{Tot} is $84.3022 - 83.7761 = 0.5261$.

The block (greenhouse) sum of squares is:

$$\frac{14.77^2 + \cdots + 14.83^2}{8} - 83.7761 = 0.0164$$

and the treatment (tomato) sum of squares is:

$$\frac{10.99^2 + \cdots + 10.72^2}{6} - 83.7761 = 0.1321$$

The variation in the results is made up of 'within cell' variation and 'between cell' variation. The 'between cell' variation includes the contributions due to variations between (in this case) tomato types and greenhouses and is easily calculated using the 'cell totals' $(1.81 + 2.03) = 3.84, \ldots, (1.88 + 1.76) = 3.64$. The calculation takes the usual form:

$$\frac{3.84^2 + \cdots + 3.64^2}{2} - 83.7761 = 0.2207$$

Subtraction of this quantity from SS_{Tot} gives the 'within cell' variation, with $bt(r - 1)$ degrees of freedom associated with the replicates. In this case this is:

$$0.5261 - 0.2207 = 0.3054$$

with 12 degrees of freedom.

We ascribe the remaining 'between cell' variation to 'lack of fit' which is associated with $(b - 1)(t - 1)$ degrees of freedom. In this case this amounts to:

$$0.2207 - 0.1321 - 0.0164 = 0.0722$$

with 6 degrees of freedom. If the 'lack of fit' sum of squares proves to be significantly large then the randomised block model, given by Equation (21.9), is too simple and something more complicated is required.

We should always test for lack of fit of the model as a preliminary to other tests, since, if the model is inadequate, there will be little point in examining it!

We can summarise the information in an ANOVA table in the usual way:

Source of variation	Degrees of freedom	Sum of squares	Mean square	F-ratio
Tomato types	3	0.1321	0.0440	1.73
Greenhouses	2	0.0164	0.0082	0.32
Lack of fit	6	0.0722	0.0120	0.47
Replications	12	0.3054	0.0255	
Total	23	0.5261		

The upper 5% point of an $F_{6,12}$-distribution is 3.00. Since 0.47 is less than this, there is no significant evidence of the model being inadequate. We can therefore consider the possible significance of the differences between the tomatoes and between the greenhouses. Neither proves to be significant at the 5% level (1.73 < 3.49 and 0.32 < 3.89). Once again Adam has not found anything of significance!

Note

◆ The contribution that we have described as 'lack of fit' is sometimes called the **interaction** effect.

Exercises 21d

1 An experiment was performed to investigate the yields (in bushels per acre) of two varieties of soy beans (Ottawa Mandarin, OM, and Blackhawk, B). Four fields were divided into quarters, two of which were assigned to OM and two to B.
Analyse the results given below to determine whether there is evidence of a significant difference between the two varieties.

Field	OM	B
1	29, 36	63, 48
2	46, 30	48, 47
3	53, 25	55, 56
4	30, 39	54, 55

2 The amount of precipitate in a chemical reaction that is formed in a ten-minute period depends upon the temperature of the reaction and the catalyst being used. The 36 members of a chemistry class take part in an experiment in which three catalysts (A, B and C) are compared at each of four temperatures. The results (grams of precipitate) are summarised below.

Catalyst	Temperature °C	
	25	30
A	2.4, 2.8, 2.6	3.1, 1.9, 3.8
B	2.5, 2.8, 2.7	3.6, 3.0, 2.9
C	2.6, 2.9, 2.8	3.6, 3.2, 3.2

Catalyst	Temperature °C	
	35	40
A	3.3, 2.7, 4.1	3.5, 3.3, 3.1
B	3.6, 3.7, 3.6	3.5, 3.9, 4.4
C	3.7, 4.0, 4.1	3.5, 4.0, 3.6

Determine whether there is significant evidence that the usual randomised block model is inadequate. Report on any effects that you find significant.

Chapter summary

◆ Results are summarised in **ANOVA tables**, with columns headed 'Source of variation', 'Degrees of freedom', 'Sum of squares', 'Mean square' $\left(= \frac{\text{S.S}}{\text{d.f.}}\right)$, and '$F$-ratio' (a ratio of mean squares).

In calculating **sums of squares** it is helpful to begin by calculating the correction factor, which is the square of the grand total, divided by the number of observations.

F-ratios are compared with the upper percentage points of an F-distribution.

Abbreviated ANOVA tables for the various designs are as follows:

◆ **Comparison of several populations:**

Source of variation	Degrees of freedom	Sum of squares
Between samples	$k - 1$	$\sum_{i=1}^{k} \frac{y_{i+}^2}{n_i} - \frac{y_{++}^2}{n}$
Within samples	$n - k$	$\sum_{i=1}^{k} \left(\sum_{j=1}^{n_i} y_{ij}^2 - \frac{y_{i+}^2}{n_i} \right)$
Total	$n - 1$	$\sum_{i=1}^{k} \sum_{j=1}^{n_i} y_{ij}^2 - \frac{y_{++}^2}{n}$

◆ **Randomised blocks:**

Source of variation	Degrees of freedom	Sum of squares
Blocks	$b - 1$	$\sum_{j=1}^{b} \frac{y_{+j}^2}{t} - \frac{y_{++}^2}{bt}$
Treatments	$t - 1$	$\sum_{i=1}^{t} \frac{y_{i+}^2}{b} - \frac{y_{++}^2}{bt}$
Residual	$(b - 1)(t - 1)$	(by subtraction)
Total	$bt - 1$	$\sum_{i=1}^{t} \sum_{j=1}^{b} y_{ij}^2 - \frac{y_{++}^2}{bt}$

◆ **Latin squares:**

Source of variation	Degrees of freedom	Sum of squares
Treatments	$m-1$	$\sum_{i=1}^{m} \dfrac{y_{i++}^2}{m} - \dfrac{y_{+++}^2}{m^2}$
Rows	$m-1$	$\sum_{j=1}^{m} \dfrac{y_{+j+}^2}{m} - \dfrac{y_{+++}^2}{m^2}$
Columns	$m-1$	$\sum_{k=1}^{m} \dfrac{y_{++k}^2}{m} - \dfrac{y_{+++}^2}{m^2}$
Residual	$(m-1)(m-2)$	(by subtraction)
Total	m^2-1	$\sum_{i=1}^{m}\sum_{j=1}^{m} y_{ijk}^2 - \dfrac{y_{+++}^2}{m^2}$

◆ With a **replicated design** in which there are r repeat values for each of the d cells of the design, there are $rd-1$ degrees of freedom associated with the total sum of squares. There is one extra row in the ANOVA table corresponding to the replications and associated with $(r-1)d$ degrees of freedom. The sums of squares for treatments, etc, are calculated as usual, but the contribution that would have been labelled 'residual' is now labelled 'lack of fit'.

Exercises 21e (Miscellaneous)

1 The following observations refer to the antler diameters (in mm) of Colorado mule deer measured at the base of the antlers. The deer were all approximately the same age (1.3 years) and the object of the experiment was to see if herds from different areas differed significantly in antler size.

Area	Diameters
A	20, 14, 9, 13, 17, 20, 21
B	22, 23, 17, 18, 21, 15, 20, 21
C	15, 20, 23, 19, 21
D	23, 17, 20, 22, 20, 15

2 An experiment was conducted to investigate three methods of drying mixed lettuce leaves after washing. Five employees tried each of three methods, with the following results (milligrams of water removed per 100 grams of leaves).

	Employee				
	1	2	3	4	5
Method A	950	787	897	850	975
Method B	857	989	918	968	909
Method C	917	872	975	930	954

Determine whether there is significant evidence, at the 5% level, of a difference in the efficiencies of the three methods.

3 Four different methods (A, B, C and D) of manufacturing metal tubes may result in tubes of different tensile strength.

Factory conditions are known to vary slightly from day to day and there may also be variation from machine to machine in the factory. A Latin square design is used, with the results (in units of tensile strength) being as follows.

Machine	Day of manufacture			
	1	2	3	4
1	A:16.6	B:16.9	C:17.4	D:17.4
2	D:17.1	C:16.7	B:19.3	A:16.6
3	B:18.4	A:17.3	D:17.5	C:19.1
4	C:17.5	D:17.0	A:16.7	B:19.3

Analyse the data for variations due to differences between methods, days and machines.

4 The following results refer to a study of the effectiveness of three organic compounds (A, B and C) against a type of virus that affects chick embryos.

Eighteen groups of eggs were used, two at each of three virus dilutions (I, II and III). The survival times are summarised below.

Dilution	Compound		
	A	B	C
I	87, 90	82, 71	72, 77
II	79, 80	73, 72	70, 66
III	77, 81	72, 68	62, 61

Analyse these data, stating your conclusions.

5 A factory is to introduce a new product which will be assembled from a number of components. Three different designs are considered and four employees are asked to consider their speed of assembly. The trial is carried out one morning and each of the four employees assembled design *A* from 8.30 a.m. to 9.30 a.m., design *B* from 10.00 a.m. to 11.00 a.m. and design *C* from 11.30 a.m. to

12.30 p.m. The number of products completed by each of the employees is shown in the following table.

Design	Employee			
	1	2	3	4
A	17	4	38	8
B	21	6	52	20
C	28	9	64	22

(a) Carry out a two factor analysis of variance and test at the 5% significance level for differences between designs and between employees. You may assume that the total sum of squares about the mean (SS_T) is 3878.9.

(b) Comment on the fact that all employees assembled the designs in the same order. Suggest a better way of carrying out the experiment.

(c) The two factor analysis assumes that the effects of design and employee may be added. Comment on the suitability of the model for these data and suggest a possible improvement.

[AEB 91]

22 Nonparametric tests

One should never put on one's best trousers to go out to battle for freedom and truth

An Enemy of the People, Henrik Ibsen

Many of the previous chapters have relied upon a specific model (e.g. binomial, Poisson, Normal) to describe the probabilistic properties of the data. In this final chapter we are more cautious in our approach and, so far as is possible, we avoid making detailed assumptions about the nature of the population being sampled. We begin by considering tests concerning a single sample (the analogue of Chapter 15), move on to paired-sample tests (Chapter 17) and finish by looking once more at correlation (Chapter 20).

The tests described make no assumptions about the underlying distribution and therefore are often called **distribution-free** tests. Without an assumed distribution there can be no parameters and hence the name we shall use is: **nonparametric tests**.

22.1 The single-sample sign test

This test is appropriate when we have no knowledge concerning the distribution of a continuous random variable, X, but are interested in hypotheses concerning the median, m, of the distribution, for example:

$$H_0: m = 23$$
$$H_1: m > 23$$

Assuming H_0, that the median is indeed equal to 23:

$$P(X > 23) = P(X < 23)$$

whereas, if H_1 is correct:

$$P(X > 23) > P(X < 23)$$

In this case an appropriate test statistic is r, the number of observations in a sample of size n that are greater than 23. Assuming H_0, the corresponding random variable, R, has a $B(n, 0.5)$ distribution, for which:

$$P(R = r) = \binom{n}{r}\left(\frac{1}{2}\right)^{n-r}\left(\frac{1}{2}\right)^{r} = \binom{n}{r}\left(\frac{1}{2}\right)^{n}$$

For the particular H_1 given above, we would reject H_0 in favour of H_1 only if r is unusually large (where the definition of 'unusually' depends on the significance level!).

Notes
- The name of this test arises because the value of r is the number of values of $x - m_0$ that have a plus **sign**, where m_0 is the median specified by the null hypothesis.
- The test is identical to a test of a proportion (Sections 15.8 and 16.4, pp. 413, 436) with $H_0: p = 0.5$. As noted in Section 16.4, in small samples there may not be any decision rule that provides an actual significance level close to the nominal significance level.
- The simplest procedure for dealing with an observation that has a value recorded as being equal to m_0 is to ignore it and to reduce n by one.

- The test can be used with non-numerical observations. For example, consumers might be asked to say whether or not a new brand of margarine was creamier than the brand that they usually bought (see Example 1 below).
- You should be familiar with any tables of the $B(n, 0.5)$ distribution that are available for examination purposes.
- Nonparametric tests often only require simple calculations, so may be used as quick (but less powerful) alternatives to parametric tests, even in situations where the latter are perfectly valid.

Example 1

A random sample of twelve people are independently asked to say which they prefer of two cheese spreads, labelled A and B. Eleven of them choose spread A and one chooses B.

Does this provide significant evidence, at a nominal 1% level, that the two spreads will not be found equally preferable by the population?

This is a two-sided situation, with the hypotheses being:

H_0: The two spreads are equally preferable.
H_1: The two spreads are not equally preferable.

Let r be the number of people that prefer spread A. Evidence against H_0 will be provided by *either* very large values of r *or* very small values of r. Denote the corresponding random variable by R. Assuming H_0, R will have a $B(12, 0.5)$ distribution for which the individual probabilities and the cumulative upper-tail probabilities are given below (to 4 d.p.).

r	$P(R = r)$	$P(R \geq r)$
12	0.0002	0.0002
11	0.0029	0.0032
10	0.0161	0.0193
9	0.0537	0.0730

We commence by determining appropriate acceptance and rejection regions. A reasonable choice is to reject H_0 if r is equal to 0, 1, 11 or 12, and to accept H_0 otherwise. The associated significance level can be determined from the table. The total probability of the four outcomes is $2(0.0002 + 0.0029) = 0.0064$. The actual significance level is therefore the 0.64% level, reasonably close to the nominal 1% level.

Since the observed value is 11, we conclude that there is significant evidence, at an actual level of 0.64%, that the population will not find the two spreads equally preferable.

Example 2

An experiment on reaction times is concerned with the effects of distracting lights on the ability of subjects to react to a buzzer. Under ordinary conditions it is known that the median reaction time is 0.22 seconds. The times taken (in hundredths of a second) by a random sample of individuals in the presence of the distracting lights are as follows:

21, 28, 31, 23, 55, 71, 27, 66, 18, 22, 63, 20, 42

Use a sign test to determine whether there is significant evidence, at a nominal 5% level, of an increase in reaction time.

Writing m as the population median reaction time in seconds when the distracting lights are being used, the two hypotheses are:

H_0: $m = 0.22$
H_1: $m > 0.22$

The variable of interest is R, the number of observations that exceed 0.22. The effective sample size is 12, since one observation was recorded as being equal to 0.22. If H_0 is correct, then R has a $B(12, 0.5)$ distribution, as in Example 1. The observed value of R is 9, and the question is whether or not this provides significant evidence to reject H_0 in favour of H_1. A simplistic approach (adopted in the answers for Exercises 22a) is as follows. First find p, the probability of obtaining the observed value, or a more extreme value, by using the $B(12, 0.5)$ distribution. If p is smaller than the desired significance level, p_0, then reject H_0 and otherwise accept H_0. In this example, since $r = 9$, $p = 0.073$ and since $p_0 = 0.05$, we accept H_0 and conclude that there is no significant evidence of an increase in reaction time.

Exercises 22a

1 Eight apples are chosen at random from a crate containing a large number of apples. The masses, in grams, are:

147, 138, 171, 142, 152, 145, 141, 143

Use a sign test, at the 5% significance level, to test whether the median mass of the apples is 150 g. State your null and alternative hypotheses.

2 The lifetimes of a random sample of Waxlong candles, measured in minutes, are:

354, 358, 348, 342, 352
335, 364, 345, 360, 341

The manufacturer claims that the median lifetime is at least 6 hours.
Use a sign test, at the 5% significance level, to test whether the manufacturer's claim is justified. State the hypotheses tested.

3 A random sample of 20 people who enter a garden centre one morning are given (in a random order) a randomly selected tomato of type X and a randomly selected tomato of type Y. Each person is asked to say which of the two tomatoes they prefer. Twelve prefer X and the remainder prefer Y.

(i) Explain why the words 'random' and 'randomly' appear so often.

(ii) Using a normal approximation, determine the appropriate tail probability associated with the observed outcome.

(iii) Calculate the corresponding exact tail probability.

(iv) What conclusions should be drawn concerning the tomatoes?

4 A manufacturer makes a large number of chains of the same type. The breaking strength, in kilograms, of each of a random sample of 12 chains was found. The results, each correct to 4 significant figures, were as follows:

1294 1310 1297 1312 1341 1315
1328 1317 1296 1321 1307 1318

Perform a sign test, at the 5% significance level, to test the hypothesis that the average breaking strength is 1300kg against the alternative that it exceeds 1300kg.   [UCLES(P)]

5 Explain briefly what is meant by the term "non-parametric" when referring to tests of significance.

A new golf course is built. The par, i.e. the score in which a professional player could expect to complete the course in good weather, is fixed at 71. A random sample of ten professional players play the course, in good weather, and record the following scores:

Player	1	2	3	4	5	6	7	8	9	10
Score	69	66	70	73	72	68	68	70	70	74

Use a sign test, at the 5% level, to decide whether the par has been fixed correctly.
[UCLES(P)]

6 State what you understand by the terms *parametric* and *non-parametric* as applied to tests of significance.

Each of a random sample of 10 students was subjected to a stimulus and the reaction time (in seconds) was measured with the following results.

Student	A	B	C	D	E
Reaction time	0.37	0.41	0.39	0.51	0.67

Student	F	G	H	I	J
Reaction time	0.41	0.36	0.45	0.62	0.59

Use a sign test to determine whether, at the 5% significance level, the average reaction time for this stimulus is 0.60 seconds. To what *average* does this test apply?   [UCLES(P)]

7 Briefly describe circumstances in which each of the following are used:

(a) parametric tests of significance,

(b) non-parametric tests of significance.

It is believed that the material from which running tracks are made has a significant effect on the times taken for athletes to run specified distances. In order to test this, 12 athletes ran on two tracks over a distance of 200 m. One track was made from synthetic material and the other from cinders. The times, in seconds, are given in the table.

Athlete	A	B	C	D
Cinder Track	27.3	26.4	26.6	25.1
Synthetic Track	26.5	26.3	24.9	25.7

Athlete	A	B	C	D
Cinder Track	26.2	27.0	26.8	27.0
Synthetic Track	26.7	24.8	26.1	26.7

Athlete	A	B	C	D
Cinder Track	25.4	26.7	25.0	24.7
Synthetic Track	24.4	24.7	24.6	24.5

By using a sign test show that, at the 10% significance level, the median times on each of the tracks could be 26 seconds.   [UCLES(P)]

Frank Wilcoxon (1892–1965) was an American who obtained his doctorate in physical chemistry from Cornell University in 1924. For the next 25 years he worked as a chemist in various of the larger chemical firms in the USA. Only for the last 7 years of his working life was he officially employed as a statistician! However, his interest in the subject dated back to 1925, when he wished to devise statistical tests of the effectiveness of various types of insecticide and fungicide. He led the research group that worked on the development of various of the pyrethrin-based insecticides, including in particular Malathion. His statistical work concentrated on devising methods of testing that were simple and easy to understand.

22.2 The Wilcoxon signed-rank test

The sign test made no assumptions concerning the underlying distribution, but the Wilcoxon test is rather more restrictive and requires the underlying distribution to be symmetric. Like the sign test, the hypotheses under test concern the median of the distribution.

For a set of observations $x_1, x_2, \ldots$, and a null hypothesis that the population has median m_0, the sign test considered only the signs of the differences $x_1 - m_0, \ldots$, whereas the Wilcoxon test takes account of the *magnitudes* of these differences. Using this extra information, the Wilcoxon test is more likely than the sign test to correctly reject a false null hypothesis (i.e. it has greater power).

The initial procedure for a two-tailed Wilcoxon signed-rank test is as follows:

1 Arrange the differences in ascending order of *absolute value*.

2 If there are any differences exactly equal to zero, then discard them, and work with a smaller sample size.

3 Suppose that there are n non-zero differences. Retain the signs of these differences, but replace their existing magnitudes with integers, so that the smallest absolute value is replaced by 1, the next smallest by 2, and so on. These replacement values are called **signed ranks**.

For example, suppose the hypotheses are:

$$H_0: m = 100$$
$$H_1: m \neq 100$$

Suppose that a random sample of 8 observations are as follows:

92.3, 57.6, 88.8, 110.5, 100.0, 181.0, 96.0, 105.7

The corresponding differences are then:

-7.7, -42.4, -11.2, 10.5, 0, 81.0, -4.0, 5.7

Discarding the zero value and rearranging the remaining values in order of *absolute* magnitude, we get:

-4.0, 5.7, -7.7, 10.5, -11.2, -42.4, 81.0

Retaining the signs and replacing the values by ranks, we get:

-1, 2, -3, 4, -5, -6, 7

Let P be the sum of the ranks corresponding to the positive differences (i.e. $P = 2 + 4 + 7 = 13$) and let Q be the sum of the ranks of the negative differences (i.e. $Q = 1 + 3 + 5 + 6 = 15$). There are two alternative (but equivalent) test statistics T and W defined by:

$$W = P - Q \tag{22.1}$$
$$T = \text{ the smaller of P and Q} \tag{22.2}$$

The calculations involving T are slightly easier, so this is the one that we use. Extensive tables of the critical values of T have been produced, but they are not really needed since, even for small values of n, a normal approximation *with a continuity correction* is very accurate. It can be shown that, for $n \geqslant 6$, the distribution:

$$N\left(\tfrac{1}{4}n(n+1), \tfrac{1}{24}n(n+1)(2n+1)\right)$$

is appropriate. The test statistic is:

$$z = \frac{\tfrac{1}{4}n(n+1) - T - \tfrac{1}{2}}{\sqrt{\tfrac{1}{24}n(n+1)(2n+1)}} \tag{22.3}$$

where the $\tfrac{1}{2}$ is the usual continuity correction. The percentage points take their usual values (e.g. 1.96 for a two-tailed test at the 5% level).

For the previous data, after discarding the zero, $n = 7$. The smaller of P and Q is P, so $T = 13$ and $z = \dfrac{14 - 13.5}{\sqrt{35}} = 0.085$. There is no reason to reject the null hypothesis.

Notes

- A check is provided by noting that $P + Q = \frac{1}{2}n(n+1)$. In the example
  $$P + Q = 13 + 15 = 28 = \frac{1}{2} \times 7 \times 8.$$
- An equivalent procedure is to assign rank 1 to the largest value, with the next largest being 2, and so on.
- If there are differences of equal magnitude then the simplest procedure is to assign to each of them the average of the ranks that they would otherwise have received. Here are some examples (using H_0: $m = 100$):

Sample size	Original observations	Signed ranks
5	98.3, 100.1, 100.1, 101.4, 103.0	−4, 1.5, 1.5, 3, 5
6	98.3, 99.9, 100.1, 101.4, 101.4, 103.0	−5, −1.5, 1.5, 3.5, 3.5, 6
5	98.3, 100.1, 101.7, 101.7, 103.0	−3, 1, 3, 3, 5
7	98.3, 99.6, 100.1, 101.7, 101.7, 101.7, 103.0	−4.5, −2, 1, 4.5, 4.5, 4.5, 7

- There are a number of tests associated with Frank Wilcoxon, so it is wise always to refer to this test using its full title: the 'Wilcoxon signed-rank test'.

Example 3

According to an examinations board, the scores on a national mathematics test were approximately symmetrically distributed about a median of 61. The scores obtained by a randomly chosen sample from Greyfriars School were as follows:

$$80, 65, 62, 61, 58, 43, 37, 38, 29$$

Use the Wilcoxon signed-rank test to determine whether the sample provides significant evidence, at the 5% level, that the median of the school results is:

(i) different from the national median,
(ii) less than the national median.

The preliminary calculations are common to the two enquiries, for both of which the null hypothesis, H_0, is that the median of the Greyfriars scores is 61. Subtracting 61 from the observations we obtain:

$$19, 4, 1, 0, -3, -18, -24, -23 -32$$

The fourth observation must be omitted from subsequent calculations. Rearranging the remainder in order of magnitude, we get:

$$1, -3, 4, -18, 19, -23 -24, -32$$

The resulting signed ranks are as follows:

$$1, -2, 3, -4, 5, -6, -7 -8$$

Thus $P = 1 + 3 + 5 = 9$ and $Q = 27$. As a check we note that $9 + 27 = 36 = \frac{1}{2} \times 8 \times 9$. The smaller of P and Q is P with value 9. In case (i) the alternative hypothesis is H_1: median $\neq 61$. For a two-tailed test at the 5% significance level, the procedure is to accept the null hypothesis unless $z > 1.96$. The value of $\frac{1}{4}n(n+1)$ is $\frac{1}{4} \times 8 \times 9 = 18$, so

$$\frac{1}{4}n(n+1) - T - \frac{1}{2} = 18 - 9.5 = 8.5 \text{ and } z = \frac{8.5}{\sqrt{51}} = 1.19. \text{ This is less than}$$

1.96, so the null hypothesis that the median is 61 is accepted.

In case (ii) the alternative hypothesis is H_1: median < 61. The test statistic z, is still equal to 1.19, but on this occasion it has to be compared with 1.645, the one-tailed 5% point of a $N(0,1)$ distribution. Since $1.19 < 1.645$ there is again no significant evidence, at the 5% level, that the school median is less than the national median.

Exercises 22b

1 Eight apples are chosen at random from a crate containing a large number of apples. The masses, in grams, are:

   147, 138, 171, 142, 152, 145, 141, 143

Use a Wilcoxon signed-rank test, at the 5% significance level, to test whether the median mass of the apples is 150 g. State your null and alternative hypotheses and any assumption made about the population distribution.

2 The lifetimes of a random sample of Waxlite candles, measured in minutes, are:

   372, 352, 335, 364, 345,
   360, 354, 358, 348, 341

The manufacturer claims that the average lifetime is at least 6 hours.
Use a Wilcoxon signed-rank test, at the 5% significance level, to test whether the manufacturer's claim is justified. State the hypotheses tested.

3 A manufacturer makes a large number of chains of the same type. The breaking strength, in kilograms, of each of a random sample of 12 chains was found. The results, each correct to 4 significant figures, were as follows:

   1294 1310 1297 1312 1341 1315
   1328 1317 1296 1321 1307 1318

Perform a Wilcoxon signed-rank test, at the 5% significance level, to test the hypothesis that the average breaking strength is 1300 kg against the alternative that it exceeds 1300 kg. State any assumption necessary for the validity of the test.

4 The time intervals between certain events are believed to be symmetrically distributed about a median of 200 minutes with the alternative being that the median is not 200. The lengths of a random sample of 12 time intervals are:

   144, 289, 312, 176, 201, 209, 103, 170, 277, 311, 190, 255

Use a Wilcoxon signed-rank test to decide whether or not to accept the null hypothesis.

5 Each of a random sample of 10 students was subjected to a stimulus and the reaction time (in seconds) was measured with the following results.

Student	A	B	C	D	E
Reaction time	0.37	0.41	0.39	0.51	0.67

Student	F	G	H	I	J
Reaction time	0.41	0.36	0.45	0.62	0.59

Use a Wilcoxon signed-rank test to determine whether, at the 5% significance level, the average reaction time for this stimulus is 0.60 seconds.   [UCLES(A)]

6 It is believed that the material from which running tracks are made has a significant effect on the times taken for athletes to run specified distances. In order to test this, 12 athletes ran on two tracks over a distance of 200 m. One track was made from synthetic material and the other from cinders. The times, in seconds, are given in the table.

Athlete	A	B	C	D
Cinder Track	27.3	26.4	26.6	25.1
Synthetic Track	26.5	26.3	24.9	25.7

Athlete	A	B	C	D
Cinder Track	26.2	27.0	26.8	27.0
Synthetic Track	26.7	24.8	26.1	26.7

Athlete	A	B	C	D
Cinder Track	25.4	26.7	25.0	24.7
Synthetic Track	24.4	24.7	24.6	24.5

Use a Wilcoxon signed-rank test, at the 10% significance level, to test whether the median times on each of the tracks could be 26 seconds.   [UCLES(A)]

22.3 The paired-sample sign test

With paired samples, the null hypothesis is that the differences between the two members of each pair come from a population having a zero median. Under this hypothesis, the second member of a pair is equally likely to be larger or smaller than the first. This is just like the single-sample sign test of

Section 22.1 (p. 580). The test statistic is R, the number of positive differences. If H_0 is true then the distribution of R is $B(n, 0.5)$, where n is the number of pairs (omitting any cases having zero difference).

If the observed value of R is either very small or very large, then (depending on whether the alternative hypothesis is one-sided or two-sided) this may lead to rejection of the null hypothesis.

Note

◆ The paired-sample sign test can also be used with non-numeric data, as illustrated in Example 5.

Example 4

In the United Kingdom the birth rate (births per year per 1000 population) is about 13. In Africa it is very much higher and, with improved health care in Africa, this is causing a population explosion. Average birth rates for 1965–70 and for 1985–90 are shown below for a random sample of African countries. Since the *same* countries are used each time, the data are paired.

	Country					
	Angola	Burundi	Congo	Egypt	Gabon	Kenya
1965–70	49.1	46.5	45.1	41.8	30.9	52.2
1985–90	47.2	45.7	44.4	36.0	38.8	53.9

	Country				
	Libya	Mali	Niger	Sudan	Togo
1965–70	49.5	51.6	49.4	47.0	44.2
1985–90	43.9	50.1	51.0	44.6	44.8

Use a sign test to test, at the 5% significance level, the null hypothesis that birth rates are just as likely to have increased or decreased between the two time periods, against the alternative that they have fallen (perhaps because of increased awareness of the problems resulting from a population explosion).

———————

There are 11 countries, so, assuming the null hypothesis, the reference distribution is $B(11, 0.5)$. The required test is one-tailed, so we need to consider only the upper tail of this distribution. Either by reference to tables of cumulative probabilities of the binomial distribution, or by direct calculation, we find P(9 or more) = 3.3%, and P(8 or more) = 11.3%. For a test at the nominal 5% level, the best choice seems to be to reject H_0 only if 9 or more of the countries show a decrease in birth rate.

Since only 7 of the countries show a decrease in birth rate, we cannot conclude that there is significant evidence of a continent-wide decrease.

Example 5

A manufacturer of soap powders has created a new powder which, it is believed, will wash clothes 'even whiter than before'. To test the powder, it uses a panel, who are presented with a box of the old powder and a box of the new powder (both boxes being disguised!). The panel are asked to try out both powders and to state which they prefer. Of 25 members of the panel, 18 preferred the new powder.
Test whether this provides significant evidence, at the 5% level, of a difference between the powders.

———

We have 18 '+ signs', with $n = 25$. The reference distribution is $B(25, 0.5)$. We need to know whether the chance of getting 18 or more '+ signs' is unusually small. If it is, then we reject the null hypothesis that there is no difference in the effectiveness of the two powders in favour of the alternative that there is a difference. We can determine the probability of 18 or more '+ signs' either by using tables (if available), or by direct calculation, or (since n is reasonably large) by using a normal approximation. The approximating normal distribution has mean np and variance $np(1 - p)$, so in this case is a $N(12.5, 6.25)$ distribution.

Using a continuity correction, the z-value is given by:

$$z = \frac{17.5 - 12.5}{\sqrt{6.25}} = 2.0$$

which is (just!) greater than 1.96. We therefore conclude that there is significant evidence of a difference between the powders – perhaps the new powder *does* wash whiter than before!

22.4 The Wilcoxon matched-pairs signed-rank test

This is simply the Wilcoxon signed-rank test applied to the differences between pairs of observations. The null hypothesis is that the two sets of observations come from populations having the same distribution, with the alternative being that they do not. As in Section 22.2 (p. 584) the test statistic is:

$$z = \frac{\frac{1}{4}n(n + 1) - T - \frac{1}{2}}{\sqrt{\frac{1}{24}n(n + 1)(2n + 1)}}$$

where the $\frac{1}{2}$ is the usual continuity correction.

Note

◆ If the distributions of two random variables X and Y are the same, then the distribution of $X - Y$ will be symmetrical (as required for the Wilcoxon signed-rank test) about zero.

Example 6

Use a Wilcoxon matched-pairs signed-rank test to re-examine the data of Example 4 (p. 587).

The changes in birth rates had magnitudes:

$$-1.9, -0.8, -0.7, -5.8, 7.9, 1.7, -5.6, -1.5, 1.6, -2.4, 0.6$$

Arranging these in order of magnitude, we have:

$$0.6, -0.7, -0.8, -1.5, 1.6, 1.7, -1.9, -2.4, -5.6, -5.8, 7.9$$

Retaining the signs, but replacing the values by ranks, we get:

$$1, -2, -3, -4, 5, 6, -7, -8, -9, -10, 11$$

Thus $P = 11 + 6 + 5 + 1 = 23$ and $Q = 43$ (giving, correctly, a total of $\frac{1}{2} \times 11 \times 12 = 66$). Thus $T = 23$ and the test statistic, z, is given by:

$$z = \frac{\frac{1}{4} \times 11 \times 12 - 23 - \frac{1}{2}}{\sqrt{\frac{1}{24} \times 11 \times 12 \times 23}} = 0.84$$

Since this value is less than 1.645 we again conclude that there is not significant evidence, at the 5% level, of a reduction in birth rate.

Notes

- If there are any differences of zero, then these are discarded and the effective sample size is reduced.
- If there are a number of differences of equal magnitude then the simplest procedure is to use averaged ranks (see Section 22.2, p. 585).

22.5 The Wilcoxon rank-sum test

This test is also known as the **Mann–Whitney** test. It is suitable for *unpaired* samples. The assumption is that the samples come from populations that have the same distribution after a translation of size k:

$$P(X < x) = P(Y < x + k)$$

for all values of x. The null hypothesis is that the two populations have the same distribution (i.e. $k = 0$).

The idea of the test is simple. Suppose that the two samples are of sizes m and n, with $m \leqslant n$. We begin by replacing the $(m + n)$ observed values by the corresponding ranks. Thus the smallest value is assigned the rank 1 and the largest value is assigned the rank $m + n$.

The quantity of interest is the sum of the ranks in the (smaller) sample of size m. This is known as the **rank-sum** and is denoted by R. Although tables of critical values of R exist, it is easier to use a normal approximation with a continuity correction. The relevant distribution is:

$$N\left(\tfrac{1}{2}(m(m + n + 1)), \quad \tfrac{1}{12}mn(m + n + 1)\right)$$

so that the test statistic is:

$$z = \frac{R - \frac{1}{2}m(m+n+1) \pm \frac{1}{2}}{\sqrt{\frac{1}{12}mn(m+n+1)}} \tag{22.4}$$

where the $\frac{1}{2}$ is the usual continuity correction *with sign chosen so as to reduce the absolute magnitude of the numerator.*

Notes

♦ It is easy to see that $E(R) = \frac{1}{2}m(m+n+1)$. We argue thus. Since the ranks range from 1 to $(m+n)$, their average size is $\frac{1}{2}(m+n+1)$. Thus the expected value of the rank of a randomly chosen observation is $\frac{1}{2}(m+n+1)$ and hence, for a sample of size m, the expected total is $\frac{1}{2}m(m+n+1)$.

♦ If there are tied values then they are usually assigned the average of the relevant ranks.

♦ The **Mann-Whitney** version of the test compares R with the minimum value that it could have. For a sample of size m, this is $\frac{1}{2}m(m+1)$. Since the varying part of the Mann-Whitney test (R) is the same as for the Wilcoxon test, the tests are equivalent.

Example 7

The marks obtained by a small random sample of statistics students were as follows:

Boys	10, 22, 42, 59, 61, 63, 65, 83, 85, 90, 93
Girls	36, 53, 54, 56, 69, 84, 88

Use a nonparametric procedure to test, at the 5% significance level, the hypothesis that the two sets of marks come from populations having the same distribution.

Boys

Girls

The data are obviously not paired, so we use the rank-sum test. The ranked values are:

Boys	1, 2, 4, 8, 9, 10, 11, 13, 15, 17, 18
Girls	3, 5, 6, 7, 12, 14, 16

For a two-tailed test at the 5% level we shall reject the null hypothesis only if $|z| > 1.96$. Working with the set of girls, m is 7 and R is $3 + 5 + \cdots + 16 = 63$. Since $\frac{1}{2}m(m+n+1) = 66.5$ and $\frac{1}{12}mn(m+n+1) = 121.9$, the continuity-corrected value of z is given by:

$$z = \frac{(63 - 66.5) + 0.5}{\sqrt{121.9}} = -0.272$$

Since $|z| < 1.96$ we accept the hypothesis that the two sets of marks came from populations having the same distribution.

Example 8

An experiment is conducted to compare two methods of using a computer. One group of randomly chosen students are asked to correct a computer file using the keyboard. A second randomly chosen group are given the same task, but the corrections are made using a mouse. The second method is expected to lead to generally faster times. The times (in seconds) are as follows:

Keyboard	73, 94, 98, 130, 144, 190, 242
Mouse	68, 75, 92, 97, 108, 124, 160, 166

Use the Wilcoxon rank-sum test to determine whether there is significant evidence, at the 5% level, that the population of 'mouse-times' has a median that is less than that of the population of 'keyboard-times'.

Keyboard

Mouse

We replace the raw data by ranks:

Keyboard	2, 5, 7, 10, 11, 14, 15
Mouse	1, 3, 4, 6, 8, 9, 12, 13

The null hypothesis is that the two populations have the same distribution, with the alternative being that the population of times taken by keyboard users has a larger median. We will reject the null hypothesis only if z exceeds 1.645.

For the 7 keyboard users the sum of the ranks is 64, compared to the expected value of 56. The variance is 74.67 and hence the test statistic, z, is given by:

$$z = \frac{(64 - 56) - 0.5}{\sqrt{74.67}} = 0.868$$

Since z does not exceed 1.645, the apparent evidence of faster times using the mouse is nevertheless not found to be significant when using the rank-sum test.

Practical _____

How well can you judge the length of an interval of time? Would you improve with practice? There are lots of possible experiments, but here is a simple one that requires people to work in pairs. The object is to judge how long is one minute. One person has a watch and announces the start of the minute, while the other has to announce when he or she thinks that the minute is up. The actual time is recorded and the judger is told. This is now repeated.

The comparison is between the accuracy of the second judgement and that of the first. Either a sign test or a Wilcoxon signed-rank test could be used, with a one-sided alternative hypothesis that anticipates an improvement.

The Wilcoxon rank-sum test could be used with the first estimates of the minute, in order to see whether there is evidence of differences in accuracy between the sexes.

Exercises 22c

1 A supermarket suspects that the average weight of Grade A melons from one supplier X is less than that for Grade A melons from supplier Y. Two random samples are taken and weighed. For 6 melons from X, the results, in kg, are:

    1.62, 1.63, 1.65, 1.67, 1.69, 1.72

   For 8 melons from Y, the results are:

    1.64, 1.66, 1.68, 1.70, 1.73, 1.74, 1.76, 1.79

   Use the Wilcoxon rank-sum test to investigate whether there is significant evidence at the 5% level to support the supermarket's suspicion. State appropriate null and alternative hypotheses.

2 A group of middle managers and a group of factory supervisors each took a management course which was followed by an aptitude test. The results were:

    middle managers
    121, 180, 122, 160, 141, 97, 212, 186

    factory supervisors
    128, 197, 181, 126, 167, 99, 147

   Use the Wilcoxon rank-sum test to determine whether these samples provide significant evidence, at the 5% level, of a difference in the two underlying populations.

3 In a traffic census drivers are asked the distance, in miles, of their current journey. The figures for a random sample of 10 drivers between 8 and 9 a.m. are:

    17.9, 7.2, 25.2, 35.9, 5.5,
    3.7, 61.7, 5.1, 2.4, 12.4

   The figures for a random sample of 6 drivers between 1 and 2 p.m. are:

    30.7, 19.4, 53.2, 72.5, 9.8, 40.7

   Test, at the 5% level whether the median distance travelled between 8 and 9 a.m. is less than the median distance travelled between 1 and 2 p.m.

4 Acorns are sown in seed compost A and, after three years, the resulting 7 oak trees have heights:

    0.571, 0.608, 0.549, 0.562,
    0.531, 0.604, 0.582

   measured in metres. Acorns are also sown in seed compost B and grown in similar circumstances. After three years the resulting 11 trees have heights:

    0.539, 0.578, 0.635, 0.592, 0.545, 0.613,
    0.620, 0.574, 0.581, 0.568, 0.617

   Use the Wilcoxon rank-sum test to find whether there is significant evidence, at the 5% level, that taller trees are produced in seed compost B. State the null and alternative hypotheses.

5 A consumers' association tests car tyres by running them on a machine until their tread depth reaches a prescribed minimum. Seven tyres of brand A were tested and the equivalent mileage, measured in thousands of miles, was measured, with results:

    34.9, 36.5, 33.1, 34.2, 37.1, 37.2, 35.7

   The corresponding results for seven tyres of brand B were:

    35.1, 37.2, 39.3, 37.4, 38.5, 33.8, 39.5

   Is there significant evidence, at the 5% level, of a difference in mileage between the two brands? State your null and alternative hypotheses.

6 The contents of a random sample of 8 pots of 'Extrafrute' jam were measured. The results, in grams, were:

    342, 354, 348, 349, 350, 347, 345, 356

   The results for a random sample of 10 pots of 'Jambo' jam were:

    340, 341, 350, 348, 342,
    350, 346, 347, 342, 344

   Carry out a rank-sum test to determine whether there is evidence, at the 5% level, that there is a difference in the contents of the two brands. State the hypotheses tested.

7 In a study of the effect of eating upon pulse rate, an investigator measured the pulse rate of 10 medical students both before and after they had eaten a substantial meal. The pulse rates, in beats/min were as follows:

Subject	1	2	3	4	5
Before	105	79	103	87	82
After	109	86	109	100	90

Subject	6	7	8	9	10
Before	78	86	79	104	101
After	90	93	90	110	100

   Using the Wilcoxon matched-pairs signed-rank test, test at the 5% significance level, the hypothesis that pulse rate is unaffected by eating against the alternative that there is a difference.

8 In order to examine the effect of temperature on the breaking strength of some chains, 10 chains were randomly selected and each was cut in half. One half, chosen at random, was tested at 15 °C and the other at 20 °C. The results were as follows.

Chain	1	2	3	4	5
15 °C	1307	1324	1321	1305	1306
20 °C	1312	1320	1318	1303	1298

Chain	6	7	8	9	10
15 °C	1304	1306	1321	1315	1301
20 °C	1297	1312	1310	1306	1289

Use the Wilcoxon matched-pairs signed rank test to determine whether there is significant evidence at the 5% level that the average breaking strength is less at the higher temperature.

[UCLES(P)]

9 A new golf course is built. The par, i.e. the score in which a professional player could expect to complete the course in good weather, is fixed at 71. A random sample of ten professional players play the course, in good weather, and record the following scores:

Player	1	2	3	4	5
Score	69	66	70	73	72

Player	6	7	8	9	10
Score	68	68	70	70	74

The same players play a second round on another day when the weather is bad. Their scores for the second round are:

Player	1	2	3	4	5
Score	71	72	72	70	76

Player	6	7	8	9	10
Score	70	65	75	73	76

Use a sign test, at the 10% level, to decide whether these figures indicate that professional players' scores are higher in bad weather.

[UCLES(P)]

10 Each of a random sample of 10 students was subjected to a stimulus and the reaction time (in seconds) was measured with the following results.

Student	A	B	C	D	E
Reaction time	0.37	0.41	0.39	0.51	0.67

Student	F	G	H	I	J
Reaction time	0.41	0.36	0.45	0.62	0.59

Five minutes after each had drunk a pint of beer the same students were given the same stimulus with the following results.

Student	A	B	C	D	E
Reaction time	0.41	0.40	0.41	0.57	0.62

Student	F	G	H	I	J
Reaction time	0.48	0.45	0.53	0.59	0.71

Explain why it is better to use Wilcoxon's signed rank test rather than the sign test to compare the results before and after drinking the beer. Determine, using a 5% significance level, whether the data show that the average reaction time increased after drinking the beer.

[UCLES(P)]

11 It is believed that the material from which running tracks are made has a significant effect on the times taken for athletes to run specified distances. In order to test this, 12 athletes ran on two tracks over a distance of 200 m. One track was made from synthetic material and the other from cinders. The times, in seconds, are given in the table.

Athlete	A	B	C	D
Cinder Track	27.3	26.4	26.6	25.1
Synthetic Track	26.5	26.3	24.9	25.7

Athlete	A	B	C	D
Cinder Track	26.2	27.0	26.8	27.0
Synthetic Track	26.7	24.8	26.1	26.7

Athlete	A	B	C	D
Cinder Track	25.4	26.7	25.0	24.7
Synthetic Track	24.4	24.7	24.6	24.5

Use a Wilcoxon matched-pairs signed rank test to show that, at the 1% significance level, there is insufficient evidence that the median time on the synthetic track is lower than that on the cinder track. State the lowest significance level at which it can be concluded that the median time on the synthetic track is lower.

[UCLES(P)]

12 Explain the circumstances under which a paired-sample non-parametric test of significance would be appropriate, while a paired-sample *t*-test would not be appropriate.

For samples of twelve pairs of items from two populations, let *n* be the number of + signs obtained using a sign test. Determine the values of *n* that result in the rejection of the null hypothesis of no difference between the populations, using a symmetric two-tail test at the (nominal) 5% significance level.

Calculate the exact tail probability associated with this nominal significance level.

A meteorological station has two rain gauges. Each gauge records, at ten-second intervals, the number of standard sized water droplets that have passed through its funnel in the interval. The numbers obtained during a randomly chosen two minute period were as follows:

Gauge 1	21	53	74	91	62	71
Gauge 2	29	75	70	96	77	103

Gauge 1	51	62	61	27	15	7
Gauge 2	76	55	59	38	14	13

Test whether these observations provide significant evidence, at the 5% level, of a difference between the numbers recorded by the gauges, using

(i) the sign test,

(ii) the Wilcoxon matched-pairs signed rank test.   [UCLES]

13 (*a*) for the case of two paired samples, explain briefly

   (i) the circumstances under which the *t*-test would be appropriate,

   (ii) the relative advantages and disadvantages of the sign test and of the Wilcoxon matched-pairs signed rank test.

   (*b*) An investigator is interested in the effects of alcohol on the accuracy of throwing of a dart. Eight volunteers each throw a single dart, aiming at the centre of the dart-board. They then each have a pint of beer and then throw the dart once more. The distances (in cm) from the centre of the dart-board are recorded in the table below.

Volunteer	1	2	3	4
Before beer	5.0	2.4	6.9	0.3
After beer	3.7	7.6	7.1	5.3

Volunteer	5	6	7	8
Before beer	4.1	3.2	2.0	4.2
After beer	4.7	1.1	2.4	4.9

(i) State explicitly suitable null and alternative hypotheses.

(ii) Using the sign test, determine the set of values of α for which your null hypothesis would be rejected at the α% significance level.

(iii) Using the Wilcoxon matched-pairs signed rank test, carry out a test of the null hypothesis at the 5% significance level, and state your conclusions.   [UCLES]

14 Two different German Vocabulary tests are given to the same A-level class of nine students and the marks gained are given in the table below.

Student number	1	2	3	4	5
Test 1	72	59	67	37	35
Test 2	64	48	46	35	47

Student number	6	7	8	9
Test 1	46	86	78	83
Test 2	45	83	74	74

Carry out two separate tests of significance, each at the 5% level, to assess whether or not the vocabulary tests are of equal difficulty, using:

(a) the sign test,

(b) the Wilcoxon test.   [UCLES(P)]

15 A new individual programme for learning keyboard skills has been devised and it is required to compare it with the one in current use. A test is carried out by selecting 9 sets of identical twins at random. One of the twins is randomly allocated to the current programme (*A*) and the other is given the new one (*B*). At the end of each programme a common examination is given, with the following results.

(continued)

Twin pair	1	2	30	4	5
A score	32	43	64	30	31
B score	30	57	60	42	47

Twin pair	6	7	8	9
A score	59	70	50	47
B Score	58	75	63	58

(i) Explain why a non-parametric test may be more appropriate than a parametric test for testing for a difference between the two programmes.

(ii) Perform two different non-parametric tests, at the 10% significance level, to test for a difference in average scores resulting from the two programmes. State the average (mean, median or mode) that is used in the hypotheses.   [UCLES(P)]

Charles Spearman (1863–1945) was responsible for bringing the use of statistical methods into Psychology and thereby changing the practice of that subject once and for all. Spearman was born in London and his initial choice of career was the army. He fought with distinction in the Burmese War and did not leave the army until 1897 when he went to Leipzig to study Psychology. After jobs in various German universities he returned to London in 1907 and was Professor at University College until retiring in 1931. His first paper on correlation appeared in 1904. He taught his students to look upon Statistics as a good servant but a bad master, and advised against collecting data in the vague hope that something would turn up!

22.6 Spearman's rank correlation coefficient, r_s

Suppose that contestants in a skating competition are independently ranked by each of two judges. With experienced judges we anticipate that contestants highly ranked by one judge will also be highly ranked by the other judge. We therefore anticipate a *positive correlation* between the two rankings. If one (or both) of the judges was assigning ranks at random, or if all the contestants were equally good, then we would anticipate nearly uncorrelated rankings.

If two judges are in complete agreement with each other, then they will assign the same rank to each contestant. Denoting the difference between the ranks awarded to contestant i by d_i, this means that Σd_i^2 would be zero. If the judges disagree, then Σd_i^2 will be greater than zero, and will increase as the extent of the disagreement increases. Spearman proposed a linear function of Σd_i^2 which had two of the properties of the correlation coefficient r (Chapter 20) in that perfect agreement was represented by the value 1 and perfect disagreement (judge 1's first choice is judge 2's last choice, and so on) was represented by the value -1. Spearman's rank correlation coefficient is usually denoted by r_s and is defined by:

$$r_s = 1 - \frac{6\Sigma d_i^2}{n(n^2 - 1)} \tag{22.5}$$

Notes

◆ A check on calculations is provided by noting that the sum of the signed differences in ranks must be zero.

◆ We have introduced r_s in the context of *two* judges. However, the same arguments would apply for the comparison of the rankings given by a single judge with true rankings when these are known.

◆ If two or more items are ranked equally then it is conventional to award the average of the corresponding ranks that could have been awarded (the so-called **tied rank**).

Example 9

There are eight contestants in a knobbly knees contest. Two judges assign the ranks shown below.

Determine the value of Spearman's rank correlation coefficient for these results.

Judge	Contestant							
	A	B	C	D	E	F	G	H
X	3	4	8	1	7	5	2	6
Y	2	1	8	3	7	5	4	6

The difference, d_1, between the two ranks for contestant A, is $(3 - 2) = 1$. The remaining differences are 3, 0, −2, 0, 0, −2 and 0 (which sum to 0, as required). Since $\Sigma d_i^2 = 18$, we get:

$$r_s = 1 - \frac{6 \times 18}{8 \times 63} = 0.786$$

There appears to be a good agreement between the judges.

Example 10

A man is asked to arrange, in order of mass, five similar boxes whose contents vary. The correct order is 1, 2, 3, 4, 5, but the order chosen by the man is 2, 1, 3, 5, 4.

Determine the value of r_s for the correlation between the two orders.

The rank differences are −1, 1, 0, −1 and 1, giving $\Sigma d_i = 0$ as required and $\Sigma d_i^2 = 4$. The value of r_s is $1 - \dfrac{6 \times 4}{5 \times 24} = 0.8$.

It appears that the man is a good judge of mass.

Example 11

A wine expert is blindfolded and asked to taste nine wines and arrange them in order of price. The correct order was A, B, C, ..., I, while the order chosen by the expert was A, (B, D), C, G, (E, F, H, I). The brackets indicate wines to which the expert assigned the same price.

Using tied ranks, determine the value of r_s as a measure of the correlation between the expert's opinion and the true order.

The expert was unable to distinguish between his 2nd and 3rd choices, so both are given rank 2.5. The final four wines are given the rank $\frac{1}{4}(6 + 7 + 8 + 9) = 7.5$. The calculations are summarised in the table below.

Wine	A	B	C	D	E	F	G	H	I	Total
True rank	1	2	3	4	5	6	7	8	9	
Judged rank	1	2.5	4	2.5	7.5	7.5	5	7.5	7.5	
d	0	−0.5	−1	1.5	−2.5	−1.5	2	0.5	1.5	0
d^2	0	0.25	1	2.25	6.25	2.25	4	0.25	2.25	18.5

The value of r_s is $1 - \dfrac{6 \times 18.5}{9 \times 80} = 0.846$. The wine expert really does appear to be an expert!

▲ ——— ▲

Calculator practice ——————————————————————————————

If you have a calculator with statistical functions then it may store 'Σx' and 'Σx^2' in accessible memories. By entering the values of d_i, you then have an easy check that $\Sigma d_i = 0$ and a quick way of obtaining Σd_i^2.

Testing the significance of r_s

According to H_0 the two sets of ranks are independent of one another. In order to understand the problems associated with testing the significance of r_s in small samples, consider the case $n = 4$. Label the objects (1, 2, 3, 4) using the rank order of one judge. If the other judge ranks the objects randomly, then (assuming no tied ranks) there are 24 possible outcomes, all equally likely. These are tabulated below, together with the resulting values of Σd_i^2.

(1,2,3,4)	0	(1,2,4,3)	2	(1,3,2,4)	2	(1,3,4,2)	6	(1,4,2,3)	6	(1,4,3,2)	8
(2,1,3,4)	2	(2,1,4,3)	4	(2,3,1,4)	6	(2,3,4,1)	12	(2,4,1,3)	10	(2,4,3,1)	14
(3,2,1,4)	8	(3,2,4,1)	14	(3,1,2,4)	6	(3,1,4,2)	10	(3,4,2,1)	18	(3,4,1,2)	16
(4,2,3,1)	18	(4,2,1,3)	14	(4,3,2,1)	20	(4,3,1,2)	18	(4,1,2,3)	12	(4,1,3,2)	14

The probability distribution of r_s is therefore:

Σd_i^2	0	2	4	6	8	10	12	14	16	18	20
r_s	1.0	0.8	0.6	0.4	0.2	0	−0.2	−0.4	−0.6	−0.8	−1.0
Prob	$\frac{1}{24}$	$\frac{3}{24}$	$\frac{1}{24}$	$\frac{4}{24}$	$\frac{2}{24}$	$\frac{2}{24}$	$\frac{2}{24}$	$\frac{4}{24}$	$\frac{1}{24}$	$\frac{3}{24}$	$\frac{1}{24}$

The distribution of r_s is evidently discrete. This causes problems when discussing significance levels. As an example, suppose (with $n = 4$) we would like to perform a one-tailed test at the 5% level. Looking at the probability distribution we can see that the outcome $r_s = 1.0$ would be judged significant, because the probability of obtaining this value or a larger value by chance is $\frac{1}{24}$, which is less than 5%. However, the outcome $r_s = 0.8$ would not be judged significant, because the probability of obtaining this value or a larger one by chance is $(\frac{3}{24} + \frac{1}{24}) = \frac{1}{6}$, which is greater than 5%.

The '5% level' is therefore a *nominal* one – the actual probability is not really '5%', but is $\frac{1}{24}$, which is slightly less. The problem of differences between actual and nominal tail probabilities was mentioned earlier (Sections 16.4 and 22.1).

Working at the nominal 1% level, when $n = 4$, no outcome would be judged significant because a sufficiently small tail probability does not exist! Since $\frac{2}{24} > 5\%$, with this small sample size we cannot achieve a significant result even at the 5% level when using a two-tailed test.

An extract from the table, given in the Appendix (p. 626), of (one-tailed) critical values for r_s is given below.

n	5%	1%	n	5%	1%	n	5%	1%	n	5%	1%
4	1.000	*	7	.714	.893	10	.564	.745	13	.484	.648
5	.900	1.000	8	.643	.833	11	.536	.709	14	.464	.626
6	.829	.943	9	.600	.783	12	.503	.678	15	.446	.604

The entries in the table are the smallest values of r_s (to 3 d.p.) which correspond to tail probabilities less than or equal to 5% (or 1%). The observed value is significant *if it is equal to, or greater than*, the value in the table. The actual significance level never exceeds the nominal value shown in the table.

Notes

◆ The entry for the nominal 1% level when $n = 4$ is an asterisk because such a small tail probability cannot be achieved in that case.

◆ In cases with tied ranks, the critical values given are conservative. The true tail probabilities may be appreciably smaller than the nominal values.

◆ For larger values of n than those given in the table in the Appendix, use the result that, for two independent rankings, $r_s\sqrt{\dfrac{n-2}{1-r_s^2}}$ approximates an observation from a t_{n-2}-distribution.

Example 12

Determine whether the values of r_s obtained in the previous three examples show significant evidence (at the 5% or 1% levels) to reject the hypothesis that the judgements were made at random. Use one-tailed tests since, in each case, we are searching for agreement rather than disagreement.

———————

For the 8 knobbly knees in Example 9, the value 0.786 exceeds the 5% point (0.643), but not the 1% point (0.833). The hypothesis that the judgements were made at random is rejected at the 5% level: the judges were evidently looking for similar knee characteristics!

For the 5 masses in Example 10, the two errors made were one error too many! The hypothesis that the judgements were made at random is not rejected at the 5% level.

The wine expert in Example 11 did indeed do well! Despite the tied ranks, the value 0.846 comfortably exceeds the nominal 1% critical value (0.783). The hypothesis that the judgements were made at random is rejected at the 1% level.

Practical ————————————————————————

How good are you at assessing differing masses? Experiment with 8 tubes of Smarties. Fill the tubes with different numbers of Smarties and then see if, after shuffling, you can rearrange the tubes in the correct order (i.e. in order of the numbers of Smarties). This is easier when the numbers differ considerably.

You can experiment to see how accurate you are at assessing mass. A Smartie typically weighs just under a gram.

Alternative table formats

Different tables present different information about the percentage points or critical values of r_s. Here are some examples:

◆ *Cambridge Elementary Mathematical Tables*, 2nd Edition, Miller and Powell (CUP)

An extract is as follows:

n	$\rho_{[.95]}$	$\rho_{[.99]}$
4	.800	
5	.800	.900
6	.771	.886

In this case the observed value is significant *if it is greater than* the value in the table (as opposed to 'equal to, or greater than'). As in the previous case, the exact significance level never exceeds the nominal value.

♦ *Elementary Statistical Tables*, Dunstan, Nix, Reynolds and Rowlands (RND Publications)

An extract is as follows:

One tail Two tail	10% 20%	5% 10%	2.5% 5%	1% 2%	0.5% 1%
4	1.0000	1.0000	1.0000	1.0000	1.0000
5	0.7000	0.9000	0.9000	1.0000	1.0000
6	0.6571	0.7714	0.8286	0.9429	0.9429

In this case 'the critical values given are those whose significance levels are nearest to the stated values'. This means that in some cases the tail probability corresponding to the tabulated value will greatly exceed the nominal tail probability.

♦ *Statistical Tables*, Murdoch and Barnes (Macmillan)

These tables have the same format as our own, but some values (for example, the 5% critical value for $n = 12$) are slightly in error.

♦ *New Cambridge Elementary Statistical Tables*, Lindley and Scott (CUP)

These tables have a very different format, using critical values of Σd_i^2 rather than of r_s. Here is an extract:

P	5	2.5	1	0.5	0.1	$\frac{1}{6}(n^3 - n)$
$n = 4$	0	–	–	–	–	10
5	2	0	0	–	–	20
6	6	4	2	0	–	35

The value of Σd_i^2 is significant at the stated level if the observed value is equal to, or less than, the value given in the table.

♦ *Understanding Statistics*, Upton and Cook (OUP)

In the Appendix (p. 626) we give separate tables of the one-tailed and two-tailed critical values. Here is an extract from the table of critical values for the two-tailed test:

n	5%	1%	n	5%	1%	n	5%	1%	n	5%	1%
4	*	*	7	.786	.929	10	.648	.794	13	.560	.703
5	1.000	*	8	.738	.881	11	.618	.755	14	.538	.679
6	.886	1.000	9	.700	.833	12	.587	.727	15	.521	.654

As before, an observed value is significant at the stated nominal level if it is equal to, or greater than, the tabulated value (which is given correct to 3 d.p.). The asterisks indicate that significance at these levels is not achievable for these sample sizes.

22.7 Using r_s for non-linear relationships

The product–moment correlation coefficient is concerned with the extent to which X and Y are *linearly* related. Suppose, however, that we had the following data:

x	1	2	3	4	5	6
y	1	8	27	64	125	216

The values *exactly* satisfy the simple relation $y = x^3$, but the product–moment correlation coefficient, r, is not equal to 1 because the relation is not linear (in fact, $r = 0.938$). By contrast, $r_s = 1$. Indeed $r_s = 1$ in all cases where y increases as x increases and $r_s = -1$ in all cases where y decreases as x increases.

Example 13

The size of a plant (y cm^2) is believed to be related to the distance of the plant from its nearest neighbour (x cm). A random sample of 12 plants is selected and their areas and nearest-neighbour distances are given in the table below.

x	35	41	86	15	23	66	27	39	91	44	52	18
y	54	58	100	50	48	75	51	60	115	62	64	30

Plot these data on a scatter diagram. Does there appear to be a relation between x and y?
Determine the value of Spearman's rank correlation coefficient for these data and determine whether that value differs from 0 at the 1% significance level using a two-tailed test.

———

The scatter diagram shows a clear relationship – as x increases, so (on the whole) y increases. The relationship might be quadratic rather than linear (especially since y is a measurement of area rather than distance).
    The two sets of ranks and their differences are as follows:

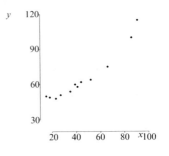

Rank of x	5	7	11	1	3	10	4	6	12	8	9	2
Rank of y	5	6	11	3	2	10	4	7	12	8	9	1
Difference	0	1	0	−2	1	0	0	−1	0	0	0	1

As required the differences in the ranks sum to zero, while $\Sigma d_i^2 = 8$. Thus:

$$r_s = 1 - \frac{6 \times 8}{12 \times 143} = 0.972$$

The 1% critical value for a two-tailed test with $n = 12$ is 0.727: there is therefore significant evidence of a relation between the two variables.

Exercises 22d

1 An expert wine taster was asked to rank 8
clarets in order of preference without knowing
the price. The results were as follows, the least
preferred being ranked 1.

Price	£2.90	£3.50	£3.80	£4.20
Taster	3	2	1	8

Price	£5.10	£5.80	£6.20	£7.10
Taster	7	6	4	5

Calculate Spearman's rank correlation
coefficient between the taster's ranking and
the price. Test the coefficient for significance
at the 5% level, stating your conclusions
clearly.   [UCLES(P)]

2 The yield (per hectare) of a crop, c, is
believed to depend on the May rainfall, m.
For nine regions records are kept of the
average values of c and m, and these are
recorded below

c	8.3	10.1	15.2	6.4	11.8
m	14.7	10.4	18.8	13.1	14.9

c	12.2	13.4	11.9	9.9
m	13.8	16.8	11.8	12.2

Calculate the value of ρ, Spearman's rank
correlation coefficient, for the above data and
determine whether it is significantly greater
than zero at the 5% level.   [UCLES(P)]

3 In a ski-jumping contest each competitor made
2 jumps. The orders of merit for the 10
competitors who completed both jumps are
shown in the table below.

Ski-jumper	A	B	C	D	E
First jump	2	9	7	4	10
Second jump	4	10	5	1	8

Ski-jumper	F	G	H	I	J
First jump	8	6	5	1	3
Second jump	9	2	7	3	6

(*a*) Calculate, to 2 decimal places, a rank
correlation coefficient for the performances
of the ski-jumpers in the two jumps.

(*b*) Using a 5% level of significance and quoting
from the tables of critical values provided,
interpret your result. State clearly your null
and alternative hypotheses.   [ULSEB(P)]

4 An expert on porcelain is asked to place 7
china bowls in date order of manufacture
assigning the rank 1 to the oldest bowl. The
actual dates of manufacture and the order
given by the expert are shown in the table
below.

Bowl	A	B	C	D
Date	1920	1857	1710	1896
Expert's order	7	3	4	6

Bowl	E	F	G
Date	1810	1690	1780
Expert's order	2	1	5

Find, to 3 decimal places, the Spearman rank
correlation coefficient between the order of
manufacture and the order given by the
expert.

Refer to one of the tables of critical values
provided to comment on the significance of
your result. State clearly the null hypothesis
which is being tested.   [ULSEB(P)]

5 A random sample of size 12 is taken from
those men who have at least one grown-up
son. The heights, to the nearest centimetre, of
the fathers and their eldest sons are given
below.

Father	190	184	183	182	179	178
Son	189	186	180	179	187	184

Father	175	174	170	168	165	164
Son	183	171	170	178	174	165

Calculate Spearman's rank correlation
coefficient between the heights of the fathers
and the sons.

It is subsequently revealed that, in copying
down the above data, the heights of two sons,
adjacent in the above list, had been accidentally
interchanged. Find, for Spearman's coefficient,
the maximum change that this mistake can
have made.   [UCLES(P)]

6 Briefly describe circumstances in which it is appropriate to use

(a) a rank correlation coefficient.

(b) the linear (product moment) correlation coefficient.

The following table gives measures of concentration span and spatial ability for 15 schoolchildren with reading difficulties.

Concentration span	15	16	17	19	20
Spatial ability	40	42	35	30	41

Concentration span	22	23	24	28	29
Spatial ability	23	22	29	31	28

Concentration span	30	31	32	34	38
Spatial ability	37	33	27	26	25

(i) Plot a scatter diagram and comment on its implication for the correlation between the two factors.

(ii) Calculate Spearman's rank correlation coefficient.

(iii) Stating any necessary assumptions about the data, test, at the 5% significance level, whether the data provide evidence of negative correlation between concentration span and spatial ability. [UCLES(P)]

7 When a dyslexic child is asked to write down two digits there is a probability p that the child will write the digits down in reverse order. In an experiment, a sequence of six digits, arranged in ascending order, is dictated to the child two digits at a time.

(i) Show that the probability that the child writes the sequence down in the correct order is $(1 - p)^3$.

(ii) Show that the least possible value of S, Spearman's rank correlation coefficient between the correct sequence and the sequence written by the child, is $\frac{29}{35}$.

(iii) Find, in terms of p, an expression for $E(S)$. [UCLES(P)]

22.8 r_s is the product–moment correlation coefficient for ranks

We will demonstrate that, when the x-values and the y-values both consist of permutations of the numbers 1 to n (as is the case for ranks), then the usual formula for r gives the same value as that for r_s.

We need the standard results that the sum of the numbers 1 to n is $\frac{1}{2}n(n + 1)$, and that the sum of their squares is $\frac{1}{6}n(n + 1)(2n + 1)$. Thus the quantities S_{xx} and S_{yy} are given by:

$$S_{xx} = S_{yy} = \frac{n(n + 1)(2n + 1)}{6} - \frac{1}{n}\left(\frac{n(n + 1)}{2}\right)^2 = \frac{n(n^2 - 1)}{12}$$

and hence:

$$\sqrt{S_{xx}S_{yy}} = \frac{n(n^2 - 1)}{12}$$

Also:

$$S_{xy} = \Sigma xy - \frac{(\Sigma x)(\Sigma y)}{n}$$

$$= \Sigma xy - \frac{n(n + 1)^2}{4}$$

while:

$$\Sigma d^2 = \Sigma(x - y)^2$$

$$= \Sigma x^2 - 2\Sigma xy + \Sigma y^2$$

$$= 2 \times \frac{n(n+1)(2n+1)}{6} - 2\Sigma xy$$

So that:

$$\Sigma xy = \frac{n(n+1)(2n+1)}{6} - \frac{\Sigma d^2}{2}$$

and hence:

$$S_{xy} = \frac{n(n+1)(2n+1)}{6} - \frac{\Sigma d^2}{2} - \frac{n(n+1)^2}{4}$$

$$= \frac{n(n^2-1)}{12} - \frac{\Sigma d^2}{2}$$

So:

$$r = \frac{1}{\sqrt{S_{xx}S_{yx}}} \times S_{xy}$$

$$= \frac{12}{n(n^2-1)} \times \left\{ \frac{n(n^2-1)}{12} - \frac{\Sigma d^2}{2} \right\}$$

$$= 1 - \frac{6\Sigma d^2}{n(n^2-1)}$$

$$= r_s$$

Calculator practice —————————————————————

The implication of this result is that if your calculator has statistical functions that include a button marked r, then, by entering the two sets of ranks as (x, y) values, the calculator will provide the value of r_s automatically. Try it!

22.9 Kendall's τ

Sir Maurice Kendall (1907–83) was the son of a Kettering publican. After failing to get a place in his local grammar school, he nevertheless won a scholarship to study mathematics at Cambridge. On graduating he initially worked at the Ministry of Agriculture, where his work involved the study of time series. Kendall's work at that time underlies the modern approach to the analysis of data of that type.

He also became interested in methods for measuring correlation and his initial paper on rank correlation was published in 1938, though 'Kendall's τ' only became widely used following the publication in 1948 of his book *Rank Correlation Methods*. This followed closely on his major work, *The Advanced Theory of Statistics*. He became Professor of Statistics at the London School of Economics in 1949, where he remained until 1961. From 1972 to 1980 he was director of the World Fertility Study. He was knighted for his services to Statistics.

Just as there are different measures of location (e.g. mean, median, mode) and spread (e.g. range, standard deviation), each of which is sensible, but each of which focuses on a slightly different aspect of the data, so there are different measures of rank correlation that represent different aspects of the rank orderings.

Kendall's idea was to examine the number of 'neighbour-swaps' needed to produce one rank ordering from another rank ordering. As an example, suppose that we have four objects, and the two orderings (A, B, D, C) and (D, B, A, C). To convert the second ordering into the first using neighbour-swaps we could proceed as follows:

> (D, B, A, C)   the second ordering
> (D, A, B, C)   swap A and B
> (A, D, B, C)   swap A and D
> (A, B, D, C)   swap B and D to give the first ordering

This is not the only possible sequence of neighbour-swaps. We could, for example, have used the sequence:

> (D, B, A, C)   the second ordering
> (B, D, A, C)   swap B and D
> (B, A, D, C)   swap A and D
> (A, B, D, C)   swap A and B to give the first ordering

Both sequences take 3 neighbour-swaps. You can experiment, but you will not be able to restore the first ordering from the second ordering using any fewer! The same number are required if we convert the first ordering into the second – simply read the previous sequences in reverse.

Denoting the minimum number of neighbour-swaps needed by Q, Kendall proposed using the measure τ, defined by:

$$\tau = 1 - \frac{4Q}{n(n-1)}$$

The symbol τ is the Greek letter 'tau' which is pronounced either like the first syllable of the word 'towel' or like the word 'tor'.

The swapping idea that underlies Kendall's τ is difficult to use with larger values of n. Fortunately, there is an easier way of counting the minimum number of neighbour-swaps! The procedure is as follows:

1 Rearrange the items in the order specified by the first ranking.
2 Write down the ranks assigned to these items by the second ranking to give a re-ordered second ranking.
3 Consider each number in the re-ordered second ranking in turn, and count how many numbers *of those to the right* are smaller than the number being considered.
4 Sum these numbers. This sum is the required minimum number of neighbour-swaps.

This is easier than it sounds! For example, consider the two orderings (A, B, D, C) and (D, B, A, C). The first item in the first ranking (A) was ranked 3 in the second ranking, while B was ranked 2, D was ranked 1 and C was ranked 4. The re-ordered second ranking is therefore (3, 2, 1, 4). The counts are as follows:

3	a count of 2 – the numbers 2 and 1
2	a count of 1 – the number 1
1	a count of 0 – the number to the right is larger
4	a count of 0 – there are no numbers to the right!
Total	$Q = 3$

Here are two more examples of re-ordered second rankings:

Ranking	(4, 1, 5, 6, 3, 2)	(1, 5, 7, 4, 3, 6, 2)
count	3, 0, 2, 2, 1, 0	0, 3, 4, 2, 1, 1, 0
Q	8	11
τ	−0.067	−0.048

An equivalent way of obtaining the sum of the counts uses a diagram. Suppose, for example, that the items A, B, C, D and E are ranked in the order C, E, A, B and D by one judge and in the order C, D, A, E and B by another judge. In a diagram, we arrange the two orderings one above the other, join the corresponding items and count the number of crossings (which is Q).

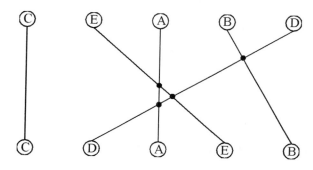

In this case $Q = 4$ and hence $\tau = 0.2$.

Without using a diagram, we could write the items down in the order given by the first judge, determine the corresponding ranks given by the second judge and hence the counts and their sum:

Item	C	E	A	B	D	Sum
First judge	1	2	3	4	5	
Second judge	1	4	3	5	2	
Counts	0	2	1	1	0	4

Alternatively (but equivalently) we could write the items down in the order specified by the second judge and proceed accordingly:

Item	C	D	A	E	B	Sum
Second judge	1	2	3	4	5	
First judge	1	5	3	2	4	
Counts	0	3	1	0	0	4

Notes

- Use letters and *not* numbers to identify the individuals being ranked. There are quite enough numbers flying about already!
- The last of the counts must always be a zero.

Testing the significance of τ

The problems associated with the discreteness of the distribution of r_s apply equally to τ. An extract from the table, given in the Appendix (p. 627), of (one-tailed) critical values for τ is given below.

n	5%	1%	n	5%	1%	n	5%	1%	n	5%	1%
4	1.000	*	7	.619	.810	10	.467	.600	13	.359	.513
5	.800	1.000	8	.571	.714	11	.418	.564	14	.363	.473
6	.733	.867	9	.500	.667	12	.394	.545	15	.333	.467

If the observed value is greater than or equal to the value in the tables (which is given correct to 3 d.p.) then the observed value is significant at the nominal (one-tailed) 5% or 1% level. The actual significance level will never be larger than the nominal one.

Notes

- Kendall's τ is sometimes denoted by r_k.
- Kendall's τ may be used with two-tailed tests as well as 1-tailed tests.
- Kendall's τ is usually smaller in magnitude than Spearman's r_s.
- As for r_s, tables of τ appear in a variety of forms and you should make sure you understand how to use the tables available in the examinations. For $n > 10$ a normal approximation suffices (using an appropriate continuity correction). For Q, the approximation uses the following distribution:

$$N\left(\tfrac{1}{4}n(n-1), \tfrac{1}{72}n(n-1)(2n+5)\right)$$

Example 14

The following data refer to the amounts of haemoglobin (in g/dl) and the numbers of red blood cells (in hundred million per cl) in samples of blood taken from mothers during labour.

Mother	A	B	C	D	E	F	G	H	I	J
Haemoglobin, x	11.7	14.2	13.7	13.5	14.6	13.8	13.9	11.4	11.6	13.6
Red blood cells, y	349	449	454	441	468	476	473	448	397	496

The hypotheses are H_0: the variables are independent of one another, and H_1: there is a relation between the variables leading to positively correlated ranks. Determine whether there is significant evidence to reject H_0 in favour of H_1 at the 5% level using (i) Spearman's r_s, (ii) Kendall's τ.

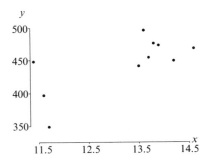

We begin with a quick look at the data, both to get an idea of the likely strength and sign of the correlation and also to check for any outlying values which might need investigation before proceeding with the calculations.

(i) To calculate Spearman's r_s we need the differences in the ranks:

Mother	A	B	C	D	E	F	G	H	I	J	Total
Haemoglobin rank	3	9	6	4	10	7	8	1	2	5	
Red blood cells rank	1	5	6	3	7	9	8	4	2	10	
d	2	4	0	1	3	-2	0	-3	0	-5	0
d^2	4	16	0	1	9	4	0	9	0	25	68

Hence $r_s = 1 - \dfrac{6 \times 68}{10 \times 99} = 0.588$. This exceeds the one-tail 5% point (0.564), so, using Spearman's coefficient we conclude that there is significant evidence for the rejection of H_0 in favour of H_1.

(ii) For Kendall's τ we again look at the ranks of the mothers. Using the haemoglobin levels, the rank order is (H, I, A, D, J, C, F, G, B, E) while for the red blood cells it is (A, I, D, H, B, C, E, G, F, J). We count the number of swaps required either using a diagram, or by re-ordering the data.

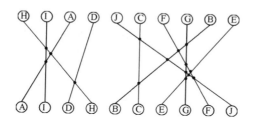

Using the diagram we can see that the mothers divide into two groups, with mothers A, D, H and I showing low values for both quantities. Although we have used straight lines in the diagram, you may find it easier to use curves, so as to avoid multiple crossings. Each crossing must involve two lines only. Here there are 15 crossings and hence $Q = 15$ and $\tau = 1 - \dfrac{4 \times 15}{10 \times 9} = \dfrac{1}{3}$.

Using the counting procedure we begin by ordering the mothers according to one of the rankings. Using the haemoglobin ranking we get:

Haemoglobin rank	1	2	3	4	5	6	7	8	9	10	Total
Mother	H	I	A	D	J	C	F	G	B	E	
Red blood cells rank	4	2	1	3	10	6	9	8	5	7	
Count	3	1	0	0	5	1	3	2	0	0	15

Thus $Q = 15$ and $\tau = \dfrac{1}{3}$, as before. Since $\dfrac{1}{3}$ does not exceed the one-tailed 5% critical value (0.467), using Kendall's τ we cannot conclude that there is significant evidence for the rejection of H_0 in favour of H_1.

Note

♦ The fact that the two test results appear to contradict one another does not mean that one test is better than the other, nor that one test is wrong. The tests are based on quantities (Σd_i^2 and Q) which are affected in different ways by perturbations of the rankings. At the 5% significance level (leaving aside the problems of discreteness), when two rankings are chosen at random, and the null hypothesis is true, each test will reject the null hypothesis on (about) 5% of occasions – but they will not usually be *exactly* the same 5%.

Practical

Here is another practical about the judgement of time. Work in pairs, with one person using an accurate watch and the other person guessing when some period of time has elapsed. Choose similar length periods.

The following sequence of time intervals (in seconds) should work well:
18, 20, 17, 15, 19, 16.
    The person being timed should judge each interval in turn without being informed how well or badly he or she has done. Calculate the rank correlation between the guesses and the true values.

Exercises 22e

1 An expert wine taster was asked to rank 8 clarets in order of preference without knowing the price. The results were as follows, the least preferred being ranked 1.

Price	£2.90	£3.50	£3.80	£4.20
Taster	3	2	1	8

Price	£5.10	£5.80	£6.20	£7.10
Taster	7	6	4	5

Calculate Kendall's rank correlation coefficient between the taster's ranking and the price.

Test the coefficient for significance at the 5% level, stating your conclusions clearly.
[UCLES(P)]

2 The yield (per hectare) of a crop, c, is believed to depend on the May rainfall, m. For nine regions records are kept of the average values of c and m, and these are recorded below.

c	8.3	10.1	15.2	6.4	11.8
m	14.7	10.4	18.8	13.1	14.9

c	12.2	13.4	11.9	9.9
m	13.8	16.8	11.8	12.2

Calculate the value of τ, Kendall's rank correlation coefficient, for the above data and determine whether it is significantly greater than zero at the 5% level. [UCLES(A)]

3 In a ski-jumping contest each competitor made 2 jumps. The orders of merit for the 10 competitors who completed both jumps are shown in the table below.

Ski-jumper	A	B	C	D	E
First jump	2	9	7	4	10
Second jump	4	10	5	1	8

Ski-jumper	F	G	H	I	J
First jump	8	6	5	1	3
Second jump	9	2	7	3	6

(a) Calculate, to 2 decimal places, Kendall's rank correlation coefficient for the performances of the ski-jumpers in the two jumps.

(b) Using a 5% level of significance and quoting from the tables of critical values provided, interpret your result. State clearly your null and alternative hypotheses. [ULSEB(A)]

4 An expert on porcelain is asked to place 7 china bowls in date order of manufacture assigning the rank 1 to the oldest bowl. The actual dates of manufacture and the order given by the expert are shown in the table below.

Bowl	A	B	C	D
Date	1920	1857	1710	1896
Expert's order	7	3	4	6

Bowl	E	F	G
Date	1810	1690	1780
Expert's order	2	1	5

Find, to 3 decimal places, the Kendall's rank correlation coefficient between the order of manufacture and the order given by the expert.

Comment on the significance of your result. State clearly the null hypothesis which is being tested. [ULSEB(A)]

5 A random sample of size 12 is taken from those men who have at least one grown-up son. The heights, to the nearest centimetre, of the fathers and their eldest sons are given below.

Father	190	184	183	182	179	178
Son	189	186	180	179	187	184

Father	175	174	170	168	165	164
Son	183	171	170	178	174	165

Calculate Kendall's rank correlation coefficient between the heights of the fathers and the sons.
(continued)

It is subsequently revealed that, in copying down the above data, the heights of two sons, adjacent in the above list, had been accidentally interchanged. Find, for Kendall's coefficient, the maximum change that this mistake can have made. [UCLES(P)]

6 The following table gives measures of concentration span and spatial ability for 15 schoolchildren with reading difficulties.

Concentration span	15	16	17	19	20
Spatial ability	40	42	35	30	41

Concentration span	22	23	24	28	29
Spatial ability	23	22	29	31	28

Concentration span	30	31	32	34	38
Spatial ability	37	33	27	26	25

Calculate Kendall's rank correlation coefficient.

Stating any necessary assumptions about the data, test, at the 5% significance level, whether the data provide evidence of negative correlation between concentration span and spatial ability.

[UCLES(P)]

7 There are eight finalists (*A* to *H*) in a music competition. After hearing them play I decide that their order of merit is *C* first, *E* second and the remainder in the order *G*, *A*, *B*, *D*, *F*, *H*. Later the judges announce that *G* has won the competition and that the final order was *G*, *C*, *A*, *D*, *E*, *B*, *H*, *F*. Determine the value of Kendall's rank correlation coefficient between my order of merit and that of the judges, and state whether it is significant at the 5% level.

[UCLES(P)]

8 Five sacks of logs are labelled *A*, *B*, *C*, *D* and *E*. The lightest sack is sack *A*, with sack *B* being the next lightest, and so on. A weight-lifter is asked to arrange the sacks in order of increasing mass. However, the weight-lifter simply arranges the three lighter sacks in a random order followed by the two heavier sacks in a random order. Find the probability that the sacks have been arranged in the correct order.

Denote by *K* the value of Kendall's coefficient of rank correlation between the weight-lifter's ordering and the correct ordering. By listing all possible orderings, or otherwise,

(i) show that $P(K = 0.4) = 0.25$,

(ii) tabulate the distribution of *K*, and find its mean and variance.

A sixth sack, *F*, is now delivered. This sack is much lighter than the others, and the weight-lifter places this sack in the correct position. Determine the new mean value of *K*. [UCLES]

9 Five brands of tea are bought from a supermarket and ranked according to their purchase prices (all distinct). A tea-taster ranks them according to her own preference. Determine how many different rankings by the tester will lead to a value of 1 or 0.8 for τ, Kendall's rank correlation coefficient. State the total number of different possible rankings by the taster.

Given that a value of 0.6 for τ may be obtained in 9 ways, evaluate $P(\tau \geqslant 0.8)$ and $P(\tau \geqslant 0.6)$, assuming that all rankings by the taster are equally probable.

The actual rankings by price and taste are given in the following table.

Tea brand	A	B	C	D	E
Price	2	1	4	3	5
Taste	3	1	5	2	4

Calculate the value of Kendall's rank correlation coefficient for this data set.

Using the results obtained above, test, at the 10% significance level, whether there is a positive correlation between price and taste.

[UCLES]

Chapter summary

♦ **The single-sample sign test:**
Based on the number of values greater than the median value specified by H_0. Assuming H_0, this number is an observation from a $B(n, 0.5)$ distribution.

♦ **The Wilcoxon signed-rank test:**
This requires a symmetric distribution. The values of differences of observations from the median specified by the null hypothesis are replaced by ranks, with the signs of the differences being retained. Denoting the sum of the positive ranks by P and of the negative ranks by Q, the statistic T is defined as the smaller of P and Q. The test statistic, z, is defined by:

$$z = \frac{\frac{1}{4}n(n+1) - T - \frac{1}{2}}{\sqrt{\frac{1}{24}n(n+1)(2n+1)}}$$

♦ **The paired-sample sign test:**
Assuming H_0: medians are same, the number of plus signs in the list of differences has a $B(n, 0.5)$ distribution.

♦ **The Wilcoxon matched-pairs signed-rank test:**
The within-pair differences are ranked in order of magnitude, with the signs being retained. The test is then equivalent to its single-sample version with hypothesised median 0.

♦ **The Wilcoxon rank-sum test:**
With samples of sizes m and n ($m \leqslant n$), denote the rank-sum for the sample of size m by R. The test statistic, z, is defined by:

$$z = \frac{R - \frac{1}{2}m(m+n+1) \pm \frac{1}{2}}{\sqrt{\frac{1}{12}mn(m+n+1)}}$$

♦ **Spearman's rank correlation coefficient, r_s:**
Let d_i be the difference between the two ranks assigned to item i. Then:

$$r_s = 1 - \frac{6\Sigma d_i^2}{n(n^2 - 1)}$$

♦ **Kendall's rank correlation coefficient, τ:**
Let Q be the minimum number of neighbour-swaps required to change one ranking into the other. Then:

$$\tau = 1 - \frac{4Q}{n(n-1)}$$

Exercises 22f (Miscellaneous)

1 The following table gives the standardised mortality rates (deaths per 1000 persons, corrected for changes in age structure) and the percentage of marriages in England and Wales that were carried out by the Church of England.

Date	1870	1875	1880
Mortality rate	21.5	21.5	19.5
Marriage %	75.6	74.2	72.2

Date	1885	1890	1895
Mortality rate	18.7	19.3	18.5
Marriage %	70.4	69.8	68.4

Date	1900	1905	1910
Mortality rate	18.2	15.3	13.4
Marriage %	66.7	63.6	61.3

(i) Determine the value of the product–moment correlation coefficient, r, for the relation between mortality rate and percentage of marriages conducted by the Church of England.

(ii) Determine the values of Spearman's rank correlation coefficient, r_s, for each of the pairs (Date, Mortality rate) and (Date, Marriage percentage).

(iii) Comment on your results.

2 A student of statistics was heard to say that a rank correlation coefficient gives an approximation to the linear (product moment) coefficient. Sketch a scatter diagram to illustrate a situation where this statement is incorrect.

It is suspected that students who finish mathematics examinations early tend to achieve better results than those who do not. The following data show the times taken (t minutes) and scores (s) of a sample of 12 randomly selected students who sat a mathematics examination.

t	25	28	32	35	40	44
s	81	59	88	79	95	86

t	47	49	52	55	59	60
s	38	44	54	48	68	73

$$[\Sigma t = 526, \ \Sigma t^2 = 24\,614, \ \Sigma s = 813,$$
$$\Sigma s^2 = 59\,001, \ \Sigma st = 34\,624]$$

(i) Calculate, for the sample, the variances of s and t. Calculate also $\Sigma(s - \bar{s})(t - \bar{t})$ and hence or otherwise calculate the linear (product moment) correlation coefficient.

(ii) Calculate Spearman's rank correlation coefficient for the sample.

Hence test, at the 5% significance level, the hypothesis that students who finish early tend to do better than those who finish later.

[UCLES]

3 It is not always possible for women to conceive naturally and one possible treatment is that of 'in vitro' fertilisation where embryos are transferred after fertilisation has taken place outside the human body. The following data give the results of such treatment at 10 randomly selected clinics, showing the numbers of embryo transfers (x) and of pregnancies (y).

No. of embryo transfers (x)	5	9	11	19	27
No. of pregnancies (y)	2	1	3	5	8

No. of embryo transfers (x)	40	52	63	80	90
No. of pregnancies (y)	10	13	14	12	9

Correct to 3 decimal places, the data are summarised by

$$\Sigma\sqrt{x} = 58.012, \ \Sigma x = 396.000, \ \Sigma\sqrt{y} = 26.184$$
$$\Sigma y = 77.000, \ \Sigma\sqrt{xy} = 171.493.$$

(i) Plot a scatter diagram of $\sqrt{y}$ against $\sqrt{x}$, and comment on the apparent relation between $\sqrt{y}$ and $\sqrt{x}$.

(ii) Calculate the linear (product moment) correlation coefficient between $\sqrt{y}$ and $\sqrt{x}$, and comment on its value.

(iii) Calculate the value of Spearman's (or Kendall's) rank correlation coefficient between $\sqrt{y}$ and $\sqrt{x}$. Stating your null and alternative hypotheses, test the significance of the coefficient, using a 1% significance level.

[UCLES]

4   Explain why it is better to use a Wilcoxon matched-pairs signed rank test, rather than a sign test, to test for a difference between two populations.

The task completion times, in minutes, for a random sample of 12 operatives using two different methods are given in the table below.

Operative	1	2	3	4	5	6
Method A	9.1	8.6	8.2	9.0	8.7	9.1
Method B	8.4	8.8	7.6	9.4	9.2	8.2

Operative	7	8	9	10	11	12
Method A	9.5	8.9	10.0	9.6	9.5	8.3
Method B	9.4	7.9	8.7	8.4	8.7	8.0

(i)   Use both tests mentioned above to test, at the 5% significance level, whether Method B results in a smaller median completion time than Method A. Comment on the results.

(ii) Test, at the 5% significance level, whether the median time for Method A is 8.35 minutes.   [UCLES]

5   Explain why it is advisable to plot a scatter diagram before interpreting a correlation coefficient calculated for a sample drawn from a bivariate distribution.

Sketch rough scatter diagrams indicating the following

(i)   a linear (product moment) coefficient close to zero but with an obvious relation between the variables,

(ii) a non-linear relation between the variables yielding a rank correlation coefficient of +1.

If a sample correlation coefficient has a value close to +1 or to −1 what further information is needed before it can be decided whether a relationship between the variables is indicated?

It is hypothesised that there is a positive correlation between the population of a country and its area. The following table gives a random sample of 13 countries with their area x, in thousand km^2, and population y, in millions.

Country	1	2	3	4	5
x	2.5	28	30	72	98
y	0.5	5	2	4	42

Country	6	7	8	9	10
x	121	128	176	239	313
y	21	16	3	14	37

Country	11	12	13
x	407	435	538
y	6	17	22

Plot a scatter diagram and comment on its implication for the hypothesis.

Calculate a suitable correlation coefficient and test its significance at the 5% level.   [UCLES]

6   The ages, in months, and the weights, in kg, of a random sample of 9 babies are shown in the table.

Baby	A	B	C	D	E
Age (x)	1	2	2	3	3
Weight (y)	4.4	5.2	5.8	6.4	6.7

Baby	F	G	H	I
Age (x)	3	4	4	5
Weight (y)	7.2	7.6	7.9	8.4

(a)   Calculate, to 3 decimal places, the product-moment correlation coefficient between weight and age for these babies. Give a brief interpretation of your result.

(b)   Find the equation $y = ax + b$, of the regression line of weight on age for this sample, giving the coefficients a and b to 3 decimal places. Interpret the meaning of your values of a and b.

(c)   Use this equation to estimate the mean weight of a baby aged 6 months.

State any reservations you have about this estimate, giving your reasons.

A boy who does not know the weights or ages of these babies is asked to list them, by guesswork, in order of increasing weight. He puts them in the order:

A C E B G D I F H

(continued)

(d) Obtain, to 3 decimal places, a rank correlation coefficient between the boy's order and the true weight order.

(e) Referring to the tables provided and using a 5% significance level, discuss any conclusions you draw from your result.

[ULSEB]

7 Explain how you used, or could have used, a correlation coefficient to analyse the results of an experiment. State briefly when it is appropriate to use a rank correlation coefficient rather than a product-moment correlation coefficient.

Seven rock samples taken from a particular locality were analysed. The percentages, C and M, of two oxides contained in each sample were recorded. The results are shown in the table.

Sample	1	2	3	4
C	0.60	0.42	0.51	0.56
M	1.06	0.72	0.94	1.04

Sample	5	6	7
C	0.31	1.04	0.80
M	0.84	1.16	1.24

Given that

$$\Sigma CM = 4.459, \quad \Sigma C^2 = 2.9278, \quad \Sigma M^2 = 7.196$$

find, to 3 decimal places, the product-moment correlation coefficient of the percentages of the two oxides.

Calculate also, to 3 decimal places, a rank correlation coefficient.

Using the tables provided state any conclusions which you draw from the value of your rank correlation coefficient. State clearly the null hypothesis being tested. [ULSEB]

8 Some children were asked to eat a variety of sweets and classify each one on the following scale:

strongly dislike/dislike/neutral/like/like very much.

This was then converted to a numerical scale 0, 1, 2, 3, 4 with 0 representing "strongly dislike". A similar method produced a score on the scale 0, 1, 2, 3 for the sweetness of each sweet assessed by each child (the sweeter the sweet the higher the score).

The following frequency distribution resulted

		liking				
		0	1	2	3	4
	0	5	2	0	0	0
sweetness	1	3	14	16	9	0
	2	8	22	42	29	37
	3	3	4	36	58	64

(a) Calculate the product moment correlation coefficient for these data. Comment briefly on the data and on the correlation coefficient.

(b) A child was asked to rank 7 sweets according to preference and sweetness with the following results:

RANKS

Sweet	A	B	C	D	E	F	G
Preference	3	4	1	2	6	5	7
Sweetness	2	3	4	1	5	6	7

Calculate Spearman's rank correlation coefficient for these data.

(c) It is suggested that the product moment correlation coefficient should be calculated for (b) and Spearman's rank correlation coefficient for (a). Comment on this suggestion. [AEB 91]

9 A machine-hire company kept records of the age, X months, and the maintenance costs, £Y, of one type of machine. The following table summarises the data for a random sample of 10 of the machines.

Machine	A	B	C	D	E
Age, x	63	12	34	81	51
Maintenance cost, y	111	25	41	181	64

Machine	F	G	H	I	J
Age, x	14	45	74	24	89
Maintenance cost, y	21	51	145	43	241

(a) Calculate, to 3 decimal places, the product moment correlation coefficient.
(You may use $\Sigma x^2 = 30\,625$, $\Sigma y^2 = 135\,481$, $\Sigma xy = 62\,412$.)

(b) Calculate, to 3 decimal places, the Spearman rank correlation coefficient.

(*continued*)

(c) For a different type of machine similar data were collected. From a large population of such machines a random sample of 10 was taken and the Spearman rank correlation coefficient, based on $\Sigma d^2 = 36$ was 0.782.

Using a 5% level of significance and quoting from the tables of critical values provided, interpret this rank correlation coefficient. Use a two-tailed test and state clearly your null and alternative hypotheses.   [ULSEB(P)]

10 The following table gives the marks obtained by 9 randomly chosen students in their German and Mathematics examinations.

Mathematics mark	78	59	65	70	90
German mark	68	37	73	65	59

Mathematics mark	42	35	502	58
German mark	55	47	42	52

It is desired to test whether there is positive correlation between marks in Mathematics and German. Discuss, giving reasons, whether a rank correlation coefficient or a linear (product moment) correlation coefficient would be more appropriate.

Calculate

(i) Spearman's rank correlation coefficient.

(ii) Kendall's rank correlation coefficient.

Show that when testing for positive correlation at the 5% level one of these coefficients provides significant evidence whereas the other does not.

Comment on the apparent contradiction.   [UCLES]

11 Within a particular branch of industry a random selection of ten firms is chosen for detailed study. For 1988 their numbers of employees (in hundreds), x, and the average annual salaries of their eight most senior staff (in thousands of pounds), y, are recorded in the table below.

x	2.2	15.3	4.7	17.3	12.3
y	29.3	33.9	28.6	33.8	34.7

x	186.8	53.4	312.6	19.1	3.7
y	43.2	38.2	47.6	37.6	29.1

Calculate Kendall's rank correlation coefficient for these data. Test, at the 2% significance level, the hypothesis that there is no correlation between the number of employees and the average salary of the eight most senior staff.

The corresponding figures for 1989 are given below, with the firms in the same order as before.

x	2.4	14.9	4.8	17.6	13.0
y	29.8	33.8	31.2	36.4	34.8

x	194.7	56.1	318.0	18.6	4.3
y	41.8	38.7	48.8	39.2	31.0

Use the Wilcoxon matched-pairs signed rank test to determine whether there is significant evidence, at the 5% level, that the number of employees in the firms in this industry increased between 1988 and 1989.

Use an appropriate t-test to determine whether there is significant evidence, at the 5% level, that the mean senior staff salary for firms in this industry increased between 1988 and 1989.   [UCLES]

12 Candidates entered for a public examination first take a 'mock' examination of the same standard which is prepared and marked by their own teachers. The marks obtained, in both the mock and public examinations, by a random sample of 20 students from a school are shown below.

Candidate	1	2	3	4	5	6	7
Mock exam.	40	80	65	50	47	71	70
Public exam.	45	77	68	61	48	75	73

Candidate	8	9	10	11	12	13	14
Mock exam.	30	15	53	42	8	35	71
Public exam.	35	25	47	49	15	36	73

Candidate	15	16	17	18	19	20
Mock exam.	63	65	29	33	79	87
Public exam.	65	68	35	39	75	88

(*continued*)

(i) It is decided to use a sign test to test at the $100\alpha\%$ level whether there is a significant difference between the marks in these two examinations. Determine the set of values for α that would lead to a conclusion that a significant difference does exist. State your hypotheses clearly.

(ii) It is believed that in the public examination 75% of all the candidates get a mark of at least 40. On this assumption, find the probability that out of 20 candidates, chosen at random from all candidates, not more than 14 get a mark of at least 40.

(iii) Explain why it would be incorrect to use the results from a particular school to test whether or not 75% of all candidates got a mark of at least 40 in the public examination. [UCLES]

13 A large number of candidates took two papers, Paper 1 and Paper 2, in the same subject. From a cursory examination of the papers a teacher concluded that Paper 2 was easier than Paper 1. The teacher then took a random sample of 8 candidates and compared their marks, which were as follows.

Candidate	1	2	3	4	5	6	7	8
Paper 1	29	49	37	61	69	61	13	45
Paper 2	66	58	82	94	38	90	26	66

Test the teacher's opinion, at the 5% level, using either the sign test or the Wilcoxon matched-pairs signed rank test, whichever you think more appropriate. Justify your choice of test.

Given that the marks on each of the two papers are approximately normally distributed, carry out a paired-sample t-test.

State in what way the results of the two tests you have carried out would be affected if

(i) each of the marks on the two papers were increased by 5,

(ii) each of the marks on Paper 1 were increased by 5. [UCLES]

14 The telephone authority is interested in the possible effect of an increase in telephone charges, which it believes may result in a reduction in the number and length of calls. In order to investigate the effect, a random sample of 10 houses is chosen and, for each house, the number of telephone calls, n, made from that house is noted. In addition, for each house, the total length x (in minutes) of a random selection of twelve of the calls is calculated. The values of n and x for the ten houses for the months before and after the increase in telephone charges are given in the table below.

House		1	2	3	4	5
Month before increase	n	121	16	20	61	27
	x	47	40	51	32	62
Month after increase	n	107	14	24	52	21
	x	43	38	39	33	51

House		6	7	8	9	10
Month before increase	n	19	14	80	31	41
	x	42	56	58	49	50
Month after increase	n	16	13	62	36	34
	x	43	51	55	44	50

All the x-values for a given month may be assumed to be observations from the same normal distribution, but the n-values are not normally distributed. Test, at the 5% significance level,

(i) the hypothesis that there has been no decrease in the number of calls made,

(ii) the hypothesis that there has been no decrease in the length of the calls made. [UCLES]

15 (a) What is meant by saying that the (product moment) correlation coefficient is independent of the scale of measurement?

(b) Ten architects each produced a design for a new building and two judges, A and B, independently awarded marks, x and y respectively, to the ten designs, as given in the table below.

Design	1	2	3	4	5
Judge A (x)	50	35	55	60	85
Judge B (y)	46	26	48	44	62

Design	6	7	8	9	10
Judge A (x)	25	65	90	45	40
Judge B (y)	28	30	60	34	42

(continued)

It is given that $\Sigma x = 550$, $\Sigma x^2 = 34\,150$, $\Sigma y = 420$, $\Sigma y^2 = 19\,080$, $\Sigma xy = 25\,020$.

Calculate the (product moment) correlation coefficient between the marks awarded by the two judges.

Calculate also Spearman's rank correlation coefficient for the data and test, at the 1% level, the hypothesis that there is no correlation between the marks awarded by the two judges.

Discuss briefly the relative merits of using these two different correlation coefficients with this particular set of data.   [UCLES]

Appendix

Cumulative probabilities for the binomial distribution

For the given values of n and p, the table gives the values of $P(X \leqslant r)$.

n	r	p									
		0.05	0.10	0.15	0.20	0.25	0.30	0.35	0.40	0.45	0.50
2	0	.9025	.8100	.7225	.6400	.5625	.4900	.4225	.3600	.3025	.2500
	1	.9975	.9900	.9775	.9600	.9375	.9100	.8775	.8400	.7975	.7500
3	0	.8574	.7290	.6141	.5120	.4219	.3430	.2746	.2160	.1664	.1250
	1	.9928	.9720	.9393	.8960	.8438	.7840	.7183	.6480	.5748	.5000
	2	.9999	.9990	.9966	.9920	.9844	.9730	.9571	.9360	.9089	.8750
4	0	.8145	.6561	.5220	.4096	.3164	.2401	.1785	.1296	.0915	.0625
	1	.9860	.9477	.8905	.8192	.7383	.6517	.5630	.4752	.3910	.3125
	2	.9995	.9963	.9880	.9728	.9492	.9163	.8735	.8208	.7585	.6785
	3		.9999	.9995	.9984	.9961	.9919	.9850	.9744	.9590	.9375
5	0	.7738	.5905	.4437	.3277	.2373	.1681	.1160	.0778	.0503	.0313
	1	.9774	.9185	.8352	.7373	.6328	.5282	.4284	.3370	.2562	.1875
	2	.9988	.9914	.9734	.9421	.8965	.8369	.7648	.6826	.5931	.5000
	3		.9995	.9978	.9933	.9844	.9692	.9460	.9130	.8688	.8125
	4			.9999	.9997	.9990	.9976	.9947	.9898	.9815	.9688
6	0	.7351	.5314	.3771	.2621	.1780	.1176	.0754	.0467	.0277	.0156
	1	.9672	.8857	.7765	.6554	.5339	.4202	.3191	.2333	.1636	.1094
	2	.9978	.9842	.9527	.9011	.8306	.7443	.6471	.5443	.4415	.3438
	3	.9999	.9987	.9941	.9830	.9624	.9295	.8826	.8208	.7447	.6563
	4		.9999	.9996	.9984	.9954	.9891	.9777	.9590	.9308	.8906
	5				.9999	.9998	.9993	.9982	.9959	.9917	.9844
7	0	.6983	.4783	.3206	.2097	.1335	.0824	.0490	.0280	.0152	.0078
	1	.9556	.8503	.7166	.5767	.4449	.3294	.2338	.1586	.1024	.0625
	2	.9962	.9743	.9262	.8520	.7564	.6471	.5323	.4199	.3164	.2266
	3	.9998	.9973	.9879	.9667	.9294	.8740	.8002	.7102	.6083	.5000
	4		.9998	.9988	.9953	.9871	.9712	.9444	.9037	.8471	.7734
	5			.9999	.9996	.9987	.9962	.9910	.9812	.9643	.9375
	6					.9999	.9998	.9994	.9984	.9963	.9922
8	0	.6634	.4305	.2725	.1678	.1001	.0576	.0319	.0168	.0084	.0039
	1	.9428	.8131	.6572	.5033	.3671	.2553	.1691	.1064	.0632	.0352
	2	.9942	.9619	.8948	.7969	.6785	.5518	.4278	.3154	.2201	.1445
	3	.9996	.9950	.9786	.9437	.8862	.8059	.7064	.5941	.4770	.3633
	4		.9996	.9971	.9896	.9727	.9420	.8939	.8263	.7396	.6367
	5			.9998	.9988	.9958	.9887	.9747	.9502	.9115	.8555
	6				.9999	.9996	.9987	.9964	.9915	.9819	.9648
	7						.9999	.9998	.9993	.9983	.9961
10	0	.5987	.3487	.1969	.1074	.0563	.0282	.0135	.0060	.0025	.0010
	1	.9139	.7361	.5443	.3758	.2440	.1493	.0860	.0464	.0233	.0107
	2	.9885	.9298	.8202	.6778	.5256	.3828	.2616	.1673	.0996	.0547
	3	.9990	.9872	.9500	.8791	.7759	.6496	.5138	.3823	.2660	.1719
	4	.9999	.9984	.9901	.9672	.9219	.8497	.7515	.6331	.5044	.3770
	5		.9999	.9986	.9936	.9803	.9527	.9051	.8338	.7384	.6230
	6			.9999	.9991	.9965	.9894	.9740	.9452	.8980	.8281
	7				.9999	.9996	.9984	.9952	.9877	.9726	.9453
	8						.9999	.9995	.9983	.9955	.9893
	9								.9999	.9997	.9990

Missing values correspond to probabilities equal to 1.000, correct to 4 decimal places.

Cumulative probabilities for the Poisson distribution

For the given values of λ, the table gives the values of $P(X \leqslant r)$.

r	λ									
	0.1	0.2	0.3	0.4	0.5	0.6	0.7	0.8	0.9	1.0
0	.9048	.8187	.7408	.6703	.6065	.5488	.4966	.4493	.4066	.3679
1	.9953	.9825	.9631	.9384	.9098	.8781	.8442	.8088	.7725	.7358
2	.9998	.9989	.9964	.9921	.9856	.9769	.9659	.9526	.9371	.9197
3		.9999	.9997	.9992	.9982	.9966	.9942	.9909	.9865	.9810
4				.9999	.9998	.9996	.9992	.9986	.9977	.9963
5							.9999	.9998	.9997	.9994
6										.9999

r	λ									
	1.2	1.4	1.6	1.8	2.0	2.5	3.0	3.5	4.0	5.0
0	.3012	.2466	.2019	.1653	.1353	.0821	.0498	.0302	.0183	.0067
1	.6626	.5918	.5249	.4628	.4060	.2873	.1991	.1359	.0916	.0404
2	.8795	.8335	.7834	.7306	.6767	.5438	.4232	.3208	.2381	.1247
3	.9662	.9463	.9212	.8913	.8571	.7576	.6472	.5366	.4335	.2650
4	.9923	.9857	.9763	.9636	.9473	.8912	.8153	.7254	.6288	.4405
5	.9985	.9968	.9940	.9896	.9834	.9580	.9161	.8576	.7851	.6160
6	.9997	.9994	.9987	.9974	.9955	.9858	.9665	.9347	.8893	.7622
7		.9999	.9997	.9994	.9989	.9958	.9881	.9733	.9489	.8666
8				.9999	.9998	.9989	.9962	.9901	.9786	.9319
9						.9997	.9989	.9967	.9919	.9682
10						.9999	.9997	.9990	.9972	.9863
11							.9999	.9997	.9991	.9945
12								.9999	.9997	.9980
13									.9999	.9993
14										.9998
15										.9999

Missing values correspond to probabilities equal to 1.0000, correct to 4 decimal places.

The normal distribution function

The table gives the values of $\Phi(Z) = \mathrm{P}(Z \leqslant z)$, where Z has a normal distribution with mean 0 and variance 1.

z	0	1	2	3	4	5	6	7	8	9	1	2	3	4	5	6	7	8	9
					$\Phi(Z)$										ADD				
0.0	.5000	.5040	.5080	.5120	.5160	.5199	.5239	.5279	.5319	.5359	4	8	12	16	20	24	28	32	36
0.1	.5398	.5438	.5478	.5517	.5557	.5596	.5636	.5675	.5714	.5753	4	8	12	16	20	24	28	32	36
0.2	.5793	.5832	.5871	.5910	.5948	.5987	.6026	.6064	.6103	.6141	4	8	12	15	19	23	27	31	35
0.3	.6179	.6217	.6255	.6293	.6331	.6368	.6406	.6443	.6480	.6517	4	7	11	15	19	22	26	30	34
0.4	.6554	.6591	.6628	.6664	.6700	.6736	.6772	.6808	.6844	.6879	4	7	11	14	18	22	25	29	32
0.5	.6915	.6950	.6985	.7019	.7054	.7088	.7123	.7157	.7190	.7224	3	7	10	14	17	20	24	27	31
0.6	.7257	.7291	.7324	.7357	.7389	.7422	.7454	.7486	.7517	.7549	3	7	10	13	16	19	23	26	29
0.7	.7580	.7611	.7642	.7673	.7704	.7734	.7764	.7794	.7823	.7852	3	6	9	12	15	18	21	24	27
0.8	.7881	.7910	.7939	.7967	.7995	.8023	.8051	.8078	.8106	.8133	3	5	8	11	14	16	19	22	25
0.9	.8159	.8186	.8212	.8238	.8264	.8289	.8315	.8340	.8365	.8389	3	5	8	10	13	15	18	20	23
1.0	.8413	.8438	.8461	.8485	.8508	.8531	.8554	.8577	.8599	.8621	2	5	7	9	12	14	16	19	21
1.1	.8643	.8665	.8686	.8708	.8729	.8749	.8770	.8790	.8810	.8830	2	4	6	8	10	12	14	16	18
1.2	.8849	.8869	.8888	.8907	.8925	.8944	.8962	.8980	.8997	.9015	2	4	6	7	9	11	13	15	17
1.3	.9032	.9049	.9066	.9082	.9099	.9115	.9131	.9147	.9162	.9177	2	3	5	6	8	10	11	13	14
1.4	.9192	.9207	.9222	.9236	.9251	.9265	.9279	.9292	.9306	.9319	1	3	4	6	7	8	10	11	13
1.5	.9332	.9345	.9357	.9370	.9382	.9394	.9406	.9418	.9429	.9441	1	2	4	5	6	7	8	10	11
1.6	.9452	.9463	.9474	.9484	.9495	.9505	.9515	.9525	.9535	.9545	1	2	3	4	5	6	7	8	9
1.7	.9554	.9564	.9573	.9582	.9591	.9599	.9608	.9616	.9625	.9633	1	2	3	4	4	5	6	7	8
1.8	.9641	.9649	.9656	.9664	.9671	.9678	.9686	.9693	.9699	.9706	1	1	2	3	4	4	5	6	7
1.9	.9713	.9719	.9726	.9732	.9738	.9744	.9750	.9756	.9761	.9767	1	1	2	2	3	4	4	5	5
2.0	.9772	.9778	.9783	.9788	.9793	.9798	.9803	.9808	.9812	.9817	0	1	1	2	2	3	3	4	4
2.1	.9821	.9826	.9830	.9834	.9838	.9842	.9846	.9850	.9854	.9857	0	1	1	2	2	3	3	4	4
2.2	.9861	.9864	.9868	.9871	.9875	.9878	.9881	.9884	.9887	.9890	0	1	1	2	2	3	3	4
2.3	.9893	.9896	.9898	.9901	.9904	.9906	.9909	.9911	.9913	.9916	0	1	1	1	2	2	2	3	3
2.4	.9918	.9920	.9922	.9924	.9927	.9929	.9931	.9932	.9934	.9936	0	0	1	1	1	1	2	2	2
2.5	.9938	.9940	.9941	.9943	.9945	.9946	.9948	.9949	.9951	.9952	0	0	0	1	1	1	1	1	1
2.6	.9953	.9955	.9956	.9957	.9958	.9960	.9961	.9962	.9963	.9964	0	0	0	0	1	1	1	1	1
2.7	.9965	.9966	.9967	.9968	.9969	.9970	.9971	.9972	.9973	.9974	0	0	0	0	0	1	1	1	1
2.8	.9974	.9975	.9976	.9977	.9977	.9978	.9979	.9979	.9980	.9981	0	0	0	0	0	0	1	1	1
2.9	.9981	.9982	.9982	.9983	.9984	.9984	.9985	.9985	.9986	.9986	0	0	0	0	0	0	0	0	0

Upper-tail percentage points of the standard normal distribution

The table gives the values of z for which $P(Z > z) = 1 - \Phi(z) = q\%$, where the distribution of Z is $N(0, 1)$.

$q(\%)$	z	$q(\%)$	z	$q(\%)$	z	$q(\%)$	z	$q(\%)$	z
50	0.000	15	1.036	2.5	1.960	1.0	2.326	0.04	3.353
45	0.126	14	1.080	2.4	1.977	0.9	2.366	0.03	3.432
40	0.253	13	1.126	2.3	1.995	0.8	2.409	0.02	3.540
35	0.385	12	1.175	2.2	2.014	0.7	2.457	0.01	3.719
30	0.524	11	1.227	2.1	2.034	0.6	2.512	$0.0^2 5$	3.891
25	0.674	10	1.282	2.0	2.054	0.5	2.576	$0.0^2 1$	4.265
24	0.706	9	1.341	1.9	2.075	0.4	2.652	$0.0^3 5$	4.417
23	0.739	8	1.405	1.8	2.097	0.3	2.748	$0.0^3 1$	4.753
22	0.772	7	1.476	1.7	2.120	0.2	2.878	$0.0^4 5$	4.892
21	0.806	6	1.555	1.6	2.144	0.1	3.090	$0.0^4 1$	5.199
20	0.842	5	1.645	1.5	2.170	0.09	3.121	$0.0^5 5$	5.327
19	0.878	4.5	1.695	1.4	2.197	0.08	3.156	$0.0^5 1$	5.612
18	0.915	4	1.751	1.3	2.226	0.07	3.195	$0.0^6 5$	5.731
17	0.954	3.5	1.812	1.2	2.257	0.06	3.239	$0.0^6 1$	5.998
16	0.994	3	1.881	1.1	2.290	0.05	3.291	$0.0^7 5$	6.109

Percentage points for the *t*-distribution

If T has a t_d-distribution then a tabulated value, t, is such that $P(T < t) = p\%$

d	75	90	95	97.5	99	99.5	99.75	99.9	99.95
					$p(\%)$				
1	1.000	3.078	6.314	12.71	31.82	63.66	127.3	318.3	636.6
2	0.816	1.886	2.920	4.303	6.965	9.925	14.09	22.33	31.60
3	0.765	1.638	2.353	3.182	4.541	5.841	7.453	10.21	12.92
4	0.741	1.533	2.132	2.776	3.747	4.604	5.598	7.173	8.610
5	0.727	1.476	2.015	2.571	3.365	4.032	4.773	5.893	6.869
6	0.718	1.440	1.943	2.447	3.143	3.707	4.317	5.208	5.959
7	0.711	1.415	1.895	2.365	2.998	3.499	4.029	4.785	5.408
8	0.706	1.397	1.860	2.306	2.896	3.355	3.833	4.501	5.041
9	0.703	1.383	1.833	2.262	2.821	3.250	3.690	4.297	4.781
10	0.700	1.372	1.812	2.228	2.764	3.169	3.581	4.144	4.587
11	0.697	1.363	1.796	2.201	2.718	3.106	3.497	4.025	4.437
12	0.695	1.356	1.782	2.179	2.681	3.055	3.428	3.930	4.318
13	0.694	1.350	1.771	2.160	2.650	3.012	3.372	3.852	4.221
14	0.692	1.345	1.761	2.145	2.624	2.977	3.326	3.787	4.140
15	0.691	1.341	1.753	2.131	2.602	2.947	3.286	3.733	4.073
16	0.690	1.337	1.746	2.120	2.583	2.921	3.252	3.686	4.015
17	0.689	1.333	1.740	2.110	2.567	2.898	3.222	3.646	3.965
18	0.688	1.330	1.734	2.101	2.552	2.878	3.197	3.610	3.922
19	0.688	1.328	1.729	2.093	2.539	2.861	3.174	3.579	3.883
20	0.687	1.325	1.725	2.086	2.528	2.845	3.153	3.552	3.850
21	0.686	1.323	1.721	2.080	2.518	2.831	3.135	3.527	3.819
22	0.686	1.321	1.717	2.074	2.508	2.819	3.119	3.505	3.792
23	0.685	1.319	1.714	2.069	2.500	2.807	3.104	3.485	3.768
24	0.685	1.318	1.711	2.064	2.492	2.797	3.091	3.467	3.745
25	0.684	1.316	1.708	2.060	2.485	2.787	3.078	3.450	3.725
26	0.684	1.315	1.706	2.056	2.479	2.779	3.067	3.435	3.707
27	0.684	1.314	1.703	2.052	2.473	2.771	3.057	3.421	3.690
28	0.683	1.313	1.701	2.048	2.467	2.763	3.047	3.408	3.674
29	0.683	1.311	1.699	2.045	2.462	2.756	3.038	3.396	3.659
30	0.683	1.310	1.697	2.042	2.457	2.750	3.030	3.385	3.646
40	0.681	1.303	1.684	2.021	2.423	2.704	2.971	3.307	3.551
60	0.679	1.296	1.671	2.000	2.390	2.660	2.915	3.232	3.460
120	0.677	1.289	1.658	1.980	2.358	2.617	2.860	3.160	3.373
∞	0.674	1.282	1.645	1.960	2.326	2.576	2.807	3.090	3.291

Percentage points for the χ^2 distribution

If X has a χ^2_d distribution, then a tabulated value, x, is such that
$P(X < x) = p\%$.

d	\multicolumn{3}{c}{Lower tail}	\multicolumn{6}{c}{Upper tail}
	0.5	2.5	5	90	95	97.5	99	99.5	99.9
1	0.0^43927	0.0^39821	0.0^23932	2.706	3.841	5.024	6.635	7.879	10.83
2	0.01003	0.05064	0.1026	4.605	5.991	7.378	9.210	10.60	13.82
3	0.07172	0.2158	0.3518	6.251	7.815	9.348	11.34	12.84	16.27
4	0.2070	0.4844	0.7107	7.779	9.488	11.14	13.28	14.86	18.47
5	0.4117	0.8312	1.145	9.236	11.07	12.83	15.09	16.75	20.52
6	0.6757	1.237	1.635	10.64	12.59	14.45	16.81	18.55	22.46
7	0.9893	1.690	2.167	12.02	14.07	16.01	18.48	20.28	24.32
8	1.344	2.180	2.733	13.36	15.51	17.53	20.09	21.95	26.12
9	1.735	2.700	3.325	14.68	16.92	19.02	21.67	23.59	27.88
10	2.156	3.247	3.940	15.99	18.31	20.48	23.21	25.19	29.59
11	2.603	3.816	4.575	17.28	19.68	21.92	24.72	26.76	31.26
12	3.074	4.404	5.226	18.55	21.03	23.34	26.22	28.30	32.91
13	3.565	5.009	5.892	19.81	22.36	24.74	27.69	29.82	34.53
14	4.075	5.629	6.571	21.06	23.68	26.12	29.14	31.32	36.12
15	4.601	6.262	7.261	22.31	25.00	27.49	30.58	32.80	37.70
16	5.142	6.908	7.962	23.54	26.30	28.85	32.00	34.27	39.25
17	5.697	7.564	8.672	24.77	27.59	30.19	33.41	35.72	40.79
18	6.265	8.231	9.390	25.99	28.87	31.53	34.81	37.16	42.31
19	6.844	8.907	10.12	27.20	30.14	32.85	36.19	38.58	43.82
20	7.434	9.591	10.85	28.41	31.41	34.17	37.57	40.00	45.31
21	8.034	10.28	11.59	29.62	32.67	35.48	38.93	41.40	46.80
22	8.643	10.98	12.34	30.81	33.92	36.78	40.29	42.80	48.27
23	9.260	11.69	13.09	32.01	35.17	38.08	41.64	44.18	49.73
24	9.886	12.40	13.85	33.20	36.42	39.36	42.98	45.56	51.18
25	10.52	13.12	14.61	34.38	37.65	40.65	44.31	46.93	52.62
30	13.79	16.79	18.49	40.26	43.77	46.98	50.89	53.67	59.70
40	20.71	24.43	26.51	51.81	55.76	59.34	63.69	66.77	73.40
50	27.99	32.36	34.76	63.17	67.50	71.42	76.15	79.49	86.66
60	35.53	40.48	43.19	74.40	79.08	83.30	88.38	91.95	99.61
70	43.28	48.76	51.74	85.53	90.53	95.02	100.4	104.2	112.3
80	51.17	57.15	60.39	96.58	101.9	106.6	112.3	116.3	124.8
90	59.20	65.65	69.13	107.6	113.1	118.1	124.1	128.3	137.2
100	67.33	74.22	77.93	118.5	124.3	129.6	135.8	140.2	149.4

For values of d greater than 100, use the result that $\sqrt{2X}$ has an approximate normal distribution with mean $\sqrt{2d-1}$ and variance 1.

Percentage points for the *F*-distribution

Upper 5% points

v							u						
	1	2	3	4	5	6	7	8	12	24	40	∞
1	161.4	199.5	215.7	224.6	230.2	234.0	236.8	238.9	243.9	249.1	251.1	254.3
2	18.51	19.00	19.16	19.25	19.30	19.33	19.35	19.37	19.41	19.45	19.47	19.50
3	10.13	9.55	9.28	9.12	9.01	8.94	8.89	8.85	8.74	8.64	8.59	8.53
4	7.71	6.94	6.59	6.39	6.26	6.16	6.09	6.04	5.91	5.77	5.72	5.63
5	6.61	5.79	5.41	5.19	5.05	4.95	4.88	4.82	4.68	4.53	4.46	4.36
6	5.99	5.14	4.76	4.53	4.39	4.28	4.21	4.15	4.00	3.84	3.77	3.67
7	5.59	4.74	4.35	4.12	3.97	3.87	3.79	3.73	3.57	3.41	3.34	3.23
8	5.32	4.46	4.07	3.84	3.69	3.58	3.50	3.44	3.28	3.12	3.04	2.93
9	5.12	4.26	3.86	3.63	3.48	3.37	3.29	3.23	3.07	2.90	2.83	2.71
10	4.96	4.10	3.71	3.48	3.33	3.22	3.14	3.07	2.91	2.74	2.66	2.54
12	4.75	3.89	3.49	3.26	3.11	3.00	2.91	2.85	2.69	2.51	2.43	2.30
15	4.54	3.68	3.29	3.06	2.90	2.79	2.71	2.64	2.48	2.29	2.20	2.07
18	4.41	3.55	3.16	2.93	2.77	2.66	2.58	2.51	2.34	2.15	2.06	1.92
20	4.35	3.49	3.10	2.87	2.71	2.60	2.51	2.45	2.28	2.08	1.99	1.84
25	4.24	3.39	2.99	2.76	2.60	2.49	2.40	2.34	2.16	1.96	1.87	1.71
30	4.17	3.32	2.92	2.69	2.53	2.42	2.33	2.27	2.09	1.89	1.79	1.62
40	4.08	3.23	2.84	2.61	2.45	2.34	2.25	2.18	2.00	1.79	1.69	1.51
60	4.00	3.15	2.76	2.53	2.37	2.25	2.17	2.10	1.92	1.70	1.59	1.39
∞	3.84	3.00	2.60	2.37	2.21	2.10	2.01	1.94	1.75	1.52	1.39	1.00

The lower 5% point of an $F_{u,v}$-distribution is the reciprocal of the upper 5% point of an $F_{v,u}$-distribution.

Upper 2.5% points

v							u						
	1	2	3	4	5	6	7	8	12	24	40	∞
1	647.8	799.5	864.2	899.6	921.8	937.1	948.2	956.7	967.7	997.2	1006	1018
2	38.51	39.00	39.17	39.25	39.30	39.33	39.36	39.37	39.41	39.46	39.47	39.50
3	17.44	16.04	15.44	15.10	14.88	14.73	14.62	14.54	14.34	14.12	14.04	13.90
4	12.22	10.65	9.98	9.60	9.36	9.20	9.07	8.98	8.75	8.51	8.41	8.26
5	10.01	8.43	7.76	7.39	7.15	6.98	6.85	6.76	6.52	6.28	6.18	6.02
6	8.81	7.26	6.60	6.23	5.99	5.82	5.70	5.60	5.37	5.12	5.01	4.85
7	8.07	6.54	5.89	5.52	5.29	5.12	4.99	4.90	4.67	4.42	4.31	4.14
8	7.57	6.06	5.42	5.05	4.82	4.65	4.53	4.43	4.20	3.95	3.84	3.67
9	7.21	5.71	5.08	4.72	4.48	4.32	4.20	4.10	3.87	3.61	3.51	3.33
10	6.94	5.46	4.83	4.47	4.24	4.07	3.95	3.85	3.62	3.37	3.26	3.08
12	6.55	5.10	4.47	4.12	3.89	3.73	3.61	3.51	3.28	3.02	2.91	2.72
15	6.20	4.77	4.15	3.80	3.58	3.41	3.29	3.20	2.96	2.70	2.59	2.40
18	5.98	4.56	3.95	3.61	3.38	3.22	3.10	3.01	2.77	2.50	2.38	2.19
20	5.87	4.46	3.86	3.51	3.29	3.13	3.01	2.91	2.68	2.41	2.29	2.09
25	5.69	4.29	3.69	3.35	3.13	2.97	2.85	2.75	2.51	2.24	2.12	1.91
30	5.57	4.18	3.59	3.25	3.03	2.87	2.75	2.65	2.41	2.14	2.01	1.79
40	5.42	4.05	3.46	3.13	2.90	2.74	2.62	2.53	2.29	2.01	1.88	1.64
60	5.29	3.93	3.34	3.01	2.79	2.63	2.51	2.41	2.17	1.88	1.74	1.48
∞	5.02	3.69	3.12	2.79	2.57	2.41	2.29	2.19	1.94	1.64	1.48	1.00

The lower 2.5% point of an $F_{u,v}$-distribution is the reciprocal of the upper 2.5% point of an $F_{v,u}$-distribution.
The table for the upper 1% points is on the next page.

Percentage points for the F-distribution (continued)

Upper 1% points

v	1	2	3	4	5	6	7	8	12	24	40	∞
1	4052	4999	5403	5625	5764	5859	5928	5981	6106	6235	6287	6366
2	98.50	99.00	99.17	99.25	99.30	99.33	99.36	99.37	99.42	99.46	99.47	99.50
3	34.12	30.82	29.46	28.71	28.24	27.91	27.67	27.49	27.05	26.60	26.41	26.13
4	21.20	18.00	16.69	15.98	15.52	15.21	14.98	14.80	14.37	13.93	13.75	13.46
5	16.26	13.27	12.06	11.39	10.97	10.67	10.46	10.29	9.89	9.47	9.29	9.02
6	13.75	10.92	9.78	9.15	8.75	8.47	8.26	8.10	7.72	7.31	7.14	6.88
7	12.25	9.55	8.45	7.85	7.46	7.19	6.99	6.84	6.47	6.07	5.91	5.65
8	11.26	8.65	7.59	7.01	6.63	6.37	6.18	6.03	5.67	5.28	5.12	4.86
9	10.56	8.02	6.99	6.42	6.06	5.80	5.61	5.47	5.11	4.73	4.57	4.31
10	10.04	7.56	6.55	5.99	5.64	5.39	5.20	5.06	4.71	4.33	4.17	3.91
12	9.33	6.93	5.95	5.41	5.06	4.82	4.64	4.50	4.16	3.78	3.62	3.36
15	8.68	6.36	5.42	4.89	4.56	4.32	4.14	4.00	3.67	3.29	3.13	2.87
18	8.29	6.01	5.09	4.58	4.25	4.01	3.84	3.71	3.37	3.00	2.84	2.57
20	8.10	5.85	4.94	4.43	4.10	3.87	3.70	3.56	3.23	2.86	2.69	2.42
25	7.77	5.57	4.68	4.18	3.85	3.63	3.46	3.32	2.99	2.62	2.45	2.17
30	7.56	5.39	4.51	4.02	3.70	3.47	3.30	3.17	2.84	2.47	2.30	2.01
40	7.31	5.18	4.31	3.83	3.51	3.29	3.12	2.99	2.66	2.29	2.11	1.80
60	7.08	4.98	4.13	3.65	3.34	3.12	2.95	2.82	2.50	2.12	1.94	1.60
∞	6.63	4.61	3.78	3.32	3.02	2.80	2.64	2.51	2.18	1.79	1.59	1.00

The lower 1% point of an $F_{u,v}$-distribution is the reciprocal of the upper 1% point of an $F_{v,u}$-distribution.

Critical values for the product–moment correlation coefficient, *r*

It is assumed that X and Y are uncorrelated and have normal distributions.

Critical values for one-tailed tests
The values in the table are the upper-tail 5% and 1% points of the distribution of *r* and hence are appropriate for two-tailed 10% and 2% tests.

n	5%	1%	n	5%	1%	n	5%	1%	n	5%	1%
4	.900	.980	13	.476	.634	22	.360	.492	40	.264	.367
5	.805	.934	14	.458	.612	23	.352	.482	50	.235	.328
6	.729	.882	15	.441	.592	24	.344	.472	60	.214	.300
7	.669	.833	16	.426	.574	25	.337	.462	70	.198	.278
8	.621	.789	17	.412	.558	26	.330	.453	80	.185	.260
9	.582	.750	18	.400	.543	27	.323	.445	90	.174	.245
10	.549	.715	19	.389	.529	28	.317	.437	100	.165	.232
11	.521	.685	20	.378	.516	29	.312	.430	110	.158	.222
12	.497	.658	21	.369	.503	30	.306	.423	120	.151	.212

Critical values for two-tailed tests
The values in the table are the upper-tail 2.5% and 0.5% points of the distribution of *r* and hence are appropriate for two-tailed 5% and 1% tests.

n	5%	1%	n	5%	1%	n	5%	1%	n	5%	1%
4	.950	.990	13	.553	.684	22	.423	.537	40	.312	.403
5	.878	.959	14	.532	.661	23	.413	.526	50	.279	.361
6	.811	.917	15	.514	.641	24	.404	.515	60	.254	.330
7	.754	.874	16	.497	.623	25	.396	.505	70	.235	.306
8	.707	.834	17	.482	.606	26	.388	.496	80	.220	.286
9	.666	.798	18	.468	.590	27	.381	.487	90	.207	.270
10	.632	.765	19	.456	.575	28	.374	.478	100	.197	.256
11	.602	.735	20	.444	.561	29	.367	.470	110	.187	.245
12	.576	.708	21	.433	.549	30	.361	.463	120	.179	.234

For values outside the range of the tables, use the fact that, assuming H_0, $r\sqrt{\dfrac{n-2}{1-r^2}}$ is an observation from a t_{n-2}-distribution.

Alternatively, use the result that $r\sqrt{n-1}$ is approximately an observation from a $N(0,1)$ distribution.

Critical values for Spearman's rank correlation coefficient, r_s

It is assumed that at least one ranking consists of a random permutation of the numbers 1 to n.

Critical values for one-tailed tests

The entries in the table are the smallest values of r_s (to 3 d.p.) which correspond to one-tail probabilities less than or equal to 5% (or 1%). The observed value is significant *if it is equal to, or greater than*, the value in the table. The exact significance level never exceeds the nominal value. The table can also be used to provide 10% and 2% critical values for two-tailed tests for r_s. The asterisk indicates that significance at this level cannot be achieved in this case.

n	5%	1%	n	5%	1%	n	5%	1%	n	5%	1%
4	1.000	*	11	.536	.709	18	.401	.550	25	.337	.466
5	.900	1.000	12	.503	.678	19	.391	.535	26	.331	.457
6	.829	.943	13	.484	.648	20	.380	.522	27	.324	.449
7	.714	.893	14	.464	.626	21	.370	.509	28	.318	.441
8	.643	.833	15	.446	.604	22	.361	.497	29	.312	.433
9	.600	.783	16	.429	.582	23	.353	.486	30	.306	.425
10	.564	.745	17	.414	.566	24	.344	.476	40	.264	.368

Critical values for two-tailed tests

The entries in the table are the smallest positive values of r_s (to 3 d.p.) which correspond to two-tail probabilities less than or equal to 5% (or 1%). The observed value is significant *if it is equal to, or greater than*, the value in the table. The exact significance level never exceeds the nominal value. The table can also be used to provide 2.5% and 0.5% critical values for one-tailed tests for r_s. The asterisks indicate that significance at the stated levels cannot be achieved in these cases.

n	5%	1%	n	5%	1%	n	5%	1%	n	5%	1%
4	*	*	11	.618	.755	18	.472	.600	25	.398	.511
5	1.000	*	12	.587	.727	19	.460	.584	26	.390	.501
6	.886	1.000	13	.560	.703	20	.447	.570	27	.383	.492
7	.786	.929	14	.538	.679	21	.436	.556	28	.375	.483
8	.738	.881	15	.521	.654	22	.425	.544	29	.368	.475
9	.700	.833	16	.503	.635	23	.416	.532	30	.362	.467
10	.648	.794	17	.488	.618	24	.407	.521	40	.313	.405

For $n > 40$, assuming H_0, r_s is approximately an observation from a normal distribution with mean 0 and variance $\dfrac{1}{n-1}$.

Critical values for Kendall's τ

It is assumed that at least one ranking consists of a random permutation of the numbers 1 to n.

Critical values for one-tailed tests

The entries in the table are the smallest positive values of τ (to 3 d.p.) which correspond to one-tail probabilities less than or equal to 5% (or 1%). The observed value is significant *if it is equal to, or greater than*, the value in the table. The exact significance level never exceeds the nominal value. The table can also be used to provide 10% and 2% critical values for two-tailed tests for τ. The asterisk indicates that significance at this level cannot be achieved in this case.

n	5%	1%	n	5%	1%	n	5%	1%	n	5%	1%
4	1.000	*	11	.418	.564	18	.294	.412	25	.240	.333
5	.800	1.000	12	.394	.545	19	.287	.392	26	.237	.329
6	.733	.867	13	.359	.513	20	.274	.379	27	.231	.322
7	.619	.810	14	.363	.473	21	.267	.371	28	.228	.312
8	.571	.714	15	.333	.467	22	.264	.359	29	.222	.310
9	.500	.667	16	.317	.433	23	.257	.352	30	.218	.301
10	.467	.600	17	.309	.426	24	.246	.341	40	.185	.256

Critical values for two-tailed tests

The entries in the table are the smallest positive values of τ (to 3 d.p.) which correspond to two-tail probabilities less than or equal to 5% (or 1%). The observed value is significant *if it is equal to, or greater than*, the value in the table. The exact significance level never exceeds the nominal value. The table can also be used to provide 2.5% and 0.5% critical values for one-tailed tests for τ. The asterisks indicate that significance at the stated levels cannot be achieved in these cases.

n	5%	1%	n	5%	1%	n	5%	1%	n	5%	1%
4	*	*	11	.491	.600	18	.346	.451	25	.287	.367
5	1.000	*	12	.455	.576	19	.333	.439	26	.280	.360
6	.867	1.000	13	.436	.564	20	.326	.421	27	.271	.356
7	.714	.905	14	.407	.516	21	.314	.410	28	.265	.344
8	.643	.786	15	.390	.505	22	.307	.394	29	.261	.340
9	.556	.722	16	.383	.483	23	.296	.391	30	.255	.333
10	.511	.644	17	.368	.471	24	.290	.377	40	.218	.285

For $n > 40$, assuming H_0, τ is approximately an observation from a normal distribution with mean 0 and variance $\dfrac{2(2n + 5)}{9n(n - 1)}$.

Factors for control chart action lines

Sample size	Factor for mean	Factors for range	
n	A_2	D_3	D_4
2	1.880	0	3.267
3	1.023	0	2.575
4	0.729	0	2.282
5	0.577	0	2.115
6	0.483	0	2.004
7	0.419	0.076	1.924
8	0.373	0.136	1.864
9	0.337	0.184	1.816
10	0.308	0.223	1.777
11	0.285	0.256	1.744
12	0.266	0.284	1.716
13	0.249	0.308	1.692
14	0.235	0.329	1.671
15	0.223	0.348	1.652

Random numbers

07552	37078	70487	39809	35705	42662
28859	92692	51960	51172	02339	94211
64473	62150	49273	29664	05698	05946
55434	20290	33414	26519	65317	47580
20131	05658	01643	17950	74442	30519
04287	26200	37224	23042	85793	50649
19631	42910	35954	88679	34461	45854
52646	83321	52538	41676	71829	00734
11107	55247	73970	67044	29864	72349
16311	04954	92332	51595	96460	77412
37057	83986	98419	76401	15412	68418
33724	28633	85953	82213	07827	48740
43737	15929	19659	52804	72335	25208
16929	84478	31341	60265	19404	27881
10131	98571	20877	34585	22353	54505
29998	48921	60361	12353	28334	84764
96525	74926	82302	97562	57805	40464
49955	60120	14557	04036	55397	54710
27936	70742	69960	69090	25800	53457
43045	75684	77671	70298	21292	27677
38782	35325	61068	64149	73456	06831
47347	47512	09263	83713	04450	31376
98561	93657	76725	55243	95540	31611
30674	43720	80477	82488	44328	55607
20293	63332	24626	56001	23528	85302

Answers

Exercises 1a

1

0	1	2	3	4	5	6
6	8	4	2	3	0	1

2

0	1	2	3
7	5	4	2

3

4	5	6	7	8
1	4	7	5	4

4

3	4	5	6	7	8	9	10	11
1	1	2	3	2	5	2	3	1

5

25	26	27	28	29	30	31	32	33	34
1	0	0	5	3	1	4	3	0	1

6

46	47	48	49	50	51	52	53	54
1	3	4	2	3	5	3	3	1

7

1	2, 8
2	1, 4, 7, 7, 7
3	1, 4, 5, 7
4	1, 2, 8
5	0 Key: 3|5 = 35

8

3	4
3	5, 5, 8
4	1, 1, 4
4	5, 6, 7, 7
5	1, 1
5	6
6	2
6	
7	3, 4 Key: 6|2 = 62

9

16	2
17	
18	2, 7
19	2, 5, 7
20	0, 4, 8, 9
21	0, 3, 3, 5, 7, 9
22	4, 9 Key: 16|2 = 162

10

1	74, 99
2	22, 44, 63, 84
3	21, 35, 62, 76, 77
4	13, 14, 24
5	02, 10 Key: 5|02 = 50.2

Exercises 1b

1 Heights in ratios $6:8:4:2:3:0:1$

2 Heights in ratios $7:5:4:2$

3 Heights in ratios $1:4:7:5:4$

4 Heights in ratios $1:1:2:3:2:5:2:3:1$

5 Heights in ratios $4:5:9:7:2:5:1:1:1$

6 (i) and (ii) ratios 1:3:5:2

   (iii) Sector angles $33°$, $98°$, $164°$, $65°$

7 (iii) Sector angles $29°$, $65°$, $95°$, $99°$, $73°$

10 (ii) Sector angles (boys) $34°$, $71°$, $102°$, $102°$, $51°$

   and (girls) $22°$, $57°$, $85°$, $95°$, $101°$

   Pie chart radii optionally in ratio $\sqrt{148}:\sqrt{114}$

12 (i) Sector angles: (1872) $311°$, $49°$, $0°$;

   (1931) $330°$, $27°$, $3°$; (1965) $319°$, $18°$, $24°$

13 (ii) Sector angles $144°$, $97°$, $45°$, $42°$, $32°$

14 (ii) Sector angles $101°$, $36°$, $32°$, $30°$, $24°$, $22°$, $21°$,

   $12°$, $9°$, $5°$, $4°$, $65°$

Exercises 1c

1 Boundaries at 749.5, etc. Frequencies: 5, 2, 9, 4

2 Boundaries at 24.95, etc.

   Frequencies: 1, 3, 5, 4, 3, 2, 2

3 Boundaries at (say) 35.0, 40.0, etc.

   Frequencies: 1, 3, 6, 4, 6, 3, 2

4 Boundaries at (say) 9.5, 19.5, etc.

   Frequencies: 3, 7, 4, 7, 5, 2, 2

5 Boundaries at 0, 19.5, 39.5, 59.5, 99.5

   Heights in ratios $4.10:17:12:3$

6 Boundaries at -0.5, 29.5, 49.5, 69.5, 99.5

   Heights in ratios $3:18:55.5:14$

7 Boundaries at 0, 1, 5, 15, ... , 100 (say)

   Heights in ratios

   $759:782:742:772:930:779:671:568:517:302:66$

8 Boundaries at 16 (say), 20, 25, 30, 35, 40, 45 (say)

   Heights in ratios $13.1:34.7:49.7:32.3:10.7:2.0$

9 Heights in ratios

   $572.4:1158:1399:1441:1141:878:553:319:83$

10 Boundaries at 0, 2, 5, 9, 13

   Heights in ratios $17.5:17:20.75:8.75$

11 Boundaries at 0, 4.5, 9.5, 14.5, 19.5, 29.5

   Heights in ratios $2.2:7:15:18:2.5$

Exercises 1d

1 (i) Plot at (4.5, 0), (14.5, 1), (24.5, 3), etc.

   (ii) Plot at (9.5, 0), (19.5, 1), ... , (69.5, 37)

2 (i) Plot at (624.5, 0), (674.5, 1), etc.

   (ii) Plot at (649.5, 0), (699.5, 1), ... , (899.5, 37)

3 Plot at (4, 0), (4, 1), (5, 1), (5, 5), (6, 5), ... , (8, 21)

4 Plot at (3, 0), (3, 1), (4, 1), (4, 2), (5, 2), ... , (11, 20)

5 Plot at (0, 0), (0, 4), (1, 4), (1, 9), (2, 9), ... , (8, 35)

6 Plot at (0, 0), (2, 35), (5, 86), (9, 169), (13, 204)

7 Plot at (0, 0), (1, 759), (5, 3888), (15, 11 305), ... , (100, 58 190)

   [Ignore round-off error]

8 Plot at (16,0), (20, 52.4), (25, 225.8), ... , (45, 699.2)

Exercises 1e

1 Time series: connected points from (0, 6.5) to (9, 9.2).
2 Scatter diagram. Broken *y*-axis.
3 Time series: connected points from (1, 25.5) to (6, 17.2). Broken *y*-axis.
4 Time series: connected points from (1, 255) to (8, 243). Broken *y*-axis.
5 Time series: production shown by connected points from (1, 238) to (12, 228). Doubly broken *y*-axis omitting sections (0, 50) and (120, 200). Exports shown on same graph with dotted line connecting points from (1, 57) to (12, 91).

Exercises 1f (Miscellaneous)

1
1	9
2	1, 2, 3, 3, 4, 4, 4, 4, 5, 6, 8
3	1, 4, 8, 9, 9, 9
4	2, 3, 3, 4, 4, 5, 6, 8
5	0, 3, 6
6	0, 1, 3
7	Key: 5|3 = 53
8	3

2
1.6	0, 9
1.7	0, 2, 6, 9
1.8	7, 8
1.9	2, 3, 3, 9
2.0	1, 8
2.1	1, 1, 2
2.2	2, 3
2.3	Key: 2.2|3 = 2.23
2.4	4

3 For deaths, a pie chart with two sectors having angles 254° and 106° is required.
4 Scatter diagram. Broken axes. Points could be labelled.
5 Use doubly broken *y*-axis omitting sections (0, 4000) and (4600, 6900).
Conviction rates: 0.635, 0.622, 0.611, 0.615, 0.607, 0.593. Show using a broken *y*-axis ranging from 0.59 to 0.64. Ideally show graph below previous one using same *x*-axis.
6 (*a*) 86°, 38°, 32°, 20°, 168°, 16°
  (*b*) 5.48 cm
7 Heights in ratios 2 : 4 : 3 : 14 : 4 : 2 : 0.5
8 Heights in ratios 4 : 15 : 25 : 9 : 6 : 2.3 : 1

Exercises 2a

1 B
2 5–8 yrs
3 25–34
5 11.21 s
6 (i) 0   (ii) 1.06

7 (i) 8   (ii) 7.4
8 4.54 m
9 0.003
10 £11.96
11 90.61 kg, 91.24 kg
12 22
13 (i) 2   (ii) 2.8
14 $\frac{1}{2}$ in
15 50
16
2	9
3	1, 1, 2
3	5, 5, 5, 7, 7, 8, 9
4	0, 1, 1, 2, 2, 3, 4
4	5, 8 Key: 3|7 = 37 Median = 39

17
3	4, 5, 5, 8
4	1, 1, 4, 5, 6, 7, 7
5	1, 1, 5, 6
6	
7	4, 5
8	2 Key: 7|5 = 75 Median = 46.5

18 (i) 31.6   (ii) 31
19 (i) 29.8   (ii) 29.5
20 (i) 9.73   (ii) 10   (iii) 10

Exercises 2b

3 (i) 36   (ii) 204   (iii) 92
4 36.5
5 1239
6 9.81
7 −2
8 15.4

Exercises 2c

Some answers may differ slightly because of different assumptions concerning the accuracy with which the original data were recorded.

1 0.39
2 1.2
3 3.97, 4
4 27.9, 27.5
5 (i) 89   (ii) 100
6 (i) 114 miles   (ii) 80–99 miles
7 (i) 16 010 miles   (ii) 15 000–15 999 miles
8 (i) 26.4   (ii) 21–25
9 $a = 2$, $b = 3$
10 £14 450

Exercises 2d

1 (i) 1008   (ii) 2.08   (iii) 1080
2 1.000 000 005, 1.000 000 005, 1.000 000 006
3 £801
4 29.78
5 (i) £54.95

6 £2.62

7 70.74

8 50.04

9 1003.1 mb

10 0.8177 mm

11 16 010 miles

12 26.37

13 5.9

14 £15 300

15 −0.4805

16 19.85

17 44

Exercises 2e

1 0.81 mm, 0.83 mm

2 48, 52

3 993 mb, 1015 mb

4 20, 35; 35.5

5 2.5, 5; 2

6 41, 55; 44.7

7 15 220 miles, 16 830 miles; 14 280 miles, 15 070 miles

8 20, 33; 19, 34

9 (a) 29   (b) (i) 8.5 min   (ii) 11.4 min

10 (a) Plot at (5,6), (10,20), (15,85), (20,148), (25,172), (30,184), (35,194), (40,200)

(b) 16.2 litres   (c) £7.64   (d) 19%

Exercises 2f

1 15, 7; $Q_1 = 6$, $Q_2 = 9.5$, $Q_3 = 13$

2 18, 11.5; $Q_1 = 12.0$, $Q_2 = 19.0$, $Q_3 = 23.5$

3 $Q_1 = 149$, $Q_2 = 180$, $Q_3 = 300$; outlier at 660

4 $Q_1 = 157$, $Q_2 = 225$, $Q_3 = 331$

5 Smokers: $Q_1 = 119$, $Q_2 = 125$, $Q_3 = 130$

Non-smokers: $Q_1 = 113$, $Q_2 = 116$, $Q_3 = 129$

6 $Q_1 = 9$, $Q_2 = 22$, $Q_3 = 36$; outlier at 104

7 21, 12.25; $Q_1 = 2.25$, $Q_2 = 5$, $Q_3 = 14.5$

8 (i) Experienced rats:

11	8
12	0, 1
12	5, 6, 7, 8, 9
13	0, 0, 1, 2
13	5, 7 Key: 12|6 = 126

Inexperienced rats:

12	6
13	4
13	5, 9
14	2, 4
14	5, 5, 7, 9
15	2, 3
15	6 Key: 14|7 = 147

(ii) Experienced: $Q_1 = 125$ s, $Q_2 = 128.5$ s, $Q_3 = 131$ s

Inexperienced: $Q_1 = 138$ s, $Q_2 = 145$ s, $Q_3 = 150$ s

9 (a) (ii) 6, 3

10 (a)

40	1, 2, 3, 4, 4, 6, 7, 7, 8, 8
50	0, 2, 2, 2, 3, 4, 6, 7, 8, 8
60	0, 2, 3, 3, 6, 6, 7, 7, 8
70	0, 0, 2, 2, 4, 4, 6, 7, 8, 8, 8
80	0, 1, 2, 5, 5, 6, 6, 7
90	3, 3, 4 Key: 50|2 = 52

(b) 66 miles, $Q_1 = 52$ miles, $Q_3 = 78$ miles

11 (i) Plot with 'step edges' at (1,21), (2,52), (3,70), (4,88), (5,96); 2, 1

(ii) mean > median

Exercises 2g

1 2.8, 1.17, 1.04

2 15.2, 4.75

3 12.0, 4.0

4 1.315 kg, 0.122 (kg)²

5 3.05

6 71.8

7 −29.0, 6.99

8 83.9, 574.8

9 1.16, 0.457

10 0.0721, 0.0721

11 −2.1, 1.45

12 1.66

13 10.5, 6.70

Exercises 2h

1 (i) $\frac{5}{3}$ (ii) 1   (iii) 2.47

2 (i) 6   (ii) $\frac{19}{3}$ (iii) 1.13

3 0.744

4 4.65

5 4, 3.39

6 36.2, 1.42

7 45.5

8 £14.65, £² 5.21

9 0.0136 mm

10 1 030 000 (miles)²

11 7.09

12 15.9

13 204, 256

14 1.685 m, 0.475

15 £14.07, 7.21

16 30 700 km, 70.9×10^6 (km)²

17 1 000 001.33 milligram, 10.07 milligram

Exercises 2i

1 (i) 0.97   (ii) −0.60   (iii) 0.63

2 0.42

3 0.41

4 −0.06

5 0.16

6 0.22

7 Quartile coefficient = 1

8 (i) 3.76, 1, 7.83, 0, 4   (ii) 0.48   (iii) 1.1   (iv) 0.5
9 0.09

Exercises 2j

1 (i) 20, 16   (ii) 20, 64
2 $y = \frac{1}{8}(5x + 100)$
3 −80, 120
4 60, 19.2
5 46, 9.92
6 50
7 56%
8 9.3
9 21.7 knots, 5.25 knots
10 1042p, 254
11 2.72 h, 1.19 h
12 1.29 m

Exercises 2k (Miscellaneous)

1 Median $= 15\frac{1}{2}$
2 2 people
3
16	2
17	Key: 16|2 = 162
18	2, 7
19	2, 5, 7
20	0, 4, 8, 9
21	0, 3, 3, 5, 7, 9
22	4, 9 median = 28.5

4 (b) 19.16, 20   (c) 0.958
5 75–84. Taking lower and upper limits as 16 and 99, heights in proportion to $1.75 : 1 : 8.38 : 20.79 : 9.57$
6 (i) −1.8 °C   (ii) −2 °C   (iii) 12.6   (iv) 3.6 °C
(v) 3.2 °C
7 (i) 10.4, 10, 2.73, 8.5, 13   (ii) 0.48   (iii) 0.33
8 (i) 33.5, 42.5, 23.4   (ii) 34.3, 12.7
9 35y 1m, 11y 3m   (i) 33y 9m, 17y 11m   (ii) 65.0%
10 (i) Plot at (11, 0), (12, 111), (13, 230), ..., (16, 700); 13.8 yrs
(ii) 13.7 yrs, 1.38 yrs
(iii) 14.1 yrs, 1.69 yrs
(iv) 14.1 yrs
11 (a) 6.5, 8, 6.5, 2.13   (b) 6.5, 4.79   (c) 6.92, 5.50
12 (i) (a) Heights in ratio
$1 : 3 : 10 : 24 : 5.5 : 2.75 : 0.5 : 0.125 : 0.25$
(b) 152.5 ml   (c) plot through (25,1), (50,4), (100,24), ..., (800,100); 125 ml
(ii) 1000.5 mm, 1.4 (mm)^2
13 (1) 106.9 cm   (ii) 109.0 cm,   (iii) 0.73
14 54.13 m   (a) 60 647.67 m^2 (b) 55.04 m
(c) 109.35 m^2
15 (*a*) £46.7, £12.4
16 Heights in ratio $6 : 15 : 8 : 3 : 1.5 : 0.5 : 0.11$
(i) 35 000   (ii) 24y 8m

17 Heights in ratio $39 : 23 : 11.5 : 4.5 : 1$
(ii) 1.11 min, 1.11 min
18 (i) 13.6 yrs, 1.37 yrs
(ii) plot through (11,0), (12,165), (13,349), (14,565), (15,796), (16,1000); about 13.7 years
(iii) 14.2 years
19 (a)
3.7	7
3.8	0, 3, 4, 4
3.8	7, 9
3.9	0, 0, 1, 1, 1, 1, 4
3.9	6, 6, 7, 7, 7, 8, 8, 9, 9
4.0	0, 1, 1, 1, 2, 2, 2, 3, 3, 3, 4
4.0	5, 5, 5, 6, 7, 7, 7, 9
4.1	1, 2, 3
4.1	6, 6, 8
4.2	0, 1 Key: 3.9|4 = 3.94

(b) 3.91 oz, 4.01 oz, 4.06 oz   (d) heights in ratio
$1 : 4 : 2 : 7 : 9 : 11 : 8 : 3 : 3 : 2$
20 (a) (i)
2	1, 4, 4
2	
3	0, 0, 0, 0, 1, 1, 1, 1, 2, 2, 3, 3, 4, 4, 4, 4, 4
3	6, 6, 6, 6, 7, 7, 8, 9
4	0, 0, 1, 1, 1, 2, 2, 2, 3, 3
4	5, 6, 6, 6, 7
5	0, 1, 1
5	5, 6, 6
6	2, 2 Key: 2|1 = 21

(ii) 37 min, $Q_1 = 32.3$ min, $Q_3 = 44.5$ min
21 Plot through (14.5,0), (19.5,22), (24.5,64), (29.5,134), (34.5,172), (39.5,188), (50.5,200); 27.1 cm, 8.8 cm; Plot through (13,0), (27,50), (32,100), (35,150), (42,200)
22 (a) At least 11 years old but under 12 (in 1984); 2 631 000 pupils in the UK in 1984 were at least 12 years old but under 14; 9 876 000
(b) 10.5 yrs
(c) 295.7, 690.0, 825.0, 877.0, 591.5, 105.0
23 (b) Heights in ratio $2 : 12 : 27 : 30 : 18 : 14 : 9 : 4 : 1$
(i) 30.6 A   (ii) 30.9 A   (iii) 2.22 A; 0.41
24 Plot through (10,0), (20,5), (30,16), (40,32), (45,51), (50,65), (60,77), (70,86), (80,92), (100,95);
median $= 44.1$ s, IQR $= 20.4$ s; mean $= 45.8$ s, SD $= 16.8$ s; Method A: limits at 18 and 64; Method B: limits at 29 and 63, % ages: 16%, 68%, 16%
25 2, 2
26 (a) (i) £6530, £1540   (ii) 22%
(c) 7%; SD $= £3800$; £7970, £3550
27 mode $= 2$, median $= 3$, mean $= 3.53$, SD $= 1.98$; 0.80 and 0.77

Exercises 3a

1 (ii) 7.3%, 3.9%, no
2 (i) 16.3 > 16.13   (ii) 16.4 < 16.51
3 Maytown (21.5% > 21.2%)

Exercises 3b

1 124 g

2 54.8%

3 £22.38

4 405 000

5 (i) 132   (ii) 125   (iii) 105

6 Jones 127, Smith 128

7 (i)

Car	Bus	Van	Motorcycle
100	100	100	100
107	106	111	90
116	111	122	89
118	113	123	83
118	118	128	81
118	113	127	68
118	113	125	63

(iii) Vans

Exercises 3c

1 (i) 121   (ii) 118

2 1.05 < 1.06: cheaper

3 (i) 112.5   (ii) 113.7

4 (ii) 122.1

Exercises 3d

Predictions will vary slightly depending upon the method used.

1

–	15.4	15.8	16.1	16.4
15.2	15.5	16.0	16.2	16.5
15.1	15.8	16.1	16.2	–

2 (i)

–	–	200.2	196.8	195.8
191.4	191.2	189.8	187.6	184.8
185.0	182.0	181.2	–	

(ii) 145, 191, 220, 187, 111

3

–	–	391.6	448.6	503.0	518.6
560.0	600.8	624.2	642.2	–	–

4 (i)

–	–	143.6	147.5
152.1	156.4	161.0	169.4
177.9	186.8	–	–

(ii) 188, 205, 240

5 (ii)

–	–	823.4	890.9
931.4	933.9	896.9	838.8
765.5	682.0	–	–

(iii) 2 540 000

6 (ii)

8.76	8.71	8.63
8.49	8.36	8.29	8.18
8.01	7.90		

(iii) 3.38

Exercises 3e (Miscellaneous)

1 (b) 19.86p   (c) 219

2 1984: 122.41, 143.18
   1990: 139.24, 131.03, 181.82

Index = 152.45
Total wage bill: 1980: £6760, 1984: £8560,
1988: £9980
1984 Index = 126.63, 1988 Index = 147.63

3 1982: 33 088, 14 008
   1986: 50 040, 39 292, 16 480
   1990: 80 620, 65 052, 47 564, 20 600
   Totals 1982: 143 284, 1990: 213 836
   Index 1990: 149.24
   121.55 ≈ 118.47, 147.75 ≈ 149.24: yes

4 (b) 62, 63, 65, 66, 67.5, 69, 69, 70, 71.5, 73, 74, 75, 75

5 Centred 4-point averages: 113, 115, 118, 121, 125,
128, 131, 134, 137, 140, 142, 143, 145, 148, 151, 155
Trend = 11 per year. (The question does not require
centred averages)

6 About 18.4, 18.6; −2.25, 16.4

7 (a) 3-point averages: 250, 273, 295, 292, 259, 261,
      254, 251, 244, 243, 241, 235, 236

(b) Tuesday of week 2

(c) Friday of week 1

(d) £147

8 350, 260

9 (iii) 12.24 per thousand

Exercises 5a

1 (i) $\frac{1}{2}$ (ii) $\frac{5}{6}$ (iii) $\frac{1}{3}$ (iv) $\frac{1}{3}$

2 (i) $\frac{2}{5}$ (ii) $\frac{4}{15}$ (iii) $\frac{2}{3}$

3 (i) $\frac{1}{4}$ (ii) $\frac{1}{13}$ (iii) $\frac{1}{52}$ (iv) $\frac{3}{13}$

4 (i) $\frac{9}{10}$ (ii) $\frac{4001}{10\,000}$ (iii) $\frac{1}{10\,000}$ (iv) $\frac{1}{10}$
   (v) $\frac{1}{100}$

5 (i) $\frac{3}{4}$ (ii) $\frac{3}{4}$

6 (i) $\frac{1}{3}$ (ii) $\frac{2}{3}$ (iii) $\frac{7}{30}$ (iv) $\frac{1}{3}$ (v) $\frac{3}{10}$
   (vi) $\frac{3}{5}$

7 (i) $\frac{1}{36}$ (ii) $\frac{1}{6}$ (iii) $\frac{1}{12}$ (iv) $\frac{1}{9}$
   (v) $\frac{1}{12}$ (vi) $\frac{1}{9}$

8 (i) $\frac{3}{14}$ (ii) $\frac{9}{14}$ (iii) $\frac{5}{14}$ (iv) 1   (v) 0

Exercises 5b

1 (i) $\frac{1}{2}$ (ii) $\frac{2}{3}$ (iii) $\frac{2}{3}$ (iv) $\frac{1}{2}$
   (v) *B* and *C*; *C* and *D*   (vi) *B*, *C*, *D*   (vii) $\frac{5}{6}$
   (viii) 0

2 (iv) *B*, *C*   (v) $\frac{7}{12}, \frac{5}{18}, \frac{1}{3}$ (vi) $\frac{5}{12}, \frac{1}{9}, 1, 0, \frac{11}{18}$

3 (a) P(*A*) = $\frac{1}{2}$, P(*B*) = $\frac{1}{4}$, P(*C*) = $\frac{5}{12}$, P(*D*) = $\frac{1}{9}$,
     P(*E*) = $\frac{1}{2}$
   (b) Exclusive: *A* and *E*, *C* and *D*; exhaustive: *A* and *E*

4 (i) 0.3   (ii) 0.1   (iii) 0.3   (iv) 0.4

5 (i) $\frac{1}{7}$ (ii) $\frac{3}{7}$ (iii) $\frac{4}{7}$

6 (i) $\frac{1}{10}$ (ii) $\frac{1}{4}$ (iii) $\frac{7}{10}$ (iv) $\frac{4}{5}$ (v) $\frac{1}{20}$

7 (i) $\frac{4}{13}$ (ii) $\frac{1}{26}$ (iii) $\frac{4}{13}$ (iv) $\frac{1}{4}$ (v) $\frac{1}{52}$
   (vi) $\frac{7}{13}$ (vii) $\frac{9}{26}$ (viii) $\frac{1}{13}$ (ix) $\frac{25}{52}$ (x) $\frac{1}{52}$
   (xi) $\frac{7}{26}$

8 (i) 0.283   (ii) 0.220   (iii) 0.660
   (iv) 0.373   (v) 0.340

Exercises 5c

1 (i) $\frac{13}{36}$ (ii) $\frac{4}{9}$ (iii) $\frac{7}{36}$ (iv) $\frac{3}{4}$
2 (i) $\frac{4}{9}$ (ii) $\frac{4}{9}$ (iii) $\frac{1}{9}$ (iv) $\frac{4}{9}$ (v) $\frac{2}{9}$
3 0.999
4 (i) 0.288   (ii) 0.352
5 (i) $\frac{1}{4}$ (ii) $\frac{17}{24}$ (iii) $\frac{1}{24}$ (iv) $\frac{1}{3}$
6 (i) $\frac{7}{25}$ (ii) $\frac{18}{25}$ (iii) $\frac{14}{75}$ (iv) $\frac{68}{75}$ (v) 0
(vi) 1
7 (i) 0.395   (ii) 0.5
8 $P(A) = \frac{1}{64}$, $P(B) = \frac{3267}{8000} = 0.408$, $P(C) = \frac{61}{125} = 0.488$,
$P(D) = \frac{613}{1600} = 0.383$

Exercises 5d

1 720   (i) $\frac{1}{5}$ (ii) $\frac{1}{3}$ (iii) $\frac{2}{3}$
2 120   (i) $\frac{2}{5}$ (ii) $\frac{3}{5}$ (iii) $\frac{7}{30}$
3 6.23×10^9 (i) $\frac{1}{156}$ (ii) $\frac{1}{78}$ (iii) $\frac{23}{156}$ (iv) $\frac{1}{2}$
4 12 600, 840   (i) $\frac{4}{5}$ (ii) $\frac{1}{45}$
5 900 900   (i) $\frac{2}{143}$ (ii) $\frac{2}{75\,075}$
6 (i) 24   (ii) 12
7 720   (i) $\frac{1}{6}$ (ii) $\frac{1}{6}$ (iii) $\frac{1}{30}$ (iv) $\frac{1}{6}$ (v) $\frac{3}{5}$
8 120   (i) $\frac{3}{10}$ (ii) $\frac{2}{5}$ (iii) $\frac{3}{5}$

Exercises 5e

1 455   (i) 225   (ii) 325
2 1.24×10^9, 2.99×10^{10}
3 540 000
4 1260
5 98
6 120
7 35
8 (i) 1680   (ii) 420
9 1320
10 1680
11 240
12 9450

Exercises 5f

1 (i) $\frac{2}{9}$ (ii) $\frac{14}{45}$ (iii) $\frac{31}{45}$
2 0.649
3 (i) $\frac{33}{112}$ (ii) $\frac{3}{14}$ (iii) $\frac{55}{112}$ (iv) $\frac{11}{56}$ (v) $\frac{1}{56}$
(vi) $\frac{1}{7}$
4 (i) $\frac{8}{33}$ (ii) $\frac{3}{11}$ (iii) $\frac{1}{11}$
5 20   (i) $\frac{1}{10}$ (ii) $\frac{3}{10}$ (iii) $\frac{3}{10}$ (iv) $\frac{3}{10}$
(v) $\frac{3}{5}$
6 (i) $\frac{12}{133}$ (ii) $\frac{4}{95}$ (iii) $\frac{33}{266}$ (iv) $\frac{13}{266}$
7 (i) $\frac{1}{99}$ (ii) $\frac{14}{33}$
8 $\frac{1}{2}$
9 (i) $\frac{1}{1140}$ (ii) $\frac{8}{19}$

Exercises 5g (Miscellaneous)

1 (a) $\frac{5}{324}$ (b) $\frac{5}{1944}$ (c) $\frac{5}{972}$ (d) $\frac{613}{648}$
(e) 0.335

3 (a) $\frac{3}{8}$ (b) $\frac{33}{40}$ (c) 0.2244   (d) 0.1671
4 (i) $\frac{1}{45}$ (ii) $\frac{9}{25}$
5 (i) $\frac{3}{320}$ (ii) $\frac{9}{320}$ (iii) $\frac{107}{295}$ (iv) $\frac{37}{295}$
6 (ii) $\frac{9}{32}$ (iii) $\frac{83}{128}$
7 (i) $\frac{1}{42}$ (ii) $\frac{11}{42}$ (iii) $\frac{1}{35}$ (iv) $\frac{4}{35}$
8 (i) $\frac{33}{323}$ (ii) $\frac{616}{1615}$ (iii) $\frac{33}{95}$
9 (i) 2600   (ii) 17 576

Exercises 6a

1 (i) $\frac{2}{7}$ (ii) $\frac{5}{7}$ (iii) $\frac{2}{3}$ (iv) $\frac{1}{3}$
2 (i) $\frac{5}{8}$ (ii) $\frac{5}{8}$ (iii) $\frac{37}{40}$ (iv) $\frac{32}{37}$ (v) $\frac{20}{37}$
(vi) 0   (vii) $\frac{5}{8}$
3 (i) $\frac{5}{12}$ (ii) $\frac{3}{4}$ (iii) $\frac{2}{3}$ (iv) $\frac{5}{11}$
4 (i) independent   (ii) independent
(iii) not independent
5 (i) $\frac{2}{3}$ (ii) $\frac{1}{2}$
6 (i) $\frac{2}{5}$ (ii) $\frac{2}{15}$ (iii) $\frac{8}{15}$ (iv) $\frac{1}{5}$
7 (a) $\frac{1}{36}$ (b) $\frac{1}{6}$ (c) $\frac{1}{3}$ (d) $\frac{11}{36}$ (e) $\frac{1}{36}$
(f) $\frac{1}{2}$
8 (a) $\frac{1458}{4096} = 0.356$ (b) $\frac{729}{4096} = 0.178$
(c) $\frac{2187}{4096} = 0.534$ (d) $\frac{730}{4096} = 0.178$
(e) $\frac{729}{730} = 0.999$
9 $\frac{21}{40}$
10 $P(A) = \frac{14}{99}$, $P(B) = \frac{14}{33}$; $\frac{5}{21}$; $\frac{5}{7}$
12 $\frac{1}{2}$, $\frac{11}{120}$, $\frac{7}{12}$, $\frac{11}{60}$
13 (i) $\frac{1}{50}$ (ii) $\frac{9}{50}$ (iii) $\frac{27}{95}$ (iv) $\frac{19}{29}$
14 (ii) $\frac{77}{95}$
15 (i) $\frac{1}{4}$ (ii) $\frac{5}{24}$ (iii) $\frac{5}{8}$ (iv) $\frac{1}{9}$
A and B not independent
16 (i) $\frac{1}{6}$ (ii) $\frac{1}{2}$ (iii) $\frac{1}{2}$ (iv) $\frac{1}{4}$
17 (i) $\frac{1}{14}$ (ii) $\frac{97}{105}$ (iii) $\frac{37}{42}$
(iv) $\frac{85}{97}$; events not independent
18 (i) $\frac{3}{28}$ (ii) $\frac{1}{2}$ (iii) $\frac{2}{3}$
19 (a) $\frac{7}{5}$ (b) $\frac{4}{15}$ (c) $\frac{4}{15}$; $\frac{119}{450}$
20 (i) $\frac{7}{10}$ (ii) $\frac{2}{7}$ (iii) $\frac{1}{3}$ (iv) $\frac{4}{9}$ (v) $\frac{3}{5}$

Exercises 6b

1 (a) 0.012   (b) 0.030   (c) 0.400
2 (a) $x = 0.2$, $P(A) = 0.4$, $P(B) = 0.5$
(c) $P(B \cap C) = 0.2$, $P(C) = 0.5$
(d) not independent
3 (a) $\frac{5x}{11}$ (b) $\frac{5}{11}x = 2\left(\frac{9}{10} - x\right)$ (e) $\frac{7}{8}$
4 (i) (a) yes   (b) no   (c) $\frac{7}{10}$, not independent
(ii) (a) $\frac{1}{4}$ (b) $\frac{1}{3}$ (c) $\frac{3}{16}$
5 (i) (a) $\frac{7}{30}$ (b) $\frac{1}{10}$
(ii) (a) 0.0106   (b) 0.000 266

Exercises 6c

1 (i) $\frac{1}{3}$ (ii) $\frac{2}{3}$ (iii) $\frac{7}{16}$ (iv) $\frac{9}{16}$
2 (i) $\frac{3}{41}$ (ii) $\frac{38}{41}$ (iii) $\frac{27}{59}$ (iv) $\frac{32}{59}$
3 $\frac{8}{11}$
4 (i) $\frac{50}{67}$, $\frac{52}{67}$ (ii) $\frac{150}{311}$

5 (a) $\frac{3}{10}$ (b) $\frac{7}{9}$

6 (a) $\frac{3}{20}$ (b) $\frac{1}{6}$ (c) $\frac{59}{60}; \frac{1}{3}$

7 $P = \dfrac{9p}{7p + 2}$

10 $\frac{17}{1000}, \frac{3}{17}$

11 (i) 0.70, 0.68   (ii) 0.28   (iii) $\frac{21}{32}$

12 (b) (i) 0.24   (ii) 0.26   (iii) 0.62   (iv) $\frac{14}{27}$

13 (c) mutually exclusive   (d) 0   (e) $\frac{1}{6}; \frac{13}{25}. \frac{3}{4}$

Exercises 6d (Miscellaneous)

1 (i) 0.18   (ii) 0.88

2 (b) (i) $\frac{1}{4}$ (ii) $\frac{1}{52}$ (iii) $\frac{1}{26}$

3 $\frac{15}{44}$; not independent; $\frac{1}{10}$

4 (b) $\frac{1}{2}, \frac{11}{12}$ (c) $\frac{6}{7}$ (d) $\frac{4}{5}$ (e) $\frac{1}{2}$

5 (i) $\frac{5}{8}$ (ii) $\frac{3}{5}$ (iii) $\frac{13}{20}$

6 (ii) 0.512   (a) 0.064   (b) 0.479   (c) 0.457

7 (i) $\frac{63}{125}$ (ii) $\frac{54}{125}$ (iii) $\frac{369}{625}$ (iv) 44

8 (a) (i) $\frac{2}{5}$ (ii) $\frac{28}{125}$ (iii) $\frac{118}{125}$ (iv) $\frac{7}{20}$

    (b) (i) C (ii) B (iii) $\bar{A}$

    (c) (i) $\frac{179}{250}$ (ii) $\frac{21}{179}$

9 (i) $\frac{5}{6}$ (ii) $\frac{13}{15}$

10 (iii) (a) $\frac{19}{96}$ (b) $\frac{19}{96}$ (c) $\frac{173}{576}$

Exercises 7a

1

x	0	1	2
P_x	$\frac{3}{28}$	$\frac{15}{28}$	$\frac{10}{28}$

2

x	0	1	2
P_x	$\frac{9}{64}$	$\frac{30}{64}$	$\frac{25}{64}$

3

x	2	3	4	5	6	7	8
P_x	$\frac{1}{12}$	$\frac{2}{12}$	$\frac{2}{12}$	$\frac{2}{12}$	$\frac{2}{12}$	$\frac{2}{12}$	$\frac{1}{12}$

4

x	0	3	10
P_x	$\frac{18}{20}$	$\frac{1}{20}$	$\frac{1}{20}$

5

x	0	1	2
P_x	$\frac{19}{34}$	$\frac{13}{34}$	$\frac{2}{34}$

6

x	1	$\frac{1}{2}$	$\frac{1}{3}$	$\frac{1}{4}$	$\frac{1}{5}$	$\frac{1}{6}$
P_x	$\frac{1}{6}$	$\frac{1}{6}$	$\frac{1}{6}$	$\frac{1}{6}$	$\frac{1}{6}$	$\frac{1}{6}$

7

x	-5	-4	-3	-2	-1	0	1	2
P_x	$\frac{1}{36}$	$\frac{2}{36}$	$\frac{3}{36}$	$\frac{4}{36}$	$\frac{5}{36}$	$\frac{6}{36}$	$\frac{5}{36}$	$\frac{4}{36}$

x	3	4	5
P_x	$\frac{3}{36}$	$\frac{2}{36}$	$\frac{1}{36}$

8

x	0	1	2	3	4	5
P_x	$\frac{3}{18}$	$\frac{5}{18}$	$\frac{4}{18}$	$\frac{3}{18}$	$\frac{2}{18}$	$\frac{1}{18}$

9

x	0.40	1.40	2.40
P_x	$\frac{1}{400}$	$\frac{38}{400}$	$\frac{361}{400}$

10 (ii), (v), (vi), (viii), (ix)

Exercises 7b

1 (i) $P_x = \frac{1}{10}$, $x = 0, 1, \ldots 9$

    (ii) No; Y: $P_1 = P_2 = \frac{111}{300}$, $P_3 = \frac{12}{300}$,

       $P_4 = \ldots = P_9 = \frac{11}{300}$

2 $P(N = n) = \frac{1}{6} \left(\frac{5}{6}\right)^{n-1}$, $n = 1, 2, \ldots$

    (i) 0.335   (ii) 0.598

3 $P_0 = \frac{5}{6}$, $P_1 = \frac{1}{6}$

4 $P_0 = P_1 = \frac{1}{2}$

6 (i) $P_n = \frac{1}{20} \left(\frac{19}{20}\right)^{n-1}$, $n = 1, 2, \ldots$

    (ii) $P_0 = \frac{19}{20}$, $P_1 = \frac{1}{20}$

Exercises 7c

1 $\frac{5}{4}$

2 $\frac{5}{4}$

3 5

4 0.65

5 $\frac{1}{2}$

6 0.408

7 0

8 $\frac{35}{18}$

9 2.30

10 $E(X) = 4.5$, $E(Y) = 2.66$

11 6

12 $\frac{1}{6}$

13 $\frac{1}{2}$

14 1

Exercises 7d

1 $\frac{55}{28}$

2 $\frac{65}{32}$

3 $\frac{169}{6}$

4 5.45

5 $\frac{21}{34}$

6 0.249

7 $\frac{35}{6}$

8 $\frac{35}{6}$

9 5.385

10 (i) $\frac{9}{4}$ (ii) $\frac{33}{4}$ (iii) $\frac{5}{4}$ (iv) $\frac{5}{4}$

11 (i) $\frac{21}{5}$ (ii) $\frac{147}{5}$ (iii) $\frac{23}{5}$ (iv) 2

Exercises 7e

1 $\frac{45}{112}$, 0.634

2 $\frac{15}{32}$, 0.685

3 $\frac{19}{6}$, 1.780

4 $\frac{2011}{400} = 5.03$, 2.24

5 $\frac{25}{68}$, 0.606

6 0.0818, 0.286

7 $\frac{35}{6}$, 2.415

8 $\frac{665}{324} = 2.05$, 1.43

9 0.095, 0.308

11 3

12 0.1, 0.9

13 $\frac{4}{7}, \frac{2}{7}, \frac{4}{35}, \frac{1}{35}; \frac{8}{5}, \frac{16}{25}$

Exercises 7f (Miscellaneous)

1 (a) $P(X = 0) = \frac{15}{32}$, $P(X = 1) = \frac{11}{32}$, $P(X = 2) = \frac{5}{32}$,
$P(X = 3) = \frac{1}{32}$
(b) $\frac{11}{16}$ (c) $\frac{4}{11}$

2 (i) $\frac{3}{10}$ (ii) 52p

3 (i) $\frac{5}{18}$ (ii) $\frac{10}{3}$

4

s	0	1	2
P_s	$\frac{3}{25}$	$\frac{20}{25}$	$\frac{2}{25}$

; mean $= \frac{24}{25}$

5 $\frac{1}{3}(2n + 1)$

6 (i) $\frac{1}{4}$ (ii) $\frac{7}{16}$ (iii) $\frac{3}{16}$

n	0	1	2	3
$P(N = n)$	$\frac{3}{16}$	$\frac{7}{16}$	$\frac{5}{16}$	$\frac{1}{16}$

$E(N) = \frac{5}{4}$

7 (i) $\frac{3}{16}$ (ii) $\frac{17}{32}$; $\mathrm{Var}(X) = 0.601$

8 (a) $\frac{1}{4}, \frac{5}{4}$ (c)

z	2	3	4	5	6
$P(Z = z)$	$\frac{1}{8}$	$\frac{1}{4}$	$\frac{1}{4}$	$\frac{1}{4}$	$\frac{1}{8}$

$4, \frac{3}{2}$

9 (a) 20 (b) 0.0387 (c) 0.265

Exercises 8a

1 (i) 18 (ii) 18 (iii) −6 (iv) 18

2 (i) 1, 4 (ii) 10, 36 (iii) −7, 36

3 $\frac{1}{4}, \frac{1}{36}$

4 (i) 81 (ii) 85 (iii) 78

5 2, 3

6 −4, 9

7 $\frac{15}{2}, \frac{5}{2}$

8 £50, £2

9 $(a, b) = \pm\left(\dfrac{1}{\sigma}, -\dfrac{\mu}{\sigma}\right)$

10 σ^2

11 $(c, d) = (3, 85)$ or $(-3, 115)$

12 $1, \frac{35}{3}$

13 10, 50

Exercises 8b

1 $E(U) = 10$, $\mathrm{Var}(U) = 12$, $E(V) = 12$, $\mathrm{Var}(V) = 7$,
$E(W) = -9$, $\mathrm{Var}(W) = 19$

2 (i) 7 (ii) 25 (iii) 0 (iv) 337
(v) $\frac{25}{12}$ (vi) 2

3 $\mu, \frac{1}{2}\sigma^2$

4 (i) 50, 73 (ii) 442

5 −20, 106

6 150, 25

Exercises 8c

1 $12, \frac{1}{9}$

2 3.5, 0.171

3 6, 0.775

4 $10 - 9p$, $\frac{81}{200}p(1 - p)$

5 $90\,\mathrm{kg}$, $0.4\,\mathrm{kg}^2$, $22\,500\,\mathrm{kg}$, $158\,\mathrm{kg}$

6 401

7 84, 252

8 $E(m) = 1$, $\mathrm{Var}(m) = \frac{2}{9}$, $P(M = 1) = \frac{13}{27}$,
$P(M = 2) = \frac{7}{27}$
$E(M) = 1$, $\mathrm{Var}(M) = \frac{14}{27}$, $E(U) = \frac{5}{3}$, $\mathrm{Var}(U) = \frac{8}{27}$

Exercises 8d

1 $\frac{1}{256}t^4(1 + 3t)^4$ (i) $\frac{27}{128}$ (ii) $7, \frac{3}{4}$

2 $P(X = 0) = P(X = 3) = \frac{1}{8}$
$P(X = 1) = P(X = 2) = \frac{3}{8}$; $\frac{3}{2}, \frac{3}{4}$

3 $P(R = -2) = \frac{1}{4}$, $P(R = 0) = \frac{1}{2}$, $P(R = 2) = \frac{1}{4}$

4 $\frac{7}{2}, \frac{35}{12}$

5 $\left(\dfrac{t^2 + 1}{2t}\right)^{10}$ (i) $\frac{15}{128}$ (ii) 0 (iii) 0 (iv) 10

7 (i) $k = e^{-1}$ (ii) e^{t-1}; 1, 1 (iii) 0.209

8 $G_X(t) = \dfrac{t(1 - t^n)}{n(1 - t)}$
$E(X) = \frac{1}{2}(n + 1)$ (i) $\frac{25}{2}$ (iv) $\frac{5}{72}$

9 (i) $a = 2$, $b = 1$ (ii) $\{3, 4, 5, \ldots\}$ (iii) 2
(iv) $\dfrac{t^6}{(2 - t^2)}, \frac{5}{64}$

10 $G_X(t) = \dfrac{1}{2 - t}$; 1, 2, $G_Y(t) = \dfrac{4k}{(2 - t)^2}$, $k = \frac{1}{4}$
(i) 3 (ii) $\frac{7}{128}$

11 $G_r(t) = \frac{1}{27}(9 + 6t + 4t^2 + 8t^4)$,
$E(X_r) = \frac{46}{27}$, $\mathrm{Var}(X_r) = 2.65$
$G_{SUM}(t) = \frac{1}{729}(9 + 6t + 4t^2 + 8t^4)^2$, 0.219;
$G_{DIFF}(t) = \frac{1}{729}(9 + 6t + 4t^2 + 8t^2)\left(9 + \dfrac{6}{t} + \dfrac{4}{t^2} + \dfrac{8}{t^4}\right)$;
0.270, 0.365

12 (i) $\dfrac{pt}{1 - (1 - p)t}$
(ii) $\dfrac{1}{p}$, $p(1 - p)^{r-1}$, where $p = \dfrac{n}{2^{n-1}}$; 6

13 (a) f_1 and f_2 (iii) $\dfrac{1}{\ln 2}$ (b) (i) $\frac{3}{8192}$ (ii) −1

14 170, 111

Exercises 8e (Miscellaneous)

1 (a) 2.8, 4.96 (b) $P_0 = 0.09$, $P_2 = 0.12$, $P_4 = 0.22$,
$P_6 = 0.24$, $P_8 = 0.17$, $P_{10} = 0.12$, $P_{12} = 0.04$, 5.6, 9.92
(c) $\frac{8}{39}$ (d) $\frac{13}{73}$, 1

2 (i) (a) $\frac{2}{15}$ (b) $\frac{2}{5}$
(ii) (a) $k = \frac{1}{100}$ (b) 3.54, 0.468 (c) 14.7, 11.7

3 (a) 0.128
(b) $P_r = (0.2)(0.8)^{r-1}$, $r = 1, 2, \ldots$, geometric
(c) 0.512; 10, 40; 0.0768

4 (i) $q = 1 - 3p$ (ii) p, $\sqrt{(3p - p^2)}$
(iii) $P(Y = -2) = p^2$, $P(Y = -1) = 2p(1 - 3p)$,
$P(Y = 0) = 1 - 6p + 13p^2$,
$P(Y = 1) = 4p(1 - 3p)$,
$P(Y = 2) = 4p^2$, $E(Y) = 2p$

5 $\mu = \frac{21}{4}$; $\frac{1}{24}, \frac{21}{2}$

6 2, 1; $P(Y = -4) = P(Y = 4) = \frac{1}{16}$,
$P(Y = -2) = P(Y = 2) = \frac{1}{4}$, $P(Y = 0) = \frac{3}{8}$; 4, 3

7 $k = \frac{1}{15}; \frac{2}{3}, \frac{34}{45}, \frac{3}{225}, \frac{4}{3}, \frac{68}{45}$

8 $£(108 - 4x)$; $x = 2$; $(£^2) \frac{245}{9}$

Exercises 9a

1 (i) $\frac{1}{27}$ (ii) $\frac{6}{27}$ (iii) $\frac{12}{27}$ (iv) $\frac{8}{27}$
(v) $\frac{8}{27}$ (vi) $\frac{12}{27}$ (vii) $\frac{6}{27}$ (viii) $\frac{1}{27}$

2 (i) $\frac{81}{256}$ (ii) $\frac{54}{256}$ (iii) $\frac{1}{256}$

3 (i) $\frac{1}{4}$ (ii) $\frac{3}{8}$; No

4 $\frac{5}{72}$ (i) No (ii) No

5 (i) 0.0001 (ii) 0.9999

6 0.336

Exercises 9b

1 (i) 0.250 (ii) 0.311 (iii) 0.393 (iv) 0.617
(v) 0.711

2 0.315

3 0.318

4 0.844, 0.156

5 0.211

6 0.0172

7 (i) 0.531 (ii) 0.984

8 0.368

9 0.408

10 0.275

Exercises 9c

1 0.015

2 4

3 (i) 0.328, 0.410, 0.205, 0.0512, 0.0064, 0.000 32
(ii) 0.0751, 0.225, 0.300, 0.234, 0.117, 0.0389

4 (i) 0.118, 0.303, 0.324, 0.185
(ii) 0.349, 0.387, 0.194, 0.057

5 0.167, 0.149, 0.158

6 $B(n_1 + n_2, p)$

7 6

Exercises 9d

1 (i) 0.942 (ii) 0.001 29

2 (i) 0.0548 (ii) 0.111 (iii) 0.633

3 (i) 0.869 (ii) 0.703

4 0.717

5 0.0171, 0.0471

6 0.135

Exercises 9e

1 10, 8

2 (i) 2, 1.6 (ii) 1, 0.9 (iii) 3, 2.5

3 $\mu = \frac{5}{3}$; $\frac{25}{18}$, 0.485

4 4.4, 0.332

5 4.2, 1.65; 0.653

6 0.197

7 0.1, 0.9; 0.007 44, 0.000 827

8 $\mu = 18.4$, $\sigma = 1.21$; 0.599, 0.982

9 15, 0.7

10 101

Exercises 9f (Miscellaneous)

1 (i) 0.187 (ii) 2

2 $\frac{23}{120}$; 27.9, 39.7, 23.5, 7.4, 1.3, 0.1, 0.0

3 $(0.9)^n$, 29

4 56

5 (a) 0.736 (b) 0.188

6 0.996

7 0.205

8 (i) 0.0081, 0.0756, 0.2646, 0.4116, 0.2401 (ii) 8.34
(iii) 8 (iv) 0.652

9 (i) 0.8 (ii) 0.180

10 (a) (i) $\frac{1}{2}$ (ii) $\frac{1}{40}$ (iii) $\frac{1}{20}$ (iv) $\frac{9}{20}$
(b) (i) 0.590 (ii) 0.009 (iii) 0.237

Exercises 10a

1 (i) 0.135 (ii) 0.271 (iii) 0.271 (iv) 0.677
(v) 0.594

2 (i) 0.986 (ii) 0.090 (iii) 0.0144

3 (i) 0.175 (ii) 0.440 (iii) 0.384

4 0.469

5 3

6 (i) 0.191 (ii) 0.658

7 0.0341

8 0.256

9 0.368

Exercises 10b

1 0.135, 0.271, 0.271, 0.180, 0.0902

2 0.175, 0.175, 0.146, 0.104

3 0.013, 0.076, 0.303, 0.607

4 2.5, 8.96

Exercises 10c

1 (i) 0.9161 (ii) 0.9665

2 (i) 0.0629 (ii) 0.0023

3 (i) 0.1128 (ii) 0.3359

4 $\lambda = 1.8$

5 $\lambda = 2.0$

6 $P(X = x) = \dfrac{e^{-1}}{x!}$, 0.0144, 0.205

Exercises 10d

1 (i) 0.433 (ii) 0.762

2 (i) 0.594 (ii) 0.857

3 0.638

4 0.953

5 (i) 0.697 (ii) 0.072

6 0.067

7 0.164; 0.2, 0.2

8 Exact: (i) 0.141 (ii) 0.228 (iii) 0.236
Approx: (i) 0.149 (ii) 0.224 (iii) 0.224

9 (i) 0.0488   (ii) 2303
10 0.677
11 (i) 0.325   (ii) 0.221; 3.5; 0.321

Exercises 10e (Miscellaneous)

1 (i) 0.310   (ii) 0.819
2 0.492
3 0.135, $\geqslant 461$
4 (i) 0.267   (ii) 0.468
5 (i) 0.222   (ii) 0.0649
6 (i) 0.819   (ii) 0.407   (iii) 0.122
7 (i) $C = \frac{1}{32}$, mode = 1, mean = $\frac{63}{32}$
    (ii) (a) 0.249   (b) 0.929   (c) 0.508; 0.542
8 0.115, 6
9 (i) 0.987   (ii) 0.514; 0.187
10 (i) $\frac{1}{21}$ (ii) $\frac{20}{9}$; 0.528
11 (a) 0.0498   (b) 0.199   (c) 0.166
12 (a) (i) 1, 4   (ii) 10,
    10   (b) (i) 0.577   (ii) 0.0404
    (iii) 0.0404   (iv) £260, £118.32
13 (a) (i) 0.670 32   (ii) 0.061 55   (iii) 0.148 13
    (b) $e^{-2n/5}$, 5   (c) (i) $\frac{3}{2}$ (d) 0.594
14 (i) 0.974   (ii) 87.9
15 (i) 0.127   (ii) 0.044
    (iii) Binomial: 0.585 or Poisson: 0.577   (iv) 113
    (v) 0.606
16 0.043
17 (i) (b) 1.5   (ii) 0.577   (iii) 0.025
18 (a) {0, 1, 2, 3, 4}   (i) $\frac{14}{55}$ (ii) $\frac{42}{55}$
    (b) (i) 0.238   (ii) 0.392
    (c) Poisson with mean 32.5

Exercises 11a

1 (i) $\frac{1}{18}$ (ii) $\frac{5}{27}$ (iii) $\frac{14}{27}$
2 (i) $\frac{3}{16}$ (ii) $\frac{7}{16}$ (iii) $\frac{1}{8}$ (iv) $\frac{1}{16}$
3 (i) $\sqrt{5}$ (ii) $\frac{7}{16}$ (iii) 0   (iv) $\frac{1}{4}$ (v) $\sqrt{5}$
4 (i) $\frac{1}{2}$ (ii) $\frac{3}{4}$
5 (i) 1   (ii) 0   (iii) $\frac{3}{4}$
6 $\frac{1}{24}$ (ii) 3   (iii) $\frac{13}{36}$
7 (i) No   (ii) Yes   (iii) No   (iv) Yes
8 12; $\frac{1}{10\,000}(600c^2 - 80c^3 + 3c^4)$; 0.916, 0.949, 0.973;
    750 gallons

Exercises 11b

1 (i) $\frac{1}{8}$
    (ii) $F(x) = \begin{cases} 0 & x \leqslant 1 \\ \frac{1}{16}(x+5)(x-1) & 1 \leqslant x \leqslant 3 \\ 1 & x \geqslant 3 \end{cases}$
    (iii) $\frac{9}{16}$ (iv) 2.12
2 (i) $\frac{1}{4}$ (ii) 0
    (iii) $F(x) = \begin{cases} 0 & x \leqslant -2 \\ \frac{1}{8}(4 - x^2) & -2 \leqslant x \leqslant 0 \\ \frac{1}{8}(x^2 + 4) & 0 \leqslant x \leqslant 2 \\ 1 & x \geqslant 2 \end{cases}$
    (iv) $\frac{1}{8}$

3 (i) $\frac{1}{4}$
    (ii) $F(x) = \begin{cases} 0 & x \leqslant 1 \\ \frac{1}{2}(x-1) & 1 \leqslant x \leqslant 2 \\ \frac{1}{4}x & 2 \leqslant x \leqslant 4 \\ 1 & x \geqslant 4 \end{cases}$
    (iii) 1.2   (iv) 3.2
4 (i) $\frac{1}{4}$
    (ii) $F(x) = \begin{cases} 0 & x \leqslant -2 \\ \frac{1}{8}(x+2)^2 & -2 \leqslant x \leqslant 0 \\ \frac{1}{16}(8 + 6x - x^2) & 0 \leqslant x \leqslant 2 \\ 1 & x \geqslant 2 \end{cases}$
    (iii) $\frac{11}{16}$ (iv) $\frac{3}{16}$
5 (i) $\frac{1}{3}$
    (ii) $F(x) = \begin{cases} 0 & x \leqslant -2 \\ \frac{1}{9}(8 + x^3) & -2 \leqslant x \leqslant 1 \\ 1 & x \geqslant 1 \end{cases}$
    (iii) -2 (iv) -1.52
6 (i) $\frac{3}{19}$
    (ii) $F(x) = \begin{cases} 0 & x \leqslant 0 \\ \frac{1}{19}\{(x+2)^3 - 8)\} & 0 \geqslant x \geqslant 1 \\ 1 & x \geqslant 1 \end{cases}$
    (iii) 0.596
7 (i) k (ii) $\frac{3}{4}$
    (iii) $F(x) = \begin{cases} 0 & x \leqslant 1 \\ \frac{1}{4}(x^3 - 3x + 2) & 1 \leqslant x \leqslant 2 \\ 1 & x \geqslant 2 \end{cases}$
8 (i) -6
    (ii) $F(x) = \begin{cases} 0 & x \leqslant 3 \\ (x-3)^2 & 3 \leqslant x \leqslant 4 \\ 1 & x \geqslant 4 \end{cases}$
    (iii) 3.5, 3.87
9 (i) $\frac{1}{9}$
    (ii) $F(x) = \begin{cases} 0 & x \leqslant 0 \\ \frac{1}{9}x & 0 \leqslant x \leqslant 1 \\ \frac{1}{9}(4x - 3) & 1 \leqslant x \leqslant 3 \\ 1 & x \geqslant 3 \end{cases}$
    (iii) $\frac{57}{40}$
10 (i) 1
    (ii) $F(x) = \begin{cases} 0 & x \leqslant 0 \\ x(2 - x) & 0 \leqslant x \leqslant 1 \\ 1 & x \geqslant 1 \end{cases}$
    (iii) 0.293
11 (i) $\frac{2}{7}$
    (ii) $F(s) = \begin{cases} 0 & s \leqslant 0 \\ \frac{1}{7}s(s + 6) & 0 \leqslant s \leqslant 1 \\ 1 & s \geqslant 1 \end{cases}$
    (iii) 0.536   (iv) 0.536
12 (i) $\frac{1}{129}$
    (ii) $F(t) = \begin{cases} 0 & t \leqslant 5 \\ \frac{1}{387}(t^3 - 125) & 5 \leqslant t \leqslant 8 \\ 1 & t \geqslant 8 \end{cases}$
    (iii) 0.461
13 (a) $a = 1.4$, $b = -0.8$ (b) 0.404, 0

Exercises 11c

1 (i) $\frac{4}{3}$ (ii) 2   (iii) $\frac{2}{9}$ (iv) $\frac{4}{9}$
2 (i) 1   (ii) $\frac{7}{15}$
3 (i) $\frac{3}{2}$ (ii) $\frac{7}{12}$

4 (i) $\frac{20}{9}$ (ii) $\frac{235}{162} = 1.45$ (iii) $\frac{5}{2}$ (iv) $\frac{11}{27}$
   (v) $\frac{302}{15} = 20.1$ (vi) $\frac{16}{27}$

5 (i) $\frac{1}{8}$ (ii) $\frac{17}{4}$ (iii) $\frac{1}{192}$ (iv) $\frac{1}{48}$

6 (i) $\frac{14}{3}$ (ii) $\frac{8}{9}$

7 (i) $\frac{4}{65}$ (ii) $\frac{844}{325}$ (iii) 0.0765   (iv) 0
   (v) 0.153

8 (i) $\frac{1}{2}$ (ii) $\frac{1}{12}$ (iii) 6   (iv) 1

9 (i) $\frac{1}{9}$ (ii) $\frac{17}{4}$ (iii) 0.482

10 $k = 2,\ \mu = 1;\ 0.325$

11 $c = \frac{1}{3},\ 0.5,\ 0.022$

Exercises 11d

1 (i) $f(x) = \begin{cases} \frac{2}{x^2} & 1 < x < 2 \\ 0 & \text{otherwise} \end{cases}$
  (ii) 1.39

2 (i) $-\frac{1}{8}$ (ii) $\frac{1}{8}$ (iii) $\frac{13}{6}$ (iv) $\frac{11}{36}$

3 (i) $\frac{1}{a^2}$ (ii) $\frac{4}{3}$
  (iii) $f(x) = \begin{cases} \frac{9}{8}\left(x - \frac{4}{3}\right) & \frac{4}{3} < x < \frac{8}{3} \\ 0 & \text{otherwise} \end{cases}$
  (iv) 2.28

4 (i) $f(x) = \begin{cases} \frac{1}{4} & 4 < x < 8 \\ 0 & \text{otherwise} \end{cases}$
  (ii) 6   (iii) 5, 7   (iv) 6

5 (i) $f(x) = \begin{cases} \frac{1}{8} & -4 < x < 0 \\ \frac{1}{8} & 4 < x < 8 \\ 0 & \text{otherwise} \end{cases}$
  (ii) 2   (iii) 0   (iv) $\frac{52}{3}$

6 (i) $a = -\frac{1}{2},\ b = \frac{1}{2}$
  (ii) $f(x) = \begin{cases} \frac{1}{4} & 0 < x < 2 \\ \frac{1}{2} & 2 < x < 3 \\ 0 & \text{otherwise} \end{cases}$
  (iii) 1, 2.5

7 (i) $a = \frac{3}{7},\ b = \frac{1}{14}$
  (ii) $f(x) = \begin{cases} x & 0 < x < 1 \\ \frac{3}{14}x^2 & 1 < x < 2 \\ 0 & \text{otherwise} \end{cases}$
  (iii) 1   (iv) 1   (v) $\frac{191}{168}$

Exercises 11e

1 (i) $f(x) = \begin{cases} \frac{1}{2} & 0 < x < 2 \\ 0 & \text{otherwise} \end{cases}$
  (ii) $F(x) = \begin{cases} 0 & x \leqslant 0 \\ \frac{1}{2}x & 0 \leqslant x \leqslant 2 \\ 1 & x \geqslant 2 \end{cases}$
  (iii) $\frac{1}{4}y,\ (0 < y < 4)$
  (iv) $g(y) = \begin{cases} \frac{1}{4} & 0 < y < 4 \\ 0 & \text{otherwise} \end{cases}$

2 (i) $a = 1,\ b = 7$ (ii) $\frac{1}{3}$

3 (i) 0.4   (ii) 0.577

4 $c = 1,\ d = 9$

5 $a = 4,\ b = 10$

6 (a) (i) 90   (ii) 2700   (b) (i) 180   (ii) 10 800

7 (a) $(h, k) = (b - a, a)$ or $(a - b, b)$
  (b) $r = \sigma\sqrt{12},\ s = \mu - \sigma\sqrt{3}$

8 $\frac{1}{4}$

9 (i) $\frac{2}{\pi}$ (ii) $\frac{2}{\pi}$ (iii) $\frac{1}{2}$ (iv) $\frac{1}{2}$

10 (i) $\frac{x^2}{a^2}$ for $x \leqslant a$, 1 for $x > a$ (iii) $\frac{2}{3}a$

11 $\frac{1}{6}h^2,\ \frac{1}{2}$

Exercises 11f

1 (i) $f(x) = \begin{cases} \frac{1}{3}e^{-x/3} & x > 0 \\ 0 & \text{otherwise} \end{cases}$
  (ii) $F(x) = \begin{cases} 0 & x \leqslant 0 \\ 1 - e^{-x/3} & x \geqslant 0 \end{cases}$
  (iii) 0.632   (iv) 0.264   (v) 0.104

2 3.48

3 (i) 0.607   (ii) 0.632

4 0.275

5 0.574

6 (i) 0.135   (ii) 0.368

7 (i) 0.472   (ii) 0.223; 0.0821

8 (i) 0.195   (ii) 0.135

Exercises 11g

1 (i) $\frac{1}{\pi(1 + x^2)}$ (iii) 1

3 (i) $\frac{1}{2}$ (ii) $\frac{1}{12}$ (iii) $\frac{1}{4}$ (iv) $\frac{1}{5}$

Exercises 11h (Miscellaneous)

1 (i) $\frac{5}{8}, \frac{19}{320}$ (ii) $\frac{5}{16}$ (iii) 0.0305   (iv) $\frac{29}{40}$

2 $\frac{3}{4}, \frac{19}{80}$
$$F(x) = \begin{cases} 0 & x \leqslant 0 \\ \frac{1}{16}x(12 - x^2) & 0 \leqslant x \leqslant 2 \\ 1 & x \geqslant 2 \end{cases}$$
0.00659

3 (i) 0.455, 3   (ii) 3.64, 4.95
  (iii) $F(x) = \begin{cases} 0 & x \leqslant 1 \\ \ln x/\ln 9 & 1 \leqslant x \leqslant 9 \\ 1 & x \geqslant 9 \end{cases}$

4 (i) $\frac{1}{4}$ (ii) 0.134
  (iii) $f(x) = \begin{cases} 2(1 - x) & 0 < x < 1 \\ 0 & \text{otherwise} \end{cases}$
  (iv) $\frac{1}{3}, \frac{8}{15}$

5 (i) $\frac{3}{4}, \frac{1}{4}$ (iii) $\frac{1}{2}$ (iv) $\frac{27}{64}$ (v) $\frac{175}{48}$

6 (a) $\frac{2}{5}$,   (b) 0.162   (c) 0.446   (d) 0.239
  (e) $\frac{1}{5}$

7 (a) $F(w) = \begin{cases} 0 & w \leqslant 0 \\ \frac{1}{3125}w^4(25 - 4w) & 0 \leqslant w \leqslant 5 \\ 1 & w \geqslant 5 \end{cases}$
  (b) 0.650   (c) 0.794   (d) 3.75

8 $a = \frac{1}{50},\ b = \frac{1}{15}$ (a) $\frac{20}{3}$ m   (b) $\frac{11}{3}$
  (c) $\frac{311}{6} = 51.8;\ \frac{9}{20}$

Exercises 12a

Note that the use of normal distribution values from a calculator may lead to slightly different values to those given, which are based on tables.

1 (i) 0.8849   (ii) 0.0359   (iii) 0.0808
   (iv) 0.2119
2 (i) 0.0113   (ii) 0.5403   (iii) 0.3848
3 (i) 0.5762   (ii) 0.1096   (iii) 0.5208
4 (i) 1.4   (ii) -0.4 (iii) -1.2 (iv) 2.6
   (v) 1.6   (vi) -1.8
5 (i) 0.6   (ii) 1.6

Exercises 12b

1 (i) 0.1587   (ii) 0.9452   (iii) 0.1151
   (iv) 0.7881
2 (i) 0.8041   (ii) 0.3413   (iii) 0.2515
   (iv) 0.0968
3 (i) 0.7881   (ii) 0.0548   (iii) 0.5434
   (iv) 0.3674
4 (i) 0.2119   (ii) 0.2347   (iii) 0.1859
   (iv) 0.8844
5 (a) 0.8849   (b) 0.2119   (c) 0.3446
   (d) 0.9641   (e) 0.2881   (f) 0.8490
   (g) 0.3087

Exercises 12c

1 (i) 0.8243   (ii) 0.1084   (iii) 0.0787
   (iv) 0.6981
2 (i) 0.000657   (ii) 0.999849   (iii) 0.00225
   (iv) 0.99758
3 (i) 0.4231   (ii) 0.6970   (iii) 0.2772
   (iv) 0.1679
4 (i) 0.7181   (ii) 0.5248   (iii) 0.1241
   (iv) 0.4783
5 (i) 0.3696   (ii) 0.6915   (iii) 0.2266
   (iv) 0.8413
6 (i) 0.1056   (ii) 0.5986   (iii) 0.6915
   (iv) 0.2266
7 63 g
8 0.0931, heavy, heavy

Exercises 12d

1 (i) 1.881   (ii) 1.645   (iii) -3.090
   (iv) -2.326 (v) 3.719   (vi) 3.090
2 (i) 29.405   (ii) 28.225   (iii) 4.550   (iv) 8.370
   (v) 38.595   (vi) 35.450
3 6.47
4 -0.037
5 24.4
6 5.62
7 $\mu = 3.08$, $\sigma = 3.66$
8 (i) 411.6 g   (ii) 432.8 g
9 6.08, $\sigma < 6.08$
10 $\mu = 2.07$ m, $\sigma = 0.458$ m

11 $\mu = 507.1$ m, $\sigma = 7.34$ m
12 $\mu = 3.62$, $\sigma^2 = 0.00256$
13 (i) 0.66   (ii) 75.7
14 $\mu = 50.15$, $\sigma = 4.00$

Exercises 12e

1 (i) 0.866   (ii) 0.710   (iii) 0.203   (iv) 0.609
2 (i) 0.500   (ii) 0.650   (iii) 0.242   (iv) 0.807
3 (i) 0.244   (ii) 0.500   (iii) 0.132   (iv) 0.411
4 (i) 0.327   (ii) 0.079   (iii) 0.186   (iv) 0.0023
5 0.977
6 0.655
7 (i) 0.217   (ii) 0.00024
8 0.074
9 (i) 0.197   (ii) 0.345
10 (i) 0.798   (ii) 0.323   (iii) 0.132   (iv) 0.228
11 (ii) 0.0802   (iii) 0.516
12 (i) 0.904   (ii) 0.952; 0.324
13 0.981   (i) 0.037   (ii) 0.0016 (assume equal size batches)
14 (a) 0.0228   (b) 0.988   (c) 0.083   (d) 0.209
15 (i) 0.252   (ii) 0.290   (iii) 0.722   (iv) 21.9
16 (i) 0.933   (ii) 0.006   (iii) 0.858
17 (i) 0.559   (ii) 0.933   (iii) 0.289
18 (i) 0.037   (ii) 0.815   (iii) 108, 1.40
19 (a) 0.036   (b) 0.020
20 197 mm; 0.292, 0.939, 0.881, 208 mm

Exercises 12f

1 $N(15, \frac{5}{14})$, 0.453
2 (i) 0.923   (ii) 0.988
3 (i) 0.094   (ii) 0.902
4 (i) 0.977   (ii) 0.977
5 193
6 65
7 (i) 0.843   (ii) 0.986   (iii) 0.998
8 (i) 0.937   (ii) 9.59
9 (i) 0.885   (ii) 207.7   (iii) 0.344   (iv) 1.10
10 0.050, 0.819
11 (a) $N\left(\mu, \frac{\sigma^2}{n}\right)$ (b) 28
12 (a) 0.159   (b) 2, 0.125   (c) 1, 0.0625; 0.103
13 (a) 0.773   (b) 0.628   (c) 2.80 kg   (d) 0.936
   (e) $N(19.0, 0.0408)$ (f) 18.7 kg to 19.4 kg
14 (ii) (a) 0.345   (b) 0.611   (iii) 0.758
   (iv) 31 mm   (v) 0.263   (vi) 97
15 (a) (i) 0.036   (ii) 0.915; 0.836, 8   (b) 0.601

Exercises 12g

1 0.114
2 0.835
3 0.036
4 0.848
5 (i) 0.113   (ii) 0.047
6 0.913, 0.016

7 0.197, 0.088

8 0.994

9 0.821

10 0.068

11 (i) 0.324 (ii) 0.608; 0.929

12 0.196, 9, 0.0024, £13 200

13 (a) $n = 10$, $p = \frac{1}{2}$ (b) (i) 0.694 (ii) 0.049; 0.0019

14 0.859 (c) 0.203 (d) 0.034

Exercises 12h

1 0.069

2 0.959

3 0.092

4 0.962

5 0.964

6 (a) 0.102 (b) 0.099; 0.749

7 (i) 0.858 (ii) 0.155 (iii) 0.22

8 (i) 0.067 (ii) 0.286 (iii) 0.739 (iv) 0.465
(v) N(29.7, 29.7), 0.442

9 (ii) (1) 0.195 (2) 0.567 (4) 0.147

10 (i) 0.404 (ii) 0.009 (iii) 0.099; 0.088

11 (i) 0.938 (ii) 0.371 (iii) 0.192 (iv) £16.40

12 (i) (A) 0.741 (B) 0.037 (iii) 100

Exercises 12i

1 (b) 178, 272

2 (a) (i) 96.2, 27.3 (ii) 113.4, 78.6 (d) 0.64

4 (i) Heights in ratios $3 : 13 : 52 : 178 : 242 : 166 : 80 : 22 : 2$

Exercises 12j (Miscellaneous)

1 0.527, 0.966

2 (a) 0.194 (b) 0.667 (c) 0.736 (d) 0.683

3 (a) 0.0064 (b) 0.189 (c) 0.189

4 0.159, 0.774, 0.067; 2.7 kg; £37.53

5 94.9%, 0.496; 2.19 μm, 0.703 μm; 3.1%

6 (i) 0.38% (ii) 98.6%;
0.008 mm; 9.508 mm, 0.007 mm

7 (a) 0.115 (b) 0.311 (c) 167.75 (d) 0.096
(e) 0.053

8 (a) 0.115 (b) 33.3 g (c) N(510, 375), 0.788
(d) 0.111

9 (a) 0.04 mm, 0.05 mm (b) 21.2% (c) 0.059
(d) 0.042 mm, (e) 5.9%

10 (i) 0.976 (ii) 0.026 (iii) 0.027

11 0.021; 0.143; 0.814

12 (i) 0.018 (ii) 507.3 (iii) 14

13 0.21; 450 g, 12.98 g

14 (i) $c = \frac{1}{10}$, $d = \frac{1}{6}$; $\frac{16}{3}$ (ii) $\frac{11}{100}$
(iii) 240, 256; 0.800

15 (*a*) 0.971 (*b*) 78 (*c*) 10

16 0.809, 0.026

17 0.692, 0.030

18 (i) 105 hr (ii) 0.105 (iii) (a) A (b) B

19 (i) 0.988, 0.606 (ii) 0.855 (iii) 0.783

20 (a) (ii) 0.90 (b) (i) 0.155 (ii) 0.497, 0.042

21 (c) 0.311 (d) 0.149
(e) 0.697 (Poisson) or, more accurately, 0.711
(binomial)

22 $\mu = 2.56$, $\sigma = 0.27$; 5.11 mm, 0.38 mm; 0.99; 0.88

23 (i) f$(x) = 1/(2a)$ for $-a < x < a$, f$(x) = 0$ otherwise
(ii) 0, $\frac{1}{3}a^2$
(iii) 0, $\frac{1}{3}na^2$, N$(0, \frac{1}{3}na^2)$
(iv) 5×10^{-8}, 2.12×10^{-9}

24 0.579, 24.6 kg, 0.897
(i) N$(25.05n, 0.0625n)$
(ii) N$(1002 - 25.05n, 0.9 - 0.0225n)$
(iii) N$(1002, 0.9 + 0.04n)$; 14

25 (i) 0.734 (ii) 29.2 cm (iii) 0.801
(iv) (1) 0.930, (2) 0.057 (v) 0.608 cm

26 £440 (a) B$(n, 0.006)$, where n is sample size
(b) Risk $= 1.35\% < 5\%$

Exercises 13b (Miscellaneous)

12 (a) 8, 5.6

Exercises 14a

1 (6.34, 10.06), (5.76, 10.64)

2 (2.55, 2.86), (2.48, 2.93)

3 (i) (10.33, 11.01) (ii) (10.27, 11.08)

4 (98.9, 111.7)

5 (i) (18 600, 19 700) (ii) (130 000 000, 138 000 000)

6 (28.4, 32.5)

7 7

8 (i) 65.6 min, 5.65 min (ii) (62.9, 68.3)

9 (i) 1.25 g (ii) $\sigma_n = 1.09$ g (iii) $\frac{s}{\sqrt{n}} = 0.196$ g;
(1000.6, 1001.9)

10 (30.4 kg, 32.4 kg)

11 (i) 9.875 ohms, 0.25 ohm^2,
(ii) (9.745 ohms, 10.005 ohms)

12 (i) 145 (ii) 225 (iii) (140.9, 149.1)

13 (259 ml, 273 ml)

14 N$\left(\mu_2 - \mu_1, \frac{\sigma_1^2}{n_1} + \frac{\sigma_2^2}{n_2}\right)$ (£1541, £1659); (£117, £283)

Exercises 14b

1 (0.823, 0.963)

2 (0.502, 0.583)

3 (i) (0.667, 0.800) (ii) (0.777, 0.905)

4 65 700

5 (0.314, 0.384)

6 (0.100, 0.188)

7 (0.260, 0.379)

Exercises 14c

1 (i) $(6.12, \infty)$ (ii) $(-\infty, 10.28)$

2 (i) $(-0.542, \infty)$ (ii) $(-\infty, 3.79)$

3 (0.731, 1)

4 (0, 0.138)

5 (0, 608.7)

6 $(115.9, \infty)$

Exercises 14d

1 (i) 0.01   (ii) 0.05   (iii) 0.725

2 (i) 0.975   (ii) 0.995   (iii) 0.90

3 (i) 0.95   (ii) 0.20   (iii) 0.15

4 (i) 1.860   (ii) 2.602   (iii) 4.773   (iv) -2.681
   (v) -2.776 (vi) 3.169

Exercises 14e

1 (i) (7.08, 9.58)   (ii) (6.53, 10.13)

2 (999.56, 1004.81)

3 (1.65, 3.09)

4 (89.9 g, 114.8 g)

5 (£66.20, £195.60)

6 128.8, 36.9, (122.6, 135.0) (in km h^{-1})

7 (78.4 N, 88.4 N)

8 (i) $(7.32, \infty)$ (ii) $(-\infty, 9.34)$

9 $(999.9, \infty)$

10 $(1.82, \infty)$

11 (i) 172.7, $s^2 = 64.57$, (167.0, 178.4)

Exercises 14f

2 $\sqrt{\frac{p(1-p)}{96}}, \sqrt{\frac{p(1-p)}{100}}$

3 $\mu = 2.5, \sigma^2 = 2.75$
   (i) $P(\bar{X} = 1) = \frac{1}{6}$, $P(\bar{X} = 2) = \frac{2}{6}$, $P(\bar{X} = 3) = \frac{2}{6}$,
   $P(\bar{X} = 4) = \frac{1}{6}$
   (ii) $P(V = 0) = \frac{1}{6}$, $P(V = 1) = \frac{3}{6}$, $P(V = 4) = \frac{2}{6}$

Exercises 14g (Miscellaneous)

1 (a) 480   (b) (8.16, 23.8), (245, 715)

2 (i) 211.0 ml   (ii) (210.1 ml, 211.9 ml)

3 (a) (9.7 min, 19.3 min)   (b) (11.8 min, 14.6 min)

4 (a) (i) (5.94 mg, 9.56 mg)   (ii) 9.27 mg
   (iii) (2.49 mg, 13.01 mg)   (b) 18
   (c) (0.068, 0.172)

5 (0.1335, 0.3165), (1009 g, 1036 g), (1026 g, 1039 g),
   (£910, £940)

6 (i) 31   (ii) (a) 0.005 53   (b) 0.194
   (c) 0.0518; 0.014

7 (a) 0.13   (b) 51y 2m, 9y 3m   (e) (0.49, 0.59)

8 (i) 4.40, 22.49   (ii) (1.46, 7.34)
   (iii) (2.05, 6.15)

Exercises 15a

1 $z = 2.18 > 1.96$: significant evidence that mean has changed.

2 $z = -2.72 < -1.645$: significant evidence that test takes less time.

3 $z = 1.32 < 2.326$: no significant evidence of increase in life.

4 $z = -1.53 > -1.881$: no significant evidence that μ is overstated.

5 $z = 1.48 < 1.645$: no significant evidence to confirm belief.

6 $z = -1.48 < -1.282$: significant evidence that mean distance is reduced.

7 $z = 1.99 < 2.326$: no significant evidence that $\mu > 1.75$.

8 $\bar{x} > \mu_0 + \dfrac{2.576\sigma}{\sqrt{n}}$ or $\bar{x} < \mu_0 - \dfrac{2.576\sigma}{\sqrt{n}}$;
   $\mu_0 > \bar{x} + \dfrac{1.645\sigma}{\sqrt{n}}$

9 $\frac{25}{3}$; 0.986; Accept $\mu = 105$ if total weight exceeds 5224 g, otherwise accept $\mu < 105$.

10 173; $z = 1.09 < 1.645$: no significant evidence that they will find more.

11 N(160,36); $z = -2.58$: significant evidence at any level not less than 1%; average weight outside the interval 160 ± 7.06 (assuming 5% level).

12 $\frac{1}{2}, \frac{1}{12}$; N(6,1); $z = -1.230 > -1.96$ (assuming 5% level): no significant evidence of difference.

13 $z = -1.86 < -1.645$: evidence sufficient.

Exercises 15b

1 $z = 2.17 < 2.326$: no significant evidence of increase.

2 $z = -4.13 < -2.576$: significant evidence mean not 70.0.

3 $z = -1.33 > -1.645$: no significant evidence to refute rumour.

4 $z = -2.13 < -1.645$: significant evidence of less time.

5 $z = -1.80 < -1.645$: significant evidence mean < 20.

6 $z = -2.03 < -1.645$: significant evidence $\mu < 80$.

Exercises 15c

1 $z = -0.79 > -1.282$: no significant evidence.

2 $z = -2.49 > -2.576$: no significant evidence.

3 $z = 1.72 > 1.645$: significant evidence proportion under-estimated.

4 $z = 1.74 > 1.645$: significant evidence $\lambda > 1$.

Exercises 15d

1 $z = 1.48 < 1.645$: no significant evidence of bias.

2 $z = -1.35 > -1.96$: no significant evidence against the claim.

3 $z = 2.35 > 2.326$: significant evidence against the claim.

4 $z = -1.07 > -1.282$: no significant evidence of decrease.

5 $z = -3.03 < -2.326$: significant evidence $p < 0.4$.

6 $z = -2.74 < -1.405$: significant evidence that support is over-estimated.

7 $z = 1.57 > 1.405$: significant evidence that proportion is understated.

8 $z = 1.99 > 1.751$: significant evidence of a greater proportion.

9 $z = 1.22 < 1.645$: no significant evidence that proportion > 0.7.

10 (i) $\frac{1}{3}$; (ii) $z = 1.833 > 1.751$: significant evidence of smoking.

Exercises 15e

The degrees of freedom, d, are given as a guide.

1 $t = -1.50 > -1.833$; $d = 9$: no significant evidence of grounds.

2 $t = -2.00 > -2.262$; $d = 9$: no significant evidence of bias.

3 $t = 2.17 > 1.895$; $d = 7$: significant evidence that mean mass > 2000 g

4 $t = 2.19 > 1.73$; $d = 19$: significant evidence that radiation level has increased.

5 (i) $\sqrt{\text{(unbiased estimate of population var.)}}$
(ii) $N(0,1)$; t with $d = n - 1$.

6 0.106   (i) $z = -1.66 > -1.96$: no significant evidence of change.
(ii) $t = -2.33 < -2.262$; $d = 9$: significant evidence of change in mean.

7 $t = 2.48 > 1.833$; $d = 9$: significant evidence of higher yield.

8 3.88%, 77.0 cl. $t = 1.82 < 2.110$; $d = 9$: no significant evidence that calibration incorrect.

Exercises 15f

CI is an abbreviation for Confidence Interval.

1 Significant evidence that mean has changed:
(a) $p = 0.029 < 0.05$
(b) CI is (460.4, 462.0); excludes 460.3.

2 Significant evidence that test takes less time:
(a) $p = 0.003 < 0.05$
(b) CI is $\mu < 5.76$; excludes 5.82.

3 No significant evidence of increase in life:
(a) $p = 0.094 > 0.01$
(b) CI is $\mu > 94.13$; contains 95.2.

4 No significant evidence of increase:
(a) $p = 0.015 > 0.01$
(b) CI is $\mu > 99.83$; contains 100.

5 Significant evidence mean not 70.0:
(a) $p = 0.00003 < 0.01$
(b) CI is (68.6, 69.7); excludes 70.0.

6 No significant evidence to refute rumour:
(a) $p = 0.184 > 0.10$
(b) CI is (905.1, 965.9); contains 960.

7 (a) 1.43, 0.656, 0.118
(b) (1.20, 1.67), 2.02 outside CI: significant evidence of a change.

8 (i) 4.31, 2.70   (ii) (3.78, 4.84)
(iii) 6 is outside CI: significant evidence against politician's claim.

9 (0.356, 0.484), 0.5 outside CI: significant evidence that proportion is not 0.5.

10 (a) (200.7 g, 209.1 g)   (b) $\frac{1}{5}$, (0.057, 0.343)
(c) 0.0038.

11 (a) 0.400, (0.208, 0.592)   (b) (494.4 g, 499.9 g)
(c) (i) 6.0 g,
(ii) (435.9 g, 473.9 g)
(d) 454 inside CI: no significant evidence against claim.

Exercises 16a

1 (i) Acceptance: $\bar{X} < 14.40$, Rejection: $\bar{X} > 14.40$
(ii) 0.281, 71.9%

2 (i) Acceptance: $\bar{Y} > 19.18$, Rejection: $\bar{Y} < 19.18$
(ii) 0.361, 63.9%

3 (i) $-6.39 < \bar{X} < -4.61$ (ii) 0.132, 86.8%

4 (i) $\bar{X} > 53.48$ (ii) 0.297, 70.3%

5 (a) Accept $\mu = 13$ since 13.21 in (12.1, 13.9)
(c) Accept $\mu = 13$ since $13.21 > 13$ (no need to calculate actual limit!)
(d) (11.9, 14.5)

6 (i) Rejection: $\bar{X} \leqslant 24$, accept $p = 0.3$ (ii) $a = 22$
(iii) 87%

Exercises 16b

1 $\ln 2$, 1; $f_0(x) = e^{-x}$ for $x > 0$ and 0 otherwise;
$f_1(x) = \dfrac{1}{(1 + x)^2}$ for $x > 0$ and 0 otherwise;
$c = 3.00$, power $= 25.0\%$

2 (i) $c = 8.42$, power $= 29.1\%$ (ii) $c = 5.86$, 0.343

3 (i) $c = 0.4$, power $= 19\%$ (ii) $c = 2$, power $= 75\%$

4 4.98%, power $= 65.6\%$

5 $\bar{x} > \dfrac{1.645}{\sqrt{n}}$; 16

Exercises 16c

1 Accept if $\bar{x}$ in (0.00401, 0.00439): approx. 0.184, 0.559, 0.885, 0.988; yes, since $\bar{x} = 0.00389$ not in acceptance region.

2 No; 16   (i) approx. 0.182   (ii) approx. 0.444
(iii) approx. 0.870

3 $c = a(1 - \alpha)$; power $= 1 - \frac{c}{r}$ for $r > c$ and 0 for $r \leqslant c$

4 $c = 3k(1 - \sqrt{\alpha})$,
power $= \{1 - \frac{c}{3\mu}\}^2$ for $\mu > \frac{c}{3}$ and 0 for $\mu \leqslant \frac{c}{3}$

5 $c = -k\ln(\alpha)$, $e^{-c/\mu}$

6 $c = -k\ln(1 - \alpha)$; $1 - e^{-c/\mu}$

7 e^{-c}

Exercises 16d

1 $H_0: p = \frac{1}{6}$, $H_1: p > \frac{1}{6}$, Acceptance region: $n \leqslant 3$;
Rejection region: $n \geqslant 4$; 7%.   (i) Accept $p = \frac{1}{6}$
(ii) Accept $p > \frac{1}{6}$

2 Acceptance region: $2 \leqslant n \leqslant 6$, 9.6%

3 (a) (ii) 0.82, (b) 5.74%

4 H_0: $p = 0.3$, H_1: $p < 0.3$; number contaminated $\leqslant 2$, actual level 3.5%; power = 40.5%; $z = -2.69 < -2.326$: significant evidence that less than 30% are contaminated.

5 (i) 7.5%   (ii) 32.3%

6 8

7 (a) H_0: $p = 0.35$, H_1: $p < 0.35$; Acceptance region: number of males $\geqslant 5$, actual level 11.8%, significant evidence that proportion is lower than usual
   (b) $z = 1.78 < 1.96$: no significant difference.

8 H_0: $\mu = 8.5$, H_1: $\mu < 8.5$; 7.44%; 81.5%; 15.0%

9 H_0: $p = 0.75$, H_1: $p > 0.75$

Exercises 16e

1 (a) (i) 0.374   (ii) 35.9   (b) 4   (c) 103.2 cm
   (d) $(1 - p_1)(1 - p_2)(1 - p_3)$

2 A: $e^{-2x}(1 + 2x + 2x^2)$, where $x = 50p$
   B: $e^{-x} + xe^{-3x}(1 + 2x + 2x^2)$

3 Missing entries: Method I: 0.999, 0.677, 0.206;
   Method II: 0.931

4 59.0, 32.8, 7.3, 0.8, 0, 0; $m = 29$, $n = 271$

5 (a) 0.987, 0.924, 0.813, 0.537, 0.254, 0.098
   (i) 0.5   (ii) 9%   (c) 0.120   (d) 0.69%

Exercises 17a

1 (i) $-1.76 < -1.645$, significant evidence that $\mu_1 < \mu_2$
   (ii) $-1.76 > -1.960$, no significant evidence that $\mu_1 \neq \mu_2$

2 $1.99 < 2.326$, no significant evidence that second mean is greater than first.

3 $3.63 > 1.645$, significant evidence that $k_A > k_B$

4 $1.64 < 1.960$, no significant evidence that writing lengths differ.

Exercises 17b

1 (i) $1.346 < 1.645$
   (ii) $1.350 < 1.645$, no significant evidence to support suspicion in either case.

2 $0.88 < 1.282$, no significant evidence that morning mean less than afternoon mean.

3 (i) $1.85 > 1.645$
   (ii) $1.86 > 1.645$, significant evidence that taller trees are produced in A in both cases.

4 $2.01 > 1.960$, significant evidence of difference in mean distances.

Exercises 17c

1 0.000 512, $2.63 > 1.812$: significant evidence of more beer from first pub.

2 $2.69 < 2.921$, no significant evidence of difference in mean weights.

3 $1.03 < 1.415$, no significant evidence that mean greater than 346 g.

$1.34 < 1.746$, no significant evidence of change in mean mass.

4 (i) $0.86 < 2.101$, no significant evidence of difference between ages.
   (ii) (0.385, 0.815)

5 133.73, $0.58 < 1.714$, no significant evidence to reject hypothesis of same mean.

6 $2.43 > 2.326$, significant evidence of increase in sales.
   $1.78 < 2.508$, a change in previous conclusion.

Exercises 17d

1 $(-0.44, 8.24)$

2 $(0.74, 7.86)$

3 $(0.14, 0.48)$

4 $(-0.079 \,\text{km}, 0.000 \,\text{km})$

5 $(-3.2 \,\text{miles}, 8.3 \,\text{miles})$

6 $(0.005 \,\text{litres}, 0.064 \,\text{litres})$

7 $(4.3 \,\text{g}, 36.5 \,\text{g})$

8 (a) $(78.0, 82.0)$ (b) $(3.2, 6.8)$

9 $(0.22 \,\text{yrs}, 0.38 \,\text{yrs})$

10 $(£5440, £7560)$

11 (a) $(23.8 \,\text{miles}, 26.3 \,\text{miles})$
   (b) $(0.9 \,\text{miles}, 4.7 \,\text{miles})$

12 (i) Means: 7.02 lb, 6.90 lb,
   variances: $13.0 \,(\text{lb})^2$, $10.4 \,(\text{lb})^2$
   (ii) $(6.4 \,\text{lb}, 7.6 \,\text{lb})$
   (iii) $(-0.8 \,\text{lb}, 1.1 \,\text{lb})$, assertion not discredited

13 (a) $13.1 \,\text{kg}^2$ (b) $(36.6 \,\text{kg}, 37.8 \,\text{kg})$
   (c) $(1.1 \,\text{kg}, 3.3 \,\text{kg})$
   (d) significant evidence of difference between the means

Exercises 17e

1 $0.79 < 1.645$, no significant evidence of difference in proportions.

2 $2.95 > 2.326$, significant evidence that proportion lower.

3 $1.99 > 1.960$, significant evidence of difference between town and region at 5% level;
   $0.46 < 1.960$, no significant evidence of difference at 5% level.

4 $(0.36, 0.50)$, $1.33 < 1.645$, no significant evidence of difference at 10% level.

Exercises 17f

1 (i) $(6.3, 18.1)$ (ii) $(9.1, 15.3)$

2 (i) (a) 2661.2   (b) $1.50 < 1.796$, no significant evidence of difference.
   (ii) (b) $2.72 > 2.201$, significant evidence of difference.

3 $1.89 > 1.833$, significant evidence in support of belief.

4 $0.46 < 2.120$, no significant evidence of difference at 5% level.
   $3.95 > 2.306$, significant evidence of difference at 5% level.

5 $1.73 < 2.101$, no significant evidence of difference.
   $2.67 > 2.262$, significant evidence of difference.

6 $0.50 < 2.571$, no significant evidence of disparity.

7 $1.81 < 2.262$, no significant evidence to reject hypothesis.

$2.64 > 1.833$, significant evidence in support of belief.

Exercises 18a

1 (i) 0.025   (ii) 0.95   (iii) 0.005   (iv) 0.990
   (v) 0.05

2 (i) 0.090   (ii) 0.045   (iii) 0.015

3 (i) 14.86   (ii) 12.83   (iii) 2.706   (iv) 7.879
   (v) 9.348

Exercises 18b

1 $1.77 < 7.82$; $d = 3$: no significant evidence that X not Poisson with mean 1.6.

2 $4.29 < 11.3$; $d = 3$: no significant evidence to reject null hypothesis.

3 $6.56 > 5.99$; $d = 2$: significant evidence sample biased.

4 $6.80 < 7.82$; $d = 3$: no significant evidence that machine not working correctly.

5 $7.33 < 13.8$; $d = 2$: no significant evidence of experiment being performed incorrectly.

6 $3.48 < 7.82$; $d = 3$: no significant evidence that sample not chosen at random.

7 $\frac{1}{6}$; $3.96 < 5.99$; $d = 2$: no significant evidence that distribution not same as for X.

8 $x_2 = 253.39$, $x_3 = 254.61$, $x_4 = 256.02$; 8 each
   (i) $6.75 < 9.49$; $d = 4$: significant evidence in support of manufacturer's claim.
   (ii) 0.800

Exercises 18c

Values may differ slightly depending upon the accuracy of calculation.

1 Pool 2 and 3: $2.61 < 9.21$; $d = 2$: no significant evidence to reject hypothesis that $X \sim B(3, 0.45)$.

2 Pool 3 and 4: $5.91 < 11.3$; $d = 3$: no significant evidence to reject hypothesis that $X \sim B(4, 0.3)$.

3 Pool 3 and $\geqslant 4$: $0.091 < 7.82$; $d = 3$: no significant evidence to reject hypothesis that $X \sim$ Poisson with mean 1.2.

4 Pool each outer pair: $5.50 < 15.1$; $d = 5$: no significant evidence to reject hypothesis that $X \sim N(20, 36)$.

5 Pool 4, 5 and 6: $3.34 < 9.49$; $d = 4$: no significant evidence to reject hypothesis that $X \sim B(6, \frac{1}{3})$.

6 Pool 3, 4 and $\geqslant 5$: $1.77 < 7.82$; $d = 3$: no significant evidence of bias.

7 Pool orange and yellow: $8.67 > 7.82$; $d = 3$: significant evidence not a random sample.

8 Pool 4 and $\geqslant 5$: $1.63 < 9.49$; $d = 4$: no significant evidence to rejecture conjecture.

9 $B(5, 0.2)$; Pool 4 and 5: $2.89 < 9.49$; $d = 4$: no significant evidence of sample not being consistent with predicted distribution.

10 $P(10 < X < 20) = 0.0606$, $P(20 < X < 30) = 0.2417$, $P(40 < X < 50) = 0.2417$, $P(50 < X < 60) = 0.0606$, $P(X \geqslant 60) = 0.0062$; Pool $(x < 20)$ also $(x \geqslant 50)$, $14.8 > 13.3$ (1% level); $d = 4$: significant evidence, at 1% level, that the times taken do not follow the stated normal distribution

11 Pool $(x < 177)$ also $(x \geqslant 183)$, $7.09 < 9.49$ (5% level) $d = 4$: no significant evidence, at 5% level, that distribution of heights is not $N(180, 3)$.

12 (i) $8.00 < 16.9$ (5% level); $d = 9$: no significant evidence, at 5% level, to reject hypothesis of randomness.
   (ii) Pool (2 and 3), also (4 or more): $X^2 = 5.91 < 7.82$ (5% level); $d = 3$: no significant evidence, at 5% level, to reject hypothesis of randomness.

Exercises 18d

Values may differ slightly depending upon the accuracy of calculation.

1 Estimated $p = 0.3$, pool 3 and 4: $9.88 > 9.21$; $d = 2$: significant evidence distribution not binomial.

2 Estimated $p = 0.15$, pool 2, 3 and 4: $0.096 < 3.84$; $d = 1$: no significant evidence to reject binomial.

3 Estimated mean $= 1$, pool 3, 4 and $\geqslant 5$: $7.49 > 5.99$; $d = 2$: significant evidence distribution not Poisson.

4 Estimated mean $= 0.5$, pool 2, 3 and $\geqslant 4$: $0.33 < 3.84$; $d = 1$: no significant evidence to reject Poisson.

5 Estimated $p = 0.1$, pool 2, 3 and $\geqslant 4$: $0.72 < 3.84$; $d = 1$: no significant evidence to reject binomial.

6 (b) $6.10 < 7.82$ (5% level); $d = 3$: no significant evidence, at 5% level, to reject normal.

7 Estimated $p = 0.4$, pool 3 and 4: $11.1 > 9.21$; $d = 2$: significant evidence fit is not good.

8 Pool 0 and 1, also 5, 6 and $\geqslant 7$: $9.55 > 7.82$; $d = 3$: significant evidence model not adequate.

9 (i) $\bar{x} = 2.00$ (ii) Pool 5, 6, 7 and $\geqslant 8$: $10.9 > 9.49$; $d = 4$: significant evidence distribution not Poisson.

10 (i) 2.46, $s^2 = 2.65$ (ii) Pool 5, 6 and $\geqslant 7$: $3.65 < 9.49$; $d = 4$: no significant evidence to reject Poisson.

11 Pool 2, 3 and $\geqslant 4$: $2.39 < 3.84$; $d = 1$: no significant evidence to reject Poisson.

12 Pool first 4 categories, also last 3: $4.91 < 16.81$; $d = 6$: no significant evidence to reject $N(\mu, 1)$.

13 (a) ($\bar{x} = 72.24$, $s = 3.403$), $10.68 > 7.82$; $d = 3$: significant evidence that model not adequate.

Exercises 18e

Values may differ slightly depending upon the accuracy of calculation.

1 $7.54 < 9.49$; $d = 4$: no significant evidence of differences.

2 $19.3 > 13.28$; $d = 4$: significant evidence of association.

3 $X_c^2 = 4.01 > 3.84$; $d = 1$: significant evidence of association.

4 Pool 16–17 and 18–24: $3.47 < 7.82$; $d = 3$: no significant evidence of differences.

5 (i) $1.94 < 5.99$ (5% level); $d = 2$: no significant evidence, at 5% level, to reject hypothesis.
   (ii) $X_c^2 = 0.121 < 3.84$; (5% level); $d = 1$: no significant evidence, at 5% level, to reject hypothesis.

6 $3.27 < 5.99$; $d = 2$: no significant evidence of association.

7 $4.78 < 5.99$; $d = 2$: no significant evidence of different preferences.

8 Combine C and D or E: $13.00 < 13.28$; $d = 4$, or combine I and II: $8.90 < 11.34$; $d = 3$: no significant evidence of association.

9 $2.01 < 5.99$; $d = 2$: no significant evidence of association.

10 $3.27 < 7.38$; $d = 2$: no significant evidence of association.

11 $17.01 > 5.99$; $d = 2$: significant evidence of association.

12 $X_c^2 = 7.40 > 3.84$; $d = 1$: significant evidence of association.

Exercises 18f

1 $I = 251 > 233$ (approx); $d = 199$: significant evidence not Poisson.

2 $I = 106.8 < 124$ (approx); $d = 99$; no significant evidence not Poisson.

3 $I = 69.4 > 55.8$ (approx); $d = 39$: significant evidence not Poisson.

4 (i) Pool $\geqslant 3$: $0.75 < 5.99$; $d = 2$: no significant evidence of departure from randomness
   (ii) $I = 67.2 < 82$ (approx); $d = 63$: no significant evidence of departure from randomness. 0.041, 8

Exercises 18g (Miscellaneous)

1 (a) $13.20 < 14.07$; $d = 7$: no significant evidence that sample not random.
   (b) *Either* Pool 2 and higher; $0.870 < 5.99$; $d = 2$, *or* Pool 1 and higher; $0.22 < 3.84$; $d = 1$: Either way no significant evidence that blue sweets are not distributed at random.
   (c) $I = 88.8 < 124$ (approx): no significant evidence that not Poisson.

2 Estimate $\lambda = \frac{5}{52}$; $1.34 < 7.82$; $d = 3$: no significant evidence to reject meteorologist's conjecture.

3 $12.5 < 14.68$; $d = 9$: no significant evidence to reject hypothesis. $z = -2.37 < -1.96$, significant evidence of fewer doubles.

4 (a) B(20,0.08)

(b) $P(X = 1) = 0.3282$, $P(X = 2) = 0.2711$, $P(X \geqslant 6) = 0.0038$
   (c) (i) 0.034   (ii) 0.932
   (d) $3.44 < 11.07$; $d = 5$: no significant evidence of results not being consistent with model.

5 (i) $E(X) = \dfrac{1}{p}$ (ii) $\bar{y} = 1.50$

Y	1	2	3	4
Frequency	133.3	44.4	14.8	4.9

Pool 4, 5, 6 and $\geqslant 7$: $2.71 < 4.61$; $d = 2$: no significant evidence that sample fails to support the belief.

6 (i) $20.94 > 9.49$; $d = 4$: significant evidence of association. Least $k = 0.1$.
   (ii) Estimated $p = \frac{11}{40}$; (0.195, 0.355)

7 (a) $14.29 < 15.51$; $d = 8$: no significant evidence of colours not being equally likely.
   (b) Pool final two rows: $4.31 < 9.49$; $d = 4$: no significant evidence of association.

8 (a) (i) 0.0283   (ii) 0.0354
   (b) $17.24 > 16.27$; $d = 3$: significant evidence of association.

Exercises 19a

1 (1.46, 15.0)
2 (38.2, 141)
3 (i) 2.18, 3.65   (ii) (0.817, 3.55)
   (iii) (1.73, 12.2) contains 4.0, no significant evidence that $\sigma^2 \neq 4.0$ (iv) (0.944, 3.42)
4 $t = 31.4\sigma^2$, $(0.551, \infty)$
5 (6.24, 44.5), 28.2
6 (i) 58.9, 8.73   (ii) (5.57,15.6)   (iii) (57.5, 60.3)
   (iv) (58.1, 59.8)

Exercises 19b

1 $35.1 < 36.42$: no significant evidence to reject $\sigma^2 = 10$
2 $10.4 > 5.63$: no significant evidence to reject $\sigma^2 = 100$
3 $2.22 > 1.735$: no significant evidence to reject $\sigma^2 = 0.3$
4 $32.9 > 16.01$: significant evidence that variance is not 1
5 $30.2 > 21.7$: significant evidence that machine has gone wrong

Exercises 19c

1 (i) 6.94   (ii) 19.25   (iii) 2.00   (iv) 2.43
2 (i) 0.0956   (ii) 0.0657   (iii) 0.276   (iv) 0.198
3 (i) 0.104, 9.60   (ii) 0.111, 5.05
   (iii) 0.0254, 5.10   (iv) 0.143, 5.99

Exercises 19d (Miscellaneous)

1 $1.34 < 5.29$: no significant evidence to reject same variance.

2 $3.36 > 2.64$: significant evidence that Sid is more variable.

3 $1.34s$, 0.1

4 (i) $(0.233, 3.44)$ (ii) $(83.0, 296)$

5 $1.51 < 3.6$ (approximately!): no significant evidence to reject variances equal; $t = 1.82 < 2.09$: no significant evidence to reject means equal.

6 (a) $0.263, 0.660$ (b) $(0.128, 0.808)$ (c) $(0.092, 1.53)$
   (d) Interval includes 1: no significant evidence to reject $\sigma_1^2 = \sigma_2^2$.
   (e) 0.263.

7 (a) $(30.5, 104.3)$
   (b) (i) $(2350, 2400)$ (ii) $(2330, 2420)$

8 (a) $1.96 < 15.44$: no significant evidence to reject $\sigma_1^2 = \sigma_3^2$
   (b) $(21.8, 143.5)$
   (c) 154.4

Exercises 20a

1 (ii) $\Sigma x = 10$, $\Sigma x^2 = 16.92$, $\Sigma y = 65.5$, $\Sigma y^2 = 609.13$, $\Sigma xy = 99.48$
   (iii) $\bar{x} = 1.25$, $\bar{y} = 8.19$, $S_{xx} = 4.42$, $S_{yy} = 72.85$, $S_{xy} = 17.61$
   (iv) $a = 3.21$, $b = 3.98$ (vi) $y = 4.01$

2 (ii) $\Sigma g = 327.3$, $\Sigma g^2 = 22\,177.55$, $\Sigma h = 263.6$ $\Sigma h^2 = 16\,626.22$, $\Sigma gh = 12\,488.07$
   (iii) $\bar{g} = 54.55$, $\bar{h} = 43.93$ (iv) $h = 67.80 - 0.437g$

3 (ii) $\Sigma w = 1750.4$, $\Sigma w^2 = 735\,651.02$, $\Sigma z = 0.0658$, $\Sigma z^2 = 0.007\,352\,36$, $\Sigma wz = -4.654\,66$
   (iii) $\bar{w} = 350.1$, $\bar{z} = 0.0132$ (iv) $w = 406.3 - 4269z$
   (vii) $w = 162.1$

4 (c) $\bar{x} = 42.417$, $\bar{y} = 61.083$, $b = 0.966$, $y - 61.1 = 0.966(x - 42.4)$
   (d) $y = 20.1 + 0.966x$ (f) $x = 33$

5 (b) $y = 20.7 + 0.961x$ (c) x about 32

Exercises 20b

1 (ii) $\Sigma x = 4$, $\Sigma x^2 = 46.36$, $\Sigma y = 42.6$, $\Sigma y^2 = 364.26$, $\Sigma xy = -71.01$
   (iii) $\bar{x} = 0.400$, $\bar{y} = 4.26$, $S_{xx} = 44.76$, $S_{yy} = 182.8$, $S_{xy} = -88.05$
   (iv) $a = 5.05$, $b = -1.97$ (vi) $D = 9.576$
   (vii) 4.48, 18.28

2 (ii) $\Sigma u = 22.5$, $\Sigma u^2 = 83.99$, $\Sigma v = 73$, $\Sigma v^2 = 859.52$, $\Sigma uv = 258.71$
   (iii) $\bar{u} = 3.21$, $\bar{v} = 10.43$, $S_{uu} = 11.67$, $S_{vv} = 98.23$, $S_{uv} = 24.07$
   (iv) $v = 3.80 + 2.06u$ (vi) 13.7
   (viii) $D = 48.59$ (ix) 1.67, 14.03

3 $y = 1.86 + 0.02r$; 1.91 km

4 $y = 12.4 + 0.6x$; 66.4

Exercises 20c

1 (ii) $y = -7.345 + 12.63x$ (iii) 36.85

2 (ii) $v = -0.459 + 0.204u$ (iii) 1.38

3 (ii) $t = 27.7 + 0.874v$ (iii) 80.1 (iv) 0.0033

4 (ii) $y = 0.243 + 0.004\,86x$ (iv) 3.74%

5 (i) $r = -90.8 + 116s$ (ii) 101 revs/min

Exercises 20d

1 $y = 14.86 + 8.06c$, 39.03 tonnes.

2 $\bar{A} = 14.875$, $\bar{T} = 17.5$; $A = 6.30 + 0.49T$; 16.1; 0.49.

3 (i) $\ln y = 1.759 + 0.763x$ (ii) 565
   (iii) $(0.524, 1.002)$

4 (ii) $\ln p = -12.47 + 0.198x$ (iii) 10.8

5 (i) $f = 11.7 + 0.226w$ (ii) 104.5
   (iii) $(0.132, 0.321)$

Exercises 20e

1 (i) $t = 3.43 + 1.24d$ (iii) 127 min
   (iv) $(0.68, 1.79)$

2 (i) $y = 2.19 + 1.95x$, $x = 14.74 + 0.252y$ (ii) 27
   (iii) $0.005\,06$

3 (i) $y = 15.8 + 0.722x$ (ii) 66 (iii) 59

4 (a) $x = 37.4 + 0.949y$; $y = 44.3 + 0.584x$
   (d) 152 mm, 185 mm

5 (b) (i) $y = 41.8 + 1.55x$ (ii) 51.1 (iii) 41.8

6 (i) $y = 1.51 + 1.29x$ (ii) 46.6 (iii) $(1.10, 1.48)$

Exercises 20f

1 (i) $c = 2.48 + 0.608m$ (ii) 0.593 (iii) 11.4

2 (i) 0.981 (ii) $y = -7.42 + 1.12x$, $x = 6.89 + 0.862y$
   (iii) 8.20 (iv) 13.9

3 (a) 0.745

4 (b) (i) 255 (ii) 0.935
   (d) $p = 2.58 + 0.879t$; 15

5 (a) 0.979 (b) $l = 13.2 + 0.770w$ (c) 0.77 mm

Exercises 20g

1 $r = -0.952$

2 (c) $h = 22.6 + 0.821a$ (d) $h = 47.2$ mg/l

3 (b) $r = -0.118 > -0.632$: no significant evidence that $\rho \neq 0$ (c) Patient 7

Exercises 20h (Miscellaneous)

1 (a) a (c) 0.682 (d) $\hat{\alpha} = 3.66$, $\hat{\beta} = 0.032$
   (f) For 0, 50, 100: 3.66, 5.23, 6.81

2 (c) $\beta = 0.906$
   (d) 5470, 6370, 7370, 8110, 8810, 9630

3 (i) $\frac{1}{2}$, no   (ii) 0, no

4 $a_0 = 9.73$, $b = 0.418$

5 (ii) $y = -2.80 + 11.5x$ (iv) 57 (v) $(8.7, 14.2)$

6 (ii) 0.781
7 (i) $\ln I = 1.65 - 2.88t$, $I_0 = 5.23$,
   $k = 2.88$ (ii) 1.24
   (iii) -0.984 (iv) 0.43
8 (b) $r_{uv} = -0.762$, $r_{ws} = -0.762$
9 (ii) 0.913   (iii) $s = -0.0366 + 0.340p$
   (iv) (0.265, 0.415)
10 (ii) 0.701   (iii) $a = 4.33$, $b = 0.564$
   (iv) (0.096, 1.033): no significant evidence that $\beta \neq 1$
11 (b) -0.952 (c) $m = 83.58 - 2.20d$ (d) 526
   (e) 2.20
12 (b) 136.55, 146.3, 154.45, 164.75, 170.5, 188.6, 184.9, 207.9, $y = -35.7 + 0.953x$
   (c) (i) 131.1   (ii) 140.7   (iii) 250.3

Exercises 21a

Note: small differences from the given answers may be due to rounding or truncation errors.

1

	D.F.	S.S.	M.S.	F.R.
Fertilisers	2	1426.95	713.48	13.7
Remainder	18	936.29	52.02	
Total	20	2363.24		

13.7 > 6.01: significant difference, at 1% level, in effects of fertilisers (B is best).

2

	D.F.	S.S.	M.S.	F.R.
Temps.	3	42.55	14.18	4.54
Remainder	16	49.98	3.12	
Total	19	92.53		

4.54 > 4.08: significant evidence, at 2.5% level, that average porosity is affected by firing temperature.

3 [It is easier to work with $z = 100(y - 1.3)$]

	D.F.	S.S.	M.S.	F.R.
Machines	5	29.42	5.88	1.42
Remainder	22	91.43	4.16	
Total	27	120.86		

1.42 < 2.66: accept, at 5% significance level, the hypothesis of no differences between the average densities of the outputs.

Exercises 21b

1

	D.F.	S.S.	M.S.	F.R.
Methods	2	6162	3081	19.8
Years	3	58	19	0.1
Remainder	6	933	155	
Total	11	7153		

19.8 > 5.14: very significant evidence, at 5% level, of a difference in the mean masses of the crops.

2

	D.F.	S.S.	M.S.	F.R.
Diets	2	10.800	5.400	4.33
Litters	4	15.924	3.981	3.19
Remainder	8	9.980	1.248	
Total	14	36.704		

(i) 3.19 < 3.84: no significant evidence, at 5% level, of differences between litters.
(ii) 4.33 < 4.46: no significant evidence, at 5% level, of differences between diets.

3

	D.F.	S.S.	M.S.	F.R.
Fats	3	680	227	6.54
Days	4	1044	261	7.53
Remainder	12	416	35	
Total	19	2140		

6.54 > 5.95: significant evidence, at 1% level, that mean amounts of fat absorbed differ.
7.53 > 5.41: significant evidence, at 1% level, that daily means vary.

Exercises 21c

1

	D.F.	S.S.	M.S.	F.R.
Doses	3	1517	505.7	20.7
Rabbits	3	659	219.6	9.0
Weeks	3	219	73.1	3.0
Remainder	6	147	24.5	
Total	15	2542		

20.7 > 9.78: highly significant evidence, at 1% level, of differences due to doses.

2

	D.F.	S.S.	M.S.	F.R.
Pesticides	4	35 266	8817	2.92
Rows	4	19 562	4891	1.62
Columns	4	19 645	4911	1.63
Remainder	12	36 268	3022	
Total	24	110 742		

2.92 < 3.26: no significant evidence, at 5% level, of differences between the effects of the treatments.

3

	D.F.	S.S.	M.S.	F.R.
Agents	3	3.182	1.061	7.3
Machines	3	0.502	0.167	1.2
Rubbers	3	25.862	8.621	59.2
Remainder	6	0.874	0.146	
Total	15	30.419		

7.3 > 6.60: significant evidence at 2.5% level, of differences due to chlorinating agents.

Exercises 21d

1

	D.F.	S.S.	M.S.	F.R.
Varieties	1	1190.25		13.6
Fields	3	43.25	14.42	0.2
Lack of fit	3	101.25	33.75	0.4
Replicates	8	699.00	87.375	
Total	15	2033.75		

$0.4 < 4.07$: no significant evidence of lack of fit, at 5% level.

$13.6 > 11.26$: significant evidence, at 1% level, of difference between varieties.

2

	D.F.	S.S.	M.S.	F.R.
Catalysts	2	0.976	0.488	2.8
Temps.	3	5.820	1.940	11.3
Lack of fit	6	0.407	0.068	0.4
Replicates	24	4.120	0.172	
Total	35	11.322		

$0.4 < 2.51$: no significant evidence of lack of fit, at 5% level.

$11.3 > 4.72$: significant evidence, at 1% level, of difference due to changes in temperature.

Exercises 21e (Miscellaneous)

1

	D.F.	S.S.	M.S.	F.R.
Areas	3	55.53	18.51	1.62
Remainder	22	252.00	11.45	
Total	25	307.54		

$1.62 < 3.05$: no significant evidence of differences, at 5% level, due to different areas.

2

	D.F.	S.S.	M.S.	F.R.
Methods	2	4593	2296	1.62
Employees	4	6809	1702	0.41
Remainder	8	33214	4152	
Total	14	44616		

$0.55 < 4.46$: no significant evidence of differences, at 5% level, between the efficiencies of the three methods.

3

	D.F.	S.S.	M.S.	F.R.
Machines	3	2.090	0.697	1.60
Days	3	2.745	0.915	2.10
Methods	3	6.095	2.032	4.67
Remainder	6	2.610	0.435	
Total	15	13.540		

$4.67 < 4.76$: no significant evidence, at 5% level, of variations due to differences between methods (or days or machines).

4 (a)

Source	D.F.	S.S.	M.S.	F.R.
Design	3	394.7	131.6	5.92
Employee	2	3350.9	1675.5	75.40
Residual	6	133.3	22.2	
Total	11	3878.9		

$5.92 > 4.76$: significant evidence, at 5% level, that there is a difference between designs;

$75.40 > 5.14$: very significant evidence, at 5% level, that there is a difference between employees.

5

	D.F.	S.S.	M.S.	F.R.
Compounds	2	635.11	317.56	27.8
Dilutions	2	291.44	145.72	12.7
Lack of fit	4	34.89	8.72	0.8
Replicates	9	103.00	11.44	
Total	17	1064.44		

$0.8 < 6.42$: no significant evidence of lack of fit.

$12.7 > 8.02$: significant evidence, at 1% level, of difference, due to dilution levels (and to compounds).

Exercises 22a

1 $p = 0.289 > 0.05$: no significant evidence that median $\neq 150$ g.

2 $p = 0.020 < 0.05$: significant evidence that claim is not justified.

3 (ii) 0.251 (iii) 0.252 (iv) equally preferable

4 $p = 0.073 > 0.05$: no significant evidence that average breaking strength > 1300 kg

5 $p = 0.343 > 0.10$: no significant evidence that par is incorrect.

6 $p = 0.109 > 0.05$: no significant evidence that median reaction time $\neq 0.60$ s

7 Cinder: $p = 0.388 > 0.10$: no significant evidence that median $\neq 26$ s

Synthetic: $p = 0.774 > 0.10$: no significant evidence that median $\neq 26$ s.

Exercises 22b

1 $z = 1.19 < 1.96$: no significant evidence that median $\neq 150$ g

2 $z = 1.72 > 1.645$: significant evidence that claim is not justified.

3 $z = 2.55 > 1.645$: significant evidence that average breaking strength > 1300 kg

4 $z = 0.75 < 1.96$: no significant evidence that median $\neq 200$

5 $z = 2.24 > 1.96$: significant evidence that average reaction time $\neq 0.60$ s.

6 Cinder track: $z = 0.59 < 1.645$: no significant evidence that median $\neq 26$ s.

Synthetic track: $z = 1.57 < 1.645$: no significant evidence that median $\neq 26$ s.

Exercises 22c

1 $z = -1.74 < -1.645$: significant evidence to support the suspicion.

2 $z = 0.06$: no significant evidence of difference.

3 $z = 1.90 > 1.645$: significant evidence that median distance for 8–9 a.m. is less.

4 $z = -1.00 > -1.645$: no significant evidence that B produces taller trees.

5 $z = -1.72 > -1.96$: no significant evidence of difference.

6 $z = 1.64 < 1.96$: no significant difference in contents.

7 $z = 2.65 > 2.576$: significant evidence that pulse rate is affected by eating.

8 $z = 1.83 > 1.645$: significant evidence that breaking strength is less.

9 $p = 0.011 < 0.10$: significant evidence that scores are higher.

10 $z = 1.83 > 1.645$: significant evidence of increase.

11 $z = 2.16 < 2.326$: no significant evidence that median time is lower; 1.6%.

12 $n \in \{0, 1, 2, 10, 11, 12\}$, $p = 3.86\%$;
  (i) $n = 8$: no significant evidence of difference;
  (ii) $z = 2.08 > 1.96$: significant evidence of difference.

13 (b) (i) H_0: medians equal, H_1: medians unequal.
  (ii) $\alpha > 28.9$ (iii) $z = 0.91 < 1.96$: no significant evidence of change.

14 (a) $p = 0.039 < 0.05$: significant evidence that tests are not of equal difficulty.
  (b) $z = 1.66 < 1.96$: no significant evidence that tests are not of equal difficulty.

15 (ii) $z = 1.90 > 1.645$; $p = 0.508 > 0.10$: no significant evidence of difference in medians.

Exercises 22d

1 $r_s = 0.452 < 0.643$: no significant evidence of positive correlation.

2 $\rho = 0.517 < 0.600$: no significant evidence of correlation.

3 (a) $r_s = 0.661$ (b) $0.661 > 0.564$: significant evidence of positive correlation.

4 $r_s = 0.714 \geqslant 0.714$: at 5% level, significant evidence of positive correlation.

5 $r_s = 0.818$; maximum change $= 0.035$

6 (ii) $r_s = -0.529$
  (iii) $-0.529 < -0.446$: significant evidence of negative correlation.

7 (iii) $1 - \frac{6}{35}p$

Exercises 22e

1 $\tau = 0.143 < 0.571$: no significant evidence of positive correlation.

2 $\tau = 0.389 < 0.500$: no significant evidence of correlation.

3 $\tau = 0.511 > 0.467$: significant evidence of positive correlation.

4 $\tau = 0.619 \geqslant 0.619$: at 5% level, significant evidence of positive correlation.

5 $\tau = 0.667$; maximum change $= 0.030$.

6 $\tau = -0.410 < -0.333$: significant evidence of negative correlation.

7 $\tau = 0.571 \geqslant 0.571$: significant evidence of positive correlation.

8 $\frac{1}{12}$ (ii) $P(K = 0.2) = P(K = 1) = \frac{1}{12}$,
$P(K = 0.4) = P(K = 0.8) = \frac{3}{12}$,
$P(K = 0.6) = \frac{4}{12}$;
$\frac{3}{5}, \frac{7}{150}; \frac{11}{15}$.

9 5, total $= 120$; $\frac{1}{24}, \frac{7}{60}, \tau = 0.600, p = \frac{7}{60} > 0.1$: no significant evidence of positive correlation.

Exercises 22f (Miscellaneous)

1 (i) 0.972   (ii) 0.979, -1

2 (i) 326.7, 129.8, -1012.5, $r = -0.410$
  (ii) $r_s = -0.448 > -0.503$: no significant evidence that early finishers do better.

3 (ii) $r = 0.875$ (iii) $r_s = 0.830 > 0.745$,
$\tau = 0.689 > 0.600$: significant evidence of positive correlation.

4 (i) Sign test: tail probability $= 0.073 > 0.05$: no significant evidence that method B is quicker than method A.
Wilcoxon test: $z = 2.16 > 1.645$: significant evidence that method B is quicker than method A.
  (ii) Sign test: tail probability $= 0.039 < 0.05$
Wilcoxon test: $z = 2.82 > 1.96$: significant evidence that median (A) is not 8.35.

5 $r_s = 0.555 > 0.484$, $\tau = 0.385 > 0.359$: significant evidence of positive correlation.

6 (a) $r = 0.972$ (b) $y = 3.50 + 1.04x$ (c) 9.7 kg
  (d) and (e) $r_s = 0.783 > 0.600$, $\tau = 0.611 > 0.500$: significant evidence, at 5% level, of positive correlation.

7 $r = 0.825$; $r_s = 0.929 > 0.786$: significant evidence, at 5% level, of correlation; $\tau = 0.810 > 0.714$: significant evidence, at 5% level, of correlation.

8 (a) $r = 0.460$ (b) $r_s = 0.750$

9 (a) $r = 0.937$ (b) $r_s = 0.976$
  (c) $0.782 > 0.564$: significant evidence, at 5% level, of positive correlation between maintenance costs and age.

10 (i) $r_s = 0.600 \geqslant 0.600$: significant evidence of positive correlation;
  (ii) $\tau = 0.333 < 0.500$: no significant evidence of positive correlation.

11 $\tau = 0.733 > 0.600$: significant evidence of non-zero correlation; $z = 1.83 > 1.645$: significant evidence that number of employees has increased; $t = 2.36 > 1.833$: significant evidence of salary increase.

12 (i) $\alpha > 0.26$ (ii) 0.383

13 Sign test: tail probability $= 0.035 < 0.05$: significant evidence to support the teacher;
Wilcoxon test: $z = 1.75 > 1.645$: significant evidence

to support the teacher; $t = 2.33 > 1.895$: significant evidence to support the teacher.

(i) None, none.

(ii) None, $t = 1.73 < 1.895$, no longer significant evidence.

14 (i) Wilcoxon test: $z = 1.83 > 1.645$: significant evidence of a decrease.

Sign test: tail probability $= 0.055 > 0.05$: no significant evidence of a decrease.

(ii) $t = 2.78 > 1.833$: significant evidence of a decrease.

15 (b) $r = 0.810$; $r_s = 0.745 \geqslant 0.745$: significant evidence of positive correlation.

Index